The Geology of North America
Volume P-3

Economic Geology, Mexico

Edited by
Guillermo P. Salas
Geologo Consultor
Asesor del Consejo de Recursos Minerales
Centro Minero Nacional
Carretera Mexico-Pachuca
Pachuca, Hgo., Mexico, D.F.

1991

Acknowledgment

Publication of this volume, one of the synthesis volumes of *The Decade of North American Geology Project* series, has been made possible by members and friends of the Geological Society of America, corporations, and government agencies through contributions to the Decade of North American Geology fund of the Geological Society of America Foundation.

Following is a list of individuals, corporations, and government agencies giving and/or pledging more than $50,000 in support of the DNAG Project:

Amoco Production Company
ARCO Exploration Company
Chevron Corporation
Cities Service Oil and Gas Company
Diamond Shamrock Exploration Corporation
Exxon Production Research Company
Getty Oil Company
Gulf Oil Exploration and Production Company
Paul V. Hoovler
Kennecott Minerals Company
Kerr McGee Corporation
Marathon Oil Company
Maxus Energy Corporation
McMoRan Oil and Gas Company
Mobil Oil Corporation
Occidental Petroleum Corporation
Pennzoil Exploration and Production Company
Phillips Petroleum Company
Shell Oil Company
Caswell Silver
Standard Oil Production Company
Oryx Energy Company (formerly Sun Exploration and Production Company
Superior Oil Company
Tenneco Oil Company
Texaco, Inc.
Union Oil Company of California
Union Pacific Corporation and its operating companies:
Union Pacific Resources Company
Union Pacific Railroad Company
Upland Industries Corporation
U.S. Department of Energy

Published by The Geological Society of America, Inc.
3300 Penrose Place, P.O. Box 9140, Boulder, Colorado 80301

Library of Congress Cataloging-in-Publication Data
Geología económica de México. English.
Economic geology, Mexico / edited by Guillermo P. Salas.
p. cm. — (The Geology of North America ; v. P-3)
Translation of Geología económica de México.
Includes bibliography references and index.
ISBN 0-8137-5213-2
1. Geology, Economic—Mexico. I. Salas, Guillermo P. II. Title.
III. Series.
QE71.G48 1986 vol. P-3
[TN28]
557 s—dc20
[553'.0972] 90-28579
CIP

Cover Photo: La Caridad Copper Mine, Sonora. See Chapter 20.

10 9 8 7 6 5 4 3 2

Contents

Preface

The Geology of North America series has been prepared to mark the Centennial of The Geological Society of America. It represents the cooperative efforts of more than 1,000 individuals from academia, state and federal agencies of many countries, and industry to prepare syntheses that are as current and authoritative as possible about the geology of the North American continent and adjacent oceanic regions.

This series is part of the Decade of North American Geology DNAG Project, which also includes eight wall maps at a scale of 1:500,000 that summarize the geology, tectonics, magnetic and gravity anomaly patterns, regional stress fields, thermal aspects, and seismicity of North America and its surroundings. Together, the synthesis volumes and maps are the first coordinated effort to integrate all available knowledge about the geology and geophysics of a crustal plate on a regional scale.

The products of the DNAG Project present the state of knowledge of the geology and geophysics of North America in the 1980s, and they point the way toward work to be done in the decades ahead.

This volume was first prepared and published in Spanish by the volume editor in 1988. Translation was ably accomplished by Dr. Cecily Petzall of Caracas, Venezuela, and redrafting or conversion to English of the text figures was done by Ms. Karen Canfield. Because of the difficulties of communication with many of the authors, the referencing and the reference lists, and some of the figure contents, do not reflect normal GSA standards. Some tabular data has not been updated from that submitted in manuscripts for the Spanish-language edition. Nevertheless, this book presents a valuable English-language synthesis of information about the geothermal, coal, and metal-mining sectors of the Economic Geology of Mexico.

Despite being confined to his bed during the last six months of his life, the late editor, G. P. Salas, worked on this book until the day he died. Special thanks go to Ing. Hugo Cortes Guzmán who helped Salas develop responses to many questions raised during translation of this book and who provided the information needed to complete the final chapters after Salas's death on June 29, 1990.

Allison R. Palmer
General Editor for the volumes
published by The Geological Society
of America.

J. O. Wheeler
General Editor for the volumes
published by the Geological
Survey of Canada.

Foreword

This volume has had a history that is quite different from that of other volumes of *The Geology of North America*. It was developed, produced, and privately printed in Mexico by Ing. G. P. Salas in 1988 as a Mexican contribution to The Decade of North American Geology, with the understanding that it would be translated into English as a part of this set. This translation was accomplished by Dr. Cecily Petzall of Caracas, Venezuela; most of the book was translated by early 1990. Nearly all figures needed some translation as well, and many required redrafting because the original material was photocopied from mimeographed company reports. This work was competently done by Ms. Karen Canfield of Louisville, Colorado.

I edited all translations and submitted numerous questions about meaning and inconsistencies to Salas along with photocopies of the translated pages and figures. With the concurrence of Salas, four small chapters were dropped by reasons of redundancy or relevance for an English-speaking audience. During 1990, as his health was failing, Salas received considerable help from Ing. Hugo Cortez Guzmán. Without that help, completion of this volume would not have been possible. The process was complicated by the loss of two packages of responses in the return mail, one of which had to be reconstituted after Salas died.

Referencing for the chapters is often not up to GSA standard. In many cases, lists of references, often to unpublished sources, were included with the chapters but not cited in the Spanish text; because they may represent the only sources of information, however difficult they may be to find, these were generally retained. In several instances, references were cited in the text, but no list was published with the chapter. Wherever possible, such lists were obtained or created. Many of the references to unpublished sources were incomplete. Because of inability even for Salas to contact many chapter authors, the missing information could not always be obtained. Thus, the reference lists represent the best that could be done under the circumstances.

This book reflects the status of information on Economic Geology of Mexico that could be obtained by Salas, exclusive of petroleum, in the early to mid-1980's. The companion volume on Petroleum Geology of Mexico has not yet been released by PEMEX (Petroleos Mexicana) for translation.

With all of the limitations mentioned above, it is hoped that this volume will provide a useful starting point for those interested in the geology of geothermal, coal, metallic, and some other non-metallic resources in Mexico.

A. R. Palmer
Boulder, Colorado,
October, 1990

The Geology of North America
Vol. P-3, Economic Geology, Mexico
The Geological Society of America, 1991

Chapter 1

Economic geology of Mexico

G. P. Salas
Consulting Geologist, Ave. Reforma No. 295, Col. Cuauhtemoc, 7th Floor, 06500 Mexico, D.F.

INTRODUCTION AND GENERAL MATTERS

Economic geology defines the application of geology, with all its specialties, to the discovery of fossil fuels, geothermal and nuclear energy sources, metallic and nonmetallic mineral deposits, water resources, and as the basis for civil works, all of which are of obvious economic and social value.

Disciplines in economic geology range from the exact sciences—advanced mathematics, physics, chemistry, geodetics, etc.—to the natural sciences such as micro- and macropaleontology, crystallography, petrology, mineralogy, volcanology, geochemistry, geochronology, and geohydrology.

The various elements of Mexico's economic geology are discussed in the chapters of this volume by outstanding Mexican geologists, whose expertise vouches for the high quality of this presentation. Their efforts are a valuable contribution to the knowledge of Mexico's nonrenewable resources.

The importance of the decision to carry out this project lies in the close interconnection of the earth sciences to the country's social and economic development. No country can develop without a solid knowledge of the necessary basic conditions—energy, water, and mineral resources—as well as the technical capacity to put them to effective use and the appropriate political stability, financial backing, and adequate educational level and general health of the population. Mexico fulfills these conditions; however, a comprehensive technical-scientific overview of its nonrenewable and water resources had not been attempted until now. The present status of the nonrenewable resources of the country, their interrelations, and the criteria guiding the present (1985) economic and social activity as well as the plans for future development are subjects for future volumes.

The significance of the earth sciences in any evaluation of nonrenewable and water resources, hence of *geology* and all its disciplines, for the country's social and economic development is shown in the following sections, which also include statistics regarding products derived from their application.

ENERGY RESOURCES

Hydrocarbons

Oil is the most economically important industry in the country, both in intrinsic value and in its multiplying industrial effects. The Mexican oil industry is one of the best examples of earth sciences applied to economic development by way of successful prospecting for oil.

Prior to 1938, exploration and regional geological mapping in Mexico had produced general geological information. Early geological sketch maps of the country dating from the beginning of the century were the basis upon which the geological study of the country was begun by foreign and Mexican geoscientists, especially Guadalupe Aguilera, Ezequiel Ordóñez, and Villarello, among others. In particular, Ordóñez' work in exploration and on the genesis, entrapment, and migration of oil led to the first commercial oil discovery in Mexico: La Pez-1 near Ciudad Ebano, San Luis Potosí, in what eventually became the Tampico oil province.

Foreign and native geoscientists succeeded in discovering the so-called Tampico basin oil fields in La Faja de Oro, Veracruz, Poza Rica, and the Tertiary light crude in salt dome flanks in the Tehuantepec Isthmus and Tabasco.

The nationalization of the oil industry, in March 1938, required replacing departed foreign and some native geoscientists with locally trained professionals, the first of whom graduated from the Universidad Nacional Autónoma de México (Mexico National Autonomous University) in 1939. Subsequently the Instituto Politécnico Nacional (National Polytechnic Institute) graduated petroleum geologists also. The nationalized oil industry that emerged after March 1938, with very few Mexican geologists and geophysicists, was thus able to confront the problem of exploring for new oil fields, the first of which was discovered in 1947.

Salas, G. P., 1991, Economic geology of Mexico, *in* Salas, G. P., ed., Economic Geology, Mexico: Boulder, Colorado, Geological Society of America, The Geology of North America, v. P-3.

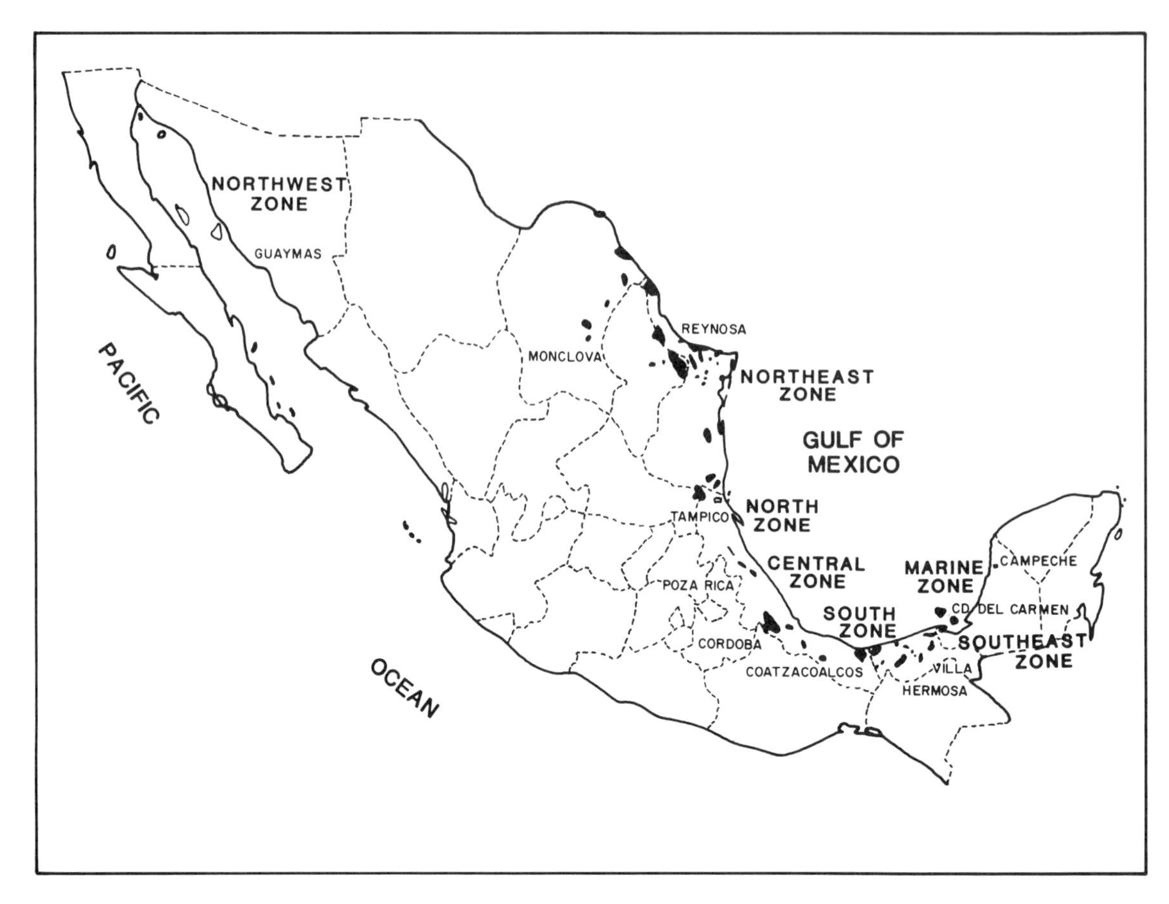

Figure 1. Present-day (1987) oil-producing areas of Mexico.

TABLE 1. HYDROCARBON ENERGY RESOURCES

Period from 1973 to 1983		
Field parties	463	
Exploratory Wells		
Producing	238	
Dry	493	
Thousands of meters drilled (to 1982 incl.)	30,041	
Fields Discovered		
Oil	125	
Gas	81	
Total Fields	206	
Production		
Oil (crude and condensate)	5,804,933,000	bbls
Gas	326,677	million m^3
Proven Reserves (1938-1983)		
Crude and condensate	72,500	million bbls.

In the period elapsed since, Petróleos Mexicanos has covered 1.7 million km^2 with geological, geophysical (seismic reflection and refraction; air and ground magnetometry; gravity, etc.), and geochemical surveys. In 1938, known oil reserves were approximately 1,276 million barrels, and production stood at 44 million barrels per year. By December 31, 1983, proved reserves were 72,500 million barrels, and total production for that year was 1,338 million barrels (reserves grew to nearly 57 times those of 1938).

Oil is the country's main source of income, one of the principal sources of employment, and the undisputed foundation of the country's industrialization process, generating 90 percent of its electric power, 4 percent (in 1985) of its gross national product, 71 percent of its exports, 55 percent of its foreign exchange, and 45 percent of the federal government's total income.

The following statistics, from the Petróleos Mexicanos Annual Report of 1983–1984, show the industry's weight in the nation's economy and consequently the importance of the earth sciences in the development of this vital Mexican industry. Table 1, showing integrated production results, illustrates that importance. Figure 1 shows the distribution of oil-producing areas in Mexico.

Electricity

Hydroelectric plants. The hydraulic resources of the river basins throughout the country are among the most important in Mexico for electric power generation. The hydroelectric development of a large number of permanent streams has resulted in the Federal Commission on Electricity's excellent system of interconnected high-tension transmission networks.

Of major importance are the Balsas and Grijalva River basins. An Integrated Development Project has set up seven power-generating dams in the Balsas River, which runs east-west across the state of Guerrero into the sea at the Guerrero-Michoacán border. "La Villita" and "El Infiernillo," both in Michoacán, have been operating for several years. The Carlos Ramírez Ulloa "El Caracol" power station is under construction in the state of Guerrero, and four others are projected for the near future.

All the dams, with or without power plants, on the Pacific slope of the Sierra Madre Occidental required geological investigation of igneous, metamorphic, and pyroclastic rocks affected by Cenozoic tectonism.

The geology of areas in the Sierra Madre Oriental, where dams are already operating and others are to be built, is very different from that of the Sierra Madre Occidental region. The dams in the states of Tamaulipas, San Luis Potosí, Veracruz, and more recently, Chiapas are founded in sedimentary rocks, mainly late Mesozoic limestones and clastic sedimentary rocks of mid–Late Cretaceous and Paleocene ages.

In all cases, the local and regional tectonics were the subjects of extensive and thorough geological and geophysical study to avoid locating the nonoverflow dam at faulted zones, to determine the permeability of the basin rock, and to locate the nearest sources of construction materials.

At present, 49 hydropower plants in Mexico have an installed capacity of 6,532,000 kW (1985; see Table 2), which saves 40,380,000 bbls of fuel per year. Foreseeable hydropower potential is 171,876,000 kW.

Geothermal energy

Geothermal energy is an important power-generating alternative, particularly in countries lacking fossil fuels. Its use is increasing rapidly as technologies for exploration, high-temperature drilling, and drilling-fluid control become more sophisticated.

Mexico has, at present, a considerable group of specialists in the location and evaluation by advanced techniques of thermal areas, as well as in developing thermal areas for immediate power production (i.e., a growing group of experts in all phases of geothermics as an alternative to fossil fuels).

As of this writing (1985), positive results of great present and future economic and strategic importance (Tables 3 and 4) have been achieved in explored areas. This does not include the hundreds of thousands of square kilometers of volcanic and tectonically highly deformed territory in Mexico where underground high temperatures may be found.

TABLE 2. POWER CAPACITY OF MEXICAN HYDROELECTRIC PLANTS

Hydropower Plants	Total Capacity (kW)	Date of Commercial Initiation	Location Municipality, State
1. Xia	170	1939	Chicomezúchil, Oaxaca
2. Bartolinas	750	1940	Tacámbaro, Michoacán
3. Jumatán	2,180	1941	Tepic, Nayarit
4.Carácuaro	120	1942	Carácuaro, Michoacán
5. Granados	940	1942	Villamar, Michoacán
6. Cointzio	480	1943	Morelia, Michoacán
7. Ixtapantongo	106,000	1944	Nvo. Sto. Tomás, México
8. Zumpimito	6,400	1944	Uruapan, Michoacán
9. Río Micos	1,052	1945	Cd. Valles, San Luis Potosí
10. Colotlipa	8,000	1946	Quechultenango, Guerrero
11. Santa Bárbara	67,575	1950	Nvo. Sto. Tomás, México
12. San Juan Viejo	228	1950	Zitácuaro, Michoacán
13. Las Minas	14,400	1951	Las Minas, Veracruz
14. Bombaná	5,240	1951	Bochil, Chiapas
15. El Encanto	10,000	1951	Tlapacoyan, Veracruz
16. Texolo II	1,600	1951	Teocelo, Veracruz
17. Tepazoico	10,880	1953	Xochitlán, Puebla
18. El Punto	960	1954	Tepic, Nayarit
19. Falcón	31,500	1954	Guerrero Garza G., Tamaulipas
20. Cóbano	52,020	1955	Gabriel Zamora, Michoacán
21. H. Martinez D'Meza	25,200	1955	V. de Allende, México
22. El Durazno	18,000	1955	Valle de Bravo, México
23. Coalcomán	488	1957	Coalcomán, Michoacán
24. Oviachic	19,200	1957	Cajeme, Sonora
25. Tingambato	135,000	1957	Otzoloapan, México
26. Mocúzari	9,600	1959	Alamos, Sonora
27. Temascal	154,080	1959	San Miguel Soyalt, Oaxaca
28. El Salto	2,975	1959	El Salto, Jalisco
29. El Fuerte	59,400	1960	El Fuerte, Sinaloa
30. Chilapan	26,000	1960	Catemaco, Veracruz
31. Tetela de Ocampo	100	1960	Tetela de Ocampo, Puebla
32. Guasuntlán	1,600	1962	Soteapan, Veracruz
33. Mazatepec	208,800	1962	Tlatlauquitepec, Puebla
34. Cupatitzio	72,450	1962	Uruapan, Michoacán
35. Tamazulapan	2,480	1962	Tamazulapan, Oaxaca
36. Gral. S. Alvarado	14,000	1963	Culiacán, Sinaloa
37. Luis M. Rojas	5,320	1963	Tonalá, Jalisco
38. El Chique	624	1964	Tabasco, Zacatecas
39. Santa Rosa	61,200	1964	Amatitán, Jalisco
40. El Novillo	135,000	1964	Soyapa, Sonora
41. La Venta	30,000	1964	La Venta, Guerrero
42. Infiernillo	1,012,000	1965	La Unión, Guerrero
43. Agustín Millán	18,900	1965	Valle de Bravo, México
44. Camilo Arriaga	18,000	1966	Cd. del Maíz, San Luis Potosí
45. J. Cecilio del Valle	21,000	1967	Metapa, Chiapas
46. Malpaso	720,000	1969	Tecpatán, Chiapas
47. La Villita	300,000	1973	M. Ocampo, Michoacán
48. La Angostura	540,000	1975	V. Carranza, Chiapas
49. Humaya	85,500	1976	Culiacán, Sinaloa

Source: Evolution of the electrical sector in Mexico (CFE, 1977).

To achieve the programmed long-range goals, it therefore seems obvious that the Federal Commission of Electricity (CFE) geotechnical specialists should continue their excellent coverage of the prospective geothermal areas with the Administration's full backing.

The Division of Geothermics of the CFE has explored 1,200,166 km^2 in the states of Baja California Sur and Norte, Sonora, Chihuahua, Jalisco, Guanajuato, México, Puebla, Michoacán, Guerrero, and Veracruz. Deep wells drilled at thermal prospects located in several of these areas have verified their future importance.

The section on geothermal energy details the regional and local geology of these major geothermal reservoirs, which are expected to afford a growing percentage of the country's power: 645 MWH are foreseen by year-end 1985.

TABLE 3. RESERVES DISCOVERED AS OF MARCH 1985

1. Proven reserves	1,340 MW
2. Probable reserves	4,600 MW
3. Possible reserves	6,000 MW
Total	11,940 MW

Coal-fired power plants

A small coal-fired plant that operated for many years in Rosita, Coahuila, preceded the Nava, Coahuila, plant, which is a few kilometers southeast of Piedras Negras, Coahuila, and is presently under construction and operation by the CFE. Chapter 10 includes a location plat of soft (noncoking) coal deposits and coal-fired power plants operating at present with coal from the Rio Escondido deposit.

Chapter 10 describes the geology of this deposit, comprising one or more noncoking coal seams in the Upper Cretaceous Olmos Formation of Coahuila (Navarro Formation of Texas), which underlies a Quaternary conglomerate in ancient terraces of the Bravo River. The variable strike of the coal (northeast to southwest in Cut I and almost north-south in Cuts II to V) allows open-cut excavation; reserves are considerable. Subsurface development is done mainly by modern longwall mining.

In the interval from 1980 to 1984, 4,387,000 metric tons of coal valued at $72,275,000 (U.S.) were sold to the CFE; by December 1984, 3,370,000 metric tons had been consumed, generating 6,066 MWH at a cost of 9¢ (U.S.)/KWH.

Nuclear energy

Not only the development of geological exploration in Mexico, but also the present world situation with respect to use and marketing of nuclear fuels must be considered under this heading.

The uranium ore boom of the 1970s peaked when the Middle East oil embargo caused oil prices to rise and forced consumer countries to restrict their energy consumption, in particular of oil-based fuels.

Plans for the year 2000 envisaged nuclear power plants as the alternative sources of energy. France, for instance, would generate 90 percent of its power through this agency; Spain, 67.5 percent; Denmark, 64 percent; Finland, 40 percent; the United States, 40 percent. This caused the price per pound of U_3O_8, "yellowcake" or concentrated fuel, to increase substantially and encouraged the search for prospective radioactive minerals for processing, and marketing to manufacturers, of the fuel cartridges that charge nuclear power plants with enriched uranium. An extensive bibliography on the use of uranium fuel for nuclear power plants in the more industrialized countries has accumulated. According to *Presence of Uramex in the development of Mexico, (Presencia de Uramex en el Desarrollo de México)* p. 63, 229 nuclear power plants were operating worldwide, and 530 more were under construction or projected. However, these programs have not achieved their objectives for various reasons, not the least being the risk of serious accidents.

Mexico's involvement in this area began in 1956 when the Comisión Nacional de la Energía Nuclear (National Nuclear Energy Commission) was established; it operated until 1971. The Instituto Nacional de la Energía Nuclear (National Nuclear Energy Institute; INEN) then took its place from 1972 to 1976. The sector was reorganized in 1979, and URAMEX (Uranio Mexicano) was created with the same former objectives of exploration, development, industrialization, and marketing of radioactive minerals, together with other unrelated obligations. For several reasons, in approximately 16 years, only 8,332 tons of in-place radioactive minerals appropriate for production were located at widely dispersed sites; neither costs nor recovery percentages were considered, since these reserves would obviously be seriously depleted if they were to be developed.

Table 5 (Presence of Uramex in the development of Mexico, p. 232–233) lists deposit localities and estimated reserves.

TABLE 4. STATISTICS

INSTALLED FIELD CAPACITY	
1. Cerro Prieto, Baja California Norte	620 MW
2. Los Azufres, Michoacán (initial)	25 MW
Present Total	645 MW
GENERATED KILOWATT HOURS (KWH)	
1. Cerro Prieto, Baja California Norte	
2. Los Azufres, Michoacán (to March 1985)	364 KWH
COST PER INSTALLED KW (JANUARY, 1985)	
1. Cerro Prieto Type, 220 Plant	$ 840.90 US
2. Los Azufres Type, 55 Plant	$ 1,218.18 US
COST PER GENERATED KWH (JANUARY, 1985)	
1. Cerro Prieto Type, 220 Plant	2.5¢ US
2. Los Azufres Type, 55 Plant	3.5¢ US

TABLE 5. NATIONAL (IN SITU) URANIUM ORE RESERVES (INEN TO DECEMBER 1976)

Locality	Tons of U_3O_8
CHIHUAHUA	
El Nopal	361.0
Margaritas-Puerto III	2,394.0
Nopal III	170.0
La Dolmita	61.2
Other	43.6
Margarita, Nopal I, II, Puerto III*	1,950.0
DURANGO	
La Preciosa	210.0
NUEVO LEON	
La Coma	997.0
Buenavista	1,256.0
El Chapote*	415.0
SONORA	
Los Amoles*	475.0
Total	8,332.8

*Inferred reserves.

HYDROLOGY

The worldwide economic importance of water is evident. Agriculture, industry, general health, power-generating processes, and life itself are impossible without it; scarcity of water pauperizes entire regions and countries. In excess, it can also be harmful, as in the case of tropical plains requiring flood control. However, knowledge of its occurrence is essential for the economic and social development of any country.

Mexico has tropical, temperate, desert, and semi-desert climates. Uniform precipitation several months a year in parts of the country feeds underground aquifers, river basins, and dams, favoring good rainfall and irrigation-dependent agriculture. However, across more than 60 percent of the territory, the climate is arid or semi-arid, with short periods of torrential rainfall that causes intense soil erosion and precludes natural—and even at times artificial—reservoiring.

Predictions are that the world population will double in the next 30 years; it could possibly be even sooner in Mexico. This underscores the need for improved water-use management, accurate precipitation measurements, knowledge of surface and ground-water accumulations in every country, catchment problems, and other information on the basis of geohydrological studies. It is therefore obvious that geohydrology is essential to the economic development of the country, at the same level of importance as all the other economic geology specialties.

Specialized government entities in Mexico have carried out outstanding geohydrological investigations of Quaternary, Cenozoic, and Mesozoic aquifers by regions and river basins, which data—not yet completed—have resulted in thousands of deep wells (see Table 6).

Statistics on the number of water wells drilled in each state are presently being compiled. Location maps and production estimates are already available. The Hydraulic Resources subdivision of the Agriculture and Hydraulic Resources Office (Subsecretaría de Recursos Hidráulicos de la Secretaría de Agricultura y Recursos Hidráulicos) estimates a yearly production in Mexico of 25 billion m^3 of water from wells drilled for irrigation, urban, and industrial use. The Water Development Division (Dirreción de Aprovechamiento del Agua) of the above organization is carrying out a nationwide inventory; counting has been completed in ten states (see Table 6). This production cannot be evaluated in terms of money; its true value lies in the fact that in its absence the Mexican economy would collapse.

As more geologically recent and older aquifers are located, it is to be expected that geohydrological studies will also produce statistics on precipitation, evaporation and evapotranspiration, percentages of runoff feeding streams and rivers and percolating down to the water table and underground aquifers, yearly volume of water loss in streams flowing to the sea, and even underground flow using radioactive tracers, all of which are needed to satisfy the growing demand for water throughout the country.

MINING INDUSTRY

Another important component of the Mexican economy is the mining industry, which produces both raw materials for domestic processing and income derived from exports of surplus minerals. Moreover, it has a multiplying effect involving the necessary means of transportation and other infrastructures, as well as energy and fuel consumption. Since the mineral deposits are usually located in remote areas, road-building for access leads to the settlement of towns, some of which have eventually become state capitals; innumerable villages and cities owe their existence to the mine workings of Colonial times.

Here again, geological activity has been vital. As mineral exploration and development advances, the technology needed to locate what is left becomes increasingly complex and costly. The requirements of modern industry also pose challenges; present-day production of consumer and capital goods demands a much wider range of metallic and nonmetallic minerals and rare earths.

More than 53 metallic and nonmetallic minerals are sold on the international market, some as uncommon as indium, osmium, palladium, and ruthenium. Many of these have not yet been located in Mexico, but investigation of their possible presence must be a future goal.

The economic importance of the geosciences in the mining industry is shown by the following integrated production and export figures from 1974 to 1983, in dollars equivalent to pesos in the corresponding years (Table 7).

TABLE 6. GEOHYDROLOGY. NATIONWIDE INVENTORY OF WATER USE (1985)

State	Area in km^2			No. of Developments		
	Total	Inventoried	Updated	Surface	Underground	Total
Aguascalientes	5,569	5,569	5,569	526	1,965	2,941
Distrito Federal	1,499	1,499	1,499	25	822	847
Durango	119,648	119,648	25,858	3,474	7,344	10,818
Guanajuato	30,589	30,589	30,589	5,877	15,731	21,608
Hidalgo	20,987	20,987	20,987	1,865	713	2,578
México	21,461	21,461	19,356	3,844	5,049	8.893
Morelos	4,941	4,941	1,800	378	758	1,136
Querétaro	11,769	11,769	11,769	1,344	1,122	2,466
Tlaxcala	3,914	3,914	754	320	545	865
Zacatecas	75,040	75,040	3,381	5,265	7,261	12,526

CONCLUSION

The foregoing has briefly summarized the topics covered in the present volume by many experienced geoscientists who have contributed significantly to the social and economic development of Mexico.

This publication is dedicated to them and to all Mexican geoscientists, in hopes that it will serve as an incentive for their continued efforts and as an essential reference for government agencies, industry leaders, investors, political decision-makers, and the general public.

TABLE 7. STATISTICS

	(in $1,000)
Metallic minerals production	12,187,979
Nonmetallic minerals production	6,273,682
Subtotal	18,461,661
Export of metallic minerals	5,792,184
Export of nonmetallic minerals	1,804,544
Subtotal	7,596,728
Grand Total	26,058,389

Printed in U.S.A.

The Geology of North America
Vol. P-3, Economic Geology, Mexico
The Geological Society of America, 1991

Chapter 2

National hydroelectric plan (1982–2000)

Hydroelectric Projects Management
Preliminary Civil and Geotechnical Engineering Office, Federal Commission on Electricity, Rio Rodano 14, 06500 Mexico, D.F.

INTRODUCTION

The importance of water resources in the context of Mexico's power system is obvious on analyzing the latest available statistics. Hydroelectric power stations generated 24.4 TWH (terawatt-hours, 10^{12} watts) in 1981, 36 percent of the overall power (almost 68 TWH) consumed that year. The history of power generation development in Mexico and of the growing importance of hydroelectricity to its present level affords a clearer understanding of the long-range planning to the year 2000 in this field.

Figure 1 shows the yearly average hydroelectric generation related to total electricity generation in Mexico and to installed capacity. Generation variations in time are due to variations in water availability, which are unpredictable. However, the hydroelectric generation and installed potential may vary, its participation in total electricity generation decreased beginning in 1970. Thus, 53 percent of the nationwide installed power in 1970 was hydroelectric and provided 57 percent of the total. In 1980, percentages were 41 and 27, respectively, that being a low-runoff year, and rose to 36 percent of the overall power generation in 1981 when precipitation was heavy.

Hydroelectric generation developed unequally to the electricity sector's overall growth, owing to the priority given at different times to the construction of thermoelectric plants. Consideration of the economic, social, operational, and conservation criteria for hydroelectric versus thermoelectric power generation shows that development of hydroelectric power is desirable.

From the economic standpoint, thermoelectric plants usually require a smaller initial investment. However, operating costs of thermoelectric plants are much higher due to the price of fuel oil. The fact that a large percentage of the most profitable hydroelectric projects has been completed already is noteworthy. Therefore, future average costs of power (kilowatt hours) will tend to increase. On the other hand, thermoelectric plants, although not wholly independent, have better possibilities of being located nearer to large consumption centers, and investment in transmission lines would therefore be lower than for hydroelectric installations.

In the social aspect, hydroelectric projects often cause difficult or eventually confrontational situations when flooding of productive lands or of more or less important communities becomes necessary, involving indemnity payments and adequate management to minimize anxiety or discontent in the affected communities. Alternatively, a dam project fosters future regional benefits that are hard to quantify, such as employment and training opportunities, and usually serves other regionally important purposes of irrigation, agriculture, and recreation. Moreover, provision must be made to minimize the usual local effects of any large-scale investment in a region, such as inflation.

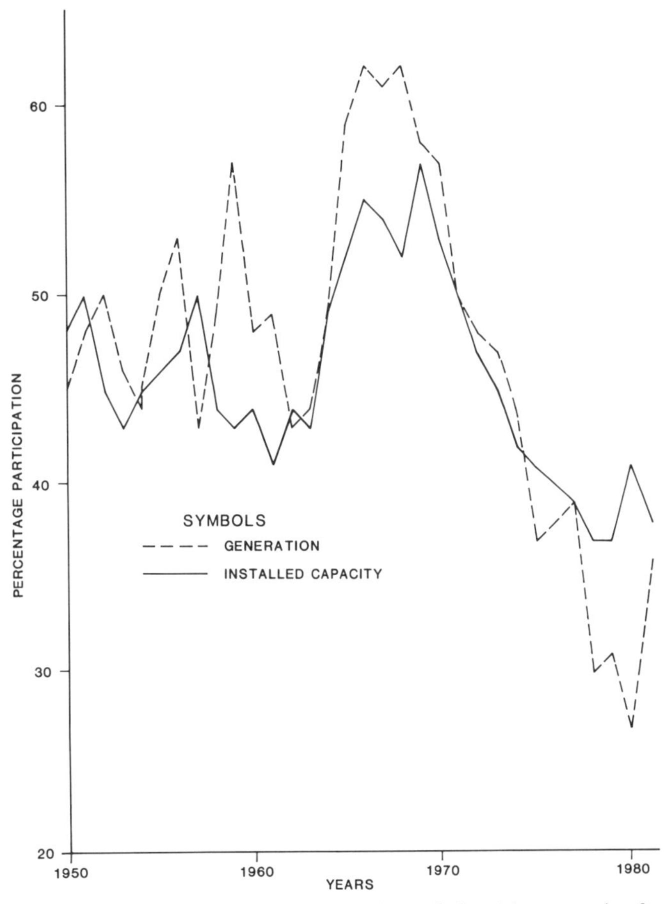

Figure 1. Hydroelectric participation in the total electricity generation for Mexico from 1950 to 1982.

Hydroelectric Projects Management, 1991, National hydroelectric plan (1982–2000), *in* Salas, G. P., ed., Economic Geology, Mexico: Boulder, Colorado, Geological Society of America, The Geology of North America, v. P-3.

In the operational aspect, hydroelectric plants are more flexible because they are the only installations capable of storing energy for use on demand and of operating readily on short notice. This makes them best suited for power generation at times of peak demand, increasing the timely value of hydroelectric kilowatt hours and allowing thermoelectric plants to operate under ideal conditions (i.e., continuously and constantly) with no fluctuations.

In multiple-use development, auxiliary construction permits different demand regimes to be compatible and to maintain operational flexibility.

As regards natural resources, hydroelectric plants run on water, a renewable resource, whereas thermoelectric generation requires fuel oil or gas derived from an important nonrenewable resource. Also, environmental pollution caused by hydroelectric plants is minimal compared to that of thermoelectric installations.

More recently, the foreign investment factor has become relevant in view of the constant increase in value of the dollar versus the peso, which inhibits the purchase of equipment or spare parts that must be paid for in foreign currency. In this sense too, hydroelectric plants have an advantage over thermoelectric—a much smaller foreign component.

Environmental protection in the future will probably be more demanding and require significant investment. Although hydroelectric plants do not normally cause pollution, they usually affect regional ecosystems. In view of the worldwide energy situation and the priority assigned to conservation of natural resources and environmental protection, it is advisable to continue using hydroelectric plants to satisfy growing demand, in spite of their higher initial investments.

In view of all the above and the possible future increases in power demand, approximately 80 TWH should be generated by hydroelectric plants by the year 2000. This involves building and putting into operation hydroelectric projects capable of generating 55.6 TWH over and above the present 24.4 TWH in the course of the next 19 years, and implies a sustained yearly growth rate of 6.4 percent for that period.

A consequence of the stagnation of hydroelectric development in the 1970s was the decline of engineering capability in that field. For this reason, although hydroelectric generation in Mexico has grown at a yearly rate of 10.5 percent in two decades, the 6.4 percent growth rate proposed for the next 19 years involves long-range planning to provide the engineering and construction capacities these projects require.

The objective for the year 2000 represents 46 percent of the hydroelectric potential identified to date. Hydrocarbons, being nonrenewable, are spoken of as reserves; hydroelectricity is spoken of in terms of yearly generation potential, since in the long run the same volume of water might be in use.

The theoretical gross hydropower generation potential is the energy that would be obtained by using all the runoff throughout the entire gradient between runoff and outlet points. The usefulness of this academic concept is that it sets a top hydropower potential, estimated in Mexico at about 500 TWH/yr. This is what would be produced if all the identified appropriate dam placement sites were developed. The hydropower potential identified to date is about 172 TWH/yr. The present feasible potential is the energy obtainable from hydropower projects of proven technical, economic, and social feasibility: 37.6 TWH/yr, including the 24.4 TWH/yr from plants operating today. The remainder (13.2 TWH/yr) would come from 15 projects whose feasibility studies have been concluded, seven of which are presently under construction. Since the proposed objective implies developing projects for a total 55.6 TWH, the magnitude of the required effort and the priority this type of project deserves are both obvious.

It should be stressed that the 80 hydropower TWH projected for the year 2000 should be thought of as a minimum. Therefore, further financial and technical analysis is needed as to the practicality of a higher objective. Once defined, the effort to meet it demands continuity, with the normal adaptations to new information. Constantly changing component prices and priorities assigned to the factors involved in project evaluations can influence decisions as to the comparative convenience of hydropower. In the past, this has caused unequal hydropower development, which if allowed to continue, would waste both human and economic resources.

PERSPECTIVES OF THE HYDROELECTRIC SUBSECTOR

POISE is the hydroelectric program envisaged in the Works and Investments Program of the Electric Sector (Programa de Obras e Inversiones del Sector Eléctrico. To analyze the consis-

TABLE 1. HYDROELECTRIC PROJECTS CONSIDERED IN THE *POISE*

Project	Installed Capacity (MW)	Annual Generation (GWH/yr)	Operation Starting Date
Caracol	570	1,275	Aug. 1984
La Amistad	66	157	Aug. 1984
Peñitas	400	1,819	Feb. 1985
Bacurato	90	275	Feb. 1985
Temascal II	240	506	May 1987
Itzantún	440	1,344	May 1987
Comedero	90	285	Apr. 1987
Picos de Guadalajara	261	436	Jan. 1988
Aguamilpa	690	2,100	Apr. 1989
San Juan Tetelcingo	540	718	Mar. 1989
Huites	300	865	Mar. 1989
La Yesca	450	1,200	Jan. 1990
Tepoa	240	743	Feb. 1990
Cajones	450	1,400	Mar. 1991
Totals	4,827	13,123	

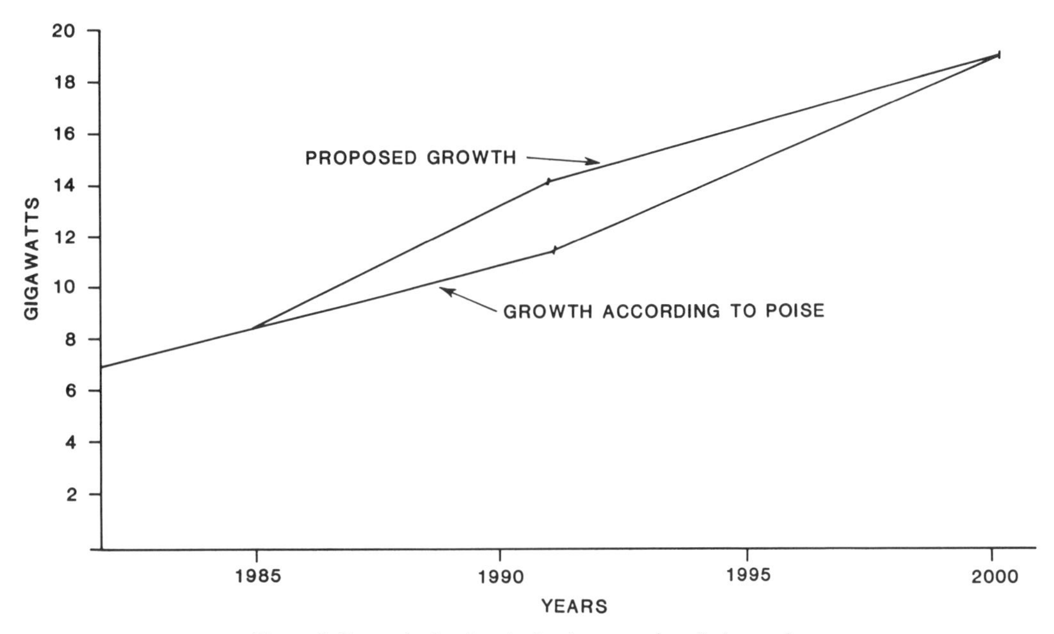

Figure 2. Future hydroelectric development: installed capacity.

tency of the POISE plans with the hydroelectric generation goal for the year 2000, the POISE hydropower development program is discussed below, according to information as of July 1982.

The POISE program contemplates a 4.3 percent annual increase in hydroelectric power generation for the 1982 to 1991 period derived from the construction of 14 major hydropower projects with a total annual generation of 13.1 TWH (see Table 1). Seven of these are presently under construction. As shown on Table 1 they will furnish an installed capacity of 2,273 MW (636 in 1984, 516 in 1985, and 1,121 in 1987) for an average annual generation of 6,600 GWH (1,500 in 1984, 2,200 in 1985, and 2,900 in 1987).

The flow of investment for these construction projects (in millions of $U.S. as of July 1982) is as follows:

Year	1982	1983	1984	1985	1986	1987
Investment	120	166	115	58	33	18

If the foreseen growth is achieved, hydropower generation will be 37.5 TWH/yr in 1991. The 80-TWH goal for the year 2000 demands completing projects capable of generating 42.5 TWH in only eight years (i.e., an average annual increase of 9.9 percent in that period).

A steady 6.4 percent annual increase throughout the 1982 to 2000 period is suggested as more advantageous than the POISE proposal (i.e., 4.3 percent from 1982 to 1991 and 9.9 percent from 1992 to 2000; see Fig. 2), mainly on the basis that either option is possible. If the 4.3 percent annual rate is kept steady until 1991, inertia would prevent a sudden sharp increase to 9.9 percent, which would be practically impossible. On the other hand, if provision is made to increase the hydroelectric growth

In either case, the total economic investment for the 1982 to 2000 period would be the same. However, the POISE proposal would involve a lower investment in the 1982 to 1991 period and a higher investment in 1992 to 2000, rather than a steady 6.4 percent growth rate throughout. The reason is that by 1991 the installed hydroelectric potential would be greater in the latter case (steady growth) than in the POISE plan; the difference in this case would be supplied by thermoelectricity that, as mentioned before, requires lower investments. Obviously, the situation would be reversed in the 1992 to 2000 period, and by the year 2000, investments in both cases would be the same (see Fig. 2).

Operating costs would be lower throughout the period in the steady-growth option, since the shadowed portion of Figure 3 would be hydroelectric, whereas if the POISE model is adopted, it would be thermoelectric generation and costlier, as noted before.

Considering the options, a steady hydroelectric growth rate throughout the entire period is proposed, which will result in modification of some of the works in the POISE plan and the addition of others.

HYDROLOGY AND GEOLOGY OF THE CHICOASEN, CHIAPAS, RESERVOIR

Hydrology

The rainfall regime has two well-defined periods: maximum precipitation—from July to November—is caused by cyclonic disturbances generated in the Gulf of Mexico, the Caribbean, and on occasion the Pacific; the dry season is from December to June. Annual average precipitation is 957 mm.

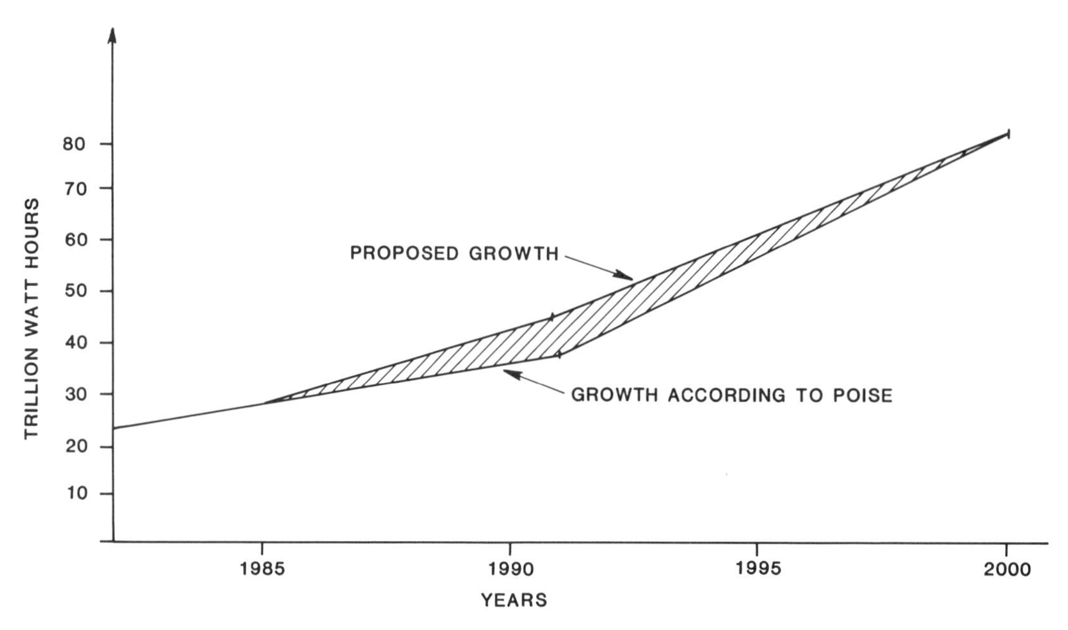

Figure 3. Future hydroelectric development: generation capability.

The Chicoasén reservoir inflow was determined by adding extracts from the La Angostura reservoir to the possible overflows from its spillway and inflow from its own basin. These results were applied to simulations of the Chicoasén reservoir operation with different parameters. The analyzed period was 23 years; runoffs from January 1952 to May 1961 were inferred, and runoffs from June 1961 to December 1974 were observed at the Chicoasén hydrometric station, located downstream from the construction site.

Geology

Physiography. Geological surveys to select the project area covered approximately 1,635 km^2 north of the town of Tuxtla Gutiérrez, Chiapas, which corresponds to the physiographic province of the high plateaus.

Location. The reservoir is set in a large Middle to Late Cretaceous limestone development forming the backbone of the Chiapas Sierra Madre. The highly folded and faulted rocks are several thousand meters thick. The course of the Chiapas Grijalva or Grande River is actually controlled by an east-west fault, which, however, does not affect the dam.

Seismic risk. In view of the proximity of the Chicoasén hydroelectric project area to one of the most seismically active regions in the world, and the presence of the Chicoasén-Malpaso and Cañada Muñiz regional faults, special care was taken to define the seismic risk. The main geologic, seismologic, and historical factors were taken into account.

The Federal Commission on Electricity has verified that none of the measurements taken at the project excavations showed abnormally high values for horizontal residual stresses, thus confirming the present-day inactivity of the nearby regional faults.

Seismologic factor. Seismic instrument records from 1900 to the present, intensities on the modified Mercalli scale recorded since 1929, direct measurements by Federal Commission on Electricity technicians, and inferences from tectonic studies based on average sea-floor seismicity were compiled and statistically analyzed by specialists of the Instituto de Ingeniería (Engineering Institute) of Mexico's National Autonomous University and of the commission itself. The results show the project area to be tectonically stable. Telluric movements, as recorded in the area, would not cause differential displacements in the project works. Recorded seismic acceleration is less than 0.15 g.

Historical factor. The physical condition of nearby buildings dating from the sixteenth century indicates that no accelerations from major earthquakes of over 0.15 g have been recorded since that time.

Geology of the dam site

The selected dam site is a 2.5-km-long canyon whose walls rise vertically from stream level (200 m above sea level) to an elevation of 330 m. Above this point the left bank slopes approximately 40° to the ridge crest at 405 m. The right bank walls are practically vertical up to 390 m. Above that point the topography undulates, with an average 5° slope to the ridge crest.

Applied geology

The foregoing geological data were extended and confirmed by 25,000 m of exploratory drilling with continuous core recovery and 3,000 m of excavations.

Printed in U.S.A.

The Geology of North America
Vol. P-3, Economic Geology, Mexico
The Geological Society of America, 1991

Chapter 3

Economic geology of the geothermal deposits of Mexico

A. Razo M., P. Reyes V., and O. Palma P.
Federal Commission on Electricity, Rio Rodano 14, 06500 Mexico, D.F.

INTRODUCTION

In a general sense, geothermal refers to the natural heat of the Earth's interior, independent of the factors involved in its surface manifestations. This heat flows directly through the rocks or is transported by fluids that rise along fractures to more or less deep zones and form the geothermal reservoirs. The heat source—the fluid—and the zone in the crust where the fluid is stored or circulates, together make up the geothermal system.

The term geothermal field implies considerations as to the feasibility of economic development. To benefit from geothermal energy implies developing the heat that underlies the Earth's surface. Energy production uses the water that comes into contact with subsurface rocks. In some areas the high heat generates steam, but in most geothermal fields the water remains in liquid form. Trapped in subsurface reservoirs, this water may be extracted by drilling wells.

ORIGIN OF GEOTHERMAL ENERGY

The distribution of temperatures over the Earth is determined by thermal flow toward the surface of caloric energy stored in the Earth's interior, plus any other type of locally produced energy. Due to the thermal conductivity of the Earth's crust, the high-temperature energy stored in the interior flows toward the surface at a rate of 1 to 2 HFU (Heat Flow Unit) measured in microcal/cm^2. Radioactive minerals in rocks, especially in granitic materials, produce another locally identified source of thermal energy. A 10-km-thick granitic crust will generate enough heat radioactively to provide a thermal flow of 0.6 HFU at the surface, a considerable amount of heat in relation to the energy radiated from the Earth's interior. Radioactively generated heat in the crust spreads over a relatively large volume, has lower temperatures, and consequently is less useful than the higher-temperature geothermal energy generated in the interior of the Earth.

The 1 to 2 HFU thermal energy flow at the Earth's surface derived from all geothermal sources by conduction from its interior is approximately 1/5,000 the solar energy flow, making it a marginal source of energy. At the surface it is not particularly significant, but it becomes so at depth, where temperatures are 500 to 1,000° C.

Convection currents in the Earth's mantle, together with volcanic activity and hydrothermal circulation within the crust, make up a transport mechanism to produce hydrothermal deposits near the surface. The crustal zone, moving in several directions by the action of convection currents in the asthenosphere, has formed oceanic and continental plates that collide or move away from each other and give rise to geologically interacting zones where plate subduction, volcanic eruptions, and mountain-building take place. Thermal energy is transported through this process from the Earth's interior to areas of the crust near the surface, where it is taken up by the underground movement of fluids and shows up at the surface as geysers, mud volcanoes, fumaroles, hot springs, sulfur springs (solfataras), etc.

Ground-water circulation at depths of several kilometers may develop a convection system in which water is heated directly or indirectly by a magmatic chamber at depths of 7 to 15 km and temperatures of 600 to 900° C. Many overheated magmatic zones are small and may cool in a few years; others may cover several cubic kilometers, remain energized for thousands of years, and give rise to major geothermal systems, which under special conditions, could become geothermal fields.

WORLDWIDE ENERGY PERSPECTIVE

Geothermal energy will not have a significant impact on worldwide energy sources in the short term. Some experts are cautiously optimistic that geothermal energy could supply 1 percent of world demand in 1990. Consequently, geothermal energy cannot be considered as an alternative to oil but rather as a complement. Oil, natural gas, and coal supplied about 70 percent of worldwide energy demand in 1982. Global production capacity of geothermally sourced power exceeded 3,700 MW (Table 1) in 1984, equivalent to that of four major power stations consuming 53 million barrels of conventional fuels per year.

Many countries seem to have geothermal resources with the potential to satisfy a large percentage of future energy demand. It

Razo M., A., Reyes V., P., and Palma P., O., 1991, Economic geology of the geothermal deposits of Mexico, *in* Salas, G. P., ed., Economic Geology, Mexico: Boulder, Colorado, Geological Society of America, The Geology of North America, v. P-3.

TABLE 1. GENERATION CAPACITY (MW) OF GEOTHERMAL PLANTS IN THE WORLD*

Country	Number of Units	June 1984	Expected 1986
1 United States	29	1,453.5	2,356.0
2 Philippines	19	781.0	1,496.0
3 Italy	42	472.1	502.1
4 Mexico	12	425.0	645.0
5 Japan	8	215.0	270.0
6 New Zealand	10	167.2	167.2
7 El Salvador	3	95.0	95.0
8 Iceland	5	41.0	41.0
9 Nicaragua	1	35.0	70.0
10 Indonesia	3	32.25	142.25
11 Kenya	2	30.0	45.0
12 Soviet Union	1	11.0	21.0
13 China	10	8.136	11.386
14 Portugal	1	3.0	3.0
15 Turkey	1	0.5	25.5
16 France	0	0.00	6.0
Total	145	3,769.686	5,896.436

*Di Pippo (1984)

has been estimated that the thermal energy stored at an underground depth of 10 km is about 3×10^{23} kilocalories, approximately equivalent to 3.5×10^{20} KWH, or to the heat provided by 4.5×10^{16} short tons of coal. This thermal energy is too spread out to be totally recoverable; however, the volumes involved are so large that even a small fraction may represent considerable value. Consequently, greater priority should be given to—and funds provided for—developing research on the interior of the globe.

By 1986, sixteen countries, including the United States, Italy, New Zealand, Japan, Iceland, the Soviet Union, Mexico, El Salvador, the Philippines, and others (Table 1), should be able to generate 5,896,436 KW in 145 generating units.

GEOTHERMAL ENERGY PROSPECTS IN MEXICO

At present, Mexico is fourth in worldwide geothermopower generation, after the United States, the Philippines, and Italy, and may improve this position in the near future owing to its numerous prospective geothermal areas. Geothermal energy may prove to be an important source of power for the country, with a potential capacity of 2,440 MW by the year 2000.

Activity in this field has developed in Mexico since the 1950s to the point that at present it involves all stages, from regional reconnaissance of thermal shows to construction and operation of geothermopower plants. The activities, programs, and costs of a nationwide development plan set up to define the country's geothermal possibilities for the 1985 to 2000 period are shown on Table 2.

TABLE 2. GEOTHERMAL EXPLORATION PROGRAM TO THE YEAR 1995
(Exploration costs in thousands of dollars as of April 1985)

Activities	1985	1986	1987	1988	1989	1990	1991	1992	1993	1994	1995	Totals
1. Geological reconnaissance in 7 states	180	190	140									510
2. Detailed geological surveys in 47 geothermal zones	370	370	410	550	550	340	340	210				3,140
3. Geochemical studies in 47 geothermal zones and in 7 states	190	190	190	230	230	140	140	80				1,310
4. Geophysical studies in 28 geothermal zones	830	1,220	1,220	1,220	1,220	1,220	1,220	1,220	1,220	830		10,920
		Geothermal Zones										
5. Drilling of 75 exploratory wells in 25 geothermal zones		1	2	2	2	3	3	3	3	3	3	
		Wells										
		3	6	6	6	9	9	9	9	9	9	
		21,750	43,510	43,510	43,510	65,260	65,260	65,260	65,260	65,260	65,260	543,870
Total Exploration Costs												559,750

Studies to date have identified 515 thermal zones in 28 states, of which field surveys done by the Federal Commission of Electricity have verified those present in 18 states. Verification of seven more is programmed for the 1985 to 1987 period (Fig. 1).

Based on the results of geostatistical exploration to select possible thermal spots and on geological reconnaissance and geochemical sampling, there could be some 30 zones in the remaining seven states, which together with the 515 already defined, add up to 545. The cost of these studies will be about $508,000 (Table 2).

Forty-five geothermal prospects have been selected for detailed geological investigation after the preliminary evaluation of the geothermal prospects in the 515 defined zones. Assuming the same proportion for the 30 zones to be located, the two additional detailed studies would make a total 47. Geological surveying will be completed in 1992 at a cost of slightly more than $3 million at 1982 prices (Table 2).

Previous experience shows that approximately 60 percent of the areas covered by detailed geological and geochemical work turns out to be suitable for geophysical investigation. On this basis, resistivity, gravity, magnetic, seismic, and thermometric studies were programmed in 28 geothermal zones, to begin in 1985 in two of the identified zones for completion by 1994, at an estimated cost of $11.5 million. Table 2 shows the program and distribution of costs.

Geochemical sampling of water and gases in the 1985 to 1992 period will complement the detailed study of the 47 geothermal prospective zones, at a cost of $1.4 million (Table 2). Various pre-feasibility studies selected 19 geothermal zones to be explored by drilling; 11 of these remained in 1985 for exploratory drilling to average depths of 2,500 m.

Assuming that 50 percent of the 28 geothermal zones slated for geophysical exploration prove of interest for exploratory drilling, the 11 areas now pending added to these would total 25. To define the existence of a geothermal resource, three wells are usually recommended for each zone, which adds up to 75 wells at

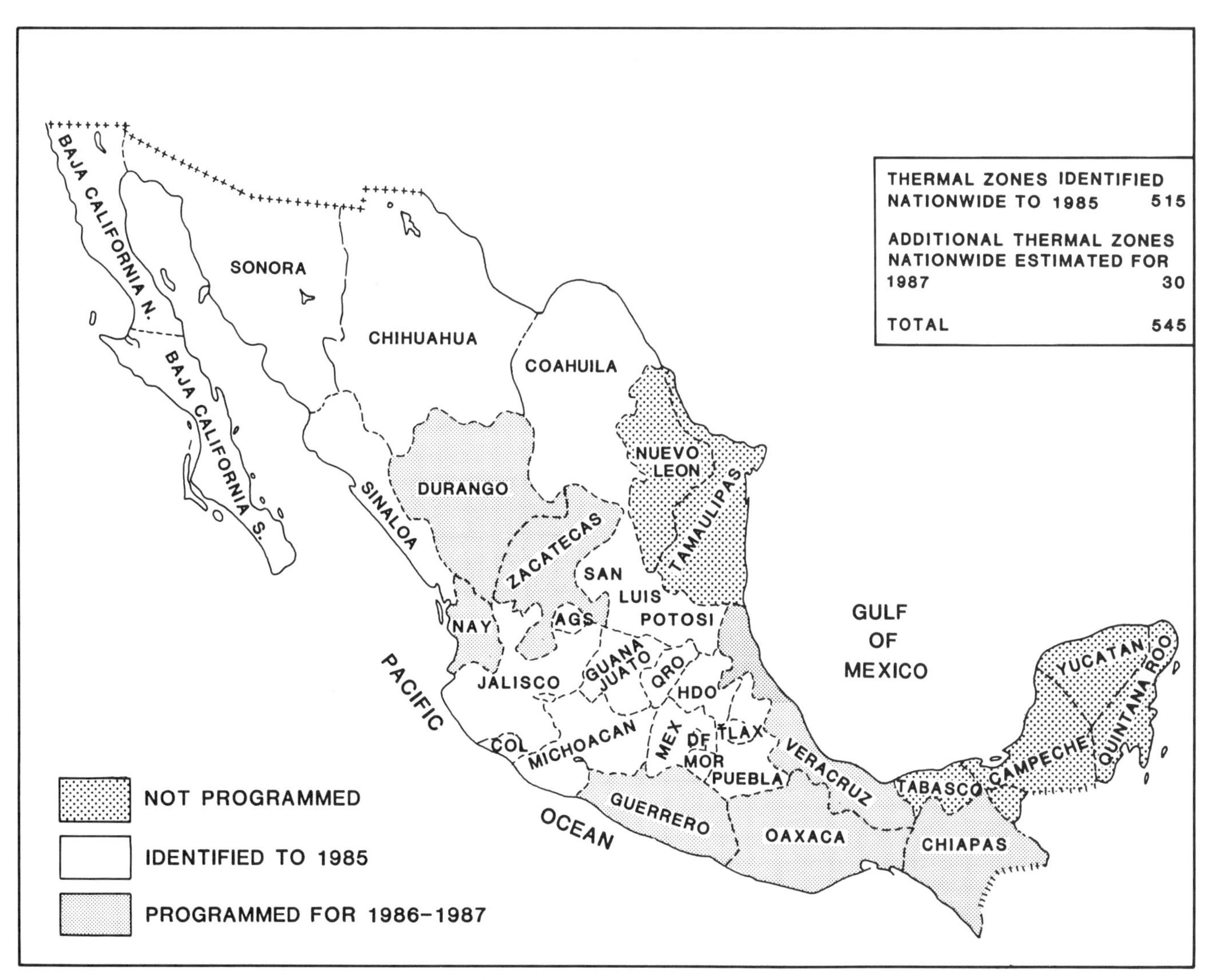

Figure 1. Identification of thermal zones in Mexico.

a cost of $544 million (Table 2). By 1996 this exploratory program will have identified all the prospective geothermal areas in the country; results to date suggest a final number of about 15.

Subsequent exploration and development drilling involves activities aimed at evaluating each well and the fields themselves to establish production capacities and reservoir development policies. Beginning with the first exploratory well in 1985, the project should continue to completion of the last producing well in the year 2000 or later. Considering that the first fields to be developed by drilling could be defined by 1986, 360 wells in the 15 geothermal fields would be needed by the year 2000 to develop a potential capacity of 1,100 MW, distributed over fields averaging 55 MW to 110 MW.

The present evaluation assumed 18 wells for a 55-MW plant built on a geothermal field in volcanic rock, including producing, failed-injection, and replacement wells. The drilling program involves a gradual increase in the number of wells drilled after 1986 and, consequently, in drilling equipment. Based on a conservatively estimated 55- to 110-MW average capacity, tentative plans are to begin design and construction of plants of this size in 1989 and operations in 1992, to be concluded in the year 2000.

Drilling, both of new fields and of developing and producing areas, normally demands geological and geochemical supervision and, in many cases, complementary and more detailed geophysical studies to direct the exploratory effort and maximize development of the geothermal fields. All of the foregoing implies that by the year 2000, 435 wells will have been drilled; an estimated 1,100 MW potential will have been verified; and approximately $6.2 million will have been invested in studies, wells, and research.

In addition to the exploration program, the completion by 1985 of plants having an installed capacity of 645 MW was planned, with another 695 MW to be incorporated into the POISE plan (see Chapter 3) by 1993. The master plan for the year 2000 would thus achieve an installed capacity of:

Built in 1985	645 MW
Proposed in the POISE for 1993	695 MW
Capacity in new fields	1,100 MW
Total installed capacity	2,440 MW

The plants proposed by the POISE are:

Los Azufres	300 MW
Cerro Prieto	220 MW
Los Humeros	125 MW
Mobile plants	50 MW
TOTAL	695 MW

BIBLIOGRAPHY

Di Pippo, R., 1984, Worldwide geothermal power development; An overview and update: Geothermal Resources Council Bulletin: v. 13, no. 1, p. 4–16.

Muffler, L.J.P., 1976, Summary of section 2; Geology, hydrology, and geothermal systems, *in* Proceedings of the 2nd United Nations Symposium on the Development and use of Geothermal Resources, May, 1975: Washington, D.C., U.S. Government Printing Office, p. xlv–li.

Palma, O., and Reyes, P., 1979, Estudio de la Zona Geotermica de Araro, Michoacán [Professional thesis]: Universidad Autonoma de Mexico, p. 4–8.

Unpublished internal reports of the Exploration Department of the Comisión Federal de Electricidad

Razo, A., 1985, Actividades de Exploración Geotérmica en Mexico: Comisión Federal de Electricidad Reporte DEX 2/85, 8 p.

Razo, A., Gonzalez, A., Hiriart, G., Gutierrez, L.C.A., Arellano, F., and Quijano, J. L., 1984, Plan Maestro para el Desarrollo Geotérmico al Ano 2000: Comisión Federal de Electricidad, Reporte DEX 2/84, 11 p.

Printed in U.S.A.

The Geology of North America
Vol. P-3, Economic Geology, Mexico
The Geological Society of America, 1991

Chapter 4

Geothermal resources and provinces in Mexico

A. Razo M., and F. Romero R.
Federal Commission on Electricity, Rio Rodano 14, 06500 Mexico, D.F.

INTRODUCTION

The large number and spectacular nature of thermal shows in the central part of the country led in the mid-1950s to the initial studies for the development of power-generating geothermal resources. By 1959 the first geothermopower plant in North America was installed in Pathé, Hidalgo, with a generating capacity of 600 KWH. Subsequently the perspective of more favorable conditions in other areas led to further investigations in Ixtlán de Los Hervores, Michoacán, and Cerro Prieto, Baja California Norte, where the largest geothermal field was discovered in 1964. Geothermoelectric generation began there in 1973, with two 37,500-MW units, and thereby the geothermal development of the country.

A census of 106 thermal spots taken from 1964 to 1965 in the central part of the country defined Los Azufres (Michoacán), La Primavera (Jalisco), and Los Humeros (Puebla) as the best geothermal prospects after Cerro Prieto. However, although some studies were done in the 1960s and early 1970s, surface exploration of Los Azufres began only in 1975 with the experience gained in Cerro Prieto, and in 1978 at Los Humeros and La Primavera. Results were favorable, and a drilling program was recommended that is presently being carried out; production has proved the presence of three major geothermal fields.

Based on the studies at Los Azufres, in the last ten years several geological, geophysical, and geochemical surveys have been carried out over 30 areas where reconnaissance reported geothermal possibilities (Table 1). The results led to deep exploratory drilling (appr. 2,500 m) in 19 geothermal areas in addition to Cerro Prieto (Baja California Norte; Table 1). Between 1975 and 1985, deep exploratory wells were drilled in ten geothermal areas, including some fields. Four of them show positive results: Cerro Prieto, Los Azufres, Los Humeros, and La Primavera; fluids detected in two others (Riíto and Guadalupe Victoria) have low enthalpy; exploration continues in three (Las Derrumbaderas, Araró, and Tulecheck); and only one (San Marcos) showed negative results (Fig. 1).

The initial explorations of the Cerro Prieto (CP) field established the presence of a reservoir in sedimentary rocks capable of producing enough steam to generate 150 MW in the CP-I power station. Later exploration to the northeast showed the reservoir extending below the average 2,500-m well-depth in the CP-II and CP-III zones and 3,500 m in CP-IV. The reservoir has enough capacity to feed two plants of 220 MW each (CP-II and CP-III), which will begin operating in 1985, and another four of 55 MW each (CP-IV) programmed to start between 1991 and 1994. Added to that of a low-pressure unit in operation since 1980, which in a second evaporation stage uses water separated from steam feeding the 150 MW of Cerro Prieto I, by 1994 the installed capacity of Cerro Prieto will be 840 MW.

Exploratory and development drilling in Los Azufres has now defined a dominantly liquid geothermal reservoir in fractured Miocene-Pliocene volcanic rocks, with an estimated energy capacity to develop over 200 MW in geothermoelectric plants. However, the field is still being explored to determine whether this potential may be increased with certainty. Five wellhead plants of 5 MW capacity each are presently operating in the field, and the construction of a 50-MW plant in the south zone, to begin operating in 1988, has been approved. Two additional plants of 55 MW each are programmed for 1990, and subsequently two more (55 MW each), depending on the reservoir response.

Explorations in Los Humeros since 1978 indicate the presence of a geothermal reservoir within a caldera in a sequence of andesites, ignimbrites, and tuffs overlying limestone and intrusive granite. The results of seven exploratory wells, through preliminary modeling of the field and a numerical simulation, have established the possibility of installing a 55-MW plant to generate power for 20 years. On this basis, three wellhead units, 5 MW each, will be installed by 1988, and construction will begin of two 55-MW plants to start operating in 1993. Exploration will continue simultaneously for more detailed evaluations of the reservoir.

The La Primavera geothermal zone lies within a caldera formed about 120 ka at the intersection of the Sierra Madre Occidental and the Mexican Volcanic Belt. The associated volcanic activity that continued until about 20 ka may have reactivated some structures of the northwest-southeast and northeast-

Razo M., A., and Romero R., F., 1991, Geothermal resources and provinces in Mexico, *in* Salas, G. P., ed., Economic Geology, Mexico: Boulder, Colorado, Geological Society of America, The Geology of North America, v. P-3.

southwest systems and fostered the development of a subsurface hydrothermal system.

Six deep exploratory wells (~2,000 m) in the Cerritos Colorados–La Azufrera area defined the presence of a geothermal reservoir in fractured Miocene-Pliocene andesites correlative with those of the Mexican Volcanic Belt basement.

Although exploration of La Primavera is still in the early stages, results to date, especially in well PR-1, indicate high-enthalpy fluids at depths of 1,800 m. This will be verified by further drilling before undertaking subsequent commercial development. Results being favorable, the installation of two wellhead units of 5 MW each will probably begin in the near future (1988). The geology, geophysics, geochemistry, drilling, evaluation, modeling, and perspectives of Cerro Prieto, Los Azufres, Los Humeros, and La Primavera are discussed in detail below.

Three exploratory wells drilled in the Mexicali Valley at the Riíto and Guadalupe Victoria thermal and geophysical anomalies indicate low-enthalpy fluids over an extensive subsurface zone in the Colorado River delta sediments. Wells ER-1B, GV-2, and Cucapá, the last drilled by PEMEX, show 150° C to 200° C temperatures at a depth of 3,000 m. These are not commercially usable for direct power generation, but will eventually be valuable when binary cycle development becomes economic.

In the Tulecheck geothermal zone the first and only deep exploratory well (several shallow wells were drilled) appears to be at the margin of a possible reservoir east or northeast of the surface thermal area. The geological and geophysical information is currently being revised to decide the location of a future 3,000-m-deep exploratory well.

Surface exploration defined three geophysical anomalies over an area of 20 km^2 in the Araró geothermal zone. The first well could not be completed in the most promising anomaly due to water and steam outflows from the casing at 33 m depth. The results of a well drilled in the second anomaly indicated a low

TABLE 1. GEOTHERMAL ZONES WHERE PREFEASIBILITY GEOLOGICAL, GEOPHYSICAL, AND GEOCHEMICAL STUDIES WERE CONDUCTED BETWEEN 1975 AND 1984

1. Los Azufres (Michoacán)*
2. Los Humeros (Puebla)*
3. La Primavera (Jalisco)*
4. Araró (Michoacán)*
5. San Bartolomé de Los Baños (Guanajuato)*
6. San Marcos (Jalisco)*
7. San Agustín del Maíz-San Juan Tararameo (Michoacán)*
8. Ixtlán de Los Hervores (Michoacán)*
9. Pathé (Hidalgo)*
10. Las Derrumbadas (Puebla)*
11. Las Tres Vírgenes (Baja California Sur)*
12. Tulecheck (Baja California Norte)*
13. Riíto-Ejido Zacatecas (Baja California Norte)*
14. Aeropuerto (Baja California Norte)
15. Las Planillas (Jalisco)*
16. Laguna Salada (Baja California Norte)*
17. Guadalupe Victoria (Baja California Norte)*
18. Nayarit (Baja California Norte)*
19. Pescadores (Baja California Norte)*
20. Culiacán (Sinola)
21. Agua Caliente de Guamuchil (Sinaloa)
22. La Ciénaga (Sinaloa)
23. Villa Corona (Jalisco)
24. Hervores de La Vega (Jalisco)
25. Agua Caliente-Buenavista (Jalisco)
26. San Agustín del Pulque (Michoacán)
27. Huandacareo (Michoacán)
28. San Sebastián (Michoacán)
29. Puroagüita (Guanajuato)
30. Comanjilla (Guanajuato)

*Geothermal areas where drilling was recommended.

TABLE 2. THERMAL SPOTS BY STATES IN MEXICO

State	Spring	Well	Fumarole	Mud Volcano	Total Spots	Geothermal Zones
Aguascalientes	-	47	-	-	47	7
Baja California Norte	18	1	1	1	21	20
Baja California Sur	7	3	4	-	14	9
Coahuila	10	2	-	-	12	12
Colima	4	-	-	-	4	4
Chiapas	2	-	2	-	4	6
Chihuahua	30	12	-	-	42	37
Durango	17	-	-	-	17	17
Guanajuato	38	147	-	-	185	31
Guerrero	6	-	-	-	6	6
Hidalgo	11	28	-	-	39	13
Jalisco	217	88	13	6	324	104
México	5	1	-	-	6	6
Michoacán	62	13	7	2	84	48
Morelos	7	13	-	-	20	7
Nayarit	23	-	-	-	23	23
Nuevo León	5	-	-	-	5	5
Oaxaca	5	-	-	-	5	5
Puebla	11	-	8	-	19	10
Querétaro	6	167	-	-	173	9
San Luis Potosí	7	66	-	-	73	19
Sinaloa	38	1	-	-	39	32
Sonora	44	36	-	-	80	53
Tabasco	2	-	-	-	2	2
Tamaulipas	2	-	-	-	2	2
Tlaxcala	1	-	-	-	1	1
Veracruz	9	-	-	-	9	9
Zacatecas	14	13	-	-	27	18
Totals	601	638	35	9	1,283	515

potential for the extension of geothermal possibilities at Araró, but did confirm their existence; the next well will be drilled in the most important surface thermal and subsurface geophysical anomalies, where temperatures of over 200° C have been geochemically estimated.

The first 2,000-m-deep well in the Las Derrumbadas geothermal zone was drilled at the northwest flank of a rhyolite dome to reach a high-temperature reservoir in deep limestones. However, it was abandoned at 804 m due to technical difficulties caused by fracturing of the volcanic rock. Future exploration is pending in either of the two anomalies that define areas of geothermal interest.

Considering that by 1985 the Cerro Prieto and Los Azufres geothermal fields should have geothermopower stations with capacity to generate 645 MW, and that by 1993 other stations will be built in these fields and at Los Humeros with a 695-MW capacity, the installed geothermal resources of Mexico for that year will be 1,440 MW. This will rise to 1,100 MW by the year 2000 if the previously described program is put into effect; the country's geothermal resource would then be 2,440 MW.

CENSUS OF THERMAL SPOTS

To define the country's geothermal prospects and their optimum development, the Federal Commission on Electricity has done several counts of thermal spots. One of the first tallies, in 1964, counted 106 thermal spots throughout the country. Later that year, and in 1965, a second tally counted a total of 116 in the central part of the country only. By 1977, 310 thermal spots had been identified. Romero and Razo (1981) gathered 338 thermal spots in their nationwide compilation; and the number rose to 434 in a subsequent count by Romero (1983b).

By 1985 integrated information from geological reconnaissance over 18 states and a bibliographic compilation of thermal spots in the remaining 13 identified 1,283 thermal shows in their various manifestations as hot springs, fumaroles, sulfur springs, mud volcanoes, water wells, or combinations. They have been grouped into 515 geothermal zones, based on assigning several to the same thermal aquifer and/or having a common origin (Fig. 1).

Table 2 shows thermal spots per state, type of surface ex-

Figure 1. Location of thermal zones in Mexico (from data compiled by H. A. Lopez and L.V.S. Rocha in 1977).

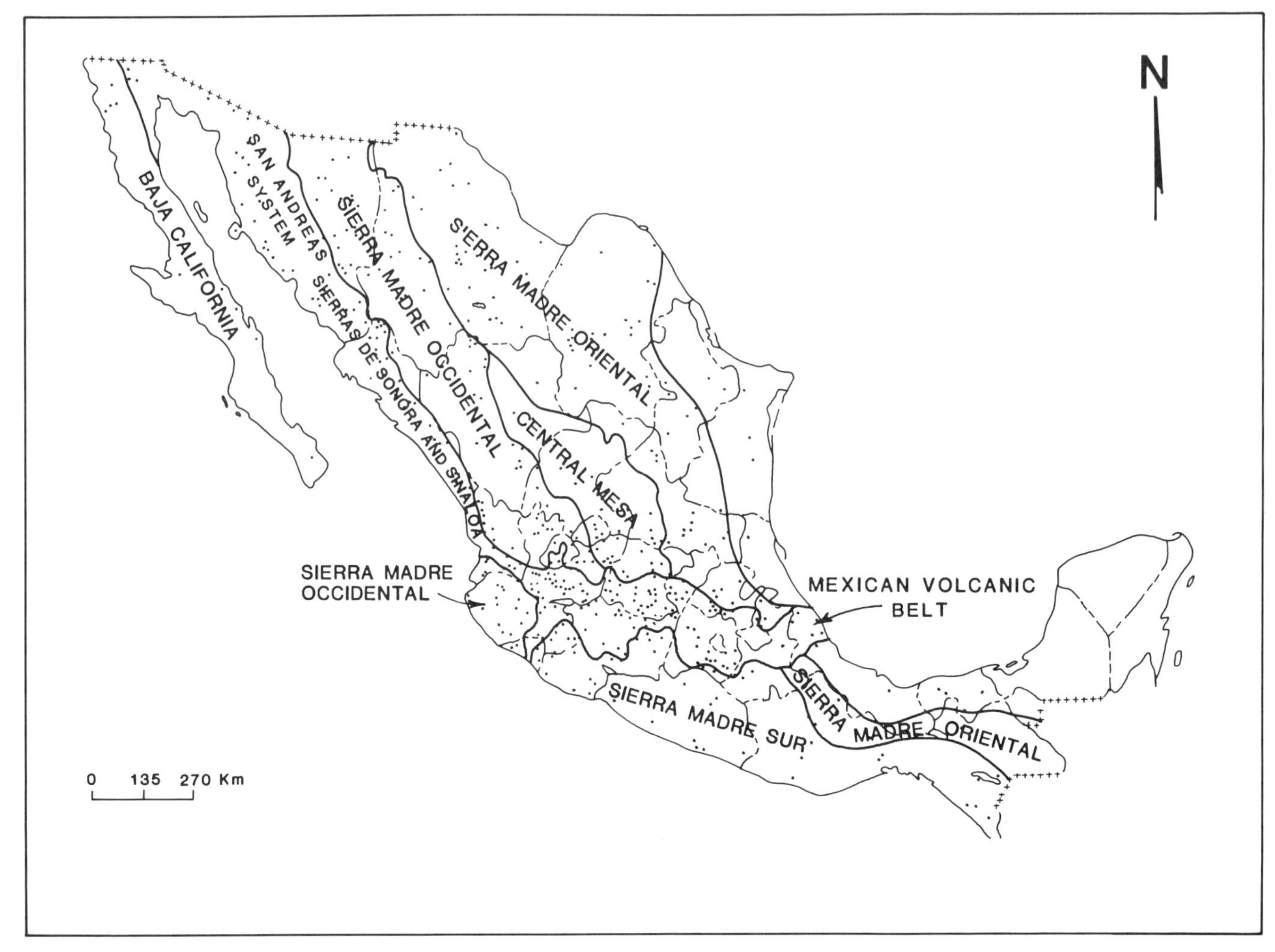

Figure 2. Geothermal provinces in Mexico (from data compiled by Veytia, 1964, 1965).

pression, and number of geothermal zones. The latter should increase as the geothermal exploration program advances to the year 2000, and will probably come to about 545. Experience proves that between these zones and the 19 drilled to 1984, 15 new geothermal fields of 55 to 110 MW capacity will probably be defined, to achieve an estimated 1,100 MW by the year 2000.

GEOTHERMAL PROVINCES

The geothermal province concept comprises the various tectonic, magmatic, and volcanic aspects of a region in which thermal activity displays similar characteristics. In the geologic history of Mexico, these events are more closely related to geothermal activity, as indicated by geological surveys in 18 states and published geological information on thermal zones in the rest of the country.

The correlation of thermal shows demonstrates that their surface manifestations often occur within defined regional geologic patterns and that exploration would profit by grouping them into geothermal provinces. Consequently, the following seven geothermal provinces have been tentatively identified on the basis of continental geologic evolution and its physiographic expression (Fig. 2): (1) Mexican Volcanic Belt, (2) Sierra Madre Occidental, (3) San Andreas System and Sonora-Sinaloa Sierras, (4) Baja California, (5) Sierra Madre Oriental, (6) Central Mesa, and (7) Sierra Madre del Sur.

Mexican Volcanic Belt geothermal province

The consolidation of geological surveys carried out for geothermal exploration purposes in the Mexican Volcanic Belt has led to a revised interpretation of some features of this geologic province. Regional mapping defines it in general as a volcanic belt spanning the country from coast to coast between 19°N and 21°N. Strictly speaking, it is an irregular belt of Pliocene-Quaternary volcanism in which rocks of diverse compositions and structural behaviors are distinguishable from those of adjacent provinces (Fig. 3).

Volcanism in the Mexican Volcanic Belt seems to have begun in the Miocene or Miocene-Pliocene; a second important

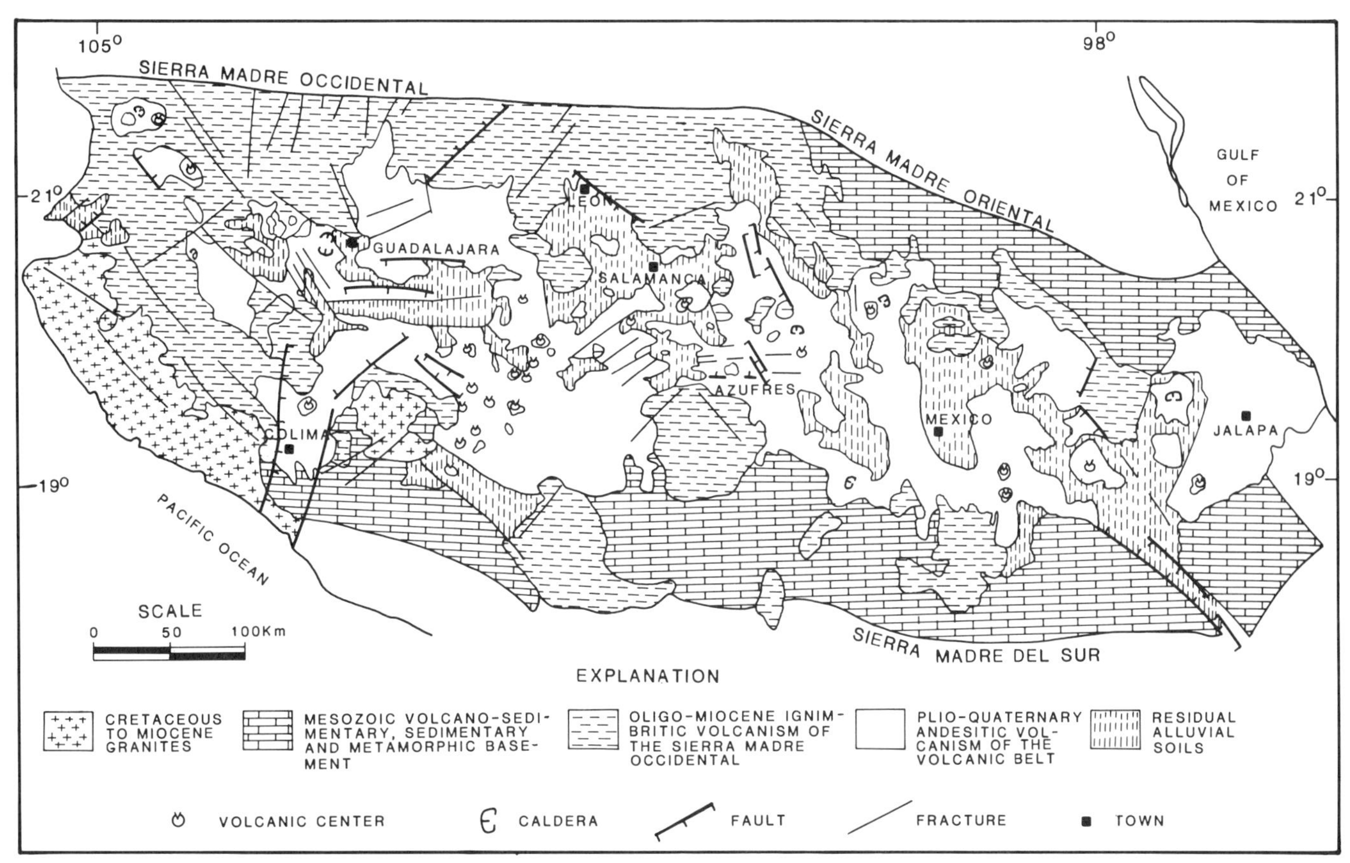

Figure 3. Lithologic units of the Mexican Volcanic Belt (from map compiled by S. Venegas, J. J. Herrera, and R. Maciel).

phase occurred in the last 4 m.y., and thus is essentially Pliocene-Quaternary. On the basis of plate tectonics, most authors hold that extension tectonism at the end of the Oligocene or beginning of the Miocene terminated the predominantly ignimbritic volcanism of the Sierra Madre Occidental and gave rise to the Mexican Volcanic Belt basic andesite series. This volcanism, probably associated with major structures, formed tectonic basins that were subsequently reactivated to produce the Pliocene-Quaternary volcanism and the northwest-southeast, north-south, and northeast-southwest structural systems that define the province.

Thermal shows in the Mexican Volcanic Belt are normally associated with systems of fractures that characterize various regions, and, where the structures are not evident, with volcanic centers related to regional patterns now masked by the volcanic products. From the west tip of Lake Chapala westward in the volcanic belt, the main structures controlling the thermal shows are northwest-southeast; south of the San Marcos Lagoon they strike north-south or northeast-southwest. Between Lake Chapala and the Sierra Nevada, in the central part of the belt, the shows predominate on structures trending east-west and northeast-southwest; in the eastern portion, they are on northwest-southeast and north-south structures (Fig. 3).

Most thermal spots in the Mexican Volcanic Belt occur in fractured andesitic volcanic rocks and appear, by their surface expression, to be more closely related to structures than to the volcanic centers. In several cases, however, they are found between acid volcanic structures or calderas. This suggests the presence of magmatic chambers near the surface as heat sources. Even in these cases, thermal activity becomes apparent in the more recent faults and fractures. Geothermal prospects are obviously greater under these geologic conditions, as experience has proved in Los Azufres (Michoacán), Los Humeros (Puebla), and La Primavera (Jalisco).

Sierra Madre Occidental geothermal province

Considering that many thermal spots in this province occur within Oligocene-Miocene volcanic rocks consisting mainly of the andesite-rhyolite-ignimbrite sequence characterizing the Sierra Madre Occidental, its limits coincide roughly with those of this physiographic province (Fig. 2). To avoid establishing subprovinces at this early stage, which future information will better define, the volcanic series occurring south of the state of Jalisco (which together with the Mesozoic intrusives forms part of the western edge of the Mexican Volcanic Belt) has been included within the Sierra Madre Occidental geothermal province.

In view of the magnitude and persistence of NNW-SSE structures in this province, it is not surprising that most of the thermal spots appear along the planes of weakness, mainly between rhyolites and ignimbrites, and even in the mid-Tertiary andesite flows and the Mesozoic and lower Tertiary intrusives of north Sonora. In the vicinity of the Mexican Volcanic Belt south of the Sierra Madre Occidental the thermal spots are controlled by north-south or nearly north-south structures.

In southwestern Jalisco the Sierra Madre Occidental structural systems change direction, and thermal activity follows mainly northeast-southwest trends in both volcanic and intrusive rocks.

San Andreas System and Sonora-Sinaloa Sierras geothermal province

The main structural systems—which control even the physiography—of the continental shelf west of the Sierra Madre Occidental, the Mexicali valley, the Sonora desert and the low-lying Sonora and Sinaloa Sierras (Fig. 2) trend NNW-SSE and NW-SE, being effects of the interaction between the American and Pacific Plates.

In Sinaloa and south Sonora, composing the southern portion of the province, thermal shows are found in Tertiary basalts, andesites, rhyolites, ignimbrites, and conglomerates, as well as over a widespread outcrop area of Cretaceous intrusives, in contrast with those of the Sierra Madre Occidental. They occur equally in rocks fractured by the older NNW-SSE, or the more recent NW-SE and N-S systems.

In the northern portion of the province, which includes the Mexicali valley and the Sonora desert, they appear within extrusive and intrusive igneous rocks, as well as in Recent alluvial deposits. The region is characterized by mainly NW-SE structures related to the transform faults that opened up the Gulf of California. These are represented in the northwest (Mexicali valley and part of the Sonora desert) by the San Andreas system, where thermal zones are conspicuous.

Baja California geothermal province

The Baja California Peninsula consists essentially of two geologic units: the northern region has large intrusive bodies and metamorphic rocks; the south is mainly Miocene andesites and intrusive, sedimentary and metamorphic rocks of various ages. The peninsula attained its present configuration in a short time (4 to 6 m.y.). Not surprisingly, its tectonic elements follow a mainly northwest-southeast trend, paralleling the transform faults that opened the Gulf of California; some of them are still active or even, as in Las Tres Vírgenes, have caused recent volcanism.

Thermal activity in this province takes place within Mesozoic intrusive and metamorphic rocks, Miocene volcanics, and even Quaternary volcanic and sedimentary rocks. In most cases it is related to the northwest-southeast structures, as elements of crustal weakness along which magmatic fluids rise or surface fluids descend. The surface fluids heat up following the anomalous gradient and subsequently crop out as thermal spots.

Sierra Madre Oriental geothermal province

This province includes geothermal zones related in some manner to the Mexican geosyncline. Many of these are found within Mesozoic sedimentary rocks, and others in volcanic and intrusive areas that locally cover or interrupt the geosyncline. Within the sedimentary rocks, thermal spots predominate in or near Cretaceous folded limestones and within the Tertiary andesites, ignimbrites, and rhyolites.

The usually low-temperature (30 to 50°C) hydrothermal flow in the sedimentary rocks appears at structures formed by the Laramide orogeny. Within the volcanic rocks they are found at variably striking tensional faults or fractures that fostered the ascent of magma in several regions, or at structures resulting from transcurrent movement, as in Chiapas.

Central Mesa geothermal province

Located between the Sierra Madre Occidental, the Sierra Madre Oriental, and the Mexican Volcanic Belt provinces, the Central Mesa includes metamorphic, sedimentary marine, and continental, volcanic, and intrusive rocks ranging in age from Triassic to Quaternary. However, thermal activity is typically that of volcanic areas, particularly where mid- and upper Tertiary acid rocks (rhyolites) are present.

Although the structures trending mainly NW-SE evidence Laramide tectonics, the thermal activity is linked to the NNW-SSE and N-S systems affecting mid-Tertiary rocks, or to the more recent (Pliocene-Quaternary) WNW-ESE and NE-SW trends. Low heating (30 to 50° C) is frequent in valley-filling alluvial and volcanic deposits such as at Salamanca, Aguas Calientes, and Villa de Reyes. Some spots seem unrelated to either tectonism or volcanism; however, heat is occasionally encountered in wells, especially when drilling through acid rock (rhyolite) underlying the sediments.

Sierra Madre del Sur geothermal province

Located in the southern part of the country, this province includes rocks ranging in age from Paleozoic to Quaternary, of metamorphic, marine, and continental sedimentary and extrusive and intrusive igneous nature, and equally diverse structural behavior. However, some systems predominate over certain areas. Thus, over a widespread region of Michoacán and Guerrero south of the Mexican Volcanic Belt structural trends are mainly northeast-southwest and north-south, the latter being more frequent in the Mesozoic marine sedimentary rocks. A few northwest-southeast alignments are also present.

Numerous intrusive and metamorphic occurrences are encountered in the states of Chiapas and Oaxaca in the Sierra Madre del Sur province. The volcanics, relatively restricted to central and southeast Oaxaca and southeast Chiapas, are late Cenozoic in age and in some cases Quaternary, as in El Tacaná. In those areas fractures trend NNW-SSE, N-S, NW-SE and—

predominantly—NE-SW. Thermal activity shows up equally within sedimentary and igneous intrusive and extrusive rocks, and even in alluvium, and no preferred structural control is clearly defined.

REFERENCES

Atwater, T., 1970, Implications of plate tectonics for the Cenozoic tectonic evolution of western North America: Geological Society of America Bulletin, v. 18, p. 3513–3536.

Salas, G. P., 1980, Carta y Provincias Metalogenéticas de la República Mexicana: Consejo de Recursos Minerales, Publication 21 E, scale 1:3,500,000.

Unpublished internal reports of the Department of Exploration, Comisión Federal de Electricidad:

Canul, D. R., and Rocha, L.V.S., 1980, Reconocimiento y evaluación de los focos termales de la parte centro y sureste del estado de Sinaloa: Report 6/80.

——, 1981, Estudio geológico del proyecto geotérmico El Chichonal, Estado de Chiapas: Report 32/81.

——, 1983, Estudio geológico con fines geotérmico de la región norte del Estado de Michoacán: Report 7/83.

Castillo, H. D., 1983, Reconocimiento y evaluación de focos termales en los estados de Jalisco y Colima: Report 20/83.

Chacón, F. M., 1982, Muestreo geoquímico preliminar de zonas termales del Estado de Durango: Report 25/83.

De la Cruz, M. V., and Hernández, Z. R., 1985, Reconocimiento geológico del proyecto geotérmico El Tacaná, Estado de Chiapas: Report 41/85.

Diaz, O. A., 1982, Reconocimiento y evaluación preliminar de los focos termales en el Estado de México: Report 33/82.

——, 1983, Evaluación geotérmica preliminar del Estado de Nayarít: Report 22/83.

Fonseca, P. H., and Ledesma, A., 1983, Posibilidades geotérmicas de Baja California y norte de Baja California Sur: Report 3/84.

Gallo, P. I., 1983, Reconocimiento y evaluación de las manifestaciones termales en la porción norte del Estado de Jalisco: Report 8/83.

Herrera, F.J.J., and Echeverría, B. R., 1983, Estudio geológico evaluativo de los recursos geotérmicos del Estado de Queretaro: Report 11/83.

——, 1983, Estudio geológico evaluativo de los recursos geotérmicos del Estado de Hidalgo: Report 12/83.

Leal, H. R., and Canual, D. R., 1981, Reconocimiento y evaluación de focos termales en el Estado de Coahuila: Report 39/81.

Leal, H. R., and Martiñón, G. H., 1981, Reconocimiento y evaluación de focos termales del Estado de Aguascalientes: Report 17/81.

Leal, H. R., and Olguín, G. A., 1981, Reconocimiento y evaluación de focos termales del Estado de Morelos: Report 63/81.

Lira, H. H., 1984, Evaluación geotérmica del Estado de Baja California Sur: Report 5/84.

López, H. A., and Rocha, L.V.S., 1979, Reconocimiento y evaluación de focos termales de la parte norte del Estado de Sinaloa: Report 10/79.

Maciel, F. R., and Ruy, A. C., 1981, Localización de los focos termales en el Estado de Chihuahua: Report 71/81.

Martiñón, G. H., 1983, Reconocimiento y evaluación de focos termales del Estado de San Luis Potosí: Report 9/83.

Quijano, L.J.L., and Gallardo, A. M., 1982, Reconocimiento y evaluación geoquímica del Estado de México: Report 45/82.

Razo, M. A., 1985, Actividades de exploración geotérmica en México: Report 2/85.

Romero, R. F., 1983a, Reconocimiento y evaluación de manifestaciones termales en el Estado de Puebla: Report 13/83.

——, 1983b, Zonas geotérmicas en la República Mexicana: Report 19/83.

Romero, R. F., and Razo, M. A., 1981, Recopilación de focos termales en la República Mexicana: Report 58/81.

Tello, H. E., and Chacón, F. M., 1983, Reconocimiento geoquímico preliminar de zonas termales del Estado de Chihuahua: Report 15/83.

Printed in U.S.A.

The Geology of North America
Vol. P-3, Economic Geology, Mexico
The Geological Society of America, 1991

Chapter 5

Main geothermal fields of Mexico; Cerro Prieto geothermal field, Baja California

A. Pelayo, A. Razo M., L.C.A. Gutiérrez N., F. Arellano G., J. M. Espinoza, and J. L. Quijano L.
Federal Commission on Electricity, Rio Rodano 14, 06500 Mexico, D.F.

INTRODUCTION

Exploration of this geothermal field was begun formally, though intermittently, in 1958; the first exploratory well was drilled in 1959, and official production of the first megawatts began in 1973. The commercial development of Cerro Prieto geothermics began with practically no previous experience. This explains the relatively long period between initial exploration, drilling, and power generation as compared with the development of similar fields in other countries and in Mexico itself.

Since 1958, but especially from the mid-sixties to the present, many studies have dealt with the Cerro Prieto geothermal field and its surroundings. The fact that more than 200 papers have been published in Mexico alone precludes even listing them here; foreign publications are at least as numerous, and unpublished reports of the Federal Commission on Electricity easily duplicate that amount as well.

Research on the Cerro Prieto field covers virtually all geoscientific disciplines: regional and local geology, structural and tectonic geology, regional and local geohydrology, subsurface geology, sedimentology, volcanology, fission-track dating, hydrothermal alteration, mineralogy, geological modeling, remote sensing, ground compaction and subsidence, determination of petrophysical properties, etc. Geophysical investigations cover resistivity, geophysical well-logging, gravity, reflection and refraction seismology, ground and aerial magnetometry, spontaneous potential, magnetotelluric analysis, microseismics, and precision gravimetry. Geochemical studies utilize geochemistry of surface shows, analysis of well fluids, geochemistry of gases, geothermometry, use of tracers, determination of radon in geothermal fluids, isotopic chemistry in fluids and minerals, and geochemical modeling. Drilling engineering covers types of pipes, cements and high-temperature drilling muds, well-completion techniques, and high-temperature hydraulics. On evaluation and development engineering, many papers deal with reinjection, temperature and pressure distribution at depth, pressure testing, mathematical simulation, and mathematical modeling. The following is an integrated review of the latest and most complete information on this field.

Geographic location and characteristics

The Cerro Prieto geothermal field lies 30 km southeast of the town of Mexicali, between 115°10′ and 115°15′W and between 32°22′ and 32°26′N on the deltaic plain of the Colorado River. The Cerro Prieto volcano, with a maximum height of 225 m, is 6 km northwest of the field (Fig. 1).

The geothermal field is located in the northeast portion of the Baja California Sierras Physiographic Province, where the Baja California Peninsula joins the Sonora desert area through the Colorado River delta.

Topographic alignments in the Mexicali Valley (La Pinta, El Centinela, Los Cucapá, El Mayor, Juárez, and San Pedro Mártir Sierras) generally trend NNW-SSE. The eastern slopes of the first four are abrupt; the western are gentler. Juárez and San Pedro Mártir show extensive high plateaus with erosion-sculpted flanks. The highest elevation, Cerro La Encantada in the Juárez Sierra, is over 3,000 m.

Regional extreme climates have strongly weathered and eroded the rocks, to the extent that most geologic features in the Mexicali Valley are masked both by sediments derived from the mountains and by Colorado River deposits. The valley drains mainly toward the Gulf of California, southeast of the Mexicali Valley. Drainage is toward the northwest to the Salton Sea only north of Cerro Prieto, and is bounded by the canal of that name.

The region lies within a great plain of Colorado River sediments, representing an extensive delta where the river flows out to the Gulf of California. Precipitation has averaged 81 to 100 mm/yr during the past 20 to 30 years.

REGIONAL GEOLOGY

The following describes the stratigraphic sequence from older to younger rocks. The terms pre- and post-batholithic have been applied to signify geologic events preceding or following the granite emplacement that climaxed in the mid-Cretaceous.

Pre-batholithic rocks. Rocks of this type are exposed northwest, west, and south of the Cerro Prieto geothermal field, in the west and southeast portions of the Sierra de Cucapá, and

Pelayo, A., Razo M., A., Gutiérrez N., L.C.A., F. Arellano G., Espinoza, J. M., and J. L. Quijano L., 1991, Main geothermal fields of Mexico; Carro Prieto geothermal field, Baja California, *in* Salas, G. P., ed., Economic Geology, Mexico: Boulder, Colorado, Geological Society of America, The Geology of North America, v. P-3.

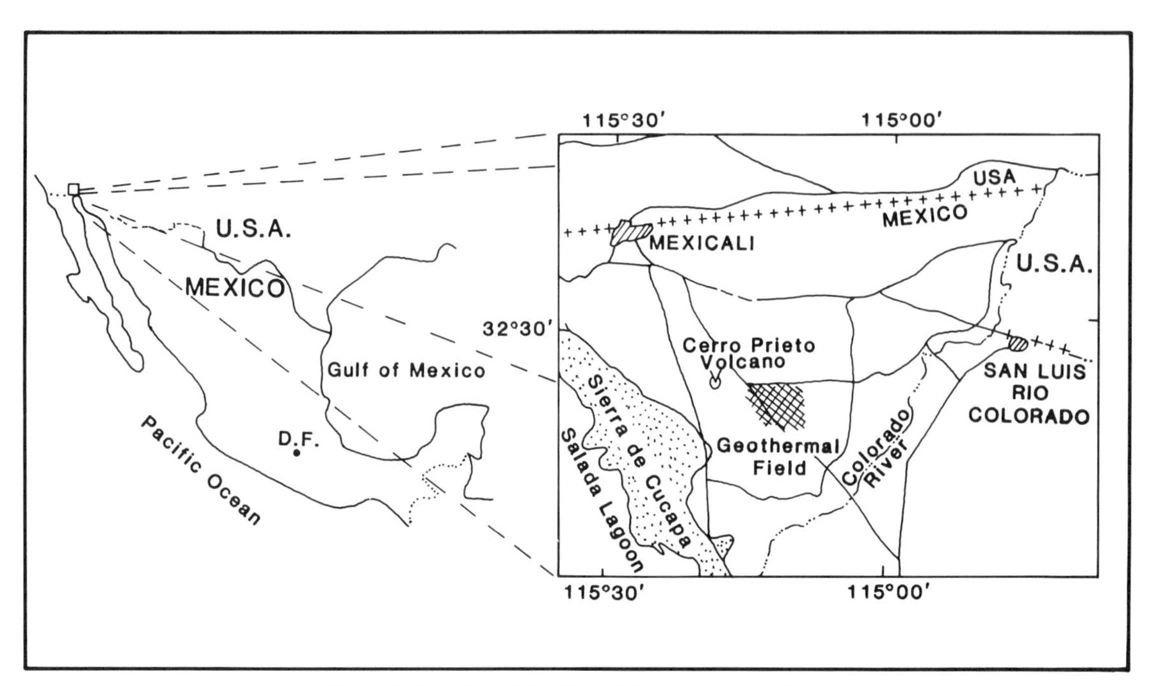

Figure 1. Location of the Cerro Prieto geothermal field.

north and southeast of the Pinta Sierra (Fig. 2). In general these rocks compose a metasedimentary complex of Paleozoic(?) limestones, sandstones, and conglomerates that are now marbles, gneisses, and schists.

Batholithic rocks. These are mainly Late Cretaceous (102 to 98 Ma) granites and granodiorites becoming locally tonalites, which form a large part of the Cucapá and El Mayor Sierras west and south of the geothermal field (Fig. 2).

Post-batholithic rocks. The post-batholithic volcanics are Miocene-Pliocene andesites, rhyolites, and dacites occurring in the Pinta Sierra, outside the area shown on Figure 2. The unit also includes Pleistocene-Holocene rhyolites to andesites, which constitute most of the Cerro Prieto volcano. Radiometric ages range from 100,000 to 50,000 years and up to 10,000 years by paleomagnetic dating.

The post-batholithic unit also includes the Mexicali valley sedimentary sequence of continental sediments derived from the Colorado River, which interfinger with alluvial deposits from the Cucapá Sierra (Fig. 2).

REGIONAL TECTONICS

Various investigations have focused on the drift of the Baja California Peninsula toward the northwest as it separates from the American Plate (Fig. 3). Thus, a subduction zone is known to have existed from the Oligocene to the Pliocene along the Baja California Sur coast, where the Farallon Plate sank beneath the American Plate. The subduction process resulted in continental-margin-type calc-alkaline volcanism (Comondú Formation).

During the Oligocene the East Pacific Rise gradually approached the continent and the subduction zone, coming into contact with the North American trough for the first time not less than 30 m.y. ago. This generated the transform faults, which became active from the contact point northward.

The contact between the East Pacific Rise and North America 10 m.y. ago was located at the southernmost level of the present Baja California Peninsula. The westward-moving American Plate caused the Farallon Plate to disappear beneath the continent. The transform faults generated by the East Pacific Rise colliding with the subduction zone gave rise to two main fault systems: the San Andreas and the Agua Blanca–San Miguel systems. Volcanic products at this stage were calc-alkaline and of extensional basalt type.

The peninsula's initial separation from the American continental mass, averaging 6 cm/yr, began some 4 to 6 m.y. ago with a proto-gulf that existed prior to the late Pliocene.

The Cerro Prieto geothermal field is presently situated within a tectonically active area, practically at the border zone between the Pacific and North American Plates, represented there by the San Andreas system, a vast area of transform faults with 5.6 cm/yr of relative movement between plates (Fig. 3).

GEOLOGY OF THE FIELD

In the vicinity of the Cerro Prieto field, igneous extrusive rocks crop out at the Cerro Prieto Volcano, which has an approximate diameter of 1,000 m and a height of 225 m above valley level. This is a conic or Vesuvian-type volcano with two superimposed volcanic centers: the northeast center contains a small crater 200 m in diameter and 60 m deep, and the southwest is a small dome. Both exhibit rhyodacitic intrusions and flows. The volcanism ranges in age from Pleistocene to Holocene, and at least five eruptive phases have been identified.

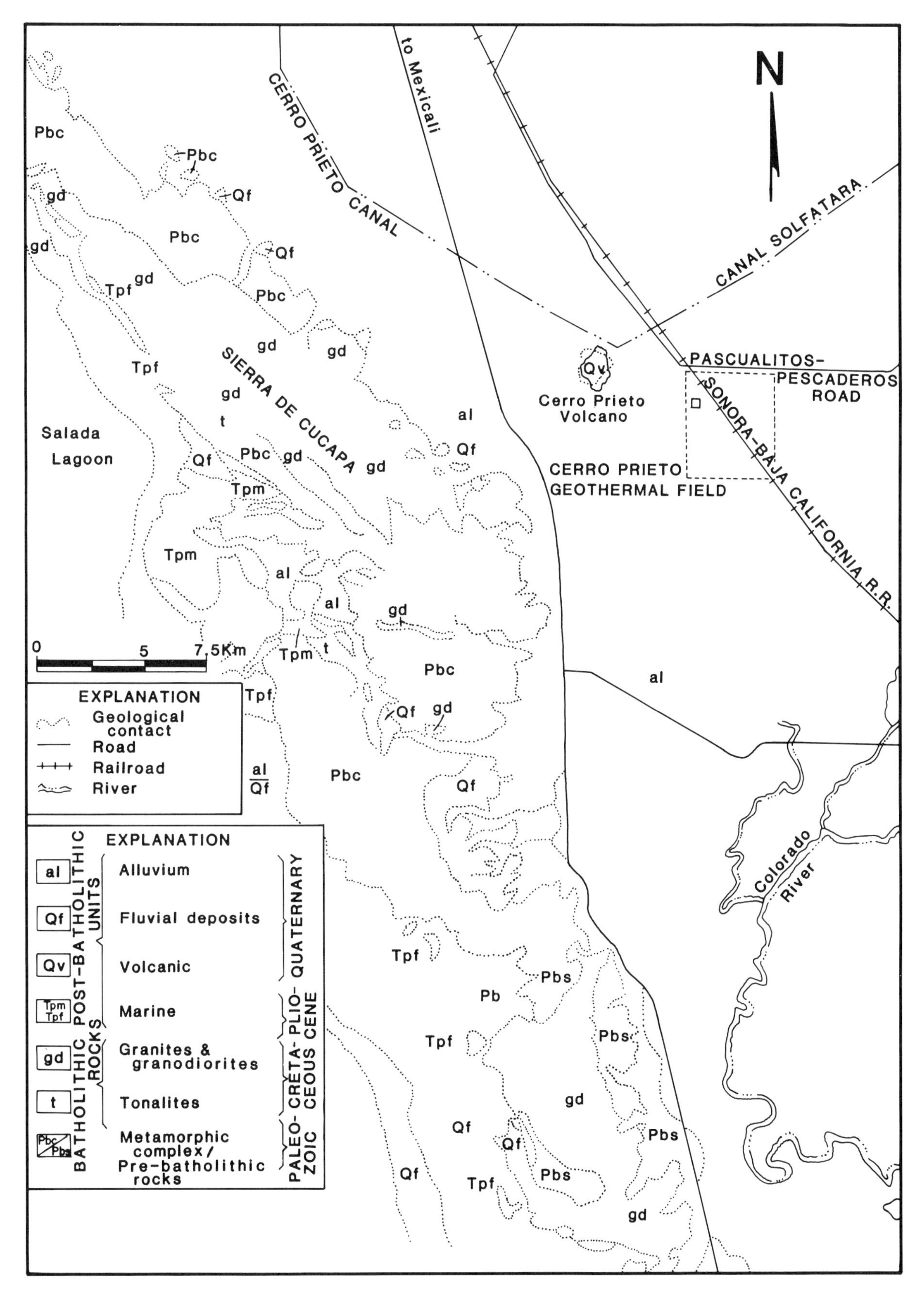

Figure 2. Regional geology of the Cerro Prieto field.

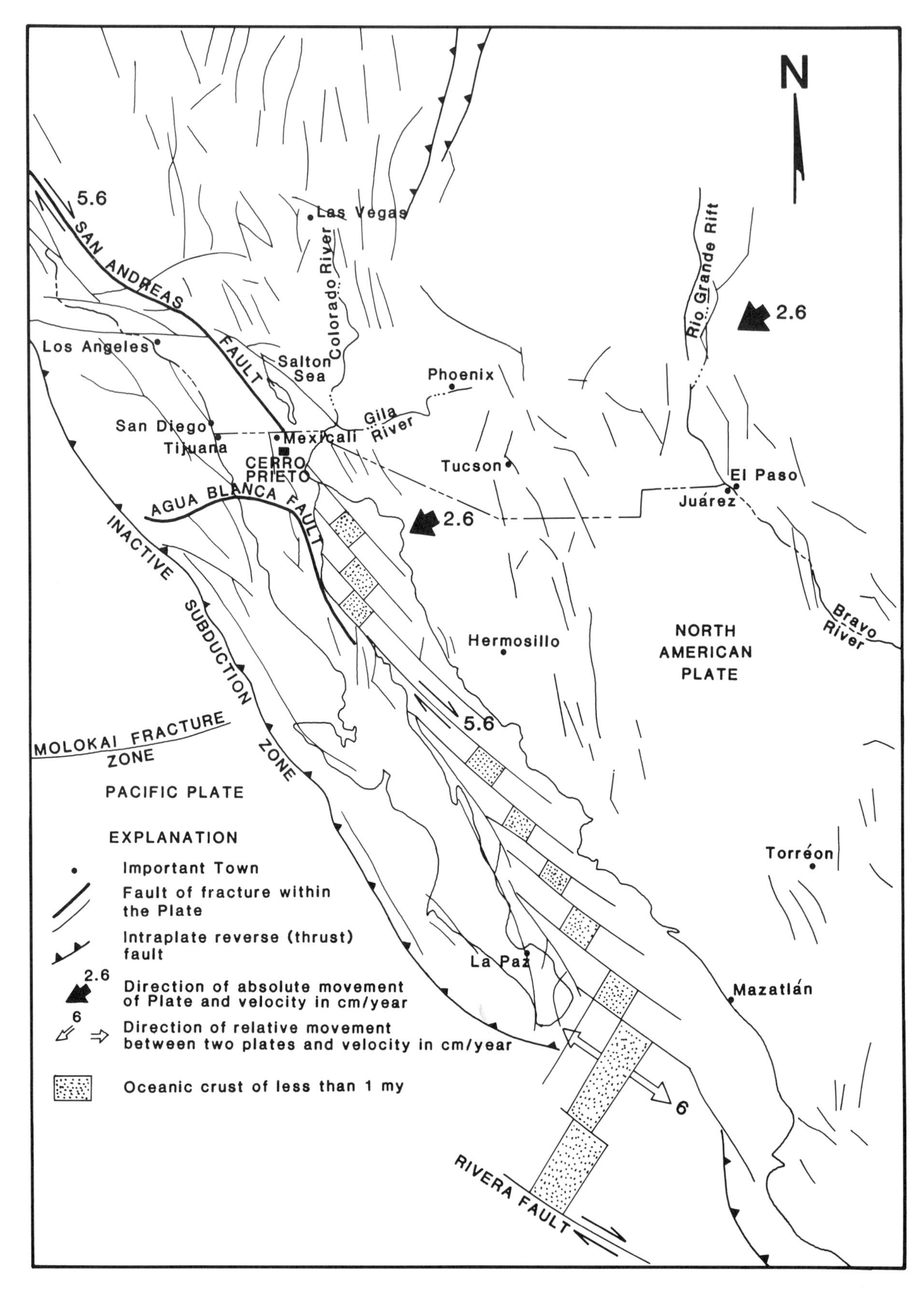

Figure 3. Regional tectonic framework of Cerro Prieto, B. C.

Recent analyses suggest that the Cerro Prieto Volcano originated by partial melting of granitic basement, rather than by differentiation of a deep-seated gabbroic or basaltic magma. The volcanic chamber was relatively small, which makes it an improbable heat source for the present geothermal system. Both the system itself and the local melting of granitic basement that produced the volcano's magmatic chamber seem rather to be effects of heat transferred through the crust by very deep-seated gabbroic intrusions.

Rare local deltaic outcrops 2 km northwest of the Cerro Prieto Volcano form promontories, ranging 2 to 15 m in height, of well-sorted, medium-grained, gray to buff, black-weathering, cross-bedded sandstones. Alternating brown and red clays and well-rounded sandstones crop out in the immediate vicinity of the Cerro Prieto Volcano.

Alluvial deposits are widespread over the region, forming the present Mexicali and Imperial (in the United States) valleys. They occur from the eastern part of the Cerro Centinela and the Cucapá–El Mayor Sierras eastward to the Altar Desert, north and northeast to the Imperial Valley and East Mesa, and south to the Gulf of California. West of the Cerro Prieto geothermal field the alluvial deposits interfinger with the Cucapá Sierra alluvial fans.

Subsurface geology

Four lithologic units have been defined in the Cerro Prieto geothermal field on the basis of well-ditch sample analyses and interpretations of the diverse geophysical surveys (Fig. 4).

Lithologic Unit D. This is the previously mentioned granitic basement emplaced in the Late Cretaceous, which crops out west of the Cucapá and El Mayor Sierras in the vicinity of the geothermal field, forming part of the Californian Batholith. Unit D gradually deepens toward the east in the subsurface of the geothermal field to over 4,000 m at Cerro Prieto III and Cerro Prieto IV. The only wells within the field that reach the basement are in the west and southwest (M-3, M-262, and M-96; for well locations, see Fig. 5).

Lithologic Unit C. Unit C is a lenticular deposit of continental-type consolidated sediments, tentatively assigned to the Tertiary (undifferentiated). The sediments were laid down in a deltaic lagoonal or estuarine-type environment and consist of alternating sandstones, siltstones, shales, slates, and argillites. The partially metamorphosed shales and siltstones are light to dark gray and occasionally black; the sandstones are mainly light gray, fine-grained, medium to well-sorted graywacke, arkose, and occasionally quartzites. Regional tectonism has produced the observed fault planes.

Unit C owes its present characteristics to geochemical and hydrothermal alteration caused by the geothermal system, together with the diagenetic circumstances of successive deposition in basins formed as a result of the prevailing tectonism. Sediment thicknesses in these buried basins are estimated in the 3,500- to 5,500-m range.

Unit C contains, among its permeable sandstone horizons, the high-temperature aquifers of the geothermal reservoir. Its thickness is estimated at 2,500 to 3,000 m, and it unconformably overlies the Unit D granitic basement.

Lithologic Unit B. Overlying Unit C is a mainly gray to dark brown mudstone (compact shale) horizon ranging in thickness from 0.0 m (at well M-189) to 800 m (well M-262). The variations in color may be due to chemical action within the deltaic depositional paleoenvironment, which is essentially the same as that of Unit C. On the basis of the virtual impermeability of the mudstones, some authors see this unit as a sealing layer covering the Unit C geothermal reservoir. However, it also contains very permeable interbedded sandstones and unconsolidated sands. The sudden subsurface thickness variations and discontinuities that are typical of high-energy environments would prevent this unit from behaving in some cases as a sealing layer, or at least as the only seal for the entire geothermal field.

Lithologic Unit A. This unit consists mainly of Quaternary poorly or nonconsolidated sediments: alternating and recurrent clays, silts, sands, and gravels. Fresh-water gastropod shells (*Amnicola*) have been found at depths of 500 and 800 m (well M-5). Semiconsolidated brown shales occur regularly in the lower portion of the unit, grading into the underlying mudstones (Unit B). As in Units B and C the depositional environment was deltaic: lagoonal or estuarine. In addition, Unit A contains andesitic and rhyodacitic lavas, which crop out at the Cerro Prieto Volcano, and sporadic basic (diabase) dikes that cut Unit C rocks and are therefore Quaternary.

ERAS	PERIOD	EPOCH	LITHOLOGY	THICKNESS	DESCRIPTION	LITHOLOGIC UNIT
CENOZOIC	QUATERNARY	PLEISTOCENE		500 to 2300m	ANDESITES, CLAY, SANDS & SCARCE GRAVELS, DIABASE DIKES, RHYODACITIC FLOWS	A
	TERTIARY			0.0 to 800m	BROWN MUDSTONES WITH INTERBEDDED BUFF SANDS & SANDSTONES	B
				~100m	BROWN SHALES & SILTSTONES, INTERBEDDED WITH BUFF SANDSTONES	C
				2500 to 3000m	GRAY TO BLACK SHALES & SILTSTONES WITH ALTERNATING WHITE & GRAYISH-WHITE SANDSTONES	
MESOZOIC	UPPER	CRETACEOUS		?	BIOTITE GRANITE	D

Figure 4. Generalized stratigraphic column of the Cerro Prieto, B. C., geothermal field (modified from De la Peña and others, 1979).

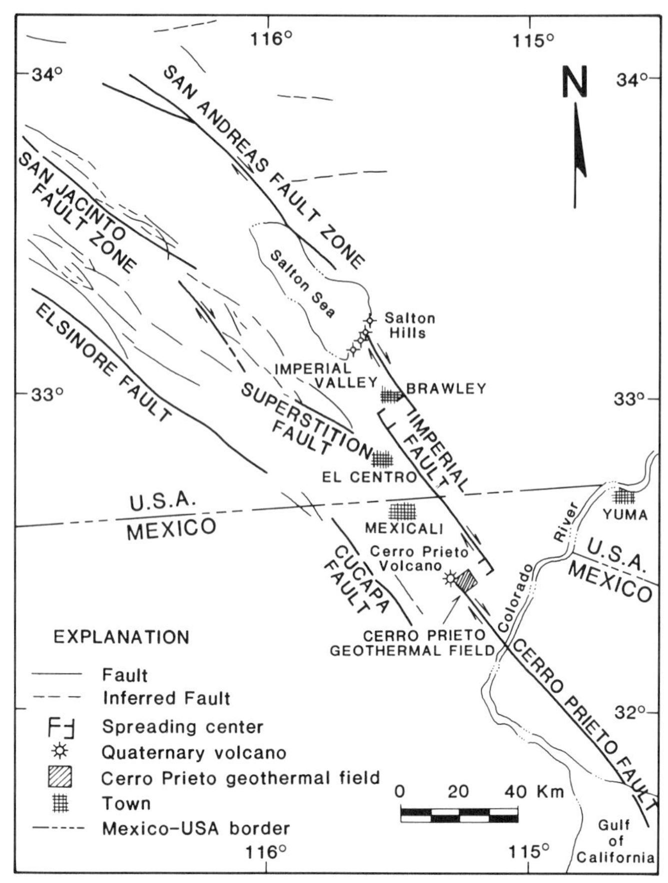

Figure 5. Regional structural elements in the Cerro Prieto, B. C., geothermal field (after Palmer and others, 1975).

Structural geology

Movement of the Pacific and North American Plates has generated stresses in the lithosphere and the consequent displacement of large blocks at the border zone where the Gulf of California is located (Fig. 3).

The northern area (Salton Sea) shows a series of small NE-SW faults. Each series, approximately 3 km long, extends from the Banning/Mission Creek strike-slip fault (part of the San Andreas system) to the northernmost end of the Brawley/Imperial strike-slip fault (Fig. 5; Palmer and others, 1975). The Cerro Prieto geothermal field lies within the San Andreas tectonic system of several generally NW-SE–trending transform faults, whence other systems, such as Cerro Prieto, branch out, also trending generally northwest-southeast. This system is parallel to the large-scale faulting such as Imperial, San Jacinto, Elsinore, etc. (Fig. 5; Palmer and others, 1975).

Geophysical surveying in the Mexicali Valley has determined the right-lateral strike-slip movement of the Cerro Prieto and Imperial transform faults that produced the generally NE-SW–trending Volcano fault system, almost perpendicular to the main NW-SE faults.

Three major faults of the Volcano fracture system—Delta, Pátzcuaro, and Hidalgo—have been identified in the subsurface, as well as other minor faults such as the Morelia fault (Fig. 6). The combined systems have apparently caused a staggered subsurface topography of rises and troughs.

The hydrothermal system

As a result of the interaction between the rocks and mineral-rich geothermal fluids, the subsurface hydrothermal system at Cerro Prieto has caused rock zoning at depth. Unit B and mainly Unit C sandstones are the most affected, since the virtually impermeable nature of the shales prevents them from interacting with the hydrothermal fluids. The appearance in the rocks of secondary minerals of lower to higher temperature of formation by action of hydrothermal fluids develops the zoning, which consists of: a montmorillonite-kaolinite zone at temperatures of 150 to 180°C and 230 to 250°C; an epidote or calc-alumina–silicate zone at 230 to 250°C and 350°C; and a biotite zone at more than 325°C. A characteristic hydrothermal mineral association, in addition to those that name it, defines each zone: quartz, calcite, K-feldspar, pyrite, prehnite, actinolite, and waikirite.

Hydrothermal minerals have replaced the original groundmass or cementation (diagenetic) minerals in the sandstones, to the extent that four lower to higher temperature mineral zones are distinguished according to the present cementation material: calcium carbonate cementation at 60 to 200°C; calcium carbonate and silica at 150 to 250°C; silica, scarce carbonate, and epidote (or transition zone) at approximately 250°C; and silica and epidote at 200 to more than 300°C. In practice, geothermal wells have proved to be good steam producers only through the horizon of silica- and epidote-cemented sandstones, a zone that has become an excellent production marker.

Several models have been proposed to explain the source of the heat. The most complete model assumes a funnel- or inverted cone–shaped gabbro or basalt intrusion, 4 km wide at the summit, at a depth of about 5 km. This intrusion presumably took place about 30,000 to 50,000 years ago northwest of Cerro Prieto I within the spreading center, or pull-apart basin assumed to exist between the Imperial and Cerro Prieto faults (Fig. 5). The model proposes a thermal plume rising toward the southwest with an approximate 45° inclination, which reached what is now Cerro Prieto I and discharged upward and toward the southwest, causing fumaroles and hot springs to rise in Volcano Lagoon (Fig. 6).

This model is somewhat deficient, however, because wells drilled near the proposed intrusion show temperatures noticeably lower than those recorded at the geothermal field, in spite of their depth (>4,000 m) and of having penetrated intrusive dikes (wells M-204 and GV-2). A more probable heat source might be high temperature conducted through the granitic basement from a much deeper zone of local magmatism than that proposed by the first model. Such a zone of magmatism, generated at the spreading or pull-apart center as a first phase in the appearance of new ocean floor within a still-continental mass, would transmit its

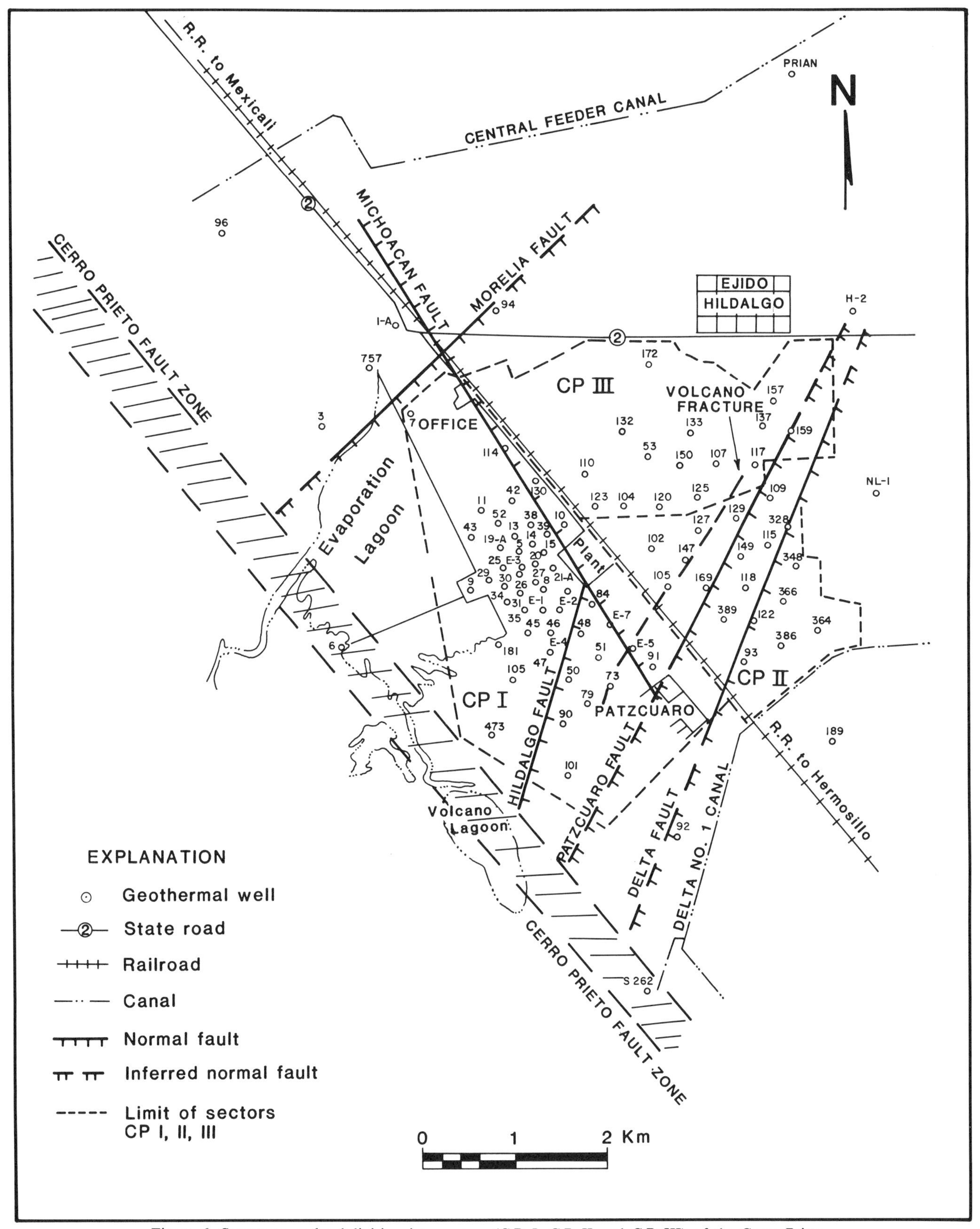

Figure 6. Structures and subdivision into sectors (C.P.-I, C.P.-II and C.P.-III) of the Cerro Prieto geothermal field (slightly modified after Fonseca, 1982).

energy through the great tensional structures that developed, such as the Volcano fault system, thus behaving as a heat source.

GEOHYDROLOGICAL CHARACTERISTICS OF CERRO PRIETO–MEXICALI VALLEY

From 1971 to 1984, both Mexican and American institutions have completed numerous studies dealing in one way or another with the geohydrologic features of the Mexicali Valley in general, and the Cerro Prieto geothermal reservoir in particular. All this information is presently being systematically integrated with updated test and measurement results, especially at Cerro Prieto. However, the main geohydrologic features of the reservoir valley are well known. Two regional aquifers are present in the Mexicali Valley: a shallow reservoir in the Unit A deltaic deposits, and a deeper one in the Tertiary shales and sandstones of Units B and C, which is that of the Cerro Prieto geothermal field. There is some interconnection between the two, because although the former is basically free water, the latter is semiconfined. Thus they are separated by an aquitard (mudstone horizon) rather than by a watertight seal, which allows some infiltration from the shallower aquifer to the lower. Some authors have suggested a subdivision of the deep aquifer, at least within the geothermal reservoir, into the Alpha and Beta aquifers, separated by a 200-m-thick shale layer. Present subsurface knowledge, however, contradicts the general presence of this subdivision, and the deep aquifer is best treated as a single unit.

The regional Unit A shallow aquifer is really a group of superimposed aquifers that behaves as one. When one is drilled for development the water from the upper and/or lower aquifers will move in time, and the boundaries of the drilled aquifer become source-boundaries. Static-level maps of wells fed by this shallow aquifer indicate that it would be naturally replenished by local infiltration at the Andrade Mesa and the Yuma Valley, northeast and east of the Mexicali Valley. From here, ground water flows generally northeast-southwest, the Colorado River acting both as replenishment source and as an underground divide for this shallow aquifer. Subsequently the underground flow of the aquifer ramifies to the northwest and southeast. The northwest branch flows through the geothermal field, south and west of Mexicali and inwards to the Imperial Valley and the Salton Sea; the southeast branch flows to the Gulf of California (Fig. 7).

The regional shallow aquifer of the Mexicali Valley has an average transmissivity of 10^{-1} m^2/s and a virtual regional storage parameter of 0.65; the infiltration coefficient in the zone is 49 percent, and local infiltration is in the range of 600 million m^3/yr. In addition, recharge throughout the Mexican-American border is estimated to be on the order of 300 million m^3/yr. Thus a net recharge of 600 to 700 million m^3/yr is estimated for the shallow aquifer.

The Cerro Prieto geothermal reservoir forms part of the regional deep aquifer, which shows less impressive features. Although the aquifer is actually larger, for estimation purposes an area of 1,225 km^2 was demarked between 32°11′ to 32°33′N and 114°45′ to 115°19′W (Fig. 8) as follows: at the base by the granitic basement (Unit D); at the summit, by the aquitard that separates it from the shallow aquifer; to the west by the Cucapá and El Mayor Sierras; south and southeast by the Gulf of California; east and northeast limits are undefined, but in the subsurface may be the buried granitic ranges; north and northeast limits are likewise undefinable, but the sedimentary rocks there seem to persist, although the basement is shallower than in the southeast, which suggests the aquifer's possible extension to the Salton Sea.

In this area, isopach mapping estimated an average thickness of almost 3,000 m for Units B and C, thinning toward San Luis–Rio Colorado (Sonora; 500 m) and thickening in the central Mexicali Valley (3,500 to 4,500 m). Applying a simple model in which natural flow through the Cerro Prieto area is assigned a hydraulic gradient of 0.008 in sediments of 7.5-md average permeability, the Law of Darcy was applied to a 3,500-m-wide flow channel corresponding to the presently known maximum northwest-southeast dimension of the field and 3,000 m deep, which is the average thickness of Unit C. Under these conditions, flow was obtained through the section of over 500 m^3/day, equal to about 6 L/s, which may be taken as the minimum natural recharge through the Cerro Prieto geothermal field in the absence of any development drilling. Obviously, once development began with the first exploratory wells in 1963, and subsequent operation of the first plant in 1973, recharge must have greatly surpassed this volume, since it is not restricted to the hypothetical 3,500-m-wide section, and also because of the altered hydraulic gradient.

The deep aquifer's water was infiltrated from the Colorado River at least 4,000 years ago and seemingly has mixed with a fraction of sea water. Present recharge comes from that same river, both by infiltration, mainly from the northeast portion of the Mexicali Valley, and by percolation from the regional shallow aquifer, especially in the west and southwest. Underground flow is slow under natural conditions, quicker along fracture zones. It will probably be greater in the northeast-southwest direction due to the larger number of development wells drilled along that trend, which follows the most favorable structures. Occasionally it would be either helped or hindered by the more regionally important northwest-southeast structures. Following the northeast-southwest trend, the underground flow should approach the heat source located northeast of the Michoacán fault, pass over it above the basement and rise to form part of the reservoir; the convection thus generated should considerably accelerate the flow velocity (Fig. 8).

The volume of sedimentary rocks in this area is 3,736 km^3, which with only 6.8 percent average porosity, results in 254 km^3 of water for the regional deep aquifer, almost 100 times the in situ volume estimated for the geothermal reservoir.

Geohydrological studies at Cerro Prieto suggest that in the Cerro Prieto I sector the geothermal reservoir interacts with the regional deep aquifer, and that development of this sector has virtually reached extraction and recharge equilibrium. If the total

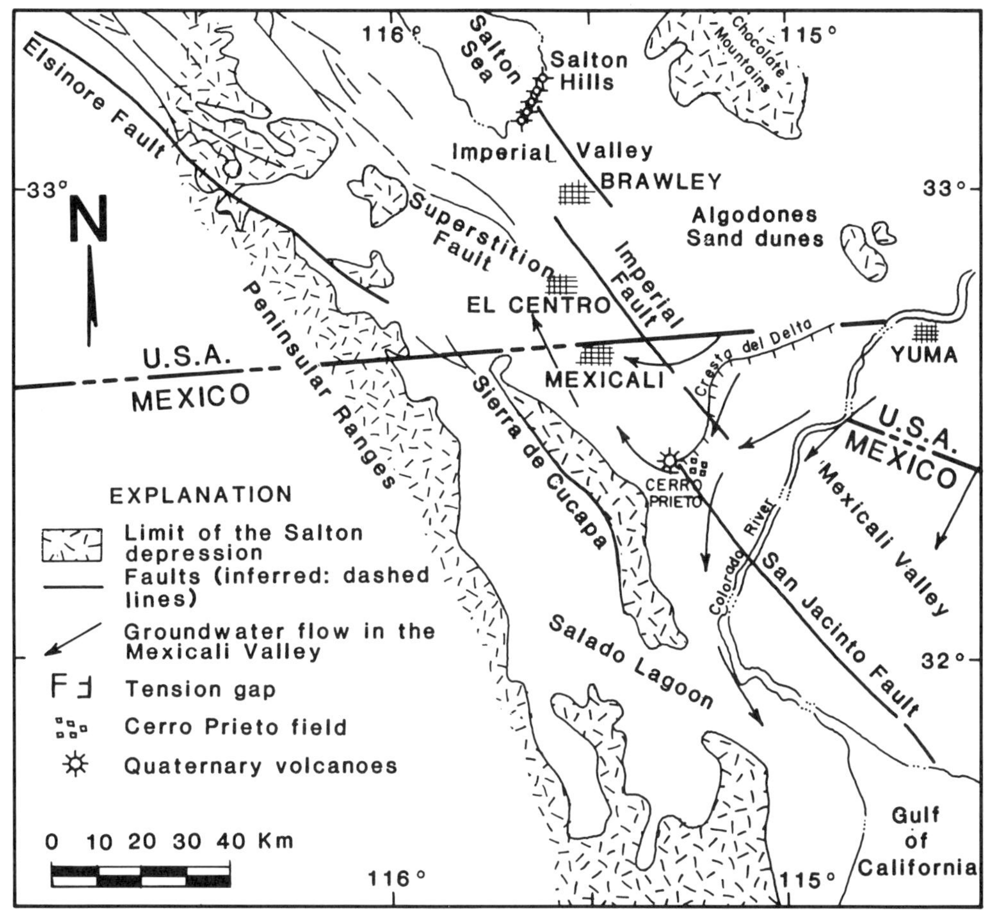

Figure 7. Probable direction of shallow aquifer subsurface flow in the Mexicali Valley (after Ramírez, 1983).

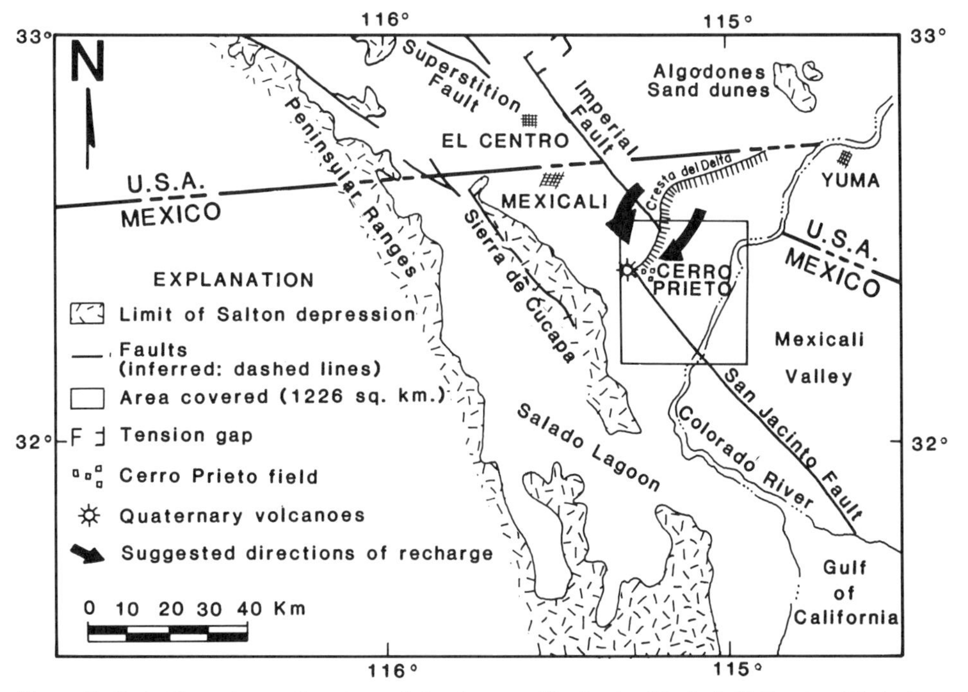

Figure 8. Probable recharge directions of the deep aquifer in the Mexicali Valley (modified after Ramírez, 1983).

combined production of Cerro Prieto I in 1983 was 4,534 ton/h of combined fluids, its recharge must be of comparable magnitude.

GEOCHEMISTRY

Geochemical studies completed during exploration and development include analyses of surface shows and of separated water and gases from wells, as well as isotopic analyses of fluids and minerals. The results have led to a better understanding of the reservoir's behavior, particularly as regards temperature variations, fluid movement, and reservoir recharge, among other aspects.

Chemical composition of surface shows

Hydrothermal shows west of the field (Fig. 9) may be grouped into five categories (Table 1):

Hot springs. Temperatures approach 90°C; pH ranges from 6.5 to 8.0, Eh from –25 to +100 mV; salinities approach 15 g/kg; low concentrations of sulfate and aluminum and moderate concentrations of silica.

Mud pots. 90 to 100°C temperatures; pH under 8.0; Eh negative; salinities typically less than 5 g/kg, although reaching at times 15 g/kg; high sulfate and low silica concentrations.

Geysers. Temperatures approaching 100°C; pH near 2; Eh positive; salinities generally less than 2 g/kg; high silica and aluminum concentrations; low Na/K indexes given in weight.

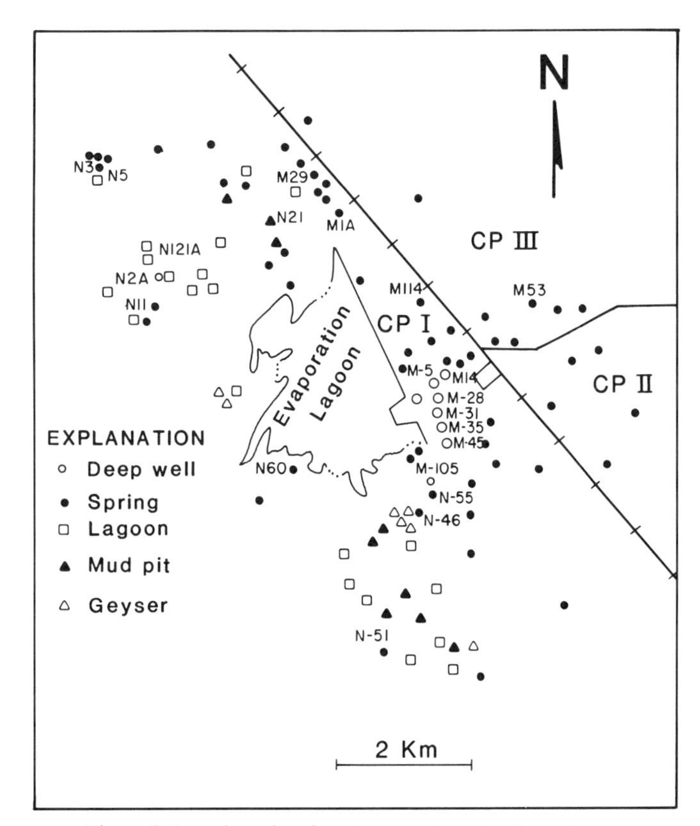

Figure 9. Location of surface thermal shows in Cerro Prieto.

TABLE 1. CHEMISTRY OF THE DIFFERENT CATEGORIES OF WATER SAMPLED IN CERRO PRIETO*

Sample Category	N29 Spring	N21 Mudpit	N46 Geyser	N12C Fumarole Condensate	N5 Lagoon
Temp. (C°)	89	100	99.5	90	22
pH	7.6	6.8	2.0	8.5	2.0
Eh (mV)	+100	-335	+310	-155	-400
SiO_2	73	30	400		200
A1	0.034		12.78	0.015	93.5
Fe	0.35	0.2	117		53
Mn	0.13	0.04	1.83	0.0036	1.88
Ca	357	38	128	12.2	2285
Mg	4.6	2	72	4	163
Na	5115	160	92	19.2	953
K	664	19	157	9	157
Li	12.4	2.3	0.06		1.5
NH_3			400		6.8
HCO_3	65	124	0	4	0
SO_4	31	307	5088	501	5203
Cl	8794	14	31	65	1258

*Concentrations in mg/kg; Valette, 1979.

Fumaroles. Two temperature intervals: approaching 45°C and 80 to 100°C; condensed fluid pH ranges from 3 to 8; Eh negative; salinities are less than 1 g/kg; sulfate and chloride concentrations are approximately 500 and 100 mg/kg, respectively.

Lagoons. Temperatures under 45°C; pH 1 to 5; Eh negative; salinities approaching 30 g/kg and high sulfate concentrations.

Chemical composition of the reservoir water

The chemical composition of the water separated from representative wells at atmospheric pressure is shown in Table 2, which also includes recorded enthalpy and temperature measurements, as well as temperatures estimated by chemical methods. Compared to sea water the Cerro Prieto brine is less saline (approximately 15 g/kg), with extremely low concentrations of magnesium, sulfate, and bicarbonates and high potassium, lithium, boron, ammonia, and silica concentrations.

Comparison of the chemical compositions of water extracted from the wells and from hot springs shows less potassium, calcium, and chlorides, and more magnesium and sulfates in the latter. These variations are ascribed to mixing of the hydrothermal water with meteoric ground water and to the rock/water interaction during the ascent of the thermal fluids.

The Na/K relation and SiO_2 content depend on the temperature and reagents in the Cerro Prieto aquifer, but remain conservative during ascent in the wells, which makes them reliable temperature indicators. Figure 10 shows Na-K-Ca temperature distributions in 1980.

Chemical composition of gases

Cerro Prieto geothermal gases are products of high-temperature reactions within the reservoir, or are introduced with the recharge water. Consequently, the gases obtained from the geothermal wells (Table 3) should reflect reservoir conditions.

As in the case of liquid-phase geothermometers, gas geothermometers have widely varying equilibrium times, which helps to gather different information from each. For instance, the Fisher-Tropsh (FT) geothermometer may be used to estimate maximum temperature in the reservoir, and the H_2S geothermometer may be used for well-bottom temperatures (Table 4).

Recent changes observed in the spatial distribution of gases within the reservoir provide evidence of meteoric water recharge of the reservoir at the west and northwest sides of the field. Degasification in the central part of the field, caused by the appearance of two phases in the reservoir, has also been detected. Noble gas measurements in some producing wells in 1977 were useful in reconstructing dynamic processes in the reservoir in terms of water origin, circulation depth, degree of steam loss, and dilution with shallow water. Positive correlations of radiogenic noble gases (He, ^{40}Ar) and atmospheric (Ne, Ar, and Kr) gas concentrations suggest that the geothermal fluids originated from meteoric water penetrating to depths of over 2,500 m (below the first boiling level), and mixed with radiogenic He and ^{40}Ar formed in the aquifer rocks. Small quantities (0 to 3 percent) of steam were subsequently lost through a Raleigh process, and a mixture with shallow cold water (0 to 30 percent) took place.

Isotopic studies

Preliminary isotopic studies of Cerro Prieto geothermal fluids and previous investigations of the Mexicali Valley ground waters (Fig. 11) suggest local recharge of the geothermal system from the area immediately to the west. The isotopic exchange of the oxygen in the water with the reservoir rock minerals at temperatures that increase with depth, and the lowering of pressure in the southeast part of the field, suggest that fluids containing less ^{18}O are invading the producing reservoir from above. The tritium and ^{14}C contents in the well fluids suggest an average age of the water that ranges from 50 to 10,000 years. The carbon and sulfur isotopic compositions agree with a magmatic origin, but a mixed organic-sedimentary origin seems more probable for the carbon and possible for the sulfur.

Inferences derived from oxygen and carbon isotopic data in calcite samples of the Cerro Prieto geothermal field depend on the type of sample. Values of $\delta^{18}O$ for the calcite in sandstones provide reliable bases for estimating reservoir stable temperatures, and the $\delta^{18}O$ values of calcite in shales can be related with the extension and spatial distribution of the underground flow. The values of $\delta^{18}O$ in calcite veins record short-lived polythermal episodes of fracture filling at temperatures that may differ from the reservoir's stable temperatures. Oxygen isotopic data in shales indicate a minimum *volumetric* water-rock relation of 2:1. This high flow is amply exceeded in sandstones, with the result that the reservoir fluids are isotopically well mixed and have relatively low salinity.

TABLE 2. CHEMICAL COMPOSITION OF WATER EXTRACTED IN CERRO PRIETO (1984)*

Well	Date	Wellhead pressure $(kg/cm^2)_g$	Extraction pressure $(kg/cm^2)_g$	Water production (ton/h)	Steam production (ton/h)	Components (in mg/l under laboratory conditions)								
						pH	Na	K	Li	Ca	Cl	HCO_3	STD	SiO_2
E-4	4/24/84	56.2	8.5	57.7	67.1	6.6	10972	3066	29	414	20299	49	36435	1002
E-6	4/24/84	12.0	8.4	108.4	45.7	7.2	10909	2959	29	509	20316	80	37130	1094
E-7	4/25/84	68.8	9.5	148.2	90.1	7.0	12319	3449	33	489	23274	47	41624	1266
M-10A	4/17/84	64.5	9.0	69.5	66.6	6.2	9856	2912	31	373	18582	76	33300	1183
M-14	4/17/84	30.6	7.3	68.3	21.1	6.8	4679	804	12	241	8335	127	14779	693
M-21A	4/23/84	7.7	7.5	19.3	13.2	7.2	5761	1263	16	256	10232	117	18414	838
M-35	4/23/84	8.8	7.7	123.4	39.1	8.1	4227	782	10	158	7503	71	13413	706
M-51	4/25/84	16.9	8.1	148.2	73.1	7.1	6100	1477	16	201	10795	100	19039	927
M-84	4/27/84	8.8	8.4	23.4	80.7	7.0	10485	2958	29	470	19797	63	35463	925
M-90	4/27/84	7.7	7.6	102.5	30.4	7.6	4562	926	12	158	7886	76	14309	708
M-91	4/27/84	9.8	7.8	85.5	57.0	7.1	10021	2704	27	373	18547	71	34003	1308
M-103	4/25/84	8.8	8.1	80.7	41.7	6.9	6104	1405	15	191	10795	92	19445	1069
M-105	4/24/84	9.1	8.6	69.4	36.7	7.3	9742	2279	25	531	17798	49	32095	930
M-114	4/30/84	7.6	7.5	92.3	24.5	7.8	7110	1357	17	432	13281	72	23270	718
M-120	4/26/84	17.9	9.5	39.7	48.0	6.7	10151	2946	30	382	18815	51	33762	1112
M-169	4/26/84	64.7	8.1	70.7	31.7	7.1	11552	3164	29	491	21462	57	39725	1266

**Note:* Data according to F. J. Bermejo, personal communication, 1984.

TABLE 3. COMPOSITION OF GASES OF FLUIDS PRODUCED IN CERRO PRIETO IN 1977 (in molar percent)

Well	Steam Fraction (X_5)	Gas/steam ($Y.10^3$)	Composition of gases (Molar %) CO_2	H_2S	H_2	CH_4	Ar	N_2	NH_3
M-5	0.287	4.46	78.86	7.63	4.66	5.18	0.014	0.60	2.28
M-8	0.284	9.24	92.02	2.59	2.33	1.51	0.0037	0.15	1.47
M-11	0.279	5.64	71.30	5.69	1.77	4.26	1.091	11.0	3.98
M-14	0.244	7.81	81.75	4.99	4.52	4.86	0.013	0.51	2.97
M-19A	0.289	5.88	82.19	7.91	2.86	3.98	0.014	0.51	2.31
M-20	0.227	9.12	91.65	2.32	1.98	1.77	0.0059	0.23	2.07
M-21A	0.273	10.1	87.93	3.17	3.97	2.95	0.0038	0.27	1.77
M-25	0.284	6.93	82.73	7.07	2.25	3.39	0.0097	0.46	3.00
M-26	0.243	4.85	80.89	8.33	4.09	4.26	0.016	0.63	1.80
M-27	0.262	9.19	86.13	3.28	4.42	3.50	0.013	0.45	1.75
M-29	0.217	5.25	82.28	4.94	2.98	4.99	0.014	0.82	2.73
M-30	0.269	4.63	82.52	5.80	4.14	4.12	0.019	0.71	2.40
M-31	0.257	6.56	80.87	5.76	5.46	5.08	0.016	0.65	2.24
M-35	0.285	7.08	91.10	2.63	2.21	2.07	0.0091	0.30	1.71
M-42	0.257	5.48	81.50	6.14	2.24	4.43	0.0083	0.49	4.58

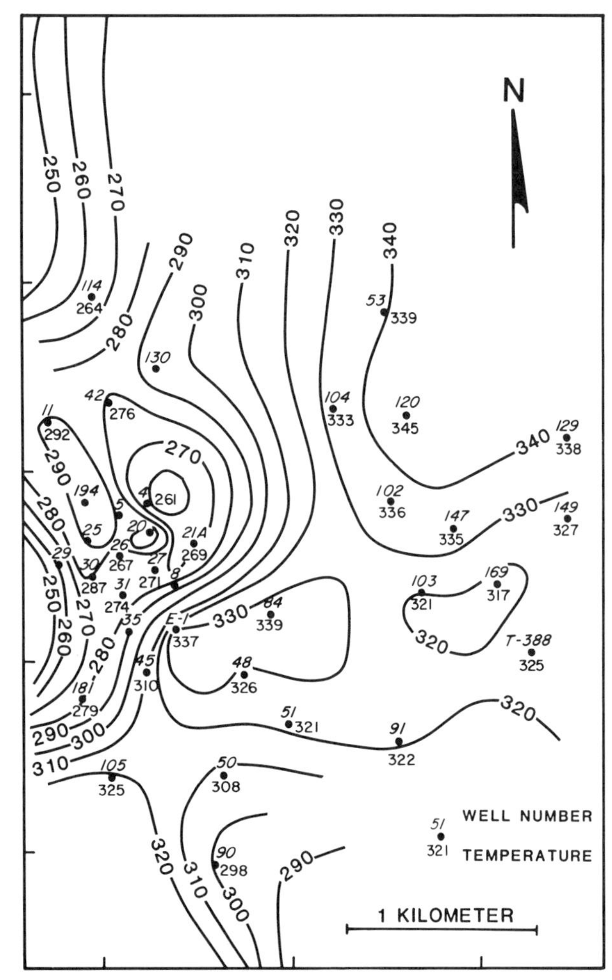

Figure 10. Distribution of temperatures (°C) obtained by Na-K-Ca geothermometry.

TABLE 4. TEMPERATURES (°C) BASED ON THE GAS AND Na-K-Ca GEOTHERMOMETERS

Well	FT*	H_2	NH_3	H_2S	NaKC
M-5	326	294	298	286	289
M-8	335	303	289	280	290
M-11	307	280	295	284	288
M-14	335	305	301	287	274
M-19 A	322	293	289	295	293
M-20	324	293	277	271	266
M-21 A	343	313	307	287	288
M-25	321	292	281	296	290
M-26	322	289	295	285	276
M-27	342	312	312	287	283
M-29	312	282	282	272	264
M-30	324	286	297	278	286
M-31	336	305	309	287	280
M-35	325	296	285	273	293
M-42	311	282	268	283	281

*Fisher-Tropsh geothermometer.

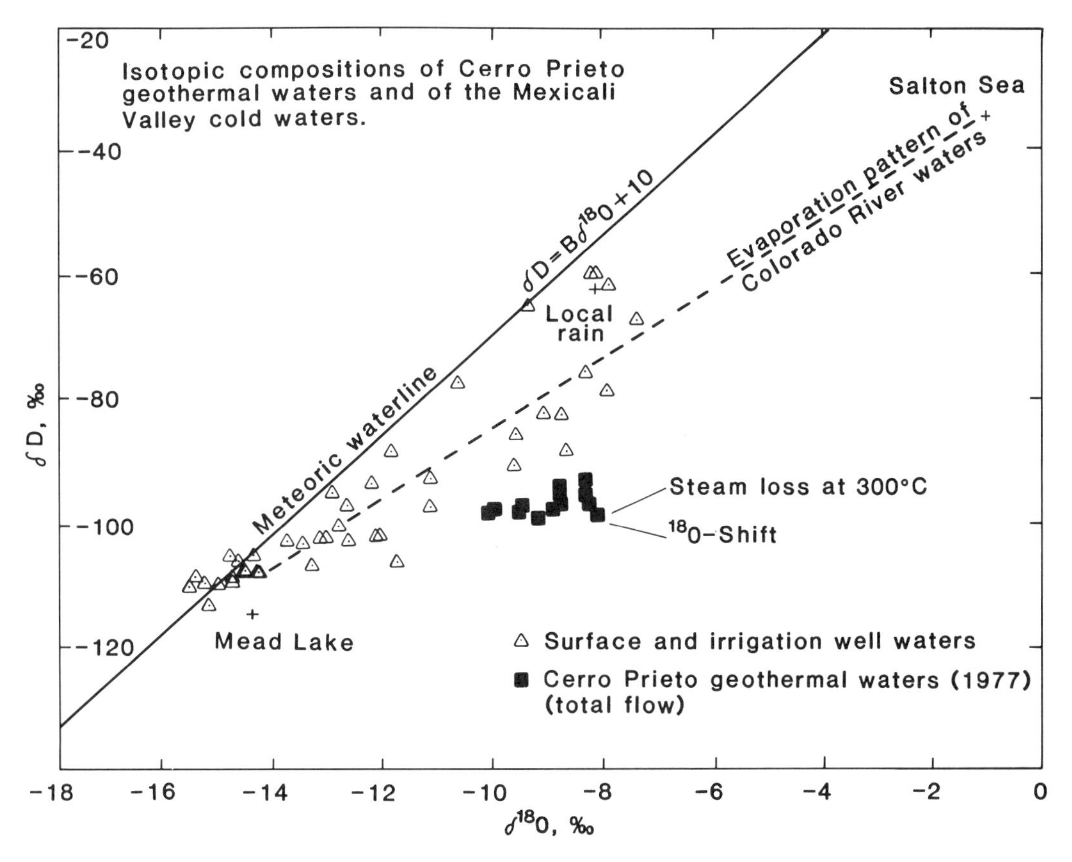

Figure 11. Deuterium and ^{18}O compositions in geothermal and surface waters.

Origin of the recharge

The chemical data indicate that changes occurring in the composition of well discharges have been mostly in the direction of a cold sweep (entry of colder water, heated on passing through an increasingly smaller volume of the hot reservoir). It is interesting to note that the opposite has not been observed (entry of hotter waters from greater depths). In the vicinity of well M-9 and possibly well M-25, isotherms show an elongated concentration of hot water surrounded by colder water (Fig. 12). Judging by the subsurface temperature distribution and the types of rocks containing the reservoir, there is a greater availability of colder water, and recharge in this area must be mainly cold. At other localities, where temperatures apparently increase uniformly with depth, recharge seems to have brought preferentially colder overlying water instead of the hotter underlying fluid.

Chemical and physical measurements indicate that, in contrast to Wairakei, New Zealand, the west portion of Cerro Prieto is not entirely closed to the entry of colder waters of low chloride content. The predominantly natural cooling of the rising hot waters is effected by mixture with cold water (dilution) rather than by boiling and steam loss. Dilution has accelerated in response to the development of the reservoir. The mixture of thermal and cold waters has been aided by the high permeability of the producing and adjacent cold-water aquifers and by the absence of a continuous low-permeability barrier between these aquifers. Although prior to development the hot water was entering the producing zone from below, this flow does not seem to have increased greatly during development, because reservoir pressures have been maintained by the entry of cold water. Recharge from below is also limited by the relatively low permeabilities resulting from mineral precipitation. Even before development began, the dilution process dominated over boiling as the main cooling mechanism of the Cerro Prieto fluid; this is shown by convective-type temperature profiles in the reservoir, the original linear relation of enthalpy and chlorides, and the small reduction in the quantity of noble gases caused by boiling.

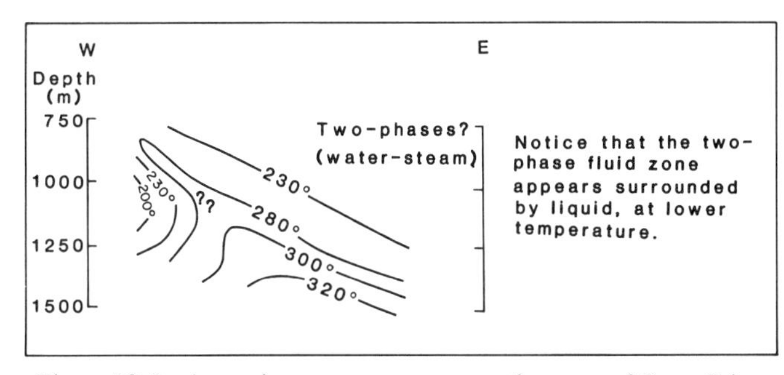

Figure 12. Isotherms in an east-west cross section west of Cerro Prieto.

Although no general boiling has occurred at Cerro Prieto to create an extensive two-phase zone, local boiling is common around the wells. This produces excess enthalpies that decrease or disappear with time, and diminishing but not disappearing silica deficiencies. This local boiling is the result of locally diminishing pressure and of intergranular (rather than through fractures) flow toward the wells. While the boiling zone expands, excess enthalpy is produced when heat from the rocks is transferred to the fluids that have been cooled by boiling. This accelerated boiling increases the concentration of dissolved silica and the deposition of quartz through evaporation around the well. As expansion of the boiling zone lessens or ceases, the rocks reach thermal equilibrium with the fluids, and the production of excess enthalpy decreases and stops. Within the now-stabilized boiling zone the cooled fluids continue to deposit quartz at a lower rate, and silica deficiency is reduced.

GEOPHYSICS

The Mexicali Valley region is covered by the continental deltaic sediments of the Colorado River, which mask the basement structures. For this reason, geophysical methods have played a significant role in exploring the Cerro Prieto geothermal field and its surroundings. Various geophysical surveys have led to better understanding of the reservoir and valley subsurface geologic features.

The first geophysical studies were seismic refraction and gravity surveys in 1962, followed by resistivity, magnetometry, gravity, spontaneous potential, magnetotelluric, seismic reflection and refraction surveys, as well as passive seismic studies and electric well-logging, which in turn includes natural potential, resistivity, gamma-ray, density, and sonic measurements. Both Mexican and American research institutions took part in these surveys. Evaluation of the area's geothermal resources continues as incoming well data are analyzed and the geophysical information is incorporated into updated reinterpretations.

The following summary of geophysical contributions to evaluation of the Cerro Prieto and general Mexicali Valley region geothermal resources contains the most relevant results. A large portion of these papers are included in the acts of the four joint CFE-DOE (Comisión Federal de Electricidad and U.S. Department of Energy) meetings that took place under a technical-scientific agreement between the governments of Mexico and the United States. Lyons and Van de Kamp (1980) and Lippman (1983) present a virtually complete list of the works carried out at Cerro Prieto.

Studies prior to commercial development

One of the first geophysical studies in the Mexicali Valley was a regional gravity survey for the purpose of clarifying the structural relation between the areas north of the Gulf of California and the Colorado River delta. The valley was covered by 400 gravity measurements to produce a topographically uncorrected Bouguer anomaly map, which showed that the mainly northwest-trending general alignment of the gravity contours is correlated with some of the major faults of the Gulf region, indicating that the Colorado River delta area is part of the Gulf of California province (Kovach and others, 1962).

Almost simultaneously (1962) the first seismic refraction survey for geothermal purposes in the Cerro Prieto area was aimed at defining the structural conditions of the basement and overlying sediments. Results of this survey suggested the basement to be about 1,500 m deep in the area of the Cerro Prieto surface shows, bounded in the west by a graben and in the east by large-scale subsidence; toward the east it appeared to deepen to over 4,500 m due to the effect of the Cerro Prieto–Michoacán fault (formerly San Jacinto).

A gravity survey covered a larger area along 18 north-south lines separated by 2-km stretches. The Bouguer anomaly map, using a density factor of 2.0 g/cc, indicated a series of blocks bounded by major faults and overlain by thick sedimentary deposits.

The combined interpretation of the above and its correlation with the geological information concluded that geothermal activity in the Cerro Prieto area is controlled by crystalline basement faulting, and that the source could be a recent intrusive a few kilometers underground, located toward the down-faulted block of the Cerro Prieto fault and east-southeast of the volcano (Fig. 13).

On the basis of refraction and surface geological information, four exploratory wells were drilled in 1964, two of which were producers (M-3 and M-5) between the 800 and 1,500 m depths. M-3 reached basement at 2,547 m and provided valuable information for geophysical interpretation; M-5 is the discovery well, having excellent temperature, pressure, and production characteristics.

Beginning in 1969, the CFE, the National Autonomous University of Mexico, and the CICESE (Centro de Investigación Científica y de Estudios Superiores de Ensenada; Ensenada Center for Scientific Research and Higher Studies), in cooperation with the University of California at Berkeley, the Pasadena Technological Institute, and the San Diego Institute of Geophysics and Planetary Physics, installed seismological stations for two months to monitor activity along the Imperial and Cerro Prieto–Michoacán faults. A considerable earthquake swarm was detected north of the Gulf of California.

The regional tectonic environment was thereby interpreted as a series of transform faults connected with five spreading centers (ridge segments), characterized by active geothermal zones, recent volcanism, earthquake swarms, and topographic depressions.

An aeromagnetic survey of the Mexicali Valley was done in July 1971 through the University of Arizona to define future exploratory prospects, the magnetic character, and the structures of the Cerro Prieto geothermal zone. The first phase of the survey covered the Cerro Prieto and Tulecheck geothermal zones with flight-line separations of 1 and 0.5 km at a barometric altitude of 305 m; a second phase was a regional survey with flight lines

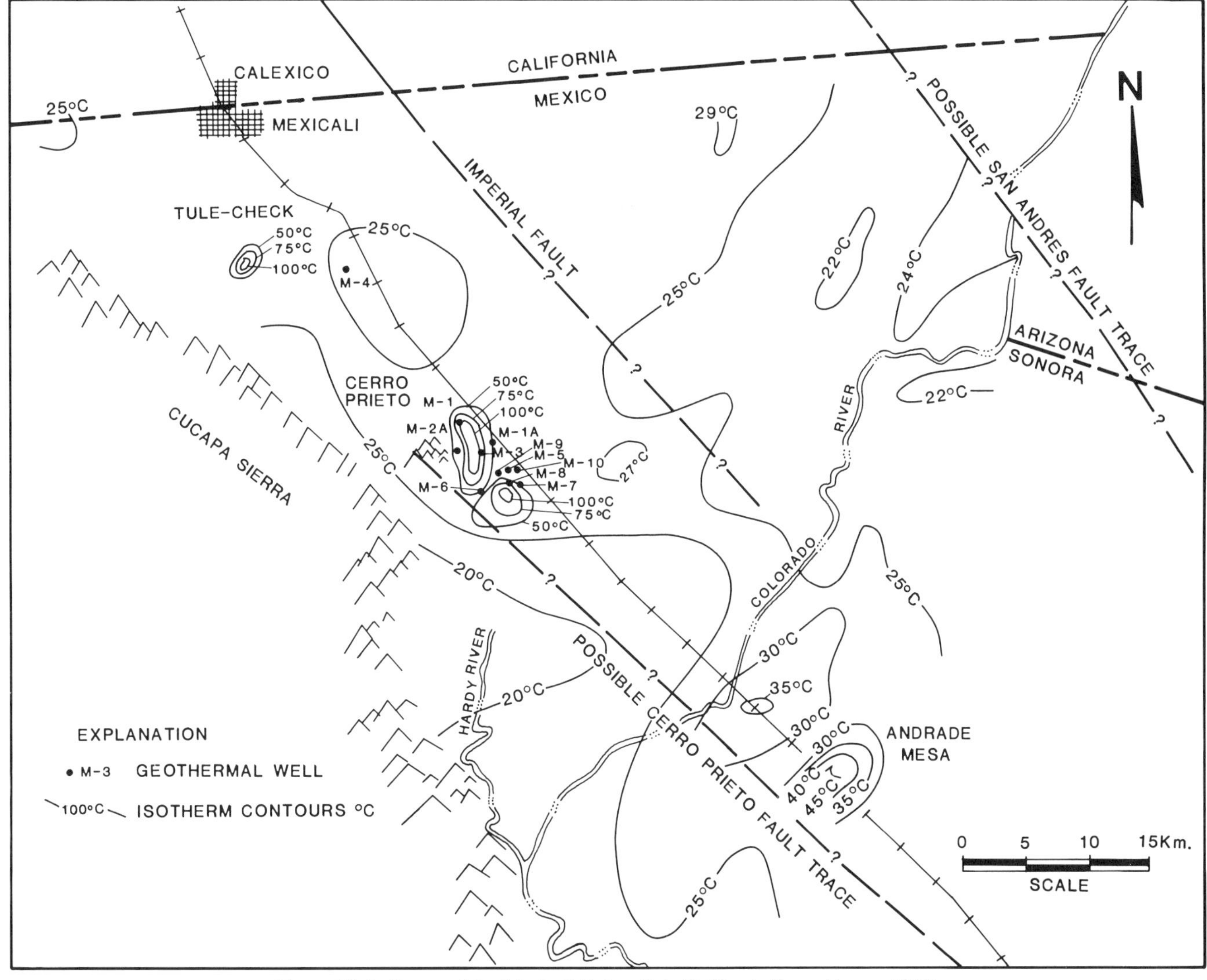

Figure 13. Isotherms of the Mexicali Valley, Baja California (from Alonso, 1964).

spaced 5 km apart at altitudes of 610 to 762 m. The total-intensity residual magnetic anomaly map showed a positive anomaly southeast of the Cerro Prieto Volcano, which was subsequently associated with the possible heat source of the geothermal activity in the region. Magnetic contours striking parallel to the regional fault pattern (northwest-southeast) indicated strike-slip and vertical faults that affected the basement. In addition, a formerly unknown probable heat source was detected in the Colorado River area, named "Panga de Abajo," and rated as a future exploratory prospect.

In 1972 the CFE began geoelectric investigations in the northwest portion of the valley to define the possible extension and geologic features of the Cerro Prieto geothermal reservoir. A first phase sought regional information on the apparent resistivity by profiling at a constant theoretical depth over an approximate area of 1,600 km^2, using the Schlumberger array of four colinear electrodes with a constant 1,000-m separation between current electrodes (AB/2 = 500 m). Data from 585 measurements indicated regionally low resistivity values (<10 ohm/m), and an area of approximately 70 km^2 surrounding the Cerro Prieto geothermal field was defined with values under 2 ohm/m; another zone of less than 2 ohm/m coincided with that of the Tulecheck thermal shows. In the second phase a detailed study covered some 40 km^2, around the Cerro Prieto I development area, using the Schlumberger array with a maximum 4-km separation of current electrodes.

Joint geophysical and geological interpretation led to exploratory drilling east of the Michoacán fault outside the initial producing area. Well M-35 was completed with producing intervals at depths of 1,080 to 1,400 m, thus extending the field toward the east and to a greater depth over approximately 10 km^2 of the development area.

Investigations following initial development

Experience, together with the results of deep exploratory and development drilling to 1972, led to the conclusion that the reservoir limits toward the north, east, and south were insufficiently defined, both horizontally and at depth, for purposes of the first geothermoelectric plant (Cerro Prieto I), and the decision was made to extend the Cerro Prieto geophysical exploration program.

On the basis of the first electrical survey results, a regional study was undertaken in 1977 over 580 km^2 in the Mexicali Valley, using the Schlumberger array of 114 vertical electrical probes. Areas selected for detailed study included the Cerro Prieto development zone, and 74 additional probes covered 208 km^2. Two low-resistivity zones were located: one was related to the development area, and the other, 15 km southeast of it, was a future exploratory drilling prospect (Fig. 14).

Simultaneously, a regional joint gravity and ground magnetometry survey followed the electrical survey lines at observation points spaced at 500- and 250-m intervals respectively. A simple Bouguer anomaly map with a Bouguer density factor of 2.0 g/cc, and a total-intensity residual magnetic map were drawn (Figs. 15 and 16). The interpretation was based on graphic and interactive trial-and-error techniques to obtain profile sections that showed horst-and-graben structures related to the Cerro Prieto and Imperial faults.

The magnetometry results defined a magnetic high centered approximately 5 km east of the geothermoelectric plant (Fig. 16). Analysis and comparison of this major anomaly with the petrography of mafic dikes recovered from well NL-1 and with other geological and geophysical information concluded that Cerro Prieto is located in a pull-apart basin. On the basis of seismic observations and Curie isotherm analysis, the summit of the main intrusive body was estimated to be approximately 3.5 km below the surface, and the present melting zone 9 to 10 km below the surface.

In a 1978 paleomagnetic survey of the rhyodacitic rocks of the Cerro Prieto Volcano, the magnetic polarity and paleopole positions suggest that volcanism began about 111 ka; the youngest volcanism to reach the surface cooled about 10 ka.

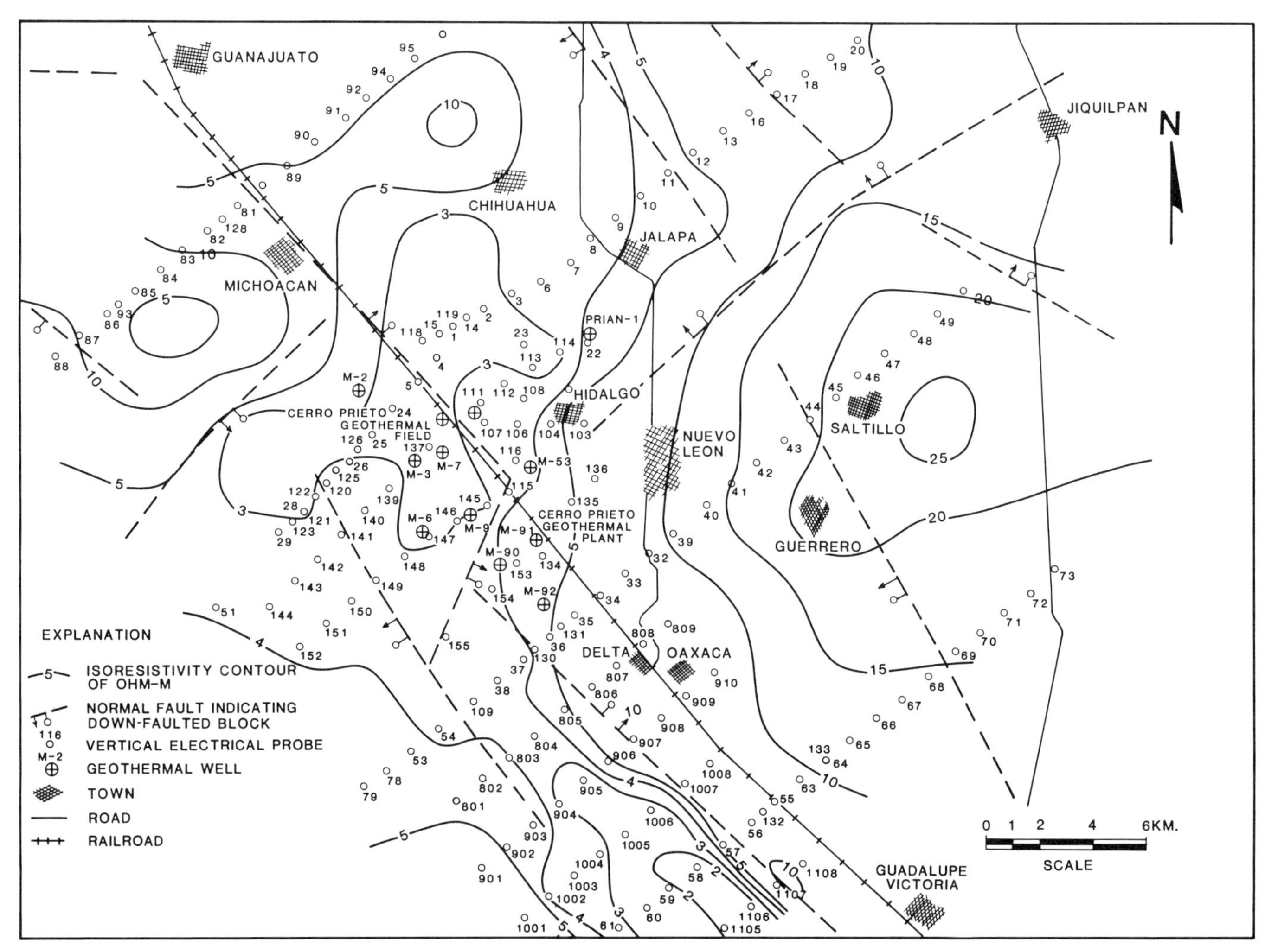

Figure 14. Apparent isoresistivity for AB/2 = 3,000 m (after Razo and others, 1976).

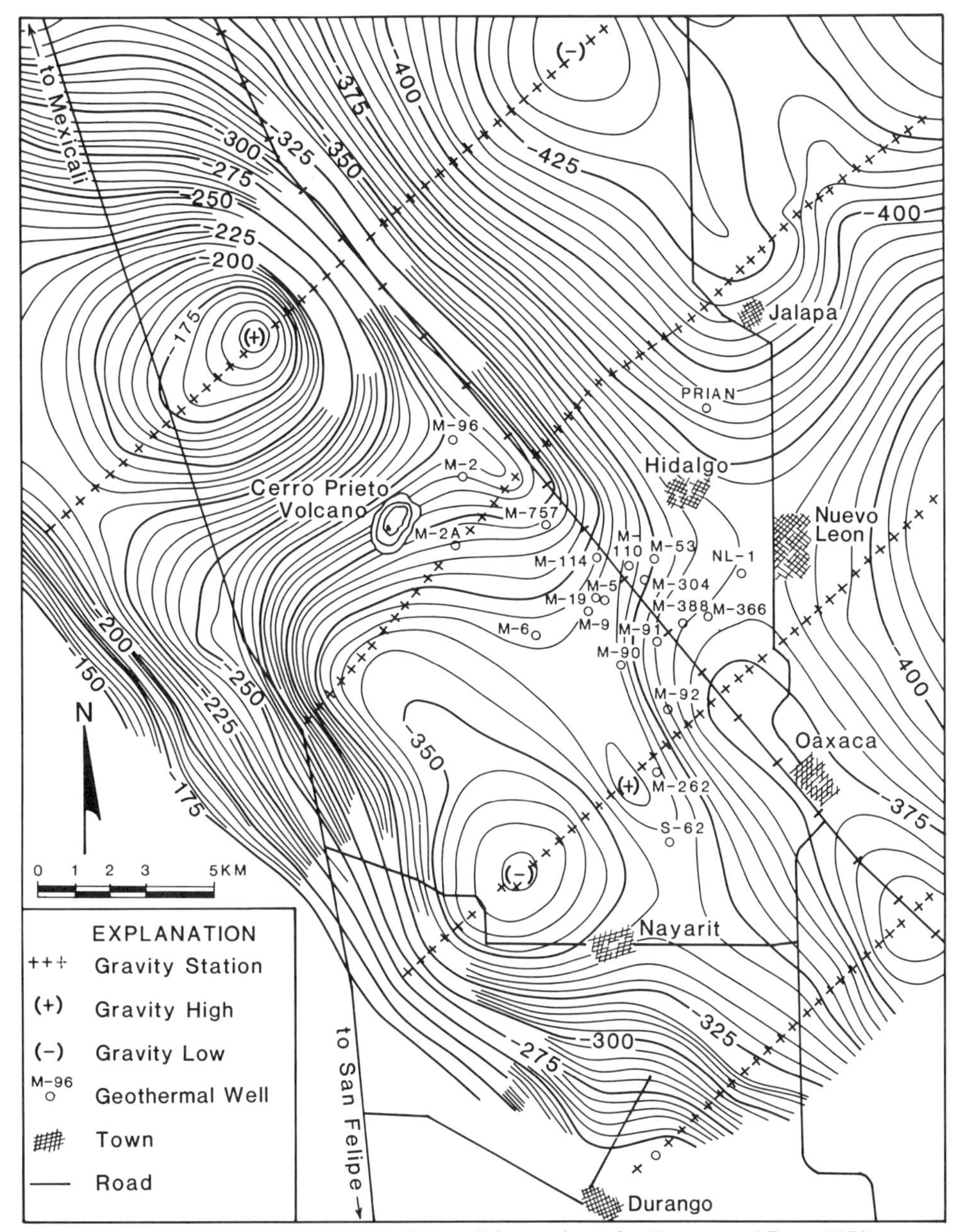

Figure 15. Bouguer anomaly map of Cerro Prieto region (after Fonseca and Razo, 1979).

As part of the joint FCE-DOE project, gravity, seismic, and resistivity surveys were programmed in 1977 for Cerro Prieto. The resistivity survey used the dipole-dipole continuous current array. Magnetotelluric and natural potential surveys were also carried out, all following lines that crossed the Cerro Prieto I plant development area (CP-I, Fig. 6).

Permanent electrodes were emplaced for future measurements, to detect the possible variations in resistivity caused by changing reservoir conditions. The bidimensional interpretation of the dipole-dipole method showed high resistivity values (4 ohm/m) in the Cerro Prieto I producing zone as compared to its surroundings (2 ohm/m) and also helped to define the deep structure in the eastern portion of the field. The models indicate that the resistive body associated with the producing zone dips 30 to 50°E to depths of over 2,000 m.

Results of the magnetotelluric method (MT) applied in 1978 to complement the resistivity surveys agree with those of the dipole-dipole and Schlumberger surveys, differing only in the seemingly clearer definition of the reservoir's deep structure.

The natural potential survey in late 1977 indicated an extensive, very long-wavelength anomaly over the geothermal field. A more detailed 1978 survey defined a bipolar anomaly passing through the center of the producing field. The source seems to correspond to a north-south–trending fault, and the anomaly was probably caused by a streaming-potential mechanism.

Four precision gravity surveys in the Cerro Prieto area from

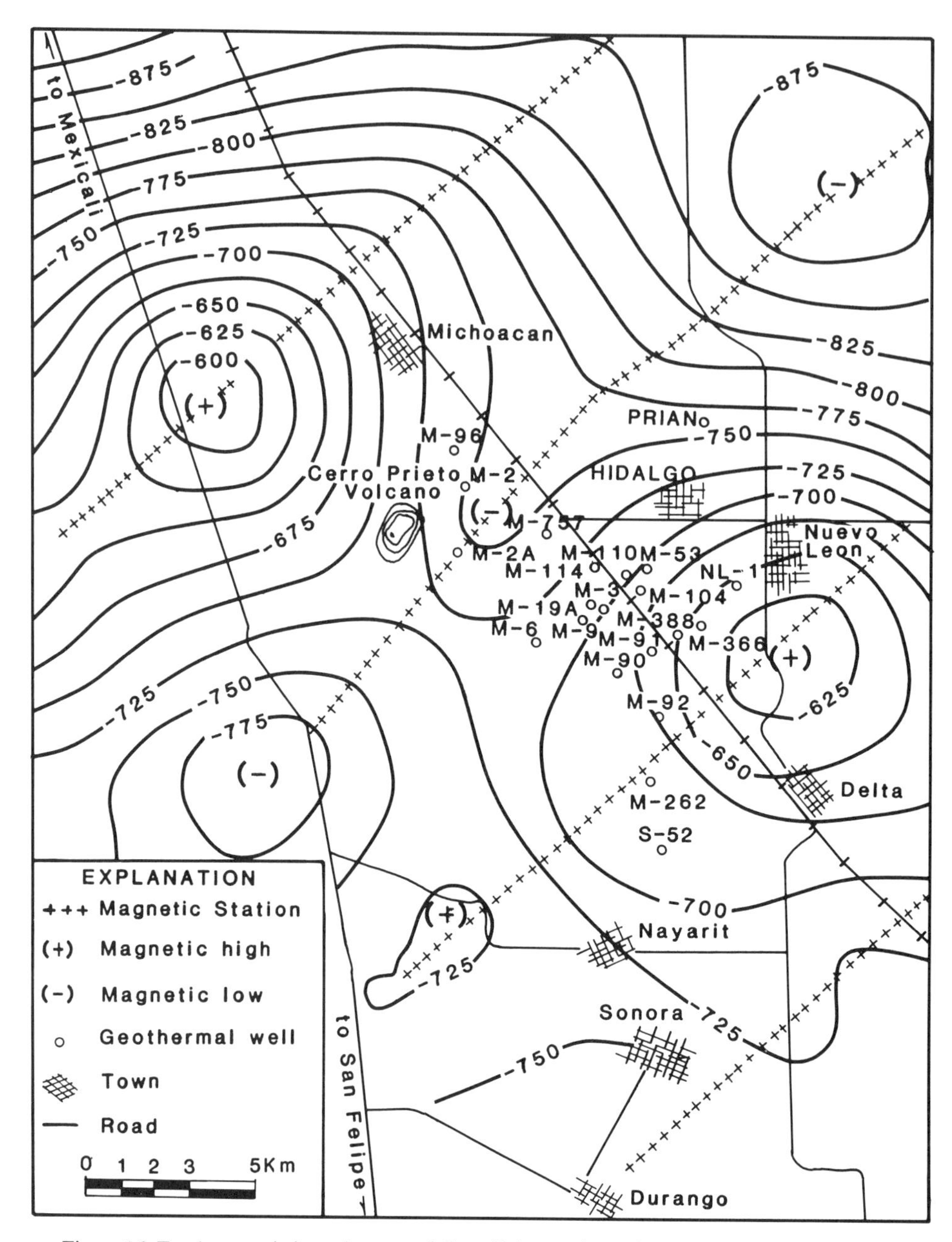

Figure 16. Total magnetic intensity map of Cerro Prieto region (after Fonseca and Razo, 1979).

1978 to 1981 investigated the subsidence possibly caused by fluid extraction; however, it proved to be associated rather with the 1979 earthquake. Gravity data were found to correlate with the results of first-order topographic surveys. A probable geothermal origin was also suggested for the subsidence, if the natural recharge of the reservoir is estimated to be less than the fluid extraction. In addition, the progressive densification of the reservoir zone was shown to be caused by mineral precipitation due to cold-water entry.

As regards seismology, from 1978 to 1981 the CFE completed approximately 400 linear km of reflection and refraction surveys, positioning the lines normal to the northwest-southeast regional structural trends. This allowed the quantification of stratal thickness and behavior, as well as structural interpretations. The Cerro Prieto field was found to extend toward the east; to the west the basement (consisting of granitic rocks, penetrated by wells M-3 and M-262) shows velocities of over 5 km/s.

The Tertiary sedimentary sequence thickness (Units B and C) was estimated at approximately 2,600 m. The uneven Quaternary sequence of varyingly compacted (semi- to nonconsolidated) continental sediments (Unit A) showed thicknesses on the order of 2,500 m.

It was concluded that in the producing zone the reflection information weakens, especially at depth beneath an interval of

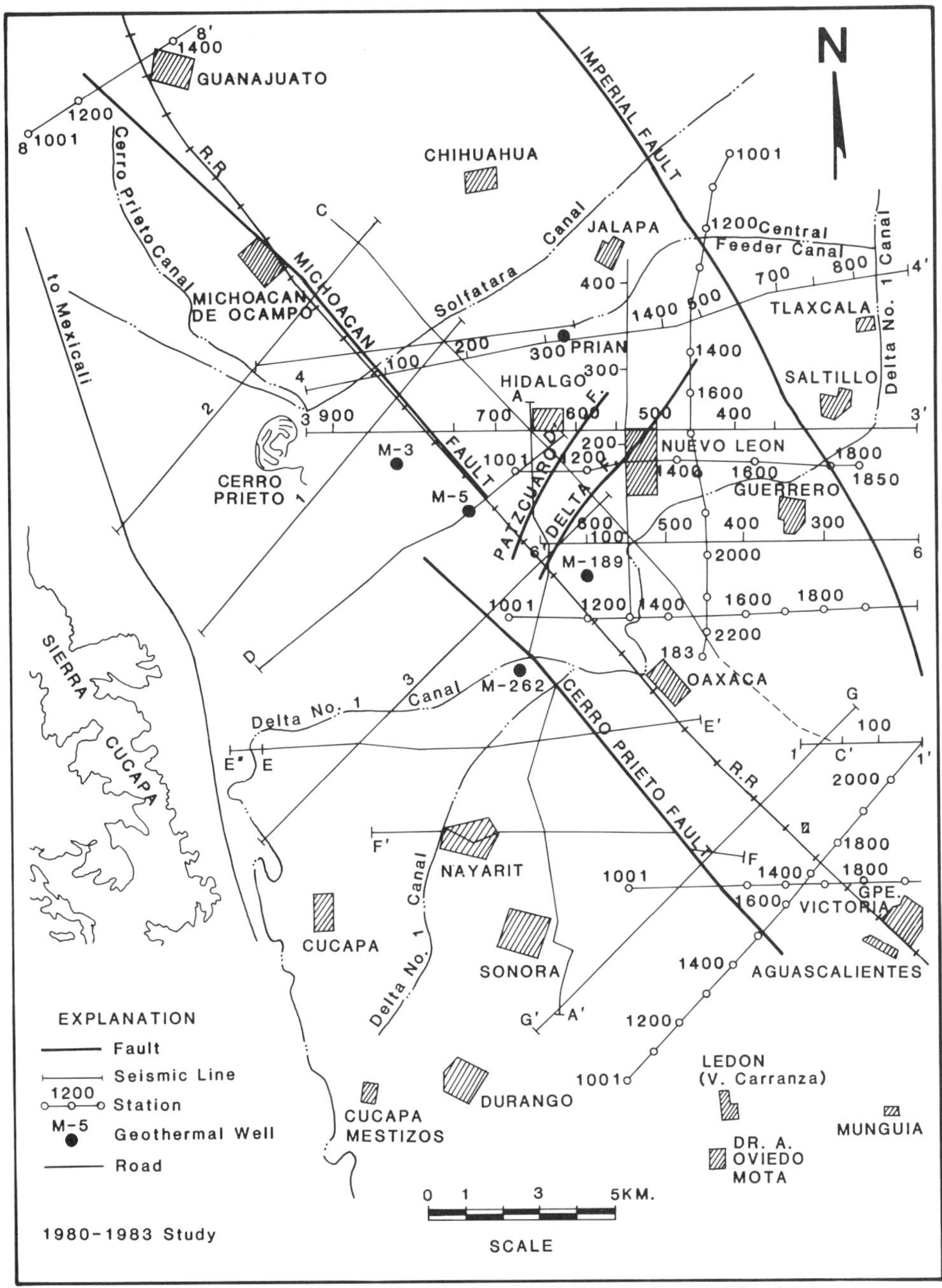

Figure 17. Seismic interpretation of the Mexicali Valley (after GYMSA-1984).

good conductors. This was also observed in the East Mesa geothermal field, where it is ascribed to the highly altered and cemented dense sediments that have been subjected to hydrothermal action. The NW-SE–trending Imperial, Michoacán, and Cerro Prieto faults were identified, as were the northeast-southwest Pátzcuaro and Delta faults (Fig. 17). These northeast-southwest structures have been at least partially confirmed in correlations of the geophysical well logs with subsurface geology (Fig. 6).

Passive seismic surveys have defined activity in the area as consisting of swarms of minor events, although some reach magnitudes of over 5.0. Microseismicity in the geothermal field and its surroundings correlates with the area's fault system, maintaining a longitudinal alignment with the Imperial and Cerro Prieto faults (Fig. 18). Other interpretations suggest that these right-laterally displaced faults are connected through an oblique fault system, which is confirmed by the presence of the northeast-southwest structures.

The seismic pattern and the composition of the fault-plane solutions suggest that tectonic deformation in the Mexicali Valley takes place mainly along two major tectonic alignments: parallel to the Imperial–Cerro Prieto transform fault system and oblique

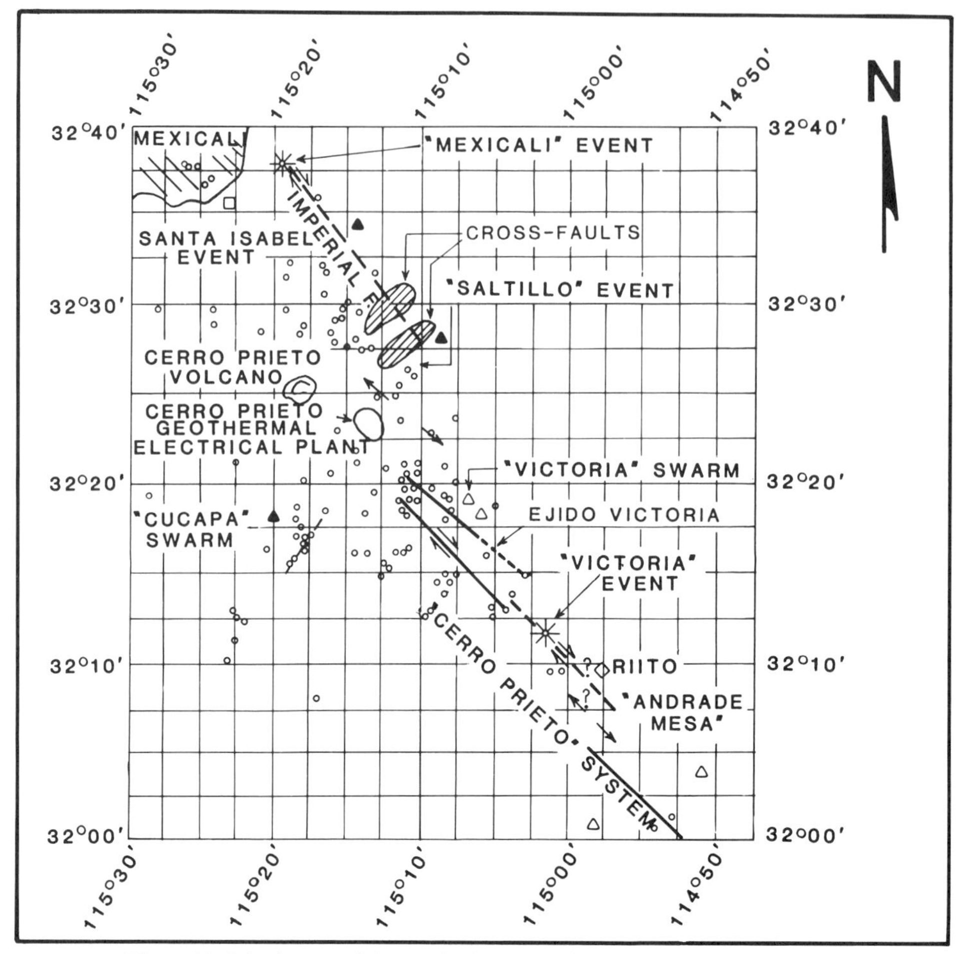

Figure 18. Seismic map of the Mexicali Valley (after CICESE, 1977 to 1980).

to it, with some normal faults. The proposed tectonic model is a graben developed between the Imperial–Cerro Prieto faults, controlled by a transform fault system, similar to that of the southern California geothermal areas of the United States.

Other studies aimed at defining microseismic activity in the geothermal reservoir area, as well as P- and S-wave attenuation and velocity in the producing zone, have established that (a) recorded seismicity in the producing area is low compared to its surroundings; (b) the first arrivals indicate right-lateral displacement of events in the field occurring along NW-SE–trending faults; (c) the scarce data suggest P-wave velocity and attenuation anomalies associated with the producing zone; (d) Vp/Vs values are high; (e) the latest survey of this type reports errors of ±0.5 km and ±1.0 km in epicenter and focal-depth determinations respectively; events tend to align along the Hidalgo fault, with a lesser tendency to group near the main producing zone; and (f) earthquake depths in this region range from 2 to 5 km.

DRILLING AT CERRO PRIETO

The Cerro Prieto geothermal field exploratory drilling has been done in several stages. Three relatively shallow wells (approximately 755 m) were drilled from 1959 to 1961 northwest of the present development area. Four deeper wells (M-3, M-5, M-6, and M-7) were drilled in 1964, two of which were producers and therefore the discovery wells for the field. Exploration and development continued in 1966 with 15 wells drilled to develop the Cerro Prieto I (CP-I) plant, which began operating in late 1973 with 75 MW of installed capacity. Development drilling has since continued, as has new exploratory drilling in the Mexicali Valley, for a total of 133 wells completed to December 1984, 103 of which are development and 30 exploratory. An additional seven wells were abandoned due to mechanical failure, drilling, or development problems; another four were drilling failures. Thus, to that date, 286,411 m have been drilled in Cerro Prieto and Mexicali Valley.

Drilling methods and procedures have evolved since 1959 owing to progressively increasing average well depths; temperatures during drilling have also risen continually, often giving rise to unfamiliar situations. Thus, the average 1,200-m depths of wells drilled in the mid-1960s increased to an average 2,500 m in the late 1970s and thence to the present average depth of 3,500 m. The deepest well drilled to date is M-204, which reached 4,125 m and was temporarily suspended in December 1984 due to mechanical problems. Figure 19 shows the evolution of the typical pipe diagram designed for Cerro Prieto wells over more than 20 years (1964 to 1982). For the purpose of solving mechanical and corrosion problems (collapses, pipe fractures, sand invasion, excessive incrustation) that arose in the early wells and required maintenance and repair costs, and which eventually amounted to a third or more of the total costs, extra-heavy pipes were used. These gave excellent results, but increased well costs considerably. In the late 1970s it was decided to use standard weight N-80 or L-80 grade casing and C-75 grade production tubing; K-55 grade was adopted for 13⅜-in and 11¾-in anchorage tubing, which is effective in controlling the recurrent corrosion problems.

The most common problems encountered in both exploratory and development drilling are (a) loss of circulation fluid, (b) formation cave-in or collapse, (c) high temperatures in the producing zone, (d) "fishing" problems, (e) circulation losses during cementation, (f) mechanical failure in casings, and (g) drilling mud or cement invasion of producing zones.

All these problems obviously escalate with increasing depth and temperatures. In particular, cave-ins seem to multiply due to the strong differential pressures generated by thermal shock at high temperatures. "Fishing" problems when recovering tools or material from the well bottom are also magnified, since conventional "fishing" tools rely heavily on electrical characteristics (free point and coupling detectors, etc.), which are distorted by high temperatures.

All of this has led to strict control over the drilling mud, a determining element. Continuous supervision of its composition, especially dissolved solids, and temperature, which must be kept within limits by means of cooling towers, has effectively aided in solving many of these problems. The drilling mud used in Cerro Prieto is typically a bentonite-base mud to which the appropriate chemical reactants are added. Materials for its preparation include barium sulfate, modified tannins, sodium lignosulfate, and caustic lignite or calcium carbonate or hydroxide, in addition to the bentonite. Temperatures over 200°C demand the use of an organic polymer to keep the mud's thixotropic properties within acceptable ranges during long periods (up to 40 hrs) of noncirculation; for instance, when pressure, temperature, or electric logs are run. Table 5 lists the required physicochemical properties of a typical Cerro Prieto drilling mud. Granular and fibrous material is used when circulation fluids are lost despite good drilling-mud control, in addition to flakes, sodium silicate, calcium chloride, and cement. When loss occurs during cementation, a polymer added to the cement usually fosters almost immediate setting.

Casing cementation has proved to be another critical element at Cerro Prieto, in particular the very frequent fluid loss while the slurry is being poured, due to changes in the pump rhythm, which generate loss-inducing excess pressures. Such

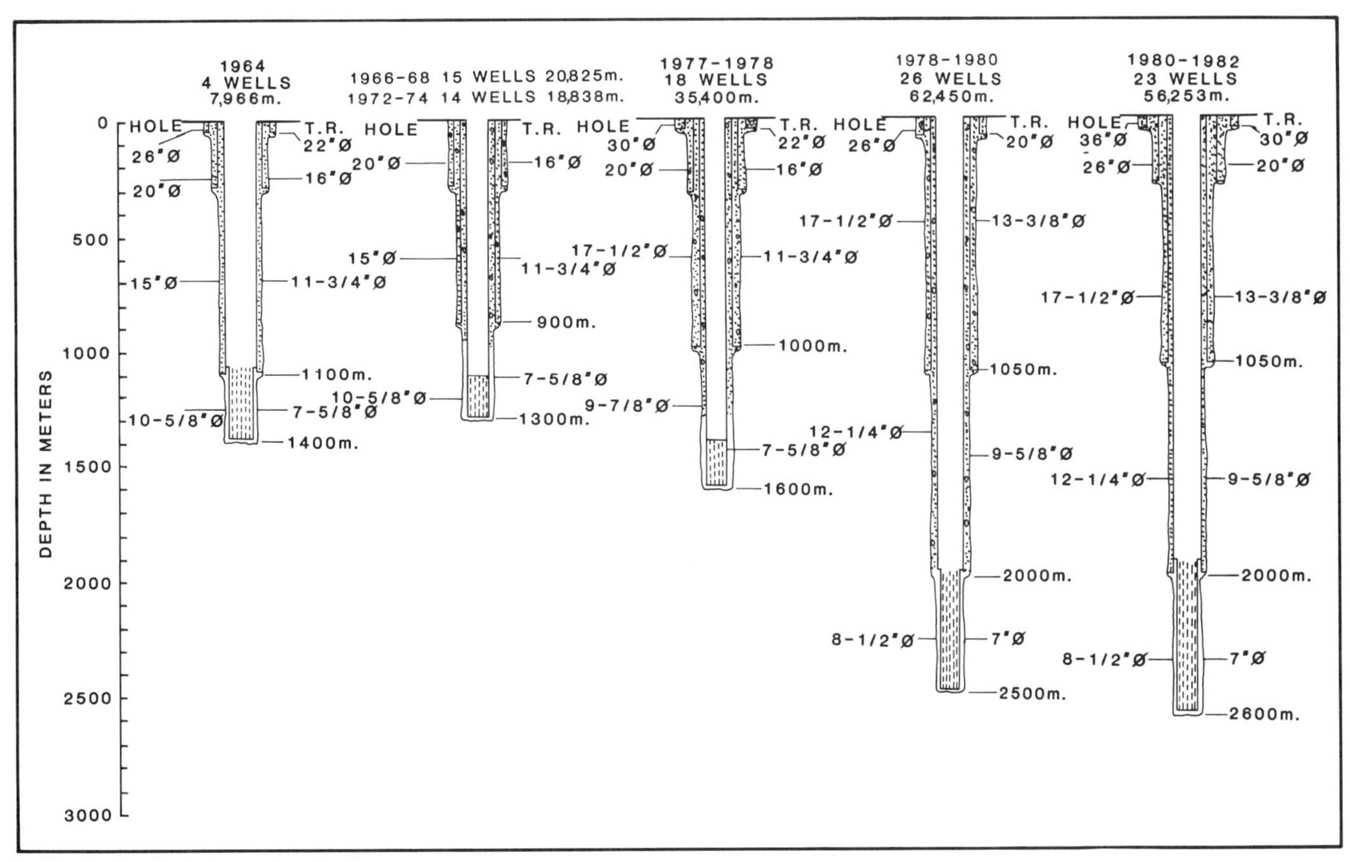

Figure 19. Prototype tubular diagrams in Cerro Prieto wells (after Dominguez, 1982).

TABLE 5. TYPICAL PHYSICOCHEMICAL PROPERTIES OF DRILLING MUD IN CERRO PRIETO

Weight	9 lb/gal
Velocity in funnel	40 s/lt
Plastic viscosity	15 cps
Yielding point	5 lb/100 ft^2
Gel	0 to 4 lb/100 ft^2 0' to 10'
Filter velicity	6 ml/30 min
Mud sealing	1 mm
pH	9
Solids	12 percent of volume
Oil	6 percent of volume
Water	82 percent volume
Sand	2 percent maximum of volume

losses cause channeling of the total absence of cement along stretches of the annular space between the casing and the formation, or between casings, which can seriously impair the well. To correct this, silica sand is applied by gravity; the sand is supplied intermittently to fill the annular space lacking cement. When drilling is resumed, the sand responds to the vibration by falling into and filling up the space.

Completion is the final stage of any geothermal well. This includes all the procedures and operations involved in the casing design to allow the reservoir-to-surface flow of geothermal fluids. Through experience a series of sometimes empirical parameters has been defined that helps to determine the proper time and conditions for successful completion of a geothermal well. These are: (a) drilling-mud entry and outlet temperatures; (b) temperature logging; (c) well-log interpretation; (d) presence of high-temperature hydrothermal minerals (epidote, biotite, chlorite); (e) type of sandstone cement (calcite or silica), and (f) lithologic features of the formation (shale color, estimated sandstone percentage, shape, and size of ditch samples).

Shale color is very important; experience has proved that as color darkens, temperatures and production problems rise. The sandstone percentage generally indicates the presence of permeable horizons, and the probabilities of good production increase with higher percentages. Finally, the ditch and core samples may indicate possible faults or fractures.

It has been found that hydrothermal conditions will be different according to the more calcareous or siliceous character of the sandstone cement, as well as the hydrothermal mineral content. Similarly, considerable differences between drilling-mud entry and outlet temperatures may indicate intervals of higher heat content at depth. The role of electric and temperature logs in defining well completion is also obvious, by reason of the subsurface geological information they provide.

Well-drilling time varies widely, depending both on depth and specific problems that may arise. For a general estimation, wells M-21A and M-150 may be taken as examples. M-21A, located nearly at the center of the CP-I area (Fig. 6), reached 1,300 m total depth in 57 days during 1973 and 1974. Figure 20 illustrates the various drilling stages; drilling time proper was 63 percent of the total period, with 37 percent taken up by pipe setting, cementing, and logging. This is the normal duration for wells of comparable depth in Cerro Prieto.

M-150, located in the northern half of the Cerro Prieto II area (CP-II, Fig. 6) was completed in 87 days in late 1978 at a final depth of 2,104 m. Figure 21 includes the progress diagram, which shows a drilling time of less than 30 percent of the total duration, more than two-thirds of which was spent logging, cementing, and other operations. In particular, it took almost 50 days, representing over half the total time, for drilling of the last 300 m of the well because of the need to solve problems such as collapses, circulation losses, and entrapments. This is a good example of the times and work rhythms involved when some of the previously mentioned difficulties arise. The initial 1,800 m were drilled in less than a month, including logging and pipe and cement settings.

Drilling practice at Cerro Prieto has shown well-hole deviation to be an undesirable alternative, because during the subsequent production process, pipes tend to erode quickly because of the effect of the solids in the fluids, which swiftly abrade the pipes at points of change of direction. Moreover, deflected pipes undergo tension-compression strains imposed by higher than normal thermal effects. Thus, although this may be a viable alternative in other geothermal fields, where fluids may carry fewer solids or thermal variations are less drastic, at Cerro Prieto it is only a last-stand attempt to save an imperiled well.

EVALUATION OF THE CERRO PRIETO FIELD

With gradually increasing knowledge of the physicochemical nature of the Cerro Prieto field, numerical evaluation of the reservoir potential became necessary. In the late 1970s the first systematic evaluation, simulation, and modeling were begun. The SHAFT (Simultaneous Heat and Fluid Transport) numerical simulator, developed by Pruess and Schroeder at Lawrence Berkeley Laboratories in 1979, was chosen for its effectiveness in relatively homogeneous geothermal fields such as Cerro Prieto.

This required integrating the relevant geological, geophysical, and geothermal data available to 1982, with the primary objective of spatially demarking the reservoir for the adequately adapted numerical program, and of providing the conceptual geothermal model together with its main average physical properties (i.e., dimensions, porosities, permeabilities, etc.).

To define the conceptual model, the horizontal limits of the reservoir at depths of 1,000, 1,300, 1,500, 1,750, 2,000, 2,500, and 3,000 m were determined; subsequent integration of these would define its vertical boundaries. Information was reliable only to the 3,000-m level, but by extrapolation, the boundary of the reservoir down to a 5,000-m depth was tentatively defined; this level was the average estimated level of the impermeable basement.

Several correlatable parameters were used to define horizontal boundaries: temperatures, lithology and Unit A/B contact,

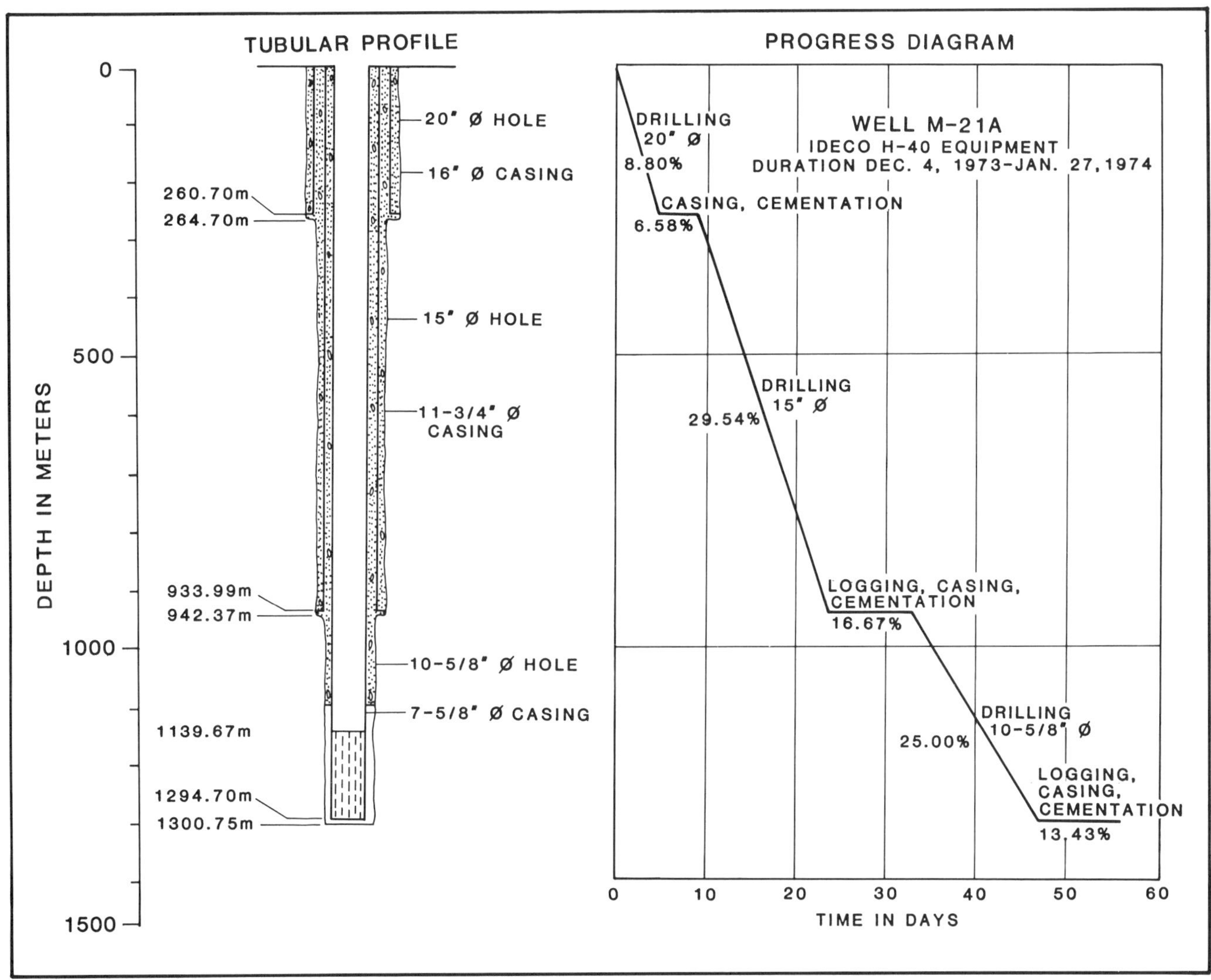

Figure 20. Progress diagram of well M-21A (after Dominguez, 1982).

increase in shale density and resistivity, depth of production pipe (slotted casing), structures, and top of the silica and epidote cement zone.

The first parameter, temperature, was the most important. Critical temperature was arbitrarily set at 250°C; that is, whatever finally resulted as a reservoir should be at temperatures of 250°C or higher. The 250°C isotherm of stabilized temperatures obtained in the wells and interpolated throughout the field for each of the abovementioned depths was thus obtained. In principle, nothing outside the area enclosed by this isotherm at the various levels would form part of the reservoir. The 150°C isotherm was likewise obtained; the area between it and 250°C was considered to be the reservoir reserve.

The second parameter, lithology, proved that the Cerro Prieto geothermal fluids are extracted from the B and C lithologic units of alternating mudstones, sandstones, shales, silty shales, and siltstones. Thus, in order for the B and C unit rocks to contain geothermal fluids, they should be enclosed by the 250°C isotherm at each level.

The boundary at the base of the nonconsolidated sediments (base of Unit A) was also drawn at each depth level. This so-called A/B contact is actually a specific depth reached in each well during drilling, at which the drill-bit velocity lessens considerably. This contact is thought to function, at least partially, as the base of a sealing layer. The A/B contact may really be either (1) the physical expression of one or several aquitards, acting as an upper vertical boundary of the geothermal reservoir; or (2) highly self-sealing zones. Statistically, producing wells begin to flow once this contact is passed. Thus, the area defined as reservoir at each level must lie below it, to ensure the existence of whatever acts as a seal. For this, the contoured A/B contact surface was obtained and its intersection with each level defined on the corresponding map, together with the isotherms.

The third parameter was the sharply increased shale density and resistivity at a certain depth, detected on the electrical well logs. This increase is related to a zone of strong hydrothermal alteration, the assumption being that a larger hydrothermal mineral deposition would have the effect, among others, of raising the density of the host rock and decreasing its conductivity. Resistivity and density increases were therefore associated with the top of the reservoir, assumed to be physically represented by a conspicuously intensified zone of hydrothermal alteration. Consequently, although logs were not run in many of the more recent wells, a hypothetical reservoir top was drawn on the basis of those increases; from this, contours corresponding to each established depth were obtained.

To delimit the fourth parameter—depth at the top of the slotted casing—contours of equal depth at the top of the slotted casing were drawn, from which each level contour was traced and correlated with those of the other parameters.

The last parameter to be considered was the top of the silica and epidote cement zone. In the field the sandstones in the producing intervals usually proved to have intergranular silica and epidote cement, the presence of epidote signaling hydrothermal conditions favorable to production. The respective contours for each level were therefore correlated with all the others. In addition, the traces of the main subsurface faults were projected on each work level at slope angles of 82°.

The boundaries of the geothermal reservoir were defined for each depth level by comparing the various parameters, as shown on Figure 22.

Similar operations and procedures at levels of 1,000, 1,300, 1,500, 1,750, 2,000, 2,500, and 3,000 m depth produced the respective areas of the reservoir, subsequently multiplied by the compensated vertical distances to obtain volumes.

The reservoir was thus structured to a depth of 3,000 m as a series of superimposed irregular prisms with a total volume of 22.9 km^3. If the data are extended to 5,000 m depth the total volume of the reservoir would be 54.2 km^3.

Similar procedures for reservoir reserve estimates resulted in 27.2 km^3 to 3,000 m depth and 67.3 km^3 to 5,000 m. The shape of the reservoir as a grouping of superimposed irregular prisms is shown on Figure 23.

Thus, the in situ reservoir is a volume of rock at a tempera-

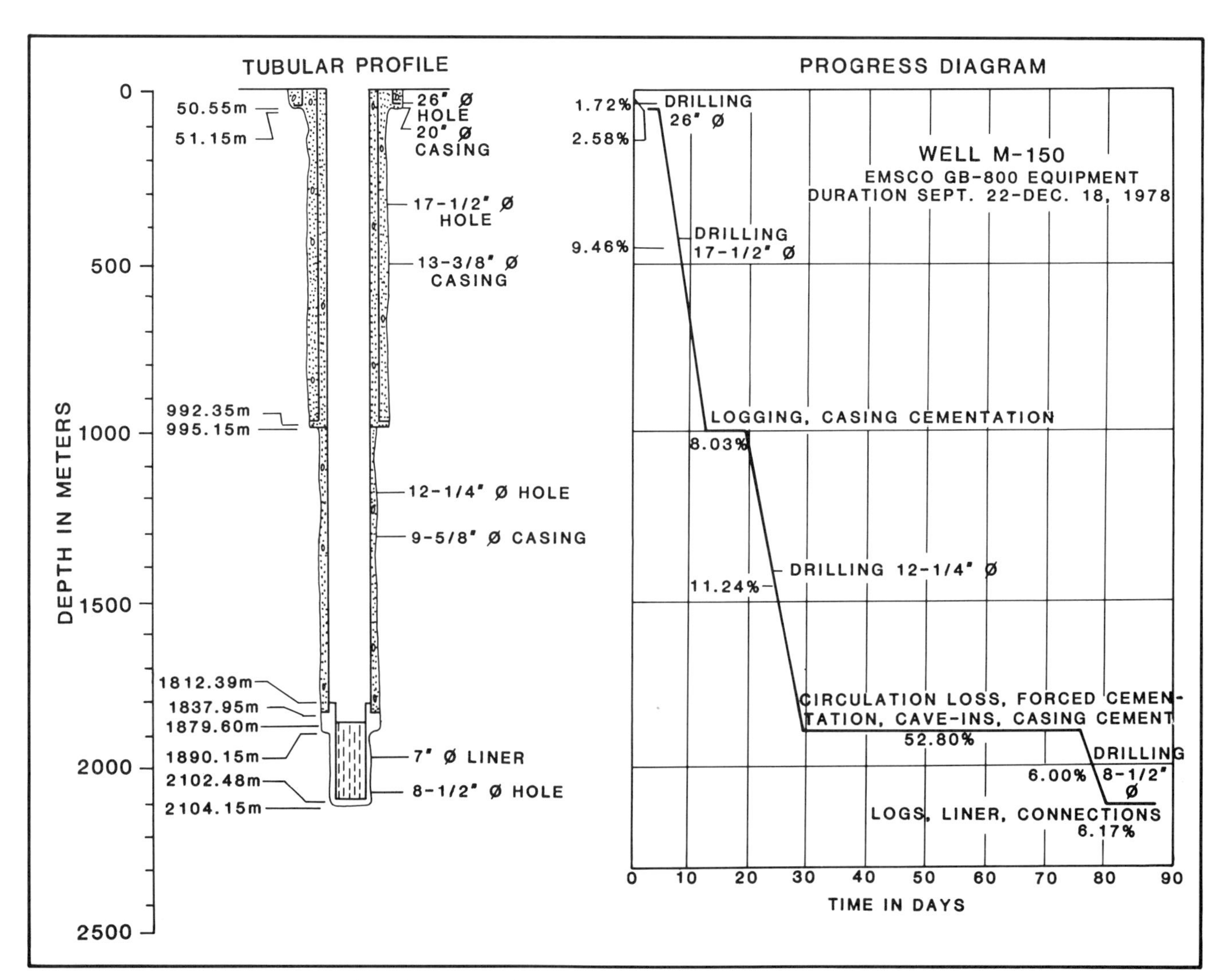

Figure 21. Progress diagram, well M-150 (after Dominguez, 1982).

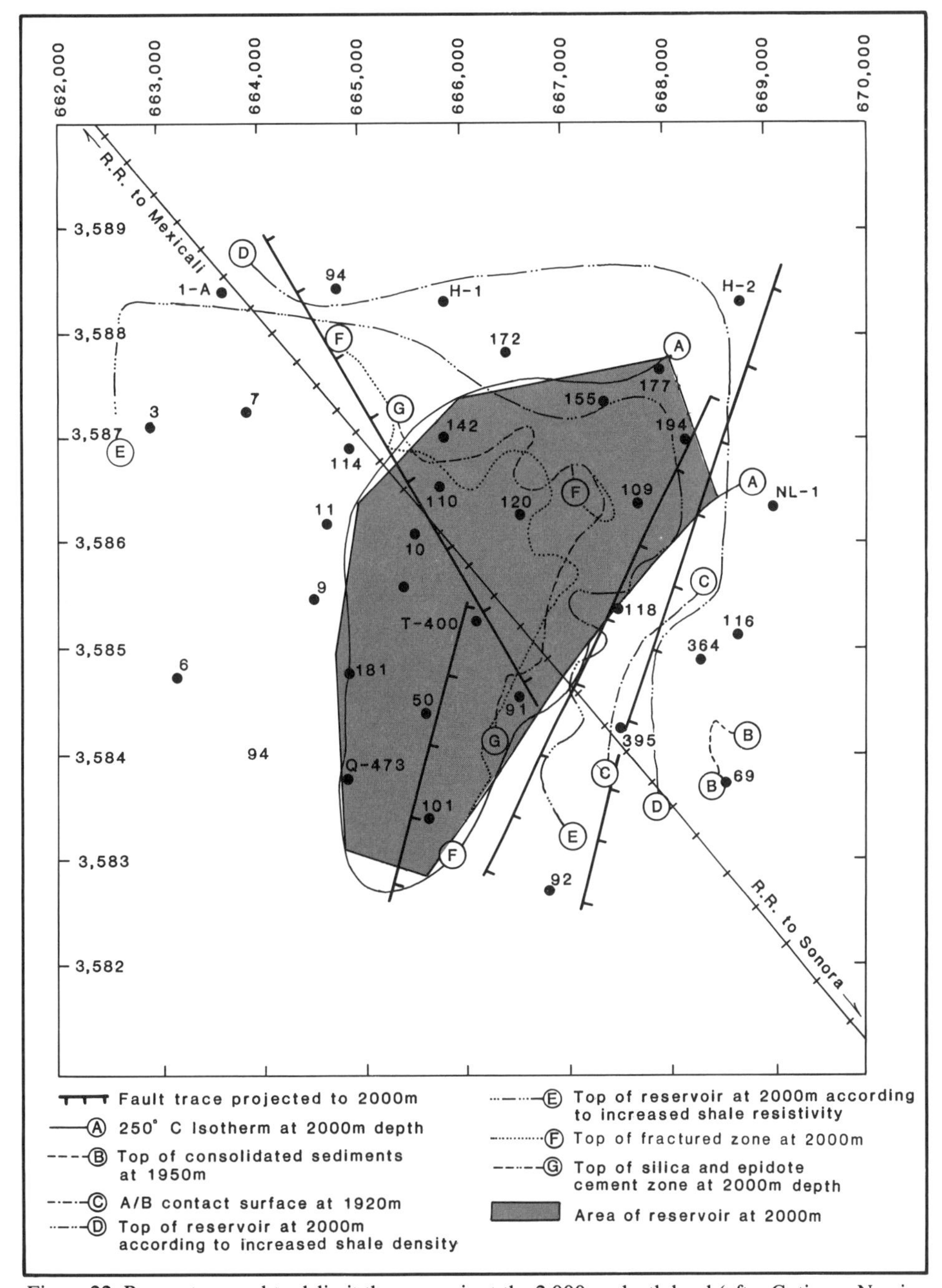

Figure 22. Parameters used to delimit the reservoir at the 2,000-m-depth level (after Gutierrez Negrin, 1982).

ture of 250°C or higher, composed of sedimentary rocks of Units B and C capable of containing geothermal fluids, showing perceptibly heightened resistivity and density values related to the geothermal activity, having silica and epidote as the main sandstone cement, topped by a sealing layer, and including a proved producing interval. The *reserve* is a cooler (between 150 and 250°C) rock satisfying the requirements for containing fluids that will feed the reservoir.

The volume of pores or empty spaces within both the reservoir and the reserve was computed to determine their fluid volumes. Taking into account that both the reservoir and the reserve consist of a heterogeneous interfingering of sandstones with shales that have low porosities and virtually no permeability, the average sandstone abundance at each level was determined on the basis of average abundances in representative Cerro Prieto well samples.

Average porosities and permeabilities were likewise defined at each level based on the results of all the analyses of Cerro Prieto sandstones done by different laboratories, the average sandstone distribution contours at each depth level, and the com-

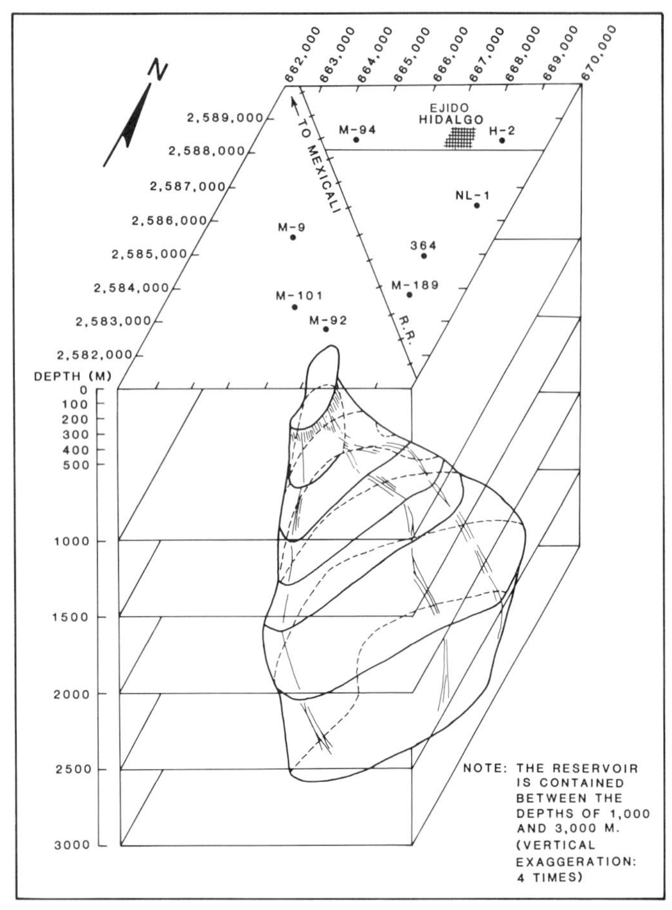

Figure 23. Idealized shape of the geothermal reservoir at Cerro Prieto, Baja California (after Gutierrez Negrin, 1982).

puted porosities and permeabilities. The partial reservoir and reserve volumes were multiplied by their total average porosities at each corresponding level, and the probable reservoir geothermal fluid volumes were thus defined. The results showed reservoir fluid volumes of 1.6 km^3 to 3,000 m depth, and 3.9 km^3 to 5,000 m depth. Reserve fluid volumes were 2.0 km^3 to 3,000 m, and 5.0 km^3 to 5,000 m. Total average porosity was 7.5 percent between 1,000 and 5,000 m, and total average permeability, 6.26 md. These data are summarized on Table 6.

The volumes cited above are the available in situ volumes for fluid occupation, not taking potential recharge into account. If only reservoir and reserve fluid volumes to the 3,000-m depth are considered, the resulting 3.6 km^3 would be capable of containing some 3,500 million tons of steam, an impressive volume in view of the fact that from 1973 to 1983 the total fluid volume extracted in Cerro Prieto I was only 298.4 million tons (Table 7). However, it should be noted that the above porosity and permeability averages are valid only for the total considered volume; locally they may either increase or decrease considerably, as proved in pressure tests.

Reservoir behavior was simulated within the limits thus defined, using a tridimensional grid with a total of 157 elements, which included the Mexicali Valley, the geothermal field, and its fault system between the 0- and 4,000-m-depth levels (Fig. 24). Data from 107 wells were compiled and grouped by elements as shown on Table 8. For the period from 1973 to 1983, the history of mixture production from the field, provided entirely by the CP-I sector, was distributed in nine elements. From 1983 onward, however, the steam needed to generate 620 MW was distributed in 16 elements, which included the first nine from CP-I, and seven CP-II and CP-III elements. As shown on Figure 24, the total simulation area was 2,250 km^2 (i.e., 45 km from east to west and 50 km from north to south).

The SHAFT reservoir behavior simulation incorporated 25 years (1973 to 1998) and required adjusting initial permeabilities and thermodynamic conditions with time. Results showed good agreement of the pressure and enthalpy patterns reported in the field from 1973 to 1983, with those computed in the nine CP-I elements for the same period. The computed steam demand also showed minimal variations compared to the actual data.

The numerical technique of the SHAFT-79 program was slightly modified to adapt to the various incidents that took place during development of the Cerro Prieto field. Pressure and temperature measurements in various wells and at different depths served as reference guides to evaluate the program estimations under fluid flow conditions.

The physical parameters of each element in the grid—porosity, permeability, density, thermal conductivity, specific heat, and compressibility—were computed on the basis of core sample laboratory tests, many of them done under pressure and temperature conditions comparable to those of the reservoir. Some data were provided by sandstone and shale percentage distributions, or from basement rocks or nonconsolidated sediments. Table 9 shows the permeability ranges, perhaps the most critical of the physical parameters.

All these values, though considerably higher than the 6-md total average obtained for the reservoir, lie within the ranges measured in core samples of the Cerro Prieto rocks. Outside the boundaries of the present reservoir, permeabilities are much higher. Nonconsolidated sediment (Unit A) permeabilities are also higher than those of the reservoir rocks.

The initial conditions of pressure and temperature were determined for each grid element on the basis of the corresponding measurements in the wells prior to the development of CP-I, and extrapolated according to the thermal and hydraulic gradients for CP-II and CP-III. The original parameters were adjusted several times to rectify differences, until the appropriate values were reached to repeat the known production history of the first ten years (1973 to 1983).

In general, the first numerical simulation results showed that the extraction regimen of fluids required to generate 620 MW can be sustained until the model date of 1998, and that in the first ten years of the field's commercial development the inflow of hot fluids from sectors CP-II and CP-III to CP-I was the chief factor in maintaining the extraction regimen. Sharp falls in pressure

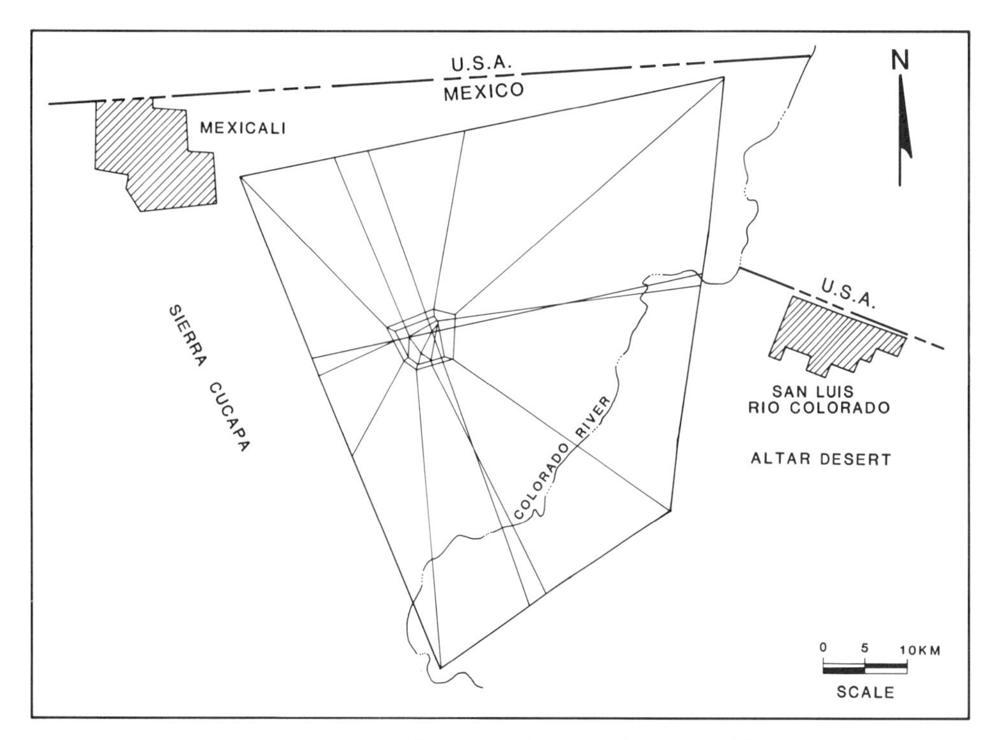

Figure 24. Total grid elements for the SHAFT-79 model.

TABLE 6. ROCK AND FLUID VOLUMES, POROSITY, AND PERMEABILITY FOR THE CERRO PRIETO RESERVOIR AND ITS RESERVE

	Rock volume (km^3)	Fluid volume (km^3)	Average porosity (%)	Average permeability (md and cm/seg)
1. Reservoir (to 3,000 m depth)	22.9144	1.61475	7.224	6.0313 md 5.8×10^{-6}
2. Reserve (to 3,000 m depth)	27.1668	2.03170	7.478	6.2596 md 6.02×10^{-6}
3. Additional reservoir (from 3,000 to 5,000 m)	31.3260	2.26300	7.224	6.0313 md 5.8×10^{-6}
4. Additional reserve (from 3,000 to 5,000 m)	40.5800	3.03480	7.478	6.2596 md 6.02×10^{-6}
5. Total reservoir (1 + 3)	54.2404	3.87775	7.224	6.0313 md 5.8×10^{-6}
6. Total reserve (2 + 4)	67.7468	5.06650	7.478	6.2596 md 6.02×10^{-6}
General average porosity			7.478	
General average permeability				6.2596 md

TABLE 7. FLUID PRODUCTION AND ELECTRIC GENERATION IN CERRO PRIETO

Year	Mixture production (ton x 10^6)	Electric generation (GMh)	Specific fluid consumption (ton/MWh)
1973	10.2	193	52.8
1974	18.7	463	40.4
1975	19.1	518	36.9
1976	22.1	579	38.2
1977	23.8	592	40.2
1978	22.0	598	36.8
1979	38.2	953	40.1
1980	33.1	915	36.2
1981	33.0	954	34.6
1982	38.7	1,263	30.6
1983	39.5	1,220	32.4
Total	298.4	8,248	

Note: Data after Mañón (1984).

TABLE 9. PERMEABILITY RANGES IN CERRO PRIETO

Modeling Zone	Horizontal Permeability	Vertical Permeability
NW of the geothermal field	30–100 md	5–10 md
Unit A (nonconsolidated sediments)	20–70 md	10–15 md
Units B and C (consolidated sediments between 2,000 and 3,000 m depth)	20–50 md	1–3 md
Unit B (between 2,000 and 4,000 m depth)	20 md	1 md

TABLE 8. GROUPING OF WELLS PER ELEMENT FOR THE SHAFT-79 MODEL

Element	Grouping of wells per element Well identification (production %)	No. of wells Total wells
1 CP-11	M-5, M-8, M-11, M-14, M-15A, M-19A, M-21A, M-28, M-27, M-29, M-31, M-35, M-38, M-42, M-25 (70%), M-30 (30%), M-39 (35%)	18 16.55
2 CP-12	M-45 (55%), M-46 (50%)	2 1.05
3 CP-13	M-50, M-48 (40%), M-51 (20%), M-90 (40%), M-101 (50%)	2.5
4 CP-21	M-34, M-43, M-130, E-3, M-25 (30%), M-30 (70%), M-39 (45%)	7 5.45
5 CP-22	M-47, M-105, M-181, M-45 (45%), M-46 (50%), E-1 (70%), E-4 (75%)	7 5.4
6 CP-23	M-84, M-48 (60%), M-51 (80%), M-79 (40%), M-90 (60%), M-101 (50%)	6 3.9
7 CP-25	M-104, M-113	2 2
8 CP-32	E-2, E-1 (30%), E-4 (25%)	3 1.55
9 CP-33	M-73, E-5, E-7, M-91, M-79 (60%)	5 4.9
10 CP-35	M-53, M-107, M-110, M-120, M-121, M-123, M-125, M-150, M-111 (25%), M-117 (70%), M-133 (75%), M-135 (45%)	12 10.15
11 CP-36	M-102, M-103, M-147, M-129, M-109 (70%), M-124 (65%)	6 5.35
12 CP-45	M-137, M-139, M-157, M-111 (75%), M-117 (30%), M-135 (55%)	8 4.6
13 CP-46	M-127, M-149, M-169, M-191, M-109 (30%), M-117 (30%), M-124 (35%)	8 4.95
14 CP-47	M-93, M-112, M-115, M-118, M-122, T-32B, T-350, T-388	8 8
15 CP-46	M-116, M-119, M-126, M-128, M-192, M-197, T-34B, T-364, T-366, T-386, T-395	11 11
16 T-201	M-114	1 1

were likewise predicted in sectors CP-II and CP-III between 1984 and 1987, tending subsequently to stabilize, possibly through overpressurization rather than by the effect of the extraction. In addition, entry of lower temperature fluids from the north was observed on simulating the initial operation of the CP-II and CP-III units. Despite their magnitude, these fluids should not greatly affect the enthalpy evolution, because higher temperature flows will also encroach from the deep zones east of the field. Finally, it will be possible to maintain well-head pressures and steam-flow values in the wells required for the CP-II and CP-III units, at least up to 1998. In CP-I, however, low pressures will preclude sustaining the present extraction rate after 1995. This has led to a recommendation that no greater volumes of mixture should be extracted than at present in this sector, and preferably they should even be reduced slightly, in order to aid recovery of the sector.

As shown above, the numerical simulation results are encouraging enough to warrant energy production of 620 MW at least until 1998. This does not imply that production cannot be maintained afterward. However, by 1998, Cerro Prieto I will have been in production for 25 years, a sufficient period to completely amortize investments in that plant.

Subsequent to the simulation, and taking into account the results of new exploratory deep wells drilled in the eastern sector of Cerro Prieto, the behavior simulation for an additional sector (CP-IV) was undertaken. The former grid of 157 elements was increased to a larger number (218) of smaller elements in the eastern sector, which would include CP-IV. The physical parameter values and initial thermodynamic conditions were maintained. Only permeabilities in the eastern sector were altered, and a deeper level—between 4,000- and 4,500-m depths—was adopted. The new simulation comprised a 30-year period, from 1973 to 2003.

An additional production of 220 MW was considered in CP-IV for staggered generation: 55 MW in June 1987, 55 MW in December 1988, 55 MW in February 1991, and a final 55 MW in November 1992. Thenceforward a total 840-MW production was simulated for Cierro Prieto, to be kept constant at least until the year 2003.

Results of this new mathematical simulation show it will be possible to extract the fluid required to produce the additional 220 MW, with bottom pressure falls in no case reaching the minimum for well-to-turbine flow. Minimum temperatures to the year 2003 will be higher than 250° C in the producing intervals considered in the simulation. Furthermore, in spite of the lower permeability values adopted, their effects on the producing intervals were negligible. In particular, the CP-IV sector showed up as capable of maintaining its assigned staggered 220-MW production, given the present distribution of pressures and enthalpies. However, production in this sector will come from depths of 3,500 to 4,500 m, and deeper drilling will be required. Otherwise, the production fluids will have low enthalpy.

To summarize, the new mathematical simulation proved that once the CP-IV sector begins production, predictable effects on CP-I, CP-II, and CP-III indicate the four sectors will be able to coexist in production, at least until the year 2003, with a combined total generation of 840 MW from 1992 onward.

SUMMARY AND CONCLUSIONS

The Cerro Prieto geothermal field is located in the boundary zone of the Pacific and North American Plates within the San Andreas fault complex, which moves at a rate of approximately 5.6 cm/yr. Metamorphic, intrusive, volcanic, marine, and continental sedimentary rocks of Paleozoic to Quaternary ages are exposed in its vicinity. Locally, the geothermal wells have penetrated sediments and sedimentary rocks, which in some cases overlie granitic basement and which have been grouped in three lithologic units: Unit A, nonconsolidated sediments; B, mudstones and sandstones; and C, shales and siltstones with interbedded sandstones. A fourth unit, D, is the Cretaceous granitic basement.

The field's major structures follow the northwest-southeast trend of the San Andreas system. However, a transverse northeast-southwest system acts as collector-distributor of underground hydrothermal fluids. The reservoir hydrothermal fluids are contained in deltaic sandstones and shales—Unit C—of undifferentiated Tertiary age.

The geothermal system is between 50,000 and 10,000 years old, and originated by heat transfer through the granitic basement from a probably very deep, tectonically caused zone of magmatism.

The Mexicali Valley has two regional aquifers: a shallow aquifer within the Unit A deltaic sediments, and a deeper aquifer within the deltaic rocks of Units B and C. The former, at present under multiple-well development for agriculture, appears to have a recharge in the range of 600 to 800 million m^3/yr. The Cerro Prieto geothermal field involves the deeper aquifer, which although less clearly defined than the upper, is believed to have supplied the reservoir zone a natural recharge of approximately 6 L/s prior to development, considering a hydraulic gradient of 0.008 and an average 7.5-md permeability. Subsurface conditions have altered with extraction, and recharge has increased considerably.

Within the 1,226-km^2 area of influence at and around the reservoir, 154 km^3 of in situ water is estimated in the aquifer. The water came from the Colorado River by infiltration during the part 4,000 years, reaching the deepest part of the geothermal reservoir chiefly from the northeast. The aquifer's underground flow probably followed a northeast-southwest direction, with some contributions from the northwest and southeast, to reach the northeast-southwest fracture system, pass through it and thereafter rise by convection to the geothermal reservoir at present under development. Part of the deeper aquifer's water was also provided by the shallower aquifer; the separation between them being an aquitard that allows some infiltration.

Chemical analyses and chemical geothermometers applied to fluids from thermal shows and wells show equilibrium temperatures of 264 to 363° C at depth in the reservoir. Isotopic analyses of well fluids indicate average ages ranging from 50,000 to 10,000 years.

Within the geologic framework of the Mexicali Valley, gravity, magnetometry, and reflection and refraction seismics were the geophysical methods contributing most to exploration of the Cerro Prieto field, by helping to define the structural and stratigraphic conditions related in some way with the geothermal reservoir. Continuous current or magnetotelluric surveys, which rely on the electric properties of the rocks, have not been successful owing to the terrain's high salinity. Seismic monitoring, precision gravity, and dipole-dipole array resistivity surveys are programmed as long-range experimental projects focusing mainly on reservoir development rhythms.

Geothermal well drilling technology at Cerro Prieto advanced significantly between completion of the first 755-m-deep wells in the 1950s, and the +4,000-m-deep wells drilled at present. The 133 wells completed to December 1984 implied drilling 286,411 m under geologic conditions affected by high temperatures, which called for special techniques and procedures in drilling and casing operations, well-pipe selection, and drilling-mud preparation.

The conceptual geothermal model of the field has combined the relevant geological, geophysical, geochemical, and production information to define the shape and boundaries of the known geothermal reservoir (with temperatures of approximately 250°C) and of its reserve (with temperatures of 150 to 250°C). The reservoir comprises a 22.9-km^3 volume, and the reserve 27.2 km^3 to the 3,000 m depth where information is reliable. By extrapolation to the 5,000 m depth, 54.3 and 67.3 km^3 volumes were obtained for reservoir and reserve, respectively. The porous volume capable of hosting geothermal fluids was 1.6 km^3 for the reservoir and 2.0 km^3 for the reserve, within the 1,000- to 3,000-m-depth interval only. If the interval from 1,000 to 5,000 m is considered, fluid volumes increase to 3.9 and 5.0 km^3 for the reservoir and reserve, respectively.

If the more restricted conditions are considered, the available geothermal fluid volume may contain some 3,500 million tons of steam, 11 times the total fluid volume extracted from the Cerro Prieto field from 1973 to 1983. The conceptual model establishes a 7.5-percent average regional porosity and a 6.26-md average regional permeability in the interval from 1,000 to 5,000 m, although local variations may be considerable.

Based on the conceptual model, the geothermal reservoir behavior was simulated twice, using the SHAFT-79 mathematical program designed by the Lawrence Berkeley Laboratory and adapted by the CFE to the particular conditions of Cerro Prieto. The first used a tridimensional grid of 157 elements comprising 2,250 km^2 of surface area to a depth of 4,000 m. The second model was patterned on a grid of 218 elements over the same area but to a 4,500-m depth in the eastern portion. Both reproduced the production history of the field between 1973 and 1983 with good approximation, confirming the field's capacity to produce 620 MW, and a total 840 MW from 1993 to at least the year 2003, combining production from the CP-I, CP-II, CP-III, and CP-IV sectors.

The simulation-predicted decreases of pressure in wells and of enthalpy in fluids were tolerable except for the CP-I sector, where it will not be feasible to sustain the present production rate after 1995. Production increase there is therefore inadvisable and should rather be decreased if possible, to improve conditions in the other sectors. The adapted SHAFT-79 program proved valid

TABLE 10. CHARACTERISTICS OF CERRO PRIETO I WELLS*

Well	Total Depth (m)	Wellhead Pressure (kg/cm^2)	Mixture Production (ton/h)	Enthalpy (KJ/kg)
E-2	1,946	21	117	1,545
E-4	1,767	65	156	1,859
E-5	1,970	12	89	1,474
E-6	3,098	13	182	1,327
E-7	2,122	69	194	1,562
M-5	1,298	7	73	1,269
M-10A	1,803	72	139	1,876
M-11	1,395	8	52	1,327
M-14	1,296	30	119	1,185
M-19A	1,448	7	193	1,273
M-21A	1,300	8	54	1,377
M-25	1,399	17	125	1,223
M-26	1,271	9	78	1,269
M-29	1,309	7	83	1,139
M-30	1,499	7	150	1,235
M-31	1,222	7	36	1,273
M-35	1,301	8	170	1,256
M-42	1,326	13	177	1,210
M-43	1,695	7	80	1,202
M-47	1,729	69	142	1,495
M-48	1,406	9	63	1,176
M-50	1,256	10	218	1,345
M-51	1,599	17	232	1,487
M-73	1,885	20	188	1,432
M-79	1,813	21	204	1,444
M-84	1,696	8	111	1,098
M-90	1,385	8	128	1,256
M-91	2,299	9	170	1,457
M-102	1,996	7	25	1,098
M-103	2,015	8	48	2,096
M-104	1,725	7	29	1,938
M-105	1,680	10	118	1,470
M-114	1,696	7	145	1,197
M-120	2,101	29	127	1,784
M-130	1,696	7	142	1,319
M-169	2,396	62	177	1,415
Total: 36 wells	60,838		4,534	

*As of March 1984; after Mañón, 1984.

TABLE 11. CHARACTERISTICS OF CERRO PRIETO II WELLS*

Well	Total Depth (m)	Wellhead Pressure (kg/cm^2)	Mixture Production (ton/h)	Enthalpy (KJ/kg)
M-93	2,561	7	266	1,334
M-116	3,041	14	477	1,400
M-119	2,956	37	473	1,431
M-122	2,822	36	377	1,555
M-128	3,012	11	452	1,418
R-364	2,920	15	512	1,427
T-366	2,981	20	520	1,515
T-386	2,657	8	245	1,436
T-395	2,775	14	401	1,334
T-400	2,445	49	449	1,651
M-115	2,603	5	187	1,404
M-118	2,664	15	550	1,502
M-127	2.504	13	232	1,478
M-124	2,436	49	517	1,574
M-147	1,908	47	546	1,833
M-149	2,395	13	229	1,538
M-169	2,396	12	350	1,515
T-328	2,695	7	224	1,562
T-348	2,895	11	438	1,595
T-388	2,570	23	448	1,495
T-350	3,117	14	461	1,510
Total: 21 wells	56,359		8,354	

*As of March 1984; after Mañón, 1984.

TABLE 12. CHARACTERISTICS OF FUTURE WELLS FOR CP-III*

Well	Total Depth (m)	Wellhead Pressure (kg/cm^2)	Combined Production (ton/h)	Enthalpy (KJ/kg)
M-109	2,396	35	451	1,574
M-110	1,996	15	521	1,449
M-117	2,495	14	512	1,683
M-120	2,101	11	272	1,737
M-125	2,315	13	454	1,570
M-191	2,499	17	539	1,677
M-133	2,450	11	346	1,590
M-137	2,504	16	494	1,555
M-139	2,494	16	500	1,694
M-150	2,104	10	164	2,043
M-157	2,621	11	390	1,700
Total: 11 wells	25,975		4,643	

*As of March 1984; after Mañón, 1984.

and useful in predicting the Cerro Prieto field behavior and is therefore of primary importance in future decision making.

Present Cerro Prieto installed capacity is 400 MW, as follows: Cerro Prieto I: 180 MW from four 37.5-MW units and one low-pressure 30-MW unit using secondary steam provided by water discarded from the other four; Cerro Prieto II: 220 MW from two 110-MW units.

Construction of Cerro Prieto III began in May 1981 with two additional 110-MW units, which should begin operating in late 1985, for a total Cerro Prieto installed capacity of 620 MW. According to the mathematical simulation results, the construction of an additional plant (Cerro Prieto IV) is feasible for a total 220 MW of installed capacity. Construction would be staggered, with four 55-MW units to be operating successively in mid-1987, late 1988, early 1991, and late 1992. Plant construction would depend on the efficient functioning of the previous plants. If the project is successful, Cerro Prieto will produce a total 840 MW.

Production and present characteristics of the CP-I wells and those programmed for CP-II and CP-III are shown on Tables 10, 11, and 12. A total of 36 wells at CP-I have an average depth of 1,690 m, an average 126 ton/h production of mixture, and a total combined 4,534 ton/h of fluid. Accumulated annual production is 39.7 million tons of fluids. Twenty-one wells have been planned and evaluated to date for the CP-II power station, with an average depth of 2,684 m, a high average production of 348 ton/h of mixture, and a complete production of 8,354 ton/h of fluid. Annual production of these 21 wells will be 73 million tons of mixture. Finally, 11 wells are programmed for the CP-III station with an average depth of 2,361 m, average mixture production of 422 ton/h per well, and a total combined production of 4,643 ton/h. Annual production will be 40.7 million tons of mixture. By 1985 the joint production of the 68 producing wells in the three geothermal field power stations will be 17,531 ton/h of mixture. There will be approximately 153 million tons of fluids yearly, representing almost 400 percent of 1983 production and over half of the entire accumulated 1973 to 1983 production (see Table 7).

The fluid production rate in Cerro Prieto from 1973 to 1983 (Fig. 25) shows an almost five-fold increase from

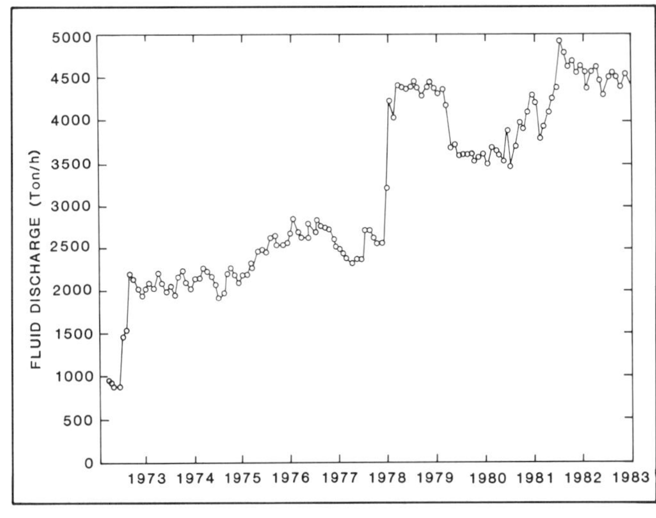

Figure 25. Fluid production curve in Cerro Prieto (after Bermejo, 1984).

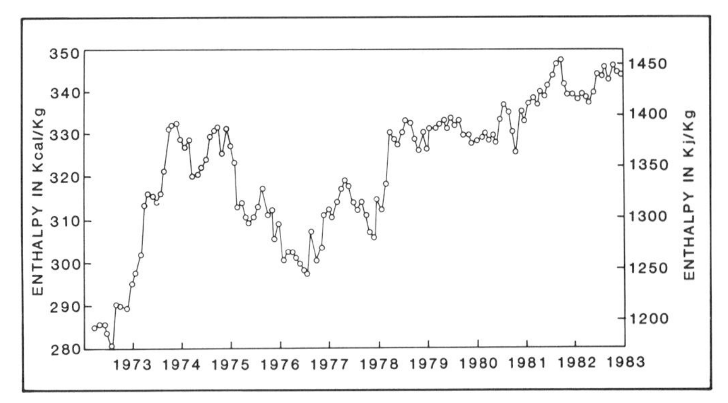

Figure 26. Average enthalpy of fluids produced in Cerro Prieto (after Bermejo, 1984).

of fluids generated by the wells increased approximately 25 percent from 1973 to 1983, despite a temporary decrease in mid-1977 (Fig. 26).

The CP-II and CP-III geothermoelectric power stations are similar to those of a conventional thermoelectric plant except for the following: the steam boiler is replaced by the geothermal reservoir and steam-producing wells; the high gas content of geothermal steam requires using direct-contact condensers; the use of geothermal steam after condensation makes it unnecessary to have an external water source to replenish losses due to evaporation in the cooling tower (Fig. 27). All this fosters high plant efficiency, of as much as 90 percent.

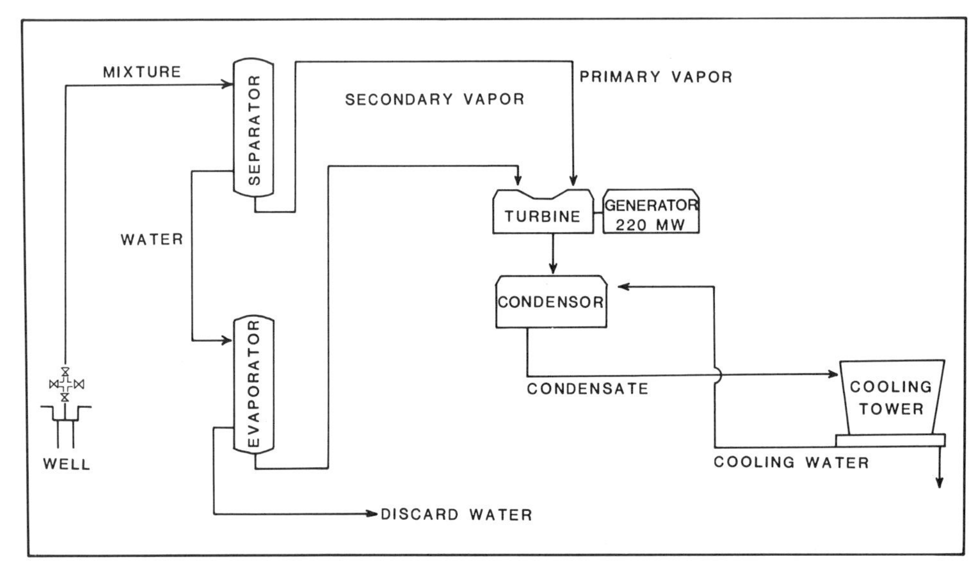

Figure 27. Flow diagram for a 220-MW geothermal power station.

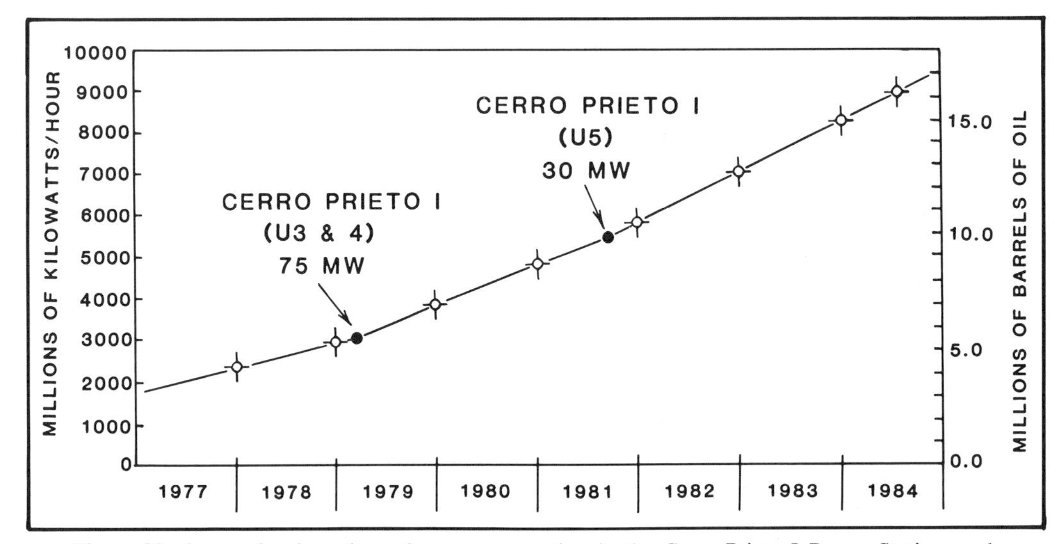

Figure 28. Accumulated geothermal energy generation in the Cerro Prieto I Power Station, and equivalent barrels of oil (after internal report of CECP).

Construction costs of such power stations vary. The CFE invested approximately 98 million dollars at June 1984 prices in building the CP-II and CP-III plants, which includes costs of the wells drilled for their operation. This figure seems high, but it should be stressed that the CP-I station alone, with its 180-MW capacity, has generated over 9 billion KWH, equivalent to some 17 million barrels of oil (Fig. 28), which at international prices for June 1984 totals some 337 million dollars. Furthermore, power generation predictions (Fig. 29) ensure complete recovery of investments in CP-II and CP-III by year-end 1985, having produced by then some 13 billion KWH, equivalent to approximately 25 million barrels of oil, which at December 1984 prices represents over 519 million dollars. If these predictions are valid and CP-IV initial production dates are kept (Fig. 29), by year-end 1995, accumulated generation will be 65 billion KWH, equivalent to almost 120 million barrels of oil and over 2.6 billion dollars at December 1984 prices. Also, part of that energy will be exported to southern California, in fulfillment of present commitments providing for annual sales in the amount of 90 million dollars at December 1984 prices.

Besides power generation, geothermal energy has many secondary applications. The most important are shown on Figure 30, and some are being put into effect in Cerro Prieto: use of

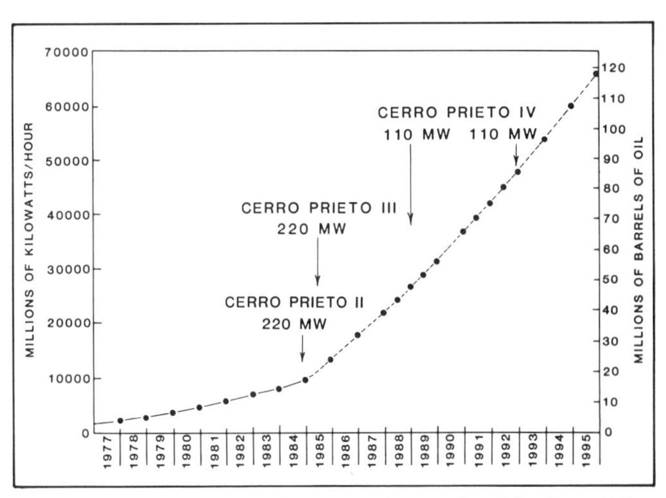

Figure 29. Geothermal electrical generation predicted in Cerro Prieto (after internal report of CECP).

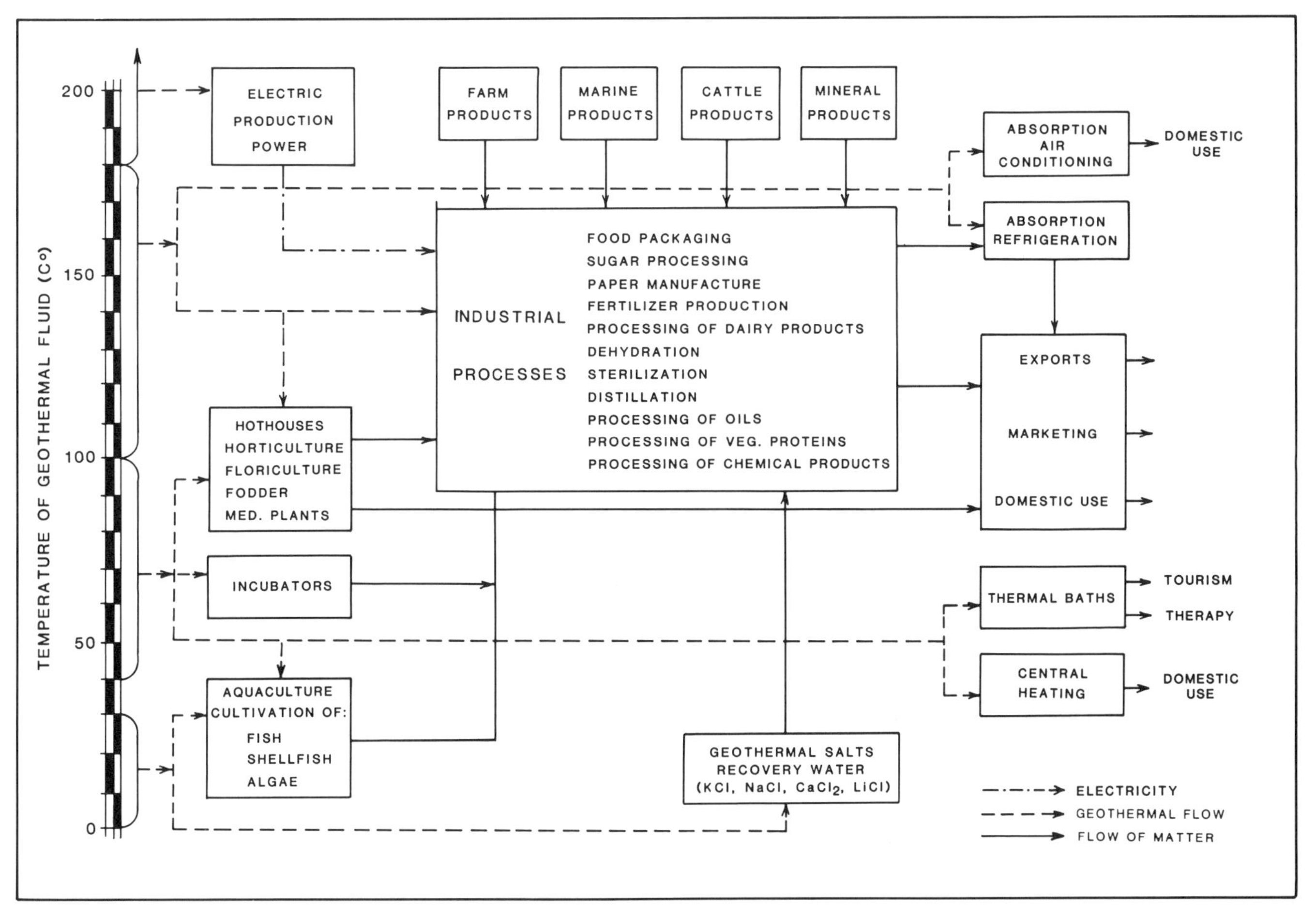

Figure 30. Potential use of geothermal energy.

discarded geothermal water in aquaculture; use of geothermal steam for air-conditioning in the main Cerro Prieto building and shortly in the CP-II and CP-III power stations; yearly production of 80,000 to 100,000 metric tons of potassium chloride, of strategic importance in nationwide agriculture and presently imported in its entirety; use of discarded geothermal steam for industrial and food production purposes; and use of low-temperature heat in farms and agricultural industries.

BIBLIOGRAPHY

Albores, A., Reyes, A., Brune, J., Gonzalez, J., Garcilazo, L., and Suarez, F., 1980, Seismicity studies in the region of the Cerro Prieto geothermal field: Geothermics, v. 9, p. 65–70.

Atwater, T., and Molnar, P., 1973, Relative motion of the Pacific and North American Plates deduced from sea-floor spreading in the Atlantic, Indian, and South Pacific Oceans, *in* Rovack, R. L., and Nur, A., eds., Proceedings of the Conference on Tectonic Problems in the San Andreas Fault System: Stanford, California, Stanford University Publications in the Geological Sciences, v. 13, p. 136–148.

Ayuso, M. A., Marquez, R., Esquer, C. A., and Negrin, G., 1984, Modeling studies in the Cerro Prieto geothermal field: Geothermal Resources Council Transactions, v. 8, p. 175–181.

Cobo, J. M., Cobo, F. J., and Bermejo, F. J., 1984, Geological model of the geothermal field at Cerro Prieto: Geothermal Resources Council Transactions, v. 8, p. 189–192.

Corwin, R. F., Morrison, H. F., Diaz, S., and Rodriquez, J., 1980, Self-potential studies at the Cerro Prieto geothermal field: Geothermics, v. 9, p. 39–47.

DeBoer, J., 1979, Paleomagnetism of the Quaternary Cerro Prieto, Crater Elegante, and Salton Buttes Volcanic Domes in the northern part of the Gulf of California rhombochasm: Actas del Segundo Simposio sobre el campo geotérmico de Cerro Prieto, Baja California, México: Mexicali, México, Comisión Federal de Electricidad and U.S. Department of Energy, Division of Thermal Energy, p. 91–98.

De La Pena, L. A., Puente, C. I., and Diaz, C. E., 1979, Modelo del campo geotérmico de Cerro Prieto: Actas del Segundo Simposio sobre el campo geotérmico de Cerro Prieto, Baja California, México, p. 303–328.

Demant, A., 1981, Plio-Quaternary volcanism of the Santa Rosalia Area, Baja California, México, *in* Ortlieb, L., and Roldan, J., eds., Geology of northwestern Mexico and southern Arizona, Field Guides and papers: México, D.F., Universidad Nacional Autonoma de México, Instituto de Geología, Estacion Regional del Noroeste, p. 295–307.

Dominquez, A. B., 1982, Perforación geotérmica en Cerro Prieto: Actas del Cuarto Simposio sobre el campo geotérmico de Cerro Prieto, Baja California, México: Guadalajara, México, Comisión Federal de Electricidad and U.S. Department of Energy, Division of Geothermal Energy, p. 217–232.

Elders, W. A., Rex, R. W., Meidav, T., Robinson, P. T., and Biehler, S., 1972, Crustal spreading in southern California: Science, v. 178, p. 15–25.

Elders, W. A., Hoagland, J. R., and Williams, A. E., 1979, Distribution of hydrothermal mineral zones in the Cerro Prieto geothermal field of Baja California, Mexico: Actas del Segundo Simposio sobre el campo geotérmico de Cerro Prieto, Baja California, México: Mexicali, México, Comisión Federal de Electricidad and U.S. Department of Energy, Division of Geothermal Energy, p. 265–284.

Fausto, J. J., Jimenez, M. E., and Esquer, I., 1981, Estado actual de los estudios sobre geoquímica hidrotermal en Cerro Prieto: Actas del Tercer Simposio sobre el campo geotermico de Cerro Prieto, Baja California, México: San Francisco, Comisión Federal de Electricidad and U.S. Department of Energy, Division of Geothermal Energy, p. 199–219.

Fitterman, D. V., and Corwin, R. F., 1982, Inversion of self-potential data from the Cerro Prieto geothermal field, Mexico: Geophysics, v. 47, p. 938–945.

Fonseca, H., and Razo, M. A., 1979, Gravity, magnetics and seismic reflection studies at the Cerro Prieto geothermal field: Actas del Segundo Simposio sobre el campo geotérmico de Cerro Preito, Baja California, México: Mexicali, México, Comisión, Federal de Electricidad and U.S. Department of Energy, Division of Geothermal Energy, p. 303–320.

Gamble, T. D., Goubau, W. M., Goldstein, N. E., Miracky, R., Stark, M., and Clarke, J., 1981, Magnetotelluric studies at Cerro Prieto: Geothermics, v. 10, p. 169–182.

Garcia, S., 1976, Geoelectric study of the Cerro Prieto geothermal area, Baja California, Mexico: San Francisco, California, Proceedings of 2nd U.N. Symposium on the Development and Use of Geothermal Resources, May 1975, p. 1003–1011.

Gastil, R. G., Phillips, R. P., and Allison, E. C., 1975, Reconnaissance geology of the state of Baja California: Geological Society of America Memoir 140, 179 p.

Goldstein, N. E., and Razo, M. A., 1980, General overview of geophysical studies at Cerro Prieto: Geothermics, v. 9, p. 1–5.

Goubau, W. M., Goldstein, N. E., and Clarke, J., 1981, Magnetotelluric studies at the Cerro Prieto geothermal field: Actas del Tercer Simposio sobre el campo geotérmico de Cerro Prieto, Baja California, México: San Francisco, Comisión Federal de Electricidad and U.S. Department of Energy, Division of Geothermal Energy, p. 357–367.

Grannel, R. B., Kroll, R. C., Wyman, R. M., and Aronstam, P. S., 1981, Precision gravity studies at Cerro Prieto; A progress report: Actas del Tercer Simposio sobre el campo geotérmico de Cerro Prieto, Baja California, México: San Francisco, Comisión Federal de Electricidad and U.S. Department of Energy, Division of Geothermal Energy, p. 329–333.

Grannel, R. B., Tarman, D. W., Clover, R. C., Leggewie, R. M., Goldstein, N. E., Chase, D. S., and Eppink, J., 1980, Precision gravity studies at Cerro Prieto: Geothermics, v. 9, p. 89–99.

Grant, M. A., Truesdell, A., and Manon, A., 1981, Production-induced boiling and cold water entry in the Cerro Prieto geothermal reservoir indicated by chemical and physical measurements: Actas del Tercer Simposio sobre el campo geotérmico de Cerro Prieto, Baja California, México: San Francisco, Comisión Federal de Electricidad and U.S. Department of Energy, Division of Geothermal Energy, p. 221–247.

Halfman, S. E., Lippman, M. J., Selwer, R., and Howard, J. H., 1984, Geological interpretation of geothermal fluid movement in Cerro Prieto field, Baja California, Mexico: American Association of Petroleum Geologists Bulletin, v. 68, p. 18–30.

Howard, J., and Apps, J. A., 1978, Geothermal resources and reservoir investigations of U.S. Bureau of Reclamation Lease Holds at East Mesa, Imperial Valley, California: University of California Lawrence Berkeley Laboratory, 5 p.

Kovack, R. L., Allen, C. R., and Press, F., 1962, Geophysical investigations in the Colorado Delta region: Journal of Geophysical Research, v. 67, p. 2485–2871.

Lippman, M. J., 1983, Overview of Cerro Prieto studies: Geothermics, v. 12, p. 265–289.

Lippman, M. J., and Manon, M. A., 1984, Status of the Cerro Prieto project: Proceedings of the 3rd Annual DOE Geothermal Program Review Meeting: San Francisco, Comisión Federal de Electricidad and U.S. Department of Energy, Division of Geothermal Energy, 21 p.

Lomnitz, C., Mooser, F., Allen, C. R., Brune, J. N., and Thatcher, W., 1970, Seismicity and tectonics of the northern Gulf of California region, Mexico; Preliminary results: Geofisica Internacional, v. 10, p. 37–48.

Lyons, D. J., and Van de Kamp, P. C., 1980, Subsurface geological and geophysical study of the Cerro Prieto geothermal field, Baja California, Mexico: University of California Lawrence Livermore Laboratory Report LBL-10540, p. 173–178.

Mañón, M. A., 1984, Recent activities at Cerro Prieto: Geothermal Resources Council Transactions, v. 8, p. 211–216.

Mañón, M. A., Sanchez, A., Fausto, J. J., Jimenez, M. E., Jacobo, A., and Esquer, I., 1978, Modelo Geoquimíco preliminar del campo geotérmico de Cerro Prieto: Actas del Primer Simposio sobre el campo geotérmico de Cerro Prieto, Baja California, México: San Diego, Comisión Federal de Electricidad and U.S. Department of Energy, Division of Geothermal Energy, p. 83–94.

Mazor, E., and Manon, M., 1978, Geochemical tracing in producing geothermal fields; A case study at Cerro Prieto: Actas del Primer Simposio sobre el campo geotérmico de Cerro Prieto, Baja California, México: San Diego, Comisión Federal de Electricidad and U.S. Department of Energy, Division of Geothermal Energy, p. 102–112.

Mazor, E., and Truesdell, A., 1981, Dynamics of a geothermal field traced by noble gases; Cerro Prieto, Mexico: Actas del Tercer Simposio sobre el campo geotérmico de Cerro Prieto, Baja California, México: San Francisco, Comisión Federal de Electricidad and U.S. Department of Energy, Division of Geothermal Energy, p. 163–177.

Nehering, N. L., and D'Amore, F., 1981, Gas chemistry and thermometry of the Cerro Prieto geothermal field: Actas del Tercer Simposio sobre el campo geotérmico de Cerro Prieto, Baja California, México: San Francisco, Comisión Federal de Electricidad and U.S. Department of Energy, Division of Geothermal Energy, p. 178–220.

Olivas, M.H.M., and Vaca, S.J.M.E., 1982, Criterios para determinar la profundidad de ademes en Cerro Prieto: Actas del Cuarto Simposio sobre el campo geotérmico de Cerro Prieto, Baja California, México: Guadalajara, México, Comisión Federal de Electricidad and U.S. Department of Energy, Division of Geothermal Energy, p. 233–244.

Palmer, T. D., Howard, J. H., and Lande, D. P., eds., 1975, Geothermal development of the Salton Trough, California and Mexico, *in* Meadows, K. F., ed., Geothermal world directory; 1975–1976 Bicentennial edition: Glendora, California, privately published, p. 280–317.

Pruess, K., Wilt, M. J., Boadvarsoon, G. S., and Goldstein, N. E., 1983, Simulation and resistivity modeling of a geothermal reservoir with water of different salinity: Geothermics, v. 12, p. 291–306.

Puente, C. E., and De La Pena, L. A., 1978, Geologia del campo geotérmico de Cerro Prieto: Actas del Primer Simposio sobre el campo geotérmico de Cerro Prieto, Baja California, México: San Diego, Comisión Federal de Electricidad and U.S. Department of Energy, Division of Geothermal Energy, p. 17–37.

Reed, M. J., 1976, Geology and hydrothermal metamorphism in the Cerro Prieto geothermal field: Proceedings of the 2nd U.N. Symposium on Development and Use of Geothermal Resources, v. I: Mexicali, México, Comisión Federal de Electricidad and U.S. Department of Energy, Division of Thermal Energy, p. 539–547.

——, 1984, Relationship between volcanism and hydrothermal activity at Cerro Prieto, Mexico: Geothermal Resources Council Transactions, v. 8, p. 217–221.

Reyes, A., and Razo, M. A., 1979, Microtectonic and potential field anomaly studies at the Cerro Prieto geothermal field: Actas del Segundo Simposio sobre el campo geotérmico de Cerro Prieto, Baja California, México: Mexicali, México, Comisión Federal de Electricidad and U.S. Department of Energy, Division of Geothermal Energy, p. 374–389.

Truesdell, A., and 6 others, 1978, Preliminary isotopic studies of fluids from the Cerro Prieto geothermal field: Actas del Primer Simposio sobre el campo geotérmico de Cerro Prieto, Baja California, México: San Diego, Comisión Federal de Electricidad and U.S. Department of Energy, Division of Geothermal Energy, p. 95–101.

Truesdell, A., Thompson, J. M., Coplen, T. B., Nehering, N. L., and Janik, C. J., 1979, The origin of the Cerro Prieto geothermal brine: Actas del Segundo Simposio sobre el campo geotérmico de Cerro Prieto, Baja California, México: Mexicali, México, Comisión Federal de Electricidad and U.S. Department of Energy, Division of Geothermal Energy, p. 241–254.

Valette, J. N., 1979, Geochemistry of the surface emissions in the Cerro Prieto geothermal field: Actas del Segundo Simposio sobre el campo geotérmico de Cerro Prieto, Baja California, México: Mexicali, México, Comisión Federal de Electricidad and U.S. Department of Energy, Division of Geothermal Energy, p. 263–280.

Valette-Silver, J. N., Thompson, J. M., and Ball, J. W., 1981, Relationship between water chemistry and sediment mineralogy in the Cerro Prieto geothermal field: Actas del Tercer Simposio sobre el campo geotérmico de Cerro Prieto, Baja California, México: San Francisco, Comisión Federal de Electricidad and U.S. Department of Energy, Division of Geothermal Energy, p. 263–280.

Von der Haar, S., and Howard, J. H., 1979, Intersecting faults and sandstone stratigraphy at the Cerro Prieto geothermal field: Actas del Segundo Simposio sobre el campo geotérmico de Cerro Prieto, Baja California, México: Mexicali, México, Comisión Federal de Electricidad and U.S. Department of Energy, Division of Geothermal Energy, p. 118–130.

Williams, A. E., and Elders, W. A., 1981, Oxygen isotope exchange in rocks and minerals from the Cerro Prieto geothermal system: Actas del Tercer Simposio sobre el campo geotérmico de Cerro Prieto, Baja California, México: San Francisco, Comisión Federal de Electricidad and U.S. Department of Energy, Division of Geothermal Energy, p. 149–157.

Wilt, M. J., Goldstein, N. E., and Razo, M. A., 1980, LBL Resistivity studies at Cerro Prieto: Geothermics, v. 9, p. 15–26.

Unpublished sources of information: Exploration Department, Comisión Federal de Electricidad

Alonso, E. H., 1964, Isotérmas del campo geotérmico de Cerro Prieto, Baja California, México: unnumbered report.

Arellano, G. F., and Razo, M. A., 1980a, Estudio de resistividad en la parte sur del Valle de Mexicali, B.C.N.: Report 4/80.

——, 1980b, Prospeccion eléctrica en el Valley de Mexicali y Campo Geotérmico de Cerro Prieto, B.C.N.: Report 32/80.

Calderon, G. A., 1958, Estudio geológico preliminar del area al sur y sureste de Mexicali, B.C., para aprovechar los recursos geotérmicos: unnumbered report.

Campos, E.J.O., Arredondo, F. J., and Vargas, L. H., 1984, Analisis de los soportes geológicos y geofísicos de pozos exploratorios en el Valley de Mexicali, B.C.N.: Report 5/84.

Diaz, C. S., and Arellano, G. F., 1979, Estudio de resistividad y potencial espontaneo en la parte sur del Valle de Mexicali, B.C.N.: Report 9/79.

Gutierrez, N.L.C.A., 1983a, Modelo geotérmico del campo de Cerro Prieto, B. C.: Report 6/83.

——, 1983b, Parámetros físicos del yacimiento en Cerro Prieto, B. C.: Report GG/9/83.

——, 1983c, Ampliación de los parámetros físicos para el campo geotérmico de Cerro Prieto, B. C.: Report GG/11/83.

Quijano, L.J.L., 1983, Origen del aqua subterranea del Valle de Mexicali y de la salmuera geotérmica de Cerro Prieto, con base en la información de isótopos ambientales: Report GQ-8-83.

Ramirez, S. G., 1983, Acuíferos y forma de recarga en el campo geotérmico de Cerro Prieto, B. C., (Recopilación bibliografica): Report GG/30/83.

——, 1985, Evaluación preliminar de la recarga del sistema geotérmico de Cerro Prieto, B. C.: Report 2/85.

Razo, M. A., and Arellano, G. F., 1978, Prospección eléctrica de la porción norte del Valle de Mexicali y campo geotérmico de Cerro Prieto, B. C.: Report 6/78.

Velasco, H. J., and Martinez, B.J.J., 1963, Levantamiento gravimétrico en la zona de Mexicali, B. C.: Bulletin 74.

Department of evaluation and deposits, Comisión Federal de Electricidad

Departmento de Evaluación y Yacimientos, 1983, Estudio de modelado en el campo geotérmico de Cerro Prieto, B. C.: Report 1383/015.

——, 1984, Comportamiento del yacimiento al integrar Cerro Prieto IV a la producción: Report 1384/024.

Executive Coordinator, Comisión Federal de Electricidad

Sanchez, R.J.J., and Terrazas, B., 1981, Hidrologia, Evaluación geohidrológica y térmica del acuífero Alfa: Report CP/1.

Theses

Barnard, F. L., 1968, Structural geology of the Sierra de los Cucapa, northeastern Baja California, Mexico, and Imperial County, California [Ph.D. thesis]: Boulder, Colorado, University of Colorado, Department of Geology, 157 p.

Secretaria de Agricultura y Recursos Hidraulicos (SARH)

Paredes, A. E., 1984, Informe preliminar sobre las condiciones geohidrologicas imperantes en el Distrito de Riego num. 014, Río Colorado, en el periodo comprendido del 9 de febrero de 1982 al 15 de febrero de 1984: Report 19.1-002.

Other reports

GEOCA, 1962, Levantamiento sismológico de refracción en la zone geotérmica de Cerro Prieto: Geólogos Consultores Asociados Report 19.2.1.2-002.

Martinez, M., 1982, Levantamiento magnetotelúrico en la area de Estacion Coahuila, Valle de Mexicali, B. C.: Baja California Institute of Science (IIE-CICESE), Report 15.1-002.

Printed in U.S.A.

The Geology of North America
Vol. P-3, Economic Geology, Mexico
The Geological Society of America, 1991

Chapter 6

Los Azufres geothermal field, Michoacán

G. H. Huitrón R., O. Palma P., H. Mendoza E., C. Sánchez H., A. Razo M., F. Arellano G., L.C.A. Gutiérrez N., and L. J. Quijano L.
Federal Commission on Electricity, Rio Rodano 14, 06500 Mexico, D.F.

INTRODUCTION

Los Azufres (Michoacán) is the second geothermal field under development in Mexico. Though still at the exploration stage, five wellhead power units are generating a total installed 25,000 KW. The geothermal zone is one of the many thermally active areas of the Geothermal Province in the central part of the country, characterized by its numerous Tertiary to Quaternary volcanic centers and high heat flow, resulting in a large concentration of thermal shows.

The Los Azufres field, 200 km west-northwest of Mexico City in northeast Michoacán (Fig. 1), covers approximately 30 km^2 of the Ciudad Hidalgo and Zinapécuaro municipalities. Hydrographically, it is centered at the drainage divide separating the closed Cuitzeo basin on the west from the Medio Balsas subbasin to the south and the Alto Lerma subbasin in the northwest, which are parts of the major hydrological pattern in the region.

The first geothermal energy evaluations in Mexico began in the 1950s at several localities. Los Azufres (formerly San Andrés) had already been recommended as an objective for geothermal exploration, but it was not until early 1975 that the Federal Commission on Electricity (CFE) undertook the first systematic geological, geophysical, and geochemical surveys to verify the area's conditions and determine the best prospects. The first exploratory phase led to a recommendation of five deep-well locations in a 25 km^2 area; these were drilled from 1976 to 1979, with encouraging results. The drilling program continued, and by 1982, five well-head turbogenerators, each with 50,000 KW capacity, had been installed. Development of the southern portion of the field to date has proved the presence of sufficient steam for a 50,000-KW plant, which will begin operation in late 1988.

Among the most relevant studies of Los Azufres and vicinity are the 1975 regional geological survey by the Geological Institute of the National Autonomous University of Mexico. This produced the first map of the area to be verified in the field. M. Palacios and F. Camacho drew up the first detailed map in 1976 and recommended deep drilling; that same year the CFE geophysical group, jointly with Electroconsult Company and an Italian group of volcanologists, carried out research, and simultaneously, Molina (1978) interpreted the preliminary geochemical information derived from analyses of the thermal shows. The geological, geochemical, and geophysical studies (the latter showing anomalies to be confirmed by deep drilling) led to the recommendation for the five exploratory wells, drilling of which began with well A-1 in the Agua Fría valley (see Fig. 5). From 1977 to 1985, A. Garfias, Z. Casarrubias, O. Rivera, and others covered the geology of 36 deep wells and produced semi-detailed mapping of the most promising geothermal areas. The geohydrological study of the area was done by F. Cedillo and others in 1981. V. de la Cruz completed detailed structural and volcanological studies in 1982. In 1985, V. Garduño and H. Martiñón undertook the structural and microstructural analysis of the field.

REGIONAL GEOLOGY

The oldest rocks in the area are part of the Mesozoic microfolded metamorphic complex of schistose rocks with interbedded calcareous horizons. The unit crops out south, east, and southeast of the area, in the Aporo, Senguio, Tuxpan, El Oro-Tlalpujahua, and Zitácuaro regions, among others (Fig. 2, Met), forming the regional basement, and is assigned to the Early Cretaceous. Southeast of the area, the metamorphic rocks are overlain by flysch-type alternating shales and sandstones assigned to the Late Cretaceous, which form the 100-km long and >50-km wide Patámbaro anticline trending NNW-SSE. The sequence was affected by the Laramide Orogeny in the early Tertiary. Overlying the sedimentary unit in the same anticline is a continental molasse-type detrital formation of conglomeratic sandstones with a high iron-oxide content, possibly corresponding to the Paleocene-Eocene Balsas Group (Fig. 2, Eoe).

Throughout the region these units are unconformably overlain at several localities by a sequence of locally brecciated and considerably weathered tuffs and andesitic-type rocks, believed to be products of Oligocene-Miocene volcanic activity (Fig. 2, Nit/Mom).

North of Tzitzio, just west of the map, middle Miocene acid tuff plateaus with brecciated horizons are found above the andesitic flows. West of Acámbaro in the Puerto de Cabras Sierra the

Huitrón R., G. H., Palma P., O., Mendoza E., H., Sánchez H., C., Razo M., A., Arellano G., F., Gutiérrez N., L.C.A., and Quijano L., L. J., 1991, Los Azufres geothermal field, Michoacán, *in* Salas, G. P., ed., Economic Geology, Mexico: Boulder, Colorado, Geological Society of America, The Geology of North America, v. P-3.

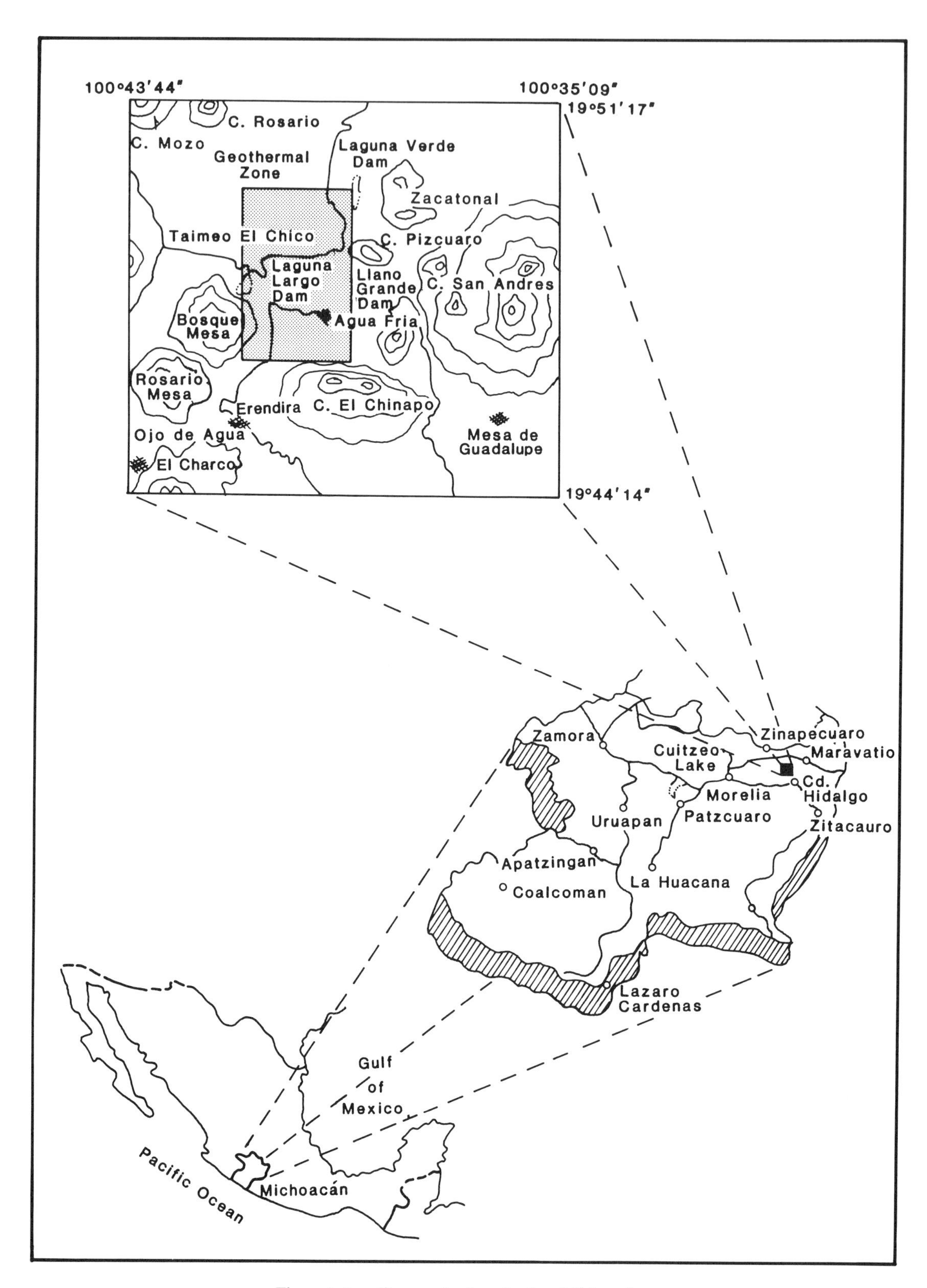

Figure 1. Locality map for Los Azufres, Michoacán.

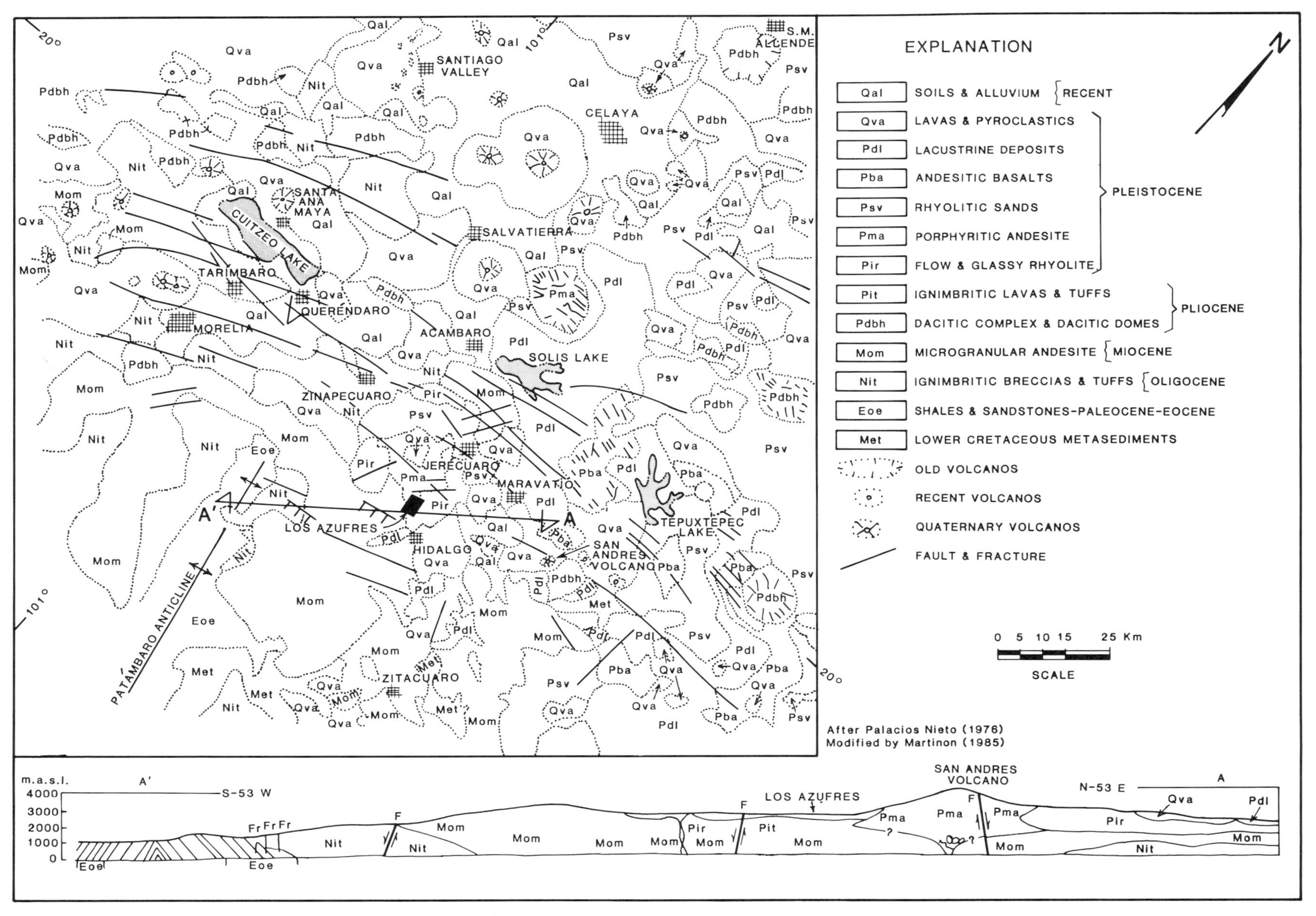

Figure 2. Regional geology of the Los Azufres area.

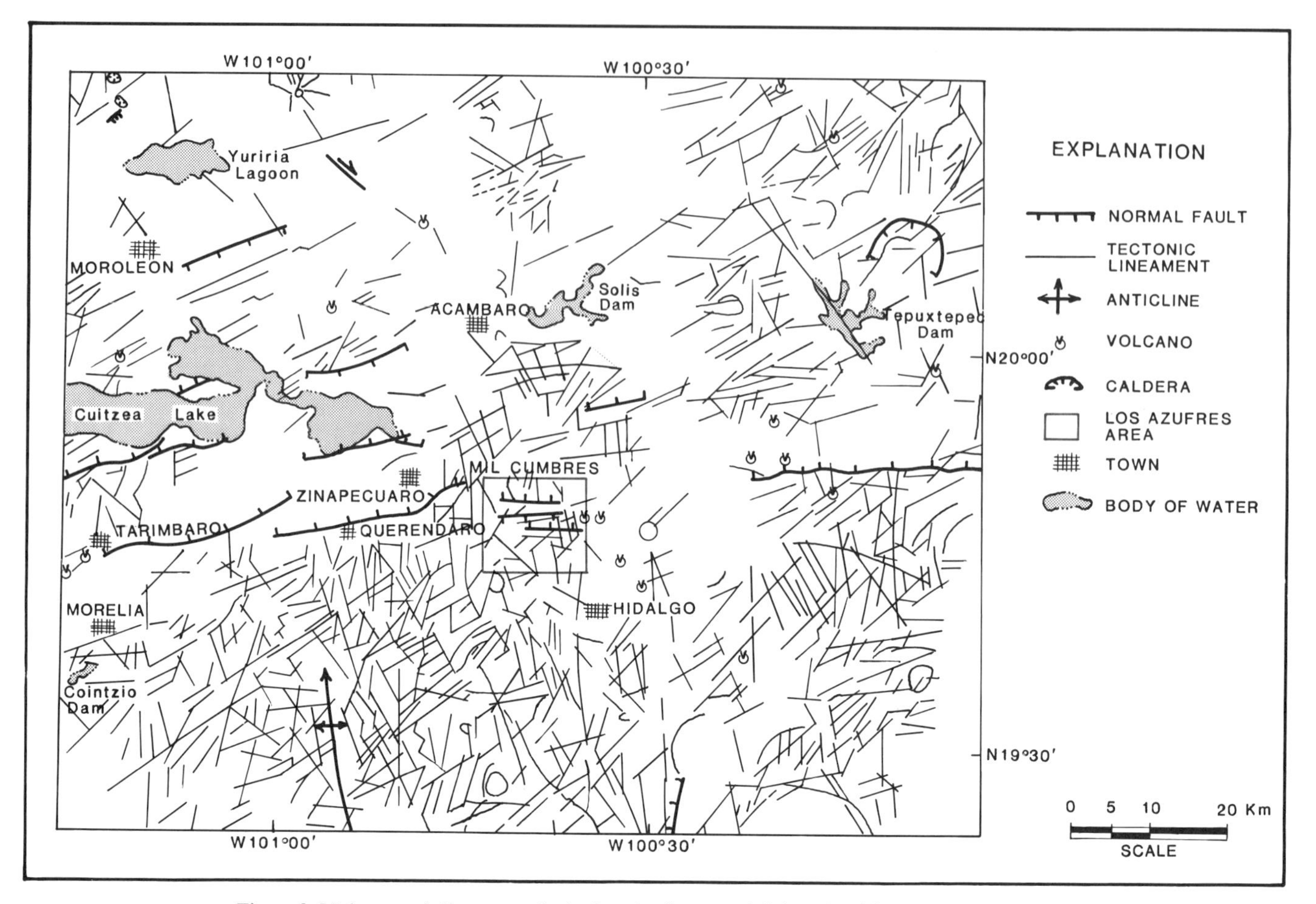

Figure 3. Main tectonic lineaments in the Los Azufres area, Michoacán. After De la Cruz (1983).

andesites are affected by a small dioritic intrusion of probable Miocene age on the basis of its field relations, although some assign it to the Pliocene.

The Los Azufres volcanic complex, of probable late Miocene to early Pliocene age (Fig. 2, Pdbh) is a series of lava flows ranging in composition from relatively basic andesite with pyroxene and olivine to dacite. The overlying Pliocene to Quaternary volcanism produced acid pyroclastic flows, andesitic lavas, rhyolitic and dacitic domes and basalt flows associated with slag cones, the latter covered locally by soils and alluvium (Fig. 2).

Tectonics

The Los Azufres area lies within the Mexican Volcanic Belt, which has undergone several tectonic episodes since possibly the Miocene, and continuing to the Quaternary. In general, Mexican Volcanic Belt tectonism is closely related to the geodynamic evolution of the Cocos, North American, and Caribbean Plates. Among other consequences, the stresses generated by these interacting plates developed a subduction zone in the Pacific known as the Acapulco Trench. The direction, depth, and length of the structures characterizing the Mexican Volcanic Belt appear to be effects of the geodynamic evolution that produced the Acapulco Trench and the Quaternary volcanic activity within it. Analysis of the plate movements leads to the conclusion that the N60°E and N7°E fault systems in the central and east portions of the Volcanic Belt are parallel and normal, respectively, to the direction of maximum compression in the Acapulco Trench.

Two structurally different systems in the Los Azufres region (Fig. 3) correspond to separate geologic provinces: the Volcanic Belt in the north and the Tertiary Balsas Basin in the south. The major, most extensive alignments (e.g., Parámbaro anticline) trend nearly north-south and occur in the Balsas Basin, together with a northeast-trending structural system. These directions and the density of the fractures are attributed to the various superimposed tectonic events that affected the province.

The Volcanic Belt structures are more recent. The main NE-SW–trending system occasionally produced structural highs (horsts) and deeps (grabens) such as Los Azufres and the easternmost portion of Cuitzeo Lake (Fig. 3). In contrast to the Balsas Basin Province, there is only one well-defined system within the belt. The north-south and northeast-southwest systems fade out toward the boundaries between the two provinces, except at specific locations where they reflect subsurface conditions.

Volcanic characteristics

The Volcanic Belt is made up of essentially andesitic stratovolcanos with pyroclastic deposits and flows ranging in composition from basalt to rhyolite, through dacites, rhyodacites, and mainly andesites. The general Pliocene-Quaternary volcanic activity of the entire Volcanic Belt covers approximately 20,000 km^2 in Michoacán State, where more than 3,000 volcanos are concentrated. The cones are mostly well preserved, allowing a good perception of the frequency of Quaternary eruptions.

The Michoacán volcanos apparently had only one primary phase of Pliocene-Quaternary activity with a minor volume of erupted lava at numerous eruptive centers. Volcanic products range from basaltic to rhyolitic; however, dacitic-rhyodacitic and rhyolitic phases are most frequent in east Michoacán (Ciudad Hidalgo sector), whereas the volcanism is more andesitic and basaltic in the more extensive western portion. Los Azufres lies in the San Andrés volcanic region to the east, composed of andesitic-dacitic (San Andrés) and rhyolitic (La Yerbabuena and El Carpintero domes) rocks.

GEOLOGY OF THE GEOTHERMAL FIELD

Exposures at Los Azufres of volcanic basic to acid calc-alkaline rocks range in age from Miocene-Pliocene to Pleistocene-Recent; considerable thicknesses of soil and alluvium are also present. The lithologic sequence from older to younger is described below (Fig. 4).

Stratigraphy

Microlitic andesite (Tmsa). This is the oldest lithologic unit within the limits of the field and forms the local basement; it is composed of andesitic flows with scarce interbedded pyroclastics, occasional basalts, porphyritic andesites, and dacitic lavas, which crop out as large irregular masses at the southernmost end of the field (Baños Eréndira) and at the northwestern flank of Cerro San Andrés in the north. The andesites are compact, vesicular, dark gray, and aphanitic, with conchoidal fracture. Some outcrops have been correlated with the Mil Cumbres and Queréndaro andesites and assigned to the Miocene-Pliocene on the basis of subsurface sample dating at 2,700 m depth (10.2 ± 0.6 Ma) and at the surface (3.1 ± 0.2 Ma).

Agua Fría rhyolite flows (Qrf). This unit unconformably overlies the andesite and consists of rhyolite domes and flows with obsidian and spherule lenses. It is exposed along the access road to the Agua Fría camp and occurs throughout the central part of the field (Fig. 4). The Agua Fría unit has two members: a lower rhyolite flow and an upper acid-brecciated tuff. Megascopically the rhyolite is pink-gray and compact, with flow texture and with spherules that are occasionally perlitic and highly weathered to kaolin. The unit ranges in age from 0.87 ± 0.02 Ma to 1.4 ± 0.4 Ma (i.e., Pleistocene).

San Andrés dacite (Qdp). The Agua Fría is unconformably overlain by the San Andrés greenish gray porphyritic dacite exposed in the easternmost part of the field (Fig. 4); the rock owes its pitted surface to dissolution of phenocrysts and inclusions of microgranular rock fragments. The unit's type locality is Cerro San Andrés, and its large, complex outcrops attain heights of as much as 700 m. K-Ar dates of 0.33 ± 0.07 Ma place it in the middle Pleistocene.

La Yerbabuena glassy rhyolite (Qrv). The La Yerbabuena glassy rhyolites make up the youngest extrusive assemblage in the geothermal field, exposed at the westernmost end where they form slightly altered circular domes. The Cerro La Yerbabuena exposures, considered typical (Fig. 4), consist of two members: glassy rhyolite below and pumiceous tuffs above. Reworked pyroclastic deposits are found around and above the domes. The unit ranges in age from 0.14 ± 0.02 to 0.30 ± 0.07 Ma (Quaternary), tending to be younger toward the west.

Tuffs and sands (Qpal). This unit comprises three types of loose or scarcely lithified material:

a. Tuffaceous-conglomeratic material formed of rhyolite flow, pumice, and obsidian fragments, exposed between San Pedro Jacuaro and Los Baños Eréndira at the southernmost end of the area.

b. Sandy tuffs of possibly explosive origin consisting of quartz crystals, fragments of pumice, glass, plagioclase, and volcanic ash with pseudo-stratification, exposed in the central-western portion of the field; this is a very friable material lacking compaction of any kind.

c. Light gray and yellowish fine material consisting of silica, organic matter, and sulfur, exposed with pseudostratification in the vicinity of the Verde and Los Azufres lagoons.

Basalts (Qvc). These Quaternary rocks crop out at the north and east ends of the field (Fig. 4) as melts and cinder and slag cones of ashes, sands, and basalt volcanic bombs.

Soils and alluvium. These occur in the field as thin layers covering and masking the other rocks; they are the weathering products of preexisting rocks and decomposed organic matter.

Structural geology

Structures at Los Azufres are crustal-type normal faults trending east-west and northeast-southwest and caused by tensional stresses; they reflect two regional systems. Two main fault systems corresponding to two different tectonic events have been defined: (1) An older northeast-southwest system affects only the oldest rocks of the area, the microlitic andesites (local basement). These exhibit two types of secondary fractures: block-fractures characterize the lower part, and horizontal slab-forming fractures the upper. Structures in this system are of minor extent (El Viejón and El Vampiro faults; Fig. 4). (2) A regional north-trending system, exposed at the Patámbaro anticline, gave rise to the northeast-southwest system and thus is penecontemporaneous and Miocene in age. This system appears clearly in the Tertiary

Balsas Basin but is very incipient in the north part of the field. A Recent east-west–trending fault system is responsible for the present landscape of horsts and grabens and staggered faults (Agua Fría, Los Azufres, Marítaro, Las Cumbres, San Alejo, etc.) that characterize the geothermal field. Most of the thermal shows occur within this system; however, since both the older and younger systems are also involved in the ascent of the fluids, some shows appear at their intersections (Fig. 4).

Subsurface geology

Present subsurface information comes from 48 wells that have penetrated five main rock types: basalts, andesites (microlitic, aphanitic, and porphyritic), pyroclastic (tuffs), dacites, and rhyolites. These make up the major lithologic units found. The first recognizable marker horizon is aphanitic andesites occurring immediately below the rhyolite flows in almost all the wells except A-1. They overlie a complex sequence of microlitic pyroxene andesites with interfingering porphyritic andesite, abundant pyroclastics, and occasional dacites. The thickest porphyritic andesites encountered in wells A-1, A-2, A-6, A-8, A-12, and A-16 (Fig. 5) correlate with a much thinner horizon in A-7 and A-26. These phenocryst-rich rocks were probably derived from cooler and more viscous lavas, or else from lavas that had flowed for very long distances.

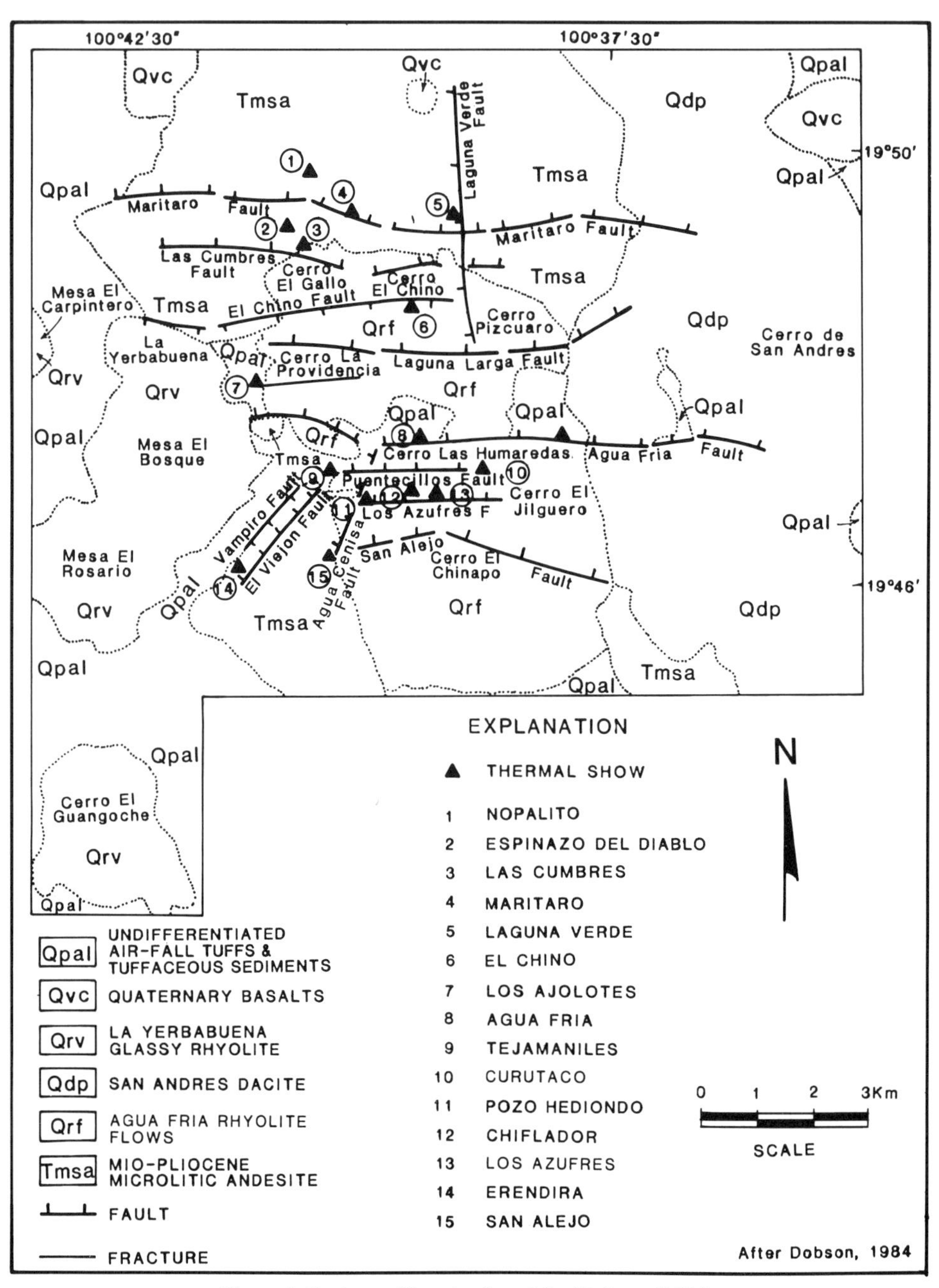

Figure 4. Geology of Los Azufres. After Dobson (1984).

At the shallower subsurface levels (<1,000 m depth), thinner rocks of basic composition occurring as melts and classified as basalts may be the products of a more recent volcanic cycle and correlative with similar rocks exposed to the north and northeast, outside of the geothermal field. At these shallower levels the dacites are more often associated with the porphyritic rather than with the fine-grained andesites, suggesting that the development of phenocrysts involves a more active magmatic differentiation. Injection well A-20 at the southernmost end of the field and actually outside its boundaries, showed the greatest abundance of dacites, which possibly indicates a different magmatic chamber south of the field.

At the deeper subsurface levels (below 1,400 m asl) the lithology varies somewhat. Dacites and porphyritic andesites are less frequent; microlitic andesites display basaltic characteristics and at times are substituted by basalts. These are encountered below 1,200 m asl in wells A-1, A-3, and A-20, among others. However, the deepest samples in wells A-1 and A-20 and at the bottom of well A-10 are dacites again, which may indicate the final acid stage of an older volcanic cycle.

Thermalism

Diverse thermal shows within the area appear mainly along structures; the most representative are Los Azufres Lagoon, Pozo Hediondo, El Currutaco, El Chiflador, Tejamaniles, Los Ajolotes, Agua Fría, San Alejo, El Chino, Marítaro, Laguna Verde, La Cumbre, and Nopalito (Fig. 4). These are steam- and gas-expelling hot springs where temperatures range from 30 to 95°C. Structural and geohydrologic conditions are particularly adequate for surface thermal activity in this area.

At several levels, isotherm contours based on well measurements (Fig. 5) were drawn to establish the relation between major structures and the highest-temperature subsurface zones. Conductive transference of heat in the field was found to take place uniformly throughout the entire reservoir rock volume. Convective transference occurs preferentially along fracture planes and lithologic contacts and develops very localized thermal plumes and lateral flows, observed in many wells, where temperature inversions take place. Intense fracturing and high flow enthalpy foster the development of steam zones, as in the Tejamaniles Module.

Figure 5 shows contours on the 225°C isotherm surface. This is the temperature considered a minimum for the geothermal reservoir to develop and defines two thermal zones at the shallower levels, which become one at depth (below 1,000 m asl). At shallow levels the 225° isotherm diminishes progressively due to the fracturing and the convective movements. The irregular shape of the isotherms clearly tends to follow structural controls imposed on one hand by faults conducting the hydrothermal fluids, and on the other by the boundary faults of the geothermal system.

In thermal show areas the rocks have undergone different forms of alteration: most extensively kaolinization, silicification, chloritization, and oxidation. In the central portion of the field, the Agua Fría rhyolites in particular are highly kaolinized.

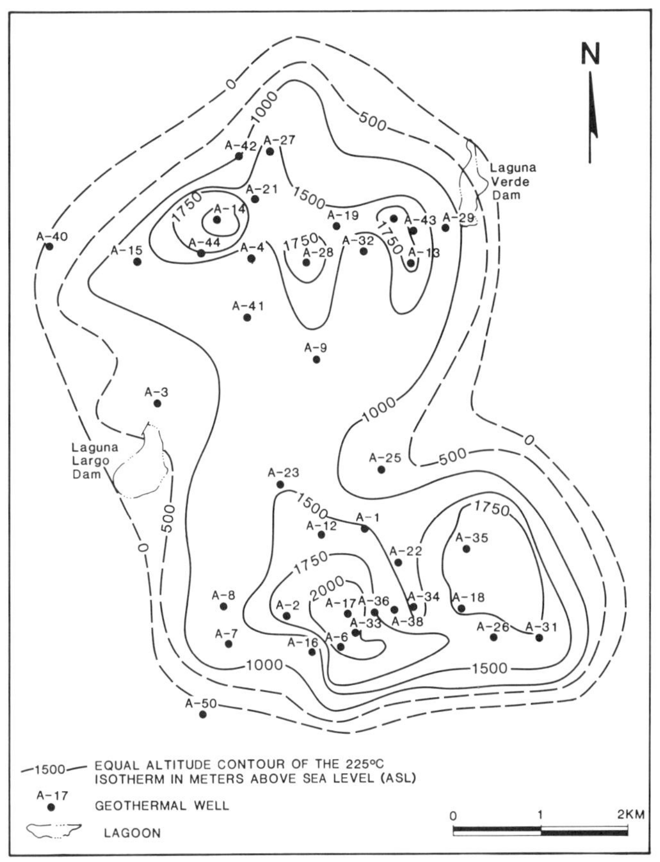

Figure 5. Contours of the 225-C isotherm, Los Azufres.

The Los Azufres subsurface geothermal system has developed a series of hydrothermal minerals as the result of geothermal fluids interacting with the host-rocks. It is difficult to establish a definite hydrothermal paragenesis in the Los Azufres subsurface, but systematic petrographic studies of ditch samples from all the wells have identified zeolites, clay minerals (montmorillonite, illite, kaolinite), chlorite, calcite, quartz, pyrite, epidote, amphiboles (tremolite-actinolite, hornblende), hematite, and oxides (leucoxene among others).

The generally irregular discontinuous mineral zoning is interpreted to result from two hypothetical superimposed geothermal systems that were spatially distinct and of different ages. However, the distribution of certain index minerals such as epidote, considered individually, agrees with that of present subsurface temperatures and with the presently producing zones of geothermal fluids, suggesting that these minerals developed in response to current conditions of the geothermal system. Estimates of the formation temperatures of minerals such as calcite, from fluid inclusions in calcite veinlets, are also similar to present temperatures measured at the same depths.

Figure 6 shows contours drawn at the top of the epidote in the Los Azufres subsurface as interpolated from well data, which indicate that at shallow levels (1,750 to 2,250 m asl) the epidote forms two zones—one in the north and one in the south—whereas at deeper levels (1,750 to 1,500 m asl) these meld into one.

The similar patterns of the epidote and isotherm contours, particularly the 225°C isotherm (Fig. 5) led to the establishment of two producing zones at Los Azufres, each with its own specific depth and fluid composition characteristics.

Geohydrologic characteristics of the field

A regional geohydrological survey over 4,700 km^2 between Los Azufres and Cuitzeo Lake (Fig. 2), comprising parts of the Cuitzeo Lake, Alto Lerma, and Medio Balsas hydrologic basins, defined the hydrologic behavior and probable relations between the aquifers and the hydrothermal system. Eleven hydrologic systems are present in the region. The aquifers are contained in alluvial and lacustrine deposits and in volcanic rocks. These aquifers are exploited by wells and water wheels in the Morelia-Queréndaro, Maravatío-Jerécuaro, and Santa Ana Maya valleys as well as on the plains bordering Cuitzeo Lake; they are untouched in the remaining valleys.

The overall area has an annual precipitation of 974 mm and average annual temperature of 16.8°C. At Los Azufres, average annual precipitation is 1,380 mm, and average annual temperature is 13°C, with 670 mm of average annual evapotranspiration.

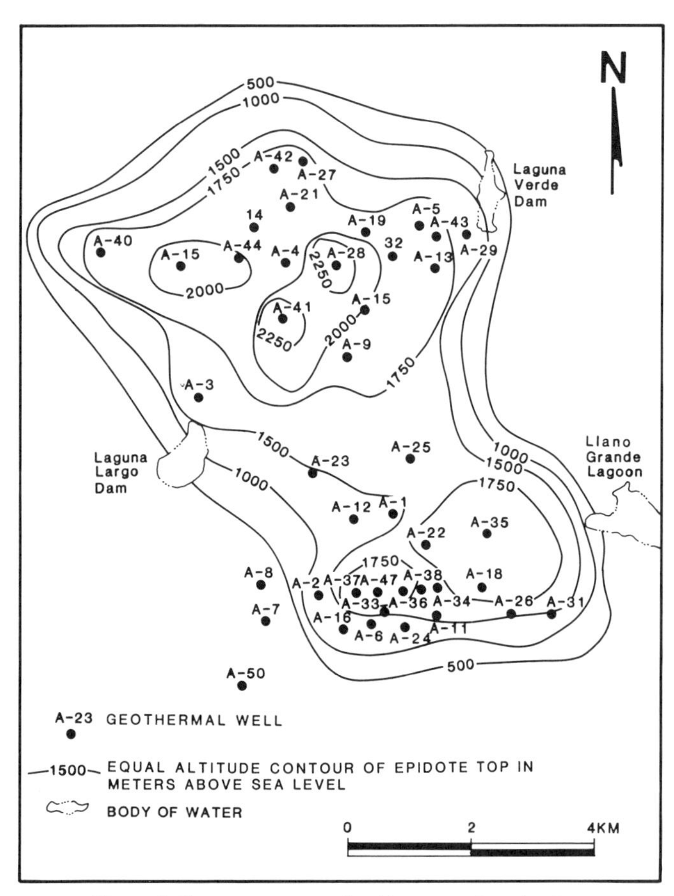

Figure 6. Contours on top of epidote zone, Los Azufres.

In the Morelia-Queréndaro Valley, aquifer recharge takes place at the foothills of the Mil Cumbres Sierra east of Zinapécuaro, west of Morelia, and west of Tarímbaro. At Los Azufres it occurs within the range itself and its surroundings. Underground flow appears to be from south to north in the Morelia area and south of the Cuitzeo area, and from southeast to northwest in the vicinity if Queréndaro.

Three aquifers were defined in the Los Azufres field: the deepest aquifer contains the geothermal reservoir. This aquifer is contained in Miocene-Pliocene andesitic rocks and connects with those at higher levels by way of faults and fractures, showing up at the surface as hot springs and fumaroles. Its recharge probably takes place by way of the structures also, through infiltration from the shallower aquifers and horizontally through the underground flow of regional aquifers connecting with that of Los Azufres.

The hydrologic equilibrium equation was applied over an area of 315 km^2, centered at the geothermal field, to define the possible Los Azufres recharge. This showed 136 million m^3/yr of infiltration, compared to 84 million m^3/yr determined theoretically. In that 315-km^2 area alone, infiltration must therefore be within that range; because the total recharge area is much greater, most probably infiltration is correspondingly greater.

Isopotential lines in Los Azufres wells drilled until 1981 established east-west and northeast-southwest flow of the hydrothermal system. Subsequent geochemical surveys show the flow also moving upward in the two hottest zones.

GEOPHYSICAL STUDIES

Geophysical investigation of the field utilized telluric, self-potential, passive seismic, and resistivity surveys carried out between 1975 and 1982 by the UNAM (Universidad Nacional Autonoma de Mexico) Geophysical Institute, CICESE (Centro de Investigación Científica y Estudios Superiores de Ensenada), and the CFE.

Telluric method

Telluric current was measured at frequencies of 8 to 0.05 Hz, assuming that responses would show low values of electric field quotient in the most conductive (or least resistive zones). Three zones were thereby defined as being of greater, average, and lesser conductivity. Measurement results show that electromagnetic wave penetration at the 8-Hz frequency, which is a probe element, varies according to the apparent resistivity (ρa) of the medium; for ρa between 5 and 10 ohm in the most conductive zone, penetration ranged from approximately 320 to 500 m, and for ρa = 20 ohm/m penetration was 1 km. Penetration was therefore taken to vary within the 320 to 1,000 m range. These results suggest a strong NNW-SSE–trending conductor at a depth of 3 to 8 km, corresponding geometrically to a dike. Subsequent information from wells and detailed geological surveys suggests, however, that if this conductor exists it is probably related to an old fracture system.

Self-potential

The self-potential survey indicated an extensive positive anomaly associated with the hydrothermal alteration zone, surrounded by a negative potential belt that widens toward the south (Fig. 7). According to flow potential theory, a flow of ions through a porous system will transport positive ion species preferentially and retain the negative; a similar situation will take place at zones of rising geothermal fluids. Thus, a positive ion-rich area would appear at the surface, whereas negative ions would concentrate where the fluids descend. Therefore, the positive anomaly would signal the most prospective geothermal area, bounded by the negative anomaly zones (Fig. 7). Both anomalies occur obviously within the geothermal zone, however, although they do define it.

Passive seismics

A passive seismic survey defined approximate values of the seismicity index together with analysis of the P- and S-wave propagation properties. Observations at five permanent seismological stations set up for a 9-month period focused on the statistical analysis of the responses to events that crossed the producing zones as compared to unknown areas of the reservoir. An east-west elongated anomalous body of low rigidity and density was interpreted in the southern zone (Fig. 8). High attenuation of shear stress waves also defined an area between the San Alejo and Tejamaniles–Los Azufres faults, very probably associated with the tensional stresses controlling these.

Although the geographic locations and depths of seismic foci are difficult to determine, they are very probably shallow, no more than 2 km deep. In the anomalous zone, for instance, the depths must be uniform due to easing of stresses by the increase or drop in pressure that characterizes geothermal reservoirs. Easing of thermal stresses within the cavities of these anomalous bodies may be associated with fluid migration toward or out of the reservoir. Also, the strongest seismic activity was found in areas surrounding stations P03, AJL, and P07, and is almost absent around stations P11 and AGF (Fig. 8).

Resistivity

The geoelectric methods used most to define resistivity anomalies related to geothermal reservoirs have been resistivity surveys. Six exploratory surveys—three of them regional—over Los Azufres defined low resistivity (10 ohm/m) anomalous zones, which were subsequently studied in detail on a total of 600 Schlumberger-type vertical electric logs. Analysis of apparent iso-resistivity and electrostratigraphic contours and profiles defined an area of values under 20 ohm/m, and within it four zones with resistivities under 10 ohm/m. These were finally grouped into two prospective geothermal zones: (1) north, in the area of wells A-27, A-40, and Cerro El Gallo; and (2) south, in the Agua Fría–Tejamaniles module (Fig. 9).

The location of the low-resistivity anomalies agrees with the geological hypothesis that the structural systems, in particular the east-west systems, foster hydrothermal fluid movement. The resistivity minima of the north and south areas are separated by a high-resistivity belt related to the scarce underground thermal activity in the area between the El Chino and Agua Fría faults (Fig. 4). The groups of low-resistivity anomalies occupy areas of 12 km^2 in the northern zone and 8 km^2 in the southern zone. Both cover the principal thermal shows and hydrothermal alteration areas.

An electrostratigraphic interpretation of Los Azufres suggests that the conductive horizon resistivities average less than 10 ohm/m, particularly approaching the faults and thermal-show areas. However, resistivities of up to 20 ohm/m are considered representative of this horizon.

In the central portion of the geothermal zone between the El Chino and Agua Fría faults, the conductive horizon with resistivities in the 10 to 20 ohm/m range is the deepest (2,500 m). Below this level, resistivities range from 55 to over 1,000 ohm/m, indicating the low permeability of the Miocene andesites that prevail in the subsurface of this area. The formations of the resistive basement underlying the conductive horizon are estimated to

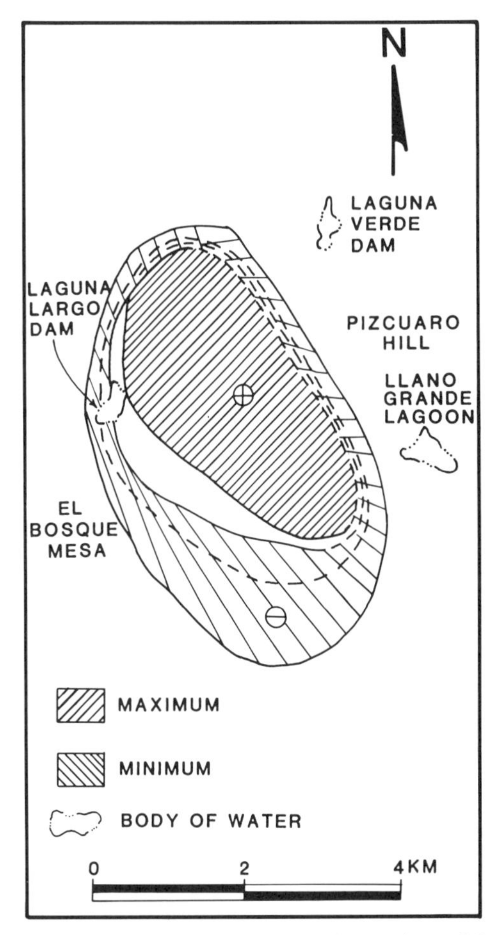

Figure 7. Anomalous self-potential zones. After Institute of Geophysics (1978).

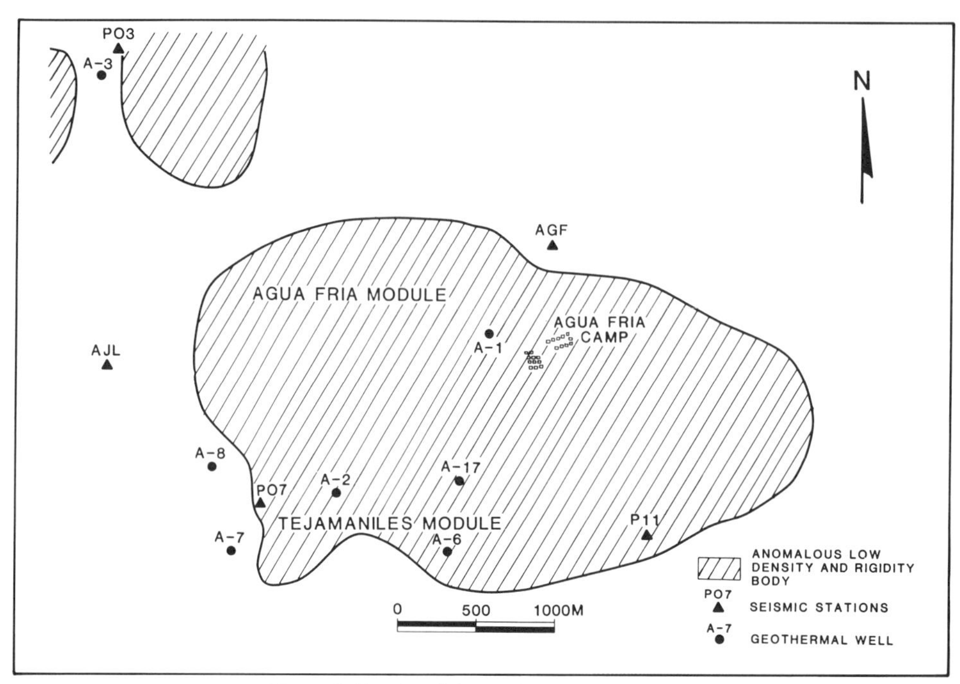

Figure 8. Results of the passive seismic survey.

lie at depths of 400 to 2,500 m; however, in some cases they could not be defined owing to their greater depth or to being saturated with geothermal fluids.

GEOCHEMICAL STUDIES

Surface thermal shows in Los Azufres were sampled for chemical analysis for several years. Two main types of water are present: sodium bicarbonate, and sulfate. The bicarbonate waters in the Eréndira and Tejamaniles areas (Fig. 4) have temperatures of 40 to 45°C, pH 6 to 7.8, and electric conductivity of 490 to 715 ohms/cm. The sulfate waters may be sodium-, calcium-, and potassium-bearing: sodium-bearing waters appear in the Los Ajolotes, Puentecillas, Laguna Larga, La Cumbre, Marítaro, Espinazo del Diablo, and Laguna Verde shows (Fig. 4) with temperatures of 40 to 88°C, pH 2 to 7.1, and electric conductivity 420 to 4,000 ohms/cm. Calcium-bearing waters are at El Currutaco and Pozo Hediondo with 80 to 88°C, pH 2.3 and 3.6, and electric conductivity of 2,225 to 6,100 ohms/cm. Potassium-bearing waters are found at San Alejo, El Chino, Nopalito, and Agua Fría with 38 to 86°C temperatures, pH 1 to 42, and electric conductivity of 300 to 1,300 ohms/cm. Table 1 shows typical compositions of bicarbonate water (Eréndira) and sulfate waters (La Cumbre).

Isotopic studies of samples from surface shows suggest that temperatures are higher in the north and lower in the south, according to the slope values of the isotopic enrichment lines. The stable-isotope content of samples in the northern zone also suggests these shows are steam discharged directly from the reservoir. In general, deuterium values (δ D^0/$_{00}$) ranged from –79.9 to –17.7 and oxygen-18 values (δ^{18}O, ^{0}/$_{00}$) from +6.91 and –10.5.

Isotopic and chemical composition of well discharges

Tables 2 and 3 show chemical compositions and specific enthalpies of fluids from wells in the northern and southern zones obtained in well samples at total discharge conditions, after waiting for the steam to dissipate; therefore, they represent the liquid phase. Bottom temperatures and specific enthalpies of the fluids produced at Los Azufres (Tables 2 and 3) show that in most cases the wells are fed a mixture of steam and liquid. This implies separation of phases in the producing layers.

Steam fractions in the feeding fluid, computed assuming an adiabatic process of field decompression in its ascent to the surface, are shown on Table 4.

A proposed geochemical model based on selective well fluid geochemical data and updated structural information uses "reactive" elements, whose concentration is a function of the aquifer water temperature, and "passive" or "conservative" elements that do not change with temperature, but help to explain concentration or dissolution in geothermal fluids on their way to the surface. The graphic representation of the steam in each well shows much greater quantities toward the south than the north (Fig. 10). This suggests some sort of steam cap in the south at shallower levels than the deeper water/steam mixture. Similarly, graphic representations of chloride contents (ppm) in waters of each well (Fig. 11) show generally homogeneous values in the northern-zone wells and widely varying, as well as minimum, values in the south; the lowest characterize the steam condensed in the shallower portion of the field.

In spite of this division of the Los Azufres field into northern and southern producing zones, the fluid feeding the wells in both appears to come from the same deep source, as shown by boron

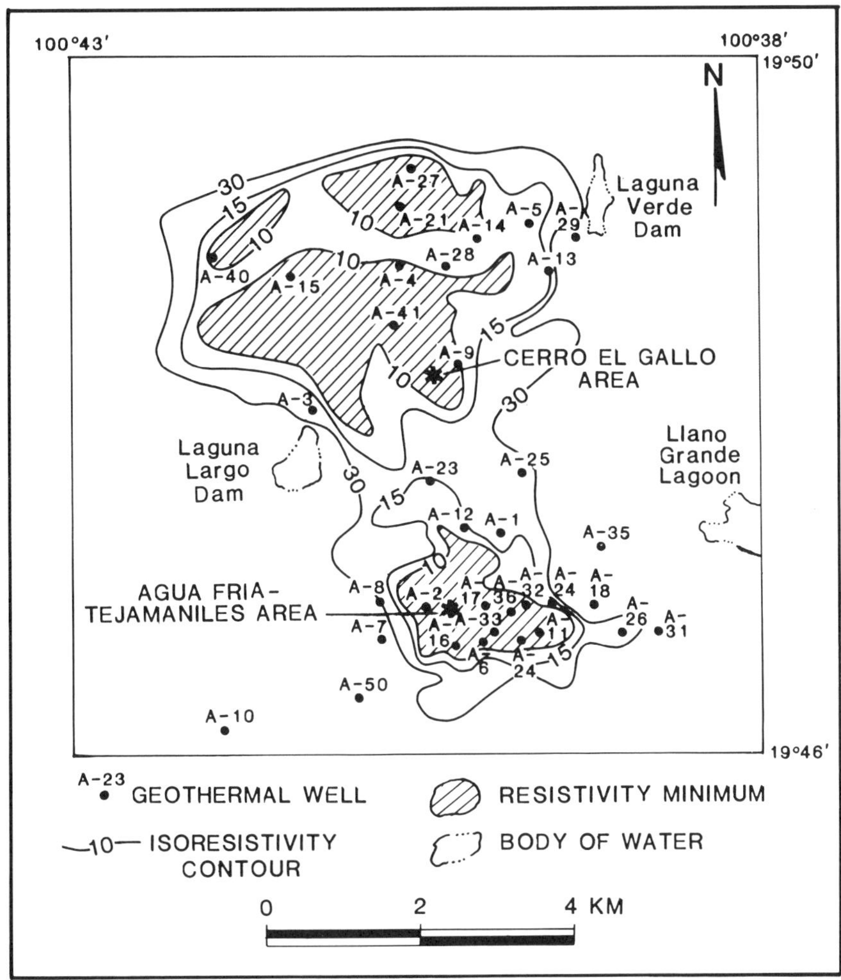

Figure 9. Apparent isoresistivities for AB/2 = 1,000 m, Los Azufres.

TABLE 1. TYPICAL COMPOSITION OF BICARBONATE- AND SULFATE-BEARING WATERS IN LOS AZUFRES, MICHOACAN

			Concentrations in ppm					
	pH	Na	K	Li	Ca	Mg	Fe	Rb
Eréndira	6.3	76	18	0.01	54	34	0.3	0.07
La Cumbre	2.0	8	6.9	0.01	1.3	8.8	63.6	0.07
	Cs	As	SiO_2	Cl	SO_4	HCO_3	NH_4	B
Eréndira	0.1	0	167	5.6	76	514	0.3	0.5
La Cumbre	0.08	0	314	31.6	4440	0	18.9	22

TABLE 2. CHEMICAL COMPOSITION OF WATER EXTRACTED FROM WELLS AT THE NORTH ZONE OF LOS AZUFRES, MICHOACAN

Well	Date	H_T (kJ/kg)	X_L (%)	Cl^-	B	SiO_2	Na	K	Ca	S.T.D.
				(Milligrams per liter)						
A-3	10/11/79	1,172	65	1,714	103	607	1,011	203	11.0	3,700
A-4	7/18/83	1,508	50	1,635	137	522	894	221	2.0	2,979
A-5	7/18/83	2,194	20	1,645	53	227	357	88	1.0	1,260
A-9	7/12/84	2,065	26	1,030	94	300	565	163	3.0	----
A-13	7/18/83	1,769	39	1,152	111	253	650	143	4.0	2,305
A-15	11/7/80	1,020	72	2,117	184	691	1,139	277	15.0	3,940
A-19	7/23/84	1,408	55	1,311	113	530	723	195	3.0	----
A-21	6/1/82	----	----	----	----	----	----	----	----	----
A-28	5/27/84	1,149	66	1,845	155	622	1,050	270	7.0	----

Notes: Data under total discharge conditions.
H_T = specific enthalpy.
X_L = percentage of liquid fraction.
S.T.D. = total dissolved solids.

TABLE 3. CHEMICAL COMPOSITION OF WATER EXTRACTED FROM WELLS IN THE SOUTH ZONE OF LOS AZUFRES, MICHOACAN

Well	Date	H_T (kJ/kg)	X_L (%)	Cl^-	B	SiO_2	Na	K	Ca	TDS
				(Milligrams per liter)						
A-1	7/10/81	1,756	40	1,779	127	578	916	248	7.0	2,690
A-2	10/18/83	1,238	62	1,829	149	589	1,026	199	9.0	2,437
A-7	4/21/80	1,149	66	2,406	120	453	1,023	144	18.0	3,816
A-8	5/9/80	1,429	54	1,502	119	626	897	188	9.0	3,423
A-16	3/18/83	1,521	50	1,689	118	452	831	137	9.0	1,953
A-18	3/18/83	1,684	43	1,210	96	347	662	131	5.0	2,579
A-22	4/15/83	1,417	55	2,117	196	626	1,104	316	5.0	4,541
A-23	4/1/83	1,516	50	1,549	127	335	847	211	5.0	3,055
A-25	3/18/83	1,374	56	1,680	132	516	900	187	5.0	3,433
A-26	2/3/82	1,529	50	1,560	134	430	825	199	6.0	3,273
A-31	10/18/83	2,193	20	715	75	190	400	77	5.0	947
A-33	10/4/83	2,559	5	188	15	43	100	17	1.0	354
A-35	8/29/83	2,532	6	237	20	52	124	26	1.0	455
A-36	7/12/84	2,215	20	678	54	159	388	85	5.0	----

Notes: Data under total discharge conditions.
H_T = specific enthalpy.
X_L = liquid fraction in percentages.
TDS = total dissolved solids.

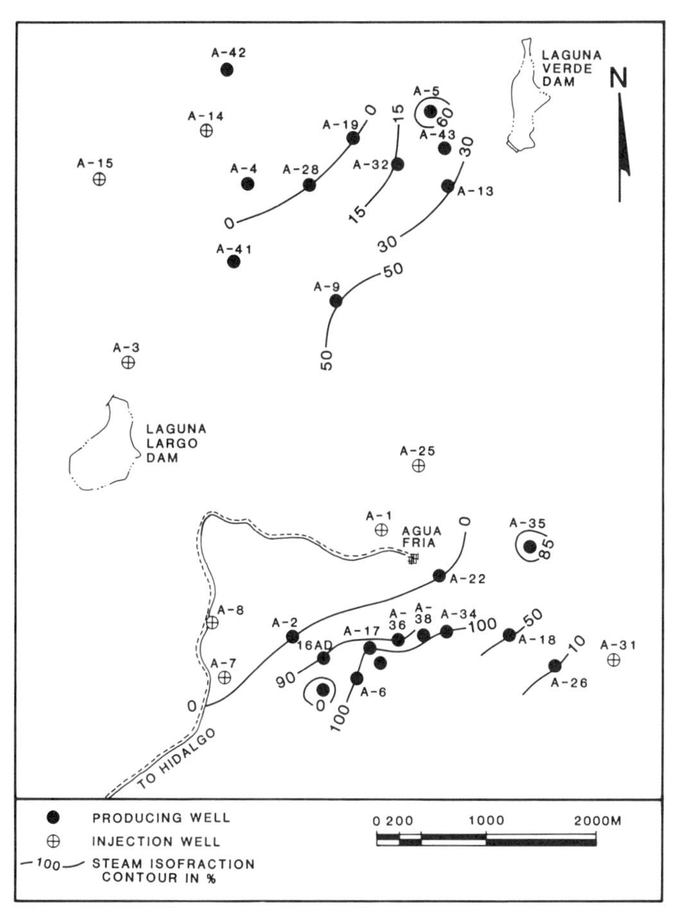

Figure 10. Well-bottom equal steam-fraction contours, in percentages, Los Azufres.

TABLE 4. BOTTOM TEMPERATURE AND STEAM FRACTION IN WELLS IN LOS AZUFRES, MICHOACAN

Well	Temperature °C	X_v (%)
A-1	320	24
A-2	281	0
A-7	248	4
A-8	290	10
A-16	264	22
A-18	284	28
A-22	328	0
A-23	310	9
A-25	289	6
A-26	306	11
A-31	280	62
A-33	269	86
A-35	288	84
A-36	294	62

Note: X_v = steam fraction.

distribution graphs; boron is the element that indicates the fluid's deep origin (Fig. 12).

EXPLORATORY AND DEVELOPMENT DRILLING

From 1976, when drilling of well A-1 began in the vicinity of the Agua Fría Camp, to the present (1985) a total 48 wells (77,094 m) have been drilled; eight of these (14,454 m) are injection wells, 12 (18,230 m) are exploratory, and 28 (44,410) are development wells. The wells are positioned—on the basis of geophysical, geological, and geochemical information—to encounter faults. Well depths vary considerably according to their location and objectives; the shallowest reached 627 m (A-17), and the deepest 3,544 m (A-44; Fig. 5). Some Los Azufres wells produce water/steam mixtures with enthalpies of 1,000 to 2,500 KJ/kg; others produce dry saturated or overheated steam with specific enthalpies of 2,262 to 2,800 KJ/kg respectively.

The southern zone differs from the northern by having abnormal pressures and severe control losses during the initial stages in almost all the wells, as a result of the strong underground fracturing. However, well A-16 was drilled there down to 2,500 m without major circulation losses, notwithstanding the very high recorded bottom temperatures (340°C). Hydraulic fracturing was therefore undertaken within the formation below the 963 m depth; initial pressure was 1,300 psi (92 kg/cm^2), and final pressure was 2,200 psi (154 kg/cm^2). The well was subsequently remeasured, with negative results; it had not connected with the nearby permeable zones in spite of the presence there of good producers.

In an attempt to locate a subsurface permeable zone in well A-16, a controlled-deviation program was set up for well A-16/D to intersect the east-west–trending Tejamaniles fault in the north (Fig. 4). A cement plug was placed in the 9⅝-in casing at 1,031 m to isolate the lower part of the well from the section where a window would be opened. A second plug was placed at 624 m, and a N61°W, 56° deviation was taken to 1,324 m where the circulation fluid was totally lost. This well was a producer. Although this is not always the case, the success of well deviation in Los Azufres makes it a future alternative for barren or poorly producing wells under certain conditions.

GEOTHERMAL MODEL OF THE FIELD

Seven parameters were used to define the shape and dimensions of the reservoir: (a) the fault system providing major secondary underground permeability; (b) temperature, with the 225°C isotherm distribution as reservoir boundary and the 150°C isotherm as the probable reserve boundary; (c) subsurface distribution of hydrothermal alteration minerals such as epidote, which indicates good geothermal conditions; (d) apparent resistivity, related to underground thermal fluid zones; (e) permeable zones detected by circulation losses during drilling; (f) hydrological information defining a deep aquifer within the Miocene-Pliocene volcanic rocks; and (g) production data records.

Two areas representing the hydrothermal system's horizontal extension were thus defined: 32.3 km^2 for the shallow levels at 2,250, 2,000, 1,750, and 1,500 m asl; and 47.2 km^2 for the deep levels of the reservoir at 1,000, 500, and 0 m asl.

In the vertical dimension the geothermal reservoir was bounded practically between the 2,250 and 500 m asl levels, and possibly could be extended to 0 m asl. The horizontal dimensions of the reservoir vary with depth, and the resulting idealized shape is an irregular pyramid tending to have several peaks toward the surface (Fig. 13).

At the shallow depths (2,250 to 1,500 m asl) the reservoir displays two independent thermal zones—northern and southern—resulting from the rise of the geothermal fluids preferentially along fault planes.

Isoresistivity contours, showing lowest resistivities where reservoir conditions are most favorable, indicate widespread shallow anomalies that fragment and shrink at depth. This is attributed to decreasing porosity and permeability in the reservoir, probably due to the sealing effect of the hydrothermal alteration, since fracturing tends to close up at depth.

Alteration mineral-content percentages in the wells average 76.5 percent of the total rock volume; small zones or "islands" where percentages are much lower are related to high-permeability zones that foster the present underground movement of the fluids; these correspond to the location of the producing wells. Alteration mineral-content percentages in the deepest wells show that the degree of alteration does not change substantially at depth. In the north, the subsurface distribution of epidote indicates an evident phase displacement between its formation temperatures and the present field isotherms, whereas in the south it coincides with present temperatures. The high alteration percentage in the rocks gives rise to self-sealing zones, which reduce the secondary permeability of the reservoir host rocks. Three zones of greater to lesser alteration intensity have been detected: the central zone at the El Chino Module, the northern zone at the Marítaro and La Cumbre Modules, and the southern zone at Tejamaniles (Fig. 4).

Reservoir and reserve volumes

Having defined the reservoir area at the several depths considered in the conceptual model, two optional volumes were computed: one takes the reservoir and its probable reserve to sea level (0 m asl) and the other, with better information, to only 500 m asl. In the first case the reservoir has a volume of 33.26 km^3, and the reserve is 46.31 km^3, for a total 79.57 km^3. In the second

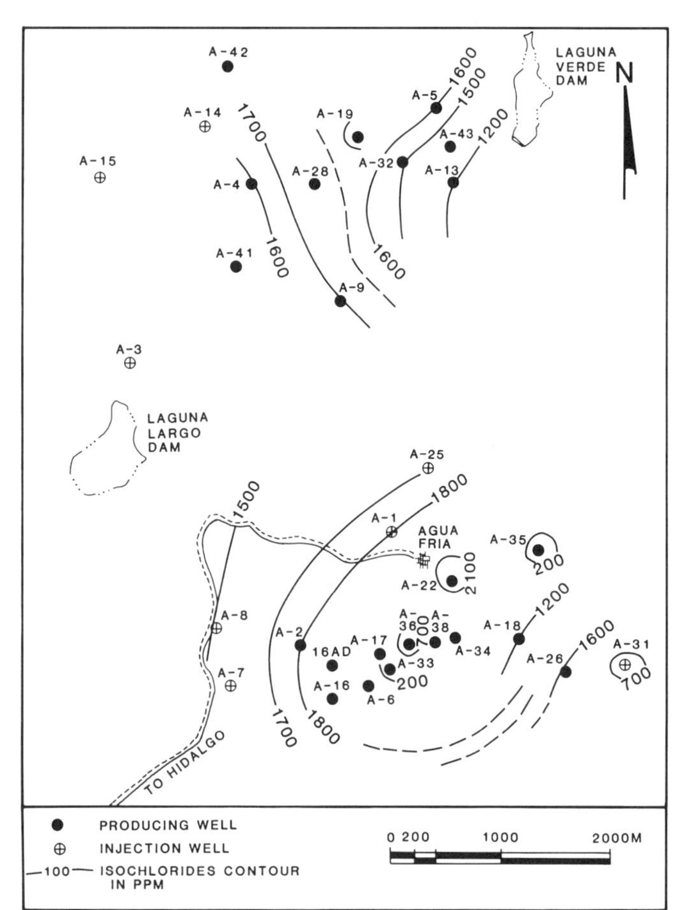

Figure 11. Contours of equal chloride content in total discharge of wells at Los Azufres.

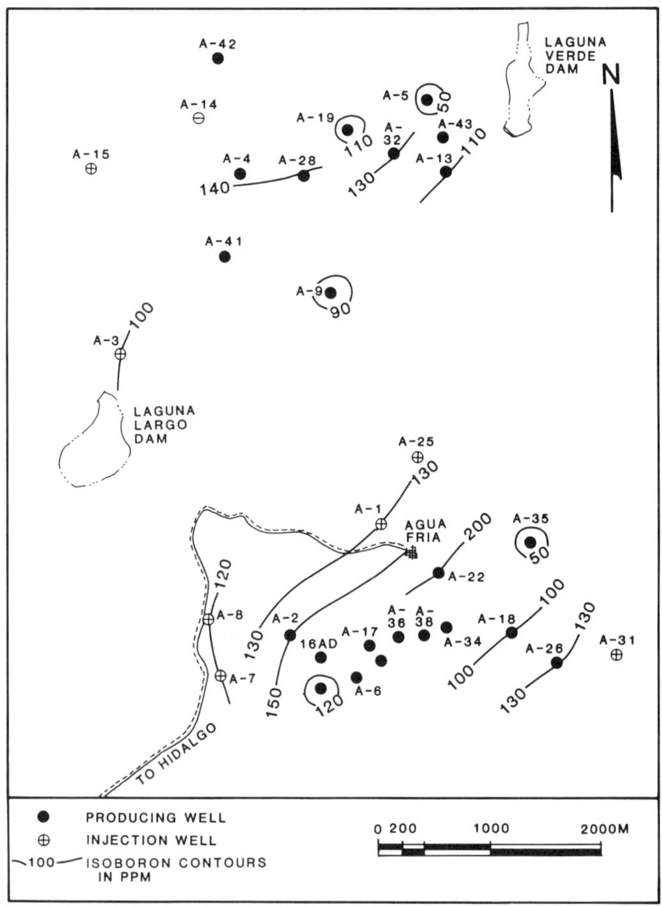

Figure 12. Contours of equal boron content in total discharge of Los Azufres wells.

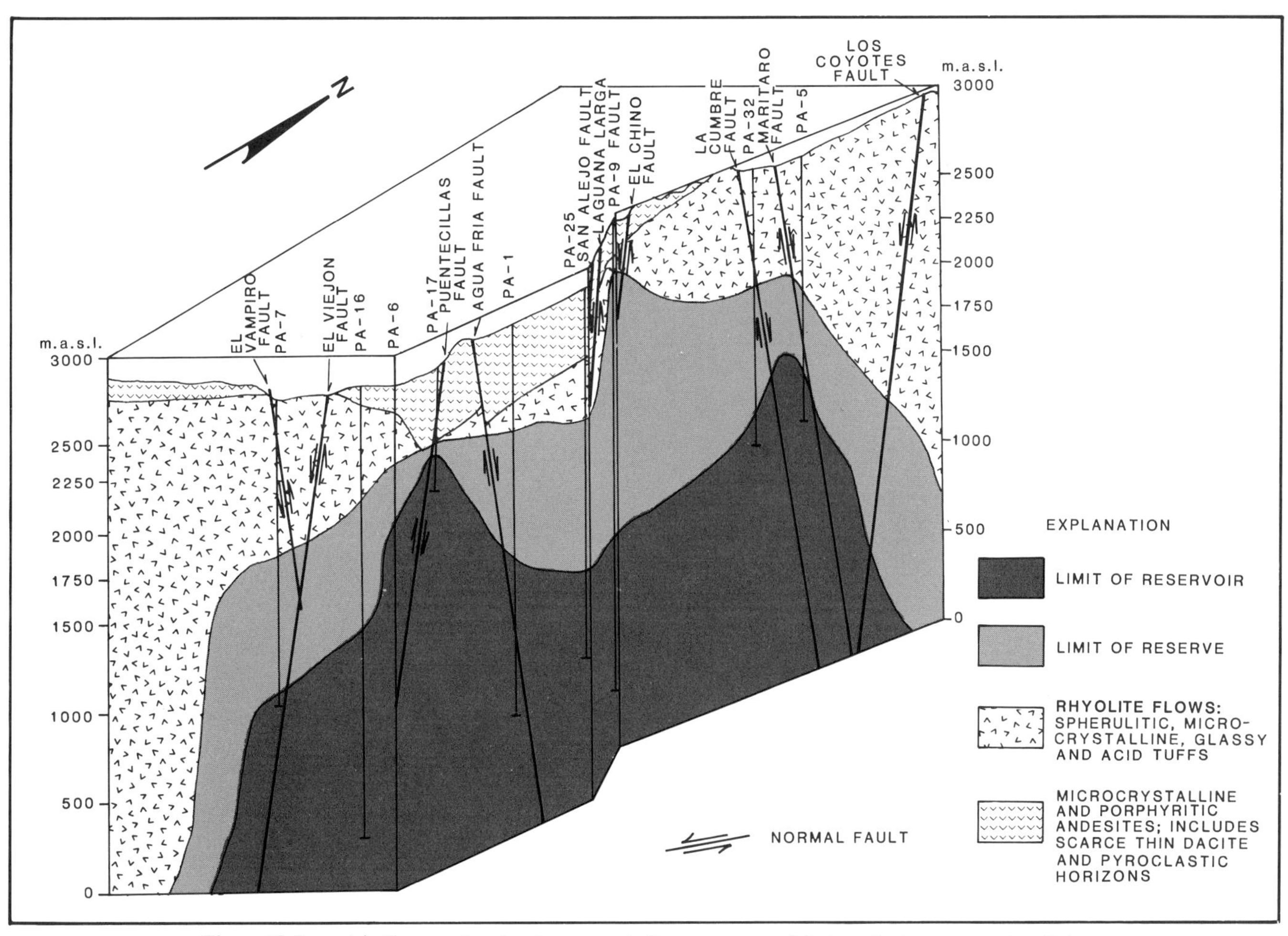

Figure 13. Isometric diagram showing the reservoir, its reserve zone, lithology, fault system, and wells in Los Azufres.

option the reservoir has a volume of 19.21 km^3, and the reserve is 35.36 km^3, for a total 54.57 km^3. These volumes correspond to rocks fulfilling the selected parameters of 225°C temperatures or more for the reservoir and 150 to 225° for the reserve.

The fluid volumes contained in these hot rocks were defined by estimating the average porosity of the total reservoir rock volume on the basis of published Tertiary andesitic fractured rock porosities affected by alteration mineral percentages recorded within the 2,250 to 500 m asl range. The presence of epidote, quartz, calcite, and pyrite is taken to indicate absence of secondary porosity in the andesites; clays, chlorite, and oxides signal higher porosities. The average frequencies of each mineral group in relation to the porosity of unaltered rocks resulted in a total weighted average porosity of 16 to 14 percent.

Fluid volumes were computed for each option, taking 14 percent as average porosity. In option 1 (to 0 m asl), reservoir fluids are 4.66 km^3 and reserve fluids 6.48 km^3, for a total 11.14 km^3. In option 2 (to 500 m asl) the reservoir fluid volume is 2.69 km^3, and the reserve is 4.95 km^3, for a total of 7.64 km^3.

Geothermal fluid volumes represent in situ water and steam without recharge effects. Assuming the more conservative alternative—2.60 km^3 of fluids in situ—, and that during development approximately one-fourth will be steam and three-fourths hot water, there will be approximately 600 million tons of steam in the reservoir. Generation of 220 MW demands 17 million tons of steam per year; Los Azufres will thus contain enough steam to generate 220 MW for 35 years.

PRESENT STATUS AND PERSPECTIVES OF THE FIELD

As mentioned above, the Los Azufres field has a known area of 32 km^2. Results of 48 exploratory and development wells define it as a dominantly liquid geothermal reservoir with zones of overheated steam. The field's production and geologic features indicate the presence of two zones, which merge at depth to form one reservoir.

Of the 28 wells drilled in the southern zone (Table 5), 14 are producers, seven are injection, three were abandoned owing to problems within the reservoir, and four were failures. Some wells produced overheated steam and others a water/steam mixture. The dry steam-producing zone is termed the "steam cap" and is

TABLE 5. PRESENT STATUS OF WELLS IN THE SOUTH ZONE OF THE LOS AZUFRES, MICHOACAN, GEOTHERMAL FIELD

Well	Characteristics	P C (M Pa-MAN)	Production at PS = 1.0 MPa abs Gv(T/H)	Ga (T/H)	Enthalpy (kJ/kg)	Observations
A-1	Hot injector	....	....	....	....	
A-2	Mixture producer	1.50	80.0	225.0	1,291.4	1979
A-6	Steam producer	1.48	28.2	0.0	2,800.0	Oct. 1984
A-7	Cold injector	....	....	....	....	
A-8	Cold Injector	....	....	....	....	
A-10	Cold injector	....	....	....	....	
A-11	Abandoned	....	....	....	....	
A-12	Failed	....	....	....	....	
A-16	Failed	....	....	....	....	
A-16D	Mixture producer	1.50	31.3	89.7	1,284.6	Sept. 1984
A-16AD	Mixture producer	1.38	26.6	12.7	2,126.4	Oct. 1984
A-17	Steam producer	1.84	57.3	0.0	2,770.0	Jun. 1984
A-18	Mixture producer	1.71	61.9	64.2	1,752.1	Jun. 1984
A-20	Failed	....	....	....	....	
A-22	Mixture producer	1.75	39.2	48.4	1,665.4	Jun. 1984
A-23	Hot injector	....	....	....	....	
A-24	Abandoned	....	....	....	....	
A-25	Hot injector	....	....	....	....	
A-26	Mixture producer	1.50	89.0	160.0	1,483.1	Dec. 1982
A-31	Hot injector	....	....	....	....	
A-33	Mixture producer	1.74	67.3	16.2	2,388.8	Jun. 1984
A-34	Steam producer	1.53	54.3	3.3	2,661.8	Jun. 1984
A-35	Mixture producer	1.32	29.2	5.4	2,463.2	Jun. 1984
A-36	Mixture producer	1.68	39.7	23.7	2,025.5	Aug. 1984
A-37	Producer	....	....	....	....	Not evaluated
A-38	Steam producer	1.71	116.6	7.1	2,661.8	Jun. 1984
A-39	Abandoned	....	....	....	....	
A-46	Drilling	....	....	....	....	
A-50	Failed	....	....	....	....	

the obvious first choice for development owing to the handier processing of the fluids.

Development of the southern zone began by installing two portable atmospheric discharge turbogenerators of 5,000 KW capacity each at wells A-6 and A-17, which, in addition to generating low-cost electricity, have led to a more precise evaluation of the field's behavior.

In the northern zone, 10 of the 20 wells drilled are producers, five injection, two are currently being studied, one was abandoned, and two were failures (Table 6). As in the southern zone, three portable 5,000-KW turbogenerators were installed in 1983 at wells A-5, A-13, and A-19, with the same objectives of amortizing investment and defining field behavior. To maintain maximum power generation capacity, A-19 was reinforced with steam from A-28 when its own steam production dropped.

A three-dimensional biphase numerical simulator capable of simulating the field behavior in detail was set up recently to evaluate mass and energy contents in Los Azufres and to predict its evolution during development. In the first stage the numerical modeling was restricted to the Tejamaniles zone for generation of 55 MW (SIMAZ-84 project). The numerical data described above were coherently integrated to feed the simulator and study three possible development scenarios that refer to open, closed, and semi-infinite boundary conditions.

As would be expected, the most critical situation corresponds to the case of a closed reservoir not receiving mass or energy from outside. Under these conditions the central zone of steam wells in Tejamaniles would quickly lose pressure. An estimated 12-bar limit would be reached in 11 years, after which the development of the reservoir would no longer be possible. However, even in this critical option the peripheral areas would still have more than enough resources to continue generating the requisite 55 MW.

In all other cases the zones under development never attain

TABLE 6. PRESENT STATUS OF WELLS IN THE NORTH ZONE OF THE LOS AZUFRES, MICHOACAN, GEOTHERMAL FIELD

Well	Characteristics	P C (M Pa-MAN)	Production at PS = 1.0 MPa abs Gv(T/H)	Ga (T/H)	Enthalpy (kJ/kg)	Observations
A-3	Cold injector	....	....	....	....	
A4	Mixture producer	1.5	56.0	144.4	1,338.8	May 1984
A-5	Mixture producer	1.5	79.0	36.3	2,145.0	May 1983
A-9	Mixture producer	1.5	56.8	32.4	2,082.4	Aug. 1984
A-13	Mixture producer	1.5	57.5	43.0	1,915.0	Nov. 1983
A-14	Hot injector	....	....	....	....	
A-15	Hot injector	....	....	....	....	
A-19	Mixture producer	1.5	18.9	37.9	1,434.0	Aug. 1984
A-21	Hot injector	....	....	....	....	
A-27	Abandoned	....	....	....	....	
A-27A	Being studied	....	....	....	....	
A-28	Mixture producer	1.5	32.3	63.3	1,444.5	Apr. 1984
A-29	Failed	....	....	....	....	
A-32	Producer	....	....	....	....	Not evaluated
A-40	Cold injector	....	....	....	....	
A-41	Steam producer	1.24	20.8	1.3	2,661.8	Mar. 1984
A-42	Mixture producer	....	....	....	....	Not evaluated
A-43	Mixture producer	....	....	....	....	Not evaluated
A-44	Failed	....	....	....	....	
A-45	Being studied	....	....	....	....	

abandonment pressures, and the results clearly show that in Tejamaniles at least 55 MW can be generated beyond the 20-yr period.

The first evaluations of the Los Azufres field's total capacity indicate sufficient capacity to install between 220 and 300 MW. On this basis, in addition to the 5 wellhead turbogenerators in operation since 1983, a 50-MW power station will begin operating in July 1988 in the Tejamaniles sector. Two more—Azufres U-1 and U-2, of 44 MW each—will begin operation in 1990. Subsequent construction of Azufres U-3 and U-4 will begin after testing reservoir behavior during the extraction of the fluid feeding units 1 and 2, which will also have 55 MW capacity. Almost three years of experience operating 5-MW back-pressure plants at Los Azufres, and—among other factors—the economic advantage of advanced generation, have improved future prospects because these are considered a good solution, particularly as a complement in the installation of larger capacitator plants.

A main attraction of these small, easily transported plants is that they may be installed shortly after completion of a well, which allows power generation to begin even before all the exploration, drilling, and production tests are concluded. The consequent savings in costs make the installation of wellhead back-pressure plants a definitive solution whenever the gas content of the steam is too high and/or the reservoir too small for capacitator stations, or a provisional solution for generating power until the field exploration and building of large stations are concluded.

This alternative (complementary installation of small wellhead plants) appears as a very attractive option in Mexico, particularly considering that they can be manufactured domestically with a high degree of local integration (more than 80 percent). The first ten will be produced during the next four years. The Los Azufres development plans therefore combine both solutions, and seven portable 5,000-MW units will be installed from 1986 to 1988 for a total installed capacity of 340 MW if the field proves able to maintain pressure during its development.

BIBLIOGRAPHY

Atwater, T., 1970, Implications of plate tectonics for the Cenozoic tectonic evolution of western North America: Geological Society of America Bulletin, v. 81, p. 3513–3536.

Campa, U.M.F., 1978, La evolución tectónica de Tierra Caliente, Guerrero: Sociedad geológico de México Bulletin, v. 39, p. 56–59.

Demant, A., and Robin, C., 1975, Las fases del vulcanismo en México, una síntesis en relacion con la evolución geodinámica desde el Cretácico: Instituto de Geología, Universidad Nacional Autonoma de México Revista 75, p. 1, p. 70–83.

Demant, A., Mauvois, R., and Silva, L., 1976, El Eje Neovolcánico Transmexicano: III Congress Latinoamericano de Geológia, no. 4, la, parte 18–Tils.

Giggenbach, W. F., 1981, Exploration and development of geothermal power in Mexico: Vienna, International Atomic Energy Agency.

Gutiérrez, N.L.C.A., and Aumento, F., 1982, The Los Azufres, Michoacán, México, geothermal field, *in* Lavigne, J., and Day, J.B.W., eds., Hydrothermal Studies: 26th International Geological Congress: Journal of Hydrology, v. 56, p. 137–162.

Mooser, F., 1969, The Mexican Volcanic Belt, structure and development; Formation of fractures by differential crustal heating: México, D. F., Pan-American Symposium on the Upper Mantle, (1968) pt. 2, p. 15–22.

—— , 1972, The Mexican Volcanic Belt, Structure and tectonics: Geofísca International, v. 12, p. 55–70.

Nieto, O. J., Delgado, A.L.A., and Demon, E. P., 1981, Relaciones petrográficas y geocronológicas del magmatismo de la Sierra Madre Occidental y el Eje Neovolcánico en Nayarit, Jalisco y Zacatecas: XIV Convenión Nacional, Asociación de Ingenieros de Minas, Metalurgistas y Geólogos de México, p. 327–362.

Unpublished sources of information

Exploration department, Comisión Federal de Electricidad

Aumento, F., and Gutiérrez, N.L.C.A., 1980, Geocronología de Los Azufres, Michoacán: Report 3/80.

—— , 1980, El Campo geotérmico de Los Azufres, Michoacán, México: Report 14/80.

Camacho, A. F., 1979, Geología de la zona geotérmica de Los Azufres, Michoacán, México: Report 6/79.

Cedillo, R. F., Silva, P. F., and Vargas, L.H.J., 1981, Estudio geohidrológico Los Azufres–Cuitzeo, Michoacán: Report 37/81.

De la Cruz, M. V., Aguilar, S. J., Ortega, G. D., and Sandoval, S.J.M., 1982, Estudio Geológico-estructural a detalle del campo geotérmico de Los Azufres, Michoacán: Report 9/82.

Garduño, M. V., and Martiñon, G. H., 1985, Analisis estructural de la zona sur del campo geotérmico de Los Azufres, Michoacán: Report 24/84.

Quijano, L.J.L., 1984, Contenido de gas de la zona sur del campo geotérmico de Los Azufres.

Razo, M. A., 1975, Resumen de los estudios de la zona geotérmica de la Sierra de San Andres, Michoacán: Report 1/75.

Venegas, S. S., Herrera, F. J., and Maciel, F. R., 1984, Algunas caracteristicas de la Faja Volcanica Mexicana y de sus recursos geotérmicos.

Executive Coordinator, Comisión Federal de Electricidad

Garfias, F. A., 1981, Estudio geológico de los pozos; A-21, A-19, A-22, A-23, A-3 y A-2 de Los Azufres, Michoacán.

Garfias, F. A., and Casarrubias, U. Z., 1985, Estudio Geologico de los pozos; A-20, A-7, A-5, A-6 y A-10 de Los Azufres, Michoacán.

—— , 1979–1981, Resumen geológico de los pozos; A-6, A-3, A-7, A-14, A-15, A-18, A-19, A-26 y A-25.

Executive Coordinator of Los Azufres, Section of control of the exploration and exploitation of the land

Izaguirre, A. O., 1983, Distribución y comportamiento de las isotermas con los resultados de los nuevos pozos perforados.

Department of Geothermia

Molina, B. R., 1978, Geoquímica de las manifesticiones del Lago de Cuitzeo y su relacion con Los Azufres, Michoacán.

Printed in U.S.A.

The Geology of North America
Vol. P-3, Economic Geology, Mexico
The Geological Society of America, 1991

Chapter 7

Los Humeros geothermal field, Puebla

F. Romero R.
Federal Commission on Electricity, Rio Rodano 15, 06500 Mexico, D.F.

INTRODUCTION

This chapter summarizes geological, geophysical, and geochemical studies of the Los Humeros thermal area and results to date of seven deep wells that confirm the zone's prospects as a commercial geothermal reservoir.

The Los Humeros area covers approximately 225 km^2 of eastern Puebla and western Veracruz States between 97°23′ to 97°31′W and 19°35′ to 19°45′N. Major towns in the vicinity are Altotonga (Veracruz) and Teziutlán (Puebla) in the north, Zacatepec (Puebla) in the south, and Perote (Veracruz) in the southeast. The Veracruz State capital of Jalapa lies 53 km to the southeast (Fig. 1).

Access to the area is by a paved road network connecting the region's main towns. Los Humeros is connected to Perote by an all-weather dirt road (paved at intervals) 32 km long. The principal access to Mexico City is Federal Highway 140 linking the town of Perote with Mexico City, Jalapa, and Veracruz. The Mexico-Perote-Jalapa-Veracruz and Mexico-Apizaco-Oriental-Teziutlán railroads crisscross the area. There is a small-plane airstrip at Jalapa.

The region has been covered by reconnaissance and detailed geological, petrographic, geophysical, geochemical, and geohydrological surveys, well-drilling and evaluation, and a preliminary modeling of the field.

Numerous internal reports by the Los Humeros Drilling Residence and the Exploration, Drilling, and Reservoir Evaluation Departments of the CFE (Comisión Federal de Electricidad) Geothermoelectric Projects Management are available.

REGIONAL GEOLOGY

The Los Humeros caldera occupies the easternmost portion of the Mexican Volcanic Belt and lies geomorphologically at the north boundary of the Libres-Oriental internal-drainage basin. This basin has been interpreted as a graben within a tension zone bounded on the north by the Teziutlán Massif (Sierras Tezompan and Chignautla); on the south by the northwest-southeast–trending Sierra Cuesta Blanca; on the east by the fault system affecting the west portion of the Sierra Pico de Orizaba–Cofre de Perote Volcanics; and in the west by the gap in the Sierra Madre Oriental (at the level of the towns of Cuyuaco, Libres, Oriental, and El Carmen; Fig. 2).

Regionally a strong tectonic influence is apparent. Within the graben are ridges of folded, thrusted, and horizontally displaced Mesozoic limestones deformed by Laramide compression, which preserves the northwest-southeast trend of the Sierra Madre Oriental. Intrusive granites and syenites emplaced in the mid-Tertiary trend in the same direction as the graben. Late Tertiary and Quaternary volcanic activity (which half buried the abovementioned ridges) resulted in a large variety of volcanic forms—domes, volcanic cones, phreatic and phreatomagmatic explosion cones, and stratovolcanos (Pico de Orizaba, Cofre de Perote and La Malinche)—that contrast with the sedimentary structures. Quaternary pyroclastic deposits and flows (tuffs and ignimbrites) filled the Libres-Oriental tectonic graben.

Characteristics of the Mexican Volcanic Belt in the Los Humeros region

The Mexican Volcanic Belt trends roughly east-west between parallels 19° and 21°N, within irregular boundaries defined by the extension of Pliocene-Quaternary volcanism. The belt is formed by large stratovolcanoes of dacitic to andesitic composition, slag cones, and some silicic volcanic centers or calderas. Andesitic rocks predominate; subordinate Pliocene-Quaternary acid and basic rocks are derived both from fissures and generally composite centers trending regionally east-west (Fig. 3).

For purposes of geothermal prospecting, and considering geographic distribution only, the Mexican Volcanic Belt may be subdivided into nine extensive generalized sectors (see Fig. 3). The Jalapa sector on the east contains the Los Humeros geothermal field.

Age. The volcanic activity took place during the past 3 m.y. (Pliocene-Quaternary). Ages of the Pico de Orizaba, Cofre de Perote, and La Malinche stratovolcanos range from 12,000 yr (La Malinche pumice) to 3.5 Ma for the Teziutlán andesites related to the Cofre de Perote volcanism; Pico de Orizaba flows

Romero R., F., 1991, Los Humeros geothermal field, Puebla, *in* Salas, G. P., ed., Economic Geology, Mexico: Boulder, Colorado, Geological Society of America, The Geology of North America, v. P-3.

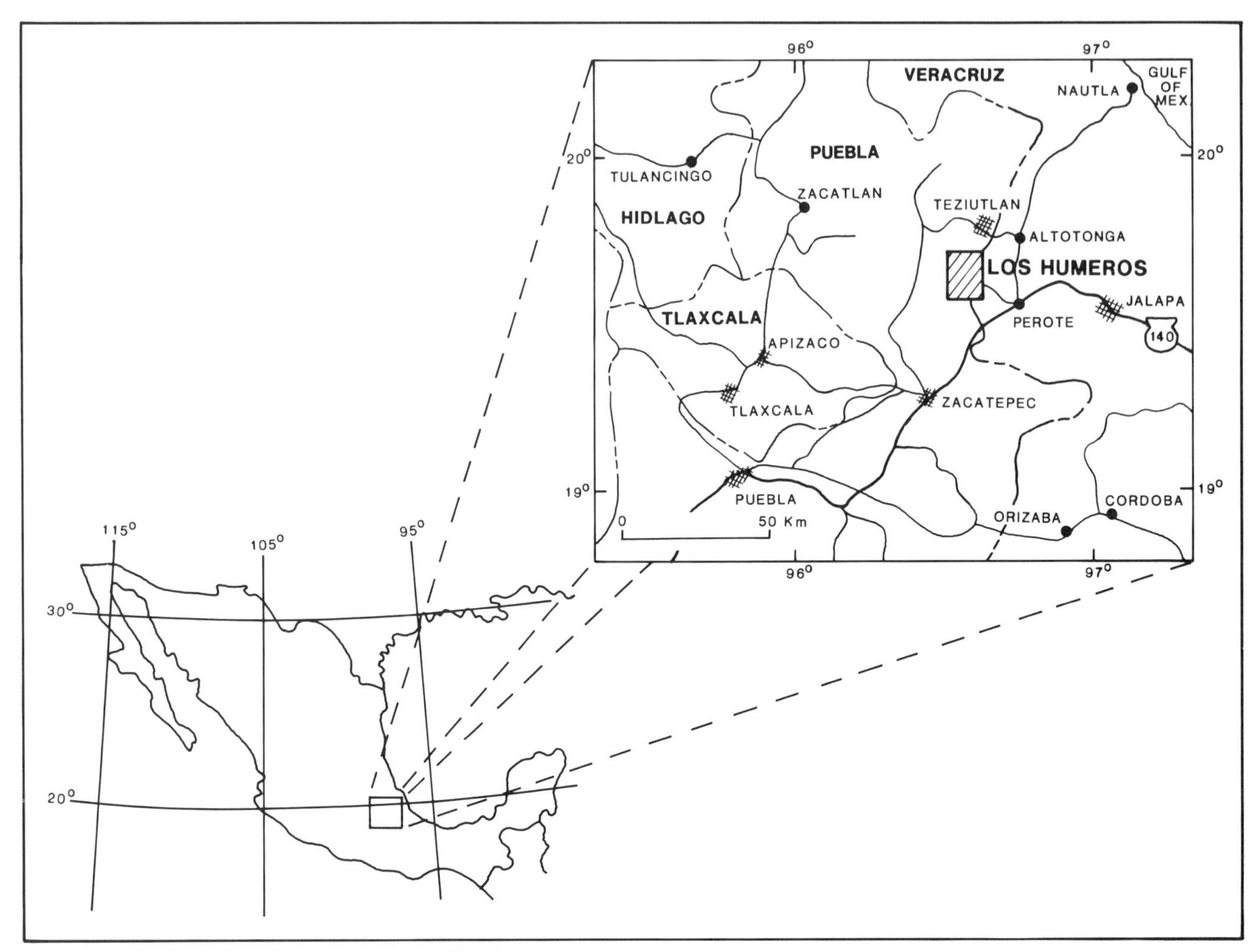

Figure 1. Location of Los Humeros, Puebla.

are less than 1 m.y. old, and flows at Cofre de Perote are 1.6 to 1.9 m.y. old.

Distribution of the volcanos. In this sector the Pico de Orizaba and Cofre de Perote stratovolcanos follow the NNW-SSE alignment of the Libres-Oriental graben. Volcanic cones and some phreatic and phreatomagmatic explosion cauldrons show NE-SW alignments also. The following types of Pliocene-Quaternary volcanism are found:

Large stratovolcanos (La Malinche, Cofre de Perote, and Pico de Orizaba), each representing volumes of more than 100 km^3 of lavas having very similar characteristics: porphyritic dacitic-andesitic rocks with abundant plagioclase and hornblende phenocrysts. Their buildup generally followed a common pattern of thick lava flows and nuées ardentes growing by superposition in lateral extensions with gentle slopes as a result of deposition of the explosive phases. When the volcanic structure grew to major dimensions, violent eruptions initially spouted pumice followed by andesitic lava and ash. At Pico de Orizaba the first eruptive phase developed an ancient volcanic structure called Sierra Negra, on top of which the present more recent cone (Citlaltépetl) arose. Glacial erosion that has affected La Malinche and Cofre de Perote has masked their volcanic history.

Monogenetic volcanos constructed of pyroclastic emissions around the conduit and small lava flows. These volcanos are mostly aligned northeast-southwest but are also east-west and northwest-southeast. Some grew at the foot or on top of the crest of the Pico de Orizaba–Cofre de Perote Range by accumulation of andesitic basalts and scoria.

Rhyolitic products. A series of rhyolitic (76 percent silica) to dacitic domes ranging in age from 0.51 to 0.24 Ma are found along the Los Humeros caldera annular fracture zone (Oyameles, Ocotepec domes, and others) and to the south (Pizarro, Las Aguilas, Cerro Pinto, and Las Derrumbadas).

Eastern boundary of the Mexican Volcanic Belt. The Sierra Pico de Orizaba–Cofre de Perote of Plio-Quaternary calc-alkaline volcanism is the easternmost volcanic activity of the Mexican Volcanic Belt; the volcanism on the slopes that drop down to the Gulf of Mexico (Naolinco-Chiconquiaco) belongs in an alkaline petrologic province related to the recent geodynamic evolution of the Gulf of Mexico.

Regional tectonics

Large bodies of igneous intrusives and metamorphic rocks form the Los Humeros regional basement. The Jurassic was a time of sea-floor spreading, giving rise to crustal instability associated with opening of the Gulf of Mexico, which caused alternating transgressive and regressive deposition. In contrast, Cretaceous sedimentation is typical of widespread transgression.

The Laramide orogenic phase of compressional deformation at the beginning of the Tertiary generated two styles of deformation in the region: north-northwest of the caldera is the basin-facies Huaycocotla anticlinorium; to the south-southeast are shelf-facies folds and thrusts. The paleogeographic boundary between them is the Libres-Oriental graben (Fig. 2). In the final stages of this phase, strike-slip faulting prompted displacements of up to 60 km of the Sierra Madre Oriental; some of these faults are basement structures.

A tensional phase (beginning of the Oligocene to the Quaternary) developed fault troughs and offsets in the Mesozoic rocks, with large volumes of lava emerging through faults and fractures. Major andesitic emissions in the upper mid-Tertiary generated the Libres-Oriental graben where the Los Humeros tectonism and volcanism subsequently took place. The first phase of this activity was a circular domal uplift approximately 40 km in diameter (as shown on satellite images) ending in the collapse of the caldera (15 km in diameter) followed by lava flows.

The regional tectonism includes three structural systems: the oldest structures are the northwest-southeast folds and thrusts of the Sierra Madre Oriental; a second northeast-southwest system of lateral faults perpendicular to the former dislocates its structures; and a third, north-south–trending, system includes the Tehuacán–Pico de Orizaba–Cofre de Perote structural alignment, which is interpreted as a right-lateral strike-slip fault antithetic to the Polochic-Motagua system of Guatemala.

Stratigraphy

Igneous (intrusive and extrusive), metamorphic, and sedimentary (marine and continental) rocks ranging in age from Paleozoic (Late Permian) to Recent, crop out in the region (Fig. 4).

Intrusive igneous (granites and granodiorites) and metamorphic (quartz and muscovite-bearing greenschists) rocks form the basal complex. Radiometric dating of the granodiorite gave a date of 246 ± 7 Ma (Late Permian). The schists are associated with green igneous rocks classified as dolerites. The basement crops out north of the caldera (Fig. 5) in the Sierras Tezompa, Chignautla, and Mazatepec and 30 km east of Los Humeros at Vega Chica and Las Minas.

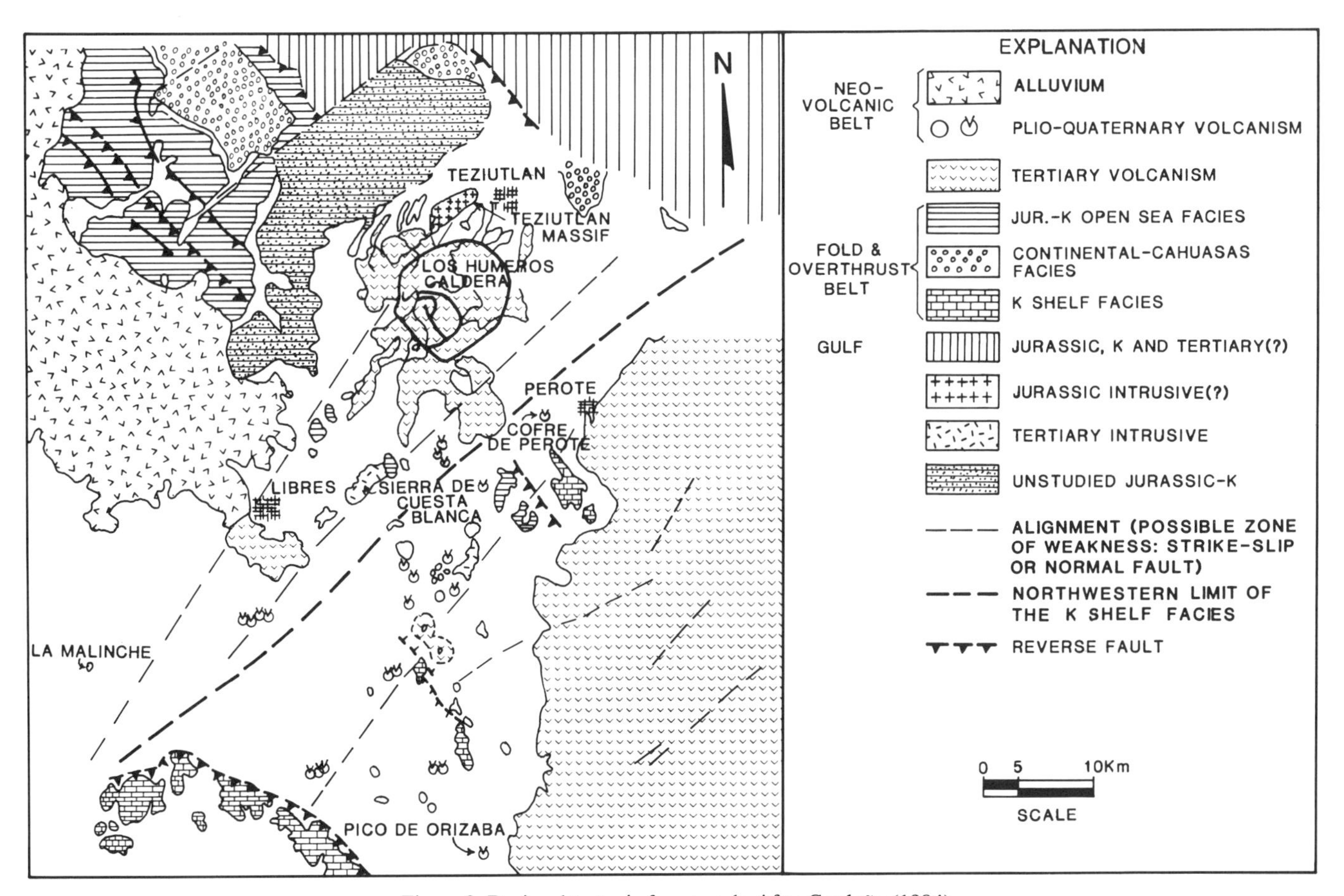

Figure 2. Regional tectonic framework. After Garduño (1984).

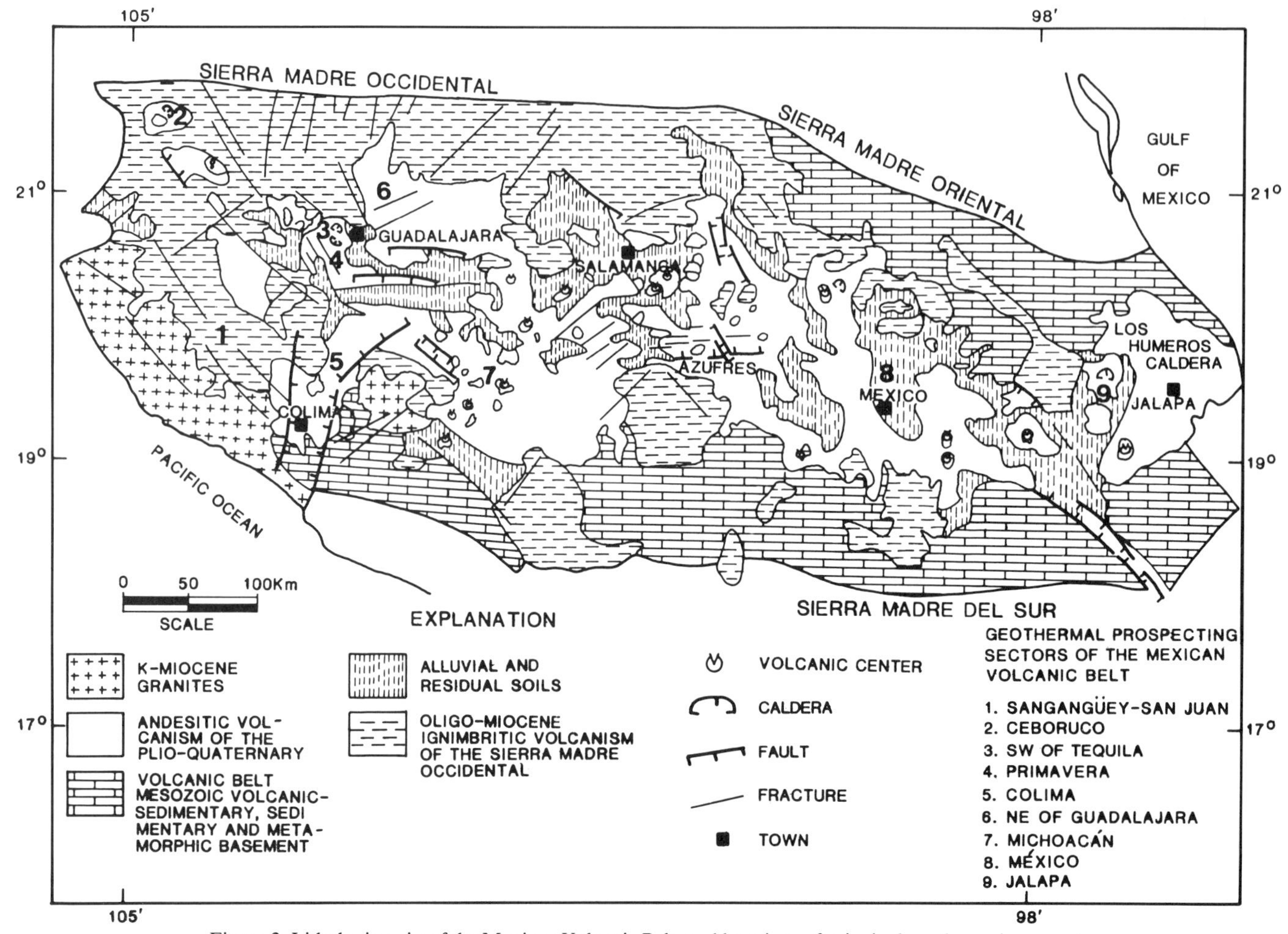

Figure 3. Lithologic units of the Mexican Volcanic Belt, and locations of principal geothermal prospecting centers. Modified from Venegas and others (1985).

A thick Mesozoic sequence of marine and continental sedimentary rocks overlies the basal complex unconformably: breccias, siltstones, sandstones, shales, conglomerates, tuffs, limestones, and marls of the following formations: Huizachal Formation (Triassic); Huaycocotla, Cahuasas, Tepexic, Santiago, Tamán, San Pedro, and Pimienta Formations (Jurassic); and (lower and upper) Tamaulipas, Orizapa, Agua Nueva, San Felipe, Guzmantla, Maltrata, Méndez, and Mezcala Formations (Cretaceous).

A series of acid rocks (granites and syenites) intruded the zones of weakness that developed in the region in the mid-Oligocene, forming stocks and associated dikes that are exposed south of the caldera (center of the Libres-Oriental Graben) and form small, northwest-southeast–trending ridges. Radiometric dates are 14.5 ± 7 Ma in the granites and 31 ± 3.7 Ma in the syenite. Skarn and marble occur as products of contact metamorphism where the intrusions invaded the Mesozoic sequence.

Extrusion of andesites and dacites, appearing as outcrops northwest and west of the caldera, began in the late Miocene. These are dated at 11 and 10.5 Ma, respectively; together they form the Alseseca Andesites volcanic unit.

After an interruption of about 5 m.y. the pre-caldera volcanism continued in the mid-Pliocene (5 ± 0.7 Ma), andesites exposed east of the caldera; late Pliocene (3.5 ± 0.3 Ma), andesite outcrops at the edge of the outer collapse area in its northern portion and in the vicinity of Teziutlán, Puebla; and other Cofre de Perote flows (~1.9 Ma) 300 m southeast of Perote (Veracruz) are known as the Teziutlán Andesites or Formation.

The caldera activity began with the intrusion of rhyolitic domes approximately 0.51 Ma and extrusion of 180 km^3 of ignimbrites (Xáltipan ignimbrites) at 0.5 Ma, which brought about the collapse of the Los Humeros caldera.

GEOLOGY OF THE LOS HUMEROS CALDERA

Volcanic rocks

Three cycles of active volcanism have been defined in the Los Humeros caldera. The first cycle, upper Tertiary (Mio-Pliocene) pre-caldera volcanism, is represented by the Alseseca and Teziutlán andesites. The second cycle, caldera volcanism, began with extrusion of rhyolite domes through annular fractures generated by the thrust of the rising magma; the subsequent emis-

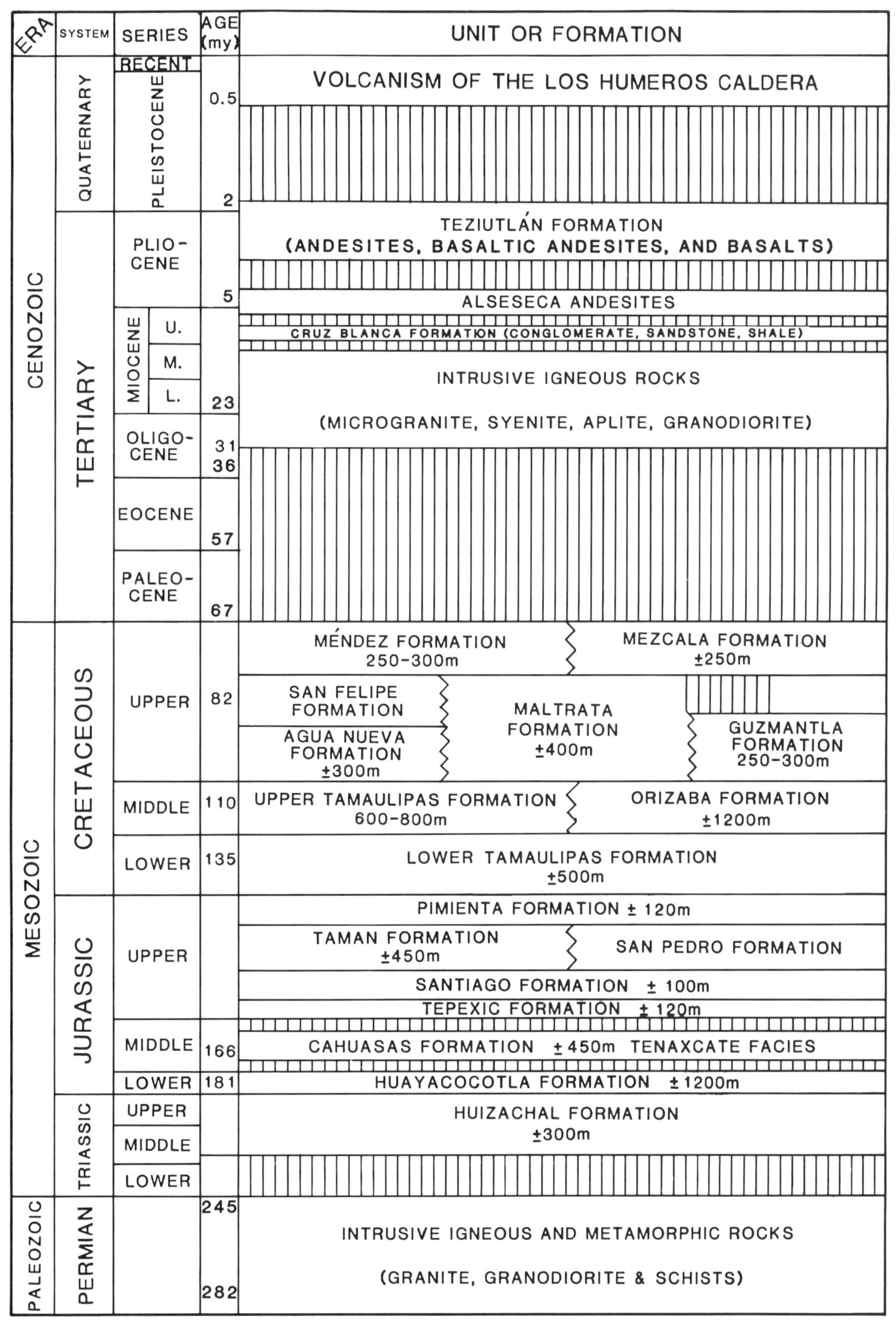

Figure 4. Lithostratigraphic units of the Los Humeros region. After Yáñez (1982).

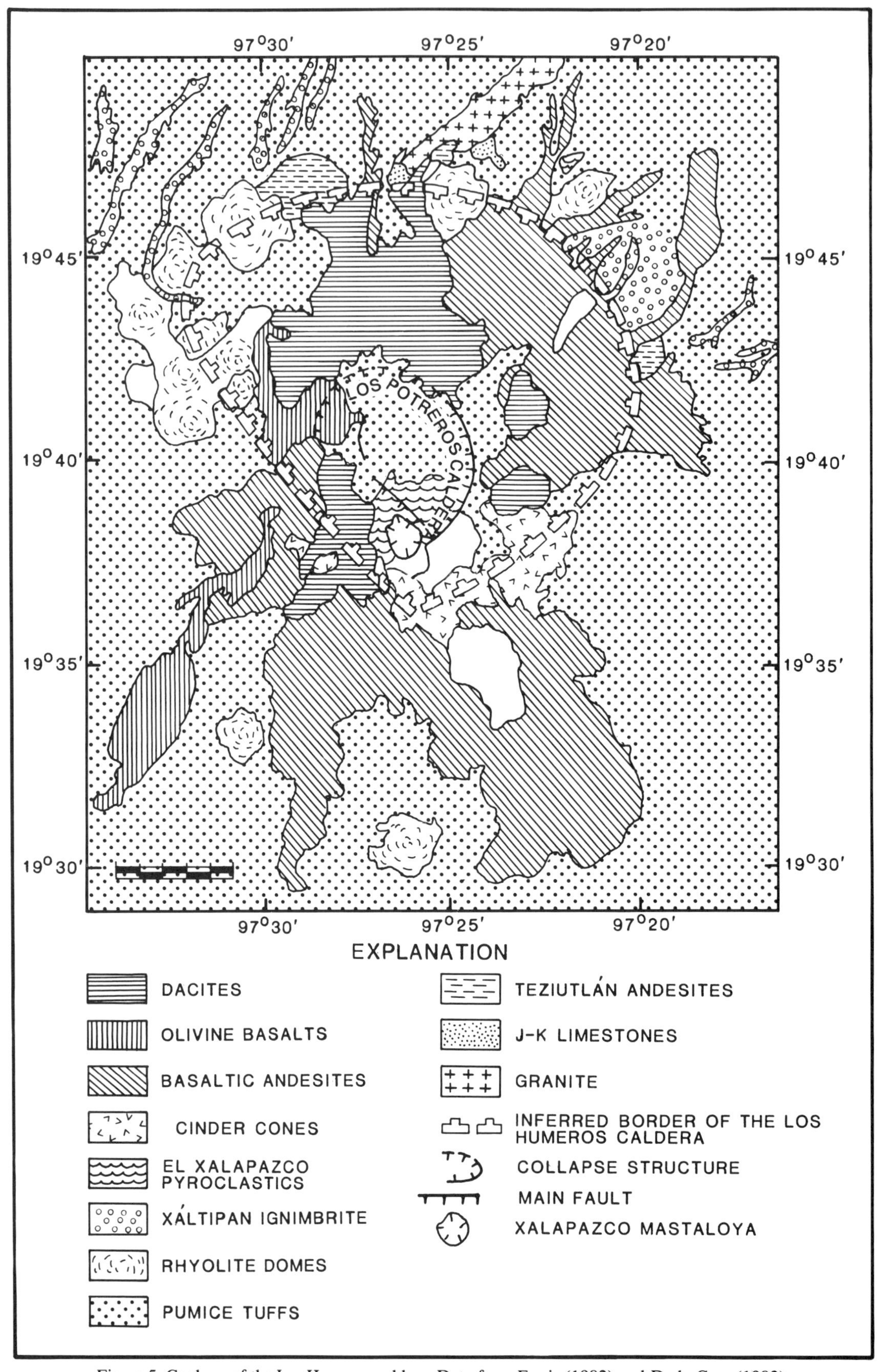

Figure 5. Geology of the Los Humeros caldera. Data from Ferriz (1982) and De la Cruz (1982).

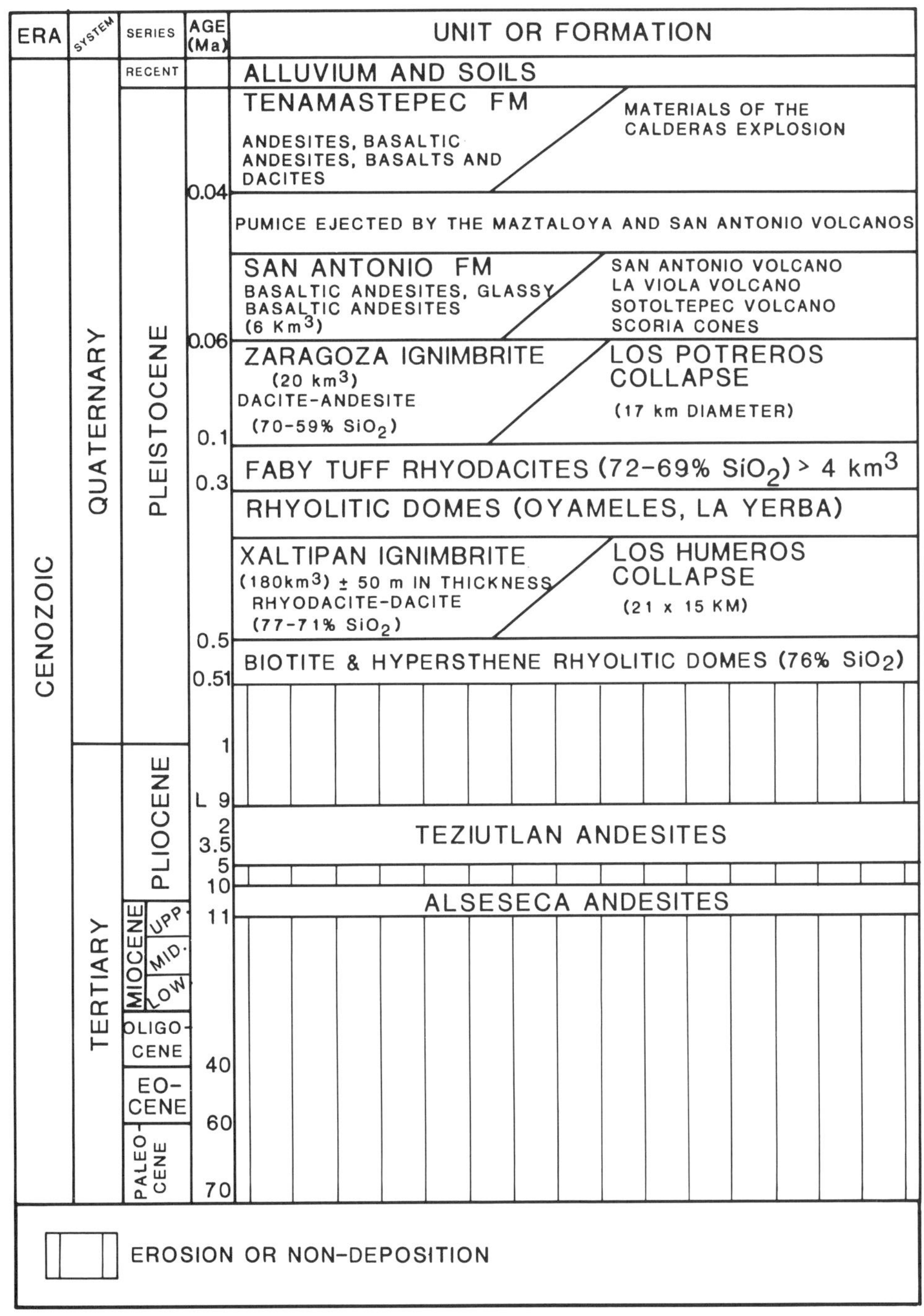

Figure 6. Pre-caldera, caldera, and post-caldera volcanic units of Los Humeros. After Ferriz (1983).

sion of 180 km^3 of rhyolitic or rhyodacitic ignimbrites (Xáltipan Ignimbrite) caused an area 21 by 15 km in diameter to collapse (Fig. 5). Resurgent domes emplaced along the caldera's annular fracture were followed by the eruption of rhyodacitic air-fall tuffs. The emission of 20 km^3 of dacitic to andesitic ignimbrites (Zaragoza Ignimbrite) generated a second caldera 10 km in diameter (Los Potreros caldera) within the Los Humeros caldera (Fig. 5). Meanwhile, an arc of volcanic scoria cones arose along the annular fracture in the southwestern part of the Los Humeros caldera, feeding widespread basaltic andesite flows that extend 10 km to the south (Fig. 5). The largest volcanic structure of this cycle was an andesitic stratovolcano that subsequently suffered an explosive event, emitting tuffs of alternating dacitic and andesitic composition, and giving rise to the Mastaloya caldera, 1.7 km in diameter (named Xalapazco-Mastaloya). The third cycle, young volcanism, includes deposition of olivine basalts, andesites, and air-fall tuffs, ending in the extrusion of dacites, the most recent of which are 40,000 years old. These units are shown on Figure 6.

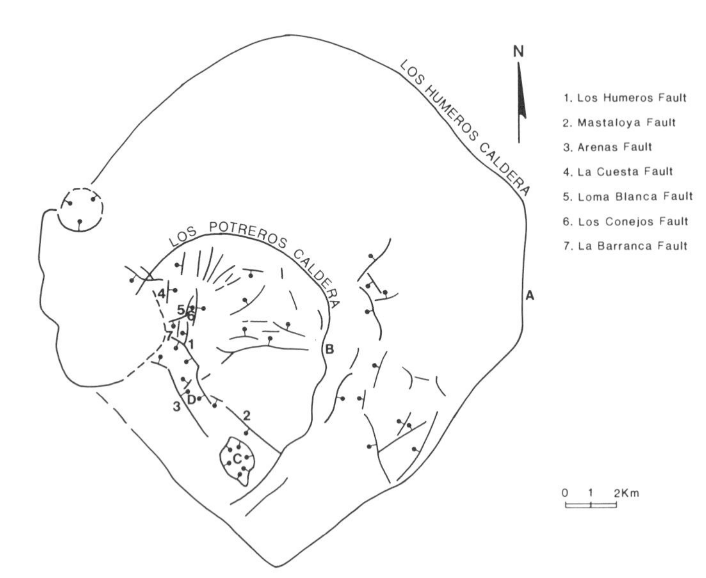

Figure 7. Structural geology of the Los Humeros caldera. The sequence of formation of the caldera structures is given by letters A to C. Graben D and the numbered faults followed the collapse of C. The bars and balls show the tilt of collapse floors interpreted from topographic expression and/or radial distribution of overflows. Modified from Torres Hernández (1985).

Structural geology

The youngest structures in the caldera are the Mastaloya and Arenas faults in the center (Fig. 7), to which the main geothermal prospects are linked. Surface hydrothermal shows are fumaroles, steaming soils, and patches of kaolinized rock and highly altered soil. The major alteration zones are associated with the Los Humeros, Loma Blanca, and La Cuesta faults (Fig. 7).

The faults with the largest displacements within the central zone are Los Humeros, Loma Blanca, La Cuesta, and Los Conejos, all of which strike north-south; the last three dip opposite to Los Humeros. La Barranca and other minor faults are associated with the main structures; the young La Barranca fault affects the caldera's final volcanic emissions.

Faults to the north of the central zone are offset toward the east. The Los Humeros fault, extending southward from Loma Blanca, dips to the west (Fig. 8). The distribution and style of the faulting suggest the presence of a major pre-caldera fault or fracture, at the longitude of Loma Blanca, controlling both fault systems. Wells H-2 and H-12, H-5 and H-7, and H-9 and H-4 (Fig. 9) exhibit major displacements of the Cretaceous sedimentary top (Fig. 8), exceeding 1,650 m between wells H-1 and H-12. Well data suggest, furthermore, that the front of the sedimentary sequence acted as an underground barrier to the reservoir and forms its western boundary (Fig. 8).

Regionally the belt of folds and thrusts affecting the Los Humeros sedimentary sequence in the subsurface is broken by transverse faults north of the caldera. Regional gravity and resistivity surveys indicate that the sedimentary structures are irregularly distributed in the caldera area, which suggests they too have been displaced by faults.

Reservoir characteristics

Seven deep wells drilled in the field to date have revealed some features of the reservoir; however, its true extension, the hydrothermal system's regional recharge, and the total thickness of the unit enclosing the reservoir are yet to be defined. At present a reservoir is known to exist trapped within Miocene-Pliocene andesitic rocks (Teziutlán Formation).

The system possibly is sealed by an average 740-m thickness of ignimbrites and tuffs emitted during the caldera's evolution. The base of the reservoir would be the Lower Cretaceous lower Tamaulipas Formation. This unit shows low permeability in the wells where it is found, suggesting it will not produce thermal fluids despite containing part of them and being a high conductor of heat, to judge by the high temperatures that have been recorded in it.

The ignimbrites and tuffs of the sealing layer include the Xáltipan and Zaragoza Ignimbrites and the Faby Tuff; these are thicker in the area of wells H-1 and H-4 and thinner (250 m) around H-2 and H-12. Alteration is mainly chloritization and oxidation with varying clay percentages; secondary minerals are calcite, pyrite, and quartz. If the alteration and mineralogy are effects of the hydrothermal system, the ignimbrites may have undergone a self-sealing process by the action of circulating hydrothermal fluids, causing their low permeability and thereby their behavior as a sealing layer. Interbedded within the ignimbrite-tuff sequence are andesite flows, at times with considerable lateral hydrothermal flow.

The andesitic sequence hosting the reservoir has been treated as one unit and named the Teziutlán Formation; however, it may include the Alseseca Andesite at its base. Within this overall grouping of porphyritic microlitic andesites, which averages 1,650 m in thickness, there are interbedded thin tuff horizons that separate several stages of lava eruptions, as well as rhyolitic, dacitic, and eventually basaltic rocks. Transmissivity values measured in these rocks at wells H-1, H-4, and H-7 were 16, 9.25, and 12.6 D-m/Cp. Andesites penetrated in wells H-2, H-5, and H-9 showed lower values and would therefore not be good producers. As would be expected, the andesites hosting the reservoir exhibit the best production conditions where they are faulted or fractured. Hence, well locations are selected adjacent to major faults at points where fracture zones presumably developed.

Andesite mineralization at producing zones includes quartz, calcite, pyrite, and epidote with subordinate mica and amphiboles, all of hydrothermal origin. Epidote characterizes the producing zone, being most abundant where production is greatest.

The lower Tamaulipas Formation, a thick sequence of limestones and clayey limestones with some chert lenses, forms the base of the reservoir. Wells H-2 and H-9 encountered 700 m of total thickness, of low permeability, and an equivalent thickness of the underlying Pimienta Formation.

Geohydrological surveys suggest that recharge of the system probably takes place through the faults shaping the caldera and its enclosed structures. A significant deep lateral recharge toward the reservoir is not excluded, lack of evidence notwithstanding.

GEOPHYSICAL SURVEYS

Among the many geophysical investigations carried out at different times by various institutions at the Los Humeros caldera, the most relevant to geothermal exploration are those discussed below.

Thermometry and thermal flow

Thermal manifestations at the Los Humeros caldera release insignificant amounts of steam through porous soils, with 50°C temperatures in scoriaceous soil, 60°C at the talus deposits of Xalapazco Mastaloya, 89°C in the pumiceous soils of the central area, and 88°C in pumice deposits near Loma Blanca (Fig. 9). Temperature measurements in boreholes approximately 3.5 m deep were recorded on temperature versus gradient graphs from five depth levels anda 530 stations to determine the thermal gradient and evaluate the heat transference regime at each locality. The following types of heat transference were defined: (a) at 70°C or more a high convective flow takes place through the soil; (b) at 40°C or more, but below 70°C, the flow is low convective; (c) at 25°C or more, but below 40°C, flow is high conductive; and (d) below 25°C, flow is low conductive.

In extensive thermal shows such as those of Los Humeros, maximum temperature measurements are insufficient for their classification as to thermal importance. Moreover the type of transference regime produced by the anomaly and its outward extension must be taken into account. This information, together with the maximum temperatures, thermal gradients, and isotherm contours, was used to rate thermal show areas in order of greater to lesser importance, as follows (Fig. 9): (a) Loma Blanca and the area north of the town of Los Humeros, characterized by complex structural control (Loma Blanca and Los Humeros faults and transverse fractures); (b) manifestations along the old road to Libres and the middle portion of the Los Humeros fault, which show a clear structural control; (c) the uplifted block of the Los Humeros fault, Xalapazco Mastaloya, and Zona Nueva, con-

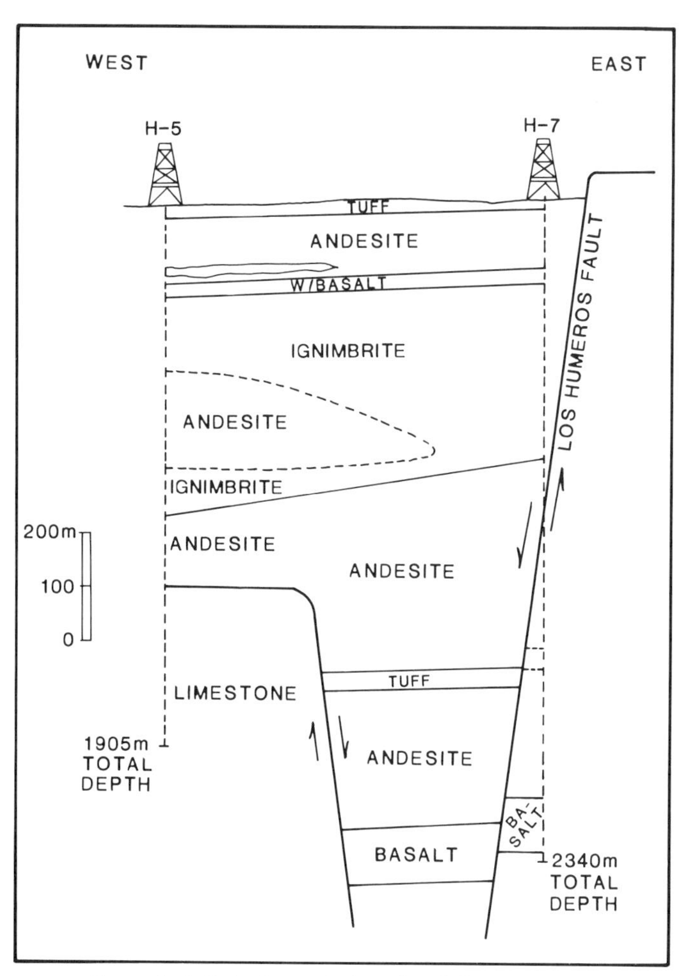

Figure 8. Lithologic correlation of wells H-5 and H-7.

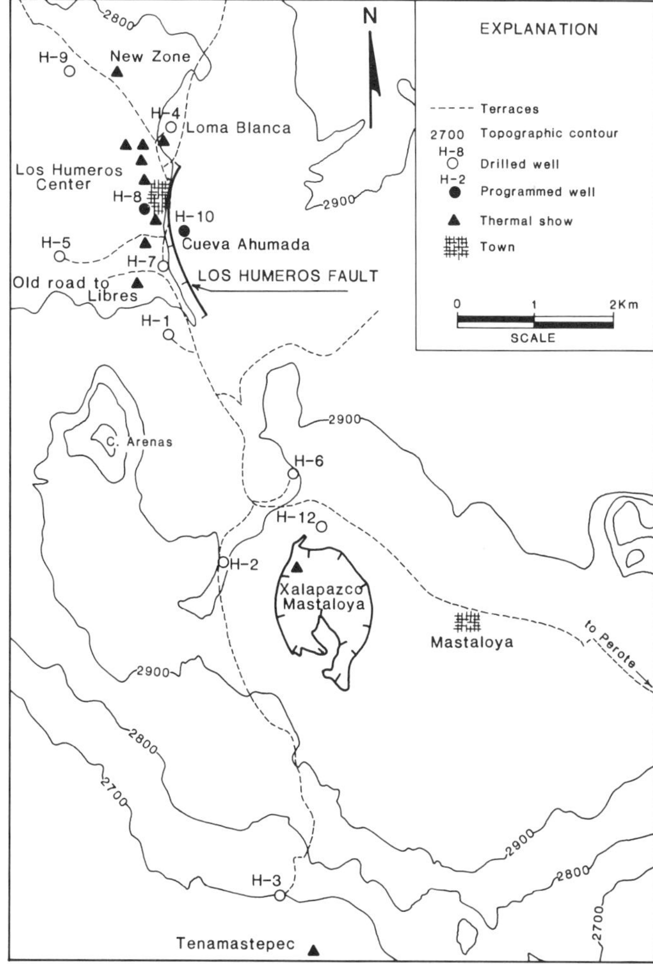

Figure 9. Location of thermal zones in Los Humeros. After García (1983).

trolled by less important faults or fractures as regards surface thermal activity; and (d) the Tenamastepec zone of slight thermal activity, limited extension, and no clear structural control.

Resistivity

A regional geoelectric survey over Los Humeros used Schlumberger-type vertical electrical probes aligned preferentially northwest to southeast. Qualitative analysis of the results indicated that strong apparent isoresistivity gradients and smallest resistivity values (230 ohm/m) are associated with thermal show areas and their related geologic structures. Quantitative interpretations identified three electrostratigraphic horizons. The first corresponds to surface volcanic rocks 20 to 400 m thick; the second to igneous rocks and conglomerates 60 to 600 m thick; and the base of this sequence was inferred to be associated with the calcareous rocks of the third horizon, showing the lowest resistivities (2 to 75 ohm/m).

After the first wells were drilled, a detailed survey, also using vertical electric probes, clearly defined the central portion of the caldera and the Xalapazco Mastaloya area (Fig. 10) as the zones of lowest resistivities (10 ohm/m) and, by correlation with drilling and other information, as the most adequate for containing the reservoir. These prospects lie at different depths, the shallowest being the central area, and the Xalapazco Mastaloya the deepest, possibly limited in the south by the collapse structure of the Los Humeros Caldera.

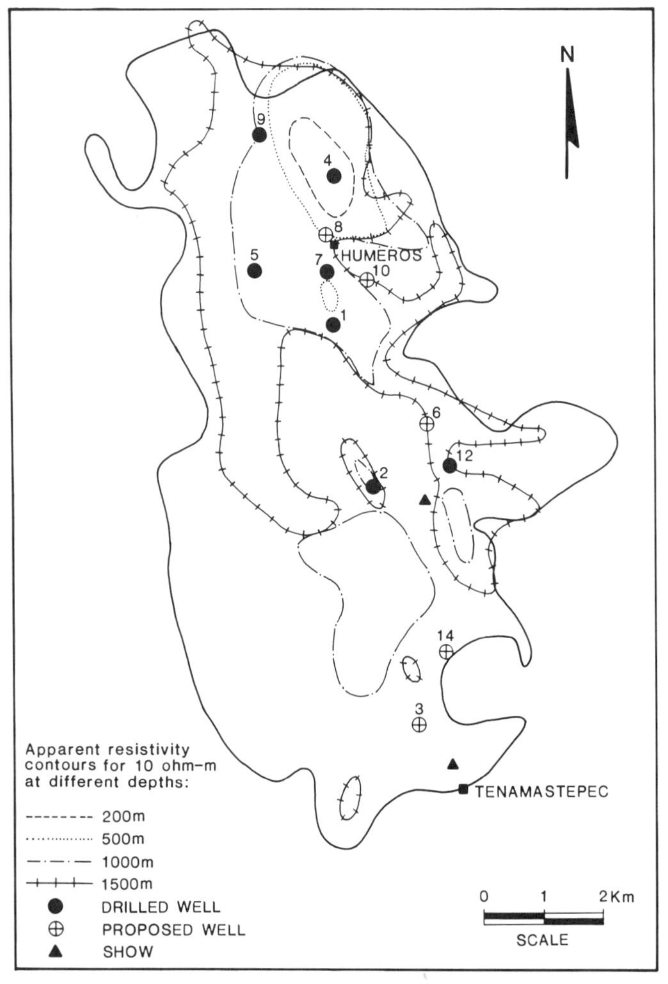

Figure 10. Integration of low-resistivity anomalies. After Arredondo Fragozo (1984).

Within the electrostratigraphic sequence of the detailed study, an outstanding conductive horizon showing the greatest thickness corresponds to andesitic and ignimbritic flows resting on top of the Lower Cretaceous limestone.

Gravity and magnetometry

An aeromagnetic survey, for the purpose of identifying geologic structures, defined the center of the caldera within the positive portion of a bipolar anomaly interpreted as the expression of a highly magnetized intrusive body at a depth of 2 km. A series of secondary anomalies was correlated with the caldera structures and with some of the faults associated with surface thermal manifestations. An anomaly linked to the Xalapazco Mastaloya was also identified.

Regional gravity data show a minimum in the caldera, apparently corresponding to its collapse. A gravity maximum in the central portion of the caldera has been attributed to a massive plug. There may be another caldera filled with volcanic material between Los Humeros and Cofre de Perote.

On the basis of Bouguer anomalies and well data, two major lithologic units are inferred within the caldera: the volcanic sequence, and the underlying calcareous series. Gravity profiles helped to define the behavior of the limestone top as well as a tabular body possibly associated with the Cerro Arenas volcanic neck.

Paleomagnetism

Several late Pliocene and Pleistocene pyroclastic outcrops within and outside the caldera were sampled and studied to define maximum emplacement temperatures. Initial remnant magnetization and demagnetization values in the rocks define them as "high-temperature" deposits emplaced at 450°C or more.

Microseismicity

A microseismicity survey used three portable stations to investigate the caldera's seismic activity and its relation to the volcanological, structural, and thermal aspects. Of a total 96 recorded events, half were classified as swarms and the rest as tremors, which are typical of active or recent volcanic structures. A zone of strongly attenuated seismic waves was defined in the central area of the caldera, possibly related to an anomalous temperature.

Telluric method

This process uses natural electromagnetic fields to deduce approximate contrasts in the electrical conductivity of geologic formations. A colineal array in two adjoining 500-m intervals at

0.05- and 8-Hz frequencies was used to map the caldera. Theoretically these frequencies probe at two different depths; penetrations are defined by the sequence of resistivities and their absolute values. The survey was carried out along two northeast to southwest lines located north of and in the center of the caldera. Responses showed marked conductivity contrasts at the different frequencies in the western portion of both lines. The northern line provided evidence of the western boundary of the caldera in that zone.

GEOCHEMISTRY

The first geochemical studies sampled and analyzed waters from waterwheels, wells, springs, and lagoons, as well as gases emitted by fumaroles and steaming soils within and south of the Los Humeros cauldera in search of regional thermal anomalies. Results showed low hydrogen sulfide and high carbon dioxide contents, and the presence of H_2 and methane, indicating high-temperature zones in the subsurface of the caldera. Average bottom temperatures computed by gas geothermometry at each manifestation were 312°C.

Geochemistry of geothermal well fluids

Tables 1 and 2 summarize the results of chemical analyses of water, gas, and condensate samples from four geothermal wells.

Water from well H-1 is sodium-bicarbonate with low salts and high boron percentages (Table 1). The low sulfate content suggests it is possibly condensed steam mixed with recently infiltrated rain waters. Liquid-phase geothermometry in H-1 shows temperatures of 256 to 270°C and 310°C for the gaseous phase (Tables 1 and 2); gas percentage was 3 percent by weight.

Well H-2 initially produced steam with some sodium-bicarbonate water (Table 1); the condensate is ammonium-bicarbonate due to the abundant carbon dioxide and ammonia in the gaseous phase. The well subsequently produced steam only. A triangular chloride-bicarbonate-boron diagram places the condensate samples from H-2 in the high-boron content region (endogenous steam zone). During initial production from H-2, liquid-phase geothermometry temperatures were 215 to 250°C. Chemical composition of gases in the fluid (Table 2) shows high ammonia concentrations, and the total gas content in the fluids was the highest of the field (20 percent by weight). Gaseous-phase thermometry showed temperatures on the order of 330°C (Table 2).

Chemistry of the fluids in well H-4 showed wider variations. pH values varied according to the steam production; fluid pH was more acid, both in the seam moisture and the condensate, when production was highest and neutralized when it dropped. Boron concentration in H-4 varied directly with steam production; ammonia (NH_4^+) concentrations increased considerably when the pH was neutralized, forming ammonium bicarbonate. Chloride appeared in condensate and steam moisture samples; when pH was acid the chloride present in the geothermal fluid appeared as hydrochloric acid gas.

The gas in H-4 shows high concentrations of hydrogen sulfide and hydrogen (Table 2). The abnormal hydrogen content probably caused chemical imbalance between the methane, car-

TABLE 1. CHEMICAL COMPOSITIONS OF EXTRACTED WATER, SAMPLED AT ATMOSPHERIC PRESSURE AND TEMPERATURES, OBTAINED BY LIQUID PHASE GEOTHERMOMETRY

(Concentrations in mg/l)

Well	Ph	Na	K	Ca	Cl	B	SiO_2	NH_4	SO_4	HCO_3	SiO_2 T°C	NaKCa T°C	NaK T°C
H-1	8.2	304	84	3	87	214	750	6	115	250	252	270	259
H-2	8.3	470	50	5	367	186	634	14	90	634	244	218	222
H-7	6.8	212	29	6	68	1645	580	10	96	60	237	215	245

TABLE 2. CHEMICAL COMPOSITION OF THE NONCONDENSABLE GASES AND TEMPERATURES OBTAINED BY GASEOUS PHASE GEOTHERMOMETRY

Well	T sep. °C	CO_2	H_2S	NH_3	H_2	CH_4	N_2 + Cl	% in wt.	T gas °C
H-1	166	952	20	15	6	6	2	3	310
H-2	180	862	9	17	9	9	94	20	330
H-4	261	784	94	0.03	119	0.4	0.14	6	422
H-7	135	841	60	69	10	10	10	2	290

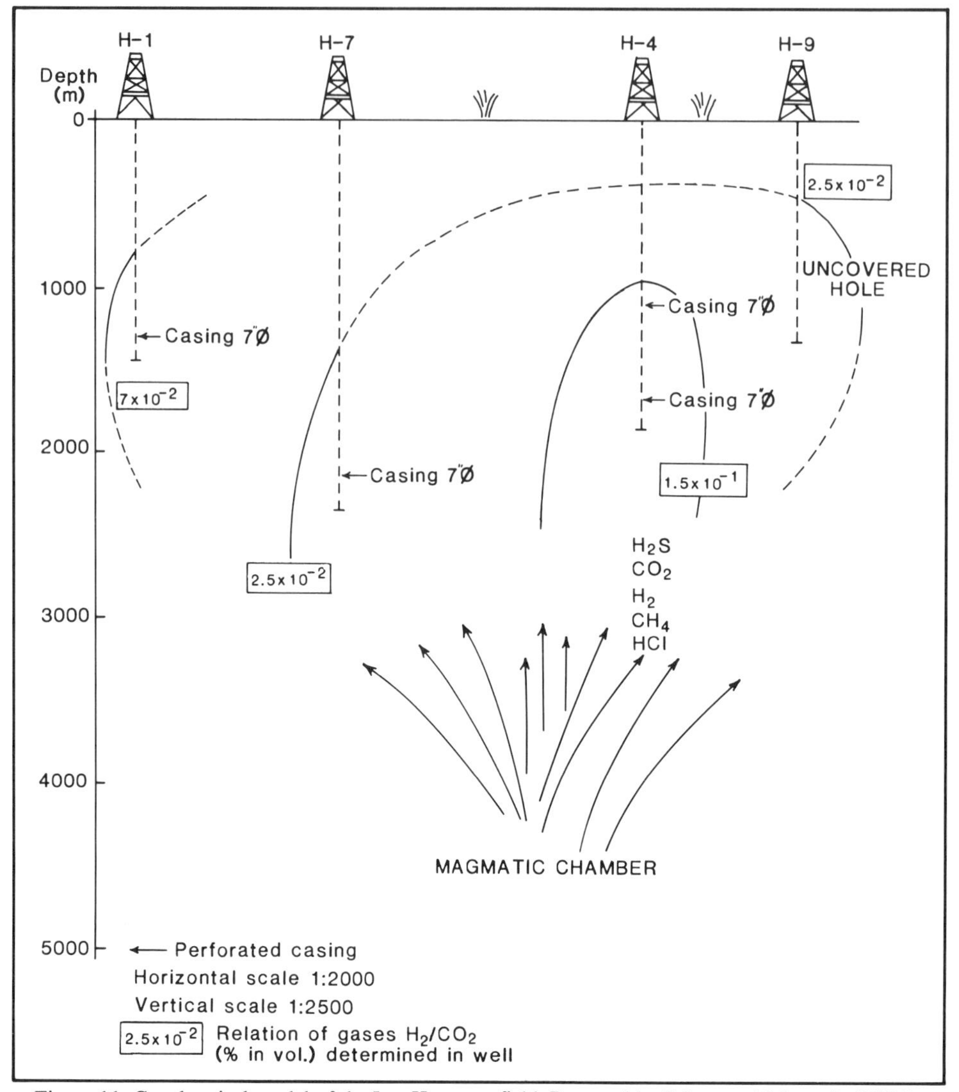

Figure 11. Geochemical model of the Los Humeros field. Prepared by López and Munguía (1984).

bon dioxide, and the steam, leading to excessively high temperature determinations when applying gas geothermometry (Table 2). This suggests that H-4 fluid is probably more closely related to the heat source. Gas was 6 percent by weight.

For its part, the scarce water separated from well H-7 was sodium-chloride with low salinity and high boron content (Table 1). Chemical analysis of the noncondensable gases showed high ammonia and hydrogen sulfide concentrations (Table 2) with 1 percent of gas by weight. Tables 1 and 2 list liquid- and gaseous-phase geothermometry temperatures.

The general chemistry of the fluids produced in the four wells indicates interaction between sedimentary and igneous rocks. The presence of sedimentary rocks is suggested by the low chlorine-boron relation in the liquid phase and the high ammonium content in the gaseous phase. Gaseous-phase geothermometry temperatures are higher than in the liquid phase, indicating that the gases come from greater depths.

Geochemical model of the field

Because most of the wells are steam producers, the preliminary geochemical model used the geochemistry of noncondensable gases, applying the hydrogen/carbon dioxide relation in volume percentage. Figure 11 shows the largest emission of endogenous gas occuring in the area of well H-4 and fanning out to those of H-7, H-9, and H-1. The concentration of hydrogen sulfide isovalues also agrees with the previous model; high (10 to 20 percent) hydrogen sulfide concentrations appear in H-4, 4 to 10 percent in H-7 and H-9, and 1 percent in H-1.

Water geochemistry in well H-1 indicates that the hydrothermal fluid in this area is probably meteoric water mixed with condensate (of high boron content) coming from a deeper geothermal zone. Accordingly the geochemical information to date indicates the field to be a relatively shallow (1,000 to 2,500 m deep) reservoir influenced by a deeper one.

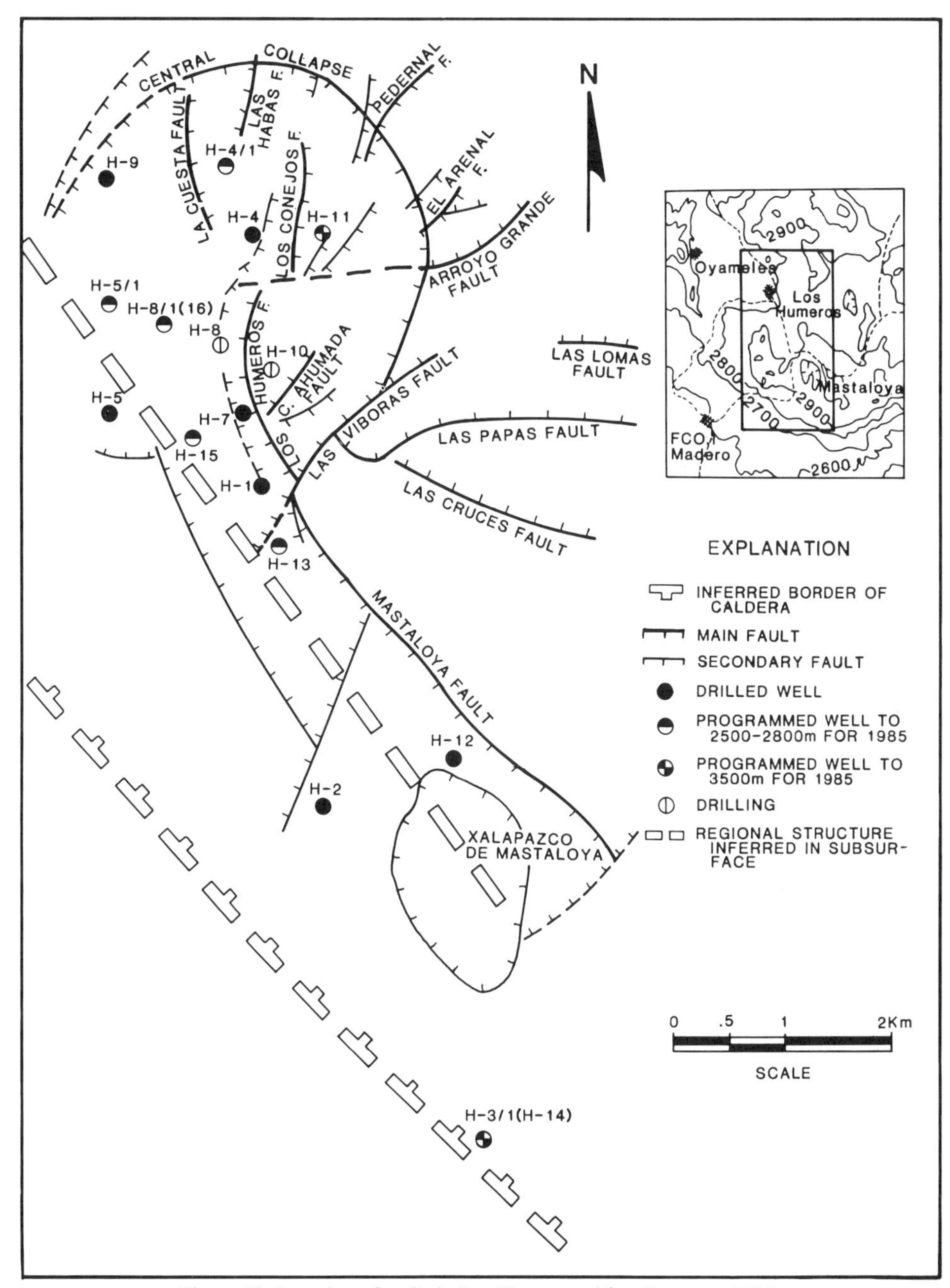

Figure 12. Location of wells in Los Humeros. After Romero (1984).

EXPLORATION AND PRODUCTION WELLS

Seven exploratory wells have been drilled in the Los Humeros geothermal area on the basis of geological, geophysical, and geochemical information (Fig. 12). Results have been as varied as their locations, depths, and objectives: two wells are producers (H-1 and H-7), two failures (H-5 and H-2), two were abandoned (H-4 and H-12) because of technical drilling problems arising from the reservoir, and one is currently under study (H-9). A summary of the relevant information follows:

Well H-1. With a total depth of 1,458 m, this well was positioned on a low-resistivity anomaly and designed to penetrate the Los Humeros fault at depth (Fig. 12). Maximum recorded temperature during drilling was 251.65°C at 1,250 m. Maximum equilibrium temperature is within the 223 to 232°C range. Maximum temperature recorded during heating was 267.3°C at 1,432 m. Production is 174 Ton/h of water-steam mixture, of which only 68.5 Ton/h is steam; the remaining 105.5 Ton/h is water. Enthalpy is 1,279 kJ/kg under atmospheric conditions. Results of the temperature, pressure, and injectivity tests suggest the main contributing zone to be at well bottom (1,457 m). Average transmissivity was on the order of 1.5×10^{-8} m^3/Pa-s (16D-m/cp). Production data at different discharge diameters established the well's capacity to generate 5 MW at a separation pressure of 0.8 MPa.

Well H-2. This well, drilled northwest of the Xalapazco de Mastaloya caldera, did not anticipate underground faulting. Total depth was 2,301 m; the section from 1,150 m to the bottom consists of predominantly calcareous sedimentary rocks (Fig. 12). The well showed high temperature (330°C) and low permeability (1.9×10^{-9} m^3/Pa-s) and is therefore a poor producer: 4 Ton/h of saturated steam was fed from zones at depths of 1,100, 1,800, and from 2,200 m to the well bottom.

Well H-4. This well was positioned on a low-resistivity anomaly at Loma Blanca, which is characterized by widespread alteration with evidence of fumarolic thermal activity (Fig. 12). Total depth was 1,880 m, seeking the possible northward extension of the Los Humeros fault; only volcanic rocks were encountered. Maximum temperature while drilling was 299°C at 1,100 m; maximum equilibrium temperature was evaluated in the 326 to 333°C range. The well produced overheated steam initially estimated at 100 Ton/h, but dropping relatively quickly to 38 Ton/h; it was believed to come from the 1,000- to 1,100-m depth interval. The transmissivity coefficient based on the injectivity test was 9.5×10^{-9} m^3/Pa-s. With a 45-Ton/h production of saturated steam at 0.8 m Pa (8 bars) pressure it can generate 4 MW. The well is presently closed owing to corrosion problems caused by the high hydrogen sulfide content.

Well H-5. Similar to H-2, this well was drilled to a maximum depth of 1,905 m for the purpose of verifying the possible connection between a low-resistivity anomaly and the underground geothermal reservoir (Fig. 12). It also penetrated limestones below the 1,555-m level, which indicates a structural high with respect to H-1. Maximum temperature of 204.5°C and equilibrium temperature of 219°C were both recorded at 1,548 m. Recorded temperatures and pressures at different dates show decreasing thermal characteristics, which indicates a restriction in the heat transference. To the present depth (1,905 m), production possibilities therefore seem slight, or else the restrictive effect may be due to damage at the well bottom while cementing.

Well H-7. The drilling objective was to define the northward extension of the reservoir discovered by H-1, seeking the Los Humeros fault and improved underground permeabilities (Fig. 12); total depth was 2,340 m. Maximum temperature was 340°C at the 2,340-m depth and the maximum equilibrium temperature was within the 334 to 340°C range. Production under atmospheric conditions is 38.39 Ton/h of steam, which is expected to generate 2 MW and 0.6 Ton/h of water. Enthalpy of the mixture is 2,608 kJ/kg. The transmissivity coefficient as per injectivity tests was 1.22×10^{-8} m^3/Pa-s.

Well H-9. The two-fold objective was to investigate the possible northward prolongation of the reservoir within the volcanic sequence and to define a possible deep reservoir in the Cahuasas Formation, which underlies the lower Tamaulipas, Pimienta, and other limestones (Fig. 12). The well was positioned near a gravity high interpreted as a subsurface uplift of the sedimentary sequence. To date, 2,500 m have been drilled; the top of the limestone was encountered at 1,733 m. A decision to continue is pending because a granitic intrusive was found underlying the limestones in well H-12 instead of the Cahuasas Formation. Maximum recorded temperature was 278°C at 2,076 m. Fluid entry thought to be saturated steam occurred at 500 to 550 m, while drilling caused loss of well control; wellhead pressures on the order of 250 psi developed on flowing through a 2-in relief line. Maximum temperature recorded at this level (500 m) was 202°C after a 24-hr rest. The interval was considered too superficial and short-lived and was subsequently cemented; drilling then continued to the present 2,500 m depth. Other significantly contributing intervals have not been detected, and a decision to drill further is pending.

Well H-12. This well was located near the Mastaloya fault on the basis of the information provided by H-1 and H-7, seeking to encounter the fault at depth and to penetrate the Cahuasas Formation at 3,500 m. The well was drilled to a final depth of 3,104 m through 2,700 m of volcanic rocks directly overlying an intrusive. Maximum temperatures recorded after a 24-hour rest were 304°C at 1,550 m, 250°C at 1,900 m, 361°C at 2,900 m, and 362°C at 2,984 m after a 14-hr rest. Possibly producing levels at 1,550-m and 1,900-m depths will be evaluated by injectivity tests, and are expected to show favorable results. Maximum-temperature zones below the 2,900-m depth are of no interest, being in intrusive granites of low permeability.

CONCEPTUAL GEOTHERMAL MODEL OF THE FIELD

The preliminary conceptual geothermal model of the reservoir was based on surface and subsurface geological, geophysical, and geochemical data together with information provided by the wells. The tridimensional demarcation and volume computation of the hydrothermal system (reservoir) and its reserve used the following parameters: isotherms, lithology, geophysical data (resistivity), alteration mineralogy, and the structural system. These values were plotted along eight preselected sections on a topographic map of the field, and contours were drawn at four preselected depth levels: 2,000, 1,500, 1,000, and 500 m asl. The reservoir and its reserve were thus demarked at each level; the respective areas and volumes at each level were also computed. Table 3 shows these figures at each level and their totals. These were the basis for an isometric tridimensional model showing the spatial distribution of the reservoir and reserve, the former between the 1,750 to 500 m asl range (Fig. 13), which is the only interval in which presently available information suggests the presence of a geothermal reservoir.

With the above volumes, the probable geothermal fluid volume was computed. Table 3 shows the volume of hot andesitic

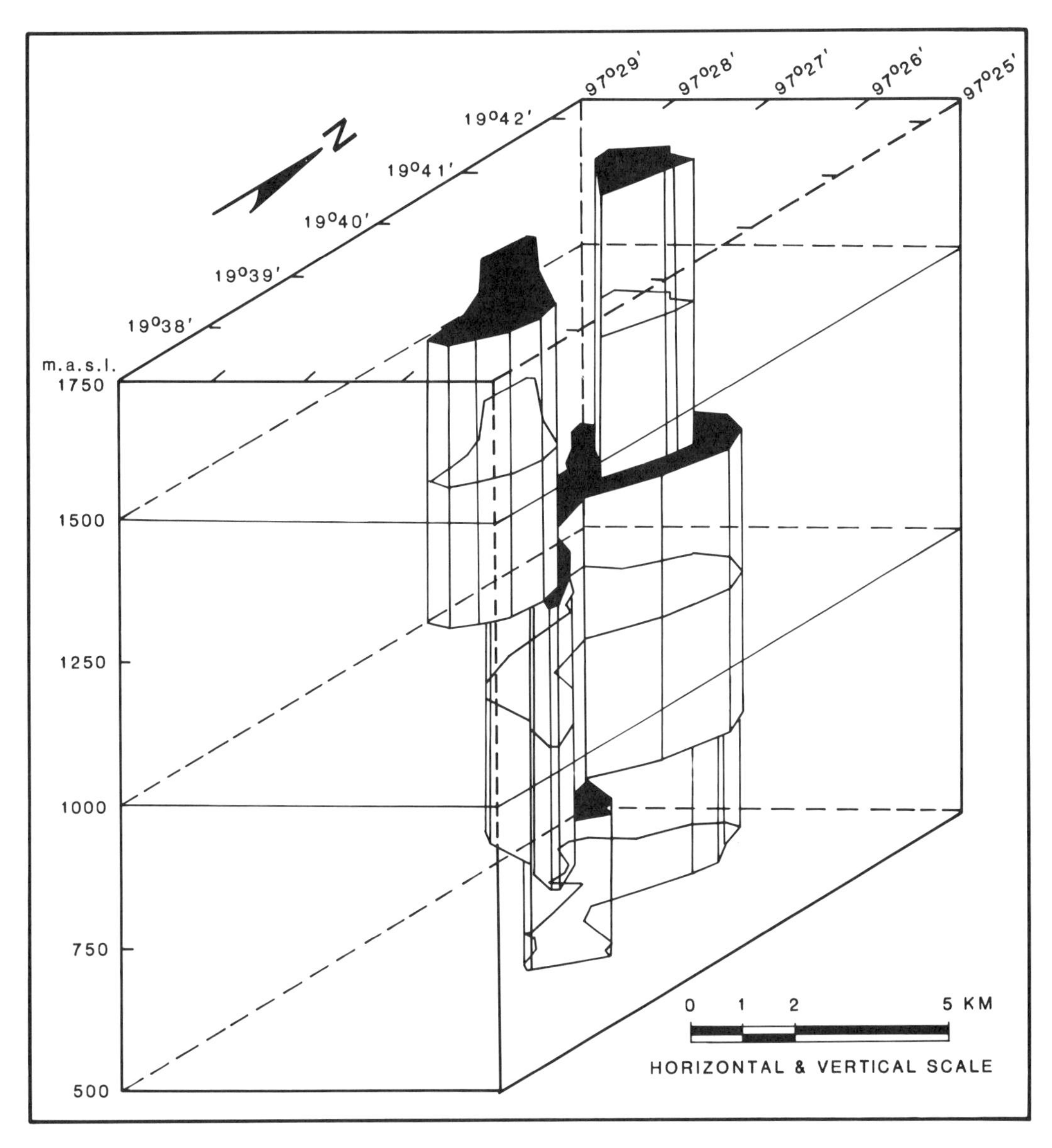

Figure 13. Isometric view of the Los Humeros reservoir. After Romero (1984).

rock (over 150°C for the reserve and over 250°C for the reservoir), which satisfies the established parameters and hence possibly contains geothermal fluids. To compute the probable volume of geothermal fluid content, the empty-space volume within reservoir and reserve that could be occupied by geothermal fluids was a requisite. This would be the absolute porosity of the rock hosting the reservoir and the reserve. Abundances of alteration minerals and total alteration in each well drilled through the host andesites were considered, as well as the pertinent literature. The average porosity, obtained by applying a specific formula, was combined with the volumes to define a probable geothermal fluid volume of 1.4 km^3 for the reservoir and 3.77 km^3 for the reserve.

Figure 14 is a block diagram of the Los Humeros geothermal field enclosing the eventual reservoir and reserve.

TABLE 3. HOT ROCK AREA AND VOLUME OF LOS HUMEROS, PUEBLA, RESERVE (>250°C)

Level (masl)	Area of the reserve (km^2)	Volume of the reserve (km^3)	Area of the additional reserve (km^2)	Volume of the additional reserve (km^3)	Area of the reservoir (km^2)	Volume of the reservoir (km^3)
2,000	4.4	1.1	....	....	....	....
1,500	13.8	6.9	3.4	1.7	5.3	2.6
1,000	19.0	9.5	11.4	5.7	10.0	5.0
500	22.5	5.6	23.4	5.8	7.9	1.9
Total	59.7	23.1	38.2	13.2	23.2	9.5

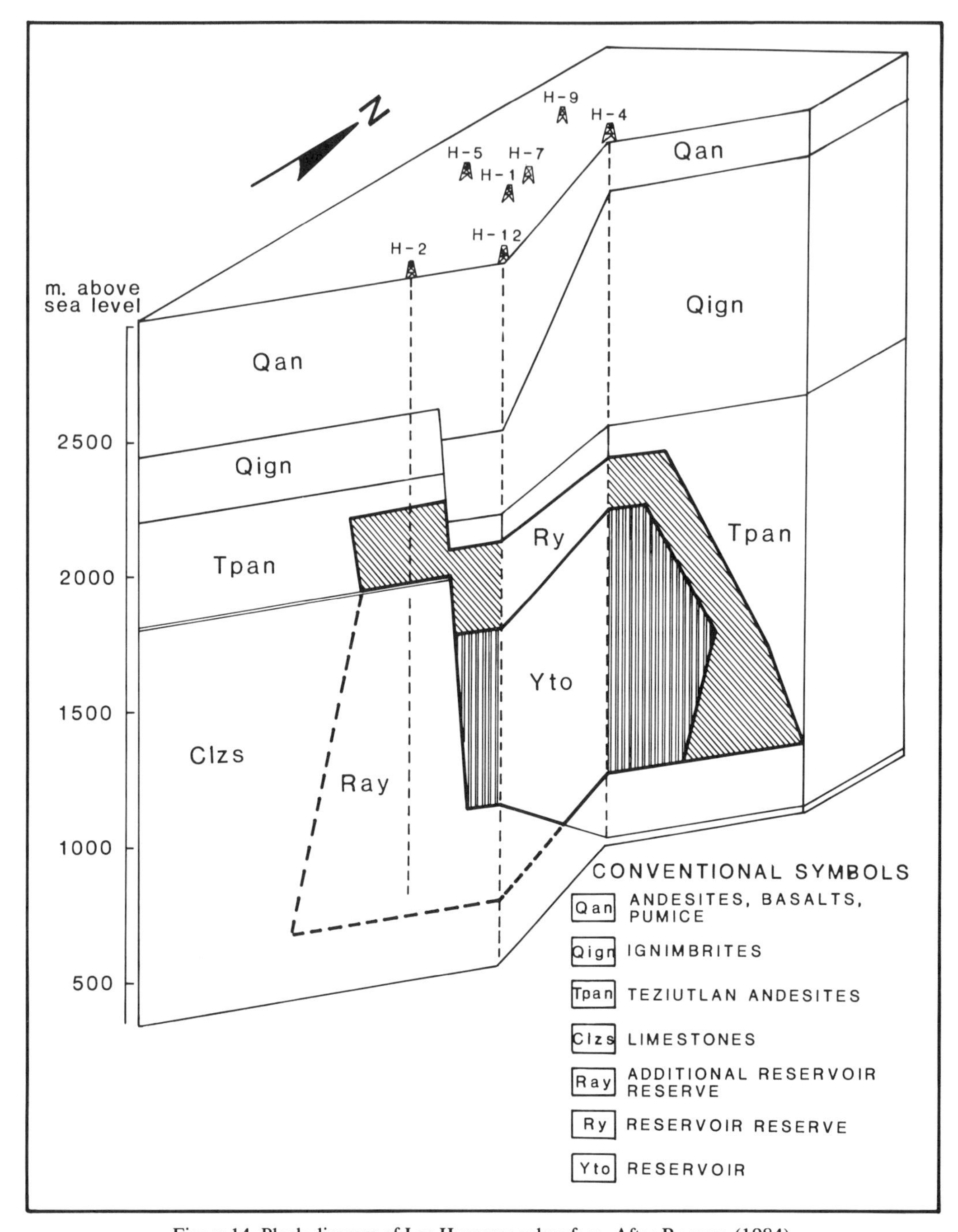

Figure 14. Block diagram of Los Humeros subsurface. After Romero (1984).

PRESENT STATUS AND PROSPECTS OF DEVELOPMENT

Of the seven wells drilled in the field to the present, H-9 and H-12 are still pending evaluation; H-1 and H-7 are rated as commercial, the first with a 170 Ton/h production of mixture and enough pressure to generate 4 MW, and the latter with 35 Ton/h of steam to generate 2 MW. H-4 was closed due to corrosion problems (H_2S); H-9 and H-12 are presently undergoing injectivity tests whose favorable results are anticipated. H-8 is presently drilling; at its present (1987) depth of 2,150 m, it shows attractive thermal features.

The numerical evaluation of the field used to compute fluid and heat reserves in the reservoir and simulate its behavior was based on the preliminary geothermal model in a development program designed to generate 55 MW for 20 years on a SHAFT (Simultaneous Heat and Fluid Transport) mathematical simulation model, which requires solving equations that describe mass and energy flow in two phases through a porous medium. The following data were considered:

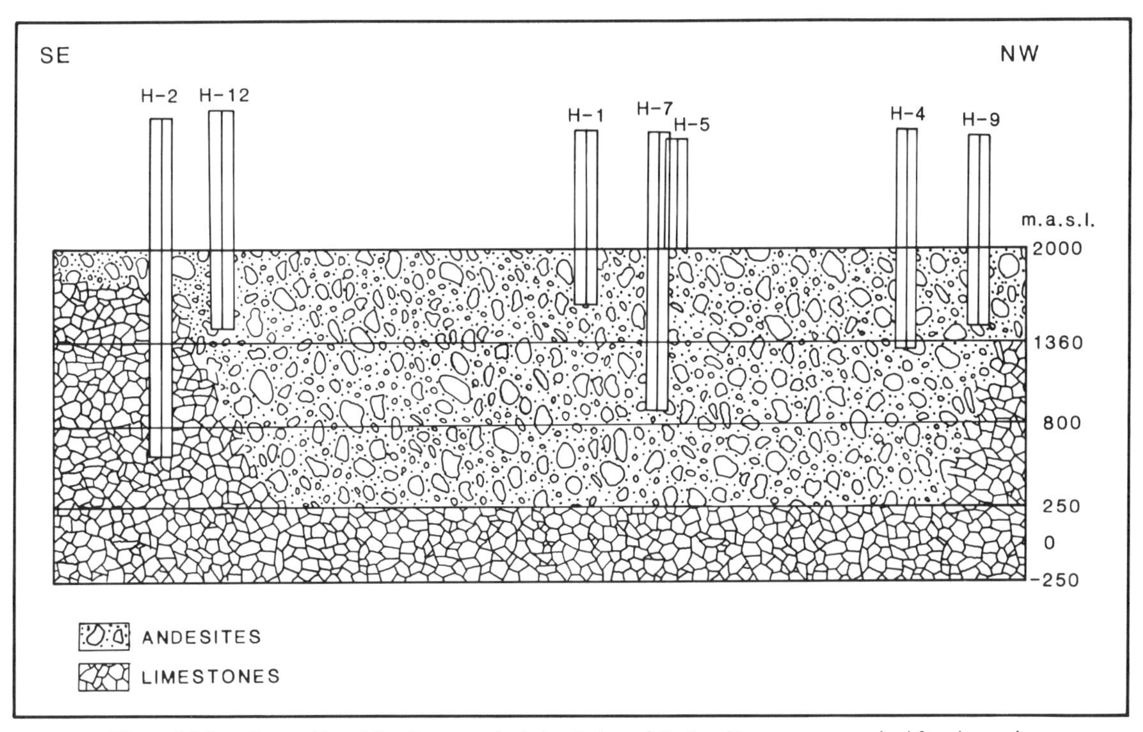

Figure 15. Levels considered for the numerical simulation of the Los Humeros reservoir. After Ascensio and others (1984).

a. *Geometry of the reservoir.* A grid based on the conceptual model of the reservoir was defined and horizontally divided into 35 elements repeated at 4 depth levels for a total of 140 elements. The selected depths were 1,360, 800, 250, and –250 m asl; the top of the first of these was placed at 2,000 m asl. The first three levels were assigned to the andesites, and the last (250 to –250 m asl) to the limestones (Fig. 15).

b. *Frontier conditions.* Based on the conceptual geothermal model, the area for simulation was divided into three parts: a geothermal reservoir; a low-permeability/high-temperature (300°C) zone; and a zone of geothermal possibilities, to be verified. The probable recharge was also defined (Fig. 16). At this point, temperatures for each level and section of elements were established.

c. *Petrophysical properties.* These were defined at each level: rock density (2.7 g/cc for andesites and 2.6 g/cc for limestones); porosity (8 percent in andesites and 2 percent in limestones); thermal conductivity (1.9 and 2.5 W/m°C for andesites and limestones, respectively), horizontal and vertical permeabilities, and specific heat for each lithologic type.

d. *Development scheme.* This was designed on the basis of a plant installation having a nominal capacity of 55 MW for 20 years, which would require approximately 10 Ton/h of steam to generate 1 MW at an initial pressure of 0.8 Mpa (8 bars). According to producing-well characteristics (H-1 and H-7) 1,260 Ton/h of geothermal fluid (350 kg/s) would be needed and are obtainable with only seven of the total 140 elements.

Simulation results show the steam needed to generate 55 MW for 20 years can be extracted with tolerable pressure drops

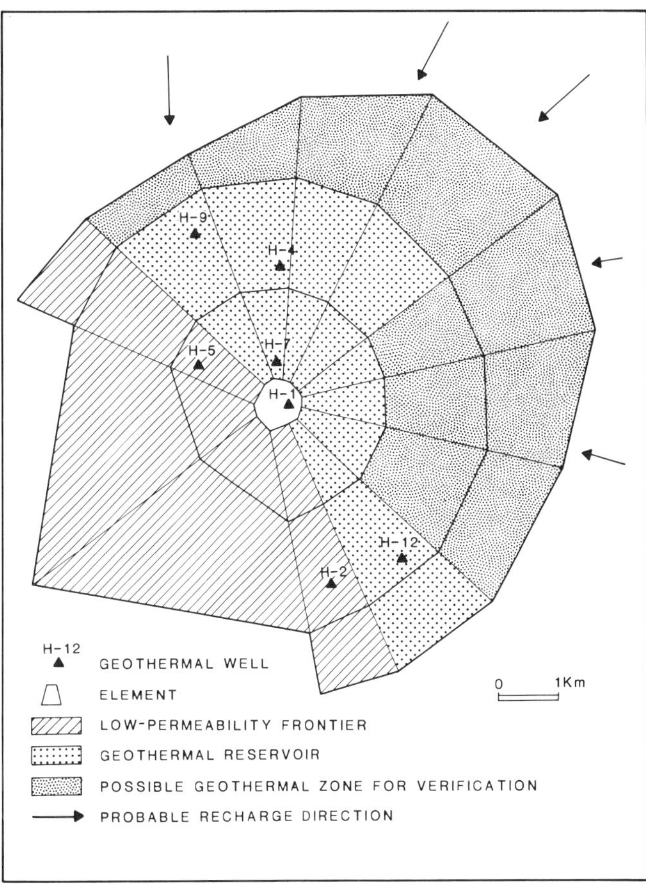

Figure 16. Reservoir frontier and recharge conditions for the Los Humeros numerical simulation. After Ascensio and others (1984).

in the simulated elements. It is therefore feasible to drill the necessary wells for production and continued exploration of the field to establish its true potential, since the 55 MW may be considered as a minimum for Los Humeros. On this basis, three wellhead units of 5,000 KW each will be installed by 1988, and construction will begin on two 55-MW plants designed to begin operating in 1993. The five-year lapse between both projects will allow the reservoir's thermodynamic conditions to evolve and more precise information to be gathered.

BIBLIOGRAPHY

Alvarez, R., 1978, Telluric, self-potential, and surface temperature profiles on Los Humeros Caldera: Geofísica Internacional, v. 17, p. 445–460.

Demant, A., 1978, Caracteristicas del Eje Neovolcánico Trans-mexicano y sus problemas de interpretation: México, D. F., Instituto de Geológia, Universidad Nacional Autonoma de México Revista, v. 2, no. 2, 5 p.

Del Rio, L. L., 1982, Búsqueda de zonas con tectonismo extreme: Teoría y una aplicación a la geotérmica; Caldera Los Humeros: Geofísica Internacional, v. 21, no. 3, 65 p.

Ferriz, H., 1982, Geologic and preliminary reservoir data on the Los Humeros geothermal system, Puebla, Mexico; Proceedings of 8th Workshop on Geothermal Reservoir Engineering, December, 1982: Stanford, California, Stanford University, SGP-TR-60, p. 19–24.

Flores, L., Sing, S. K., and Urrutia, J., 1978, Aeromagnetic survey of Los Humeros Caldera, Mexico: Geofísica Internacional, v. 17, p. 415–428.

Pone, L., and Rodriquez, C., 1978, Microearthquake activity associated to Los Humeros Caldera, Mexico; Preliminary Survey: Geofísica Internacional, v. 17, no. 4, p. 461–478.

Urrutia, F., 1983, Paleomagnetic estimation of emplacement temperature of pyroclastic deposits; Preliminary study of Caldera de Los Humeros and Alchichica Crater: Geofísica Internacional, v. 23, no. 3, p. 277–292.

Venegas, S. S., Maciela, F. R., and Herrera, F.J.J., 1984, Algunas characterísticas de la faja volcánica mexicana y de sus recursos geotérmicos: Geofísica Internacional, v. 24, p. 1–47.

Unpublished internal reports of the Comisión Federal de Electricidad

Arredondo, R., 1978, Estudio geoeléctrico de detalle en la zona geotérmica de Los Humeros, Puebla, Morelia, Michoacán: Report 67/82.

Ascencio, C. F., and 11 others, 1984, Evaluación numérica preliminar del campo geotérmico de Los Humeros, Puebla: Report 5.5.8-002.

Cedillo, R. F., 1984, Estudios geohidrológicos en la área de Los Humeros–Derrumbadas, Morelia, Michoacán [abs.]: Report 17/84.

De La Cruz, M. V., 1983, Estudio geológico a detalle de la zona geotérmica de Los Humeros, Puebla: Report 10/83.

Departamento de Evaluación y Yacimientos, 1983, Informe preliminar sobre el campo geotérmica Los Humeros, Puebla: Report 1383-007.

Guarduño, M.V.H., and Romero, R. F., 1984, Análisis geológico estructural preliminar del campo geotérmico Los Humeros, Puebla: Report 14/84.

Garduño, M.V.H., Romero, R. F., and Torres, H. R., 1985, Estudio estructural y de teledetección en la región de Perote, Veracruz: Report 12/85.

García, E.G.H., 1983, Estudio termométrico a 3.5 m de profundidad en la zona geotérmica de Los Humeros: Report 23/83.

Gutiérrez, N.L.C.A., 1982a, Petrografía del Pozo H-2, campo geotérmico Los Humeros, Puebla: Report 22/82.

—— , 1982b, Litología y zoneamiento hidrothermal en los pozos H-1 and H2 del campo geotérmico Los Humeros, Puebla: Report 23/82.

—— , 1983, Litología y zoneamiento hidrotermal en el Pozo H-4 del campo geotérmico Los Humeros, Puebla: Report 55/82.

López, M.J.M., and Munguía, B., 1984, Pozo H-4, comportamiento geoquímico del fluido: Report 5.4-004.

Medina, M. M., 1984, Evaluación del Pozo H-7 de campo Los Humeros, Puebla: Report 5.4-005.

Palacios, M.L.M., and García, V. M., 1981, Informe geofísico del proyecto geotérmica Los Humeros–Derrumbadas: Report 5.2-002.

Pano, A. A., 1975, Bosquejo estratigráfico, estructural y tectonico del área de Teziutlán y áreas circunvecinas: PEMEX VI Excursión Geológica, Report IGPR-107.

Torres, H. R., 1983, Campo geotérmico Los Humeros, Puebla: Report R.P.H.-001.

—— , 1984, Geología estructural de Los Humeros, Puebla: Report R.P.H.-005.

Viggiano, G.J.C., 1984a, Columna petrográfica del Pozo H-7, campo geotérmico Los Humeros, Puebla: Report 5.1.2-004.

Yáñez, G. C., and García, 1982, Exploración de la región geotérmica Los Humeros–Las Derrumbadas, estados de Puebla y Veracruz: Report 5.1.2-03a.

Unpublished theses

Robin, C., 1982, Relations volcanologiques-magmatologiques-geodynamiques, application en passage entre volcanismes alcalin et andestique dans le sud mexicain: Clermont, France, Universidad de Clermont Ferrand II.

Tarango, O. G., 1967, Estudio geológico petrolero del área Río Laxaxalpa–Teziutlán, Estado de Puebla [professional thesis]: Mexico, D. F., Escualo Superior de Ingeniera y Arquitectura, p. 11–46.

Printed in U.S.A.

The Geology of North America
Vol. P-3, Economic Geology, Mexico
The Geological Society of America, 1991

Chapter 8

La Primavera geothermal field, Jalisco

Saúl Vengas S., Germán Ramírez S., Carlos Romero G., Pablo Reyes V., Antonio Razo M., Luis C. A. Gutiérrez N., Francisco Arellano G., and José Perezyera y Z.
Federal Commission on Electricity, Rio Rodano 14, 06500 Mexico, D.F.

INTRODUCTION

Preliminary exploration to evaluate the La Primavera (Jalisco) geothermal prospect began in the 1960s with photogeological interpretation and geochemical surveys of the surface shows. In the late 1960s the (then) Office of Geothermal Exploration of the Comisión Federal de Electricidad (CFE) drew up a wide-ranging geological project to explore the central portion of Jalisco State. Results indicated La Primavera to be the most favorable of several geothermal prospects by reason of its geologic structural conditions and the ages and compositions of its rocks. Detailed geological, geochemical, geophysical, and geohydrological surveys were carried out to select exploratory well locations. Five geothermal wells (PR-1, PR-2, PR-4, PR-5, and RC-1) were drilled in a first phase between January 1980 and August 1982. Subsequently, PR-1 was deepened, and drilling began on PR-8 in 1984.

The Sierra La Primavera lies 15 km west of the city of Guadalajara (Fig. 1) between longitudes 103°28′ and 103°38′W and latitudes 20°32′ and 20°43′N, in a volcanic area of fumaroles and hot springs. The region forms part of a series of valleys, basins, and block mountains of mainly mid-Tertiary to Pliocene-Quaternary volcanic rocks. Drainage is parallel and dendritic; in the rainy season, intermittent streams carry water to the valleys and basins draining toward the Río Grande de Santiago in the north and the Río Ameca in the west, and thence to the Pacific Ocean.

The Sierra La Primavera stands out as a dome-like promontory of volcanic rocks in a caldera-type semicircular array forming the natural water divide of the area. The principal streams (arroyos) are el Caracol, Llano Grande, Hondo, and Cerritos Colorados (see Fig. 4 for locations).

REGIONAL GEOLOGY

In the state of Jalisco in western Mexico, the Sierra Madre Occidental ignimbritic formations are covered by extensive layers of Mexican Neovolcanic Belt basaltic-andesitic rocks. Plate tectonics show the existence of three major structures in the Pacific that form a triple junction: the East Pacific Rise, the Rivera fracture zone, and the Mesoamerican Trench. The relation of the Mexican Neovolcanic Belt to these tectonic elements remains to be explained, but in its northwestern part within the continent, three major structures form the east-west Chapala, northwest-southeast Tepic-Chapala, and north-south Colima grabens, possibly reflecting the structural arrangement at the Pacific coast (Fig. 1). The intersection of these three major structures south of Guadalajara may represent an incipient triple junction in the continent. Other data seem to support this inference: the Quaternary volcanism within the Tepic-Chapala and Colima grabens, and in particular the silicic peralkaline rocks in the Sierra La Primavera, are characteristic of continental rift areas.

The oldest rocks in the region belong to an Early Cretaceous intrusion exposed at Sierra La Laja and Sierra Tapalapa south of La Primavera (Fig. 1). This dioritic intrusive unconformably underlies a Cretaceous sequence of marine sedimentary rocks: limestones, shales, and sandstones approximately 500 m thick. In Sierra Tapalapa these rocks form a north-south–trending anticline whose axis crops out again at Sierra La Laja.

The basic volcanism unconformably underlying the sequence in Sierra Tapalapa (Fig. 1) began in the Oligocene, possibly in response to subduction of the Farallon Plate beneath the American Plate. In general it consists of andesitic rocks with interbedded pyroclastics, tuffs, breccias, and agglomerates, also seen at the bottom of the Río Grande de Santiago canyon and northwest of the town of San Francisco Tesistán.

Oligocene-Miocene volcanic activity gave rise to major effusions of ignimbrites, pyroclastics, basaltic and rhyolitic lavas and domes, forming what is known as the Sierra Madre Occidental. In the neighborhood of La Primavera this volcanism occurs in the Sierra Tapalapa, and in the Río Grande de Santiago canyon (Fig. 1). The highly fractured state of these rocks can be seen clearly northwest of Sierra La Primavera in the vicinity of the Santa Rosa nonoverflow dam.

Sierra Madre Occidental volcanism ceased approximately 20 Ma owing to consumption of the Farallon Plate beneath the

Venegas S., S., Ramírez S., G., Romero G., C., Reyes V., P., Razo M., A., Gutiérrez N., L.C.A., Arellano G., F., and Perezyera y Z., J., 1991, La Primavera geothermal field, Jalisco, *in* Salas, G. P., ed., Economic Geology, Mexico: Boulder, Colorado, Geological Society of America, The Geology of North America, v. P-3.

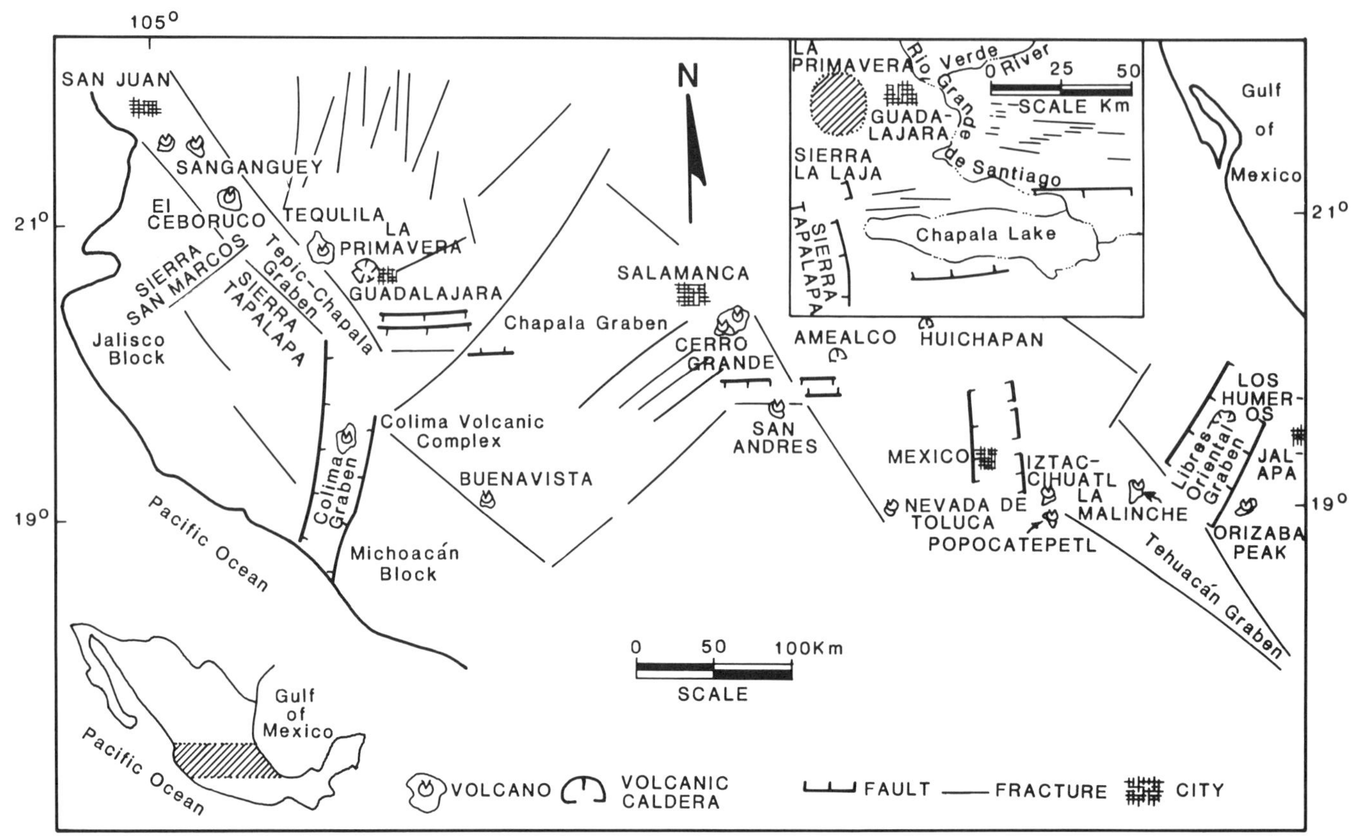

Figure 1. Location and volcanic-structural context of the La Primavera geothermal field. After Gutiérrez (1981a).

American Plate. The consequent collision of the East Pacific Rise with the American Plate brought about the reorientation of Pacific plate tectonics. Thus, at the end of the Miocene and beginning of the Pliocene the entire Sierra Madre Occidental region was covered by widespread andesitic and basaltic fissure flows together with pyroclastics, basaltic ignimbrites, and scarce rhyolite flows. This volcanism possibly responded to extensional tectonics developed by effects of the tectonic reorientation in the Pacific and represents the initial Mexican Neovolcanic Belt magmatic activity, characterized mainly by its crosswise position with respect to the other physiographic provinces of Mexico. The type localities of this fissure volcanism are found in the region of Sierra San Marcos, and parts of Sierra Tapalapa (Fig. 1).

The second stage of intrusions in the region may correspond to the possible effects of the Tertiary Laramide orogeny. These are mostly quartz monzonites affecting rocks of the Sierras San Marcos, Tapalapa, and La Laja.

Pliocene-Quaternary extensional tectonics generated a series of grabens in response to the separation of the Baja California Peninsula from the continent; in time these were occupied by Lake Chapala and a number of other small lakes. These grabens were subsequently filled with continental sediments of mainly volcanic composition, which are still accumulating today.

The principal volcanism of the Mexican Neovolcanic Belt began at the end of the Pliocene in response to subduction of the Cocos Plate beneath the American Plate. Widespread basic volcanism within the regional Tepic-Chapala graben in the Quaternary began with fissure emissions such as the Mesa Santa Rosa, subsequently shifting to the central activity of the Tequila (Jalisco), El Ceboruco, San Juan, and Sangangüey (Nayarit) volcanic complexes (Fig. 1), as well as several cinder cones of basaltic composition. Acid magmatic activity in the region developed rhyolitic domes to the north and west-southwest of La Primavera, and the La Primavera caldera complex, itself, related to the probable migration of the magmatic chamber, which gave rise to the Tequila Volcano.

GEOLOGY OF THE LA PRIMAVERA CALDERA

Sierra La Primavera is an upper Pleistocene volcanic complex of comagmatic (rhyolitic-peralkaline) lavas, sand flows, and pumice pyroclastics, with interbedded lacustrine and fluvial sediments (Fig. 2). The lavas erupted along two arcs, forming a series of domes, which have been mapped as two separate units on the basis of their topographic expression. Petrographic analyses indicate two rhyolitic units: porphyritic and aphanitic. The older porphyritic lavas have a 5 to 15 percent phenocryst content (Na-sanidine, quartz, ferrohedenbergite, and fayalite) and are more peralkaline than the Recent aphanitic lavas.

Radiometric dating in La Primavera established a sequence

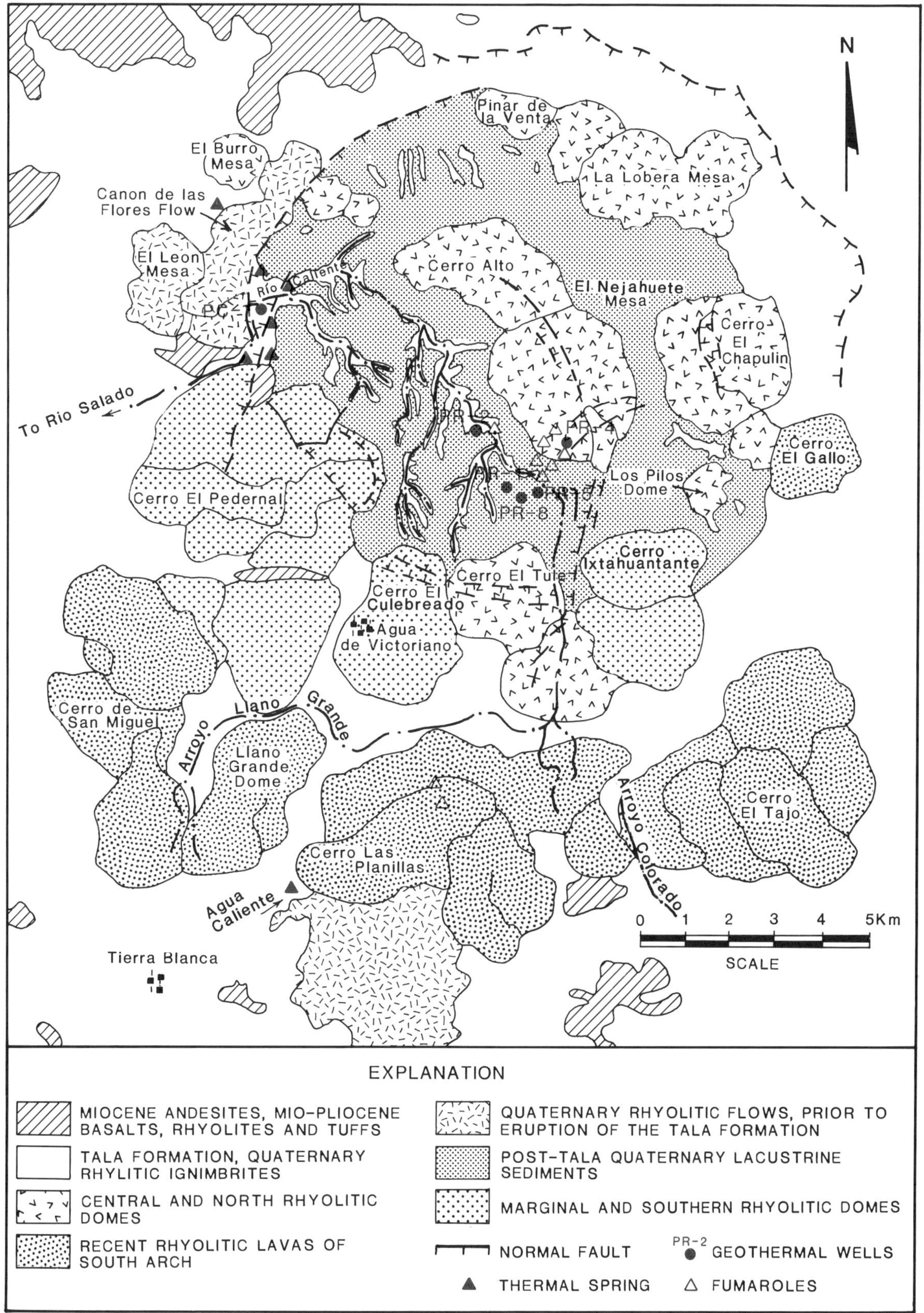

Figure 2. Geology of the La Primavera geothermal field. Slightly modified after Mahood (1980).

of superimposed dome-shaped eruptive events in a semicircular array similar to that of a volcanic caldera, generated about 100,000 years ago by rhyolitic lava flows through an arched fracture that formed domes of approximately 1 to 4 km in diameter. A second group of porphyritic lava domes developed in the south 80,000 to 60,000 years ago through semicircular fractures of the caldera ring. These lavas exhibit a slight change in their texture, which is intermediate between that of the older porphyritic units and most of the Recent aphanitic units. The aphanitic rhyolites were emplaced 50,000 or 60,000 years ago south of the preceding lavas, forming a volcanic arc at the periphery of the central caldera ring. The lenticularly shaped flows from these domes are 3 to 4 km long. The most recent of these is a 30,000-yr-old rhyolitic dome 2 km in diameter that tends to be less peralkaline. A complex series of pyroclastic flows with lacustrine and fluvial deposits lies within the first dome-bounded caldera ring.

The composition of the volcanic products of the La Primavera complex appears to reflect geochemical evolution in the upper portion of a magmatic chamber beneath the caldera.

Pre-Tala rhyolitic lavas

The earliest rhyolitic lavas of the Sierra La Primavera, erupted about 120,000 B.P., crop out in only two areas, although they may exist beneath younger units. One outcrop area north and northwest of the map area comprises the Río Salado porphyritic dome and the Cañón de Las Flores aphanitic flow, underlying the Tala Tuff (Fig. 2). At Cañón de Las Flores and La Primavera spas in this same area, the Tala Tuff is seen to overlie the Mesa El León porphyritic dome, possibly in relation to an event occurring between the above flows.

Another pre-Tala lava is exposed south of the Sierra La Primavera in the Cerro Las Planillas area where scattered outcrops of a 120,000-yr-old aphanitic dome, probably predating the Cerro Las Planillas, cover about 2 km^2 (Fig. 2).

Tala Tuff

At 95,000 B.P., an estimated 20 km^3 of pyroclastic flow filled the valleys in the vicinity of La Primavera, covering about 700 km^2. This has been named the Tala Tuff, with its type locality west of Sierra La Primavera. The Tala Tuff is an ignimbritic flow of rhyolitic composition with high silica content, and contains several flows that may be grouped into three units.

A white aphanitic pumice that makes up almost 90 percent of the tuff volume occurs as a thick intra-caldera flow surrounded by laminar flows in the neighboring valleys. The intra-caldera part of this unit is exposed in the scarps of the Río Caliente fault west of the Río Caliente Spa and Hotel, where the tuff appears partially welded. The laminar flows are not welded and in many places are glassy. Over several km^2 east of the town of Tala the tuff surface shows polygonal fractures and conduits that indicate the passage of a fumarole phase. Large-scale fumarole activity possibly developed through heating of ground waters by the laminar flows. These flows appear to have followed three directions: toward the west, forming a gently sloping plain from the range to beyond Tala; toward the northeast, possibly extending beneath the suburbs of Guadalajara; and toward the south, where outcrops are scarce.

The second Tala Tuff unit is a 10-m-thick layer of fine white aphanitic pumice and coarser gray pumice containing isolated quartz and sanidine phenocrysts. The fluid and compressed shape of these pumices suggests a mixture of two magmas. The third unit, some 10 m thick, consists of white pumice fragments with scattered quartz and sanidine fragments in a fine-grained, light pink groundmass. These last two units are present only in the central part of the Sierra, which suggests that the caldera collapsed after the first unit was formed. Radiometric dates place the Tala Tuff eruption at about 95,000 to 96,000 B.P.

In the development of the caldera, the roof of the magmatic chamber probably did not collapse as a piston, since no radial fractures are observed at La Primavera. Along the entire circumference the collapse must have taken place as a series of small staggered faults, forming a basin subsequently filled with pyroclastics and lacustrine sediments.

Domes formed about 95,000 B.P.

Deposition in the center of the caldera continued for a short period after the last Tala Tuff pumice was laid down, giving rise to fine-grained sediments (Fig. 2). A series of domes developed over these deposits along two parallel arcs: at the northeast edge of the caldera, possibly along a fracture underlying the Pinar de la Venta, Mesa La Lobera, Cerro El Chapulín, and Cerro Ixtahuantante domes; and at its center, forming the Cerro Alto, Mesa El Nejahuete, and Cerro El Tule domes.

The basal contact of the domes with the lacustrine sediments is a breccia possibly developed by the flow itself, since it diminishes toward the dome edges. This contact generally appears when the fine-grained deposits thin out, approximately 10 m above a giant pumice horizon. Eventually the dome lavas are seen to overlie the surface of the lacustrine deposits, as well as some pumice flows covering the sediments. At the contact, the lavas consist of dense glass, scarce spherulites, and devitrification possibly caused by rapid cooling in the lake waters.

Lavas erupted about 75,000 B.P.

A new group of domes appears to have erupted about 75,000 B.P. at the southeast edge of the caldera, producing porphyritic and aphanitic lavas containing an approximate total of 3 km^3 of magma. K/Ar dates for these domes are validated by their position within the lacustrine stratigraphic sequence. At Arroyo Hondo the 75,000-yr-old Cerro El Culebreado dome is seen overlying the 95,000 B.P. Cerro El Tule dome. The basal contacts of these younger domes are less well exposed than those of the older, being covered by lush vegetation. However, contacts may be located approximately in Arroyo Ixtahuantante, at Cerro El Culebreado, and at the edges of the Cerro El Pedernal aphanitic flow.

Uplift of the caldera

The younger 75,000-yr-old lavas are covered by only 10 to 20 m of pyroclastics. These reworked, fine-grained, quickly falling pumice deposits exhibit the characteristics of subaerial deposits only in the high parts of the domes. Subsequent regional uplift produced the present topography of the Sierra.

The strongest deformation triggered by the uplift took place outside the caldera, because the lacustrine sediments there dip radially 10 to 20°, whereas toward the center, the dips are almost horizontal or 2° at most. Faulting in the central part appears in association with the uplift, and many of the faults scarcely affect the sediments. This suggests a general uplift of the caldera as a whole, except at its edges.

Lavas of the southeast arc

The uplift was probably caused by the displacement and rise of the magmatic chamber toward the southeastern part of the caldera about 60,000 B.P., which fostered the flow of 7 km^3 of magma as lavas and domes of mostly aphyric obsidian that make up the volcanic arc southeast of the caldera. The lavas flowed radially away from the center of the Sierra, whereas the 75,000- and 95,000-yr-old domes show no marked directional flow. This suggests that the uplift occurred after eruption of the younger domes but prior to the emplacement of the southeast-arc lavas (i.e., between 75,000 and 60,000 B.P.; Fig. 2).

The extrusion of these southeast-arc domes was preceded by the eruption of aphanitic pumice chemically related to the 60,000-yr-old lavas. This occurs in the caldera lake, crowning the lacustrine deposits. Likewise, south of the caldera a small pyroclastic deposit is chemically identical to the southeast-arc lavas; it extends over approximately 2.5 km^2 and forms hills north of Tierra Blanca.

The southeast-arc lavas tend to become younger toward the east. Thus the Cerro San Miguel central flow underlies the Llano Grande, and the Arroyo Colorado is covered by the eastern flank of the Cerro Las Planillas. The Cerro El Tajo and Cerro El Gallo domes, the outwardmost and youngest of the arc, dated radiometrically at 20,000 and 30,000 B.P., respectively, are the last expressions of volcanism in La Primavera and mark the end of the eruptive cycle. In addition to the circular features that gave rise to the caldera (Fig. 2) other faults and fractures trending north-south, northwest-southeast, northeast-southwest, and east-west reflect regional tectonics (Fig. 1). All occur within the volcanic complex, none of them extending beyond the Sierra La Primavera to the older rocks. This implies that whereas regional tectonism controlled the location of the caldera complex, the fracturing within it was due to local adjustments compensating for changes in the level of the underlying magmatic chamber. However, although the faults and fractures are local, they evidently respond to regional patterns and were possibly caused by residual effects of these patterns on the subsurface rocks of the caldera. The surface thermal manifestations are, of course, controlled by these structures.

HYDROGEOLOGY

According to meteorological information from 19 climatological stations over a 20-yr period (1960 to 1979), the region has average annual temperature and precipitation of 20°C and 957 mm, respectively. True evapotranspiration is 617 mm, infiltration coefficient is 28 percent, and a potential volume of 5.9 million m^3 of water infiltrates yearly in an area of 61 km^2.

The Río Caliente, an Ameca tributary in the western part of La Primavera, flows from several thermal and cold springs. Gauging records show that hot-spring flows tend to be constant; hence a good part of the flow must come from interaction between the shallow aquifer and one or more deeper ones.

From the local hydrogeological viewpoint, La Primavera is at the water divide area of three shallow aquifers: Tesistán-Atemajac-Toluquilla, Ahualulco–Río Salado–Ameca, and San Marcos. However, the geothermal area is mainly the east portion of the second of these. The northwest-southeast–trending regional structural system partially controls the superficial underground flow. Since those structures are results of the various tectonic elements of western Mexico, they may rightly be thought to influence the deep regional aquifer.

Geohydrologic characteristics of the rocks exposed at La Primavera and vicinity show that the Quaternary volcanic rocks—basalts and andesites of the Tequila Volcano flows, pyroclastics and the rhyolitic domes derived from the caldera—are permeable, forming the main recharge areas for the shallow aquifers. The Tertiary volcanic rocks allow meteoric waters to infiltrate and may develop low-yield aquifers, except for the Miocene-Pliocene andesites, which may develop major aquifers. However, it is possible that the deeper underground flow in these rocks is controlled mainly by fracturing and not necessarily associated with any rock type.

At La Privavera the shallow aquifer is recharged in the main part of Sierra La Primavera by rainwater infiltration and by the deeper thermal aquifers appearing at the surface in the Caliente River and vicinity. Ground waters flow generally westward; they leave the caldera and descend the Río Salado Valley to the Río Tala and onward toward the Río Ameca Valley.

The presence of a deep aquifer related to a geothermal system in La Primavera is indicated by the hot springs and fumaroles in the area. This deep regional aquifer is contained in Miocene-Pliocene andesitic rocks exposed southeast of the caldera and becoming deeper toward it, as shown by wells PR-1 and RC-1 where the andesitic rocks were encountered at 1,080 and 1,200 m depths, respectively. This fosters an underground northwestward flow.

GEOPHYSICAL EXPLORATION

Diverse geophysical surveys have covered the Caldera La Primavera: UNAM (Universidad Nacional Autonoma de México) gravity and magnetometry surveys in 1970; CFE Geothermal Exploration Office magnetometry, thermometry, natural potential, and resistivity surveys in 1978 to 1980; and a CICESE (centro de Investigación Científica de Estudias Superiores de Ensenada) microseismicity study in 1980. The integrated geophysical, geological, and geochemical results defined the geothermally prospective areas as the center of the Caldera La Primavera (Cerritos Colorados–La Azufrera), the vicinity of the Las Planillas dome, and the western edge of the caldera (Río Caliente) (see Figs. 1 and 4). Apparent resistivity anomalies of under 20 ohm/m for semielectrode spacings between 2,000 and 3,000 m were detected in all three zones, which also show good agreement with surface thermometry anomalies and with microearthquake swarms. The most salient of the three was at the center of the caldera (Fig. 3).

Most significant of all were the resistivity results, where they were strongly related to the geothermal aspects of high temperatures, salinity, and porosity, or their combinations. The apparent resistivity anomalies show a remarkable alignment with the major structural systems, especially the northwest-southeast trend; the extension and interconnection of resistive minima, as electrode spacing widened when contouring apparent resistivities, is also extraordinary.

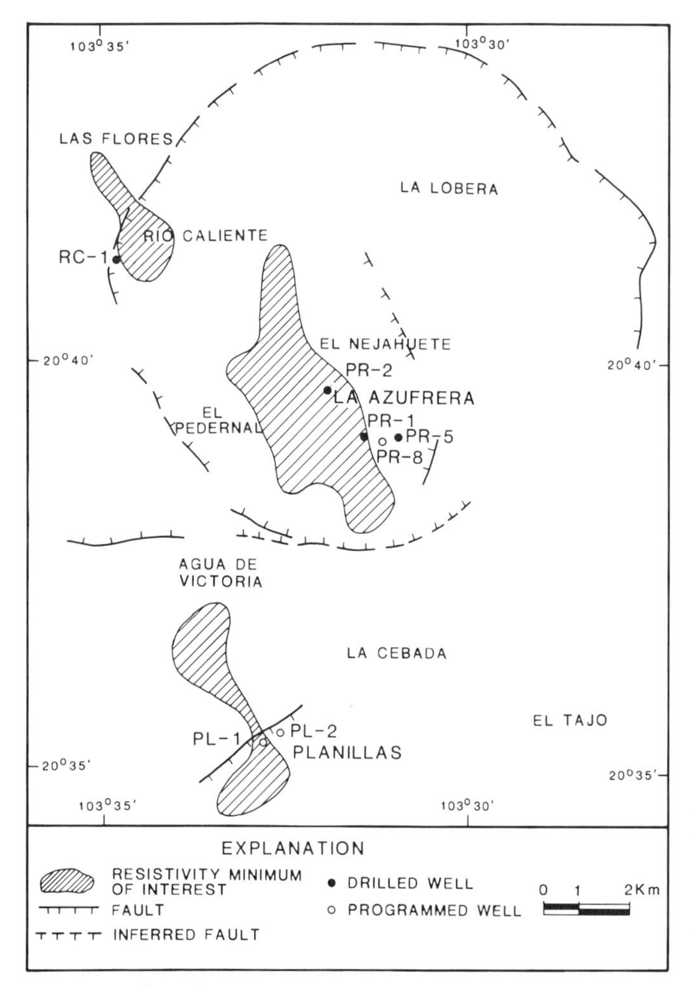

Figure 3. Resistivity minimums at La Primavera.

One of the least resistive anomalies (10 ohm/m), in the Cerritos Colorados–La Azufrera area, seems to extend at greater depth southwestward to the Cerro El Culebreado area at the edge of the caldera. The anomaly detected in the vicinity of the Las Planillas rhyolitic dome is of interest in that the low resistivity may be related to the hydrothermal system, which is associated with the northeast-southwest fracturing. The low resistivity anomaly in the Río Caliente area, where well RC-1 was later drilled with negative results, is influenced by the surface thermal fluids coming from the central part of the caldera. These strongly saturate the rocks in that area and therefore have lower resistivities, unrelated to significant subsurface temperatures.

Drilling in Cerritos Colorados–La Azufrera proved the existence of 230 to 300°C temperatures at 1,000 to 2,000 m depths, in correspondence with a low resistivity anomaly of 7 km^2.

Thermometric results at depths of 0.5 to 1.0 m defined several surface anomalies, of which La Azufrera–Cerritos Colorados and Las Planillas were the most significant, with maximum temperatures on the order of 90°C. Other hot spring areas—La Primavera, Río Caliente, and Agua Caliente—have temperatures of about 65°C. The La Azufrera–Cerritos Colorados isotherms in Figure 4 illustrate the detailed work done over each zone. Correlation of the isotherm contours with the geology indicates that the most highly thermal areas are closely related to the presence of structures.

The UNAM regional gravity and magnetometry surveys (1970), and the CFE vertical component magnetometry, concluded that the Caldera La Primavera collapse zone is apparently related to a magnetic low (Fig. 5) and to a Bouguer Anomaly minimum. The anomaly gradient decreases from northeast to southwest (Fig. 6) and develops a northwest-southeast–elongated anomaly coinciding with the alignment of the resistivity minimums and again shows the relevance of the northwest-southeast structural system.

The natural potential survey covered mainly the central part of the caldera (Cerritos Colorados–La Azufrera area) and part of the Río Caliente zone. In general the anomalies show positive areas directly or approximately above surface thermal shows: values of up to +280 mV at La Azufrera–Cerritos Colorados, and higher values (+380 mV) at Río Caliente and vicinity over a larger area and with an apparent north-south alignment. These values are interpreted as reflections of the thermal activity in both areas, and of the hydrologic system, in addition, at Río Caliente (Fig. 7).

Microseismicity data show that the most intense and frequent events occur in the central (Cerritos Colorados) and southwest parts of the caldera.

GEOCHEMICAL EXPLORATION

Thermal activity in the La Primavera area shows up as springs at Río Caliente and Agua Caliente (southwest of Las Planillas; Fig. 2). Approximate flow at Río Caliente is 400 L/sec at 65°C and less at Agua Caliente with a temperature of 63°C. Waters in both are high in sodium bicarbonate, and have chloride concentrations of 100 to 150 mg/L (Table 1). Cationic-composition geothermometers indicate equilibrium temperatures of about 175°C in these waters. Several fumarole areas occur in the south-central part of the caldera with sulfur deposits in some cases; CO_2 and H_2S are the most abundant gases, with lesser amounts of N_2, H_2, and CH_4.

The chemical composition of waters extracted at atmospheric pressure from well PR-1 (Table 1) differs considerably from that of the springs because these springs serve only as discharge for the Tala Tuff mesothermal aquifer, which owes its thermalism to partial admixture of rising geothermal fluids. For this reason, geothermometer temperatures do not represent the deep geothermal reservoir.

Table 2 shows chemical compositions of gases in well PR-1. Methane (CH_4) and hydrogen (H_2) concentrations are characteristic of magmatic differentiation. The nitrogen (N_2) values and carbon dioxide/hydrogen sulfide (CO_2/H_2S) ratio are those of a high-temperature geothermal zone. Certain gases such as carbon dioxide, methane, or hydrogen, which were used to evaluate subsurface temperatures, show these to range from 240 to 300°C.

DRILLING

Drilling in the La Primavera field began on January 4, 1980; five wells (PR-1, RC-1, PR-2, PR-4, and PR-5) had been drilled by 1984 for a total 7,605 m, and a sixth was in progress.

Well PR-1, drilled on a low-resistivity anomaly in the Cerritos Colorados–La Azufrera area, aimed at encountering geologically inferred structures. Drilling, done by stages, was completed at 1,226 m; casings were set as follows: 20-in diameter at 40 m; 13⅜-in diameter at 233 m; 9⅝-in diameter at 670 m; 7-in diameter at 600 to 659 m; and 7-in slotted diameter at 659 to 1,159 m. Initial production conditions were unfavorable in spite of reaching a maximum 284°C at 1,150 m; the well was therefore deepened to 1,822 m and completed with a slotted 4½-in diameter casing in the 1,430- to 1,822-m interval. Maximum measured well-bottom temperature was then 292°C, production conditions

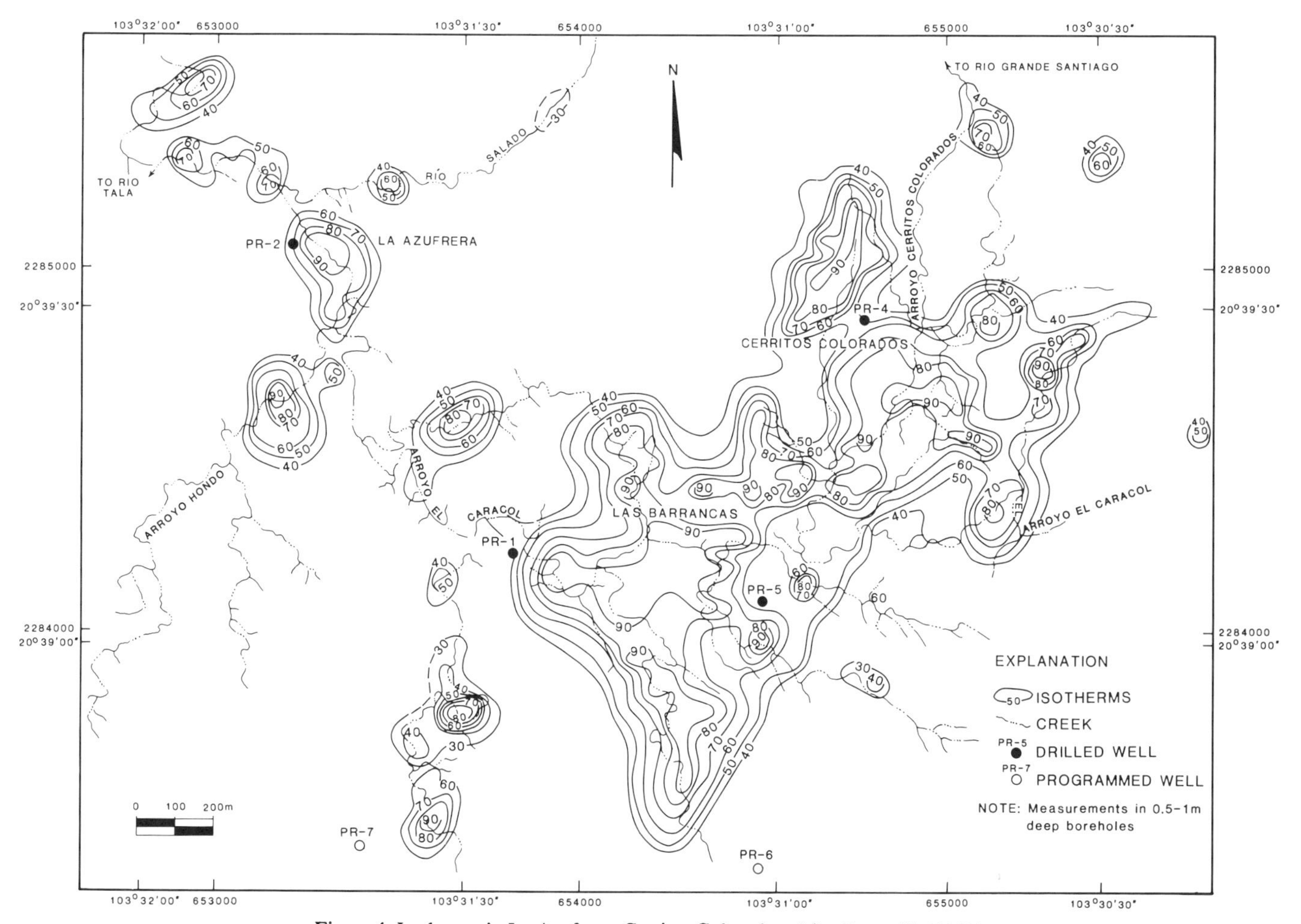

Figure 4. Isotherms in La Azufrera–Cerritos Colorados. After Reyes V. (1981).

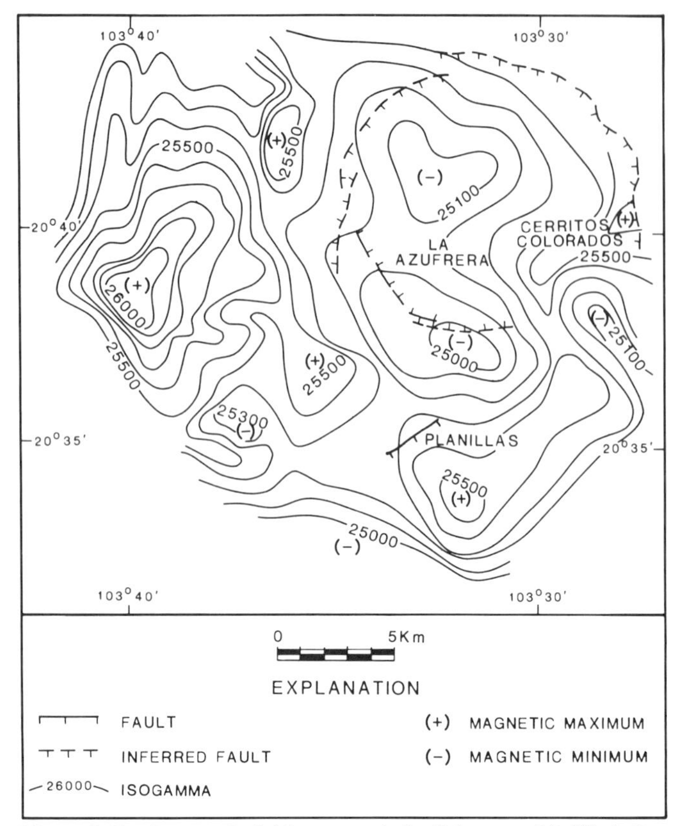

Figure 5. Total magnetic field contours in La Primavera. Modified after Del Castillo (1970).

TABLE 1. CHEMICAL COMPOSITION OF TWO REPRESENTATIVE SPRINGS IN THE LA PRIMAVERA AREA AND OF WATERS FROM WELL PR-1 AT ATMOSPHERIC PRESSURE (0.85 bar)

Mg/l	Río Caliente	Agua Caliente	Pozo PR-1
Na^+	291.0	287.0	760.0
K^+	12.0	20.0	191.0
Ca^{++}	5.7	2.4	4.2
Mg^{++}	0.3	0.3	0.05
Li^+	1.1	1.2	7.8
Rb^+	0.03	<0.02	1.2
Cs^+	0.02	0.02	1.1
Cl^-	105.0	143.0	1,260.0
SO^{2-}	27.0	9.0	59.0
HCO_3^-	453.0	337	151.0
SiO_2	225.0	178.0	1,068.0
B	17.0	17.0	169.0
pH	7.0	7.6	8.37

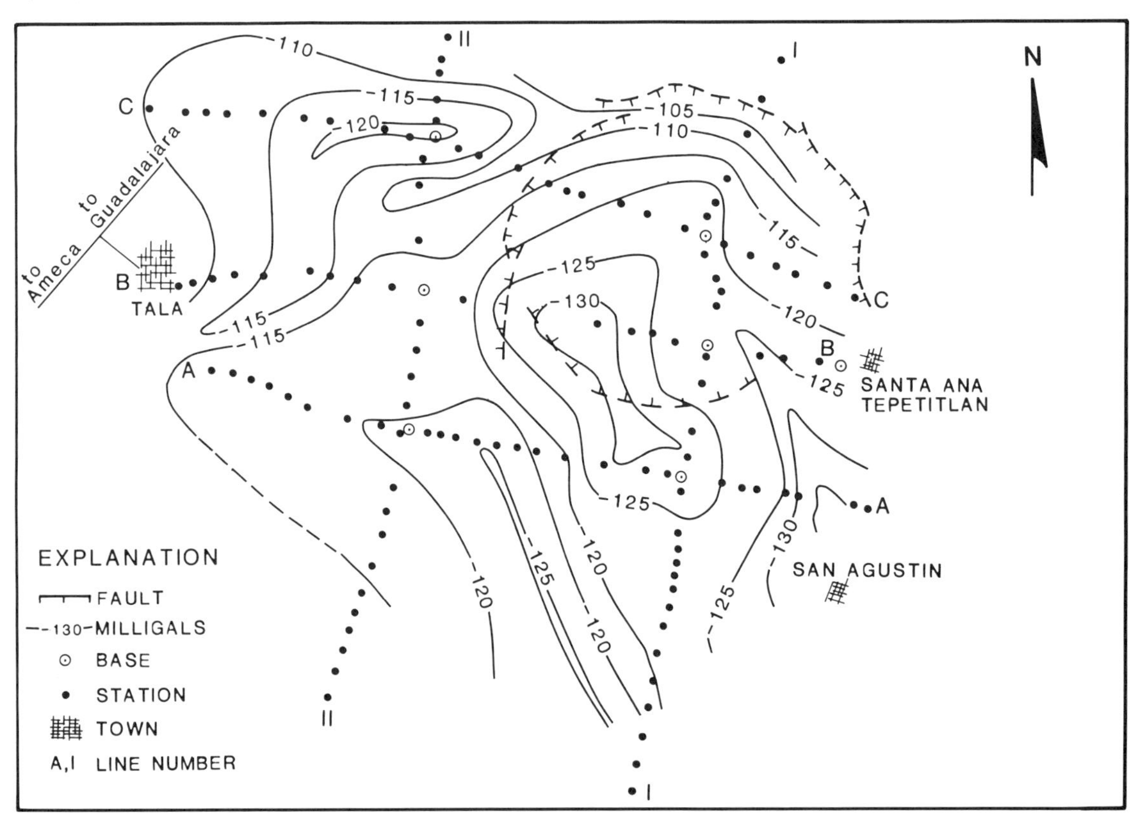

Figure 6. Bouguer Anomaly contours in La Primavera. After Del Castillo (1970).

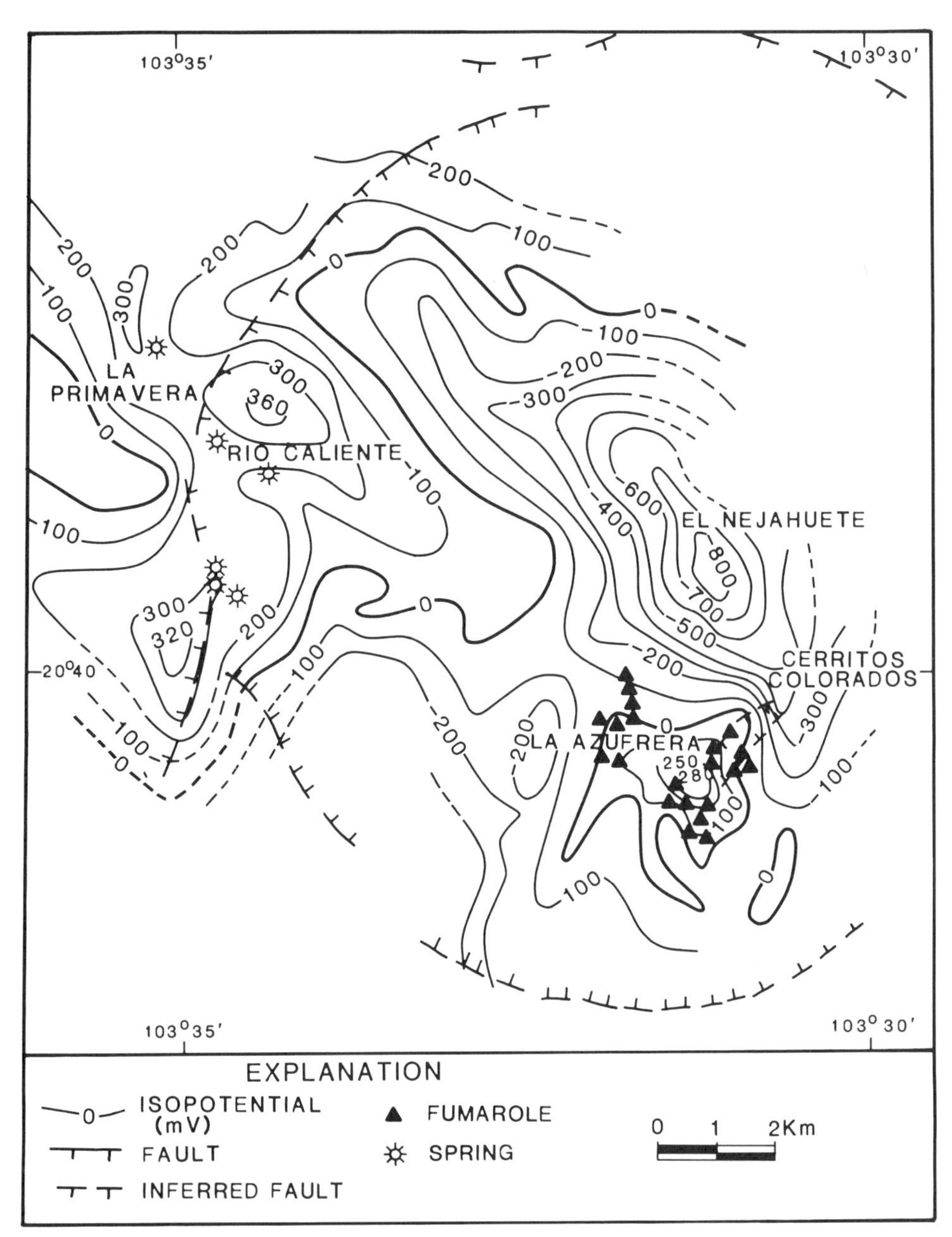

Figure 7. Natural-potential anomalies in La Primavera. After Romero (1981).

TABLE 2. CHEMICAL COMPOSITION OF GASES FROM WELL PR-1
(in mmol/mol)

E kJ/kg	Ps bars	Gas total	CO_2	H_2S	NH_3	H_2	Ar	N_2	CH_4
1,918	7.75	25.6	979	11.4	2.01	1.94	0.02	1.03	4.35

improved, and the well now produces 60 Ton/h of steam—enough to generate approximately 5 MW.

Well RC-1 was located on the Río Caliente low-resistivity anomaly near the northeast margin of the caldera, seeking to intercept the collapse structure at depth. Final depth was 1,900 m; well-indurated rock made it unnecessary to set larger diameter casing, and a casing of only 13¾-in diameter was set at 197 m. Maximum recorded temperature was 99°C at 1,800 m, and the rock was practically impermeable throughout. The conclusion was that the well had encountered the nonpermeable marginal part of the hydrothermal system and that most of the thermal flow takes place within the superficial aquifer near the surface.

Well PR-2, drilled on the same PR-1 low-resistivity anomaly near the La Azufrera thermal zone, sought to encounter a northwest-southeast and northeast-southwest structure at depth for purposes of increasing permeability in the geothermal reservoir. Final depth was 2,000 m, and circulation losses at the Tala Tuff interval were constant. A maximum temperature of 305°C was recorded during drilling at the 1,950-m depth. Permeability decreased with depth, and the well was completed with casing as follows: 20-in diameter at 47 m; 13⅜-in diameter at 263 m; 9⅝-in diameter at 998 m; blind 7-in diameter at 1,567 m; and slotted 7-in diameter at 1,567 to 1,995 m. The well produced 22 Ton/h of mixture, of which 9 Ton/h was steam, 8 Ton/h water, and 5 Ton/h gas.

Well PR-4 looked for increased subsurface permeability within the Cerritos Colorados graben near a fumarole zone and at the north edge of the low-resistivity anomaly. The well was drilled to 668 m and abandoned due to circulation loss of drilling fluid, especially at the Tala Tuff interval; maximum temperature of 82°C was measured at 500 m.

Well PR-5, aimed to encounter the PR-1 reservoir (284°C at 1,200 m) and planned for a total depth of 2,000 m, was drilled 750 m northeast of PR-1 in a highly fractured area with some surface thermal activity. It was completed at only 1,215 m due to circulation losses at the Tala Tuff interval. Maximum temperature during drilling was 262°C at the 1,050-m depth, although 280°C temperatures were measured on subsequent development. Casings were set as follows: 30-in diameter at 27 m; 20-in diameter at 325 m; 13⅜-in diameter at 473 m; 9⅝-in diameter at 571 m; 7-in diameter was hung from 524 to 1,213 m and set from 524 to 874 m. Initial production was 151 Ton/h of mixture, 35 Ton/h of which were steam and the remainder water. Subsequent casing and setting failures rendered the well unproductive.

All the wells encountered a similar lithologic sequence: thin lacustrine sediments, a varying thickness of the Tala Tuff and alternating andesitic flows (predominant), and pyroclastic deposits, with occasional rhyolitic and basaltic flows.

HYDROTHERMAL MINERALOGY

Ditch-sample thin sections from all five wells were petrographically studied at 4-m intervals. The following alteration minerals were identified and counted: oxides, hematite, calcite, chlorite, clay minerals, zeolites, sericite, pyrite, leucoxene, quartz, and epidote. Significant quantities of epidote (more than 10 percent of rock volume) were found only in a very restricted interval of well PR-2, associated with the trace of the La Azufrera fault plane. The other minerals appeared in practically all the wells.

The relatively closely spaced wells in the south-central part of the caldera (PR-1, PR—2, PR-4, and PR-5) and well RC-1 at the western boundary were correlated using the alteration minerals; except for PR-2 they show very similar hydrothermal mineralogies with no outstanding features. However, correlation of the alteration levels in RC-1 with those to the southeast showed that several of the levels, especially those with zeolites and pyrite, tend to dip and deepen toward RC-1. The fact that both low-temperature (zeolite group except for wairakite) and moderate-temperature minerals (pyrite, with its specific association in La Primavera) deepen abruptly toward the western edge of the caldera, suggests that the probable heat source and the reservoir itself are at a much greater depth in that direction than in the south-central part. This deepening happens because the geothermal system develops a zonation in the host rocks outward from the heat source to the periphery, wherein secondary mineral associations characterize different temperatures, pressures, and fluid activity. Not only do the zeolites and pyrite follow the observed trend but, to a lesser degree, so do the calcite and quartz. This leads to the conclusion that the western edge of the caldera where RC-1 was drilled has not experienced temperatures of more than 150 to 175°C and therefore is not a future drilling prospect, at least to 2,000 m depth. Although temperatures in the past must have been higher than now, they never reached those of the wells in the south-central part of the caldera, at least above 2,000 m depth.

On the other hand, secondary minerals in the south-central wells indicate relatively moderate temperatures—no higher than 250°C—in every case. Indeed, in PR-1, the highest-temperature association that was encountered was calcite-pyrite-chlorite with insignificant traces of epidote; this implies that its final depth was probably insufficient, barely entering a marginal area of the geothermal reservoir. Stable highly thermal conditions would probably be found at greater depth.

In summary, mineralogical associations in the La Primavera field wells are characteristic of active geothermal regimes such as Cerro Prieto (Baja California), Los Azufres (Michoacán), Los Humeros (Puebla), and other areas of the world. This implies the presence of an active geothermal system in the caldera at greater depth than drilled to date.

WELL PRODUCTION AND EVALUATION

As previously mentioned, five deep exploratory wells were drilled at La Primavera, of which only PR-1 is commercial after deepening, PR-2, presently flowing through 1-in tubing, will require deepening to improve production. RC-1 and PR-4 are shut down without flow lines, and PR-5, initially productive, remains shut down. PR-8 is presently being drilled (1984).

To evaluate power generation capacities, bottom temperatures, pressures, and flows were repeatedly measured in each well, as well as production at the surface in some cases. The following summarizes the general reservoir features.

Initially, well PR-1 showed the highest temperature under static conditions (drilling and heating stages): 284°C between 1,150 m and 1,226 m depth. Under dynamic (flow) conditions, however, the temperature dropped to 203°C, due perhaps to a slight thermal fluid invasion of the reservoir, different physicochemical characteristics, and a high carbon dioxide (CO_2) content. Further drilling to 1,822 m increased the initial 37 Ton/h production of mixture to 60 Ton/h, the first indication of a commercial geothermal reservoir with temperatures approaching 300°C below the 1,800 m depth.

PR-2, producing from the 1.567- to 1.995-m interval, recorded the highest bottom temperature (304°C) and saturation conditions. PR-5 showed subcooled liquid between 1,100 and 1,215 m and saturation conditions from 1,100 m to the surface. The subcooling coincides with fluid entry (400 L/min) recorded in Spinner tests.

Correlation of the formation equilibrium temperatures evaluated for each well showed PR-5 to have the best thermal characteristics, at lesser depths than PR-2, PR-1, and RC-1. This suggests that subsurface thermal conditions tend to improve southeastward in the direction of PR-5 within the caldera.

Initial production from PR-5 was higher (35 Ton/h of steam) than in PR-1 to 1,226 m and PR-2, although with a lower steam-water mixture enthalpy. This subsequently decreased by stages until the well ceased to flow.

Testing results of PR-5 and the high water flow (114 Ton/h) led to the conclusion that the well was affected by waters of low thermal content invading the production interval.

Initial development of PR-1 to 1,226 m and PR-2 showed generation capacities of 2 MW and 1 MW, respectively, at 0.6 Mpa (6 bars). PR-2 behaved erratically, possibly due to high carbon dioxide gas content and/or mixture of fluids of different physicochemical characteristics.

The conclusion is that thermodynamic and production characteristics of the reservoir will improve at depth, and future wells should be drilled to 3,000-m depths.

Well production to date has been affected mainly by: (a) invading waters of low thermal content—more so in PR-5, which initially showed a higher production rate (35 Ton/h of steam) than PR-1 and PR-2 and subsequently ceased to flow; (b) the low transmissivity of the producing rocks; and (c) in PR-2 also the large emission of carbon dioxide gas.

CONCLUSIONS AND PRESENT STATUS OF THE FIELD

The La Primavera geothermal field lies at the intersection of the Sierra Madre Occidental and Mexican Volcanic Belt volcanic provinces and structurally at the intersection of the east-west Chapala and northwest-southeast Tepic-Chapala grabens. The field is within a Pleistocene caldera structure generated by a series of acid volcanic events beginning 100,000 B.P., the most recent evidence being barely 30,000 years old. Volcanic products are of peralkaline rhyolitic type with very large quantities of silica (more than 75 percent) and include domes, flows, ignimbrites, and tuffs. The center of the caldera contains lacustrine deposits.

The principal surface structures are annular or circular in shape, caused by caldera collapses and warping by the rising magmatic chamber. Faults and fractures trending northwest-southeast, north-south, and northeast-southwest within the caldera reflect regional subsurface structures. The surface fumaroles and springs appear to be associated with these structures, in particular the northwest-southeast–trending fumaroles.

The hydrothermal reservoir is in Miocene-Pliocene andesites correlative with those at the base of the Mexican Volcanic Belt, which enclose a deep regional aquifer flowing probably from southeast to northwest and controlled by similarly trending regional structures.

The geothermal system gave rise to strong secondary mineralization of the subsurface rocks characterized by alteration minerals that are typical of geothermal fields: quartz, calcite, zeolites, chlorite, pyrite, and small amounts of epidote.

Geochemical, mineralogical, and production test results indicate that the wells in the south-central part of the caldera reached only a shallow marginal sector of the reservoir and that better thermodynamic and production characteristics may be expected at greater depth.

Of the five wells drilled to date (PR-1, PR-2, PR-4, PR-5, and RC-1), only RC-1, at the western edge of the caldera, may be rated a failure, with temperatures no higher than 100°C in almost 2,000 m. PR-4 is really an abandoned well that could not drill through the Tala Tuff thickness owing to drilling fluid circulation losses and the impossibility of setting casings to the programmed final depth. PR-2 and PR-5 fell short of reaching zones of high reservoir pressure and/or improved permeability in spite of encountering production temperatures, subsequently diminished in PR-5 by the invasion of cooler waters.

Present prospects of the La Primavera geothermal field are encouraging, as shown by PR-1, which on deepening a further 596 m found good permeability, high pressures, and excellent temperatures. PR-8, drilling at present (1984), is expected to confirm the presence of a commercial reservoir in this field.

BIBLIOGRAPHY

Del Castillo, G. L., Márquez, C. R., and Sandoval, O.J.H., 1971, Anomalías magnetométricas y gravimétricas regionales y su relación con la geológia del área geotérmica de La Primavera: Instituto de Geofisica, Universidad Nacional Autonoma de México, Jalisco Investigación 1050.

Mahood, G. A., 1980, Geological evolution of a Pleistocene rhyolitic center; Sierra La Primavera, Jalisco, Mexico: Journal of Volcanology and Geothermal Resources, v. 8, p. 199–230.

McEvilly, T. V., Mahood, G. A., Majer, E. L., Schechter, B., Truesdell, A. H., 1978, Seismological/Geological field study of the Sierra La Primavera geothermal system: Berkeley, University of California preliminary report.

Unpublished sources of information

Exploration Department, Comisión Federal de Electricidad

Ballina, L.H.R., 1982, Estudio geoeléctrico de resistivadad de Las Planillas, Jalisco: Report 61/82.

Casarrubias, U. Z., and Venegas, S. S., 1981, Informe del Pozo RC-1.

Gutiérrez, N.L.C.A., 1981a, Petrografía y mineralogía de los pozos PR-1 y RC-1 de La Primavera, Jalisco: Report 1/81.

——, 1981b, Litología y mineralogía secundaria del Pozo PR-2, La Primavera, Jalisco: Report 19/81.

——, 1982, Litología y mineralogía de alteración del Pozo PR-4 del campo geotérmico de La Primavera, Jalisco: Report 7/82.

Ramírez, S. G., 1982, Hidrología superficial y subterránea de las zonas geotérmicas La Primavera–San Marcos–Hervores de la Vega, Jalisco: Report 19/82.

Ramírez, S. G., and Mata, V. M., 1982, Hidrogeoquímica de la zona La Primavera–San Marcos–Hervores de la Vega, Jalisco.

Ramírez, E. G., Casco, R. J., and Mata, V. M., 1982, Hidrogeología regional de la zona geotérmica La Primavera–San Marcos–Hervores de la Vega, Jalisco: Report 14/82.

Razo, M. A., Ruiz, O. R., Godoy, J. G., and Leal, H. R., 1978, Reporte geológico preliminar del área de La Primavera, Jalisco: Report 8/78.

Reyes, V., P., 1982, Estudio geoeléctrico de detalle en el área geotérmica La Azufrera–Cerritos Colorados, Jalisco: Report 57/82.

Romero, G. C., 1981, Estudios geológicos y geofísicos en el área geotérmica de La Primavera, Jalisco: Report 5/81.

Romero, G. C., Reyes, V. P., and Venegas, S. S., 1979, Resultados preliminares de la exploración geoelétrica et el área Cerritos Colorados–La Azufrera de la zona geotérmica de La Primavera, Jalisco: Report 7/79.

Venegas, S. S., 1981, Geología de la Sierra de La Primavera, Estado de Jalisco: Report 52/81.

Venegas, S. S., and Casarrubias, U. Z., 1980, Informe preliminar de la geología del área Río Calienta–Aqua Brava, de la zona geotérmica de La Primavera, Jalisco: Report 13/80.

Venegas, S. S., and Ruy, A. C., 1981, Estudio geológico regional de Las Planillas en el Estado de Jalisco: Report 47/81.

Venegas, S. S., Romero, G. C., Reyes, V. P., and Palma, S. H., 1979, Informe preliminar de la geología del área La Azufrera–Cerritos Colorados de la zona geotérmica de La Primavera, Jalisco: Report 8/79.

Institute of Electrical Investigations, Universidad Nacional Autonoma de México

Mercado, G. S., 1971, Zona geotérmica de La Primavera, Jalisco, y zonas geotérmicas circundantes, Estudio geotermoquímico.

Neovolcanic Axis Division

Superintendencia de Ingeniería Química, 1980, Geoquímica preliminar del campo geotérmico de La Primavera, Jalisco.

Printed in U.S.A.

The Geology of North America
Vol. P-3, Economic Geology, Mexico
The Geological Society of America, 1991

Chapter 9

The Fuentes-Río Escondido coal basin, Coahuila

F. Verdugo D. and C. Ariciaga M.
Federal Commission on Electricity (MICARE), Rio Aodano 14, 06500 Mexico, D.F.

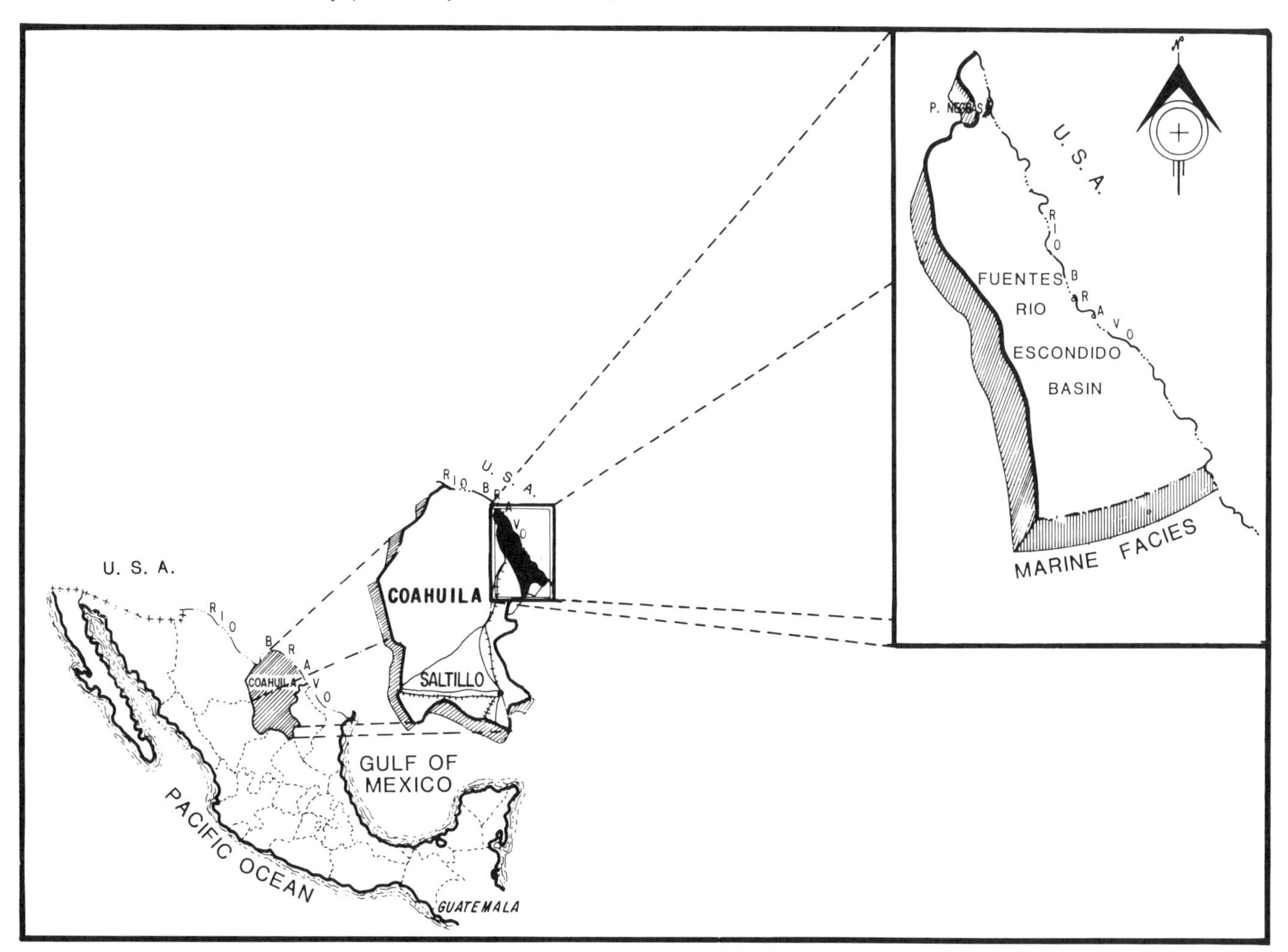

Figure 1. Location of the Fuentes–Río Escondido basin.

INTRODUCTION

Coal resources have been known to exist in the northeastern Coahuila region (Fig. 1) since the last century when there was small-scale mining across the border in Texas at Eagle Pass (Maverick County), St. Thomas District (Webb County), and San Carlos (Presidio County), among others (Evans, 1974). With the aim of diversifying energy sources for power generation, the CFE (Comisión Federal de Electricidad) under Manuel Moreno Torres launched the coal development program in 1960; since then, six geological exploration programs have been carried out over what is now known as the Fuentes–Río Escondido Basin (Fig. 2) as follows: I. 1960 to 1961; II. 1963 to 1964; III. 1967 to 1968; IV. 1973 to 1976; V. 1977 to 1979; VI. 1980.

By 1961, 12 million tons of thermal coal had been evaluated for the Venustiano Carranza Thermoelectric Project in Nava

Verdugo D., F., and Ariciaga M., C., 1991, The Fuentes–Río Escondido coal basin, Coahuila, *in* Salas, G. P., ed., Economic Geology, Mexico: Boulder, Colorado, Geological Society of America, The Geology of North America, v. P-3.

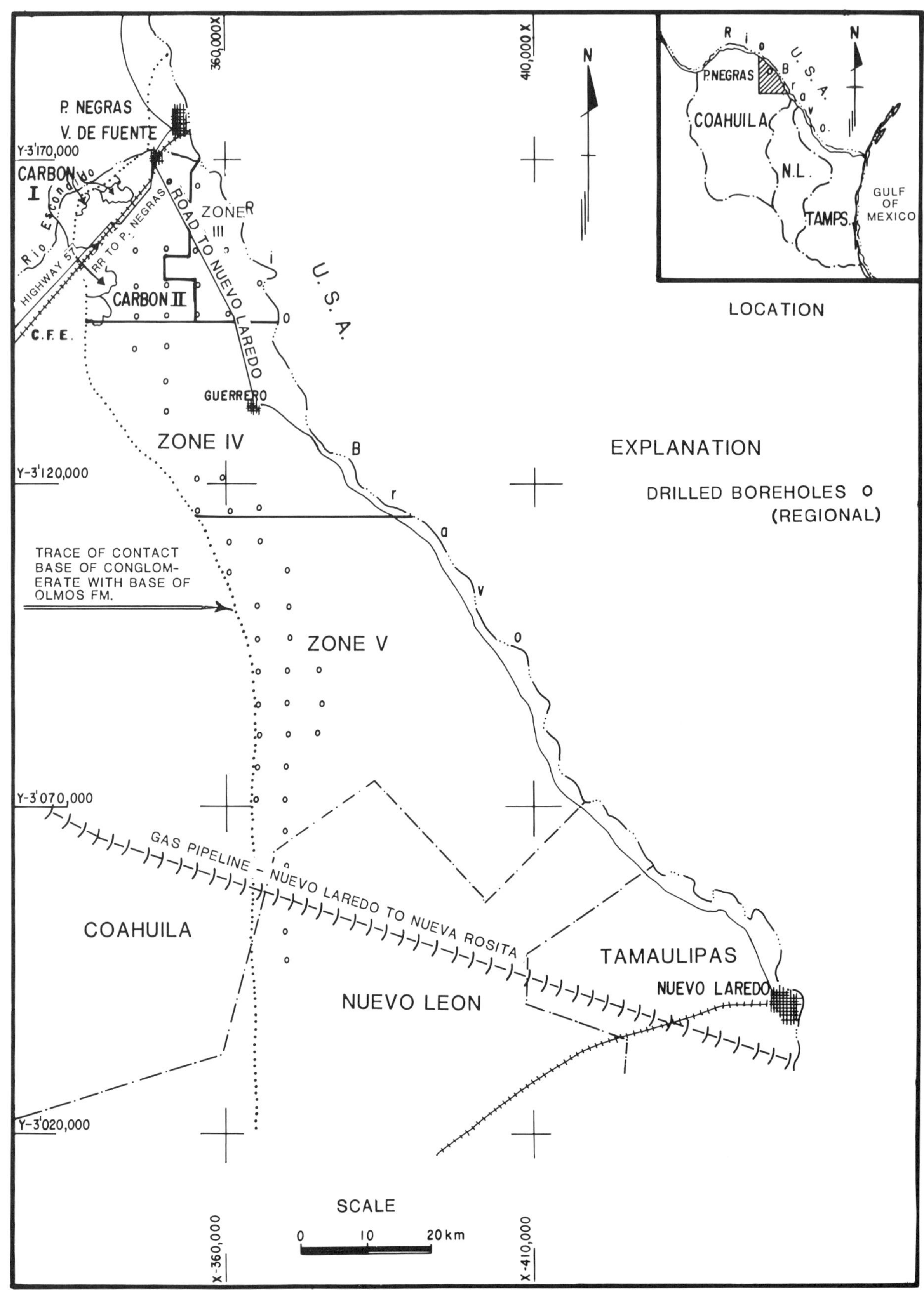

Figure 2. Location of projects within the Fuentes–Río Escondido region.

(Coahuila), which started operations in 1964 with an installed capacity of 37.5 MW, consuming 140,000 tons/year of coal from the Río Escondido Mine developed by the CFE. By 1976, 192 million tons of evaluated probable reserves justified the José López Portillo Thermoelectric Power Station Project, with 1,200 MW of installed capacity in four 300-MW units fed by a total 12,000 tons of coal per day for 30 years. The thermoelectric station is located 25 km south of the town of Piedras Negras in Coahuila (Fig. 2). To date an evaluated 600 million tons of thermal coal represents 45 percent of the tonnage required to satisfy demand to the year 2,000, based on the installed capacity of 18,683 MW for the NOINE sector as of this date (1985) (Figs. 3 and 4).

GENERAL INFORMATION

Geographic location

The Fuentes–Río Escondido coal basin covers 6,000 km^2 in the northeast portion of Coahuila State, between 100°15′ to 101°00′W and 28°15′ to 29°00′N; the town of Piedras Negras lies in its northern part (Fig. 1). Access to Piedras Negras is by the Piedras Negras–Saltillo railroad, Highway 57 (Piedras Negras to México, D.F.), and Federal Highway 2 (Ciudad Acuña to Matamoros) (Fig. 2). It also has a local airport with a paved landing strip, telegraph and telephone systems, and it communicates with Eagle Pass, Texas, by land (road and railroad) and air.

Climate and vegetation

The region has a dry, hot climate with humid, cold winters; extreme temperatures are 45° C in summer and –8° C in winter.

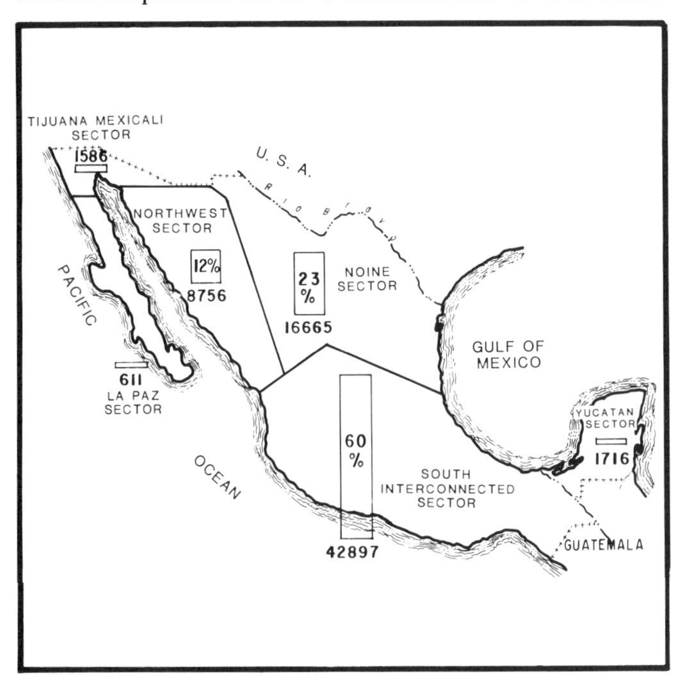

Figure 3. Regional distribution of demand, in megawatts, to the year 2000.

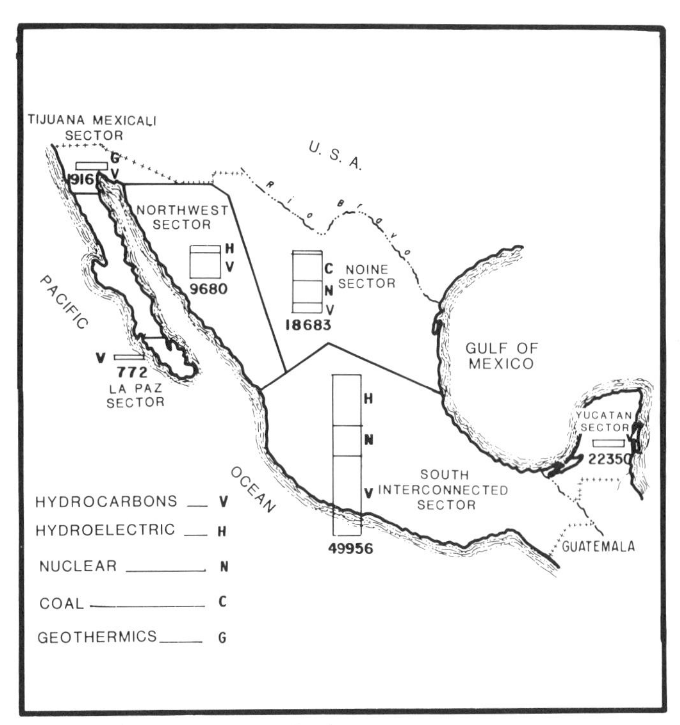

Figure 4. Installed capacity, in megawatts, to the year 2000.

In the Koeppen classification this is a dry steppe-type climate with rainy winters; average annual precipitation is 550 mm, and average annual temperature is 25.8° C. Altitude is 200 to 250 m above sea level; the distribution of soil types determines varieties and densities of the typical steppe-type herbaceous plants, bushes and cacti, the commonest of which are:

	Common name (Spanish)	**Scientific name**
HERBACEOUS	Gobernadora	*Larrea divaricata*
	Cenizo	*Tucophyllum texanus*
	Candelilla	*Euphorbia antisyphilitica*
BUSHES	Hiedras	*Rhus exima*
	Mezquite	*Prosopis juliflora*
	Uña de gato	*Acacia greggy*
	Huizache	*Acacia farnesiana*
CACTI	Nopal	*Opuntia ficusindica*
	Lechuguilla	*Agave lechuguilla*
	Biznaga	*Echiva cactus*

REGIONAL GEOLOGY

Physiography and geomorphology

The coal basin lies within the Bravo Basin (Cuenca del Bravo) subprovince of the Gulf of Mexico Coastal Plain Physiographic Province (Figs. 5 and 6). The Bravo Basin may be divided into Eastern, Central, and Western zones. The coal basin area is in the Western zone, which consists of Upper Cretaceous,

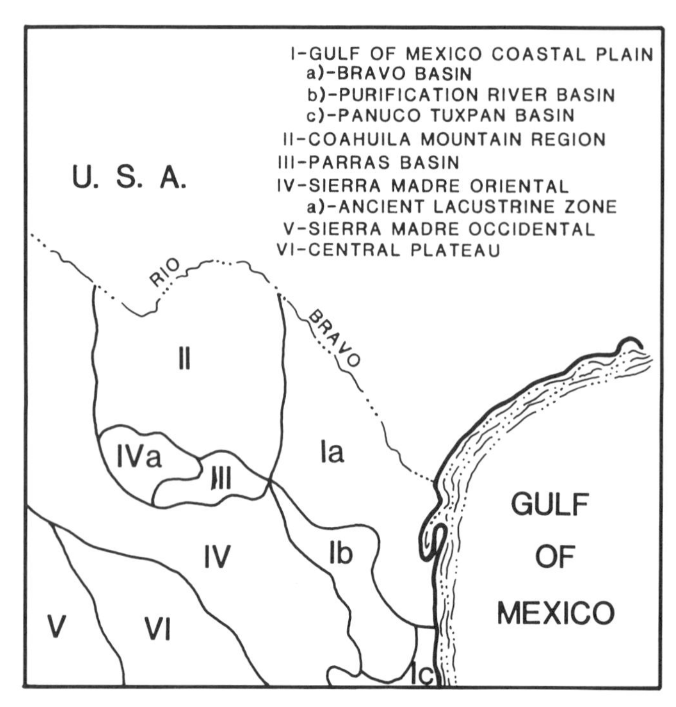

Figure 5. Physiographic provinces of northeastern Mexico.

gently dipping limestones, shales, and sandstones, producing a slightly undulating topography. Toward the west a northwest-southeast–trending anticline, predominantly in limestones, is known as the Burro Range (Serranía del Burro), which extends from the Río Bravo (Río Grande) at the Coahuila-Chihuahua state line to the towns of Zaragoza and Allende.

Considering the degree of erosion and the fluvial system, characterized by the meandering Río Bravo, the region is geomorphologically mature. Present-day relief is the result of mainly exogenous, both depositional and erosional, processes and, to a lesser extent, of endogenous events.

Stratigraphy

The Fuentes–Río Escondido Basin lithologic units range from Upper Cretaceous to Recent in age and have been studied in ditch and core samples from wells drilled by both the CFE and PEMEX (Petroleos Mexicanas). The sequence is mainly sedimentary: limestones, sandstones, shales, coal, and conglomerates of the Austin, Upson, San Miguel, Olmos, Escondido, and Sabinas-Reynosa Formations (Fig. 7).

Deposition within the basin is genetically associated with a lobate-type delta in the north, which is elongate in the south. Three main facies have been defined in the Cretaceous rocks (Fig. 8); the *Upson Formation (Campanian),* typically a prodelta facies; the *San Miguel Formation (Upper Campanian),* showing evidence of deltaic-front facies; and the *Olmos Formation (lower Maastrichtian),* of deltaic-plain character, having the appropriate elements for the development of coal seams at its base. These are strictly *diachronous* units by reason of their origin in a prograding deltaic system (Fig. 9).

The stratigraphy of the coal-bearing area has been established and correlated with the regional stratigraphy of the Bravo Basin by means of sedimentological studies (sequential analysis, primary structures, petrography, etc.) of drill core samples. The San Miguel, Olmos, and Sabinas-Reynosa Formations were identified. The second of these, belonging to the deltaic plain of a system that prevailed in Upper Cretaceous time (Campanian-Maastrichtian), is the unit of interest in coal exploration.

Austin Formation (Coniacian-Santonian)

The limestone outcrops conformably overlying the Eagle Ford Formation were originally named "Austin Chalk" by Shumard (1860). Separation of the two in the field is difficult; however, the Austin limestones exhibit no laminated structures and weather in different-sized ovoid fragments rather than in slabs as does the Eagle Ford.

Lithology and thickness. The Austin Formation is typically a gypsiferous limestone; east of Smith County in the east Texas oilfields it is mainly sandy. The Fuentes–Río Escondido Basin sequence consists of alternating thinly bedded limestones and dark gray calcareous shales. The formation weathers in greenish tones and with medium-bedded to laminated stratification. In CFE wells it is 300 m thick.

Contacts. The contacts are conformable with the underlying Eagle Ford and overlying Upson Formations.

Age and correlation. On the basis of its fossil content—*Inoceramus undulatiplicatus* Romer, *Durania austinensis,* and *Bawlites* sp.—the age of the formation is Coniacian-Santonian; it correlates with the San Felipe Formation exposed in the Tampico-Misantla sedimentary basin.

Upson Formation (lower Campanian)

Defined by Dumble (1892), with its type locality in Maverick County, Texas, this unit is the more clayey base of the Taylor Group, a dark gray or greenish gray clay usually weathering in yellowish tones and locally with small gypsum laminations and crystals.

Lithology and thickness. The unit is found from southeast Texas to south of the Adjuntas and Monclova basins in the state of Coahuila. In the present area, outcrops are found north-northwest of the town of Piedras Blancas; they consist of light to dark gray shales with small calcareous siltstone lenses and a few sandstone lenses toward the top. The shales are highly organic in some zones, and some layers are bioturbated by organic borings; fossils are gastropods, pelecypods, fish remains, foraminifera, and dinoflagellates. At the type locality, the unit is 180 m thick; outcrops near Eagle Pass are 170 m thick, which is also the thickness reported in the Fuentes–Río Escondido basin.

Contacts. This unit is conformable above the Austin and beneath the San Miguel Formations; on the basis of its strati-

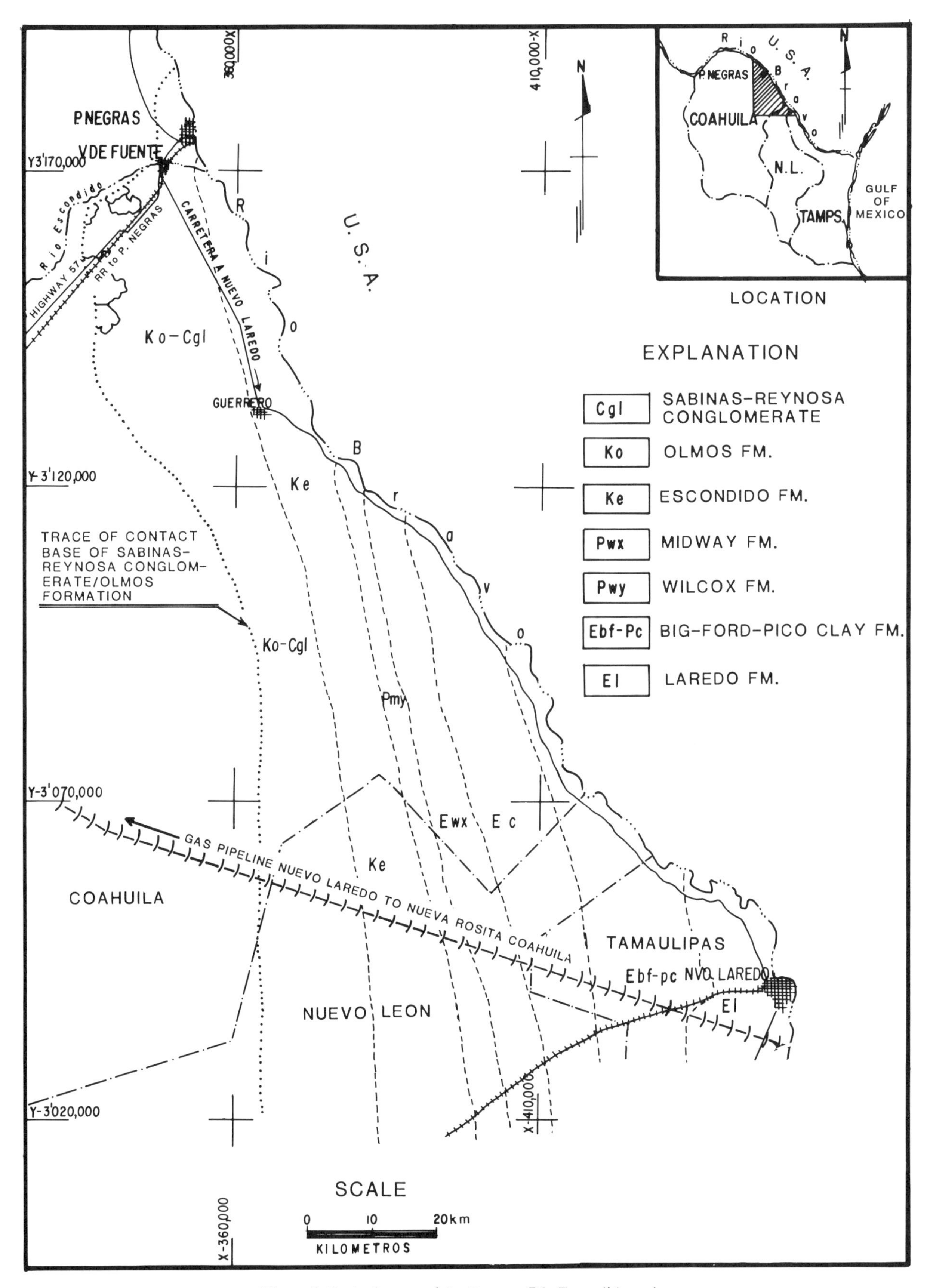

Figure 6. Geologic map of the Fuentes–Río Escondido region.

graphic relations to the latter, it is assigned to the prodelta facies of the ancient constructional high deltaic system defined in the basin.

Age and correlation. Its foraminiferal content and stratigraphic position define the unit's age as lower Campanian, equivalent to the Taylor Group of Texas and time-correlative to the lower Méndez Formation of the Tampico-Misantla basin in Mexico.

San Miguel Formation (upper Campanian)

This formation was defined by Dumble (1892) in the old San Miguel Ranch north of Eagle Pass, Texas. In Mexico it is found in the composite Parras–La Popa basin in Monclova, and in the Fuentes–Río Escondido basin where it crops out north and northwest of Piedras Negras.

Lithology and thickness. This unit was originally divided into five unnamed members at its type locality; the five, however, are not always individually correlatable: (a) fossiliferous concretions, (b) compact siltstones, (c) white sandstones, (d) nonstratified siltstone, and (e) laminated sandstone. These subunits have not been differentiated in the present area, and almost certainly no correlation is possible with members established elsewhere, since the formation is a deltaic-front facies. Locally it exhibits the character of channel-mouth bars and deposits formed by rupture of distributary channels: thickening of the deltaic-front sandstones, cross-bedding, clay intraclasts, and traces of transported organic matter of widely varying thickness and characteristics. Interbedded sandstones and shales are common at the base; toward the top, very thick sandstones interfinger with scarce shales. In some cases shales are absent, and the section consists of 20 m or more of quartz sandstone.

Fucoids or organic borings, including types of *Ophiomorpha,* are found in the formation, possibly caused by crustaceans inhabiting normally saline shallow waters. These structures appear in some cores and indicate marginal-littoral depositional environments. In the United States, this unit is explored for tar sands; in Sabala and Maverick Counties, Texas, 45 km from the town of Eagle Pass, Exxon has a project for in situ oil development from a 15-m-thick interval within this stratigraphic unit.

Contacts. This unit is conformable over the Upson and underneath the Olmos Formations; all three units are diachronous.

Age and correlation. Dumble (1892) assigned this unit the same age as the lower Taylor Group (upper Campanian), but also correlated it with the Navarro Group glauconite beds, based on his identification of *Exogyra ponderosa* and *Exogyra costata* (lower Maastrichtian) in the unit. Vanderpool (1930) prefers to assign the San Miguel Formation to the lower Maastrichtian on the basis of its faunas. In the present area it has been assigned to the upper Campanian without, however, giving it a fixed age due to its deltaic and diachronous nature, comparable to the Upson and Olmos Formations. It is time correlative with the Difunta Group of the Parras–La Popa basin, the Taylor Group in Texas, and the lower Méndez Formation of the Tampico-Misantla basin.

EON	ERA	PERIOD	EPOCH	AGE IN M. Y.	NORTH OF THE BURRO. RANGE	CARMEN RANGE	FUENTES-RIO ESCONDIDO SUB-BASIN	COLOMBIA-SAN IGNACIO SUB-BASIN
PHANEROZOIC	CENOZOIC	QUATERNARY			ALLUVIUM		ALLUVIUM	ALLUVIUM
		TERTIARY NEOCENE	PLIOCENE	2	SABINAS REYNOSA CONGL.		SABINAS REYNOSA CONGL.	UVALDE
			MIOCENE	5				
			OLIGOCENE	25				
		PALEOGENE	EOCENE					LAREDO
								PICOCLAY
								BIGFORD
								CARRIZO
								WILCOX
			PALEOCENE	56				MIDWAY
	MESOZOIC	CRETACEOUS	UPPER	66 MAASTRICHTIAN	ESCONDIDO FM.		ESCONDIDO FM.	
					OLMOS FM.		OLMOS FM.	
				69 CAMPANIAN	SAN MIGUEL FM.	AGUJA FM.	SAN MIGUEL FM.	
					UPSON FM.	TERLINGA FM.	UPSON FM.	
				77 SANTONIAN 83 CONIACIAN	AUSTIN FM.	AUSTIN FM.	AUSTIN FM.	
				90 TURONIAN	EAGLE FORD FM.	EAGLE FORD FM.	EAGLE FORD FM.	
				95 CENOMANIAN	BUDA FM.	BUDA FM.	BUDA FM.	
			LOWER	100 UPPER ALBIAN	DEL RIO FM.	DEL RIO FM.	DEL RIO FM.	
				110	GEORGETOWN FM.	BURRO REEF COMPLEX	GEORGETOWN FM.	

Figure 7. Stratigraphic correlation chart.

Olmos Formation (lower Maastrichtian)

The beds of the Olmos Formation were named "Coal Series" by E. T. Dumble (1892). Stephenson (1918) described them at Olmos Station in Maverick County and Olmos Creek, its type locality. Schmitz (1885), Bose and Cavins (1927), Evans (1974), Smith (1970), Maxwell (1962), McBride and others (1975), and others studied the distribution and nature of the Olmos. The Upper Cretaceous coal originated in a swamp-deltaic environment with laterally and vertically changing facies as a high-constructional delta—in which facies progradation and basin subsidence are constant.

Lithology and thickness. Stephenson (1918) described finely stratified sandy shales and greenish gray shales, with fine- to coarse-grained, thinly bedded, greenish gray sandstones at irregular intervals, containing coal and lignite seams toward the base of the formation. The Olmos strata cannot be individually correlated owing to their origin (deltaic plain) and lenticular nature. In the present area the formation consists of interbedded

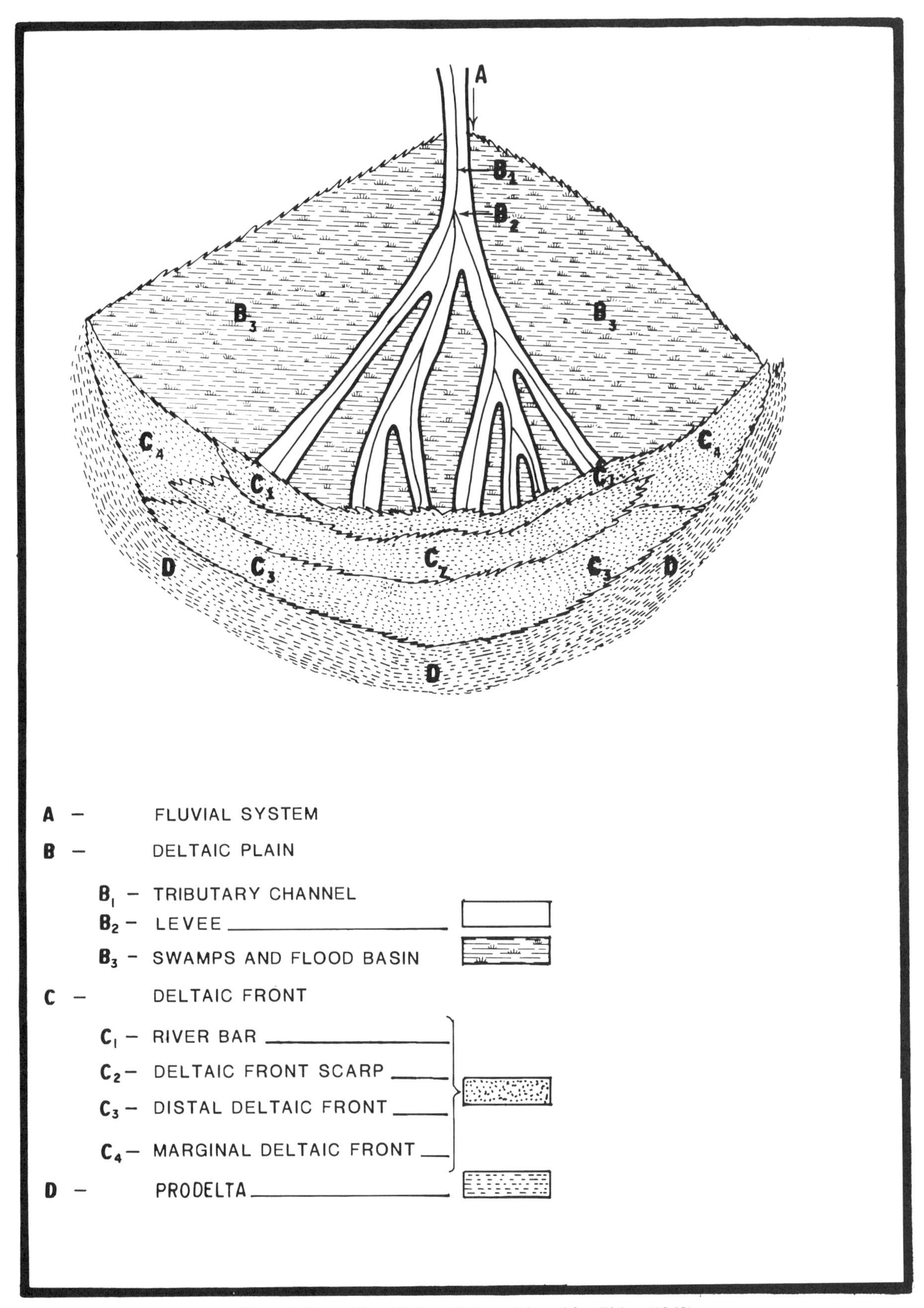

Figure 8. Depositional facies of lobate deltas. After Fisher (1969).

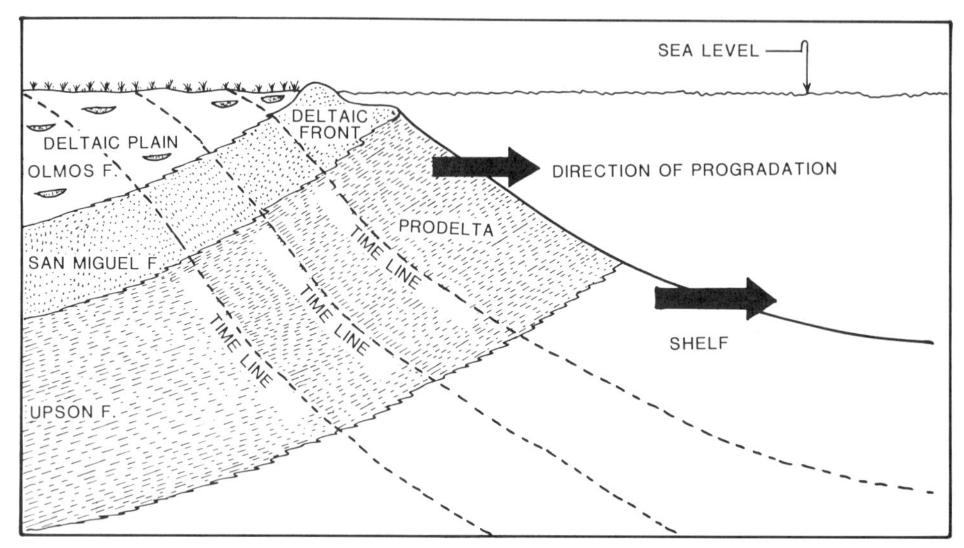

Figure 9. Progradation of a constructional deltaic system (after Scruton, 1960), showing stratigraphic relations of the principal formations of the Fuentes–Escondido coal basin.

shales and siltstones, cross-bedded and current-rippled sandstones, occasional carbonaceous shales, kaolinitic clays, and coal seams. Its lower part is composed of deltaic plain (marshes, mangrove swamps) deposits: shales having high contents of organic matter and coal seams, together with thin shales typical of bay environments and distributary-channel sandy sediments. Fluvial plain sediments predominate in the upper part: fluvial sandstones and overbank shales of the transition zone between channels and flood plains. The formation is approximately 100 m thick, ranging from a few meters to 150 m.

Contacts. This unit is conformable above the San Miguel and beneath the Escondido Formations.

Age and correlation. Fossils in the formation are cephalopods (*Sphenodiscus* sp.), pelecypods, *Exogyra costata,* and gastropod remains, corresponding to the lower Maastrichtian. The formation is equivalent in time to the lower Navarro Group and the upper Méndez Formation of the Tampico-Misantla sedimentary basin.

Escondido Formation (upper Maastrichtian)

This formation was named by Dumble (1892) with its type locality near where the Río Escondido flows into the Río Bravo, about 4 km south of Piedras Negras. The upper part of the formation crops out at Caballeros Creek, 52 km from Piedras Negras. Stephenson (1918) mentions the Cretaceous-Tertiary contact on the American side of the Río Bravo.

Lithology and thickness. The Escondido Formation is 180 to 250 m thick in Maverick County, Texas. Excellent outcrops of this highly resistant unit are found north and east of Eagle Pass where the base of the formation is exposed. In Mexico it occurs in the Adjuntas, Sabinas, and Fuentes–Río Escondido basins. The formation is composed of fossiliferous shales, mudstones, sandy siltstones with fucoids, and numerous sandstone layers.

The following fossils have been identified in outcrop and at the type locality: *Sphenodiscus pleurisepta, Casidulus* sp., *Exogyra costata, Ostrea glabra,* crustaceans, pelecypods, and gastropods, which place it in a neritic environment.

Contacts. This unit is conformable above the Olmos and beneath the Paleocene Midway Group; in outcrop areas it is partially covered by the Miocene-Pliocene Sabinas-Reynosa Conglomerate.

Age and correlation. Based on its faunal and floral content and stratigraphic position it is assigned to the late Maastrichtian, equivalent to the upper Navarro Group in Texas and to the Difunta Group of the Parras–La Popa basin.

Midway Group (Eocene)

This unit is composed mainly of calcareous siltstones and sandy shales, impure limestone, disseminated glauconite, and mollusk concentrations that include ammonites (*Sphenodiscus* sp.) for which a Paleocene age has been defined. Since the Cretaceous-Tertiary contact has been used to mark the eastern boundary of the Río Escondido Project, these sediments are not included in the stratigraphic column for the present area. The boundary was set arbitrarily for exploration purposes, and the coal seams may continue at depth below the 600 m level, which would extend the commercial prospects for the project.

Sabinas-Reynosa Conglomerate (Miocene-Pliocene)

According to Humphrey (1956), this unit includes conglomeratic deposits of the Sabinas coal region; in the Sabinas basin it crops out along the Sabinas River. The approximately

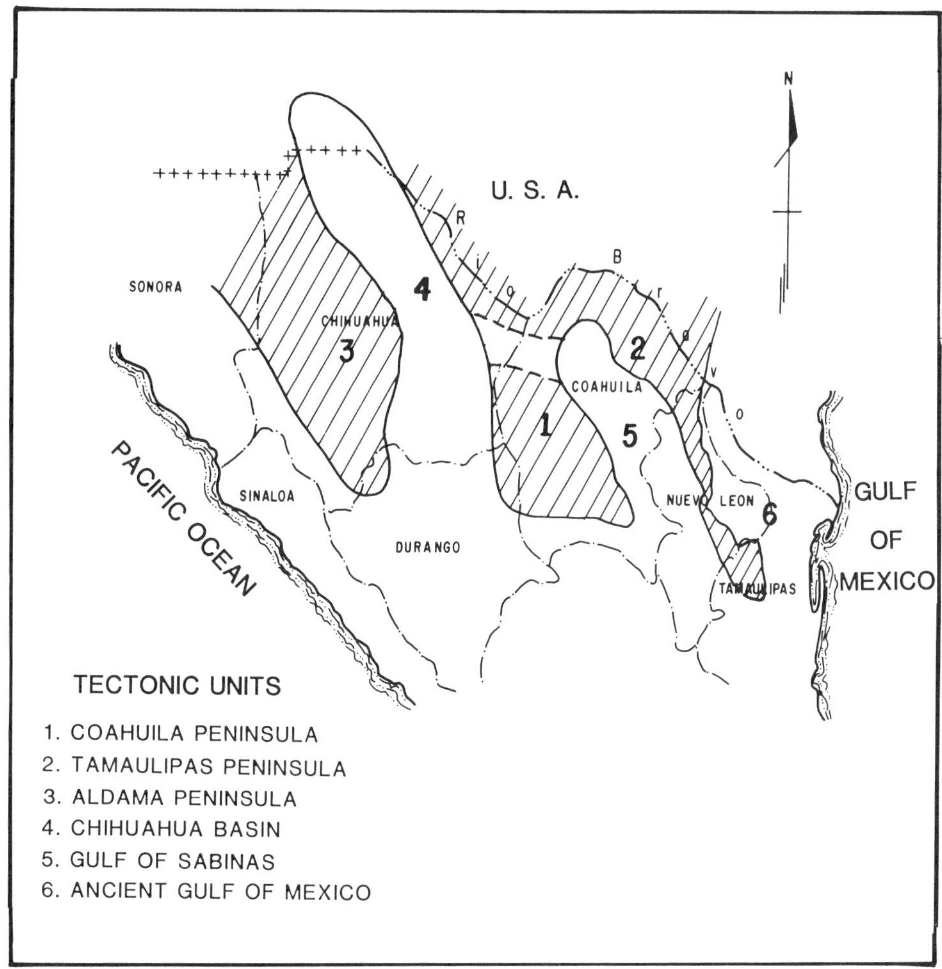

Figure 10. Upper Jurassic paleogeography of northern Mexico. After Humphrey and Diaz (1956).

30-m-thick unit consists of reworked limestone fragments, 0.05 to 1 m in size, cemented by calcium carbonate, unconformably overlying the previously described formations. It is exposed toward the west in the present area where it consists of variably sized limestone detritus, sands, and "caliche"; the unit is the principal aquifer in the region due to its porosity and permeability, wide extent, and shallowness in most of the exploratory wells.

The unit is assigned to the Miocene-Pliocene on the basis of its lithostratigraphic position. It is capped by a variable thickness (1 to 15 m) of "caliche" generated by carbonate precipitation. The residual soils are Recent alluvium at the top of the stratigraphic column in the present area. The possible use of "caliche" in the manufacture of cement increases the economic potential of this formation.

Regional paleogeographic framework

In the course of the geologic evolution of this region, major geologic events gave rise in the Late Jurassic to the formation of two paleogeographic elements: the Coahuila and Tamaulipas Peninsulas (Fig. 10). At the beginning of the Cretaceous (Neocomian) the emerged landmasses were exposed to erosion, and thick sediments accumulated in the existing basins, represented by the Taraises and Cupido Formations (Neocomian-Aptian). At the end of the Neocomian the seas had covered the exposed areas except for the southern part of the Coahuila Peninsula; simultaneous slow subsidence caused the waters to become deeper, and deposition of the calcareous Cupido and La Peña Formations continued until Albian time (Fig. 11). Late Cretaceous epeirogenic movements shallowed the seas, and the calcareous-clayey sediments (shales and limestones) of the Eagle Ford Formation were laid down continuously until Coniacian-Santonian time.

A series of uplifts from the Cenomanian onward caused marine regression, laying bare extensive coastal plains in northern and central Coahuila where large deltaic systems developed and gave rise to the Coahuila coal deposits. Such deltaic systems extended approximately from Piedro Negras, Coahuila, and Eagle Pass, Texas, in the north, to Monclova in the south, Lampazos in the east and Cuatrociénagas in the west. The deposits are believed to have come from the west (Figs. 12, 13).

At the beginning of the Campanian, the prodelta facies of a deltaic system began to develop in what is now known as the

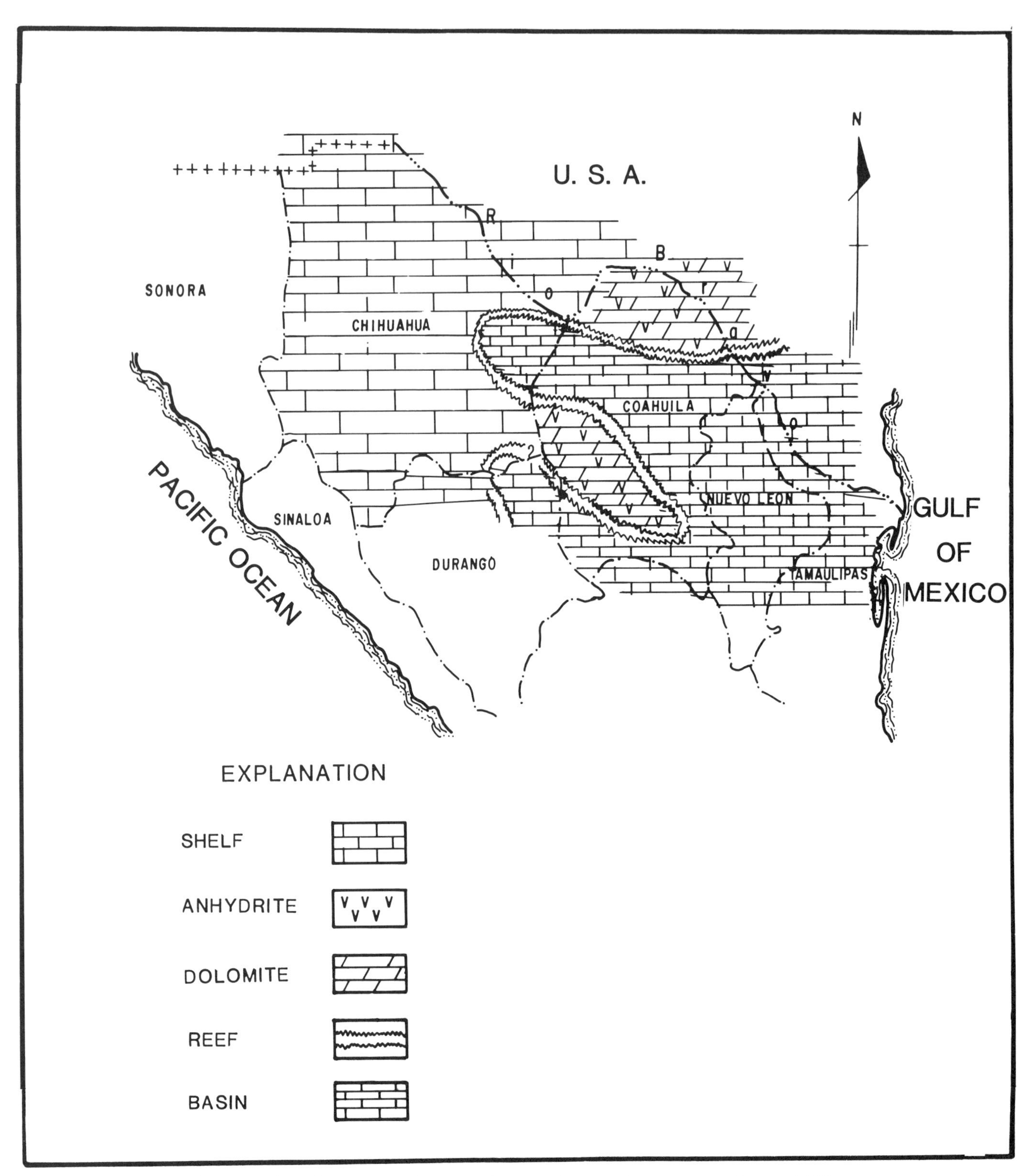

Figure 11. Albian-Cenomanian paleogeography of northern Mexico. After Lopez Ramos (1969).

Fuentes–Río Escondido basin with the deposition of fine-grained sediments of the Upson Formation. Regression continued, and the San Miguel Formation was laid down at the end of the Campanian in a shore or deltaic front environment (Fig. 12). The seas regressed, leaving clean sands over most of the area. The Olmos Formation (deltaic-plain deposits) was laid down at the beginning of the Maastrichtian. These three formations belong to the deltaic system prevailing in the area and are considered as diachronous units.

The sedimentological characteristics of the Olmos Formation change upward from its original deltaic nature, with fluctuating neritic and continental deposition, to sandstones and silts derived from emerged landmasses; these were probably laid down in late Maastrichtian time and are represented by the Escondido Formation. At this time, stronger epeirogenic movements heralded the beginning of the Laramide Orogeny, which continued until the Paleocene-Eocene, modifying the older depositional structures and generating the forces that uplifted, folded, and faulted the Lower and Upper Cretaceous sediments at the western edge of the area. Subsequently, at the beginning of the Tertiary and until the Pliocene, the area remained exposed to erosion, and the Sabinas-Reynosa Conglomerate was laid down unconformably on the previous units.

Structural geology

The dominant structure of the Fuentes–Río Escondido basin is a 2 to 3° east-dipping homocline whose northwestern portion has been eroded, causing erosional unconformity between the Sabinas-Reynosa Conglomerate and the Cretaceous units (Fig. 5). The Cretaceous units are gently folded; fault and fracture systems in the Olmos Formation are evidence of Laramide tectonic activity.

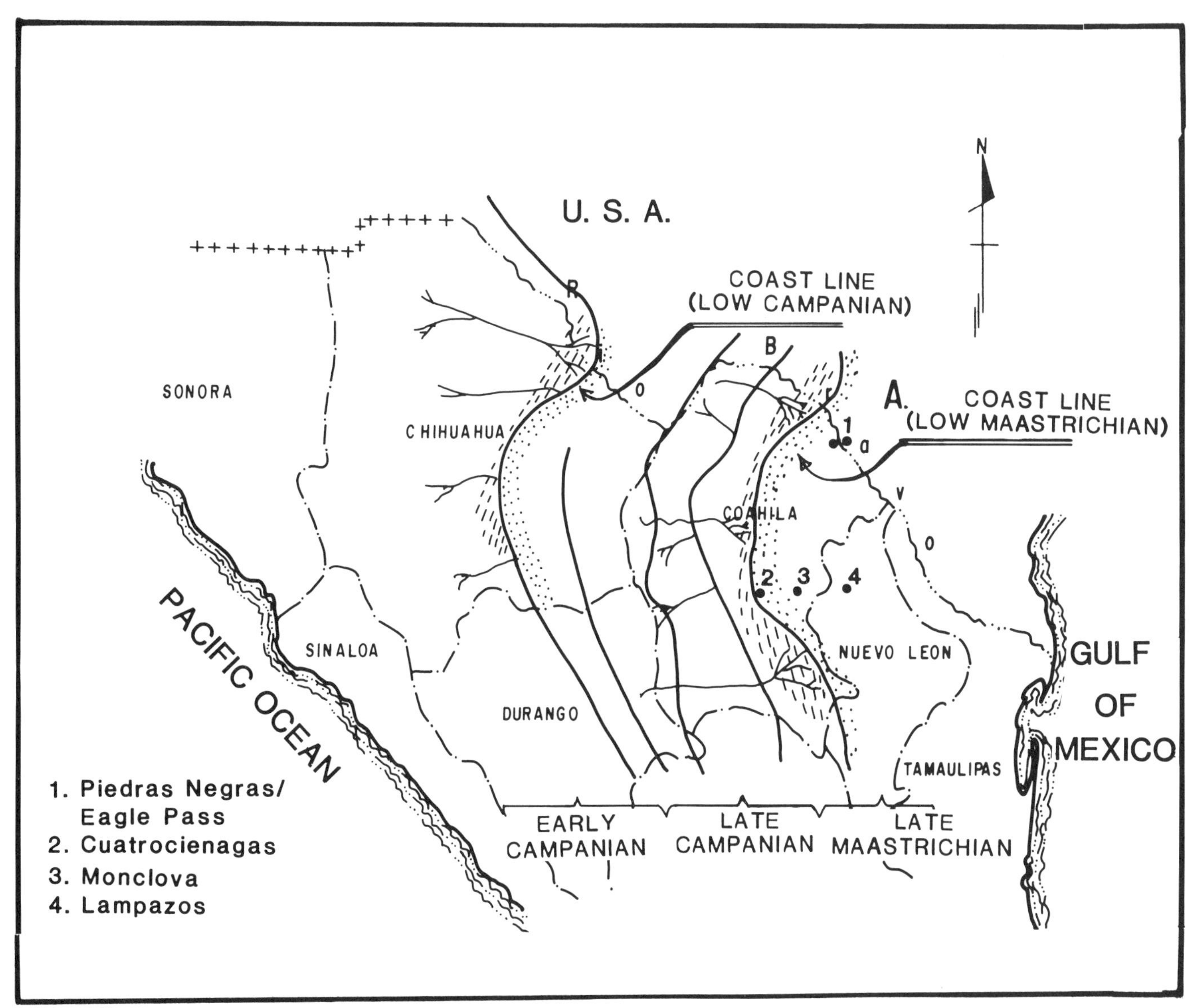

Figure 12. Upper Cretaceous paleogeography of northern Mexico. After Weide (1967).

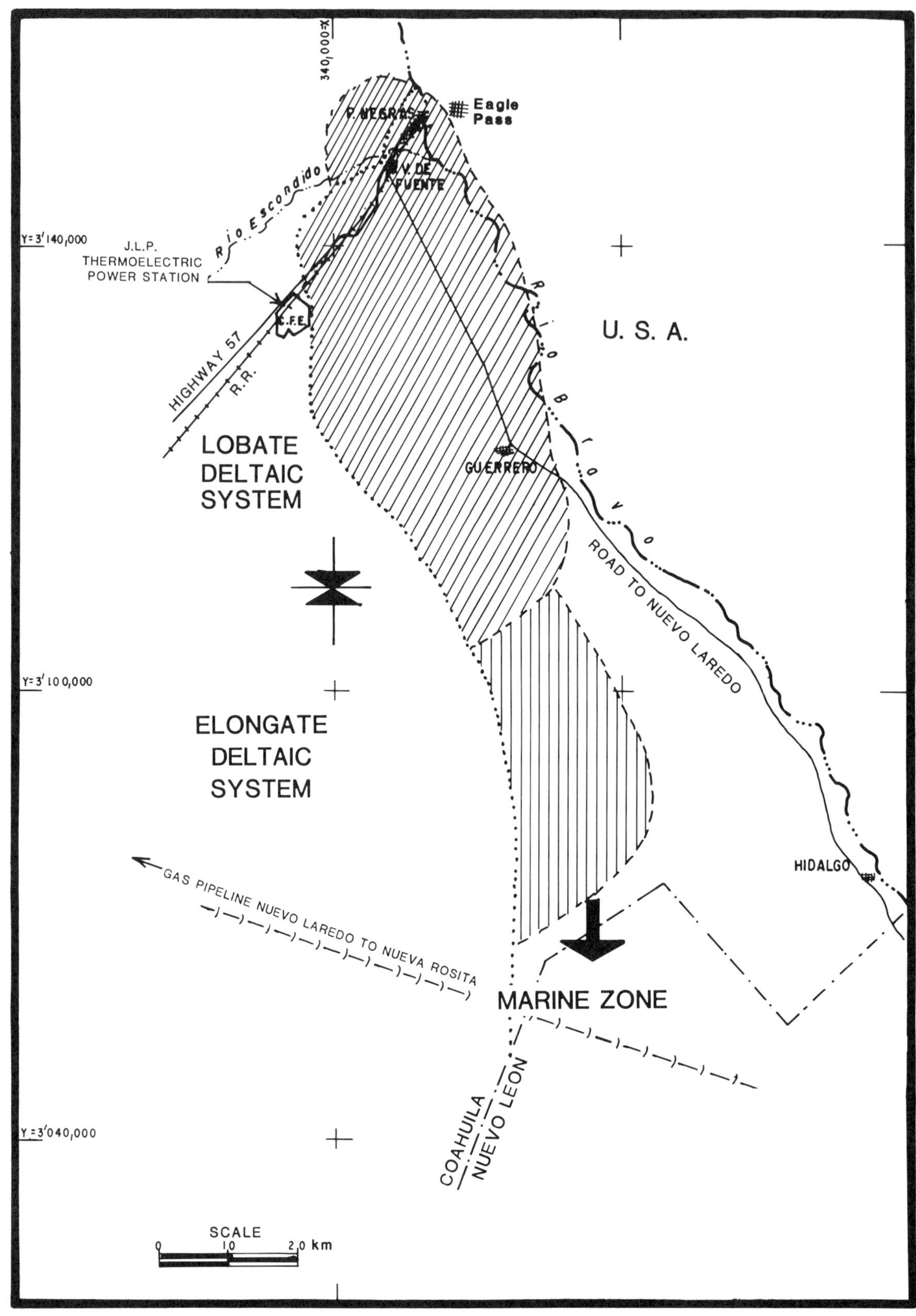

Figure 13. Deltaic systems in the Fuentes–Río Escondido basin.

LOCAL GEOLOGY

Geomorphology

In Lobeck's (1939) classification, the Río Escondido Project area is in a geomorphologically rejuvenated late-mature region in which subparallel streams drain down steep slopes. A fracture system at times controlling the drainage, allows hill systems to develop in the calcareous Sabinas-Reynosa Conglomerate, which overlies the stratigraphic section unconformably.

Stratigraphy

As mentioned above, the economically important units are of constructional deltaic character (i.e., erosional and depositional factors prevail over the destructive effects of wave action). Facies that represent such systems are defined, from the sea toward the shore, as the prodelta, the deltaic front, and the deltaic plain facies. Peat deposits accumulated in the deltaic plain facies and evolved into coal due to the effects of heat and pressure. The sedimentological facies have been identified in the Fuentes–Río Escondido area where the prodelta is represented by the Upson Formation, the deltaic front by the San Miguel, and the deltaic plain by the Olmos deposits.

According to McBride and others (1975), the units in the depositional sequence are diachronous; hence, continuous coal horizons are of different ages. For evaluation purposes, this means that the continuity of these coal horizons is determined by the lobate character of the plain and by the seaward-advancing delta. However, structural interpretation has defined contemporaneous small, local deeps, which also restricted the coal beds.

The lithology of the units corresponds in sequence to a regression gradually leading to coarser deposits toward the top of the section. Initial deposition is thus represented by the development of the deltaic plain where mainly fluvial sediments shaped the drainage, bounding areas of clayey and marsh deposits in which peat accumulated. This facies consists locally of two megasequences, the lower of which contains up to eight coal-bearing sequences separated by paleo-channel, dike, flood-plain, and channel-rupture subfacies, described lithologically as follows:

Ideal depositional sequence:

7.	Coal	C
6.	Carbonaceous shales	Cs
5.	Shales	Sh
4.	Siltstones	St
3.	Sandstones with intercalations	Si
2.	Fine-grained sandstones	Fs
1.	Medium-grained sandstones	Ms

Although the sequences do not always show all the units, the sequential order is preserved (Fig. 14).

The second megasequence conformably overlies the above, separated by a diastem, which at times is an erosional surface.

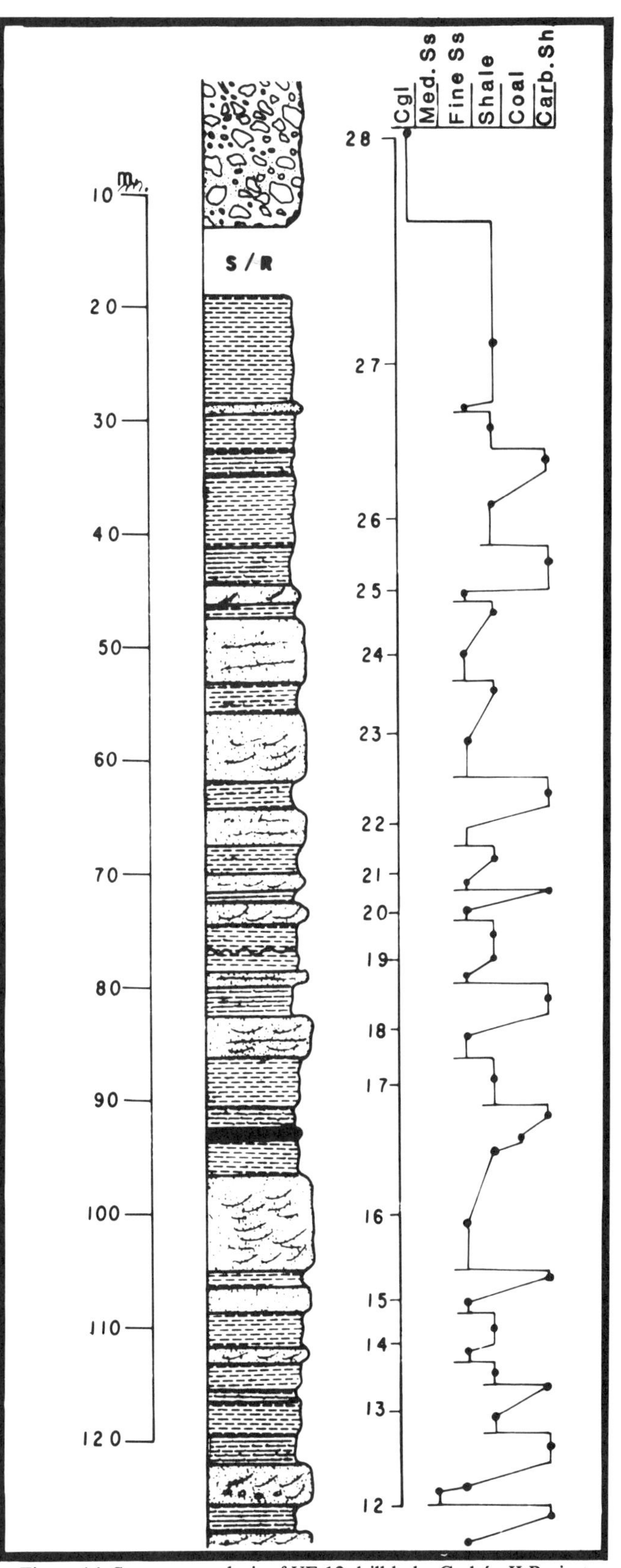

Figure 14. Sequence analysis of VF-12 drill hole, Carbón II Project.

This sequence developed in the course of a gradual transgression, during which the Escondido Formation was finally laid down. Its economic importance is secondary because the destructive aspect of the transgression precluded the formation of extensive deltaic plains where coal seams could develop. The coal deposits found occasionally within the unit are of allochthonous or mixed nature. Fine-grained and clayey deposits increasing gradually toward the top signal subsidence and the progressive advance of the shoreline toward the continent.

Structural geology

Structural elements identified in vertical-probe exploration by the CFE led to the conclusion that the regionally east-dipping homocline is broken by gentle undulations roughly paralleling the direction of the dip. Interpreted groups of fractures associated with normal faulting allow structural correlation of the coal-bearing levels. Vertical displacement, ranging from few to a maximum of 30 m, are principally cumulative (i.e., fault systems are staggered and it becomes difficult to locate the faults precisely between probes).

Table 1 shows preferred fracture or fault and fold strikes defined by stereographic analysis on polar projection of bedding planes determined in triangulation of the coal-bearing horizon of the Carbón II Project (Figs. 15, 16).

TABLE 1. PREFERRED STRIKES OF STRUCTURAL ELEMENTS, CARBON II PROJECT

Structural Element	Section I	Section II	Section III	Section IV	Section VI	Section VII
Fractures and/or Faults	N05°E N05°W N-5°	N15°W N45°W N15°E	N45°E N50°W N70°W	N65°W N61°E N85°E	E-W N30°W N48°	N25°E N22°W
Fold axis	N50°E	N80°E	N-S N20°E N40°W	N35°E N10°E	N05°W	N22°W

COAL RESERVES

Origin

The relation of the original peat deposit to the source organic vegetable matter defines coals as autochthonous or allochthonous, according to whether the vegetable matter grows, dies, and accumulates in the same place or is transported and deposited elsewhere. Petrographic analyses of the Río Escondido coals classify them in the first category, although some allochthonous traits, such as the presence of coal seams in the upper Olmos Formation megasequence, have been detected. The original peat deposits probably resulted from a combination of both processes, in view of the predominantly clastic nature of the deltaic system. The thermal evolution of the peats was a consequence mainly of basin subsidence, although subsequent tectonism caused the coal to mature heterogeneously, as indicated by anomalously high reflectance values in areas of relatively strong tectonism (faults and folds).

Classification and characterization

In Alpern's (1980) classification of solid fossil fuels, the Río Escondido coals are rated mainly as bituminous, with average reflectance value R = 0.58. Although some samples have been classified as lignites, they are essentially vitric (average vitrinite content 78.35) with a facies of mixed character (average mineral matter 33.2 percent). The corresponding washability is 31.5 percent (Obregón, 1981).

The following physicochemical characteristics were defined: air-drying loss; moisture at 103 to 110°C; natural bed (or equilibrium) moisture; volatile matter; fixed carbon; ash; sulfur; true density; free-swelling index; gross calorific value; elemental analysis comprising carbon, hydrogen, oxygen, nitrogen, sulfur; mineral analysis of ash to determine aluminum, phosphorus, titanium, silica, iron, calcium, magnesium, sodium, SO_3, and chlorine; pyritic and organic sulfur and as sulfide; grindability index, ash-fusion temperatures under oxidizing and reducing atmospheres; equivalent silicon content; acid/base ratio; slagging and fouling factors, etc. (Table 2).

The above characteristics were obtained under strict laboratory conditions applying certified standards issued by the American Society of Testing and Materials (ASTM), the International Standards Organization, the National Bureau of Standards (USA), the Bureau of Mines (USA), the National Coal Board (UK), and specialized bibliographies. In the ASTM classification, these results define the Río Escondido as a bituminous, high-volatile C coal.

Evaluation of reserves

Historically coal reserves evaluations have been computed by stages according to diverse standards. The CFE divided the project into five exploration zones (Fig. 2; Table 3). According to the degree of the available knowledge on the above areas, in situ reserves are classified as proven, probable, and possible, following U.S. Bureau of Mines and U.S. Geological Survey criteria.

Drillhole density was initially fortuitous, particularly in the Carbón I Project. Subsequent geostatistical studies laid the groundwork for a north-south and east-west oriented control grid, on which reserves could be classified as proven (drillhole spacing 1 km or less), probable (drillhole spacing 2 km), or possible (drillhole spacing 4 km).

Quality of the coal has been defined according to its use for heating or power-generation purposes; ash-content percentages, maximum caloric potential, moisture, volatile matter, and density are therefore factors to be considered for evaluation purposes.

These parameters have been plotted on statistical graphs to determine their interrelations and predict their values in terms of development planning (see Fig. 17 and Table 4).

In late 1977, auxiliary data from geophysical well logs were added—conventional electric, stratigraphic, lateral, neutron, caliper-density, sonic, and temperature logs—affording information on the physical and chemical characteristics of the stratigraphic column and the coal horizons. In connection with the latter, such parameters as ash content, caloric potential, and density were obtained from the geophysical logs by applying statistical analysis to the laboratory test results. Aquifer potential, as regards both water quality and aquifer thickness, was also established.

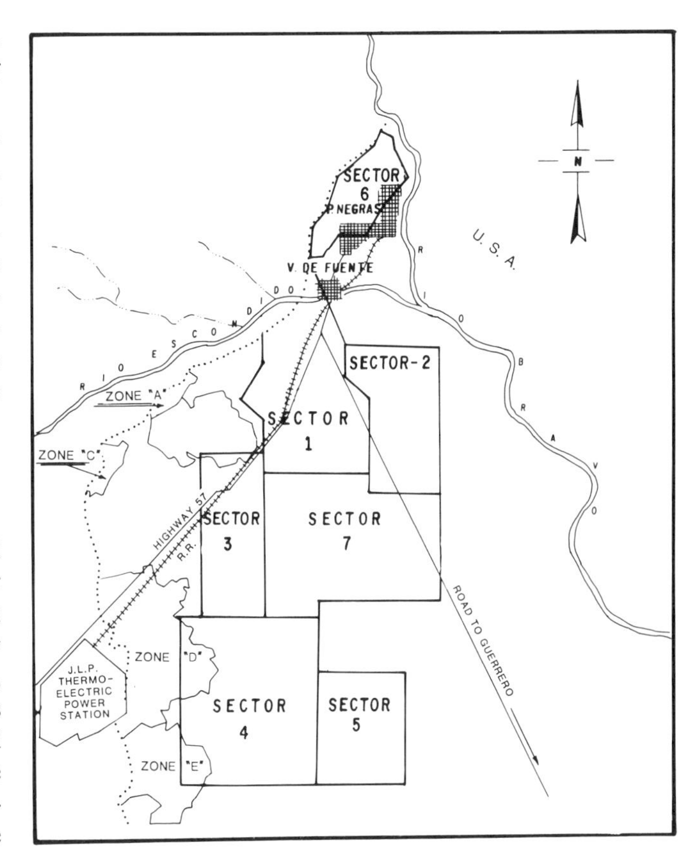

Figure 15. Map of Carbón II Project sectors from which data for Figures 16 and 17 were obtained.

Development

MICARE (Minera Carbonífera Río Escondido), a quasi state-owned company with a CFE participation, was created for the purpose of developing the coal deposits by means of both open-pit and underground mining (Fig. 18). An exploitability depth of 50 to 60 m was established for the open-pit operations, depending on the coal/steriles ratio (Figs. 19, 20). The coal is extracted with hydraulic excavators, power shovels, front-end loaders, and all-terrain dump trucks. Underground mining is done by the longwall method with self-advancing jacks along rectangular 200 by 1,200 m panels. Future mining operations will be affected by gas and oil reservoirs associated with the paleochannels, and with the San Miguel Formation below the 250-m-depth level, requiring previous degasification.

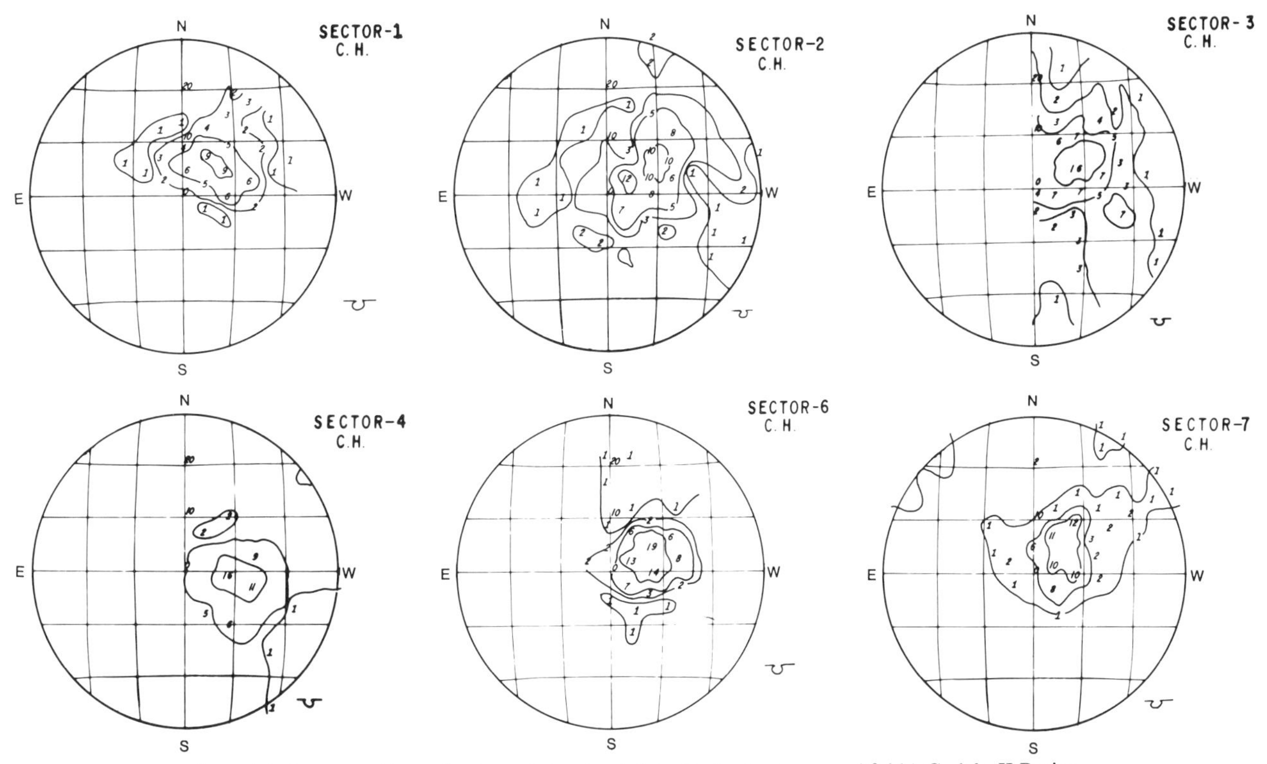

Figure 16. Stereographic projections of structural data (faults, fractures, and folds) Carbón II Project.

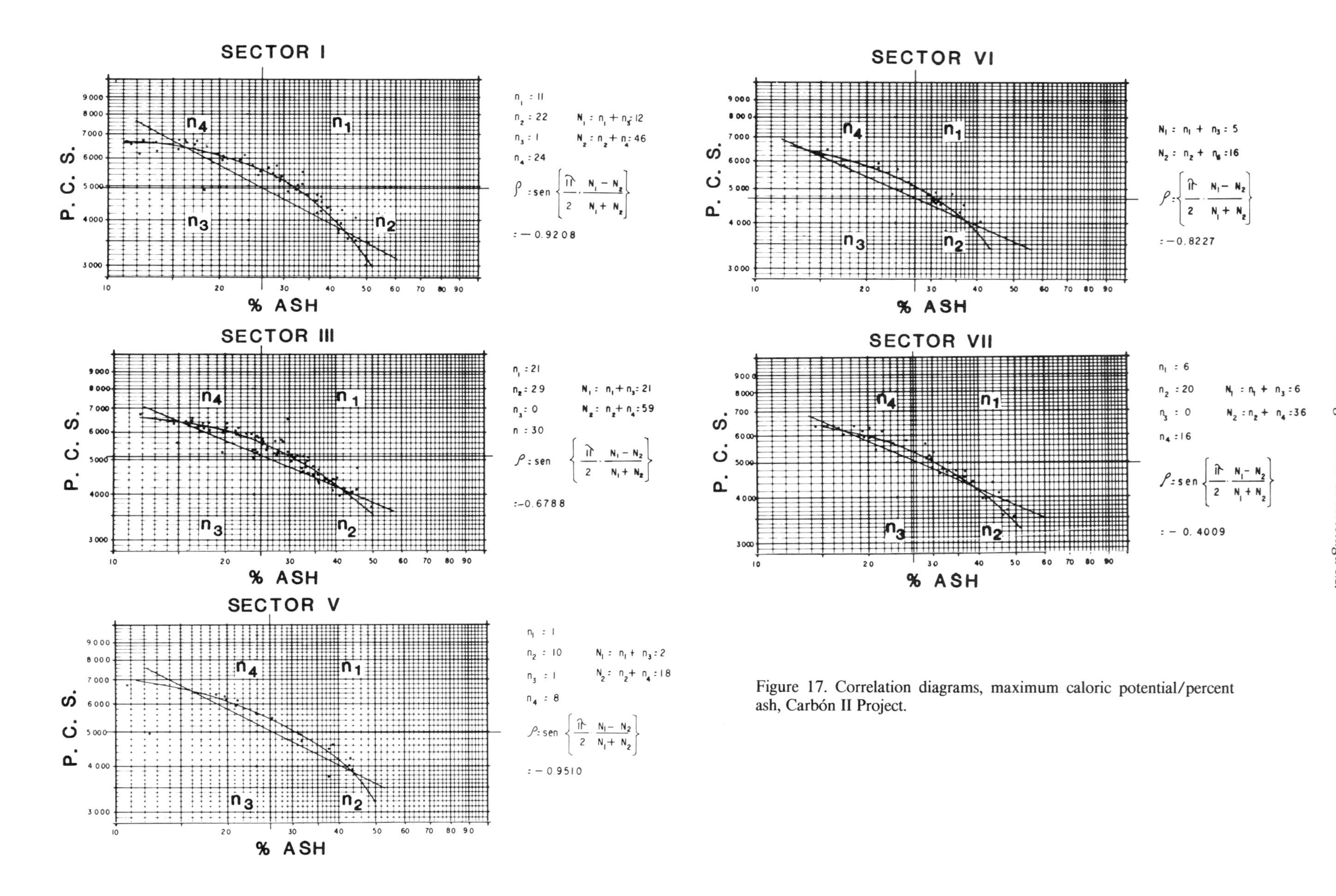

Figure 17. Correlation diagrams, maximum caloric potential/percent ash, Carbón II Project.

TABLE 2. PHYSICO-CHEMICAL VALUES OF RIO ESCONDIDO COALS

Coal		**Ash**	
Primary analysis*	wt. %	Mineral analysis	wt. %
Total moisture	4.16	Iron (Fe_2O_3)	4.34
Volatile matter	30.50	Calcium (CaO)	0.31
Ash	33.27	Magnesium (MgO)	1.01
Fixed carbon	32.07	Sodium (Na_2O)	0.83
Gross caloric value (cal/gr)	4,581.0	Potassium (K_2O)	0.80
Net caloric value (cal/gr)	4,406.6	Silicon (SiO_2)	63.03
		Titanium (TiO_2)	0.77
Elemental Analysis	**wt. %**	Phosphorus (P_2O_5)	0.09
Hydrogen	3.49	Sulfur (SO_3)	0.25
Carbon	47.20		
Nitrogen	1.23	**Fusion and Oxidizing**	**°C**
Oxygen	9.59	Initial deformation	1,540
Sulfur	1.01	Softening temperature	1,540
Chlorine	0.05	Hemispheric temperature	1,540
Total moisture	4.16	Fluid temperature	1,540
Ash	33.27		
Sulfate sulfur	0.05	**Type of Ash**	Eastern
Pyritic sulfur	0.39		
Organic sulfur	0.82		
Complementary Determinations and Concepts			
Equilibrium moisture	6.0	Silicon ratio	0.92
Free-swelling index	1	Base/acid ratio	0.08
Grindability index†	53	Fouling index	0.090
T250°C	1,608	Slagging factor	0.033

*Corresponding to moist base in situ seam
†Significant difference in hardness of materialll
Dry ash furnaces F.P. +1,427°C
Furnaces to remove liquid ash F.P. 1,204°C

TABLE 3. DEGREE OF KNOWLEDGE FOR DETERMINATION OF RESERVES SHOWN IN TABLE 4

	Area (km^2)	Degree of Knowledge	Number of Wells
Carbon I Project	74.27	Evaluation	736
Carbon II Project	276.00	Evaluation	475
Zone III	300.00	Detail	37
Zone IV	900.00	Evaluation-regional	130
Zone V	4,450.00	Regional	10

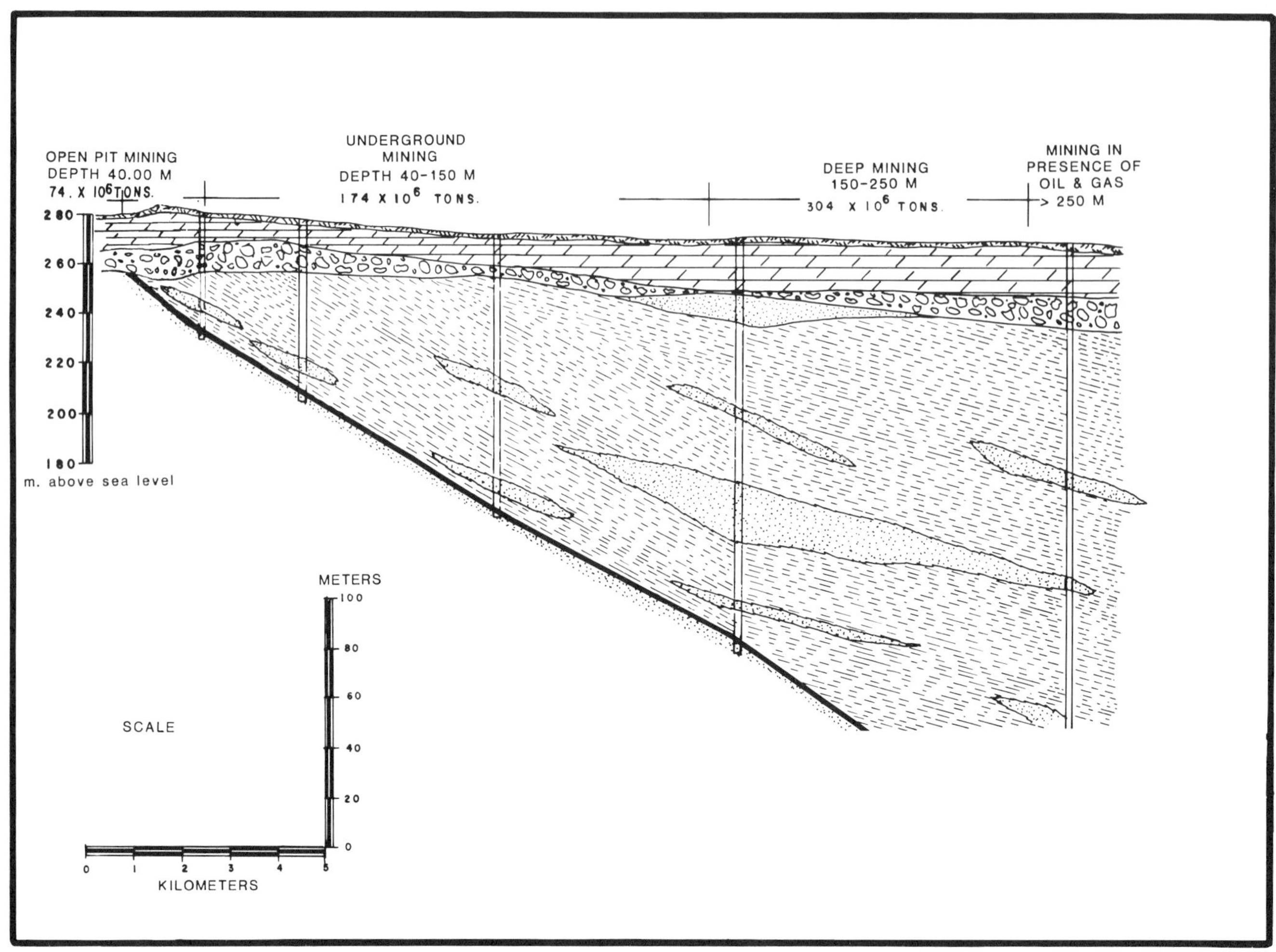

Figure 18. Idealized geologic cross section, Fuentes–Río Escondido basin.

Present capability of MICARE consists of two underground and one open-pit mining operations: Mine I, produces 6,000 tons/day on three longwall faces, and Mine II is presently under development; the first longwall face will begin extraction in 1985 and the other two in 1986 for a production of 6,000 tons/day. Cut I is operating to a maximum depth of 60 m, as per the coal/steriles ratio, producing 2,600 tons/day.

Utilization

The noncoking characteristics of the Río Escondido coal allow its use for power generation by combustion at the power station itself. The installed 1,200-MW capacity requires 3,000 tons of coal per day per unit, having 4,500 cal/kg of average gross calorific value. Specific consumption under optimum conditions is 0.55 kg of coal per kw/h produced.

The integrated use of the coal reserves has led to studies defining the possible utilization of coal ash in the construction industry in the near future. Other studies concluded that the Río Escondido coal can be used in coke mixtures. A washing plant, presently at the design stage, will ensure supply of coal to the power plant with a constant 30 percent ash content, thereby optimizing the efficiency of the thermoelectric complex.

TABLE 4. EVALUATION OF RESERVES IN THE FUENTES–RIO ESCONDIDO BASIN

	Category of Reserves (millions of tons)				
Project	Proved	Probable	Possible	Geo. reconn.	Total
Carbón I	203	--	--	--	203
Carbón II	297	--	--	--	297
Zone III	--	36	56	200	292
Zone IV	100	24	--	200	324
Zone V	--	--	--	100	100
Total	600	60	56	500	1,216

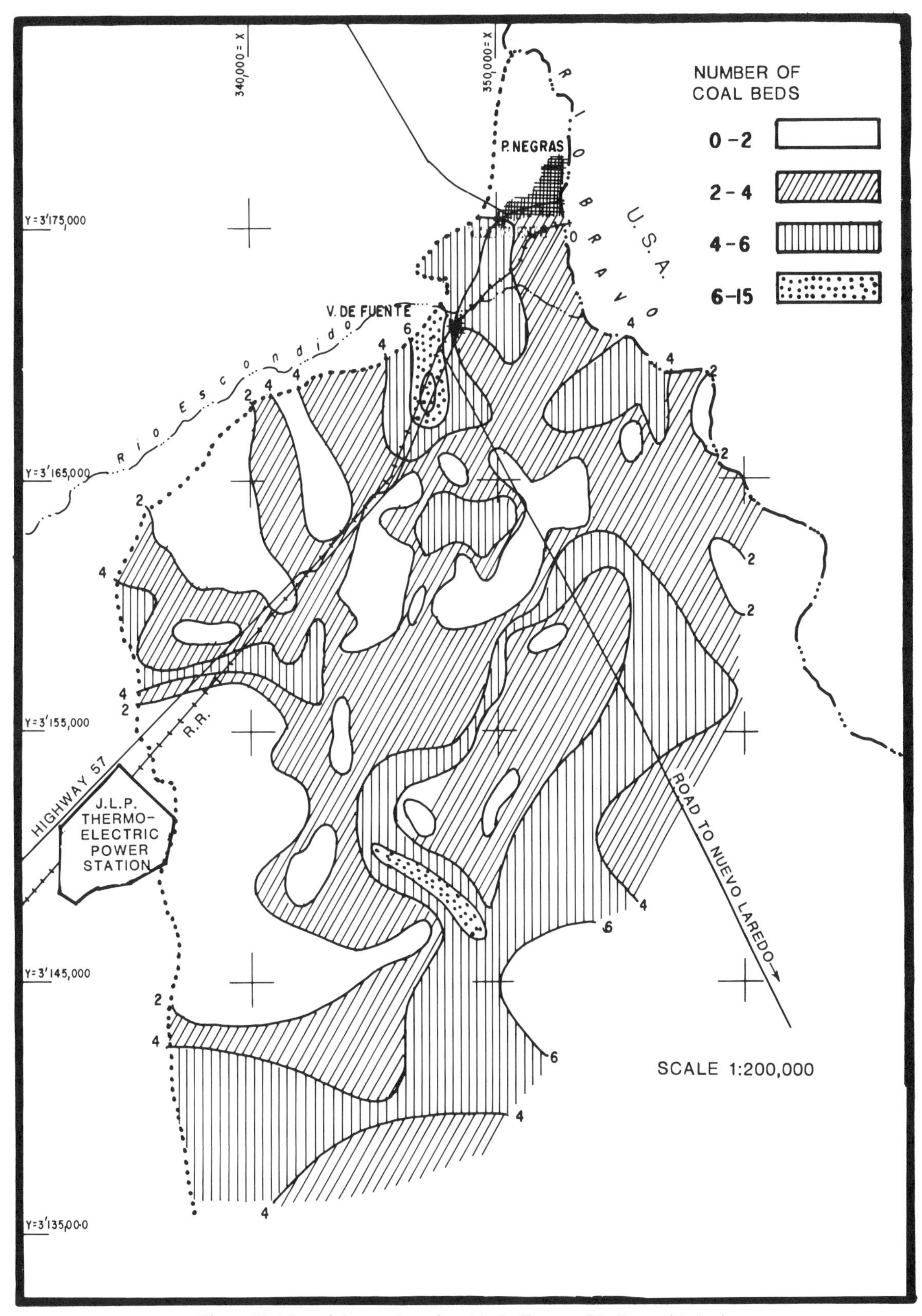

Figure 19. Map of the number of coal beds, Fuentes–Río Escondido basin.

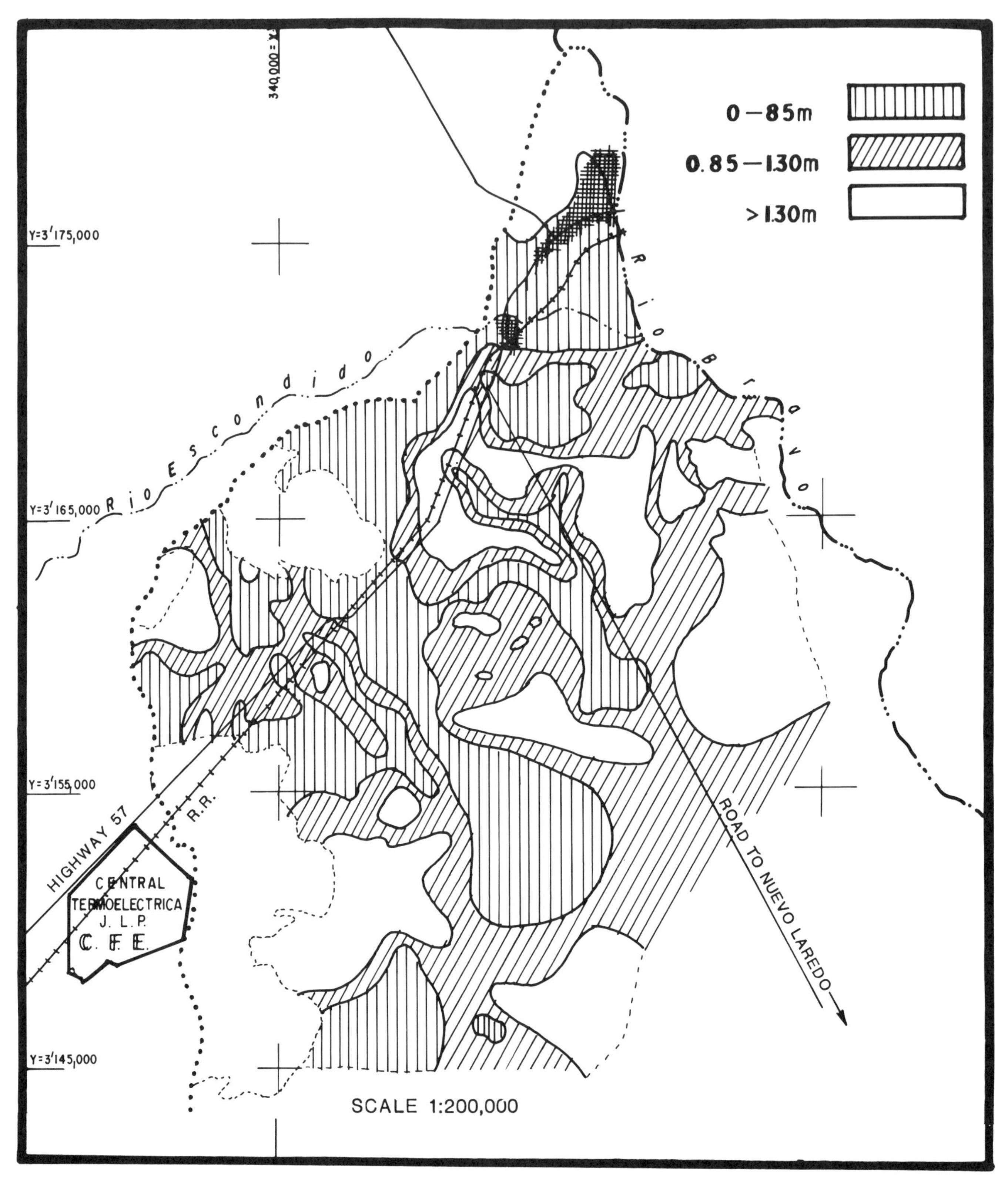

Figure 20. Map of total coal thicknesses, Fuentes–Río Escondido basin.

TABLE 5. HYDROCARBON POTENTIAL OF THE RIO ESCONDIDO COAL FIELDS

Unit	Reserves Category	Area (km^2)	Average bed Thickness (m)	Porosity (%)	Water Saturation (%)	Oil Reserves*
San Miguel Formation Sandstones	Possible	70	6	6.5 to 27.8	1 to 0.47	287.9

*Million barrels (36°API)

OTHER RESOURCES

As geological information became available in the course of evaluating the coal reserves, other resources that can be utilized simultaneously with the development of the coal have been defined.

Hydrocarbons

During exploration for the Project III coal in 1979, hydrocarbons were accidentally encountered at shallow depths (250 to 500 m). This led to definition of the geometry of these reservoirs simultaneously with the coal-exploration activities. The hydrocarbons are stratigraphically trapped in the San Miguel Formation and eventually in the Olmos Formation paleochannel facies. By 1982, drillholes ED-256, ED-259, ED-280, ED-284, ED-285A, ED-288, ED-289, ED-292, and ED-295, all of which had oil shows, were transferred to PEMEX. To date, 11 wells have been drilled, three of which produce an average 15 bbls/day.

An interinstitutional technical support project has been set up to explore for oil and coal in the basin. At present, PEMEX carries out a 16-well program in the areas of Projects III and IV. The possible hydrocarbon potential of the Río Escondido reservoir is shown in Table 5 (Rodríguez, 1984).

Tar sands, classified as heavy hydrocarbons, are known to be under development by Exxon in Maverick County, Texas, 45 km east of Piedras Negras, Coahuila, where 3×10^9 barrels of −2° API crude have been evaluated and are being developed using FAST (Fracture-Assisted Steam Flood Technology). In this process, steam is injected through pilot wells, after previously increasing the permeability by inducing fracturing through sand and water injection. The project comprises two injection wells, four producers, and three observation wells. The required steam can be generated by combustion of coal extracted from the same basin.

Caliche

This is a calcium carbonate–rich deposit formed in soils of semiarid regions, where capillary action and evaporation allow vertical flow of moisture, and carbonate precipitates toward the surface. In the present area, irregular masses of caliche develop in the upper part of the Sabinas-Reynosa Conglomerate. Data from 112 coal exploration drillholes allowed the evaluation of coal reserves that could be open-pit mined to 40-m depths at the Tajo-I and Tajo-II (Cuts I and II) sites. Thickness and distribution of the conglomerate caliche overburden was defined at the same time at both localities, and outlined an additional mining subproduct. ASTM standards were applied to characterize the deposit; the following parameters were determined by wet method laboratory analyses of the samples:

Total moisture:	0.71%
Density:	2.6 %
Loss by ignition:	35 %
Insoluble residue:	13.2 %
Calcium carbonate:	77 %
Magnesium carbonate:	1.7 %
Iron oxide:	0.57%
Aluminum oxide:	2.6 %
Silicon oxide:	17.2 %
Sulfur dioxide:	0.27%

Caliche-conglomerate reserves are 121 million tons at Tajo-I and 163 million tons at Tajo-II, for a total of 284 million tons of material. Cement is produced industrially by flotation, which enriches the CaO content of the caliche. The coal can be used in the clinker-baking process.

Fly ash

The principal additional components required for Portland cement manufacture are silica, alumina, and iron, applied in specific proportions by adding shale to the limestone when cement is made by conventional methods. At Río Escondido the CFE obtains fly ash in the coal combustion process for power generation at the José López Portillo Power Plant. Once the plant becomes fully operational it will produce 1.2 million tons of ash per year from combustion of 12,000 tons of coal per day.

The combustion residue may be classified according to its nature as fly ash, bottom ash, and furnace slag; 75 percent of the total residue volume will be fly ash, constituting a production of 900,000 tons. Bottom ash and furnace slag make up the remaining 25 percent: 300,000 tons/year.

According to its production volume and granulometry the chemical composition of the fly ash is as follows:

SiO_2	63.2 %	MgO	1.04%
Al_2O_3	23.5 %	K_2O	0.09%
Fe_2O_3	5.6 %	Na_2O	0.74%
CaO	1.38%	SO_3	1.35%

In general, the main components of a conventional cement production process are high calcium carbonate–limestone and clays with 3 to 8 percent alumina, 17 to 25 percent silica, and 2 to 6 percent iron.

GEOHYDROLOGY

The Río Escondido Project area is enclosed within the Acuña-Laredo hydrologic subbasin of the Rio Bravo Basin. The basin's stratigraphic section contains as yet unevaluated major aquifers in Lower Cretaceous limestones and at shallow levels in the Sabinas-Reynosa Conglomerate. Exploratory drilling for coal encountered the San Miguel, Olmos, Escondido, and Sabinas-Reynosa Conglomerate Formations. The three former were found to be impermeable or nonaquiferous units. They unconformably underly the Sabinas-Reynosa Formation, a regionally important aquifer consisting of conglomerates, gravels, and caliche with an average 30-m thickness, which is found at less than 40 m depths and is saturated with calcium-bicarbonate waters. The aquifer is in a granular deposit of varying distribution and petrophysical character. This generates anisotropic conditions and consequently widely variable hydrodynamic factors that make it difficult to evaluate the aquifer capacity.

Another major aquifer has been identified in limestones of the Lower Cretaceous Salmon Peak and McNight Formations. In the coal basin these units occur at levels 1,000 m below the coal zones and crop out in the Burro Range foothills east of the area. Important water volumes can probably be extracted from this aquifer.

Geohydrological study of the area is presently being undertaken by the CFE to evaluate both aquifers. The data will also help to eliminate infiltration problems during the construction and development of the mines and to determine the possible effects of contaminating agents on the regional ground waters.

REFERENCES CITED

Alpern, B., 1980, Essai de classification des combustibles fossiles solides: Documents Techniques CdF, no. 3, p. 195–210.

Bose, E., and Cavins, O. A., 1927, The Cretaceous and Tertiary of southern Texas and Northern Mexico: Texas University Bulletin 2748, p. 7–142.

Dumble, E. T., 1892, Report on the brown coal and lignite of Texas; Character, formation, occurrence, and fuel uses: Austin, Texas, Bureau of Economic Geology, 243 p.

Evans, T. J., 1974, Bituminous coal in Texas: Austin, Texas, Bureau of Economic Geology Handbook 4, 65 p.

Humphrey, W. E., 1956, Tectonic framework of northeast Mexico: Gulf Coast Association of Geological Societies Transactions, v. 6, p. 25–35.

Lobeck, A. K., 1939, Geomorphology, an introduction to the study of landscapes: McGraw Hill Book Company, Inc., 731 p.

Maxwell, R. A., 1962, Mineral resources of south Texas: Austin, Bureau of Economic Geology, University of Texas Report of Investigations, no. 43, 140 p.

McBride, E. F., and others, 1975, Deltaic and associated deposits of Difunta Group (Late Cretaceous to Paleocene), Parras and La Popa basins, northeastern Mexico, *in* Broussard, M. L., ed., Deltas models for exploration: Houston Geological Society, p. 485–522.

Obregón, A. L., 1981, Genesis y clasificación del Carbon: Mexico D.F., IV Reunion Nacional de Geológia y Geotérmia, p. 18.

Rodriguez, M. J-M., 1983, Presencia de Hidrocarburos en zonas de carbon en la Sub-Cuenca de Fuentes Río Escondido: Piedras Negras, Coahuila, México, Symposium Latinoamericano del Carbon, p. 9.

Schmitz, E. J., 1885, Geology and Mineral resources of the Rio Grande region in Texas and Coahuila: American Institute of Mining Engineering Transactions 13, p. 388–405.

Shumard, B. G., 1860, Observations on the Cretaceous strata of Texas: St. Louis Academy of Sciences Transactions, p. 582–590.

Smith, C. I., 1970, Lower Cretaceous stratigraphy, northern Coahuila, Mexico: Austin, University of Texas Report of Investigations, no. 65, 101 p.

Stephenson, L. W., 1918, A contribution to the geology of northeastern Texas and southern Oklahoma: U.S. Geological Survey Professional Paper 120, p. 129–163.

Vanderpool, H. C., 1930, Cretaceous Section of Maverick County, Texas: Journal of Paleontology, v. 4, p. 252–258.

Printed in U.S.A.

The Geology of North America
Vol. P-3, Economic Geology, Mexico
The Geological Society of America, 1991

Chapter 10

Summary of exploration and development at Río Escondido

F. Verdugo D., and C. Ariciaga M.
Department of Coal Studies, Comisión Federal de Electricidad, Río Rodano 14, 06500 México, D.F.

Minera Carbonífera Río Escondido, S. A. (MICARE) is a mostly state-owned concern established August 2, 1977, with the main objective of coal development for power generation. Present coal reserves assigned to the company are located in the Fuentes–Río Escondido basin in the northeastern part of Coahuila, and the Colombia basin in the northern part of Nuevo León; reserves in the latter, however, have not been completely evaluated. The present chapter refers exclusively to the Carbon I and Carbon II projects in the Fuentes–Río Escondido basin.

The notion of creating MICARE originated after the discovery of coal potential in the vicinity of Nava, Coahuila, during the 1960s. The properties of this coal (sub-bituminous, long-flame) preclude its use as coke for the steel industry; the possibility was then considered of using it as fuel for power generation. The first coal-fueled power plant in Mexico began to operate in 1964 with coal from a small underground mine northwest of the present Mine I facilities (Fig. 1). The venture was successful and inspired exploration over a more extensive area, as well as the conception of a vast project capable of generating approximately 10 percent of the entire power demand in Mexico by the year 2000.

After eight years of activity and large exploration programs, carried out on the one hand by Estudios Carboníferos del Noreste (Northeast Coal Studies), a dependency of the Federal Commission on Electricity, and by the MICARE Geological Department on the other, a large volume of reserves has been quantified, which will allow the operation of two coal-power plants for the next 30 years. Tables 1 and 2 show volumes computed for each of the project areas shown on Figure 1.

The exploration program and evaluations for this project are probably among the most important accomplishments in the field of earth sciences of Mexico during the past few years. Interdisciplinary groups participating in the project performed a variety of studies, from palynology of sediments associated with the coal deposits, to geostatistical analysis of well data, including, of course, geophysical information, to define the origin of the deposits as well as to establish norms and procedures for the most efficient exploration and reliable calculations of reserves. In brief, it may be said that the project has been the result of applying all the geological disciplines and the most modern techniques.

With regard to cost of power generation, Table 3 is self-explanatory.

TABLE 1. CALCULATION OF RECOVERABLE COAL FROM CARBON I AREAS (FIG. 1)

Zone	Area (hectares)	Volume Calculation Dec. 1984	Identified Reserves (tons)	Recovery Factor (%)	Expected Recovery (tons)
Mine I	1,085	30,857,224	25,197,038	79.1	19,930,857
Mine II	1,166	43,686,568	34,619,608	82	28,388,078
Mine III	1,163	30,322,957	30,322,957	79.1	23,985,458
Cut III	278	9,238,100	9,238,100	95	8,776,195
Cut IV	325	8,594,917	8,594,917	95	8,165,171
Mine IV	1,335	36,455,746	30,702,289	79.1	24,285,510
Cut V	289	6,679,806	6,679,806	95	6,345,816
Cut I	329	11,250,447	11,250,447	95	10,687,925
Cut II	514	15,692,184	15,692,184	95	14,907,575
Total	6,484	192,777,949	172,297,346		145,472,585

TABLE 2. CALCULATION OF RECOVERABLE COAL FROM CARBON II AREAS (FIG. 1)

Zone	Area (hectares)	Geological Reserves (Dec. 1983)	Recovery Factor (%)	Expected Recovery (tons)*
Sector 1	1,041	22,666,526	65.0	14,732,242
Sector 2	813	21,114,552	65.0	13,724,459
Sector 3	636	14,959,374	65.0	9,723,593
Sector 4	1,869	48,587,291	65.0	31,581,739
Sector 5	200	4,377,353†	65.0	2,845,280
Sector 7	646	15,114,333†	65.0	9,824,317
Sector 6	1,323	27,581,145	95.0	26,202,088
"ED–V"	....	72,000,000	65.0	46,800,000
Total		226,400,574		155,433,718

*Recovery factor was applied.
†Reserves as of 12/79.

Verdugo D., F., and Ariciaga M., C., 1991, Summary of exploration and development at Río Escondido, *in* Salas, G. P., ed., Economic Geology, Mexico: Boulder, Colorado, Geological Society of America, The Geology of North America, v. P-3.

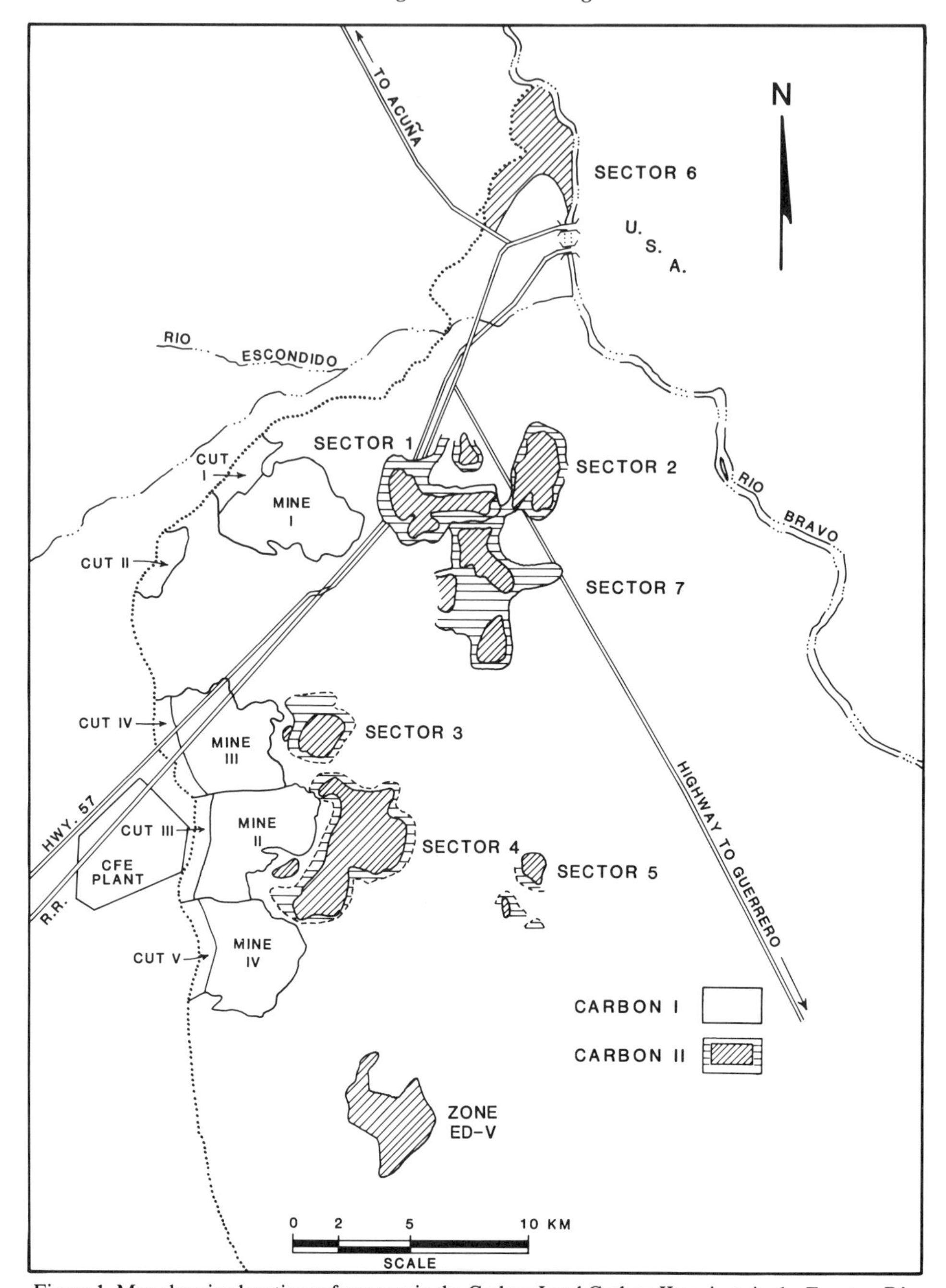

Figure 1. Map showing locations of reserves in the Carbon I and Carbon II projects in the Fuentes–Río Escondido basin.

TABLE 3. SALES OF COAL TO CFE

Year	Sales (thousands of tons)	Approximate Value of Sales (billions of pesos)
1980	316	1,600
1981	756	550
1982	601	675
1983	1,154	5,000
1984	1,560	9,550
Total	4,387	16,000*

*Approximate dollar equivalent $150 million.

Printed in U.S.A.

The Geology of North America
Vol. P-3, Economic Geology, Mexico
The Geological Society of America, 1991

Chapter 11

Geology and reserves of coal deposits in Mexico

E. Flores Galicia
Consejo de Recursos Minerales, Centro Minero Nacional, Carretera México-Pachuca, Pachuca, Hgo., Mexico

SABINAS DISTRICT: MONCLOVA (COAHUILA)

INTRODUCTION

The Coahuila Coal District (Sabinas area) is located in the east-central part of the state of Coahuila between 26°45′ and 27°70′N, and 100°15′ and 102°00′W in the Coahuila Basin and Range physiographic province. The area covers approximately 30,000 km^2, and the existence of coal has been established in subbasins of widely varying dimensions: La Esperanzas, Sabinas, Saltillito-Lampacitos, San Salvador, Castaños, Las Adjuntas, Monclova, and San Patricio (Fig. 1). The Río Escondido coal basin, to the northeast, is discussed in Chapter 9 of this volume.

The region is one of extensive plains and bolsons within which the coal-bearing subbasins occupy synclinal structures separated by steeply dipping anticlinal limestone ranges. Climate and vegetation are semidesert; as typical for northern Mexico, seasonal temperature differences are pronounced. The dentritic, generally south-southeast drainage is controlled by Ríos Alamo, Salado, Monclova, and San Juan, draining into the southeast-flowing Río Sabinas.

Access is by Federal Highway 57 (Mexico to Piedras Negras), which runs across the area from south to north, and by other secondary roads. The Mexican National Railway has a general transport rail running parallel to Highway 57, as well as numerous branches and stations at extraction centers north of the area.

HISTORICAL GEOLOGY

Jurassic

Deposition probably began in the Callovian, when the paleogeographic elements developed that defined the morphology and influenced deposition throughout almost the entire history of the Sabinas Gulf. These elements include (Fig. 2): (1) the Tamaulipas peninsula, a large unit of Paleozoic or older deformed gneisses and granites marking the north coast of the Gulf; (2) Coahuila Island marks the south coast and consists of the same type of material; locally, (Delicias area) slightly deformed Carboniferous and Permian sediments are also found; (3) La Mula Island, a small north-south–trending unit, half-blocked communication with the open seas toward the southeast.

Upper mid-Jurassic deposition was controlled by these elements. Thus, Oxfordian clastic sediments were laid down at the foothill areas of the high plateaus that were their source (La Gloria Formation); deposition continued toward the coastline as a complex of oolitic bars (Zuloaga Group, in part) and as evaporites on the shelf area (Olvido Formation). This arrangement continued during the Kimmeridgian and the Tithonian with no major changes: terrigenous deposits were laid down at the margins (La Casita Group), and fine sands east-southeastward in the direction of the open seas (La Caja and La Pimienta Formations).

Cretaceous

The same paleogeographic elements controlled the entire Lower Cretaceous deposition, which is therefore very similar to that of the Jurassic. At the northern edge of Coahuila Island, continental arkose sediments derived from it continued to accumulate (San Marcos Formation), and lithologically similar deposits occur to the south of Tamaulipas Island (Hosston Formation). The eastern and southern margins of the Gulf were closed by a thick reef (Cupido reef complex) of mostly oysters and rudists; thus, behind the barrier and within the Sabinas Gulf, lithologically very variable beach sediments accumulated. These have been subdivided into a multitude of generally thin and areally restricted formations (Menchaca, Barril Viejo, Padilla, La Mula, and La Virgen Formations). Fine basin sediments (lower Tamaulipas and Taraises Formations) occur in front of the reef and offshore. The reef and associated facies deposits ceased accumulating in the lower Aptian. Filling of the Sabinas Gulf was probably complete at that time. Then, maximum transgression (lower Aptian to lower Cenomanian) deposited calcareous units, the most extensive of the entire Cretaceous (La Peña, Aurora, and Kiamichi Formations); the latter contains restricted local reef developments (Monclova Limestone).

A general west to east regression began in the Turonian and makes up at least the northern half of the Mexican Geosyncline;

Flores Galicia, E., 1991, Geology and reserves of coal deposits in Mexico, *in* Salas, G. P., ed., Economic Geology, Mexico: Boulder, Colorado, Geological Society of America, The Geology of North America, v. P-3.

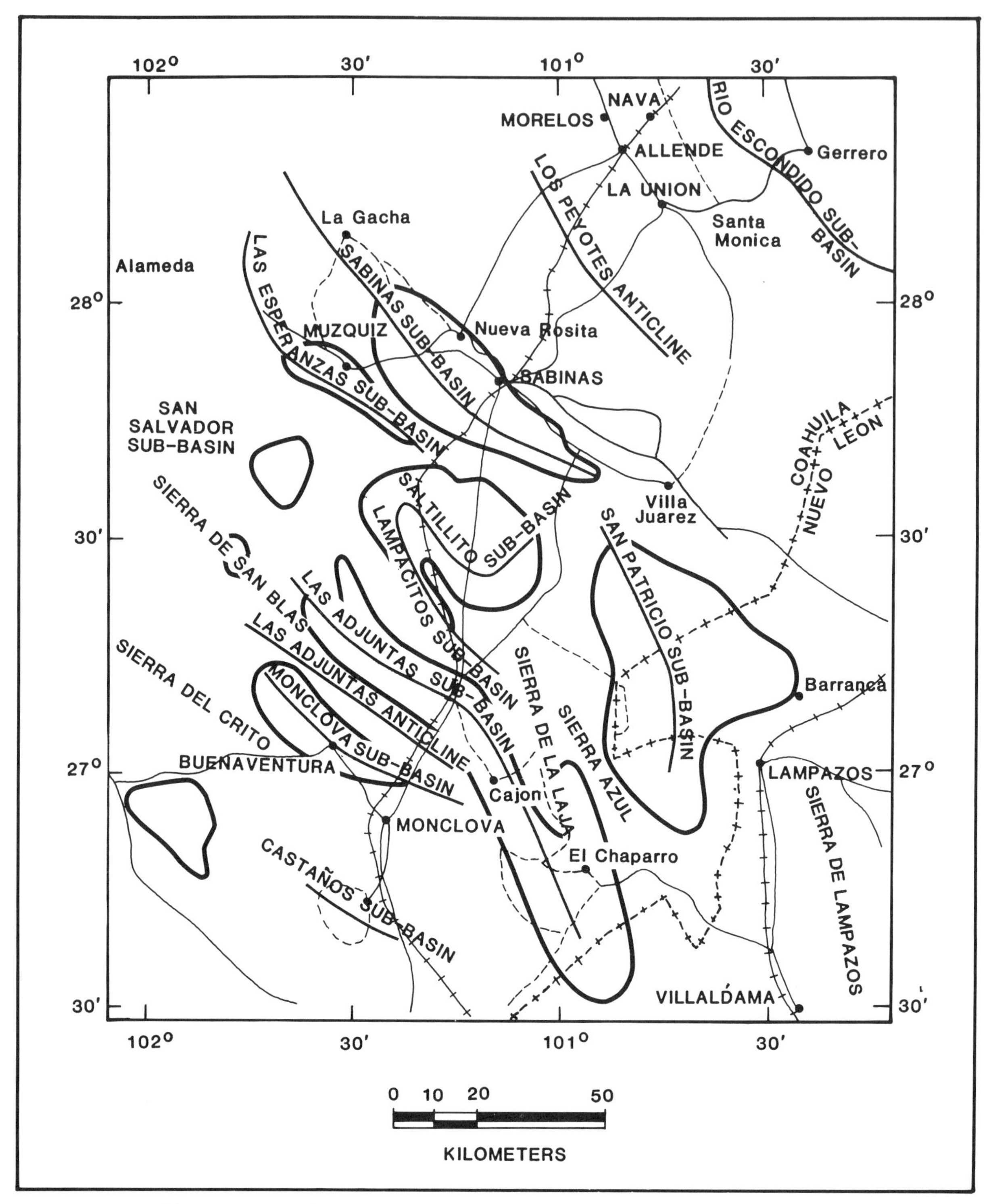

Figure 1. Map showing distribution of coal subbasins in the Sabinas-Monclova coal district.

this improtant episode culminated in the formation of coal deposits. Initially only an increased content of finer sediments within the Turonian, Coniacian, and Santonian calcareous formations (Eagle Ford and Austin Formations) is seen. In time the regression became widespread, and consequently so did the fine-grained deposits. These may be interpreted as the most basinward deposits of an advancing prograding delta (Upson Formation).

Fine-grained deposits became dominant at the end of the Campanian when the units become markedly diachronous; a fairly continuous sandstone extending over the entire area is taken

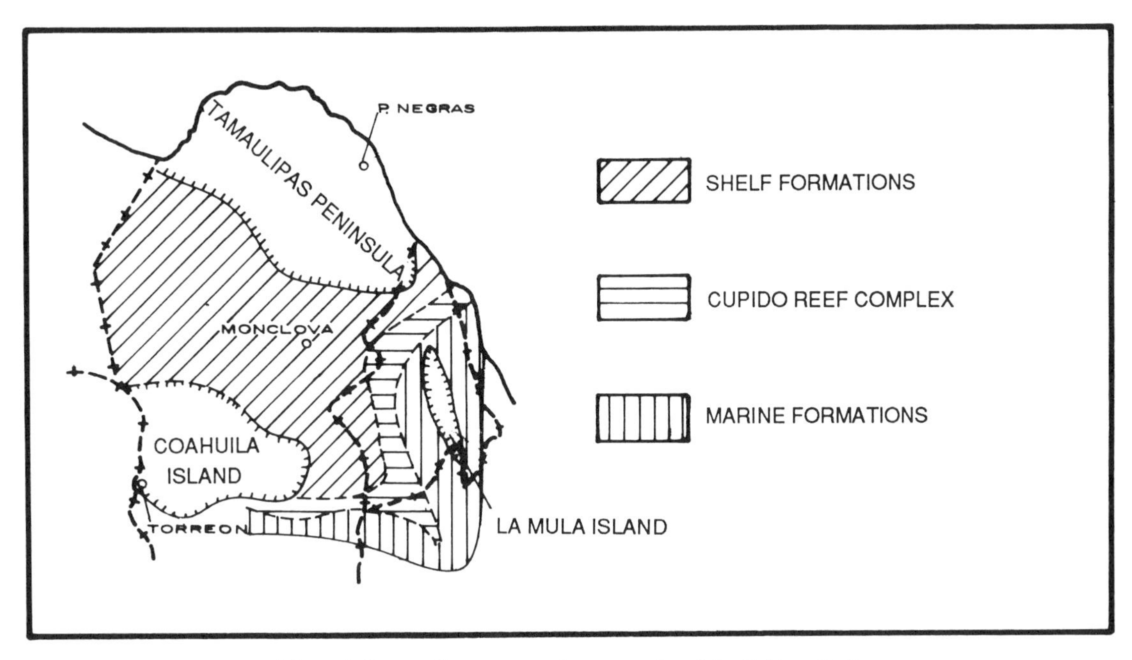

Figure 2. Lower Cretaceous paleogeography of the Sabinas Gulf.

to be the frontal deposits of the deltaic complex (San Miguel Formation). Directly overlying these deposits, a 10- to 40-m-thick sequence containing the coal seams in question is considered to be typical of a "wave-dominant" delta plain (base of Olmos Formation). The lobes of this deltaic complex lack continuity; a great many of them trend northwest-southeast, are laterally restricted, and are connected by strand plain and perhaps lagoonal deposits. This is important, because the morphology and dimensions of the lobes control the coal thicknesses. Thus, the thickest and most uniform seams (averaging 1 or 2 m in thickness) are found above the lobes, with the lower parts of the seams immediately behind the sands at the top of the San Miguel Formation; there is normally one seam to a lobe. In contrast, frequent marine invasions fostered discontinuous vegetation in areas between the lobes; consequently, the vertically numerous coal seams rarely attain thicknesses of more than 1 m.

Three major paleogeographic areas may be defined within the Coahuila Coal District at the deltaic lobe level (Fig. 3) from west to east. In the first of these, the lobes are only slightly developed, and continental facies largely predominate. A good lobe development is found in the central part, where marine and continental conditions are evenly balanced; almost all of the coal reserves are located there. Toward the east, strong subsidence led to repeated marine incursions onto the upper shelf area of the lobes and therefore largely disallowed coal development. Above this interval, the remainder of the Olmos Formation is continental and contains fluvial facies (Fig. 3).

Mesozoic deposition ended in late Maastrichtian time with renewed marine transgression from the east-northeast. Shallow beach bar and small reef deposits (Escondido Formation) overlie the Olmos continental deposits with marked regional unconformity (Fig. 4), signalling the generalized seesaw movements that heralded the Laramide orogeny.

Tertiary

Subsequent to the folding, lesser developed lacustrine environments occupied the center of the bolson areas where clays, gypsum, and travertine (San Buena Formation) accumulated during the Tertiary; conglomerates and calcareous sands were laid down at the foothill margins of the limestone ranges.

Quaternary

During this period, rivers reworked and partly mobilized the Tertiary lacustrine deposits; some volcanism (basalt flows) took place toward the east-central part of the Sabinas subbasin.

STRATIGRAPHY

The Coahuila Coal District geographic area is closely related to the Sabinas Gulf, a major paleogeographic feature. The following descriptions of coal deposits are given in the context of their geologic framework, and in stratigraphic order.

Cretaceous

The general Cretaceous sequence of the Sabinas Gulf area is well known from many publications and is therefore excluded here, except for the Upper Cretaceous units that are closely related to the coal deposits.

San Miguel Formation. At its type-locality, the old San Miguel Ranch on the Río Grande, north of Eagle Pass, Texas, the

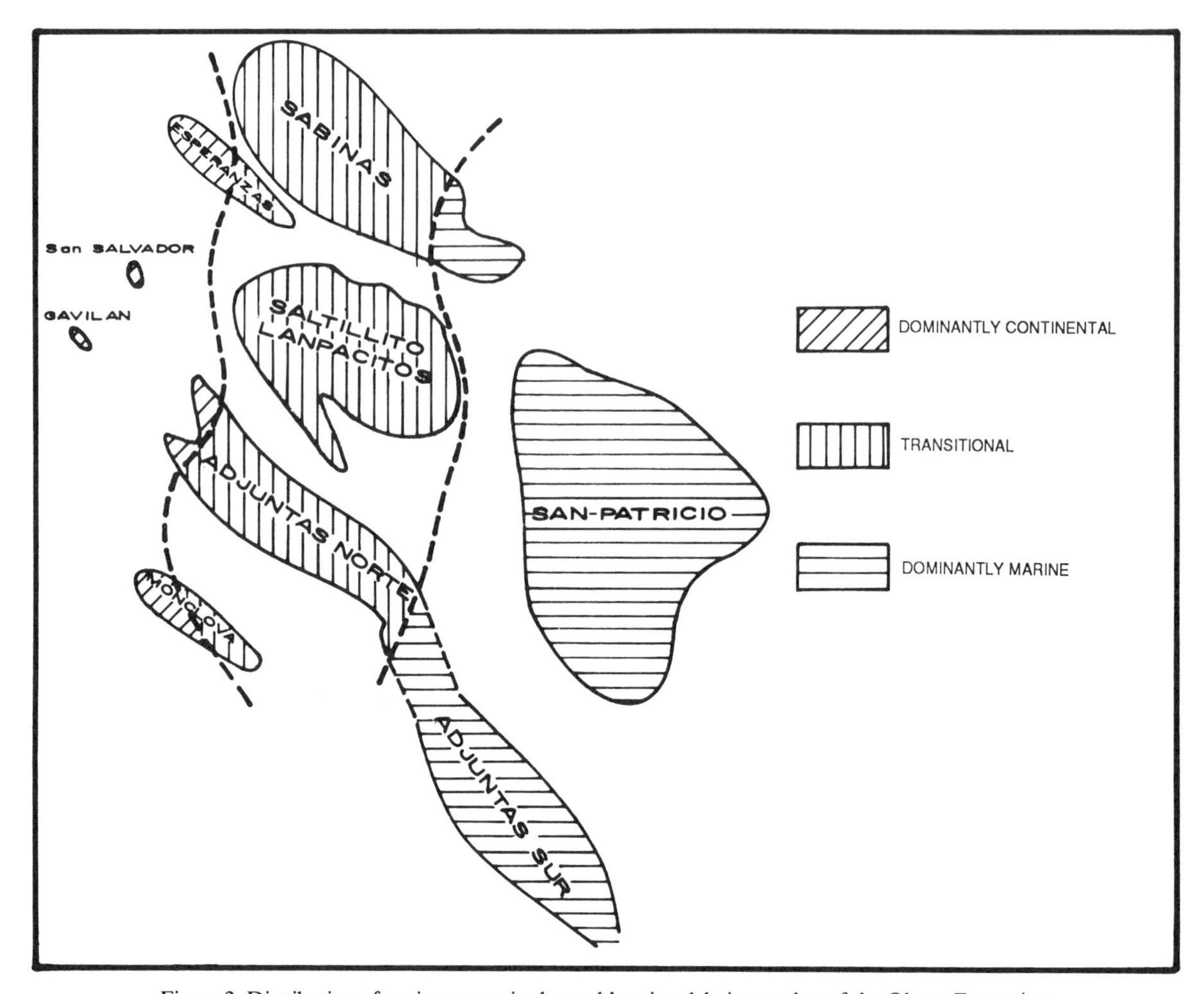

Figure 3. Distribution of environments in the coal-bearing deltaic member of the Olmos Formation.

San Miguel Formation crops out in the Las Vacas Canyon southeast of the Don Martín Dam, the El Cedral Canyon, and along the slopes of the Sierras La Rata, Ovallos, Sardinas, El Cristo, and La Gloria. The formation has been divided into five lithologic members from base to top as follows:

Member No. 1. Fossiliferous concretions at the base grade upward to mudstones; limestone lenses and thin beds occur near the base; fossils are found in both the clayey and the calcareous portions. At El Cedral it is 30 m thick.

Member No. 2. Consists of siltstone beds 0.05 m thick, some of which are unstratified. The total thickness of the member is 45 m.

Member No. 3. A basal calcareous sandstone is followed by unstratified siltstone with some clayey sandstone interbedded with massive siltstone; the unit changes gradually upward to siltstone with some sandstone, and then to siltstone with very fine sand; thickness of these beds varies from 0.02 m to 1 m; their resistance to erosion allows them to be found locally in outcrop. The unit is 68 m thick.

Member No. 4. Massive siltstones, generally hidden owing to their poor resistance to erosion. Thickness is 87 m.

Member No. 5. Light gray, fine- to coarse-grained, poorly sorted sandstone varying laterally to unstratified siltstone. Diamond-bit cores show the perfectly defined separation of the interbedded light gray sandstone and dark siltstone in layers 5 to 10 cm thick. These are topped by a massive, light gray, coarse-grained sandstone bed, locally containing a 10- to 20-cm-thick coal seam. Total thickness of the member is 30 m.

Average thickness of the San Miguel Formation is 260 m, measured at El Cedral (Robeck and others, 1956). It conformably overlies the Upson Formation and underlies the Olmos Formation, also conformably. *Ostrea saltillensis, Sphenodiscus* Meek, and *Coahuilites* Böse have been identified, and an uppermost Campanian age has been assigned to the unit on the basis of its stratigraphic position (Fig. 5).

Olmos Formation. This formation crops out at Arroyo El Saúz south of the Don Martín Dam, at El Cedral, and in all the coal region subbasins; outcrops occur also in the Ojinaga Basin and San Pedro Corralitos (Chihuahua), usually covered by alluvium.

Lithologically, the Olmos Formation consists of alternating sandstones, siltstones, shales, and coal seams at the base, in cyclic

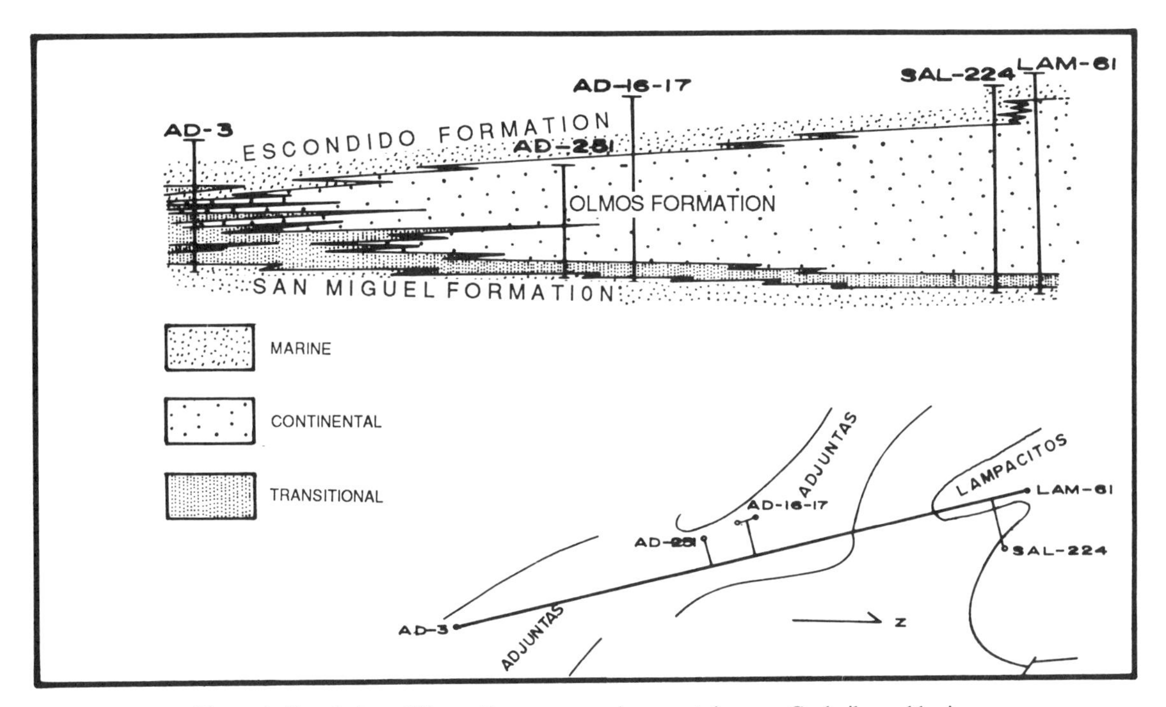

Figure 4. Correlation of Upper Cretaceous environments in some Coahuila coal basins.

sequences that are usually incomplete or truncated (Fig. 6). The facies is interpreted basically as that of a point bar fluvial sequence with fills and erosional paleochannels accompanied by sand deposition, and a channel lag of soft shale pebbles and lesser amounts of wood and bones. Upward the cycle lacks channels due to loss of hydrodynamic competence, and current-rippled fine sandstones and silts develop. Finally, dewatering of the shales, ponding, and soil development took place on the flood plain, as shown by the occurrence of a number of whitish calcareous nodule levels.

The lower part of the sedimentary section is shalier, indicating lengthier flood conditions by reason of the more gently sloping fluvial plain. Green and red colors predominate in the shale intervals upward in the section and point to a progressively dryer climate and consequently diminished possibilities for the preservation of vegetable matter.

The coal zone, located at the base of the Olmos Formation, is represented by light to dark gray shales with layers of carbonaceous black shale, interbedded with light greenish gray siltstones and lesser volumes of fine- to medium-grained clayey sandstone. The interbedded coal seams range from a few centimeters to 1 m or more in thickness. Among them is a level of fire-resistant clay providing high radioactive values. Thicknesses of up to 280 m have been reported.

The Olmos Formation overlies the San Miguel Formation conformably and underlies the Escondido Formation unconformably. Specimens of the ammonite *Sphenodiscus* sp. and *Exogyra costata* and some gastropods have been identified; it is assigned to the Maastrichtian. The formation is equivalent to the lower part of the Navarro Group of Texas and northeast Mexico and the upper Méndez Formation of the Tampico-Misantla Basin.

Escondido Formation. The first thick, massive, hard sandstone overlying the Olmos Formation marks the basal unit of the Escondido Formation. The contact is well exposed at the hills in the eastern part of Eagle Pass, Texas. The formation includes dark clays and marls interbedded with more or less thick sandstones, limestones, and resistant fossiliferous layers. Gomez and Evaristo (1983) considered the base of the Escondido Formation to be at the top of greenish gray shales with mottled reddish and purplish layers that characterize the top of the Olmos Formation. The formation crops out mostly at Mesa Cartujanos northeast of Lampazos (Nuevo León) and in elongated lenses within the valleys following the synclinal structures.

EPOCH	AGE	TEXAS AND NE MEXICO	COAHUILA
TER-TIARY	PLIOCENE	SABINAS CONGLOMERATE	SABINAS CONGLOMERATE
LATE CRETACEOUS	MAASTRICHTIAN	NAVARRO GROUP	ESCONDIDO FORMATION
			OLMOS FORMATION
	CAMPANIAN	TAYLOR GROUP	SAN MIGUEL FORMATION
			UPSON CLAY

Figure 5. Correlation of the Cretaceous formations of northeastern Mexico.

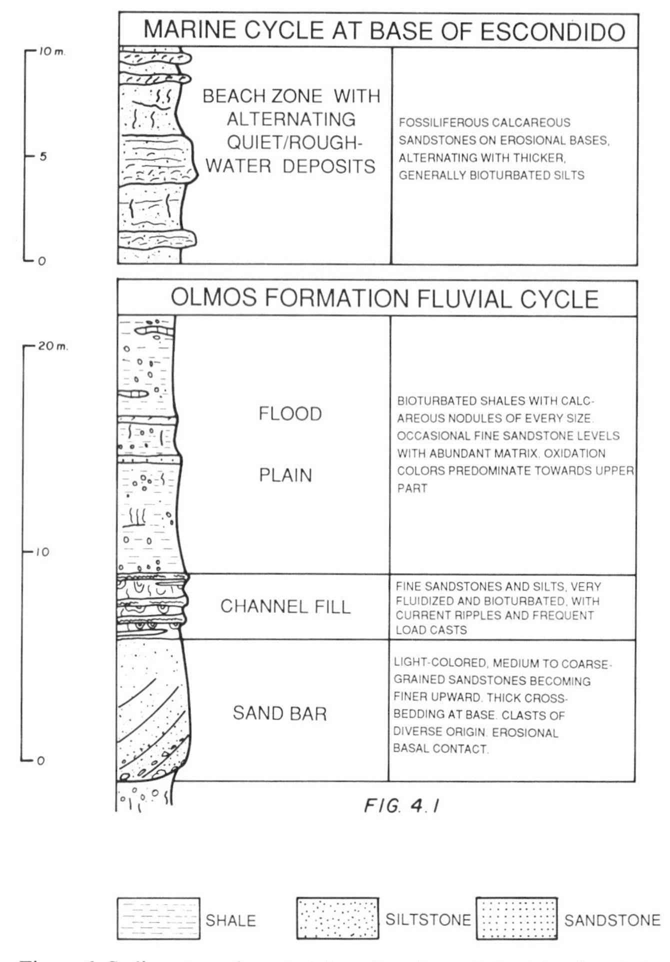

Figure 6. Sedimentary characteristics of marine and fluvial cycles of the Escondido and Olmos Formations.

The base of the Escondido Formation consists of greenish gray and purplish gray shales with friable reddish brown horizons. This sequence is very similar to the top of the Olmos Formation, differing only in the typically reddish color caused by oxidation in the latter. The section continues upward with purplish gray and reddish brown siltstones and shales alternating with medium-grained, poorly sorted greenish gray sandstones. A 40-m-thick sandstone crowning the section was thought to represent the top of the Escondido Formation, until an exploratory drillhole (AD-116) encountered an overlying greenish gray shale now assumed to be the top; this shale is usually not exposed because of its weak resistance to erosion. Both gradual and erosional contacts are frequently observed in this formation, which is 395 m thick.

Tertiary

Sabinas Conglomerate. This unit includes conglomeratic deposits of the Sabinas area, which crop out over the entire Coahuila coal province, developing fans and terraces. The unit consists of limestone, sandstone, and intrusive and extrusive rock gravels and pebbles cemented by calcium carbonate and clay. It is up to 30 m in thickness and unconformably overlies the Austin, Upson, San Miguel, and Olmos Formations. It has been assigned to the Pliocene and correlated with the Santa Inés Formation exposed in Durango. In Texas it is termed the Uvalde Gravel.

San Buena Limestone. A travertine interval cut by exploration drilling in the Adjuntas subbasin has been named the San Buena Limestone; exposures have been found in Sierra La Bartola between Monclova and Frontera Cañon de San Marcos; outcrops have also been reported south of Cuatro Ciénagas. This is a very porous light-colored limestone (travertine) representing a shallow fresh-water environment where erosion was remarkably intense; in Sierra La Bartola, it ranges in thickness from 6 to 42 m. Locally it conformably overlies the Sabinas Conglomerate; otherwise, it is unconformable above the Escondido, Olmos, and San Miguel Formations. In some outcrops it is found in contact with Eagle Ford altered shales.

The San Buena Limestone is considered to be contemporaneous and correlative with the Providencia limestones reported from the Linares area. On the basis of its unconformable stratigraphic position it has been assigned to the Pliocene.

Quaternary

Alluvium. Quaternary alluvial beds of recent conglomerates, gravels, sands, silts and clays, soils, and caliche presently undergoing calcareous cementation, occur throughout the entire Sabinas-Monclova Coal District. These deposits are probably fluviolacustrine sediments accumulated in temporary lakes and swamps developed in the bolson topography that prevailed at the beginning of the Quaternary.

COAL DEPOSITS

The coal deposits are interbedded with carbonaceous shales, siltstones, and sandstones within synclinal structures developed in the sedimentary rocks. Their distribution is extremely irregular, since they appear in very lenticular layers. The structural behavior of the coal seams is closely related to the syncline that contains them and is described here as observed in the various subbasins of the Sabinas-Monclova Coal District.

The coal deposits of the Sabinas subbasin consist of a series of seams a few centimeters to 1.8 m thick, interbedded with shales, siltstones, and sandstones in the so-called "coal zone" at the base of the Olmos Formation, which ranges in thickness from 13 m at the southeasternmost end to 27 to 30 m in the northwest. The number of known seams ranges from a maximum 15 to a reported minimum of two, possibly due in such cases to having been eroded and replaced by sandstone. This leads to the assumption that in these instances the drill encountered ancient distributary paleochannels.

The most persistent deposit is the double seam, containing the thickest and economically most prospective intervals. The double seam increases in lenticularity and sediment content in the northwest portion of the basin, where facies changes therefore seem to have been more constant and more prevailingly continental.

The double seam typically shows an intercalated, fire-resistant clay (tonstein), megascopically appearing as an ochre-yellow shale or siltstone a few centimeters to 30 cm thick. Petrographic studies report a pyroclastically structured clay composed of rock fragments, montmorillonite aggregates, devitrification products, and scarce quartz and plagioclase crystals reworked from the older vitreous matrix, and classify it as a lithic tuff of dacitic composition (tonalite-dacite series). This double seam is a very useful stratigraphic marker for correlation of the commercial coal seams.

In the Saltillito-Lampacitos, Adjuntas, and Monclova subbasins the coal seams display the same features as in the Sabinas subbasin, except for their diminishing continuity and thickness from north to south and the consequent thinning of the commercially significant interval, as shown in the chapter on reserves. The remaining subbasins, such as Esperanzas, San Salvador, and El Gavilan, show similarity in their coal seam distribution, although they are geomorphologically different due to the continental environments prevailing in these areas. Finally, the southeast Adjuntas and San Patricio subbasins are of predominantly marine facies, in which coal seams are thin and commercially nonprospective.

STRUCTURAL GEOLOGY

Deformation during the Laramide orogeny in the Sabinas Gulf was caused by compression arising from the convergence of geoanticlinal areas developed by the ancient paleogeographic elements. Thus, in the north the Tamaulipeco Geoanticline follows the margins of the Tamaulipas platform and in the south the Coahuila Geoanticline is restricted to the island of Coahuila. The convergence on both areas—which have thick basements and thin cover—of the stresses transmitted by the positive areas, produced tighter folding and general compression, in the range of 35 percent, developing extensive elongated northwest-trending anticlines and synclines. Within the synclinal areas, the following subbasins (Fig. 1) make up the Sabinas-Monclova Coal District:

Sabinas subbasin

The south flank is somewhat steep, owing to the greater influence of the Las Rusias anticline on this syncline; more variable dips occur along its southwest and southeast flanks. Its north end is the least-dipping of all the structures described to this point, due to the very gentle dips of the Domico de Los Peyotes anticline.

Saltillito subbasin

The southwestern flank of this syncline dips 10 to 12° away from the Baluarte anticline. Its northwestern flank dips gently; the Las Rusias anticline, which is asymmetrical and verges northeastward, had little influence on this syncline.

Lampacitos subbasin

This syncline is controlled on its southwestern flank by the Ovallos anticline, with 22 to 45° dips at its southern limit. Its northeastern flank dips 15 to 18° due to the weak effect of the northwest-plunging axis of the Baluarte anticline. The separation between this subbasin and the Saltillito is unclear.

Las Adjuntas subbasin

This extensive syncline is controlled by the Ovallos and La Rata anticlines, which form its northern flank and by the Santa Gertrudis and La Gloria anticlines on its southern flank. The strong southwestward divergence of the northern anticlines caused steep, locally overturned dips on the northeastern flank; the dip of the southeastern flank, in contrast, is much gentler (8 to 12°) thanks to the weak influence of the Santa Gertrudis anticline; the syncline is thus asymmetrical and trends northwest.

Monclova subbasin

This is a northwest-trending asymmetrical syncline controlled on its northeastern flank by the Santa Gertrudis anticline; dips are 18 to 22° southwest. The *Sacramento anticline* limits its southwestern flanks with 8 to 16° dips to the northeast. The syncline developed in sedimentary rocks that host the coal seams, exhibiting subordinate folding, pinch-outs, and very local faulting.

Southeast Adjuntas subbasin

This structure was identified after locating the Olmos–San Miguel contact southeast of the Adjuntas Basin and a few dirty coal outcrops, such as the 80-cm-thick interval reported at Arroyo La Colorada on the southeastern flank of Sierra La Gloria and approximately 6 km south and 12 km east of Rancho La Cruz. Two lesser outcrops of 10-cm-thick coal seams occur at Ojo de Agua. All of the above are located on the north-northwest flank of the subbasin.

The structure lies between the two large anticlines of the Sierra La Gloria to the southwest and Sierra Pájaros Azules to the northwest. Reverse faulting brings rocks of the Aurora and Escondido Formations into contact. Intrusive igneous rocks (Cerro Panuco) and granodioritic dikes affect the sedimentary rocks.

Environments in this area were predominantly marine and transitional toward the south-southeast, indicating that the deltaic facies progressively decreased in this part of the syncline and coal-bearing conditions are consequently minimal.

Esperanzas subbasin

This is a northwest-trending syncline approximately 32.5 km long and 2.5 km wide, limited on its steeply dipping (up to

TABLE 1. COMPUTED RESERVE VOLUMES (IN METRIC TONS) IN THE SABINAS MONCLOVA DISTRICT*

Subbasin	Thickness		
	>80 cm	>100 cm	>120 cm
Sabinas	238,913,000	210,337,000	167,845,000
Adjuntas	152,626,000	117,569,000	83,861,000
Saltillito Lampacitos	42,291,000	30,903,000	24,074,000
Monclova	16,455,000	11,494,000	4,945,000
Totals	450,285,000	370,303,000	280,725,000

*Computed by Mineral Resources Council.

50°) southwestern flank by the Sierra Santa Rosa. The 8- to 11°-dipping northeastern end is limited by the Sabinas subbasin.

San Salvador sub-basin

This relatively small secondary syncline, 6 km long by 3 km wide, is bordered by the Sierras Múzquiz, Ovallos, and San Jerónimo.

Castaños subbasin

This is a generally northwest-trending asymmetrical syncline covering approximately 72 km^2, limited by the Múzquiz anticline in the north, the Sierras Nadadores and Sacramento in the south, Sierra Tulillo in the east, and Sierra Sardinas in the west.

San Patricio subbasin

The prominent Pájaros Azules anticline limits this subbasin on the northwest. Dips are steeper on the northwest flank, commonly attaining up to 87°; at some localities the beds are vertical and even overturned. Large faults are present in the south. Toward the southwest, between the Alamos and San Felipe Canyons, this northwest-trending asymmetrical syncline is overturned, and a reverse fault separates it from the Cartujanos Mesa.

RESERVES

Evaluation of the Sabinas-Monclova Coal District has been carried out by various private companies and government organizations applying different criteria for the purpose, which makes it difficult to establish the basin's coal potential with any degree of certainty (Table 1). Several exploration concessions in the Sabinas and Saltillito subbasins have evaluated 1.40-m-thick coal seams, considered as the minimum thickness for commercial development. However, the Mineral Resources Council (Consejo de Recursos Minerales) has evaluated 0.80- to 1.40-m-thick coal seams and classified coal reserves as proven, probable, and possible. This has not been the case in some areas explored by other organizations (Tables 2 and 3).

TABLE 2. RESERVES OF THE SABINAS–MONCLOVA DISTRICT*

Subbasin	Concession†	Category of Reserves	Thickness (m)	Depth (m)	Reserves (10^6 tons)
			More than...		
	A.H.M.S.A.	Proven	1.40	0 to 350	461.2
	I.M.M.S.A.	Proven	1.40	0 to 350	92.0
Sabinas	C.R.M.	Proven	0.80	0 to 500	238.9
	C.F.M.	Proven	1.40	0 to 350	30.0
	Other companies	Proven	1.40	0 to 350	28.0
	C.R.M.	Proven	0.80	0 to 500	27.9
Saltillito	A.H.M.S.A.	Proven	1.40	0 to 350	108.0
	Sicartsa	Proven	1.40	0 to 350	25.6
San Patricio	Hullera Mex.	Proven	1.40	0 to 350	60.9
Lampacitos	C.R.M.	Proven	0.80	0 to 500	14.4
Adjuntas	C.R.M.	Proven	0.80	0 to 500	152.6
Monclova	C.R.M.	Proven	0.80	0 to 360	16.4
All		Proven	0.80		60.0
All		Possible	0.80		60.0
All		Resources			1,180.0

*To January 1985.
†A.H.M.S.A. = Altos Hornos de México, S.A.; I.M.M.S.A. = Industrial Minera Mexicana, S.A.; C.R.M. = Consejo de Recursos Minerales; C.F.M. = Comision de Fomento Minero; Sicartsa = Sicartsa Siderurgica Carbonifera Las Truchas, S.A.

TABLE 3. TOTAL RESERVES SABINAS–MONCLOVA DISTRICT

Category of Reserves	Reserves (10^6 tons)
Proven	1,255.9
Probable	60.0
Possible	60.0
Resources	1,180.0
Total	2,555.9

According to their physicochemical characteristics, the coal in the Sabinas and Saltillito Basins is a bituminous (1.15 percent reflectance), vitric (69.36 percent vitrinite), coal (14.4 percent ash content); in the ASTM classification it is a high-volatile A bituminous coal (Tables 4 and 5).

TABLE 4. AVERAGE ELEMENTAL ANALYSES BY BLOCKS

Sub-basin	Block or Body	Elemental Analysis					Group
		Volatiles (%)	Fixed Carbon (%)	Ash (%)	Sulfur (%)	Water Content (%)	
	A-1	17.34	46.86	33.52	0.88	1.40	1
Sabinas	A	16.95	45.99	37.05	1.19	0.30	1
	B	16.63	43.99	39.91	0.92	0.45	1
	A-1	19.73	50.59	25.87	1.94	1.40	1
Garcia	B-1	16.07	43.68	36.20	1.95	2.11	1
	B-2	15.24	43.48	38.24	1.17	1.33	1
Adjuntas							
	A-1	14.90	48.03	31.99	1.72	1.38	1
Abasolo	A-2	17.02	42.80	39.91	1.63	1.45	1
	B-1	16.50	39.21	41.78	1.34	1.11	2
Saltillito	Lampacitos Area	18.99	44.57	32.98	1.06	2.31	1
	San Alberto Area	20.19	40.89	36.34	0.97	2.08	1
Lampacitos	Mota Corona Area	19.66	47.40	30.99	0.83	1.47	1
	San Francisco Area	13.75	40.90	42.28	2.87	0.90	1
Monclova	San Buena Area	12.65	35.57	50.50	0.73	1.55	2

TABLE 5. AVERAGE ELEMENTAL ANALYSES BY SUBBASINS

Subbasin	Elemental Analysis				
	Volatiles (%)	Fixed Carbon (%)	Ash (%)	Sulfur (%)	Water Content (%)
Sabinas	16.68	44.52	38.92	0.98	0.68
Adjuntas	15.80	44.39	36.68	1.46	1.44
Saltillito Lampacitos	19.54	45.38	32.55	0.93	1.84
Monclova	13.21	38.66	44.36	2.13	1.09

COLOMBIA-SAN IGNACIO BASIN (COAHUILA, NUEVO LEÓN, AND TAMAULIPAS)

This basin includes parts of the states of Coahuila, Nuevo León, and Tamaulipas. The greater part of it lies within the Burgos sedimentary basin in the northern portion of the Gulf of Mexico Coastal Plain (Fig. 7). The area is served by the Monterrey–Nuevo Laredo and Ciudad Anahuac–Nuevo Laredo highways, and the Piedras Negras–Reynosa road, as well as the México–Monterrey–Nuevo Laredo railroad and an international airport at Nuevo Laredo, Tamaulipas. Many unpaved trails and pathways criss-cross the area and are passable in the dry season.

PHYSIOGRAPHY

The gently rolling topography and low-lying hills may be divided into two zones: (a) a zone of very low relief in the eastern part developed mainly on clays, sands, and shales; (b) a somewhat higher zone of slightly rounded or abrupt relief in the west, developed mainly on sands.

The principal hydrographic network is the Río Bravo Basin where numerous generally southeast-trending tributaries and streams converge and drain into the Gulf of Mexico.

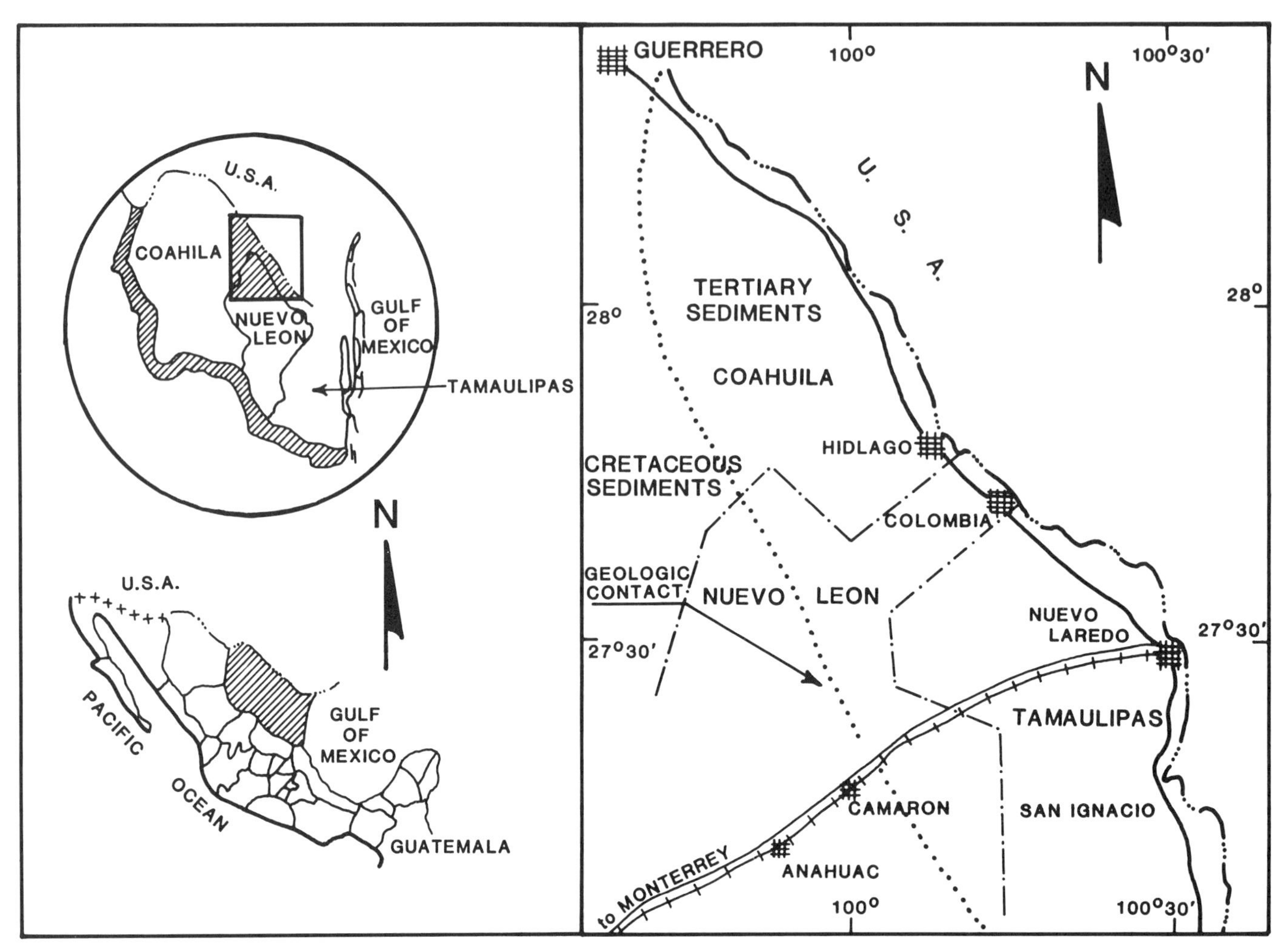

Figure 7. Location of the Tertiary basin in the states of Coahuila, Nuevo León, and Tamaulipas.

HISTORICAL GEOLOGY

The Paleocene to Recent sediments forming the basin crop out as generally north-northwest–trending bands in which the stratigraphically older beds occur in the west and become younger in the east (Fig. 8). The deposits are the result of numerous oscillating transgressions and regressions caused by subsidence and deposition, but with a coastline that always migrated east toward the present Gulf of Mexico shores. The sediments are shales and sandstones varying in composition according to their depositional environment; moreover, they were deformed and compacted, adjusting to the basin itself and giving rise to pulses and fluctuations of the coastline, which led to their interfingering.

STRATIGRAPHY AND SEDIMENTATION

The Eocene formations that constitute the major rock units of the Colombia–San Ignacio Basin (Fig. 9) are described below.

Wilcox Formation

This formation is exposed along the Río Bravo, 95 km northwest of Nuevo Laredo. It is an alternation of thinly bedded shales, of banded aspect in outcrop, and clayey sandstones; intercalated thick cross-bedded micaceous sandstones and gray and red clays are also present. Lignite is found in the lower part of the formation. A maximum thickness of 1,300 m was recorded in the Vanquerías area. The formation is assigned a lower Eocene age and it is correlated with the Wilcox Formation of south Texas. Its depositional environment was probably one of shallow waters possibly in a lagoon environment.

Carrizo Formation

This formation is considered as the basal part of the Claiborne Group, although some authors place it within the Wilcox Group. Outcrops form a belt 1.5 to 6.5 km wide from the Río

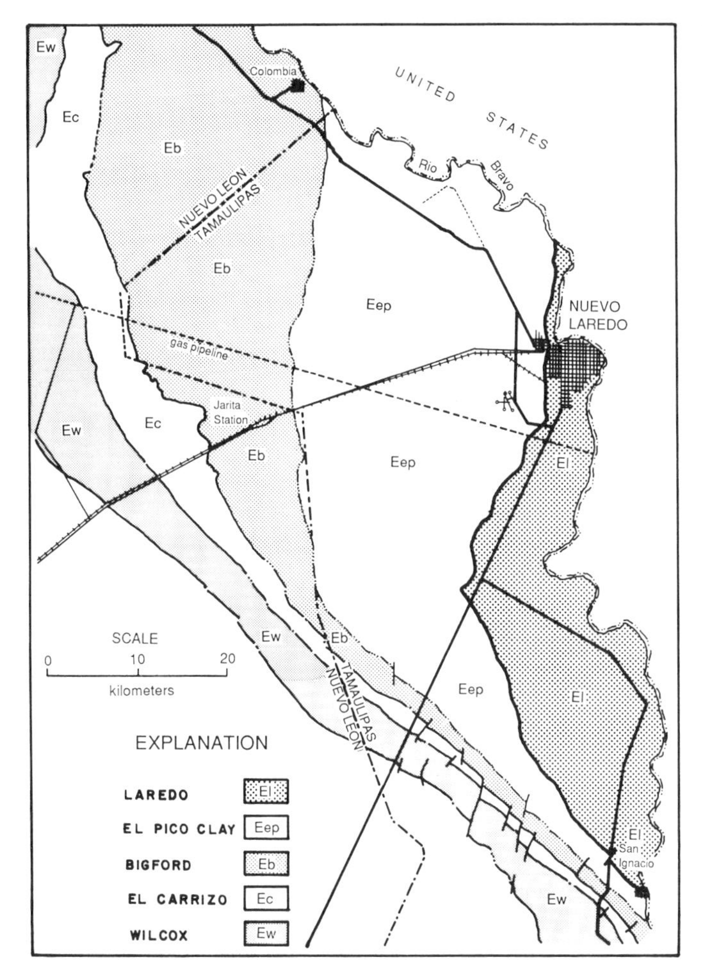

Figure 8. Geological sketch map of the Eocene formations of the Colombia–San Ignacio basin.

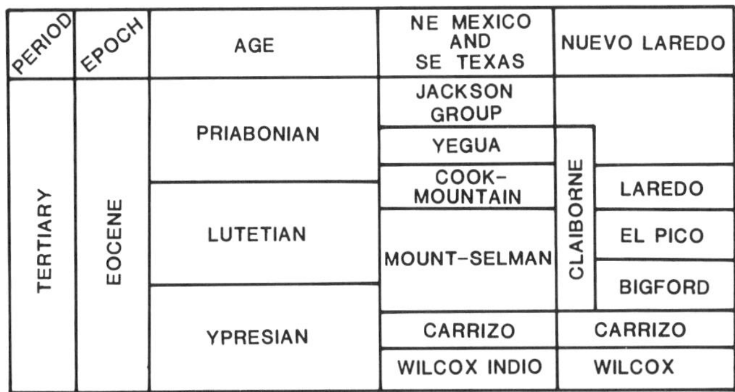

Figure 9. Correlation of the Eocene formations of northeastern Mexico and southern Texas.

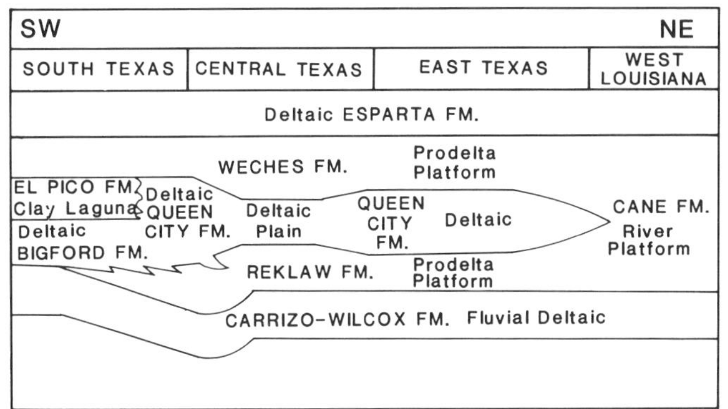

Figure 10. Regional facies changes in the Eocene formations of the Gulf Coastal Plain.

Bravo in the north to San Ignacio, and then continue to the southeast. Although essentially arenaceous, the Carrizo Formation contains numerous clays and shales, as well as interbedded thin, massive, laminated, ferruginous sandstones and some calcareous concretions. In the Río Bravo area its thickness ranges from 40 to 130 m, according to several authors. A thickness of 374 m was measured west of Hidalgo, Coahuila, increasing to 393 m along the Monterrey-Laredo highway and decreasing toward the south in the San Ignacio–El Saúz area to 315 m. The unit represents a fluvial depositional system, although coastal sedimentary characteristics are found west of Hidalgo (Coahuila).

Bigford Formation

The Bigford Formation sandstones of south Texas belong to deltaic environments of a high-destructive meander belt and piled up coastal bars (Fig 10). The formation consists of sandstones, thinly bedded shales and fossiliferous siltstones, and a few coal beds. The shales are brown and gray, gypsiferous, and make up almost 25 percent of the formation. The San Pedro seam near the top of the unit is one of the cannel coal seams known in the area. Thickness ranges from 130 to 245 m.

El Pico Clay Formation

This unit is named after El Pico hill in Webb County, Texas, where it is considered as a post-Bigford member of the Claiborne Group. The El Pico Clay crops out along the banks of the Río Bravo from the vicinity of Arroyo El Ebanito to the Miguel Alemán common, Tamaulipas. The outcrops trend southward and subsequently bend toward the southeast. Outcrop width decreases considerably from the Río Bravo to the south in the neighborhood of San Ignacio.

The formation ranges in thickness from 180 to 360 m and consists mainly of partly gypsiferous gray, green, and brown shales, fine-grained sandstones, and brown and gray siltstones in thin beds. Several coal lenses and seams are present, the most important being the Santo Tomás Seam, a cannel coal at the base of the formation.

The sediments of this unit appear to represent lagoon muds genetically associated with high-destructive deltas developed inland (i.e., northwest of the coastal bars; (Fig. 10).

Laredo Formation

This name has been used for sediments described by others in the Río Grande Embayment as Cook Mountain and Sparta to emphasize the fact that the middle Eocene lithologies and faunas in the embayment differ from those of central and east Texas. The formation is exposed for 140 km along the Río Bravo and trends almost parallel to it.

More than 50 percent of the formation is sandstone. Two thick massive sandstone members with abundant glauconite, which gives them a greenish color, are separated by an intermediate, basically shale member. The unit's thickness is 230 m in Webb County, Texas, and almost 400 m at San Ignacio.

The Laredo Formation is equivalent to the Sparta in east Texas and represents a high-constructive delta having coastal bar sands, lagoon, and prodelta muds as main facies.

STRUCTURAL GEOLOGY

The coal seams belong to or are associated with a lagoonal facies resulting from the deposition of the middle part of the Claiborne Group (Bigford, El Pico Clay): sands interbedded with clayey shales, traces of sulfur and coal seams, suggesting a subcontinental environment (i.e., shallow waters), which indicates that seas were static during this period. The overlying Yegua-Jackson sediments in adjacent parts of Texas represent lagoon facies of the coastal barrier bars system. The entire stratigraphic sequence forms a large, gently northeastward dipping (5 to 11°) homocline.

COAL DEPOSITS

The principal commercial coal deposits of Texas occur in the lower portion of the Wilcox Group at the Bigford/Pico Clay contact and also at the Yegua Formation/Jackson Group contact. The same coal seams are present in northwestern Mexico, represented by the San Pedro seam within the Bigford Formation and the Santo Tomás within the El Pico Clay, which are the most prospective. This is a cannel coal derived from marine algae.

TABLE 6. RESERVE FIGURES AS OF JANUARY 1984

Category	Million Tons	Metric Tons of Coal Equivalent x 10^3
Proved	76	65.0
Probable	3	2.6
Possible	30	25.7
Inferred	63	54.0
Resources	80	68.6
Totals	252	215.9

RESERVES

Reserves volumes have been computed by several methods; in the present case, geostatistics was applied (Table 6). Results obtained in this basin are restricted to areas of coal at shallow depths with average seam thickness of 80 cm; underground mining is presently under consideration. Development has focused on coal seams 0 to 40 m deep for open-pit mining and on coal at depths no greater than 80 m; increased potential may be possible once the evaluation at depth has been carried out. The coal in this basin is characterized mainly by its low ash content and caloric value of 6,000 Kcal/kg.

LA MIXTECA COAL BASIN, OAXACA

INTRODUCTION

The La Mixteca basin occupies the northwest part of the state of Oaxaca and extends to the neighboring states of Puebla and Guerrero, comprising three important coal areas: Tezoatlán, Mixtepec, and Tlaxiaco (Figure 11). The Tlaxiaco area, 12 km northwest of the village of the same name, is subdivided into the Allende, Rancho General, and Las Huertas sectors. The undivided Tezoatlán area is 16 km southeast of the village of that name, and the undivided Mixtepec area is 25 km south of Tezoatlán. Access is mainly by the Cristóbal Colón road, Highway 190, to km 132 at the junction with Highway 125 leading to the town of Tlaxiaco; from there an unpaved road continues to Mixtepec with various branch-offs that communicate to the entire area. This is an extensive basin covering several hundred square kilometers; however, the areas that have been studied in detail barely reach 110 km^2.

PHYSIOGRAPHY

The basin lies in the Oaxaca Highland subprovince, separated from the Balsas-Mezcala subprovince to the south by the Acatlán metamorphic uplift; both subprovinces belong to the major Sierra Madre del Sur physiographic province. The region is characterized by mountain ranges reaching altitudes of 2,500 m. Intermittent streams feed the hydrographic basin of the Río Numi and its tributaries, the Tlaxiaco and Santa Catarina, which join the Atoyac and drain into the Río Balsas. Drainage is dendritic, controlled by structures and lithologic types.

HISTORICAL GEOLOGY

The sediments of the La Mixteca coal basin have undergone the effects of at least three orogenies, the first of which most

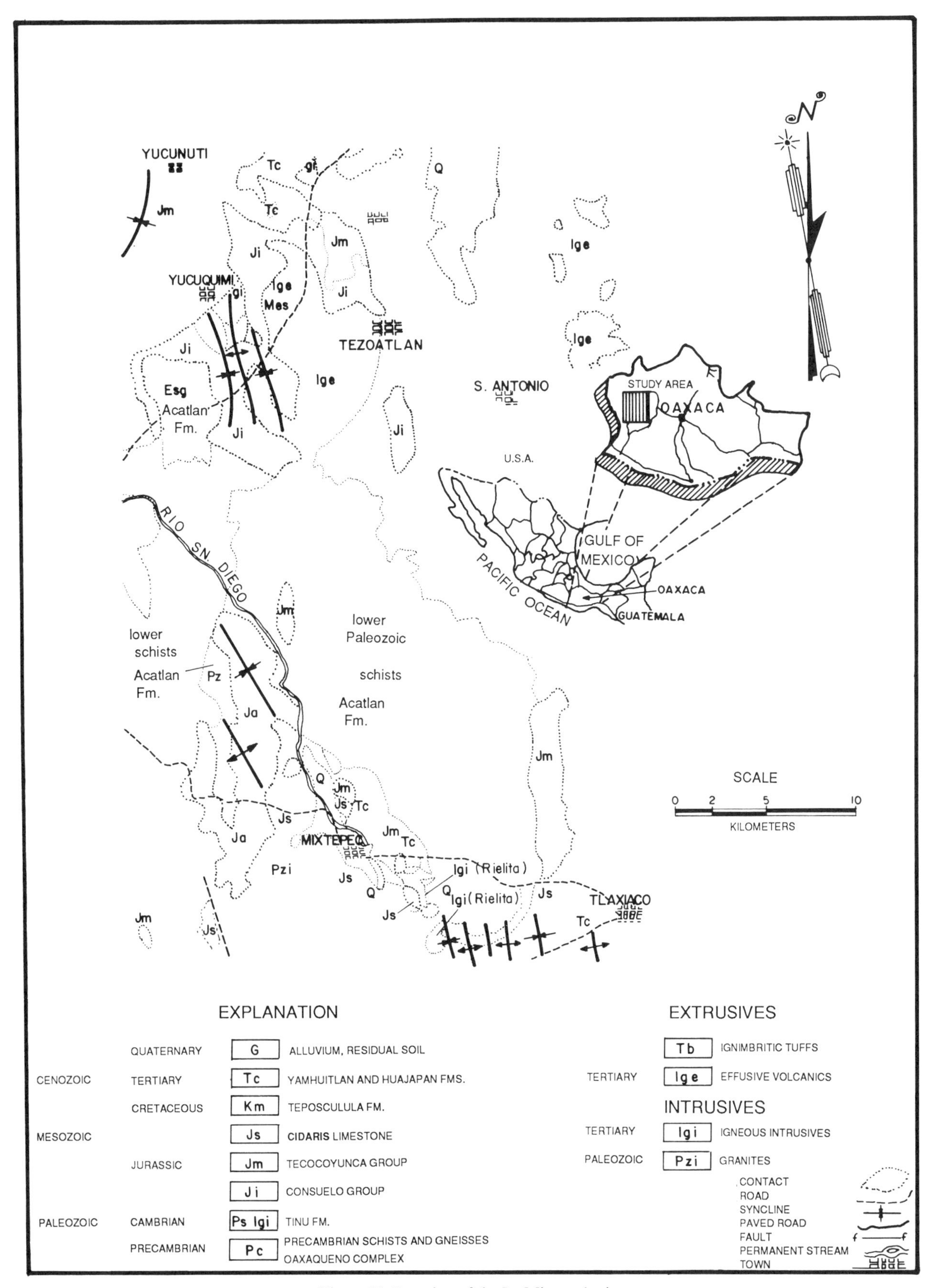

Figure 11. Location of the La Mixteca basin.

	SYSTEM	SERIES	STAGE	STRATIGRAPHIC UNIT
CENOZOIC	QUATERNARY		RECENT	ALUVIUM
			PLEISTOCENE	HUAJUAPAN
			PLIOCENE	YANHUITLAN
	TERTIARY		MIOCENE	
			OLIGOCENE	BALSAS GROUP CONGL.
			EOCENE	
			PALEOCENE	
MESOZOIC	CRETACEOUS	LATE	MAASTRICHTIAN	YUTANDACHI
			CAMPANIAN	
			SANTONIAN	
			CONIACIAN	
			TURONIAN	
		MIDDLE	CENOMANIAN	TEPOSCOLULA
			ALBIAN	
		EARLY	APTIAN	SAN ISIDRO
			BARREMIAN	IGNEOUS INTRUSIVE ROCKS
			HAUTERIVIAN	
			VALANGINIAN	
			BERRIASIAN	
	JURASSIC	LATE	PORTLANDIAN (TITHONIAN)	SABINAL GROUP
			KIMMERIDGIAN	
			OXFORDIAN	CIDARIS LS.
		MIDDLE	CALLOVIAN	TECONCOYUCA GROUP: YUCUNUTI, OTATERA, SIMOR, TABERNA, ZORILLO
			BATHONIAN	
			BAJOSIAN	
			AALENIAN	CONSUELO GROUP: CUALAC CGL.
		EARLY	TOARCIAN	ROSARIO
			PLIENSBACHIAN	
			SINEMURIAN	
			HETTANGIAN	
	TRIASSIC	LATE	RHAETIAN	
			NORIAN	
			CARNIAN	
PALEOZOIC			PERMO-CARB.	ACATLAN SCHISTS
			CAMB.-ORD.	
PRE-CAMBRIAN				GNEISS

Figure 12. Stratigraphy of the La Mixteca coal basin, Tlaxiaco, Oaxaca.

certainly altered positive clayey terrains to the metamorphic complex of the Acatlán schists, dated at 510 ± 60 Ma (Pb/alpha)—somewhere in the Early Paleozoic. The absence of unmetamorphosed Paleozoic sediments is understandable, because either the continent that emerged following orogenesis underwent extensive erosion, or most of the Paleozoic terrains were metamorphosed during orogenesis to form the Acatlán Schists. The same possibly occurred with the Triassic and Liassic sediments at a later date.

The interlayered marine/nonmarine character of the mid-Jurassic column, which shows ferns and marine fossils, may be explained as the product of marine regressions and transgressions that led to alternating or cyclically successive swamp and marine environments. The Upper Jurassic (Callovian and Oxfordian) is represented by calcareous sediments with abundant sedentary marine macrofossils typical of interior shelf environments; these give way gradually to basin deposits: marly and clayey beds with ammonites ranging in age from the Kimmeridgian to the Hauterivian.

The Mesozoic sedimentary column as a whole exhibits continuous deposition, interrupted only by the intra-Cretaceous orogeny between the Hauterivian and the Albian. The interruption is visible at the contact between the Sabinal Group and the San Isidro Formation (Fig. 12) where the Barremian and the Aptian are absent.[1] Subsequent deposition took place in littoral environments: reddish clastics with giant *Nerinea* are transitional to calcareous beds (Teposcolula Limestone) containing miliolids and rudists, which define an interior shelf environment of Cenomanian age, possibly extending to the Turonian. These interior shelf sediments are followed by basin deposits: shales and marls with *Inoceramus* and *Globotruncana,* index species of Santonian, Campanian, and Maestrichtian age. Immediately afterward the Laramide orogeny caused emergence, bringing about the definitive separation of the Atlantic and Pacific Oceans. Cenozoic deposition was continental and accompanied by strong volcanic activity during the Miocene.

STRATIGRAPHY

The lithologic units range in age from the Paleozoic to the Recent. The oldest rocks are green sericitic schists, the "Acatlán Schists," which Salas (1949) tentatively considered pre-Mesozoic. Unconformably above these schists are unmetamorphosed rocks of Aalenian to Oxfordian age, approximately 3,200 m thick, that include the upper part of the Consuelo Group (Cualac Conglomerate)[2], the Zorrillo, Taberna, Otatera, and Yucuñuti Formations of the Tecocoyunca Group, and the *Cidaris* Limestone.

[1]The inconsistency between the text here, later discussions of the Sabinal/San Isidro relations, and Figure 12 was never resolved. ARP.

[2]Inconsistency between the text here and Figure 12 was never resolved. ARP.

Paleozoic

Acatlán Formation. Salas (1949) used the term Acatlán Schist to refer to schist exposures surrounding the town of Acatlán, Puebla, and extended it to include all the metamorphic rocks underlying the entire Oaxaca sedimentary basin. Later, the name Acatlán Formation was applied only to the folded low-rank metamorphic exposures in the area of Acatlán. The unit is widespread in Acatlán and the Tlaxiaco region. It consists of quartz, sericite, and chlorite schists and occasional reddish-weathering green phyllites. It is distinguished from the other formations by its metamorphic rank, milky-white quartz veins and pockets of white mica. The unit is strongly folded and faulted, making thickness measurements impossible.

Mid-Jurassic

Cualac Conglomerate. Originally named Cualac Quartzite in the vicinity of Cualac, Guerrero, the unit was later named Cualac Conglomerate by Erben (1956) who described it near Mixtepec. It occurs extensively from north to south and throughout the Tlaxiaco Basin; in Mixtepec it borders the Acatlán Schist.

This is a conglomerate of milky quartz and schist fragments, 0.5 to 5 cm in size, in a silica groundmass; at times it appears as a true quartzite, occasionally exhibiting abundant ovoid purple-colored concretions. The unit forms sharp ridges that are readily identified in the field and on aerial photographs. It is 12 to 80 m thick in the Mixtepec region; as much as 120 m have been measured in the vicinity of Tlaxiaco. However, in the Barrio Séptimo Gorge and at the junction of the Ríos Tlaxiaco and Ocotepec it is barely 1.90 to 2 m thick, thus proving its lenticular nature.

The formation is regionally conformable with the Rosario Formation below and the Zorrillo Formation above; however, in the vicinity of Mixtepec and Tlaxiaco it unconformably overlies the Acatlán Schist directly. It is assigned an upper Aalenian and possibly lower Bajocian age on the basis of its stratigraphic position (Erben, 1956). It is nonmarine; features such as subrounded pebbles and cross-bedding indicate its clearly continental origin and its detrital constituents are directly derived from the metamorphic basement rocks.

Mid- to Late Jurassic, Tecocoyunca Group

This group was first defined as Tecocoyunca Beds; Erben (1956) raised it to group status on the basis of its great thickness and diverse lithologies. It includes five mid-Jurassic and partly Upper Jurassic continental and marine formations.

Zorrillo Formation. The type-locality of this formation is in the Tezoatlán region, Oaxaca. The unit is 20 to 80 m thick, and is composed chiefly of thinly bedded, fine- to medium-grained, occasionally cross-bedded sandstones and siltstones. At the lower levels the sandstone is conglomeratic, with abundant milky quartz pebbles containing reddish-colored calcareous siltstone concretions. It overlies the Cualac Conglomerate and underlies the Taberna Formation, conformably in both cases; the Zorrillo/Taberna contact is transitional. It is assigned to the mid-Bajocian. No faunas are present, but the unit contains an abundant fossil flora, defining its depositional environment as clearly continental.

Taberna Formation. The type locality of this formation is also in the Tezoatlán region, Oaxaca. The formation is composed of dark shales and sands, light brown-gray-weathering dark calcareous shales; it contains abundant fossiliferous concretions, gray fine-grained quartziferous sandstones and shales with spheroidal concretions that show fossil borings. It is transitional with the overlying Simón Formation. Its age is late Bajocian and it was laid down in a marine environment.

Simón Formation. The type locality for this formation is also in the Tezoatlán region. It consists of medium-bedded, yellowish, medium- to coarse-grained sandstones and fine-grained quartzose sandstones. A conglomerate of milky quartz pebbles in quartzose groundmass is also present, besides carbonaceous shales, coal seams, and some dark limestone layers. The unit is 80 to 100 m thick. It is transitional with the overlying Otatera Formation and is assigned a mid-Bathonian age (Erben, 1956). The formation is barren except for land plants and coal seams (i.e., it is of continental origin).

Otatera Formation. The type locality for this formation is also in the Tezoatlán region. The unit is composed of light brown, thinly bedded, occasionally cross-bedded, fine- to coarse-grained sandstones, shales, and siltstones. The shales are dark, with red hematitic concretions. Some calcareous layers are *Ostrea* coquinas. Thickness is 50 to 70 m. The formation is conformable with the overlying Yucuñuti Formation. Frequent pelecypods (*Eocallista* Imlay and Alencaster) and gastropods (*Phasianella* sp.) indicate an approximate early Callovian age. The Callovian was a period of limey-clayey and limey-sandy deposition in persistent shallow seas.

Yucuñuti Formation. The type locality of this formation is east of Yucuñuti, Oaxaca. Dark shales containing light brown calcareous concretions alternate with light-colored fine-grained sandstones, marls, and limestones; a sandy limestone horizon is fossiliferous. It is conformable with the ovelrying Cidaris Limestone Formation and is assigned a late Callovian age. Its depositional environment was shallow marine.

Upper Jurassic

Cidaris *Limestone Formation.* According to Cárdenas (1966) the best exposures of this formation are at Cerro de La Isleta, near Mixtepec. There, a series of 40- to 80-cm-thick brown oolitic limestones with iron oxide nodules is followed by thin- to medium-bedded gray shales and nodular marls containing an abundant and varied fauna, especially large, thick-shelled *Exogyras* and scarce *Parathyridina.* The secton ends in a thickly bedded gray pure limestone with black chert nodules and abundant *Parathyridina* and *Cidaris* spines.

In the Mixtepec region it has an average thickness of 86 m,

although it is 95 m thick at Cerro de La Isleta; however, at the Río Tlaxiaco localities, Las Vacas, Tlaxiaco-Cuquila junction, and Río Sabinal, thicknesses of 200 to 650 m are found. The lower contact is placed at the first appearance of *Parathyridina* which coincides with the first oolitic limestone beds; it underlies the Sabinal Formation conformably, and is assigned to the Oxfordian. Its sediments define a depositional environment of relatively quiet, stable open seas.

Upper Jurassic–Lower Cretaceous

Sabinal Formation. This formation consists of black thin limestones interbedded with bituminous shales. It is conformable beneath the San Isidro Formation and is assigned a Late Jurassic to Early Cretaceous (Kimmeridgian-Barremian) age, equivalent to the Kimmeridgian *Idoceras* beds in Mazapil, Zacatecas. It was laid down in a marine environment.

Lower Cretaceous

San Isidro Formation. This formation is a clastic unit exposed at the Río San Isidro southwest of Tlaxiaco, at kms 79 and 80 of the Tlaxiaco-Pinotepa National road, and on the Ríos Ocotepec, Tablas, and San Isidro, the last of which is the type locality. It is an alternating association of clastic bodies in which thick-bedded (0.80 to 1.50 m), fine- to coarse-grained, occasionally conglomeratic, buff-colored sandstones are predominant, together with medium-bedded nodular shales and red and green siltstones. A transitional layer in the upper part of the unit includes shales, sandstones, marls, clayey limestone, and dolomite with abundant gastropod and *Exogyra* faunas. It is conformable with the overlying Teposcolula Formation and is assigned to the Aptian. Depositional conditions varied from neritic to bathyal in a relatively quiet sea.

Mid-Cretaceous

Teposcolula Limestone Formation. Salas (1949) used the term Teposcolula Limestone for the calcareous sequence exposed in the immediate vicinity of San Pedro Teposcolula, Oaxaca. It occurs extensively in the Tlaxiaco region and is almost exclusively composed of a succession of whitish-weathering, thick-bedded (0.80 to 1.50 m), gray limestone layers, some of which are dolomitized and others contain rudists, with black chert lenses and nodules. The unit is easily identified in the field, where it develops sharp topographic highs and karst relief. Total thickness is estimated in 1,060 m. Its upper contact is unknown in the Tlaxiaco region, where the overlying rocks have been eroded, but in other localities it shows a conformable transitional contact with a sequence of marls and nodular shales. It is of Albian to Cenomanian age and represents a marine depositional environment.

Tertiary

Huajuapán and Yanhuitlán Formations. These formations were named by Salas (1949) to define continental sediments, igneous intrusive and extrusive rocks, and pyroclastics. The Huajuapán is distinguished from the Yanhuitlán Formation in that it contains scarce red beds, and consists mostly of brown, gray, and slightly yellow igneous rocks. In the Yanhuitlán Formation the red tones of thinly bedded (2 to 60 cm) montmorillonitic clays and subarkosic silts prevail. The Yanhuitlán overlies the *Cidaris* Limestone with angular unconformity near Tlaxiaco. Although both units have been assigned to the Tertiary, they are presently being studied in detail in view of their widely differing sedimentary features. According to the existing lithologies, the formations are ascribed a continental origin.

Recent

The youngest materials in the region are sandy gravels and clays derived from erosion of the preexisting lithologic units; these were transported and deposited on surfaces presently under cultivation, mainly in the Tlaxiaco, Cuquila, and Mixtepec Valleys.

STRUCTURAL GEOLOGY

Mid- and late Precambrian metamorphic rocks and late Precambrian sedimentary rocks make up the La Mixteca Coal Basin basement, which includes tightly folded schists and gneisses that have been affected by numerous faults, giving rise to small northeast-trending anticlines and synclines. The extensive distribution, great thickness, and mineralogy of the schists and gneisses indicate widespread regional metamorphism of preexisting granitic and sedimentary rocks.

Subsequent strong tectonism also played a significant role, affecting the basin sediments containing the coal seams. The Oaxaca sedimentary basin must have undergone several orogenies. According to Salas (1949), however, the La Mixteca was subjected only to slight readjustments that caused faulting and fracturing and gave rise to small blocks of lesser magnitude.

COAL DEPOSITS

The coal seams range from bituminous to semianthracites and occur at several geologic levels in the lower part of the mid-Jurassic Rosario, Zorrillo, and Simón Formations: alternating sandstones, shales, and siltstones indicate probable deltaic environments. Coal seam thicknesses range from a few centimeters to 3 m; most are very lenticular. As a result of the depositional conditions and of subsequent orogenic processes, a good many coal seams show high impurity contents; moreover, strong tectonism has deformed and folded them to a considerable degree. These factors, in addition to the influence of numerous igneous intrusions and extrusions, have caused the coal seams and their host sediments to occur as isolated blocks separated by numerous faults, which restricts their continuity and complicates the possibilities of commercial development.

As many as five coal seams of prospective thickness have been located in the Tlaxiaco area: one at the base of the Zorrillo Formation and the rest in the central portion of the Simón Formation; dips are variable to 45°. In the Tezoatlán area the Rosario, Zorrillo, and Simón Formations all contain coal seams. Finally, coal seams in the strongly tectonized Mixtepec area occur in the Zorrillo and Simón Formations, dipping even more steeply than in the other areas.

RESERVES

The Tlaxiaco and Tezoatlán areas have been partially evaluated, and the Mixtepec area is at the regional exploration stage. The 19.5 million tons of coal evaluated in Tlaxiaco are distributed as follows: Allende, 14,837,924; Rancho General, 1,886,385; Las Huertas, 2,573,880; and Barrio Séptimo, 280,160, on the basis of 300-m depths and more than 0.80-m thicknesses. In the Tezoatlán area, 13 million tons of proved, 20 million tons of probable, and 30 million tons of possible reserves have been evaluated. The almost vertical dips of the coal seams in the strongly tectonized Mixtepec area and the lack of subsurface information allow only an estimate of up to 30 million tons of reserves (Table 7).

TABLE 7. GENERAL RESERVES OF THE LA MIXTECA BASIN

Area	Category of reserves and resources (tons x 10^6)					
	Proven	Probable	Possible	Inferred	Resources	Total
Tlaxiaco	19.5				20	39.5
Tezoatlán	13	20	30		30	93
Mixtepec					30	30
Total						162.5

Barranca Basin, Sonora

INTRODUCTION

The Barranca Basin is located in the mid-central part of the state of Sonora, within the Sonora Desert physiographic province, 90 km southeast of Hermosillo, at 28°15′ to 29°00′N, 109°30′W (Fig. 13). A 107-km-long unpaved road connects San Marcial—the most important community in the region—with Hermosillo, from which a 96-km-long road, also unpaved, leads to Guaymas harbor. The coal deposits occupy an almost north-south–trending belt over approximately 100 km^2.

PHYSIOGRAPHY

A series of almost north-south–trending parallel hills and rises characterizes the area; Sierra Lista Blanca and Cerro El Salto and Cerro La Guásima are some of the highest. Intermittent tributaries of the Río Mátape and an intermittent stream carrying water only in the rainy season form a more or less rectangular drainage system; the relief indicates an advanced mature stage of fluvial erosion.

HISTORICAL GEOLOGY

The geologic record of the area begins in the Paleozoic, a period of thick limestone deposition in shallow temperate seas. The seas regressed possibly at the beginning of the Mesozoic until the end of the Triassic, during which time the limestones underwent erosion. In the Late Triassic and Early Jurassic the area was part of a transitional brackish-water zone of frequent successive transgression and regression, represented by the interbedded sandstones, shales, siltstones, coal seams, and a few conglomerates of the Barranca Formation. At the end of the Jurassic or during the Cretaceous the San Marcial Formation was laid down. The final retreat of the sea took place in the Cretaceous; since then the region has formed part of the continent. Volcanism became predominant at the beginning of the Tertiary, and thick lavas, tuffs, and agglomerates (Lista Blanca Formation) covered a large part of the region. Mid-Tertiary granitic intrusions caused fracturing, faulting, and tilting that generated mountain blocks, which appear to be hanging in the granitic mass. An example is the Sierra La Lista Blanca and the block formed by the hills from Agua Salada to Cerro Colorado.

Renewed volcanic activity at the end of the Tertiary produced the lavas, agglomerates, and conglomerates of the Baucarit Formation, covering some of the preexisting rocks (for example, Mesa de La Sanguijuela). The present-day topography is the result of rejuvenation in the Quaternary prior to the onset of the current erosion cycle.

STRATIGRAPHY

The surface outcrops range from Paleozoic to Cenozoic in age. The oldest rocks in the region are marbles unconformably underlying the Upper Triassic and Lower Jurassic Barranca Formation, which contains the coal and graphite seams. Metamorphic rocks of mainly sedimentary origin covering extensive areas have been tentatively assigned to the Cretaceous and named the San Marcial Formation. Overlying these is the Lista Blanca Formation of possibly Lower Tertiary age. A mid-Tertiary granitic batholith forms the basement for all the above rocks, as well as for the Upper Tertiary Baucarit Formation.

Paleozoic

Paleozoic limestones, sandstones, shales, and barite beds occur northeast of San Javier, in the Sierra Cobachi. They are dated by their fossil content as Ordovician and Permian.

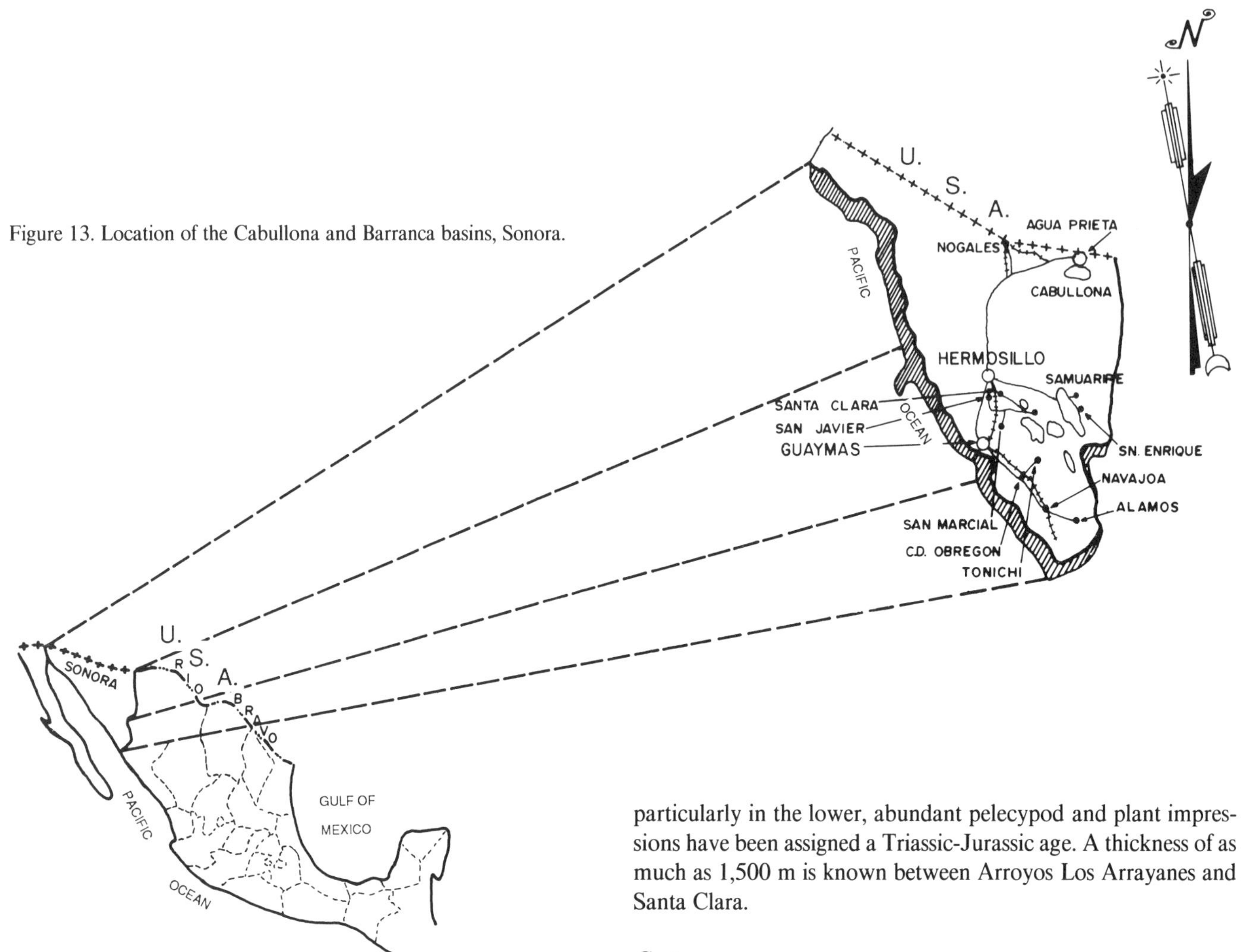

Figure 13. Location of the Cabullona and Barranca basins, Sonora.

Upper Triassic and Lower Jurassic

Barranca Formation. This formation consists of slightly metamorphosed clastic and clayey sediments with intercalated coal and graphite. The Barranca Formation is well exposed along an almost north-south–trending belt about 5 km long, extending from 1 km north of San Marcial to Cerro Colorado. The clastic sedimentary rocks, of transitional continental to marine depositional environment, have been divided roughly into lower, middle, and upper units. The lower and upper units consist mainly of quartzitic sandstones with intercalated thin shale layers. The middle unit is made up of alternating sandstones and shales and includes black carbonaceous shales with coal seams; however, coal beds enclosed in coarse-grained sandstones are also found. The contacts between these units are gradational (i.e., not clearly defined). Pesquera and others (1960) divide the formation into only two members: the lower, of interbedded carbonaceous shales and siltstones, quartzites, some conglomerates, coal, and graphite; the upper, somewhat similar to the above, but with predominant quartzites and conglomerates and without coal or graphite seams. Both members contain fossil flora and fauna; particularly in the lower, abundant pelecypod and plant impressions have been assigned a Triassic-Jurassic age. A thickness of as much as 1,500 m is known between Arroyos Los Arrayanes and Santa Clara.

Cretaceous

The unconformity between the Barranca Formation and the overlying metamorphosed Cretaceous rocks is clearly defined by a conglomerate bed, at times 12 m thick, characterized by its Barranca-derived fragments in a groundmass of mainly sedimentary origin. The metamorphosed rocks, tentatively assigned to the Cretaceous under the name of San Marcial Formation, are of mainly sedimentary origin and in very few cases igneous. They overlie the conglomerate described above and exhibit marked stratification, trending N23°W and dipping 30 to 50°W. They are found covering the east flank of the Barranca Formation.

Cenozoic

Tertiary rocks overlying folded Cretaceous and older units consist mainly of andesitic and rhyolitic tuff flows and some basalt overflows, coarse clastic deposits, and intrusive rocks of mainly granitic, granodioritic, and dacitic composition, which affect the entire described sequence. The youngest Tertiary rocks (Baucarit Formation) vary from bedded sandstones to conglomerates.

Quaternary

Alluvium and terrace deposits: unconsolidated gravels, sands, and silts form the recent materials, accumulated mostly in low-lying areas and stream beds.

STRUCTURAL GEOLOGY

The rocks of the area are highly folded, faulted, and intruded by large plutonic bodies. Several periods of deformation are recorded by unconformities between the Cretaceous rocks and the volcanics at the onset of the Tertiary and between these and the Baucarit Formation. There is little evidence of the age of the first movements. The latest developed generally north-northwest structural trends.

The Barranca Formation occurs as large displaced remnant masses of sedimentary rocks widely separated from one another. Structures within the Barranca are very complex due to the effects of the diverse regional and local upheavals and igneous intrusions, which have caused radiating normal faults and the progressive metamorphism of the anthracitic coal beds to amorphous graphite in the vicinity of the intrusive contacts. The generally very complex structure may be considered as northwest- to northeast-trending, with monoclinal dips of 10 to 45° to the northeast, northwest, and southwest. Up to 30 coal levels, from a few centimeters to 5 m thick, are present in the Barranca Formation.

COAL DEPOSITS

The coal deposits within the lower fine carbonaceous member of the Barranca Formation occur as interbedded, somewhat lenticular seams thinning out in the carbonaceous shales, or wedging out between quartzite beds or between shale and quartzite. At least four prospective coal seams have been discovered by direct or drilling exploration; others detected to date are too lenticular and thin for commercial development purposes. Seam continuity is highly broken up by multiple faults of diverse magnitude and type that cause characteristic staggering in the direction of both strike and dip and must be taken into account in development plans.

The two main types of coal in the San Marcial area show a common physicochemical character, but individual features. In general, the coal is black, hard, compact, shiny, with subconchoidal or cubic fracture, and specific gravity of approximately 1.9 gr/cm^3. The most noticeable difference between the two types of coal is that one tends to fracture easily in more or less cubic fragments and is recovered entirely during drilling, whereas the other is harder to break, fissile, and oilier to the touch.

Chemical analyses of the San Marcial coal are as follows: moisture 8 percent; sulfur 0.37 percent; hydrogen 1.60 percent; nitrogen 0.36 percent; oxygen 7.8 percent; calorific value 11,500 BTU, 6,300 (cal); ash fusion temperature 1,280°; fuel interrelation 18.6 percent; fixed carbon (dry basis and free of mineral matter) 95.3 percent. According to these values and taking maximum and minimum variations of their main contents into account, these coals are perfectly classified as anthracite, ranging from meta-anthracite to semianthracite and including anthracite.

The coal deposits developed in the transitional brackish-water environment of an extensive basin spreading throughout the central and eastern parts of the state of Sonora. Subsequent coalification processes produced bituminous coal that metamorphosed to anthracitic coal or graphite, principally when granitic and other igneous rocks intruded the Barranca Formation (Fig. 14).

RESERVES

Reserve evaluations carried out by the Mineral Resources Council, focused mainly on the San Marcial areas, and two smaller nearby areas (Fig. 13). The results of these evaluations are shown on two smaller nearby areas (Fig. 13). The results of these evaluations are shown on Table 8.

TABLE 8. RESERVE EVALUATIONS FOR THE SAN MARCIAL AREA

Area	Category of reserves and resources* (tons x 10^6)					
	Proved	Probable	Possible	Inferred	Resources	Total
Santa Clara	30	230		50,000	10,000	60,260
San Marcial	4,000	9,450	18,240			31,690
San Enrique	725					725
Remainder of Basin					50,000	50,000
Total	4,755	9,680	18,240	50,000	60,000	142,675

*Carried out by Mineral Resources Council.

CABULLONA BASIN, SONORA

INTRODUCTION

The Cabullona Basin (Fig. 13) lies in northeastern Sonora State within the Agua Prieta, Naco, and Fronteras townships and is divided into two main areas known as Cabullona and El Encino. From Hermosillo toward Agua Prieta, a paved highway runs for 18 km to Nacozari, where a 20-km unpaved trail turns east to Cabullona Station. The basin covers 2,850 km^2.

PHYSIOGRAPHY

The area belongs to the northernmost Sierra Madre Occidental physiographic province, represented there by north- and northwest-trending ranges separated by wide valleys. The area is drained by intermittent tributaries of the Río Agua Prieta, joining the Río Cabullona of the Yaqui hydrographic basin at Cabullona Station.

STRATIGRAPHY

Precambrian to Recent lithologic units occur in the area. Sedimentary rocks are predominant with subordinate igneous outcrops. Of main interest are the Cintura and Snake Ridge Formations of the Bisbee and Cabullona Groups, respectively, which contain the region's coal deposits.

Cintura Formation

This is a sequence of sandstones, siltstones, and shales with conglomerate lenses and calcareous layers. Sandstones in the lower part are fine- to coarse-grained, buff, purplish, and red in color; cross-bedding occurs in the upper part of the unit. Coal seams are present in the middle section interbedded with shales and/or sandstones; fossiliferous horizons containing *Trigonia, Turritella,* and *Ostrea* are found at these same levels. The unit is assigned to the Lower Cretaceous and has an estimated thickness of 2,800 m. It is conformable above the Mural Limestone and beneath the Snake Ridge Formation.

Snake Ridge Formation

This formation consists of a conglomerate of limestone and schist fragments with its type locality in the Cabullona Basin, Sonora. It consists of intercalated sandy conglomerates, conglomeratic sandstones, coarse-grained sandstones, and shales in beds of medium thickness. Two levels of carbonaceous shales toward the base grade laterally to coal seams. The unit, approximately 650 m thick, is conformable with the overlying Camas Sandstone Formation; locally, as in Rancho Magallanes, it is found in fault contact over the Glance Conglomerate Formation.

HISTORICAL GEOLOGY AND TECTONIC HISTORY

The Precambrian and Paleozoic rocks were affected by Permo-Triassic Cordilleran orogenesis, which uplifted and folded the central part of Sonora; subsequently these rocks were covered by Triassic and Jurassic sediments. The Triassic and Jurassic rocks in turn were folded and affected by large, mainly northwest-trending faults. Strong plutonic activity west of Sonora at the beginning of the Cretaceous caused the emergence of acid batholiths and stocks. A major part of the widespread transgression over northern and northeastern Sonora in the Neocomian resulted in the extensive Cabullona Basin, with more than 4,000 m of Upper Cretaceous sediments overlapping the eroded Paleozoic and earlier Mesozoic surfaces. At the beginning of the Tertiary and during the Laramide orogeny, east-west movements of another orogenic period affected both the Sierra Madre Occidental and east Sonora, folding the preexisting rocks; reverse faulting took place west of the Cabullona region.

The Cabullona and El Encino areas, 25 km apart, are connected by an overturned anticline. The Morita, Mural, and Cintura Formations outcropping farther south of El Encino, some 24 km from the town of Frontera, are also involved in an overturned anticline.

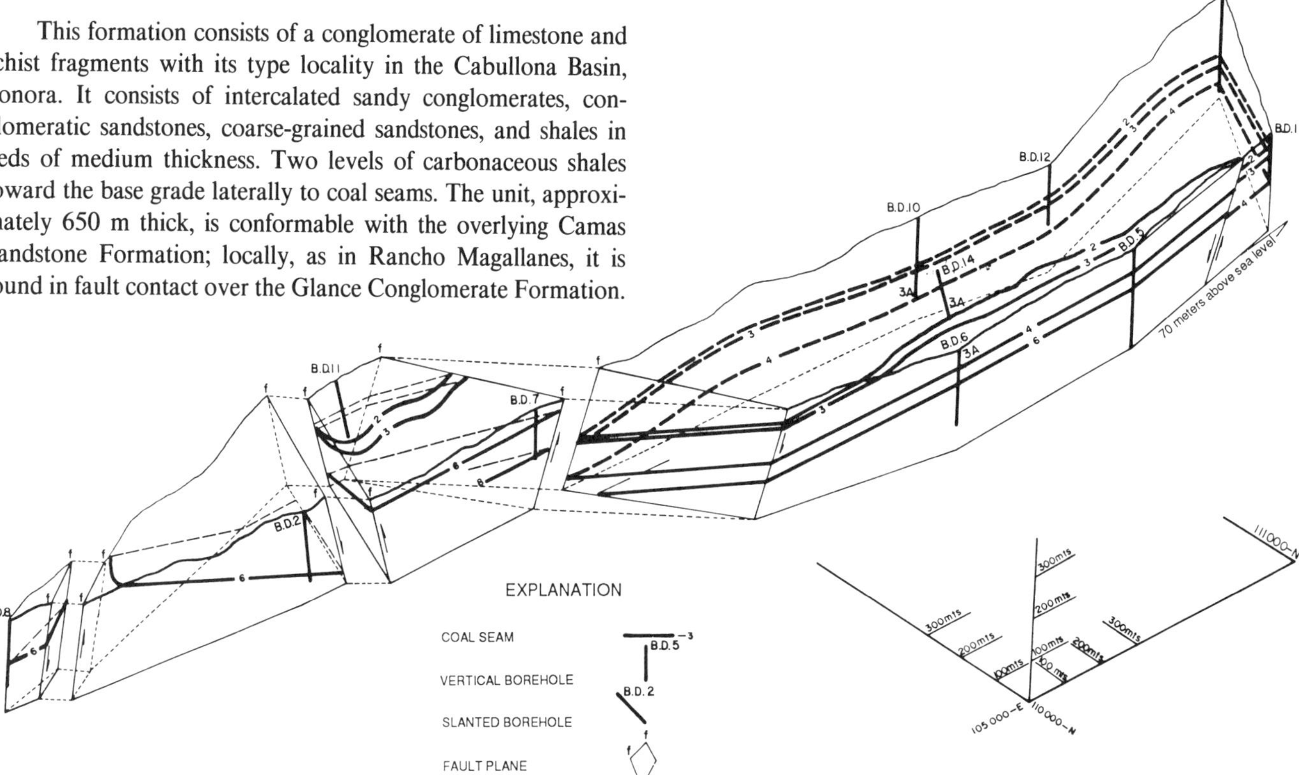

Figure 14. Block diagram showing correlation of coal seams in the San Marcial region, Sonora.

COAL DEPOSITS

The Cabullona area, with a lateral extension of 13 km,contains the San Marcos, Las Fragüitas, Santa Rosa, La Aguja, and San Juan sectors, where several coal seams are present. Thirty-eight layers of coal, carbonaceous material, or alteration products have been located in the San Marcos sector, the most prospective in the area. Eleven coal seams have been found in the El Encino sector.

Bustillo (1963) mentions parallel coal seams spaced 10 and 20 m apart, interbedded with N70°W-trending and N60°E-dipping shales and sandstones, 50 cm to 1.5 m thick, and extending along strike an average 10 km; the seam is encased in shales and is the thickest and most consistent.

The coal seams in both the San Marcos and El Encino sectors occur within the Cintura Formation. Locally the rocks are strongly folded and there may be seam repetition; however, this is difficult to judge because much of the area is covered by alluvium. Chemical analyses are as follows: moisture 3.76 percent; volatile matter 9.92 percent; sulfur 0.00 percent; fixed carbon 67.45 percent; ash 18.86 percent; calorific value 9,055 BTU = 5,034 cal. According to these values the Cabullona coal is bituminous, derived from organic sediments deposited in a transitional brackish-water basin environment and subsequently metamorphosed to coal.

RESERVES

Exploration of the Cabullona Basin is in the early stages. Therefore the chemical-petrographic characterization of the coal and reserves evaluation are premature because the coal seams are known only in outcrop. On the basis of geological studies to date, the possible resources in the basin can only be estimated; a possible volume of 80×10^6 tons may be distributed as follows:

San Marcos area: 13 km of outcrops; seven coal seams, average thickness 1.00 m each; 500-m continuity at depth; density 1.5; approximately 67×10^6 tons.

El Encino area: 6 km of outcrops; three coal seams; average thickness 1 m each; 500 m continuity at depth; density 1.5; approximately 12×10^6 tons; which adds up to a total 80 million tons of resources in the basin.

SAN PEDRO CORRALITOS BASIN, CHIHUAHUA

INTRODUCTION

This area is located within the Peña Blanca Ranch, 50 km northeast of the town of Nuevo Casas Grandes, Chihuahua, in the neighborhood of the San Pedro Corralitos railway station (Fig. 15). The San Pedro Corralitos prospect is accessed by an earthworks paralleling the railroad track for a distance of 50 km from Nuevo Casas Grandes to San Pedro Corralitos Station. From that point a turnoff to the southeast for 2 km reaches Peña Blanca Ranch. The synclinal structure covers approximately 75 km^2 at the level of the Olmos Formation.

PHYSIOGRAPHY

The San Pedro Corralitos basin forms part of the Northern Central Mesa physiographic province where isolated mountain blocks are separated by extensive plains over a large, almost desert surface. These extensive plains have an altitude of 1,200 to 1,400 m; the Sierras La Escondida and Capulín are both about 2,300 m in elevation; the principal range elevations fluctuate between 2,000 and 4,000 m. The drainage is parallel on the internal watershed in the state of Chihuahua and dendritic where coal seams are present; its ramifications multiply on the slopes of the Sierra La Escondida. Much of the water infiltrates underground through the porous soils and bolsons of the plains or is gathered in small, very short-lived lakes.

HISTORICAL GEOLOGY

The area is a mature sedimentary block consisting of a small chain of scattered, almost flat-topped hills owing their shapes to a considerably eroded lava flow. This block is strongly intruded and sustained by numerous andesitic sills and dikes.

STRATIGRAPHY

The Lower to Upper Cretaceous marine sedimentary rocks in the area belong to the Aurora Formation, attributed to the Albian on the basis of its stratigraphic position, and to the Cenomanian-Turonian Eagle Ford Formation.

In Arroyo Real del Viejo (Fig. 16), Flores G. and Gómez (1982) measured 74 m of stratigraphic section and drew the following conclusions:

1. Alcántara D. and Camacho (1977) incorrectly grouped these rocks within the Upson Formation; however, both the Upson and overlying units are present. Figure 17A shows the stratigraphy proposed by Flores G. and Gómez (1982): the coal seams are placed within the Olmos Formation, and the underlying sandstone is considered as the upper part of the San Miguel Formation.
2. The exposures correspond to a deltaic transition series consisting—from bottom to top—of deltaic slope, deltaic front, and deltaic plain (Fig. 17A).
3. These sequences and their depositional environments correlate perfectly with those of the Sabina-Monclova Coal District, and therefore the conditions for the formation of coal are similar. Upper Olmos Formation sediments are also exposed there and in Coahuila, confirming their fluvial character by the appearance of red levels toward the top of the unit.

STRUCTURAL GEOLOGY

The San Pedro Corralitos area—a sedimentary block sustained and preserved from erosion by ramifications of the igneous complexes and the Tertiary volcanism—is situated between an intrusive igneous complex (Sierra Capulín) and large intrusions in the northeasternmost Sierra La Escondida. The 10-km-long block is structurally similar to an east-southeast–trending overturned syncline (Fig. 18) with a steep, slightly overturned (80 to 89° northeast-dipping) northern flank. The more gentle southern flank dips 16 to 40° to the northeast. The block is bordered on the southeast by igneous intrusive and effusive rocks (Sierra La Escondida) and on the northwest by a fault(?) that makes it disappear suddenly in alluvium. The southern flank seems to be limited by a series of dikes and faults producing displacements in this part of the syncline. The northern flank appears to be limited by the prolongation of a series of folds in Lower Cretaceous(?) sedimentary rocks; abundant dispersed dikes and sills are present throughout most of the area.

COAL RESOURCES

Two seams, each with several associated coal levels, have been named Seams One and Two from bottom to top and are described below.

Seam One appears 12 m below the top of the San Miguel Formation with a thickness of 76 cm and a 4-cm-thick shale intercalated at 30 cm from its base. Although somewhat weathered, it is vitrinitic coal and therefore high-volatile. This high vitrain content makes it very similar to the Coahuila coals and is additional proof of the identical conditions of formation in both areas. Two other thin coal levels appear toward the top, and at 0.5 m a 15-cm-thick, very altered top layer has been identified as refractory clay. No evidence has been found of possible tectonism, which might have altered the coal's true thickness; the seam exhibits cleats that indicate only moderate tectonic pressures.

Seam Two is located 18.5 m from the top of the San Miguel Formation. Alcántara D. and Camacho (1977) mention a seam thickness of 1.80 m at this level; however, Flores G. and Gómez (1982) report a measured thickness of 30 cm, with a thinner 10-cm horizon higher in the section. Tectonism here is stronger, especially at the top, but is considered insufficient to have altered the true thickness to any considerable degree. Sandstone levels present at the top level of this seam may possibly affect and erode the coal over medium distances.

Table 9 confirms the suspected qualities of the coal. Considering that the samples are from stream-bed coal outcrops, it may be assumed that at depth the fixed carbon content will be higher,

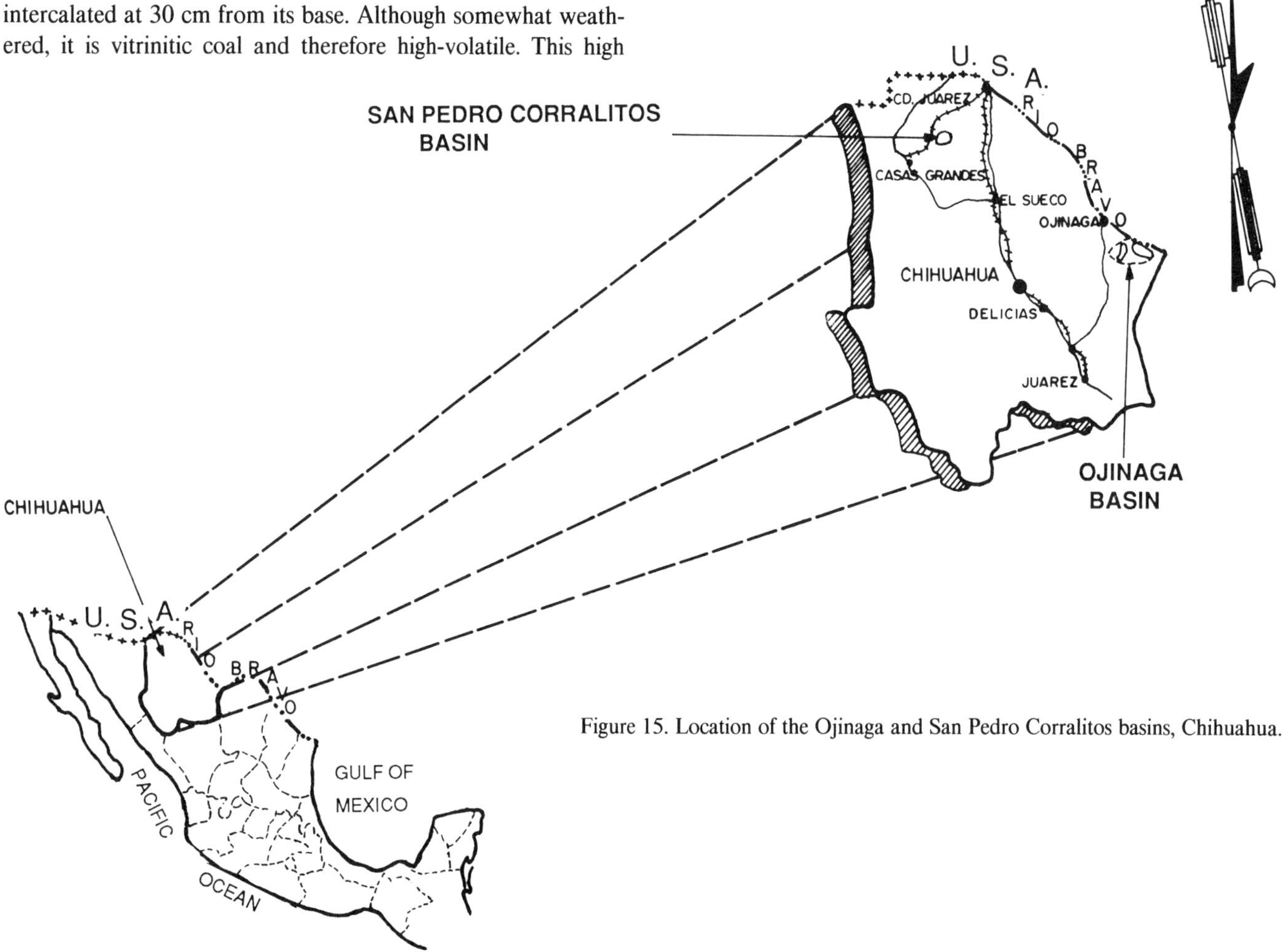

Figure 15. Location of the Ojinaga and San Pedro Corralitos basins, Chihuahua.

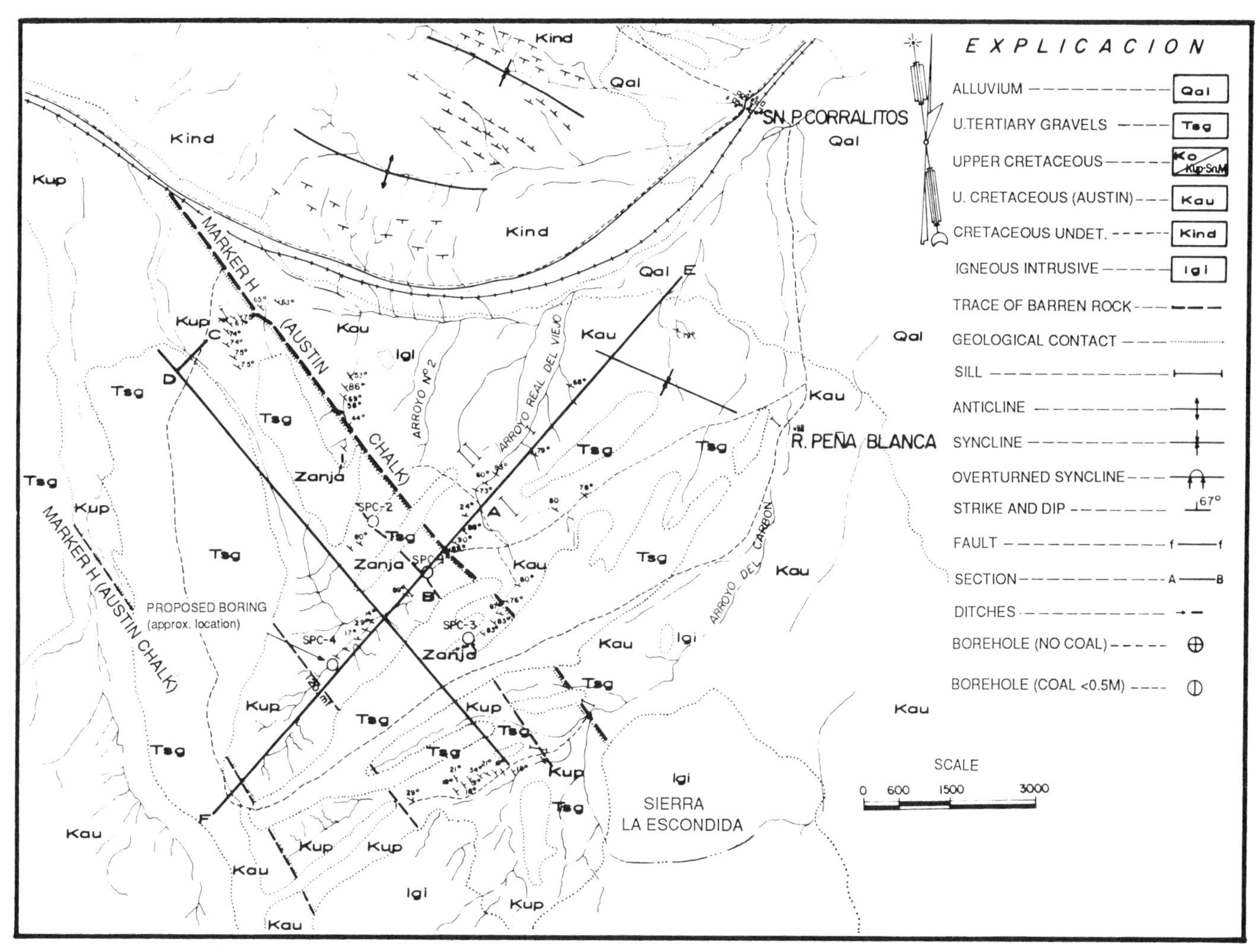

Figure 16. Geological map of the San Pedro Corralitos area, Nuevo Casas Grandes Municipality, Chihuahua.

volatile matter will show a moderate increase, and ash content will be lower. Sulfur content will probably be higher, because part of it forms soluble sulfates at the surface.

RESERVES

Prospects in this area are minimal, because (a) thickness has been extrapolated on the basis of scarce information and not verified directly; (b) the 25 to 35° dips and numerous dikes would make mechanized development extremely difficult, besides which great dpeths—of as much as 600 m in the vicinity of the fold hinge—would be quickly encountered perpendicular to the dip at short distances; and (c) reserves would be minimal for a modern production unit, with very few possibilities of increase. The San Pedro Corralitos area has no infrastructure and although relatively close to the railroad, there is no possible consumption center for this coal in the entire region. Therefore, the estimated 5,800,000 tons of coal must be categorized as inferred reserves with a maximum coal seam thickness of 0.70 m.

TABLE 9. SAN PEDRO CORRALITOS COAL ANALYSIS

Sample	Thickness (m)	Volatiles (%)	Fixed Carbon (%)	Ash (%)	Sulfur (%)	Moisture (%)
M-1-A	0.34	22.8	21.3	55.9	0.23	16.4
M-1-B	0.42	33.6	24.6	41.8	0.31	26.9
M-2-A	0.30	23.87	36.23	39.9	0.49	11.3

Float and Sink

Sample	Weight Percent	Fixed Carbon (%)	Ash (%)	Sulfur (%)
M-1-A (F)*	9.4	36.80	26.20	0.42
M-1-A (S)*	90.6		54.40	0.27
M-1-B (F)*	13.2	41.4	22.3	0.37
M-1-B (S)*	86.8		41.2	0.28
M-2-A (F)*	47.0	57.20	17.20	0.68
M-2-A (S)*	53.0		53.70	0.36

*(F) = float; (S) = sink.

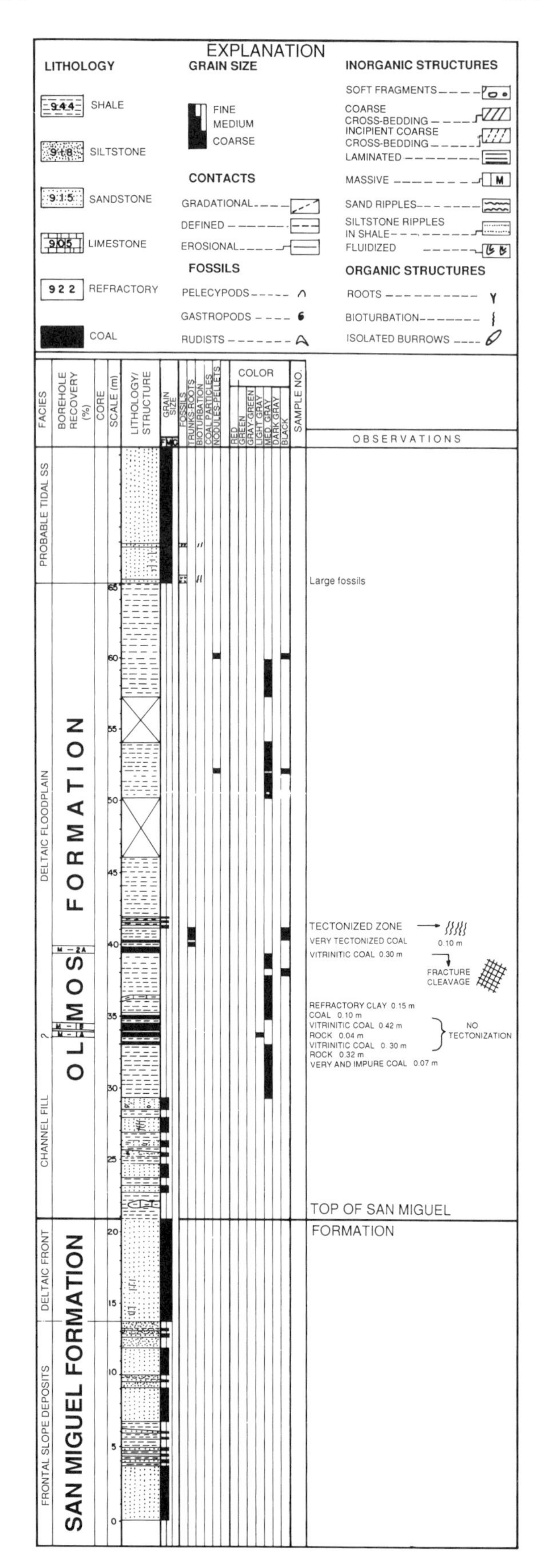

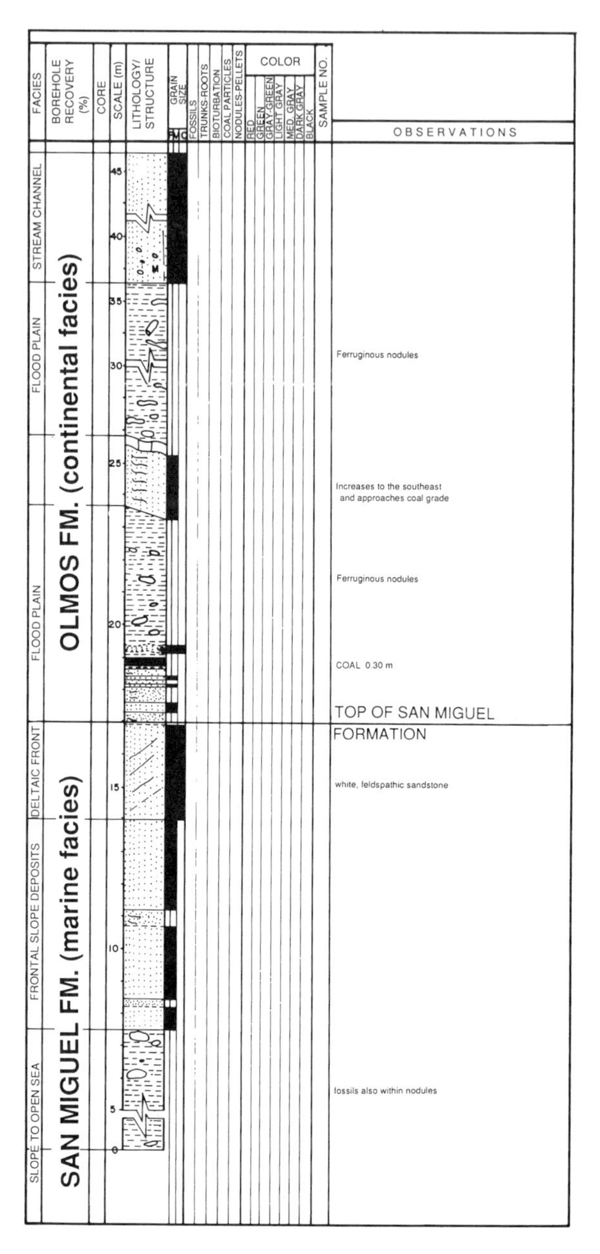

Figure 17. Stratigraphic columns for: A. San Pedro Corralitos basin (Arroyo Real del Viejo); B. Ojinaga basin.

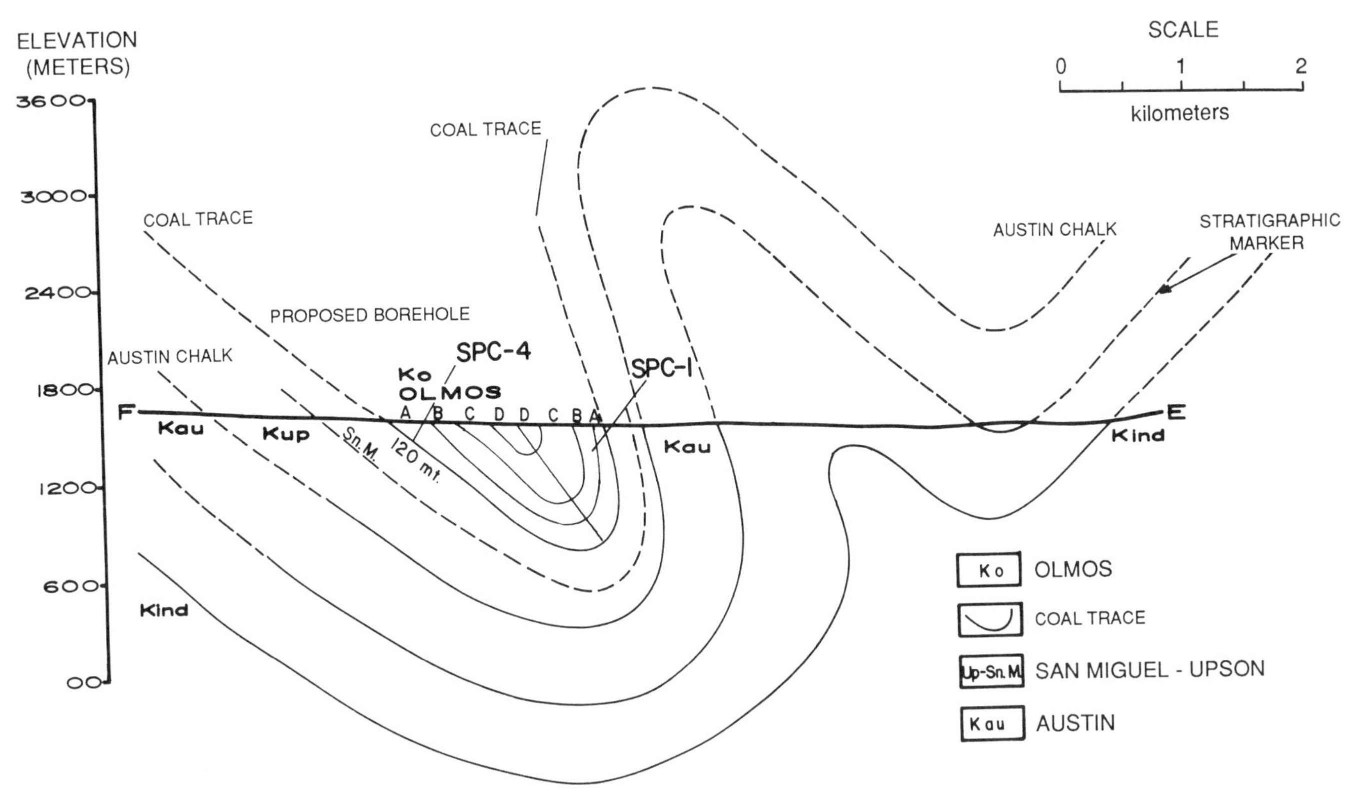

Figure 18. Structural cross section of the San Pedro Corralitos coal basin.

OJINAGA BASIN, CHIHUAHUA

INTRODUCTION

This basin is located in northeastern Chihuahua State (Fig. 15), approximately 30 km S45°E of the village of Ojinaga, at geographic coordinates 29°10′N, 104°10′W. Main access is by road from Chihuahua to Ojinaga, as well as the railroad between the two. Several dry-season transit earthworks lead to the work area from Ojinaga. The area occupies approximately 150 km^2 parallel to the Río Bravo.

PHYSIOGRAPHY

The region is part of the Basin and Range Province; topographic elevations range from 900 to 1,500 m. The ranges trend mainly northwest with some north-south exceptions. The drainage belongs to the Gulf of Mexico watershed. The principal stream is the Río Conchos, flowing into the Río Bravo near the town of Ojinaga. Drainage at the coal outcrops is dendritic; ramifications increase at the foothills of the Sierras El Mulato–San José and Rica.

STRATIGRAPHY

Several Upper Cretaceous formations are present in the region, but only the marine and continental facies (Aguja Formation) are described because the coal seam occurs at the transition from one to the other.

The most complete stratigraphic section is exposed in the Rancho San José area: two mappable units of Maastrichtian age with a combined thickness of approximately 400 m.

Marine facies

At the base, this facies consists of alternating dark gray shales and deformed sandstones with inclusions of calcareous nodules and abundant *Exogyra.* A gradational sequence of fine- to coarse-grained deformed sandstones with abundant fossils (*Ostrea,* gastropods, pelecypods, etc.) appears toward the upper part, topping the marine facies and serving as a stratigraphic marker. The entire column is intruded by sills up to 0.90 m thick. The unit is 60 m thick.

Continental facies

This facies is represented at the base by 0.35 to 0.50 m of dirty coal and a clay facies with calcareous-ferruginous concretions. A silty-sandy (stratified sandstone) sequence appears toward the top with abundant calcareous-ferruginous concretions, sills, dikes, dikelets, and dinosaur bones. The section is crowned by 8 m of basic igneous rock cover. At Block A, 48 m of stratigraphic section were measured, where exposures are generally good (Flores G. and Gómez L., 1982). The following conclusions were drawn:

Figure 19. Coal localities of Mexico.

1. Without denying the validity of the terms Continental Aguja and Marine Aguja used by several authors in the Ojinaga area for the rocks containing the coal seams, usage of the San Miguel and Olmos Formations is proposed for them here on the basis of their perfect lithologic identity with the type sections of those units in the Sabinas-Monclova Coal District. Also, a standardized Upper Cretaceous nomenclature in northern Mexico would be more helpful in regional exploration projects. The contact defined by Flores G. and Gómez L. (1982) in the area is shown on the lithologic section of Figure 17B.

2. The sedimentary features at the top of the San Miguel Formation (Marine Aguja) and base of the Olmos Formation (Continental Aguja) clearly defines a deltaic transition series from bottom to top: deltaic slope, deltaic front, and deltaic plain, each with several facies.

3. This series and its depositional environments are absolutely comparable to the San Pedro Corralitos and Sabinas-Monclova Coal District sequences, indicating a progradation of Upper Cretaceous deltas throughout northern Mexico, derived from sources in the northwest.

4. The conditions for the formation of coal in this area are therefore similar to those at San Pedro Corralitos and the Sabinas Basin; they differ from the latter only in that the continental facies predominates, and thus the balanced marine and continental conditions determining the accumulation of coal deposits are absent.

STRUCTURAL GEOLOGY

Tertiary volcanism and tectonism separated the sedimentary block into two similarly trending adjacent parts of different structural behavior.

Block A

This block is located at the northwestern slope of Sierra Rica and consists of a small, slightly arched, 3-km long, north-northwest–trending chain of hills in sediments of the Taylor-Aguja transition (Maastrichtian), dipping 20°SW in the south and up to 45°SW in the north. A reverse fault limits the block in the south-southeast and disappears suddenly in alluvium toward the southeast; the north-northwest limit is also an apparently hidden fault.

Block B

This block, between the northeastern flank of Sierra San José and the southwestern flank of Sierra Mulato, consists of continental and marine sediments of the Aguja Formation dipping 13° and 10°SE; it is less tectonized in the north.

COAL DEPOSITS

The coal seams occur in synclinal structures as regularly distributed, very lenticular sedimentary layers, interbedded with carbonaceous shales, siltstones, and sandstones. The structural behavior of the coal is closely related with the syncline containing it. In the Ojinaga Basin, in the Mina El Muerto area, and along the outcrop, only one coal level is visible in the field; it occurs 2 m above the top of the San Miguel Formation and extends over a distance of some 3 km. In general, the outcrop is poor, within a shale interval bounded by sandstones at top and bottom. The cover and residues at the old mine workings in the northernmost part of Block A prevented measuring the coal thickness reported by Del Arenal (1964). Thirty centimeters of vitrinite coal were defined within the coal level about 200 m toward the south of the old mine. From this point on, only small coal threads are found within carbonaceous shales.

RESERVES

On the basis of studies to date, no economically attractive coal thickness has been defined. Moreover, the area is strongly tectonized, and it does not seem probable that the Ojinaga Basin (Blocks A, B, and C) is prospective. However, on the basis of paleogeographic considerations inferred from knowledge of the Sabinas Basin, there is a slight chance that a west-northwest–trending coal lens may be present beneath the cover between seams A and B. Eventual reserves must be classified as geologic resources with a volume of 23,000,000 tons.

TABLE 10. COAL POTENTIAL FOR USE IN THE STEEL INDUSTRY

Metallurgical Potential	Coking Coals (x 10^6)
Proved reserves	1,255.9
Probable reserves	60.0
Possible reserves	60.0
Geologic resources	1,180.0
Total	2,555.9

TABLE 11. COAL POTENTIAL FOR ENERGY SUPPLY

Energy Potential	Thermal Coal (x 10^6)
Proved reserves	713.2
Probable reserves	92.6
Possible reserves	134.2
Inferred reserves	313.0
Geologic resources	623.0
Total	1,876.0

TABLE 12. PHYSICO-CHEMICAL CHARACTERISTICS OF MEXICAN COALS

Basin	Sub-Basin Area or Sector	Total Moisture	Proximate Analysis Ash	Volatile Matter*	Fixed Carbon†	Elemental Analysis C	H	O	N	S	FSI§	g/cm³	Caloric Value (Cal/g)**	Ash Fusion Temp, °C	Petrographic Analysis Vitrinite	Inertinite	Exinite	Reflectance	Alpern, 1981	Classification ASTM
Fuentes	Rio Escondido	4.16	33.27	48.75	51.25	75.51	5.58	15.33	1.97	1.61	1	1.6	71.23	1,540	80	15	5	0.56	Bituminous (0.56) Vitric (180) Mixed (33.27)	High Volatile Bituminous (C)
Sabinas	Sabinas	5.3	14.45	36.02	63.98	74.0	4.3	19.75	1.1	0.86	8	-	87.01	1,540	69.36	21.64	9.00	1.15	Bituminous (1.15) Vitric (69.36) Coal (14.45)	High Volatile Bituminous (A)
Sabinas	Saltillito	6.3	15.0	34.11	65.89	73.5	4.4	20.10	0.80	1.30	7	-	86.09	1,500	79.89	14.95	5.16	1.26	Bituminous (1.26) Vitric (79.89) Mixed (38.47)	Medium Volatile Bituminous (A)
Sabinas	Esperanzas	4.5	36	25.34	74.66	....	....	....	....	....	7 1/2	....	....	....	92.4	6.84	0.22	1.37	Bituminous (1.37) Vitric (92.94) Mixed (36)	Medium Volatile Bituminous
Colombia San Ignacio		3.32	18.69	56.65	43.35	78.78	6.75	10.14	1.65	2.68	1 1/2	1.37	78.40	1,338	78.37 52.62	3.57 3.74	18.06 13.64	0.45 0.45	Lignite (0.45) Vitric (78.37) Coal (18.69) Lignite (0.45) Liptic (52.62) Coal (18.69)	High Volatile Bituminous (A) High Volatile Bituminous (A)
Mixteca†	Tlaxiaco	0.1 3.8	28.87 54.61	10.78 29.18	89.22 70.82						0–7	1.5	80.32 87.07		98.97 100.00	....	0.55 1.03	0.92 1.98	Bituminous (1.79) Vitric (100) Mixed (28.87–54.61)	Low Volatile Bituminous
Mixteca† (Tezoatlan)	Plaza Lobos Plancha el Consuelo San Juan Viejo	1.2 0.76 1.38	42.48 36.46 37.96	9.5 10.74 16.22	90.5 89.26 83.22					0.70 0.50 0.45	1 1		79.54 85.59							Semi-Anthracite Semi-Anthracite Bituminous Low in Volatiles
Ojinaga**		12.99	14.93	41.84	58.14							1.58	52.39		97.7	2.93		0.50 0.76	Bituminous (0.50–0.76) Vitric (97.07) Coal (14.93)	Sub-Bituminous (C)
Barranca	Santa Clara	10.5	6.0	8.4	91.6	97.3	0.6	1.4	0.5	0.2			78.89	1,338				4.27	Anthracite	Anthracite

*Ash-free dry base.
†Point values—not basin averages.
§Free-swelling Index.
**Ash-free wet base.

TABLE 13. MEXICAN COAL POTENTIAL*

Basin	State	Reserves and Resources (millions of metric tons)					
		Proved	Probable	Possible	Inferred	Geologic Resources	Totals
Sabinas	Coahuila	1,255.9	60	60		1,180	2,555.9
Río Escondido	Coahuila	600	60	56	200	300	1,216
De la Mixteca	Oaxaca	32.5	20	30	....	80	1,625
Colombia–San Ignacio	Nuevo León, Tamaulipas	76	3	30	63	80	252
Barranca	Sonora	4.7	9.6	18.2	50	60	142.5
Cabullona	Sonora	....	....	....	....	80	80
Ojinaga	Chihuahua	....	....	....	....	23	90
Total		1,969.1	152.6	194.2	313	1,803	4,431.9

*To January 1985.

OTHER REPORTED COAL OCCURRENCES IN MEXICO

From the very beginning of coal mining in Mexico, industrial attention focused on the Coahuila basins. Consequently, no importance was attached to the systematic exploration for evaluation purposes of the many evidences of coal deposits in the rest of the country. Nevertheless coal, lignite, and peat localities have been reported. Few have been studied geologically or to define their origin; none have been quantified or evaluated.

Brief reconnaissance studies have located coal manifestations, outside of the commercially known basins, in the states of Chihuahua, Guerrero, Hidalgo, México, Michoacán, Oaxaca, Nuevo León, Puebla, San Luis Potosí, Tamaulipas, Veracruz, and Durango; some are shown on Figure 19. The best known deposits are of paludal or deltaic origin: lenticular facies of Jurassic, Cretaceous, and Eocene or younger marine or continental formations, all of which are worth investigating.

Examples of Jurassic coal are the Tlapa (Guerrero), Tejaluca, Ahuatlán, and Tecomatlán (Puebla) deposits. Cretaceous coals, which have not been studied, are Mina del Muerto, between San Carlos and Ojinaga, and the Corralitos area northeast of Nuevo Casas Grandes (Chihuahua), Bolsón de Mapimí (Durango), Temexalaco, Cuxcatlán and Xilitla (San Luis Potosí), C. Villa Pánuco, Tempoal, and Tacasneque (Veracruz). Numerous shows of Eocene or younger subbituminous coal of paludal origin are found at Yahualica and Chicontepec (Veracruz), Zacualtipan (Hidalgo, possibly Miocene), and Tamanzunchale (San Luis Potosí). In addition, there are a good number of lignite and/or peat deposits of lagoonal origin, such as Tehuichila, Zacualtipan, and San Miguel Ocaxichitlán (Hidalgo), Chalco (state of México), and others, possibly of Miocene age, in the highlands.

Almost all of these localities and many others have been the subject of geological reconnaissance; mines have been worked in very few. A few hundred tons were extracted in some cases at the beginning of the century or just after the revolution. For example, coal was extracted for Fundidora de Monterrey in Colombia (Nuevo León); the "La Salvadora" mine was developed in 1874 in Tecomatlán (Puebla). Also at Peña Ayuquila (Veracruz), coal from "El Cristo" Mine near Tempoal (Veracruz) was extracted and shipped down the Río Pánuco.

CONCLUSIONS

In the geologic evolution of the country, conditions for the formation of the present-day coal deposits prevailed during three main periods: the Upper Triassic–Middle Jurassic (200 to 170 Ma), the Upper Cretaceous (75 to 65 Ma), and the Eocene (50 to 40 Ma). The first of these is represented by the Sonora coal in the northwest and the Oaxaca deposits in the south. Most of the Mexican coal is Upper Cretaceous, in the Sabinas, Fuentes–Río Escondido (Coahuila), Ojinaga and San Pedro Corralitos (Chihuahua), and Cabullona (Sonora) subbasins. Eocene coal and lignites occur in the Colombia–San Ignacio region. Other localities of coal, lignite, and peat occurrence in the country are poorly known, not having been investigated at depth or considered as poor prospects.

According to present-day knowledge of Mexican coal reserves and characteristics (Tables 10 to 13) the coal potential of the country may be classified according to its use in the steel industry (coking coal; Table 10) or for energy supply (thermal coal; Table 11). Coal potential given in Tables 10, 11, and 13 is in metric tons. Lack of information as to the chemical properties of the coals prevents converting these figures to universal thermal units in some areas.

REFERENCES CITED

Alpern, B., 1981, Pour une classification synthetique universelle des combustibles solides: Bulletin Centres de Recherches Exploration et Production Elf-Aquitaine, v. 5, no. 1, p. 271–290.

Burckhardt, C., 1927, Cefalopodos del Jurasico Medio de Oaxaca y Guerrero: Mexico, Instituto de Geologia, Boletín 47, 108 p.

Cárdenas V., J., 1966, Contribución al conocimiento de la Mixteca Oaxaquena: Mineria y Metalurgia, v. 38, p. 15–107.

Del Arenal, R., 1964, Estudio geologico para localizacion de yacimientos de carbon en el area de Ojinaga, San Carlos, Estado de Chihuahua, Mexico: Asociación Mexicana de Geologos Petroleros Boletín.

Erben, H. K., 1956, Estratigrafia a lo largo de la carretera entre Mexico, D.F., y Tlaxiaco, Oaxaca, con particular referencia a ciertas areas de los Estados de Pueblo, Guerrero, y Oaxaca: México, D.F., 20th International Geological Congress Excursion A-12, p. 11–36.

Pesquera V., R., and Carbonell C., M., 1960, Geologia y exploración de los depositos de carbon de la region de San Marchial, Estado de Sonora: Consejo Recursos Naturales no Renovables Boletín 59, 41 p.

Robeck, R. C., Pesquera V., R., and Arrendondo S., U., 1956, Geología y depositos de carbon de la region de Sabinas, Estado de Coahuila: México, D.F., 20th International Geological Congress, 190 p.

Salas, G. P., 1949, Bosquejo geologico de la cuenca sedimentaria de Oaxaca: Asociación Mexicana de Geólogos Petroleros Boletín, v. 1, p. 79–156.

Unpublished sources of information

Alcántara D., J., and Comacho, R., 1977, Informe final proyecto carbon San Pedro Corralitos, Chihuahua: Final report unpublished.

Bustillo S., G., 1963, Deposito de carbon en la Sierra Cabullona, Municipio de Agua Prieta, Sonora: Mexico, D.F., Consejo de Recursos Minerales, Archives Micrografia.

Flores G., E., and Gómez L., F., 1982, Informe de la visita realizada y sintesis del conocimiento de los prospectos "San Pedro Corralitos y Ojinaga, Chihuahua" con posibilidades de actuación a futuros: Consejo de Recursos Minerales, unpublished.

Gomez L., F., and Evaristo F., G., 1983, Resultado obtenido en la exploración por reservas de carbon en la Subcuenca de Lampacitos y extreme sur de la Saltillito: Consejo de Recursos Minerales, unpublished.

Printed in U.S.A.

The Geology of North America
Vol. P-3, Economic Geology, Mexico
The Geological Society of America, 1991

Chapter 12

Geology of uranium deposits in Mexico

Guillermo P. Salas
Geologo Consultor, Asesor del Consejo de Recursos Minerales, Centro Minero Nacional, Carretera México-Pachuca, Pachuca, Hgo., México, D.F.
Fernando Castillo Nieto
Comisión de Fomento Minero, Puente de Tecamachalco No. 26, Col. Lomas de Chapultepec, 11000 México, D.F.

NUCLEAR ENERGY

Until 1977 it was believed that the worldwide energy problem would have to be solved by considerably increasing nuclear energy generation (Rich and others, 1977). In 1986 this is no longer the case. Mexico's 77 billion barrels of proved oil and gas reserves allow postponement of the decision to depend on nuclear energy generation to any important degree.

Prior to 1970, nuclear power plant development took the lead over coal- and oil-fueled power plants in countries, such as the Netherlands, England, Norway, and others, whose other resources were insufficient to satisfy their greater energy demand. In Mexico, the National Nuclear Energy commission was created in 1955—more for the strategic purpose of diversifying rather than developing new energy sources—after classifying all radioactive minerals as national reserves. The National Institute of Nuclear Energy replaced the commission in 1972 and was in turn replaced in 1979 by Uranio Mexicano (URAMEX), which disappeared in 1985. The Board of Mineral Resources (Consejo de Recursos Minerales) inherited the uranium exploration activities, and the Mining Development Commission (Comisión de Fomento Minero) took over the development and marketing aspects.

GEOLOGY OF URANIUM DEPOSITS

Initially, the source of the uranium in economically attractive deposits posed an enigma. However, Rich and others (1977) published the combined findings of hydrothermal deposit researchers, and the close geochemical similarity between hydrothermal and sandstone-hosted deposits is now known. The assumption is that hexavalent uranium is transported in oxidizing solutions and becomes tetravalent on contact with reducing agents such as coal, carbonaceous clay or sand, and/or oil sands, which causes it to be deposited.

The geology of the known Mexican uranium deposits is of two main types: the first occurs in igneous, especially acid extrusive, rocks such as rhyolites and rhyodacites, or in felsic intrusives in contact with sedimentary rocks such as limestones. The other type appears in sedimentary rocks, particularly mid- or upper Tertiary sandstones, but also in Mesozoic limestones.

Rich and others (1977) demonstrated that uranium tends to become enriched in the final stages of magmatic differentiation, generally coinciding with the presence of simultaneously deposited potassium and quartz. Alkaline and granitic rocks contain an average of 2 to 4 ppm of uranium, but felsic rocks, and pegmatites in particular, contain more than 10 ppm; in some cases (bostonite dikes), uranium content reaches 100 ppm. Ultrabasic rocks, on the other hand, are always uranium poor.

Hydrothermal uranium worldwide occurs in seams within fractures filled with calcareous material and occasionally with quartz; uranium also fills all sorts of holes caused by premineralization. Fractures may be micro or macro, short or long (at times over 1 km in length), and very shallow or deep-seated. The hydrothermal minerals are generally secondary minerals derived from pitchblende; in Mexico, pitchblende has been reported at some localities in the state of Chihuahua. In the Placer de Guadalupe and El Aire area (Chihuahua), carnotite veins occur in limestone fractures and in the granodiorite, causing the fractures (González Reina, 1956). Also, the Telixtlahuaca (Oaxaca) pegmatites contain allanite and betafite veins with thorium oxides and some niobium.

URANIUM LOCALITIES

The principal Mexican hydrothermal deposits studied by the Nuclear Energy National Commission, and afterward by URAMEX, are vein deposits in the states of Chihuahua, Sonora, Durango, and others. At the secondary or transported "El Chapote," "La Coma," and "Buenavista" deposits within sedimentary rocks in the state of Nuevo León, uranium minerals are hosted in sandstones of the Oligocene Frío No Marino Formation. The varied geology of these deposits is summarized in Table 1 and in the following paragraphs. Figure 1 shows the principal uraniferous localities in Mexico.

Salas, G. P., and Castillo N., F., 1991, Geology of uranium deposits in Mexico, *in* Salas, G. P., ed., Economic Geology, Mexico: Boulder, Colorado, Geological Society of America, The Geology of North America, v. P-3.

TABLE 1. IN SITU URANIUM RESERVES OF MEXICO*

Locality	U_3O_8 (tons)	
Chihuahua:		2,789
Margaritas	1,224	
Puerto III	630	
El Nopal	361	
Domitila	61	
Nopal III	40	
Others: Laguna del Cuervo, Laguna del Diablo, Sierra de Gómez, Puertos I, II, IV, V, and VIII, El Calvario, La Gloria, etc.	473	
Nuevo León:		5,075
La Coma	1,318	
Buenavista	1,256	
El Chapote	843	
Diana	940	
Peñoles	191	
Presita	185	
Dos Estados	169	
Trancas	130	
Santa Fe	43	
Sonora:		1,664
Los Amoles	900	
Others: Santa Rosalía, Noche Buena, Picacho, Luz del Cobre, Santa Rosa, Moctezuma, Huasabas, Granaditas, etc.	764	
Durango:		1,267
Coneto	664	
La Preciosa	434	
El Mezquite	90	
Others	79	
Oaxaca:		696
Santa Catarina Tayata	600	
San Juan Mixtepec	96	
	Total:	11,491

*URAMEX, May 1983.

Chihuahua

Most of the uranium localities in Chihuahua are in the Sierra Peña Blanca, about 60 km north of the town of Chihuahua. The Sierra Peña Blanca is characterized by its strongly faulted, fractured, and eroded sharp topography at a geomorphologically advanced youthful stage. Paleozoic limestones, shales, and sandstones are overlain unconformably by mainly calcareous Cretaceous rocks; the entire sedimentary sequence is covered by Tertiary volcanic rocks of acid composition. El Nopal mine is an open-pit development in a crystalline rhyolitic ignimbrite, the Nopal Formation, which exhibits reddish brown colors in the mineralized area. El Nopal or the Nopal I deposit contains an estimated 175,000 tons of ore, averaging 0.2066 percent U_3O_8.

The Las Margaritas deposit in the central portion of the Sierra Peña Blanca is hosted in a rhyolitic ignimbrite, the Escuadra Formation, overlying the Nopal Formation within a topographic depression covered by talus debris. Fractures filled with uranium minerals and iron oxides indicate the presence of the deposit. Topsoil stripping revealed the ignimbrite formation to be highly fractured and uranium-mineralized; on occasion the uranium-filled fractures show the appearance of chimneys. 1,234,000 tons of ore averaging 0.0992 percent U_3O_8 are estimated. In the Margaritas deposit the uranium is associated with molybdenum in an estimated volume of 5,308,000 tons of ore averaging 0.15 percent Mo, which represents 7,962 tons of molybdenum. In neighboring areas, an estimated 4,022,000 tons of ore averaging 0.095 percent Mo represent an additional 3,821 tons of molybdenum. When URAMEX ceased its activities in 1983, the recovery and development costs of the molybdenum were under study.

The Puerto III deposit in the central part of the Sierra Peña Blanca is hosted at the contact of the Nopal Formation with the overlying Escuadra Formation. The mineralization here occurs both filling fractures and disseminated within a seam that follows the contact. About 569,000 tons of ore with an average 0.1107 percent U_3O_8 are estimated for this deposit; underground mining development has begun on two sloping fronts.

The above three deposits have similarities in their uranium mineralization (uranophane, carnotite, and metayuyamunite); pitchblende occurs at depth in "El Nopal." These deposits make up the Mining Metallurgic Production Project of Peña Blanca, which began with the construction of a conventional acid leaching plant that was never completed.

Nuevo León

In Nuevo León, the La Coma, Buenavista, El Chapote, Peñoles, and Trancas deposits are located in the Burgos Basin bounded on the north by the Río Bravo. Paleocene to Recent sedimentary sequences crop out in parallel belts: younger toward the east and aging westward. The sediments consist of alternating marine shales and sandstones: mainly sandstones, sandy shales, volcanic ash, and carbonaceous layers of continental and paludal

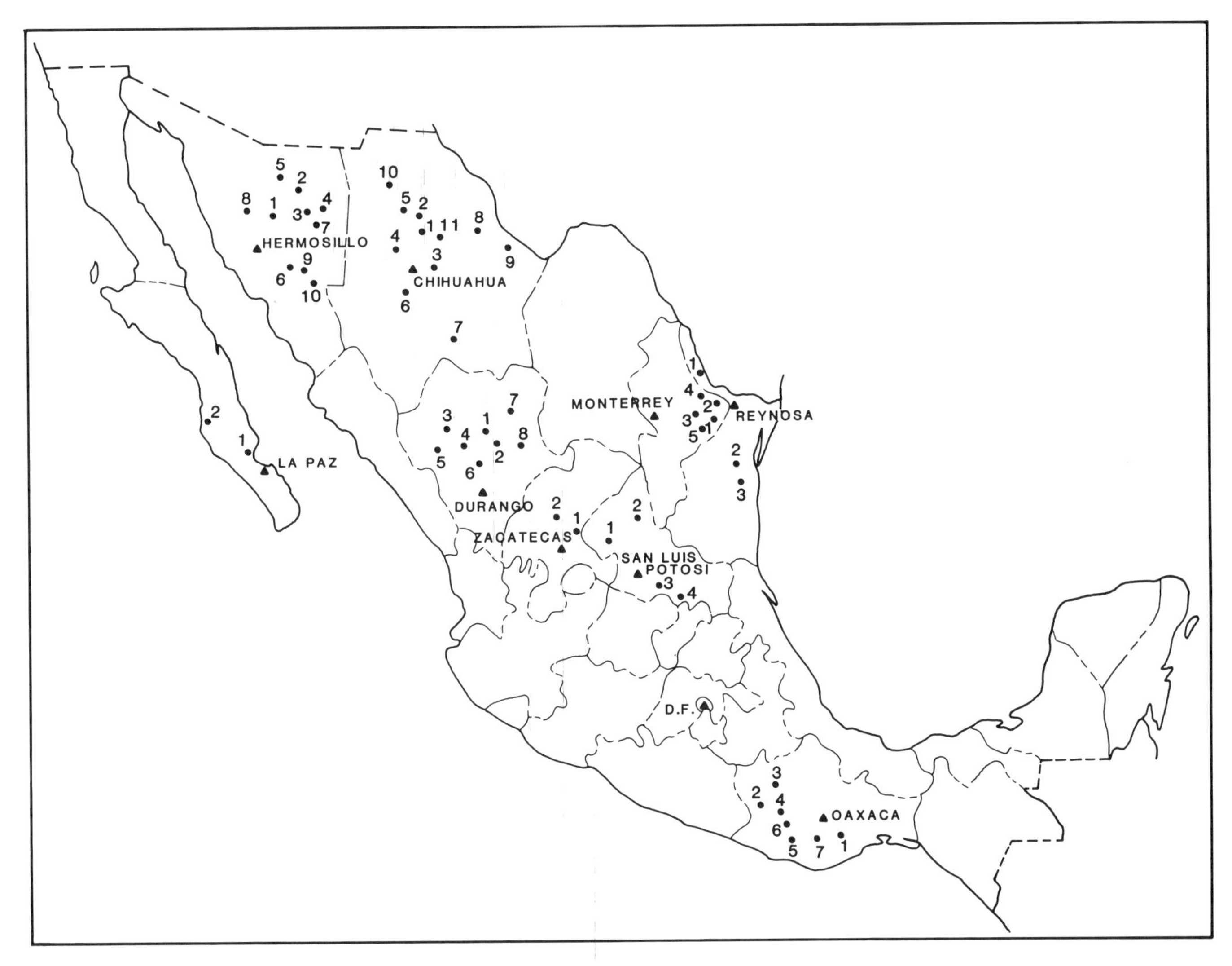

Figure 1. Principal uraniferous localities in México.

SONORA:
1. Los Amoles-Martín
2. Granaditas
3. Moctezuma
4. Huasabas
5. Santa Rosalía-Picacho
6. San Javier
7. Los Caballos
8. Nochebuena
9. Luz del Cobre
10. Yécora

CHIHUAHUA:
1. Sierra Peña Blanca
2. Laguna del Cuervo
3. La Gloria–La Mesa
4. Majalca
5. Cerros Colorados
6. Pastorias
7. Adargas
8. Ojinaga
9. Sotolar
10. Los Arados
11. Sierra de Gómez

OAXACA:
1. Los Cantiles
2. San Juan Mixtepec
3. Tezoatlán–El Pipi
4. Boca de Perro
5. Magdalena-Peñasco
6. Santa Catarina Tayata
7. Ocotlán Taviche

TAMAULIPAS:
1. Díaz Ordaz
2. Burgos-Méndez
3. Cruillas

ZACATECAS:
1. Villa de Cos-Bañón
2. Tetillas

DURANGO:
1. El Mezquite
2. Buenavista
3. Sierra de Coneto
4. Sierra de San Francisco
5. Santiago Papasquiaro
6. Chupaderos-Gramón
7. La Merced
8. La Preciosa

NUEVO LEON:
1. El Chapote-Diana
2. Santa Fe–Dos Estados
3. La Coma
4. Buenavista
5. Presita-Trancas-Peñoles

BAJA CALIFORNIA SUR:
1. San Juan de la Costa
2. Santo Domingo

SAN LUIS POTOSI:
1. Peñón Blanco–Salinas
2. Wadley–Sierra de Catorce
3. Río Verde
4. El Realito

nature. The La Coma, Buenavista, El Chapote, Peñoles, and Trancas deposits are contained in the Oligocene Frío No Marino Formation, a continuous belt of clayey sediments and semiconsolidated sands underlying a flint and limestone pebble conglomerate known as the Norma Conglomerate.

In the La Coma area, 30-m-thick sandstones contain the uranium mineralization and are covered by clayey and shaley beds alternating with tuffaceous material, overlain in turn by the Norma Conglomerate. The uranium horizon ranges from 55 to 92 cm in thickness, dips practically from the surface to depths of over 100 m, and occurs at the base of the sandy Frío No Marino Formation, in contact with fine shales of possibly late Eocene age. Exploratory drilling has defined it along a belt 30 km long and 2 to 3 km wide. About 753,000 tons of ore averaging 0.175 percent U_3O_8 have been estimated at La Coma, and 785,000 tons averaging 0.160 percent at Buenavista. These deposits contain the same uranium minerals (mainly ianthinite and coffinite); fragments of organic material act as the reducing element in the fixation of the uranium. The presence of coffinite indicates a higher degree of radiometric equilibrium. Consequently, these deposits are older than the uraninite mineralization characterizing the younger middle-upper Oligocene, lower Miocene, and Pliocene horizons of the neighboring Texas deposits.

Development of the La Coma and Buenavista deposits (included in the La Coma projection project) may be accomplished by in situ leaching, or by traditional opencut mining to maximum depths of 40 m combined with underground mining at greater depth, depending on the price of the mineral. The choice will depend on the results of in situ leaching tests, this being the most convenient process if the sandstones have good permeability. Traditional mining would be considered only if the sandstones have been made impermeable by clay sealing up the intergranular spaces.

Sonora

In Sonora, the Los Amoles deposit is about 90 km northeast of the town of Hermosillo, located in the western foothills of the Sierra Aconchi, forming part of the initial mountain ranges of the Sierra Madre Occidental. The geomorphology of the area is characterized by a series of hills emerging abruptly as isolated cliffs that consist mostly of igneous rocks, which predominate over the metamorphic and sedimentary rocks. The Sierra Aconchi is a large granite batholith in contact with medium-grade metamorphic (silicified volcanic) rocks resting as float over the granite intrusive, followed by a trachyandesite and andesite sequence of probably Lower Cretaceous age.

The Los Amoles uranium deposit is emplaced in a fracture system within the trachyandesite, which has been strongly altered by hydrothermal solutions (formation of chlorite, sericite, and kaolin, and very silicified). The dynamic effects of the metamorphism have developed complex reticulate fracture systems filled with radioactive minerals and extending to the igneous intrusive-metamorphic-extrusive contact zones. Sediments at the surface are alluvium deposited by streams descending the range toward the Río Cucurpe or Río San Miguel.

The Los Amoles deposit forms an elongated lens in which 1,915,000 tons of ore averaging 0.047 percent U_3O_8 have been estimated. The uranium mineralization is mainly secondary (uranospilite, zippeite, and gummite); primary pitchblende appears occasionally. The occurrence of the uranium oxide as disseminated impregnations and scabs in a multitude of small fractures resembling a stockwork permits opencut mining and subsequent heap leaching; the uranium solutions can then be treated by ionic exchange to obtain uranium concentrates.

All of the above deposits were covered by the three URAMEX uranium production projects: Peña Blanca, La Coma, and Los Amoles. Of these, the La Coma feasibility study could not be carried out due to metallurgical problems; consequently, development plans were not defined. The Peña Blanca and Los Amoles Projects began construction at both mine and plant levels; initially they were almost marginally commercial. At present, feasibility studies must be updated, taking into account both the lower price of uranium and the higher mining and development costs, which would make these projects noncommercial. On the other hand, if purchase of uranium becomes necessary, dependence on foreign supply could be avoided by continuing these projects.

Durango

Outstanding among several uranium manifestations in the state of Durango is the Sierra de Coneto, where tuffaceous beds in Cenozoic volcanic felsic rocks contain abundant fracture-filling and disseminated secondary uranium mineralization, principally at Montosa, Pinito, and Perla. Exploratory drilling has revealed widespread occurrence of uranium in extensive, though low-grade (200 ppm) deposits. La Preciosa is an exploratory underground mine where uranium mineralization occurs in silicified and kaolinized breccias produced by intrusive rocks affecting Cretaceous calcareous and clayey sediments and causing local brecciation. About 723,330 tons of ore averaging 0.06 percent U_3O_8 have been estimated. Exploration is presently suspended. El Mezquite is another exploratory underground mine in a Tertiary volcanic rhyolitic sequence containing fracture-filling and disseminated secondary mineralization within a roughly tabular mass in a tuffaceous layer; 180,000 tons with 0.05 percent U_3O_8 have been estimated.

Oaxaca

Uranium mineralization at Santa Catarina Tayata in western Oaxaca State is associated with tuffs and conglomerates of possible Cenozoic age, near the contact with Jurassic silicified sandstones, which in turn underlie Cretaceous limestones. The mineralization appears to be the result of remobilization of minerals by Cenozoic intrusive dikes and intermediate volcanism. The exploratory project comprises four known deposits in which

850,000 tons with 0.07 percent U_3O_8 have been evaluated. At San Juan Mixtepec the first drillholes penetrated uranium ore at the contact between Jurassic schists and older gneisses. Exploration of these localities was suspended when URAMEX was closed.

Baja California Sur

Also worth mentioning is the presence of uranium ore in phosphatic rock at San Juan de la Costa and Santo Domingo in Baja California Sur, presently used in the production of phosphoric acid for fertilizers. According to the organization in charge of developing these deposits, Roca Fosfórica Mexicana (ROFOMEX), there are 1,500 million tons of phosphate rock, in which 151,000 tons of U_3O_8 have been estimated (Table 2).

The only economical method of recovering the uranium at present is by use of the phosphoric acid that results from dissolving the rock with sulfuric acid, a process used by Fertilizantes Mexicanos (FERTIMEX) to produce fertilizer. URAMEX, together with FERTIMEX, initiated the UF—1 Project with the objective of obtaining clean phosphoric acid by removal of the radioactive pollutant, for the dual purpose of producing noncontaminated fertilizer and obtaining an uranium-rich solution, from which a concentrate would subsequently be derived. The Depa-Topo process offered by the JEN-SENER Spanish group was selected by bidding; their plans were to build a plant next to the FERTIMEX facility at Pajaritos, Veracruz, with an installed capacity of 375,000 tons of P_2O_5 per year and 150 to 160 tons of U_3O_8 contained in the yearly production of phosphoric acid. Suspension of the project in 1982, due to lack of appropriation, left it at the conceptual and basic engineering stages.

TABLE 2. ESTIMATES OF URANIUM IN BAJA CALIFORNIA SUR

Deposit	Phosphate rock (tons)	U_3O_8 (%)	U_3O_8 content (tons)
San Juan de la Costa	50,000,000	0.012	6,000
Santo Domingo	1,450,000,000	0.010	145,000
Total	1,500,000,000		151,000

REFERENCES CITED

Gashing, J., 1980, Development of uranium exploration models for the prospector-consultant system: U.S. Geological Survey Office of Resource Analysis, SRI Project Publication 7856, p. 142–147.

Rich, R. A., Holland, H. D., and Petersen, U., 1977, Hydrothermal uranium deposits, *in* Developments in economic geology, v. 6: Amsterdam, Elsevier, 264 p.

Viscaino, M. F., 1980, Presencia de URAMEX en el Desarrollo de Mexico: Méico City, Comisión de la Communicacion de URAMEX, p. 211–235.

URAMEX, 1980–1983, unpublished reports.

Printed in U.S.A.

The Geology of North America
Vol. P-3, Economic Geology, Mexico
The Geological Society of America, 1991

Chapter 13

Geohydrology

Guillermo P. Salas
Geologo Consultor, Asesor del Consejo de Recursos Minerales, Centro Minero Nacional, Carretera México-Pachuca, Pachuca, Hgo., México, D.F.
C. García H.
Secretaria de Agricultura y Recursos Hidraulicos, Insurgentes Sur, No. 476, Col. Roma, 06760 México, D.F.

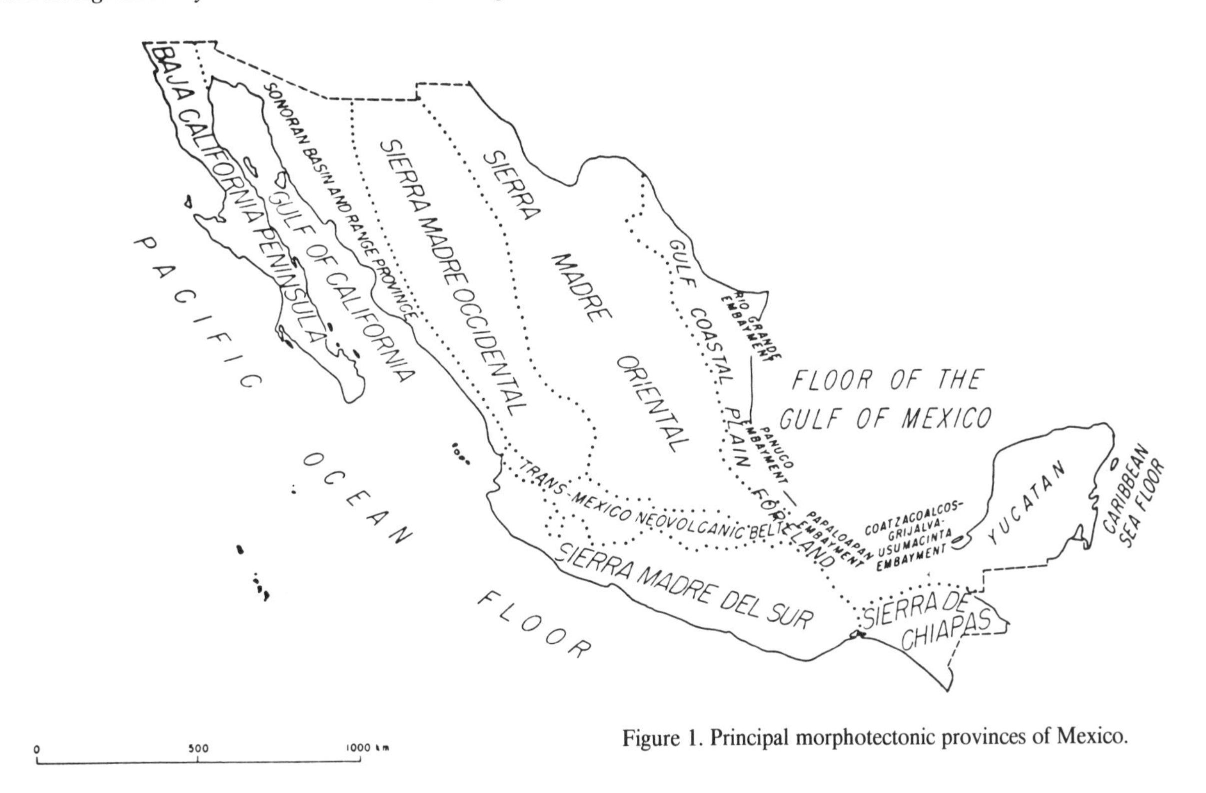

Figure 1. Principal morphotectonic provinces of Mexico.

Mexico's continental and peninsular area of approximately 2,000,000 km^2 between 15°N and 32°30′N comprises dry-climate latitudes north of the Tropic of Cancer, so that about 60 percent of the northern part of the country is desert or semidesert. The main physiographic elements are shown on Figure 1.

The Sierra Madre Occidental and Sierra Madre del Sur are formed principally by a 2,000-m thickness of interbedded extrusive volcanic and pyroclastic rocks, apparently generated by early and mid-Tertiary volcanic emissions through extensive fissures and northwest-trending faults, and middle to late Tertiary granodioritic igneous intrusions. This volcanic sequence overlies a Precambrian metamorphic basement in portions of the states of Sonora, Oaxaca, and Chiapas. Strong fluvial and tectonic erosion has deeply carved both mountain ranges, giving rise to a large number of geohydrologically important river basins (Fig. 2).

The products of erosion that fill the valleys with fine- and coarse-grained clastics hundreds of meters thick (on occasion over 1,000 m), contain Tertiary and Quaternary aquifers of considerable importance in the economy of the north and west portions of the country (Figs. 3 and 4).

The stratigraphic section of the Sierra Madre Oriental, another physiographic element of structural and hydrological control, consists of more than 5,000 m of marine platform deposits: Cretaceous and Jurassic limestones interbedded with substantial thicknesses (tens and, at times, hundreds of meters) of shales and marls. The sequence includes extensive fossiliferous reef

Salas, G. P., and García H., C., 1991, Geohydrology, *in* Salas, G. P., ed., Economic Geology, Mexico: Boulder, Colorado, Geological Society of America, The Geology of North America, v. P-3.

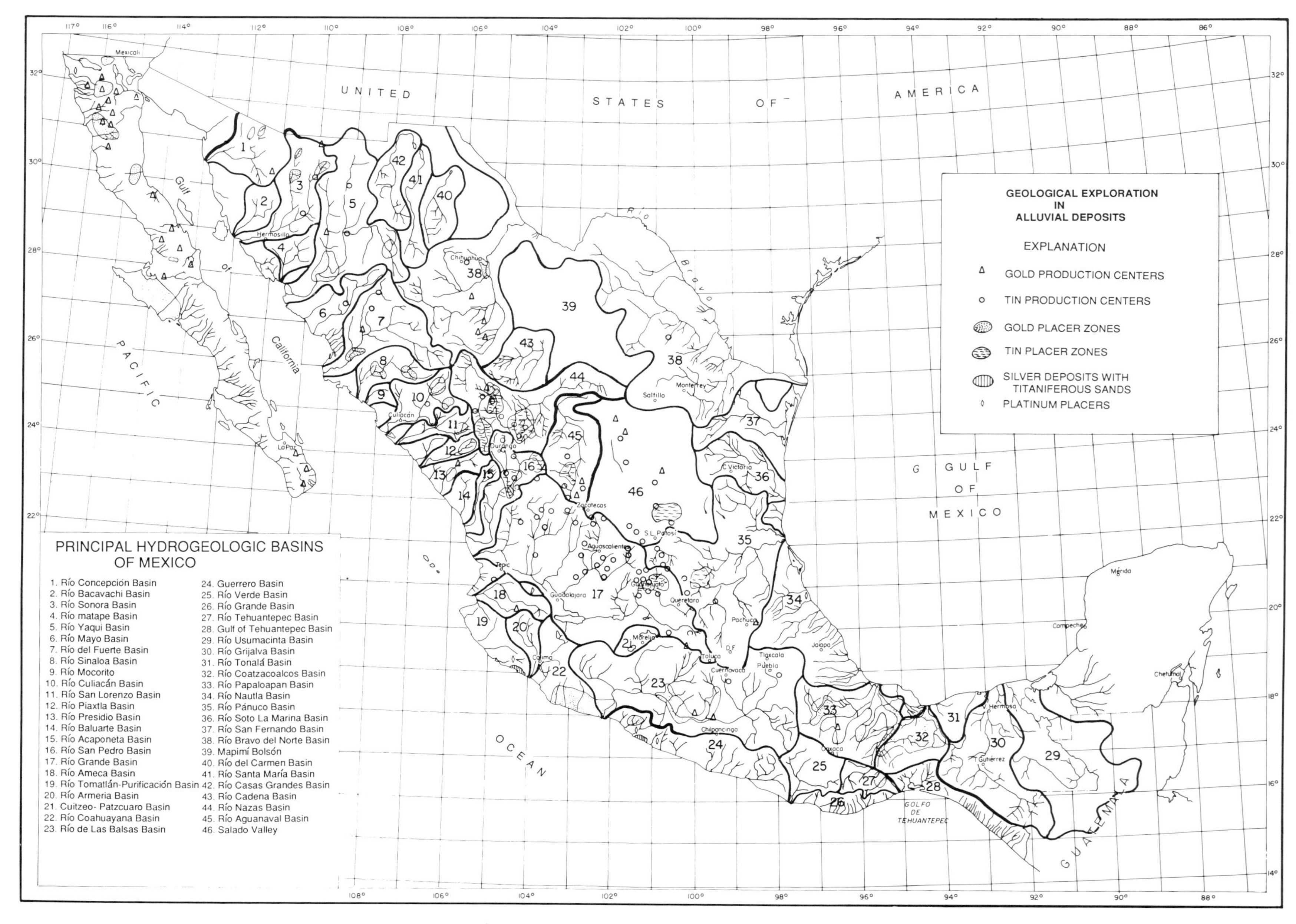

Figure 2. Principal hydrogeologic basins of Mexico.

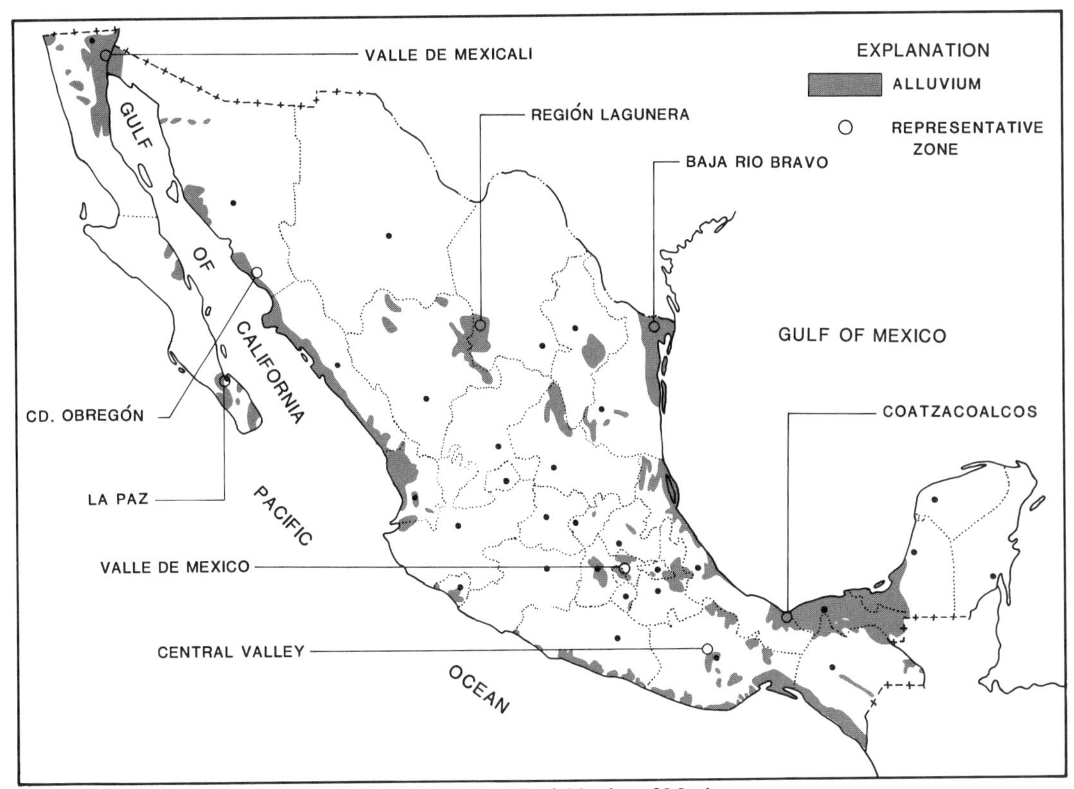

Figure 3. Major alluvial basins of Mexico.

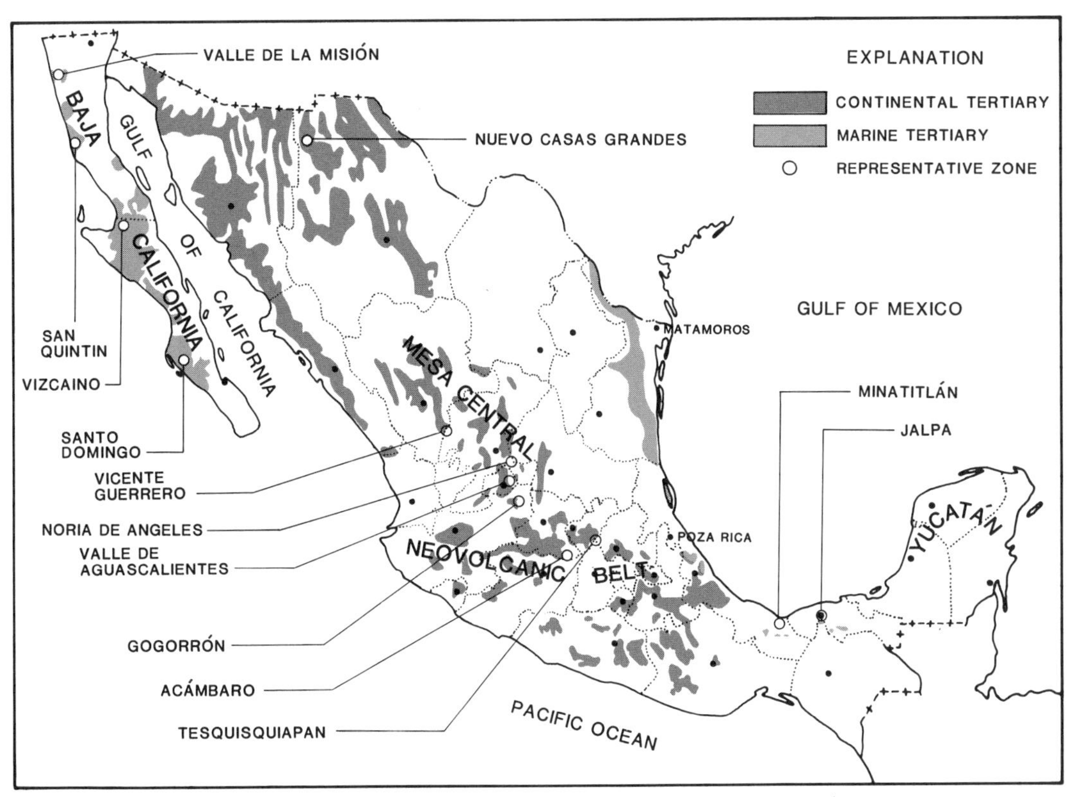

Figure 4. Distribution of major Tertiary siliciclastic aquifers of Mexico.

limestones, at times thousands of kilometers in length, which eventually became excellent porous and permeable host rocks for oil and gas (depending on their geologic position) and for water. Examples include El Abra Limestone in the Tampico-Veracruz region, the Aurora Limestone in the states of Coahuila and Chihuahua farther north and northwest, El Doctor Limestone in Querétaro farther southeast, the mixed-facies Tamabra Limestone at Poza Rica in Tabasco, and the Sierra Madre Limestone in Chiapas. Special attention has been focused on these due to their vast lateral extension within the Great Mexican Geosyncline and their importance as aquifer sources (Fig. 5).

TERTIARY CLASTIC SEDIMENTS

Clastic sediments are present in all the Eocene to Pleistocene sedimentary basins and consist of marine and continental sandstones and conglomerates, which in the appropriate geologic and structural settings, usually exhibit the porosities and permeabilities of good aquifers.

The Tertiary clastic sediments are especially found overlying the Mesozoic section, both conformably and unconformably. Marine regressions and invasions during the Tertiary gave rise to unconformities and the deposition of extensive conglomerates. Pyroclastic deposits are scarce, since volcanism during those periods was restricted. However, the porous beds, either calcareous or clastic, which are also permeable, make good aquifers.

Tertiary clastic sediments are particularly noteworthy in the Gulf of Mexico Coastal Plain valleys along the entire Gulf Coast, and also at the Mesa Central in outcrop areas of these Oligocene and Miocene sandstone and conglomerate formations (Fig. 4).

QUATERNARY CLASTIC SEDIMENTS

These clastic sediments, products of the erosion of pre-Quaternary formations, cover extensive areas all over the endorheic and open valleys in the highlands, mountain slopes, and coastal plains, and they develop important aquifers—under the appropriate conditions—by accumulating large volumes of percolated surface rainwater. In outcrop they give rise to springs of varying magnitudes, otherwise yielding generally "fresh" water for purposes of agriculture, industry, and cattle-raising.

Physiographically, the intermontane, coastal, and highland-plains basins are the best sources of water, which is extracted by water wheels and shallow wells over widespread areas throughout the country, including the Baja California del Norte desert plains and the central Peninsula east and southeast of the Sebastian Vizcaino Peninsula (Fig. 4).

Figure 6 shows the over- and underdeveloped geohydrologic zones of Mexico, together with the areas considered as balances. Villareal and others (1978) presents an excellent description of the aquifers of Mexico.

THE TRANS-MEXICO NEOVOLCANIC BELT

This geologic structure crosses the continent from west to east for almost 1,100 km between latitudes 19° and 21°, with an average width of about 220 km (Fig. 1). The belt comprises three geotectonic elements.

1. Andesitic, dacitic, and rhyolitic magmatism, 23 to 5 Ma in the north, and 18 to 5 Ma in the south, outcropping rarely and mostly covered by younger volcanics. This has therefore been poorly studied, and its limits remain undefined.

2. Younger magmatism, 2.5 Ma to Recent in the north, and 1.6 Ma to Recent in the south, consisting of associated basalts and andesites-dacites in the first case and basalts, andesites, and basic andesites in the second. Neogene basalt fields are the outstanding feature distinguishing the Neovolcanic Belt and can be traced throughout its entire extension.

3. These two volcanic elements are cut by grabens, generally oblique to the trend of the Neovolcanic Belt, that are filled by tuffs, lava flows, and lacustrine sediments; most of these troughs were closed-lake basins originally.

The Neovolcanic Belt is the most densely populated area in the country; ground waters have been overdeveloped, and most of the aquifers are at present under a permanent ban of further exploration in order to prevent their total depletion.

In order of importance, the aquifers under development are: (1) granular sediments filling the tectonic troughs; (2) basalts and their associated pyroclastics; and (3) fractured andesites and dacites, which are the sources for the tectonic trough fills.

THE CHIHUAHUA AND COAHUILA BOLSONS

These bolsons are the continuation of the Río Grande rift in northeastern Mexico, which gives rise to troughs aligned north-south and filled with fluvial sediments. These troughs are distributed from Ciudad Juarez to Torreon, except for the La Laguna area, which trends east-west. The lacustrine sediments in the troughs are fed by aquifers from the underlying reef limestones and fractured sandstones that have enormous recharge areas in the surrounding mountains. A very minor percentage of the recharge to these granular aquifers comes from infiltrated rainfall or suface runoff, except for La Laguna, which receives discharge from the Nazas River.

Tables in Villareal and others (1978) show catchment figures in an excellent statistical study; Figure 6 of Villareal and others (1978) shows the zones that have been studied by several methods and at various levels.

We consider that, although much remains to be done in Mexican geohydrology, the specialists of the last decades have made substantial progress in the evaluation of a considerable portion of the country's geohydrologic potential.

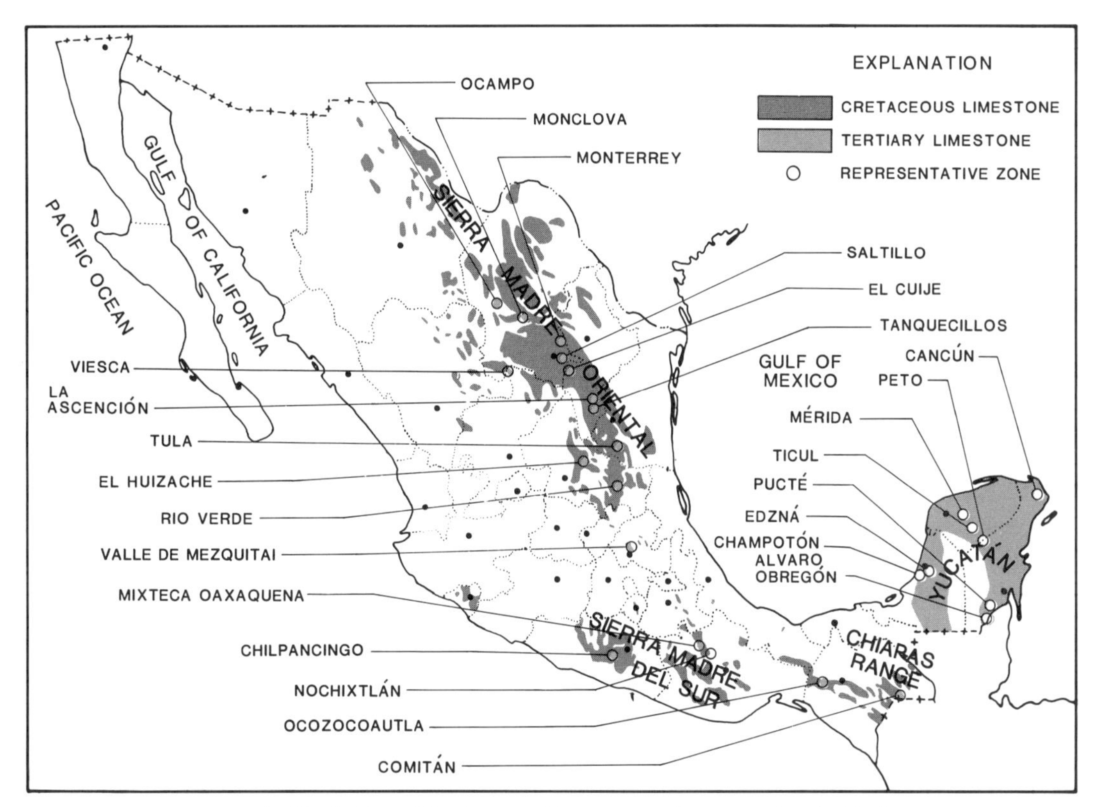

Figure 5. Distribution of Tertiary and Cretaceous carbonate aquifers in Mexico.

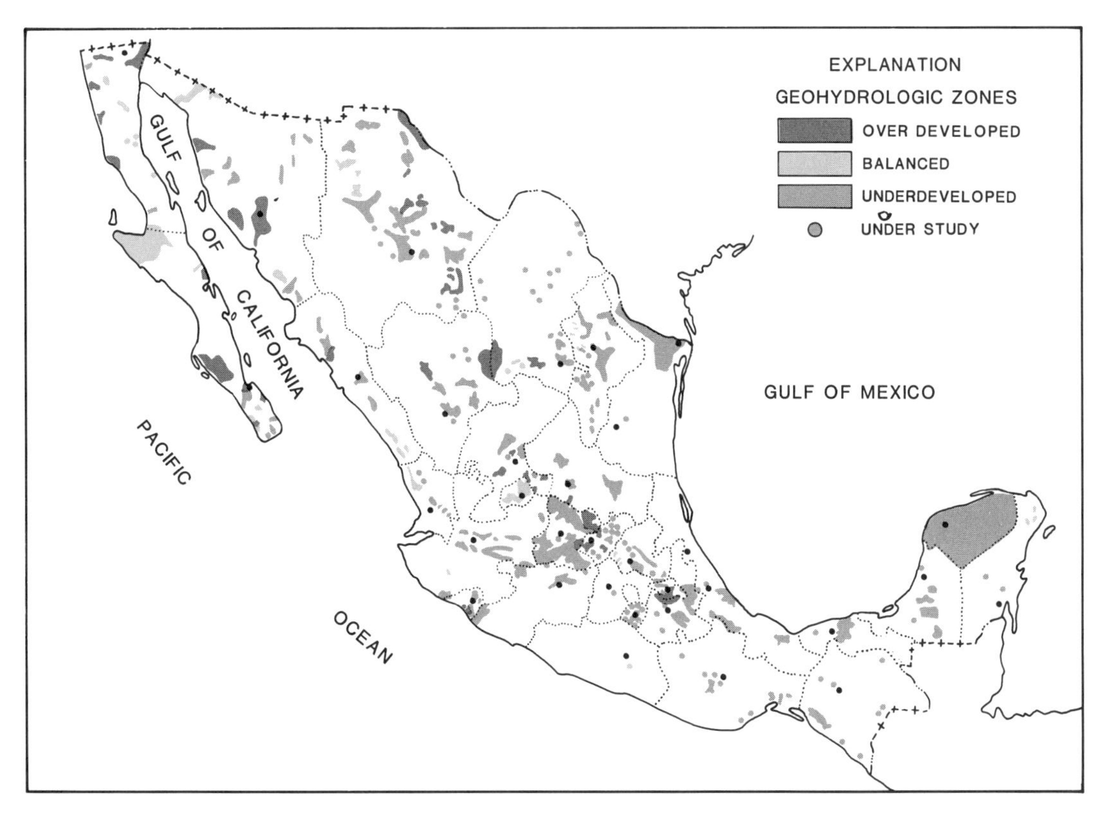

Figure 6. Map showing extent of studies and utilization of Mexican aquifers.

FLUVIAL CONTROL

It is to be hoped that these studies cover the geology not only of the underground aquifers, but also—as in fact is being done—that of the river basins for engineering construction purposes, such as the large dams in the states of Chiapas, Veracruz, Sinaloa, Guerrero, Michoacán, Sonora, Nuevo León, Tamaulipas, and others. The country's economic progress demands such a policy, since these projects not only control floods but also in many cases generate electric power and increase farm productivity through canal irrigation.

These are the reasons to assign as great a priority to geohydrology as to the earth sciences applied in the production of fuels and minerals.

REFERENCE CITED

Villareal, G. A., and others, 1978, Atlas Geohidrologico, V. 1: Mexico City, Banco Nacional de Información Geohidrologica.

Printed in U.S.A.

The Geology of North America
Vol. P-3, Economic Geology, Mexico
The Geological Society of America, 1991

Chapter 14

Metallic and nonmetallic mines; Introduction to the geology of the Metallogenic Provinces

Guillermo P. Salas
Geologo Consultor, Asesor del Consejo de Recursos Minerales, Centro Minero Nacional, Carretera México-Pachuca, Pachuca, Hgo., México, D.F.

Figure 1. Metallogenic Provinces of Mexico.

METALLOGENIC PROVINCES IN MEXICO

The Metallogenic Map and Provinces of the Republic of Mexico (Salas, 1976, 1980) was the end result of prolonged work to which many contributed: Genaro González Reyna, Ruben Pesquera Velásquez, Germán Arriaga García, Moisés Martínez Manrique, Hugo Cortés Guzmán, and many other members of the Mineral Resources Council (Consejo de Recursos Minerales) and of the Geological Institute (Instituto Geologico, UNAM). The publication stressed the importance in metallogenesis of the geochronological epochs during which the mineral deposits originated and their correlation with petrographic, metallogenic, and orogenic provinces.

There is no specific definition of a metallogenic province. Many authors share basic concepts; more recently, others, using more advanced and sophisticated techniques, have found geo-

Salas, G. P., 1991, Metallic and nonmetallic mines; Introduction to the geology of the Metallogenic Provinces, *in* Salas, G. P., ed., Economic Geology, Mexico: Boulder, Colorado, Geological Society of America, The Geology of North America, v. P-3.

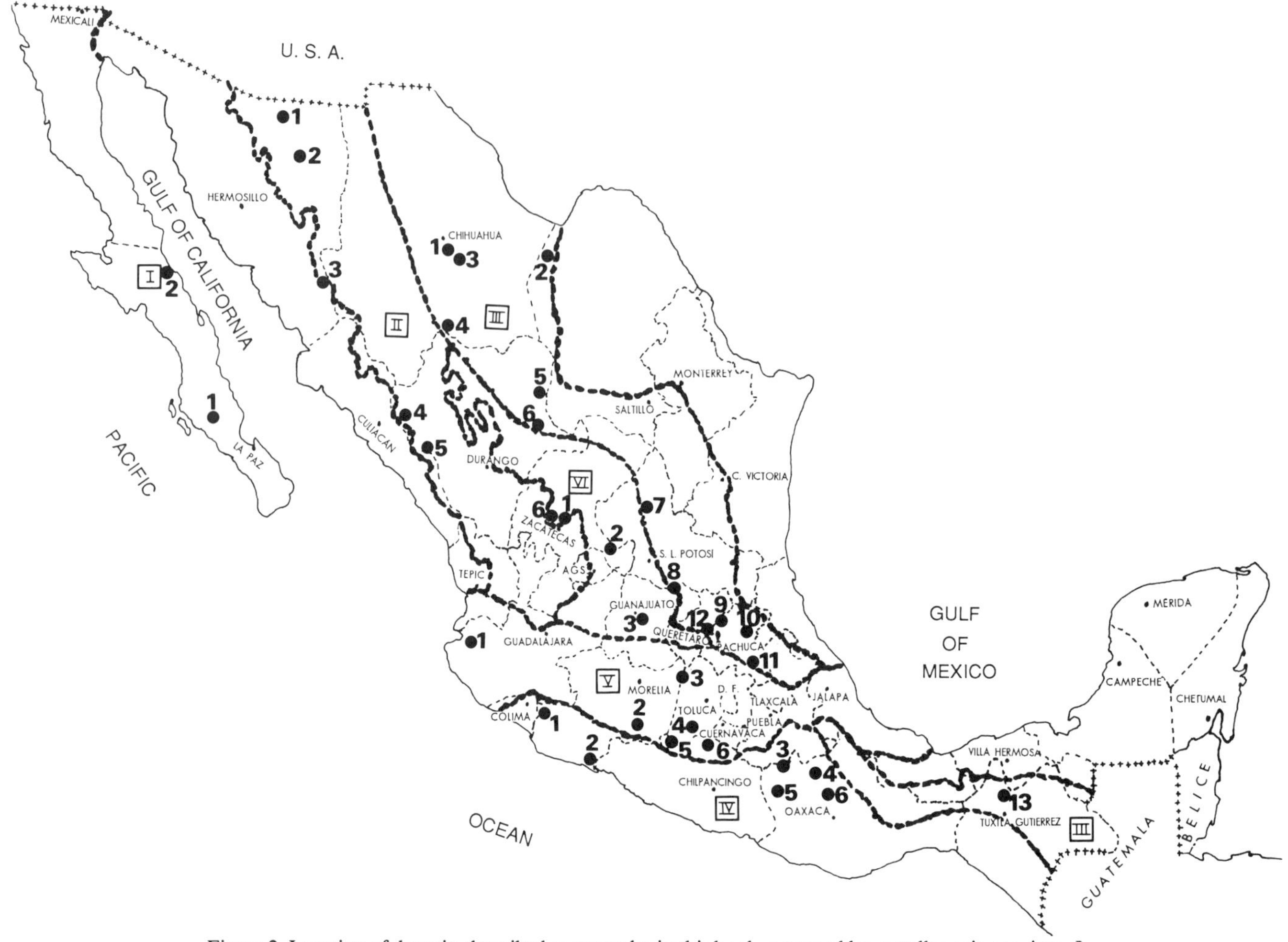

Figure 2. Location of deposits described as examples in this book, arranged by metallogenic province. I. Baja California: 1, San Hilario; 2, El Boleo. II. Sierra Madre Occidental: 1, Cananea, Sonora; 2, La Caridad, Sonora; 3, Gochicc, Sonora; 4, Topia, Durango; 5, San Dimas–Tayoltita, Durango; 6, San Martín, Zacatecas. III. Sierra Madre Oriental: 1, Santa Eulalia, Chihuahua; 2, La Encantada, Coahuila; 3, Naica, Chihuahua; 4, San Francisco del Oro, Chihuahua; 5, Anatares, Durango; 6, Verdaleña, Durango; 7, Charcas, San Luis Potosí; 8, El Realito, Guanajuato; 9, La Negra, Querétaro; 10, Zimapán, Hidalgo; 11, Molango, Hidalgo; 12, Pachuca-Real del Monte, Hidalgo; 13, Santa Fe, Chiapas. IV. Sierra Madre del Sur: 1, La Minita, Michoacán; 2, Las Truchas, Michoacán; 3, Concepción-Papalo, Oaxaca; 4, Huitzo, Oaxaca; 5, Tejocotes, Oaxaca; 6, Telixtlahuaca, Oaxaca. V. Transmexican Neovolcanic Belt: 1, Cuale, Jalisco; 2, Inguarán, Michoacán; 3, El Oro–Tlalpujahua, Michoacán; 4, Zacualpan, México; 5, Tizapa, México; 6, Taxco, Guerrero. VI. Mesa Central: 1, Fresnillo, Zacatecas; 2, Real de Angeles, Zacatecas; 3, Guanajuato, Guanajuato.

chemical or metalliferous consanguinity in certain deposits located within certain petrographic provinces.

This chapter is a first attempt to subdivide the Mexican territory into what the writer calls Metallogenic Provinces, albeit coinciding with orographic provinces (Fig. 1).

The physiographic or orographic features of Mexico are the direct result of the continent's tectonic history. The frequency with which mineral deposits occur along these features appears to demonstrate the author's belief in the close association between diastrophic movements and metallogenic processes.

Such an association has been long studied. De Launay (1913) and possibly others before him attempted to associate mineral deposits with certain geographic regions. Lindgren (1933), sharing De Launay's ideas, presented a concept of metallogenic provinces and epochs. Turneaure's discussion (1955) is probably the most adjusted to mining exploration needs with regard to metallogenic epochs and provinces. Billingsley and Locke (1941) advanced more sophisticated ideas, and almost simultaneously, Warren and Thompson (1945) searched for trace elements in western Canadian sphalerites.

On the basis of complex geochemical methods, Burnham (1959) attempted to draw metallogenic province maps in the southwestern United States on which various types of mineral deposits were plotted in what were termed Metallogenic Provinces, aligned in three separate belts and consistent with regard to both orientation trends and position of significant tectonic features. He concluded (Burnham, 1959, p. 69) that . . . "besides Metallogenic Provinces for a specific metal or group of metals, determining regional productivity appears to consist of lesser features aligned along belts that lie within, and are essentially defined by, the larger metallogenic belts. The limits of the metallogenic belts cannot be precisely defined; consequently there is no basis on which to define these limits or prevent the extension of these Metallogenic Provinces."

Perhaps more important are the conclusions as to the connection between mineral deposits and tectonic features, as well as to the period when the deposits were formed and that of the diastrophic deformation.

Burnham (1959) continues, "1) Metallogenic belts are very extensive; 2) they are not time-related with the associated tectonic features where the deposits occur; 3) there are relatively smaller segments containing mineral deposits and representing up to four Metallogenic epochs quite separated in time; 4) another conclusion is that mineral deposits are not especially associated with any particular type of host-rock, since the same rock types are present both within and outside of the mineralized belts."

Burnham analyzed chalcopyrite samples for trace elements such as antimony, arsenic, bismuth, cadmium, cobalt, germanium, indium, molybdenum, nickel, etc.; he also analyzed sphalerite samples for the same elements plus galium and thallium.

Worth mentioning is that Burnham's study included samples from the Pachuca and Guanajuato mining districts in Mexico, as well as from the states of Baja California, Chihuahua, Durango, San Luis Potosí, Sonora, and Zacatecas. Relative to this, Burnham (1959) shows eight maps of mineral concentration "zones" that reflect some degree of structural control in Mexico. However, although these have somewhat similar geographic orientations, they do not show any specific orographic or tectonic control. This may be the result of insufficient sampling to cover the entire territory for a better statistical interpretation of the structural control. Nothing has been published in Mexico to date in this field.

For the present it is more important to attempt the country's subdivision into six Metallogenic Provinces with relation to geologic and diastrophic features, in which a certain frequency of mineral deposits can be detected.

The following chapters describe each of these provinces (Fig. 2) in a preliminary tentative classification, with the expectation that future researchers will further refine this proposal.

REFERENCES CITED

Billingsley, P., and Locke, A., 1941, Structure of ore districts in the continental framework: American Institute of Mining and Metallurgical Engineers, v. 144, p. 9–59.

Burnham, W. C., 1959, Metallogenic Provinces of the southwestern United States and northern Mexico: New Mexico, Bureau of Mines and Mineral Resources Bulletin 65, 73 p.

De Launay, L., 1913, Gites Minéraux et Métalliferes: Paris and Liège, Librarie Polytechnique.

González Reyna, J., 1956, Riqueza Minera y Yacimientos Minerales de México: Mexico City, 22nd International Geological Congress, 15 maps.

Lindgren, W., 1933, Mineral deposits: New York, McGraw Hill Book Company, Inc., 930 p.

Salas, G. P., 1976, Contribution of Mexico to the Metallogenetic Chart of North America: Geological Society of America Map and Chart Series MC-13, scale 1:2,000,000.

—— 1980, Cartay Provincias Metalogeneticas de la Republica Mexicana: Consejo de Recursos Minerales, Bulletin 21-E, 2nd edition.

Turneaure, F. S., 1955, Metallogenetic Provinces and Epochs, *in* Bateman, A. M., ed., Economic Geology: New York, John Wiley and Sons, p. 38–89.

Warren, H. V., and Thompson, R. M., 1945, Sphalerites from western Canada: Economic Geology, v. 40, p. 309–335.

Printed in U.S.A.

The Geology of North America
Vol. P-3, Economic Geology, Mexico
The Geological Society of America, 1991

Chapter 15

Baja California Peninsula Metallogenic Province

Guillermo P. Salas
Geologo Consultor, Asesor del Consejo de Recursos Minerales, Centro Minero Nacional, Carretera México-Pachuca, Pachuca, Hgo., México, D.F.

INTRODUCTION

The present-day Gulf of California apparently began to take shape during the Miocene (11 to 25 Ma). According to Gastil and others (1972), the Baja California Peninsula started forming in mid-Cretaceous time. Subduction caused andesitic volcanism from Arizona to south of Sinaloa. At the end of the Cretaceous, the peninsula began to move as a block toward the northwest, and at the start of the Tertiary, it separated from the continent, opening the proto–Gulf of California. The present morphology of the peninsula began to develop between 4 and 6 Ma. There is clear evidence (Miocene sediments in the Island of Tiburon and others) of the existence of the incipient Gulf before the Miocene, but not in its present form.

Wisser (1954) believes that the geologic history of the Baja California Peninsula is similar to that of the southwestern Great Basin and, in its western part, to the Sierra Nevada in the United States. Whatever the origin of the peninsula, I agree with Wisser about the close resemblance between northwestern Sonora and the Great Basin of the western United States, and the relation between Baja California and the California Sierra Nevada (Salas, 1980).

Typical mines in Baja California Norte (Fig. 1) are: El Arco (copper), El Fenomeno and La Oliva (tungsten), and multiple gold and some iron prospects.

Typical mines in Baja California Sur (Fig. 2) are: Santa Rosalia, El Boleo (copper); El Triunfo (silver); Lucifer (manganese); San Juan de la Costa, San Hilario, Santa Rita, Santo Domingo, Tembabiche (phosphorites); and El Tigre (chromium).

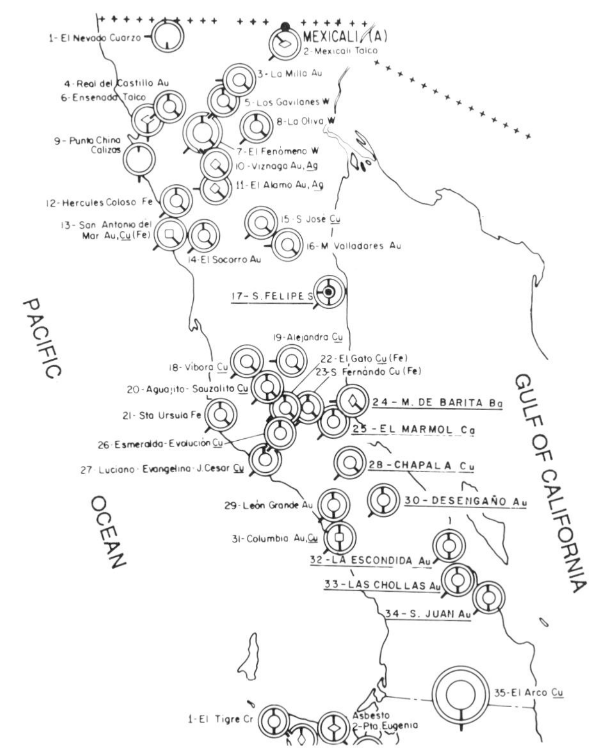

Figure 1. Metallogenic map of the north half of the Baja California Peninsula Metallogenic Province (Salas, 1976, 1980).

Salas, G. P., 1991, Baja California Peninsula Metallogenic Province, *in* Salas, G. P., ed., Economic Geology, Mexico: Boulder, Colorado, Geological Society of America, The Geology of North America, v. P-3.

Figure 2. Metallogenic map of the south half of the Baja California Peninsula Metallogenic Province (Salas, 1976, 1980).

REFERENCES CITED

Gastil, G., Phillips, R. P., and Torres, R. R., 1972, The reconstruction of Mesozoic California, *in* Tectonics-Tectonique, Toronto: 24th International Geological Congress, Proceedings, sec. 3, p. 217–229.

Salas, G. P., 1976, Contribution of Mexico to the Metallogenetic Chart of North America: Geological Society of America Map and Chart Series MC-13, scale 1:2,000,000.

——, 1980, Cartay Provincias Metalogeneticas de la Republica Mexicana: Consejo de Recursos Minerales, Bulletin 21-E, 2nd edition.

Wisser, E. H., 1954, Geology and ore deposits of Baja California: Economic Geology, v. 49, p. 44–76.

Printed in U.S.A.

The Geology of North America
Vol. P-3, Economic Geology, Mexico
The Geological Society of America, 1991

Chapter 16

Summary of structural and stratigraphic data on the Monterrey Formation outcrops of the San Hilario area, Baja California Sur

J. Ojeda R.
Consejo de Recursos Minerales, Centro Minero Nacional, Carretera México-Pachuca, Pachuca, Hgo., Mexico, D.F.

INTRODUCTION

This chapter summarizes the results of the writer's work on the paleogeography and facies changes of the Monterrey Formation in the San Hilario area of Baja California, as part of the exploration for phosphate rock prospects carried out in the area by the Mineral Resources Council (Consejo de Recursos Minerales).

Field work from March 1975 to January 1976 covered detailed mapping on 1:10,000-scale vertical airphotos of the Monterrey Formation outcrop belt, an area 70 km long and 4 to 5 km wide, together with the upper and lower portions of the respectively under- and overlying Tepetates and San Isidro Formations.

Additional information includes six cross sections along the principal stream beds, eight lithologic columns (scale 1:100), and a stratigraphic correlation table of these columns (1:2,000). As an aid to correlation and future detailed geological work, a collection of Polaroid photographs describes virtually all the Monterrey outcrops exposed at the streambeds, focusing on the section's geoeconomic features. Part of this material is included here.

Location

The San Hilario area (Fig. 1) lies 70 km north-northwest of La Paz, capital of the state of Baja California Sur, between 110°50′ to 111°20′ W and 24°10′ to 24°40′ N.

Access

The Transpeninsular La Paz–Tijuana Highway runs through the southeastern half of the area, bordering the Monterrey Formation outcrop between kms 70 (El Aguajito) and 114 (Las Pocitas), after which it turns sharply west, whereas the Monterrey outcrop continues northwest.

Several NNE-trending unpaved roads, most of them improved by the Council for transportation of machinery used in phosphate exploration, join the ranches and commons, providing access throughout the entire area.

REGIONAL GEOMORPHIC LANDSCAPE

The desolate but singularly beautiful desert landscape comprises low-lying chains of hills that gradually evolves to mesas or hills of medium elevation and culminates to the north-northeast to form the southwestern flank of the Sierra La Giganta.

The section embracing the middle to upper parts of the Tepetates Formation and the overlying Monterrey, San Isidro, and Commondú Formations ranges between 100 and 300 m above sea level.

Streams, cliffs, and erosional processes

The permanently dry, southwestward-meandering main riverbeds are covered by sandy mounds and flanked by banks of marine sediments overlain with gravels and conglomerates. These banks or "cliffs" ("cantiles" in the local usage) (Fig. 2) usually consist of loosely consolidated sediments, and can be as high as 70 to 80 m. The cantiles are the products of erosion by three typical characteristics of desert areas: (1) sudden torrential rains, following several years of drought, causing erosion of large volumes of detritus; (2) pronounced differences between day and night temperatures; and (3) strong air currents generated by the latter.

The persistence of the air-temperature factor brings about the self-erosion of the cantiles: extensive longitudinal fractures appearing at the tops of the marginal walls gradually make them break off and subdivide into large blocks, which eventually slide downslope at the slightest wind. This material accumulates, together with the gravels and conglomerates, until torrential rainfall carries it downstream to form either small but very extensive, or locally widely meandering, alluvial aggradational plains.

Ojeda, R., J., 1991, Summary of structural and stratigraphic data on the Monterrey Formation outcrops of the San Hilario area, Baja California Sur, *in* Salas, G. P., ed., Economic Geology, Mexico: Boulder, Colorado, Geological Society of America, The Geology of North America, v. P-3.

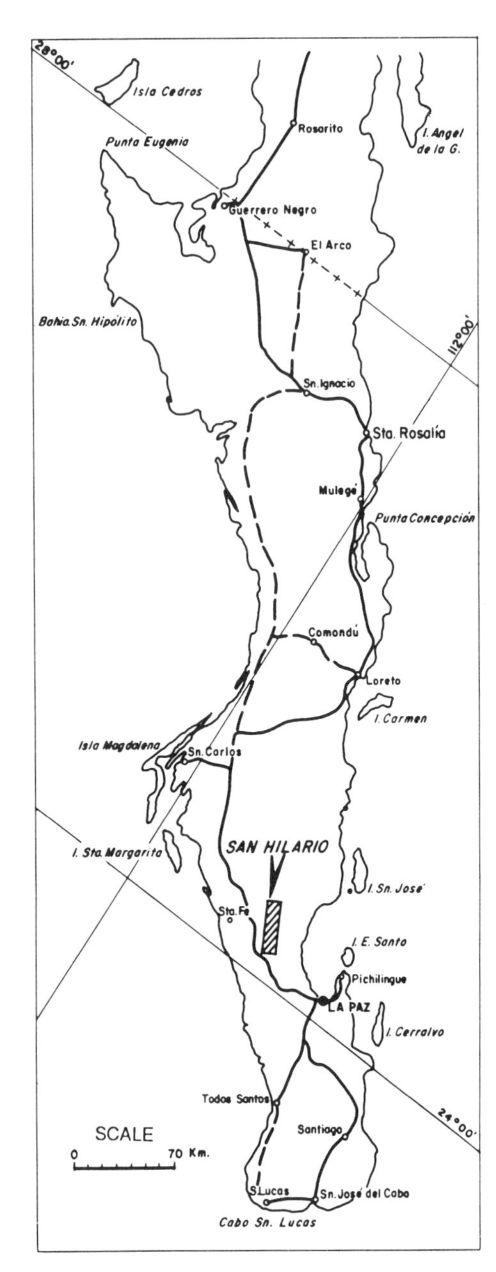

Figure 1. Index map of the San Hilario area.

Oases

At middle or upper heights upstream from the main riverbeds, occasional springs form small oases around which there are usually two or three shanties occupied by some of the few inhabitants of the area, who are otherwise spread out over the intermontane depressions or in small hamlets at some of the aggradational valleys or along the Transpeninsular Highway.

Climate

During most of the year, but particularly in spring and summer, maximum temperatures rise to 40 to 45°C between 10 a.m. and 4 p.m., gradually decrease to tolerable levels around sundown, reach the freezing point at 2 to 3 a.m., and begin to climb slowly until sunrise, when the cycle begins again. Occasional clouds are quickly dispersed by the air currents.

Vegetation

The nightly cold temperatures condense water vapor rising from the marine air currents that come in from the Pacific Ocean or the Gulf of Cortés. Dew drenches the entire area, fostering the existence of a wide variety of cacti, drought-resistant trees, and thorn bushes that make up the desert flora of the region.

REGIONAL GEOLOGICAL OUTLINE

In the San Hilario area the Monterrey Formation forms part of a gently northeast-dipping Tertiary marine sedimentary sequence on the southwestern flank of a wide northwest-trending syncline.

This sequence (Fig. 3) is interrupted at intervals by the latent tectonism of the peninsular block, which evolved gradually from relatively deep to very shallow marine conditions (Paleocene to Eocene and lower Miocene Tepetates and Monterrey Formations), continued under alternately marine and continental environments (lower to middle Miocene San Isidro Formation), and came to an end with predominantly continental deposition interrupted by final volcanism (upper Miocene Comondú Formation).

Widespread lavas and pyroclastic products covered the southern half of the peninsular block in the upper Miocene. Strong continental erosion at the end of this period caused widespread distribution of the volcanic detritus. The subsequent transgressive and regressive Pliocene-Pleistocene seas reworked and redeposited these materials, forming the thick gravel and conglomerate cover, which together with the marine-continental sediments, still accompanies the final emergence of the peninsular block.

Erosion has since continued shaping the mesas, alluvial aggradational plains, and cantiles that characterize the region.

Monterrey Formation

The predominantly clayey and diatomaceous Monterrey Formation contains abundant microfossils, particularly foraminifera, and is almost entirely pervaded by gypsum.

These obvious general features cloak several processes of deposition and alteration—silicification, dissolution and evaporation, or chemical precipitation—which reflect environmental conditions and changes during deposition.

The environmental conditions and variations are the largely inherited direct results of the previous paleogeographic and paleotectonic framework, briefly described below.

Paleogeographic and paleotectonic framework. The oldest underlying Tertiary unit, the Tepetates Formation, represents the clay, silt, and sand facies of a marine transgression in

Figure 2. Typical cliff exposures. a, San Isidro Formation; b, Monterrey Formation; c, Tepetates Formation; d, Typical aspect of erosion on Monterrey Formation.

early Eocene time, caused by the subsiding western portion of the peninsular block (Fig. 4a). The sandy facies in the middle and upper parts exhibits extensive cross-bedding (Fig. 2c, and the predominantly clayey facies at the top includes abundant gypsum. The former indicates large-scale coastal currents and deltaic-type deposition; the latter, arid environment, partial or total evaporation, and lacustrine conditions.

Emergence at the end of the Eocene initiated a period of erosion, local folding, and some volcanism emplaced during the pre-Miocene–post-Tepetates interval, building up a low-lying central area, which shortly thereafter (early Miocene) became an important structural sea-floor feature.

Subsidence in the lower Miocene was minor in comparison with that of the Eocene, as shown by the thickness of the Monterrey Formation and the very slight angular unconformity that separates it from the Tepetates.

These conditions caused deposition of the Monterrey in a shallow basin of restricted circulation and limited extension. The Monterrey sea transgressed from west to east, first covering the western flanks of the peninsula and leaving a few comparatively smaller positive areas toward the west. Later it spread toward the east-southeast to connect with the proto-Gulf embayment through the La Paz isthmus (Fig. 4b).

Merging of the cold Pacific Ocean waters and the warm proto-Gulf sea took place in a slightly submerged, low-lying central area, with the Bahía Magdalena and Bahía Vizcaíno blocks jutting out as island on the west, and the reefs left isolated from the land masses toward the east.

This connection remained permanently subject to emergence or subsidence brought about by tectonic forces, probably related to the isostatic adjustment of the proto-Gulf, acting on the southern half of the peninsula as a whole, or independently on several or each of the emerged blocks to the east and west.

The arid weather conditions that prevailed over the Eocene positive areas, added to the gradual emergence of the peninsula at the end of the lower Miocene and a weak but persistent volcanic ashfall from the east-northeast (probably related to Tertiary volcanism of the northwestern Sonora region) make up the general environment and depositional conditions of the Monterrey Formation.

Lithologic types and depositional processes. The Monterrey Formation includes a large variety of lithologic types:

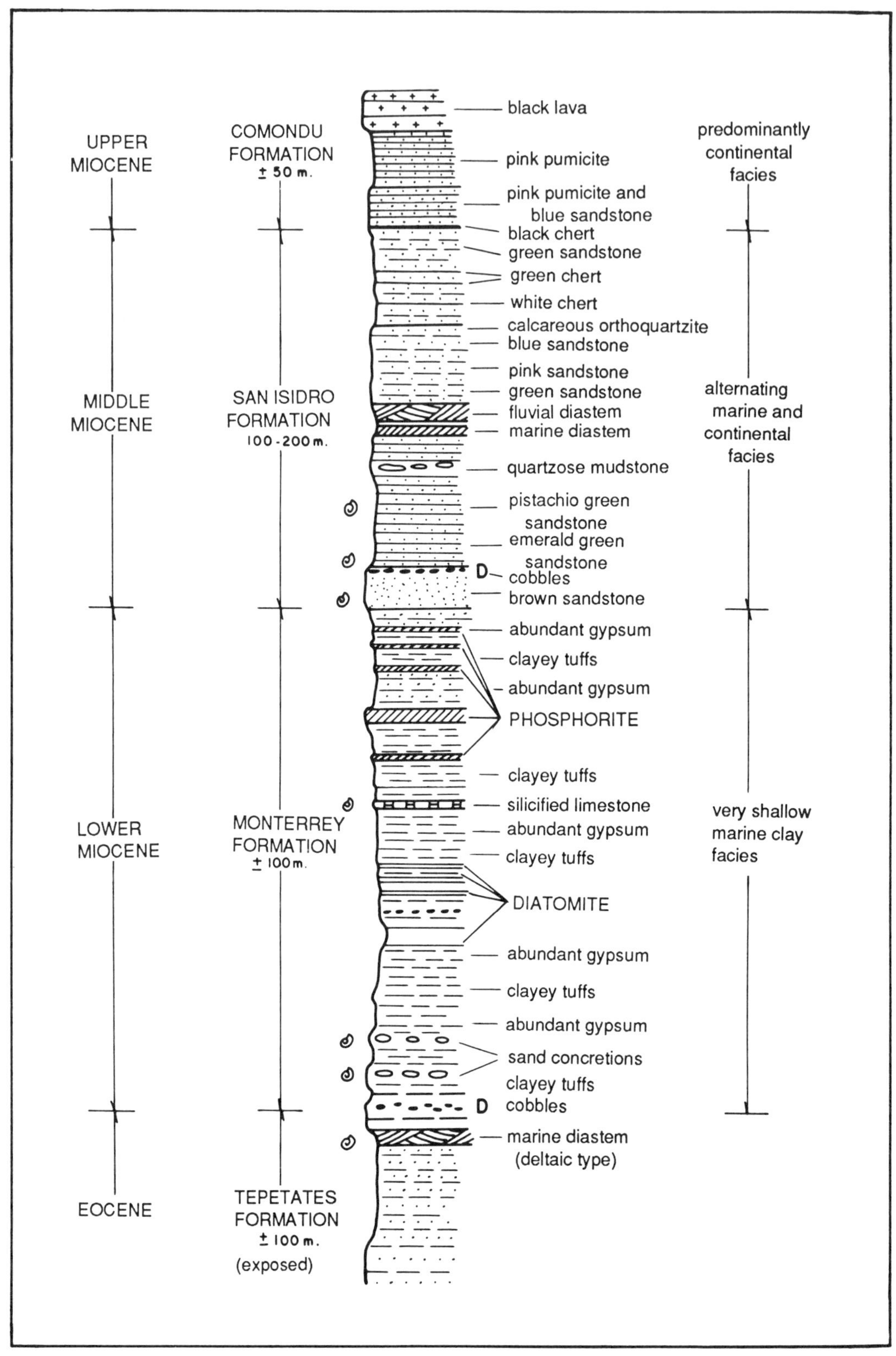

Figure 3. Schematic lithologic column of the rocks exposed at San Hilario.

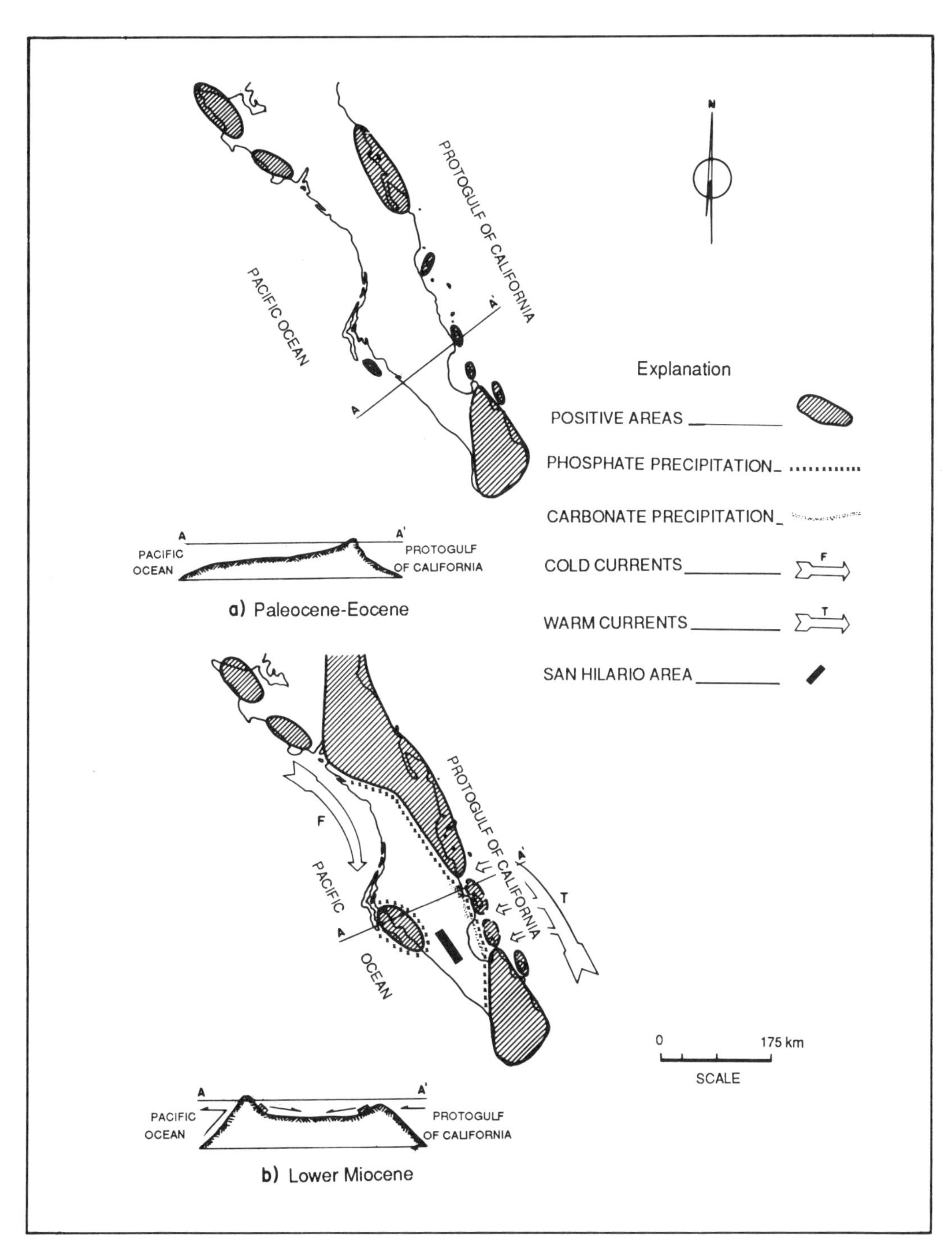

Figure 4. Paleogeographic maps. a, Paleocene-Eocene; b, lower Miocene.

clayey tuffs, silicified diatomaceous green shales, ubiquitous gypsum, and several diatomite and phosphorite seams and layers. All of these were products of two processes: a clastic process involving mainly the terrigenous and volcanic-sedimentary material, and a nonclastic, principally chemical and biochemical, calcareous sedimentation. These processes were simultaneous and gave rise to a mixed, or hybrid, depositional environment where alteration products evolved mainly through silicification, metasomatic replacement, dissolution, and evaporation or chemical precipitation.

Deposition followed the physical laws of clastic processes, but at the same time embraced nonclastic organic material (abundant planktonic and benthonic microorganisms) and gypsum derived from the arid, partially lagoonal depositional environment.

Clastic processes dominate in the physical features of the resulting sediments; however, the gradual or irregular and intermittent addition of nonclastic materials and their alteration products resulted in lenticular distribution. Consequently, many parts of the section preserving evidence of clastic processes after emergence are relatively homogeneous or slightly lenticular, especially when they have undergone silicification. On the other hand, others that show some degree of lateral and longitudinal homogeneity, become increasingly lenticular as more nonclastic—calcareous, phosphatic, or diatomaceous—material is introduced. In such cases the sedimentary mixture gradually thickens or thins, subdivides or rejoins, grades or is suddenly cut off, and occupies higher or lower stratigraphic levels.

This is therefore, a mixed or hybrid sedimentary process in which the type, volume, and distribution of the sediments laid down in a particular area during a specific geologic interval vary according to the predominant process (clastic or nonclastic). This relation depends on the environmental conditions, but in the end the greater volume of the clastic materials surpasses and dilutes the products of the nonclastic processes.

Under these conditions it becomes extremely difficult to find correlatable stratigraphic horizons. However, at San Hilario, two horizons combine clastic and nonclastic features and may be used for correlation: the main phosphorite seam (Fig. 5), and a silicified limestone 20 or 30 m underneath it. The limestone horizon, locally known as "Lajas Palo Verde," is more lenticular, and rises or sinks moderately in stratigraphic level; it appears at Arroyo Campamento and continues indefinitely toward the northwest.

Most representative stratigraphic section. The most representative section of the Monterrey Formation (Fig. 6) is exposed at Arroyo San Hilario where it is 110 m thick and consists mainly of silty, clayey or sandy, yellowish brown, sometimes slightly greenish or whitish material: mostly clayey tuffs or hybrid tuffs (term introduced by the Petrographic Laboratory of the Comisión de Fomento Minero) derived from continuous or intermittent ashfall throughout the deposition of the Monterrey. The section also includes numerous seams or layers that range in thickness from centimeters to several meters and consist of green, silicified, and locally porcelainized diatomaceous shales at different stratigraphic levels.

Figure 5. Most representative outcrop of the main phosphorite seam and the middle to upper part of the Monterrey Formation (Arroyo San Hilario).

Next in stratigraphic importance is the abovementioned "Palo Verde" silicified limestone, together with several economically significant phosphorite and diatomite or diatomaceous seams and layers, described further in greater detail. The silicified limestone is a product of metasomatic replacement between calcareous material of chemical and biochemical origin (calcareous ooze) contained both in sea waters and planktonic foraminifera, soluble silica from the siliceous ooze derived from volcanic ashfall, and siliceous tests of diatoms forming the diatomite or diatomaceous material in the section. This is a milky-white, sometimes slightly brownish, very silicified metasomatite with subconchoidal fracture, at times with a remarkably abundant foraminiferal content (in other cases absent or scarce due to dissolution by unsaturated waters), thinly interbedded with diatomaceous material (Cerro Palo Verde) that forms a persistent seam ranging from 30 to 50 cm in thickness. The rock develops a blackish brown film and breaks down into slabs upon weathering.

Several levels of sandy concretions appear toward the base and top of the section, gently crossing bedding planes for considerable distances along strike and dip until they disappear abruptly. These are ovoid, flattened or subrounded concretions, ellipsoidal in section, at times very fossiliferous, with abundant mollusk shells (particularly *Turritellas*) in some cases, and swarms of worm burrows in others.

According to Dr. Stanley Riggs of East Carolina University (personal communication, 1986), who has specialized for many years in the mechanics of shallow-marine deposition, these concretion horizons at different stratigraphic levels owe their existence to meandering submarine currents that carry sand and mollusk or other invertebrate shells and deposit them along the way at the curves. Wave action and sea-level fluctuations cause

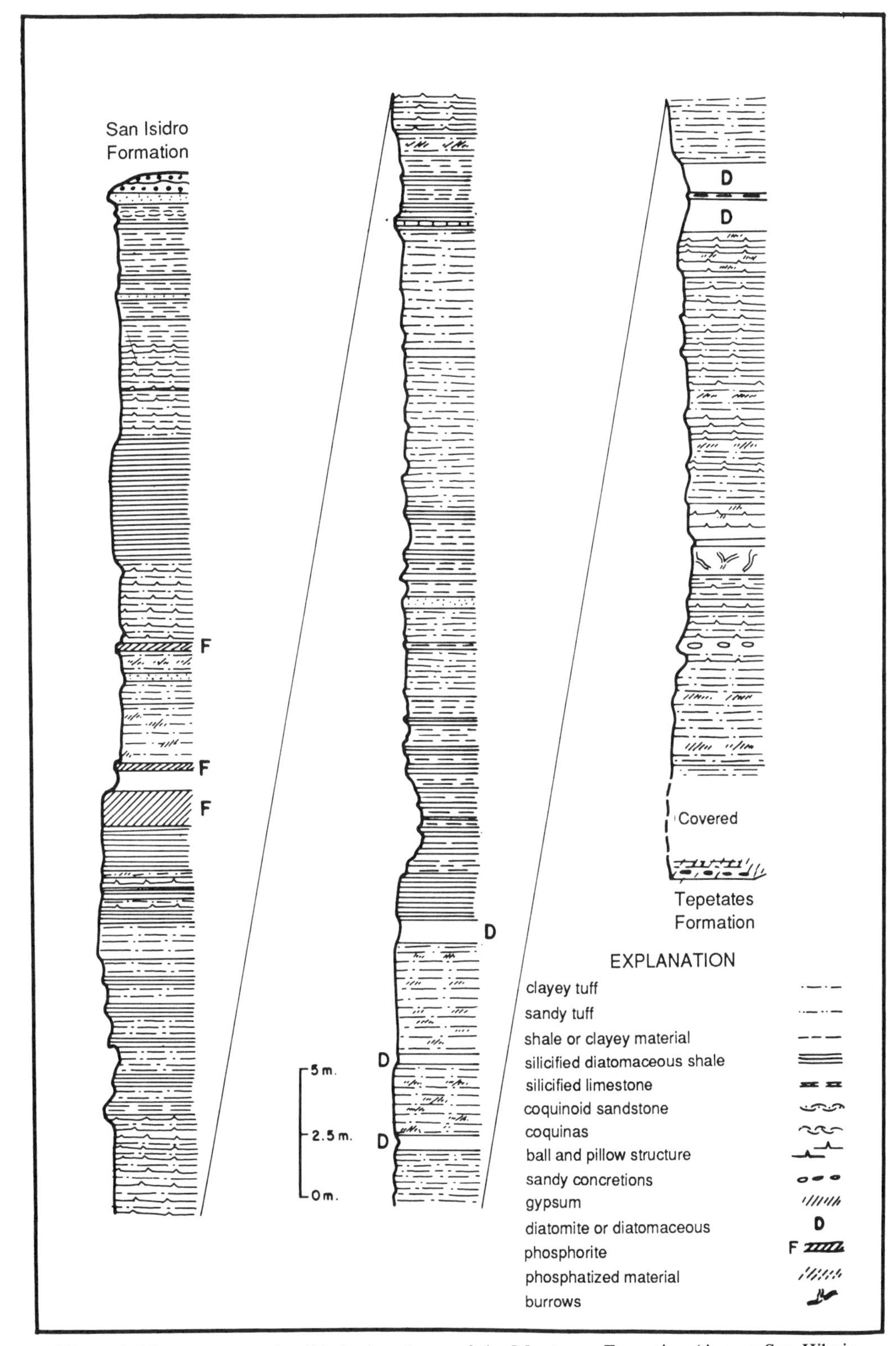

Figure 6. Most representative lithologic column of the Monterrey Formation (Arroyo San Hilario, Piedra Parada Ranch).

the interruption of these currents, whose entrained contents are then buried under new loads of sediment, until the process is renewed after a geologically short lapse. The overburden compresses the sandy fill of the buried currents, resulting in the flattened ovoid shape of the sandy concretions. The worm burrows are caused by the abundant phytophagous organisms inhabiting a shallow-marine environment of relatively stagnant or restricted circulating waters.

The section is entirely pervaded by gypsum inherited from the upper portion of the Tepetates Formation and the partly lagoonal conditions that prevailed off and on during the deposition of the Monterrey. Gypsum occurs both micro- and macroscopically and is ubiquitous; it appears microscopically as interstitial gypsum in some fine-grained sections, and macroscopically in all types of sediments, fractures, and bedding planes in the section except the Palo Verde silicified limestone. Such a distribution suggests: (1) an intimate clay-anhydrite association possibly derived from removal both of sediments on the sea floor due to circulating currents, and of sediments in the surrounding positive areas; (2) hydration of the anhydrite and its recrystallization as gypsum, which may take place even under subaqueous conditions.

The above would explain not only the ubiquitous presence of the gypsum but also the expansion or swelling, as well as the micro- and macrofracturing shown by most of the Monterrey sediments.

Facies of the Monterrey Formation

Distribution of the different Monterrey lithologic types, whatever their origin (clastic or nonclastic), develops stratigraphic facies. Most have both academic (genetic or stratigraphic) and economic significance, but two are particularly attractive economically.

Phosphatic facies. The phosphatic facies is represented by two phosphorite seams: a main lower (Fig. 7) and a secondary upper (Fig. 8) level, as well as numerous variably sized lenses and lenticles. Most of this material occurs in the middle to upper parts of the section within an approximately 30-m-thick stratigraphic interval, but frequent lenses and lenticles are restricted toward the top of the lower half. In addition, the phosphatic facies occurs: (1) at the Monterrey-Tepetates contact as a phosphatized conglomerate a few centimeters to 2 m thick, of polished quartzitic pebbles and cobbles, marine vertebrate bones, and occasional shark teeth; and (2) at the southeast half of the area, in outcropping lenticular seams of partially phosphatized tuffaceous material, 20 to 30 m below the main seam, and in a lenticular phosphorite seam 1 to 2 m thick, appearing locally at the base of the San Isidro Formation 6 to 7 m above its contact with the Monterrey.

Figure 7. Main phosphorite seam (A) overlying Monterrey Formation clays, siltstones, and tuffaceous sandstones.

Figure 8. Upper phosphorite seam (A) unconformably overlying the Monterrey Formation clays, siltstones, and tuffaceous sandstones. Note angular erosional unconformity. An oligomictic conglomerate occurs at the contact.

Main seam. This seam occurs as a practically continuous outcrop over approximately 70 km between th El Aguajito and La Suerte Ranches (Fig. 9); the relatively few interruptions are caused by small lateral or normal faults that mask or disconnect the phosphorite outcrop for intervals of as much as hundreds of meters.

The seam is somewhat lenticular, ranging from 60 or 70 cm to 1 or 1.2 m (occasionally more) in thickness. One possibly significant feature in reconstructing the phosphorite's depositional environment is the frequent presence of marine vertebrate bone fragments at the base of the seam, either dispersed or forming lenticular beds up to 15 or 20 cm thick (Cantil de La Vuelta). Among these are femur, rib, vertebrae, and occasionally jaw fragments, all partially or wholly phosphatized, as well as occasional shark teeth.

The phosphorite groundmass is very siliceous, making the seam resistant to erosion, and consequently it stands out from the bedrock, which is composed usually of quickly eroding silicified green shales thinly interbedded with clays or gypsum. In general, the silica content ranges from 40 to 50 or 60 percent, and the phosphate (P_2O_5) ranges from 9 to 10 or 12 percent throughout the entire outcrop.

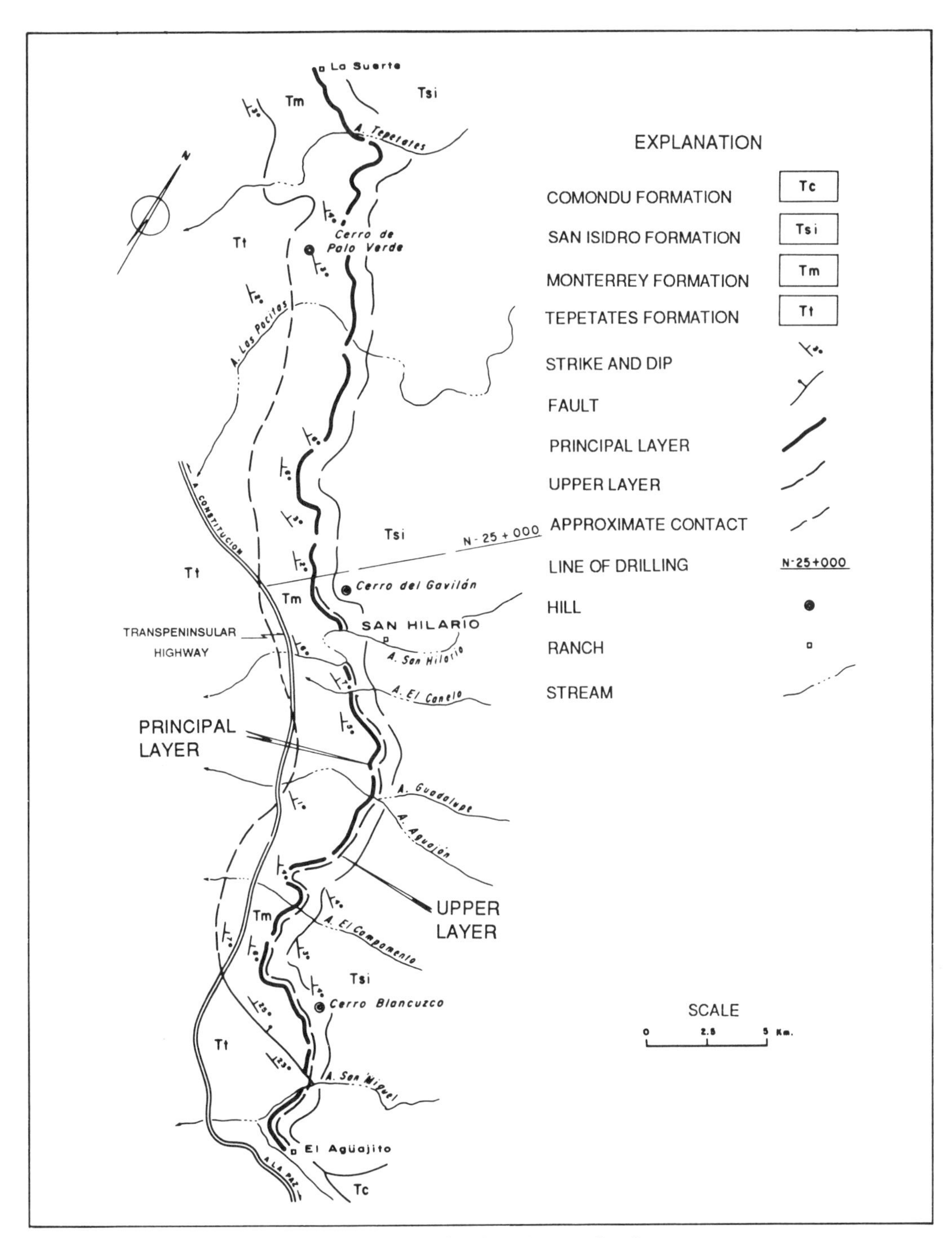

Figure 9. Outcrop map of main and upper phosphate seams.

Upper seam. The upper seam is more lenticular than the lower and extends for 25 km along the strike of the phosphorite outcrop between the western flank of Cerro El Gavilán and Arroyo San Miguel, appearing first at a lateral fault that brings it into direct contact over the lower seam for 1 or 2 km, after which they gradually separate.

The interval between the seams varies: 30 to 40 cm at Arroyo San Hilario, 4 to 5 m at Arroyo Guadalupe, 3 to 4 m at Arroyo El Campamento, and less than 70 cm to 1.5 m at Arroyo San Miguel. The thickness of the upper seam ranges from 1.20 m at the western flank of Cerro El Gavilán to 70 cm at Arroyo San Hilario, 1.50 m at Arroyo Guadalupe, and 30 to 50 cm at Arroyo El Campamento.

The outstanding feature of the upper seam is that from the Arroyo Canelo onward, it overlies an oligomictic conglomerate, a few to 70 cm thick, of polished quartzitic cobbles and pebbles as

TABLE 1. SEAMS, BEDS, AND LENSES ENCLOSING THE PHOSPHATIC FACIES OF THE MONTERREY FORMATION, SAN HILARIO, BAJA CALIFORNIA SUR

Category or type	Thickness and variability ranges (m)	Specific Characteristics	Approximated P_2O_5 (%)	Persistence or degree of continuity along strike
Lenticle	0.05 to 0.06 (occasionally 0.10 to 0.20)	Highly silicic "pisolitic" type (38.3 percent or more)	22 to 24	The entire phosphorite outcrop (~50 km).
Lenticule	0.15 to 0.20 (occasionally thickens to 0.30)	Predominantly calc. matrix with some silica and gypsum	15 to 18	May thin out and disappear suddenly over distances of hundreds of meters.
Lentil	0.30 to 0.50 (occasionally thickens to 0.60)	Calc. matrix but may contain abundant silica and gypsum	14 to 16	Variable from hundreds of meters to several (5 to 6) km.
Lens	0.30 to 0.50 to 0.70	Calc. matrix but may contain abundant silica and gypsum	13 to 15	Persists intermittently over distances of as much as 15 to 20 km but may thin or disappear abruptly.
Upper seam	0.70 to 1.20 to 1.50	Predominantly calc. matrix with some silica and gypsum, conglomeratic base and bioclastic facies	15 to 16	25 km, after which it subdivides and lenticularizes.
Lower seam	0.60 to 1.20	Predominantly silic. matrix (40 to 50 percent) contains some carbonates and gypsum, bioclastic facies at the base	9 to 12	Outcrops along the entire length of the deposit know to the writer.
Tuffaceous material	0.50 to 0.70 to 1.50	Very lenticular tuffaceous matrix with altered biotite specks	4 to 7	Hundreds of meters, but very lenticular.

Note: In addition to the above, a lenticular phosphorite seam ranging 0.70 to 1.50 meters in thickness that contains 14 to 16 percent P_2O_5 appears locally at the base of the San Isidro Formation.

well as more frequently occurring and abundant marine vertebrate bone fragments and shark teeth.

Silica percentages in this seam are considerably lower than in the lower. At the same time P_2O_5 contents are higher; two samples gathered by the writer, at the western flank of Cerro El Gavilán and at Arroyo Guadalupe, averaged 28.4 percent and 17.5 percent P_2O_5.

Lenses and lenticles. The widely diverse distribution and sizing of lenses and lenticles in the phosphate facies necessitates the use of a very specific terminology to describe them adequately, since their occurrence—as in the case of the upper and lower seams—is directly related both to the phosphatization process and to their commercial potential.

They are here classified according to thickness and degree of continuity along strike as lenses ("lentes"), lentils ("lentejon"), lenticules ("lenticulas"), and lenticles ("lentecillas"; Table 1). Generally the degree of along-strike persistence of these lenses or lenticles varies in direct proportion to their thickness and inversely to their P_2O_5 content. Thicknesses decrease upward in the section as the silica content becomes lower.

Perhaps the most obvious exception to this rule is a persistent highly siliceous (38.3 percent+) lenticle, 0.05 to 0.10 m thick, often of "pisolitic" type, that invariably follows the phosphorite outcrop along most the entire length of the deposit, although it is masked at times by erosional detritus. This lenticle is always located at the top of the main seam, at intervals ranging from a few to 20 or 30 m, and on occasion may be 15 to 20 cm thick. Its P_2O_5 content, usually 22 to 24 percent, is the highest of any in the phosphatic facies.

Frequency. The frequency of the visible occurrence of lenticles in different areas (since many are discovered during drilling) may be *gregarious* or *cumulative* (4 to 5 lenses and numerous lenticles), *limited* (2 to 3 lenses and some lenticles), *precarious* or *sporadic* (1 to 2 lenses and occasional lenticles), and *solitary* (1 lens and occasionally 1 or 2 lenticles).

Distribution is gregarious or cumulative between the Tepetates Ranch and Arroyo Las Pocitas, sporadic or solitary between Matanzas and Cerro Colorado, limited from the latter to Arroyos San Hilario and Guadalupe, sporadic or solitary between Arroyos El Campamento and San Miguel, and finally gregarious again from the latter to El Aguajito Ranch.

In the *gregarious* or *cumulative* distribution between Tepetates Ranch and Arroyo Las Pocitas, groups of lenses or lenticles *above* the main seam gradually thicken, becoming poorer in silica at the same time; this gives rise to commercially attractive concentrations (17.89 percent average P_2O_5 content). However, they practically disappear from Cerro El Gavilan to Arroyo El Campamento, a distance along which the upper seam is particularly conspicuous. It seems probable that the seam developed at the expense of the lenticles due to subaqueous removal caused by circulating currents, as suggested by (1) the change in facies, (2) the angular erosional unconformity at the base of the

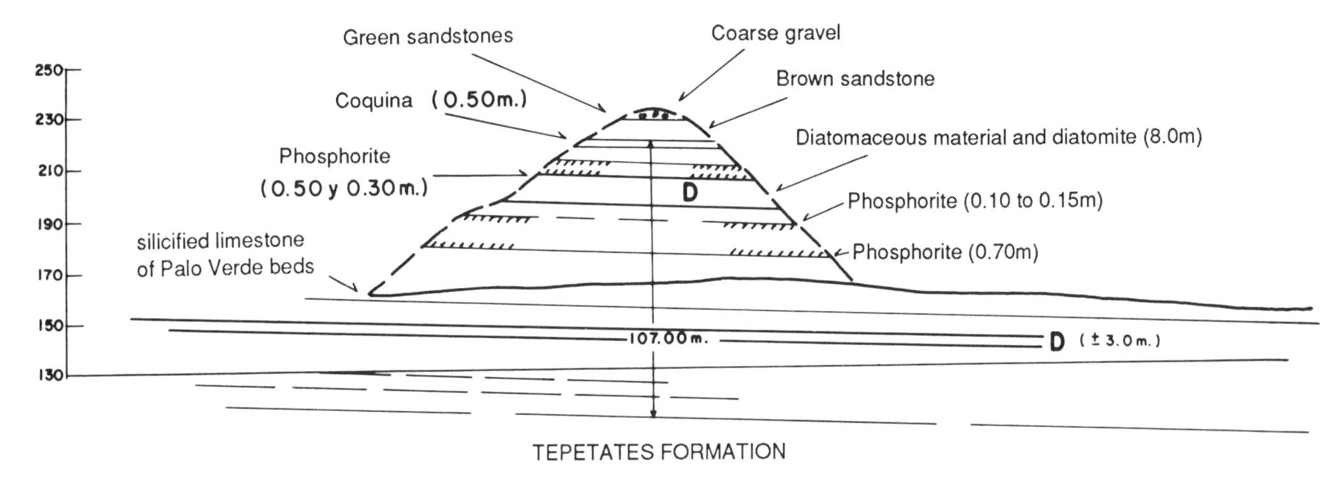

Figure 10. Cross section, looking northwest, of the Monterrey Formation exposed at Cerro de Palo Verde.

upper seam, (3) the latter's higher P_2O_5 percentage as compared to the lower seam, and (4) the oligomictic conglomerate at the base.

In the *gregarious* or *cumulative* distribution between Arroyo San Miguel and El Aguajito Ranch, a distance over which the Palo Verde silicified limestone is notably absent, lens and lenticle groupings occur *underneath* the main seam, and the phosphorite beds thin out gradually toward the southeast and disappear into thick intervals of tuffaceous sandstones with finely interbedded light brown pumicites and thin, dark gray volcanic ash beds. This occurrence is probably uneconomic; as discussed further on, an excess of volcanic material was found to be unfavorable to the distribution of both phosphorite and diatomite.

Diatomite facies. Previously mentioned analyses indicate that this is a low-grade phosphorite, difficult to extract economically because of its thickness and overburden. A possible solution would be to develop other economically attractive materials within the section, either separately in some areas or jointly with the phosphorite in others. A description of the diatomite or diatomaceous material in the Monterrey Formation is therefore included here.

Similar to the main phosphorite seam, the diatomite facies persists along strike for the entire length (70 km) of the Monterrey outcrop; however, its stratigraphic level may rise or sink gradually and it may—also gradually—subdivide, thicken, or become thinner along both strike and dip. The facies is represented by lenticular layers of practically pure diatomite enclosed in matte white diatomaceous material that grades to slightly brownish as it intermingles with host sediments. It is as compact as kaolin in some cases, may show subconchoidal fracture or be very porous, and is considerably resistant to weathering but easily eroded. It is very well exposed in the cliffs along the main streambeds that dissect the Monterrey; in the intermediate areas a thick gravel and conglomerate cover masks the outcrops. However, it may be seen clearly at Cerros Palo Verde and Blancuzco, respectively 20 and 25 km northwest and southeast of San Hilario.

At Arroyo San Hilario the stratigraphic interval enclosing the diatomaceous material is approximately 12 m thick and includes a very persistent seam, 3 m thick, and 3 layers—0.5, 0.4, and 1 m thick, respectively—that thin slightly downdip. This part of the section is 50 m below the main phosphorite seam.

At the La Vuelta cliff of Arroyo Las Pocitas, some 16 to 17 km north-northwest of San Hilario, the abovementioned seam thickens to 4 m but includes a 1-m interval of silicified green shales and finely interbedded tuffaceous-clayey material. At this point a 50- to 10-cm-thick phosphorite bed directly overlying the diatomite is evidence of the close relation between the two.

At Cerro de Palo Verde, standing out from the surrounding landscape 2 to 3 km northwest of the La Vuelta cliff, the diatomite facies subdivides into two seams, 8 and 3 m thick (Fig. 10), that appear respectively 18 m above and 30 m below the main phosphorite seam. The base of the upper seam is finely interbedded with silicified limestone.

The diatomaceous material attains maximum thickness (approximately 20 m) at Cerro Blancuzco, the only locality in the area where it includes beds, as much as 15 to 20 cm thick, of dark gray volcanic ash and light brown pumicite, and also where the material appears only 7 to 8 m below the main phosphorite seam. At Cerro Blancuzco the large volume of volcanic ash absorbed by the Monterrey depositional environment fostered the local accumulation of diatomaceous material, which however, gradually thins out toward the southeast as tuffs and tuffaceous material increase in the section, and practically disappears in the El Aguajito area. Here, the excessive volcanic material not only hindered diatomite deposition but also affected the main phosphorite seam,

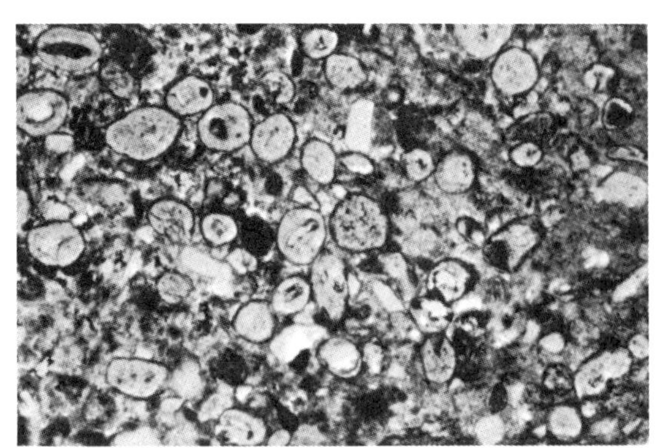

Figure 11. Photomicrograph of oolites and collophane nodules in a clayey groundmass with quartz and feldspar fragments (phosphorite at San Hilario).

which at some places is only 30 to 40 cm thick; at the same time a significant proportion of the tuffs and tuffaceous material was phosphatized.

Phosphorite-diatomite genetic environment. Many environmental and depositional factors converge in the phosphorite and diatomite genesis, as proved by the distribution of these materials at San Hilario. Rising cold marine currents loaded with nutrients (containing nitrogen and phosphorus) overflowed the Tepetates sedimentary threshold and spread over a sea-floor relief of highs and lows. They not only caused precipitation of carbonate and phosphate (Fig. 11) by modifying the temperature and pH of the partially stagnant waters in the embayment, but also carried diatom tests from northern latitudes; these accumulated on the sea floor together with diatoms that reproduced explosively in an acid medium charged with the soluble silica continually contributed by volcanic ashfall.

The subaqueous precipitation of calcareous ooze caused by the abundant planktonic foraminifera combined with that of the siliceous ooze produced by the tuffs and explosively reproducing diatoms, and with the replacement of carbonate by phosphate and soluble silica, to form the silicified limestone-diatomite-phosphorite deposits. The sudden temperature changes that may have taken place locally by merging of the cold Pacific and warm proto-Gulf waters, or a "red tide" generated by an explosive development of dinoflagellates (implicit in rising marine currents), would have caused massive extinction of the marine vertebrates, whose remains make up the bioclastic facies.

The joint presence of clayey tuffs, silicified green diatomaceous shales, silicified limestone, ubiquitous gypsum, phosphorites, diatomite, and marine vertebrate remains is not only explainable, but even logical, when a mixed sedimentary process, environmental conditions, and the Monterrey paleogeographic and paleotectonic framework are taken into account.

STRUCTURE

The structure of the Monterrey outcrop is simple: a wide syncline that trends 30° northwest on the average and dips an average of 3° northeast. Significant faulting appears toward the southeasternmost part of the area (El Aguajito); otherwise (over approximately 50 km from Tepetates Ranch to Arroyo El Campamento) only rare, small lateral or normal faults interrupt or locally mask the phosphorite outcrop (Fig. 12).

Syngenetic lenticularity

This structural simplicity hides a syngenetic lenticularity that may be almost imperceptible in the partly or wholly silicified parts of the section where the clastic process was dominant, or it may be strongly marked, irregular, and inconsistent where mostly nonclastic, chemical, or biochemical processes were involved. This deformation affected practically the entire section, to the extent that truly representative strikes and dips of the sedimentary blocks are very rare. This factor, together with the low structural dip, required both the use of indirect methods for the construction of cross sections and use of the main phosphorite seam as reference because this is the only more or less reliable structural correlation level in the whole section.

In spite of the generalized syngenetic lenticularity, it is nevertheless evident that in the end the clastic process dominated over the nonclastic, since even the calcareous, phosphatic, or diatomaceous materials that were most affected by that deformation exhibit some features that allow structural control within reasonable limits. This is particularly the case in sediments that were partially or entirely silicified in the process, such as the main phosphorite seam and, to a much lesser but immediate degree, the Palo Verde slabs and some levels of similar composition or coquinoid material. The degree of approximation imposed by the sedimentary conditions affects stratigraphic and structural control of the various Monterrey lithologies. Consequently, both the cross sections and lithologic columns shown here (Figs. 13 and 14) provide only approximate thicknesses and facies changes.

CONCLUSIONS

Of academic interest

1. The Monterrey Formation was laid down in a shallow embayment or paleo-bay of restricted circulation under arid weather conditions.

2. Deposition took place in a mixed or hybrid environment by way of two simultaneous processes: clastic, and nonclastic chemical and biochemical.

3. The processes generated alteration products mainly by

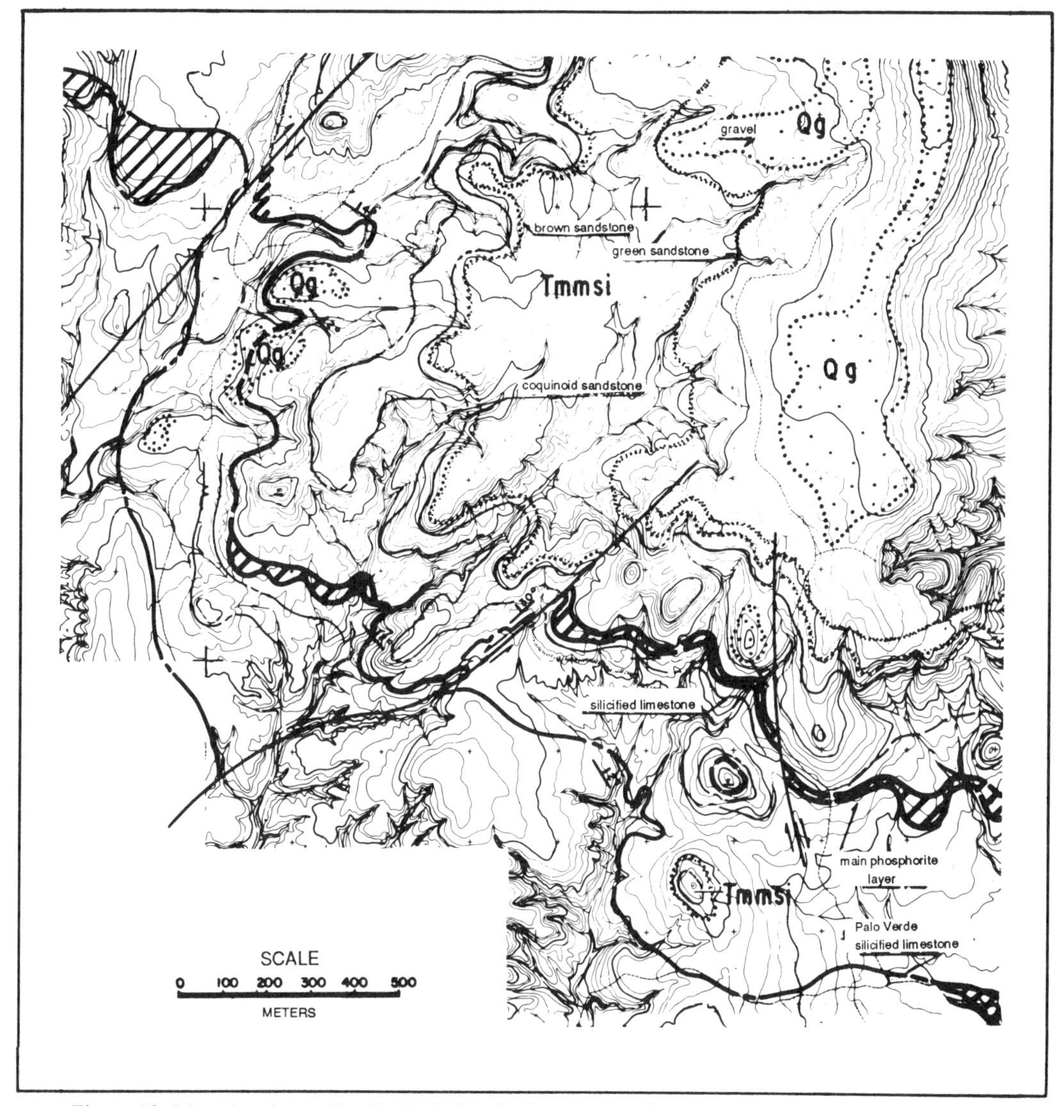

Figure 12. Map showing strike-slip faults locally interrupting the continuity of the main phosphorite seam on the north flank of Cerro El Gavilan.

silicification, metasomatic replacement, dissolution, and evaporation or chemical precipitation.

4. The clastic process predominated in the resulting deposition, enhanced by silicification; however, the gradual or irregular and intermittent introduction of nonclastic materials and their alteration products lenticularized their distribution.

5. The phosphorite is a joint product of both clastic and nonclastic processes in a specific environment, as reflected by its distribution.

6. Phosphorite distribution depends on the course followed by the environment of precipitation that was caused by rising marine currents that overflowed the Tepetates sedimentary threshold and spread over a sea-floor topography of highs and lows.

7. Migration of this environment was controlled *mainly* by the changes in temperature that arose from the merging of the cold Pacific and warm proto-Gulf waters.

Of interest for prospecting

To establish the environmental and depositional conditions of the phosphorite in greater detail, the relief of the original depositional basin must be reconstructed, and the direction of the circulating currents, together with the effects of the sea-floor relief on the migration of the environment of precipitation, must be defined. To this effect the following is recommended:

a. Detailed sampling of the already described and measured stratigraphic sections for detailed chemical, petrographic, sedimentological, and micropaleontological analyses.

b. Deep drilling with core recovery downdip at 10, 15,

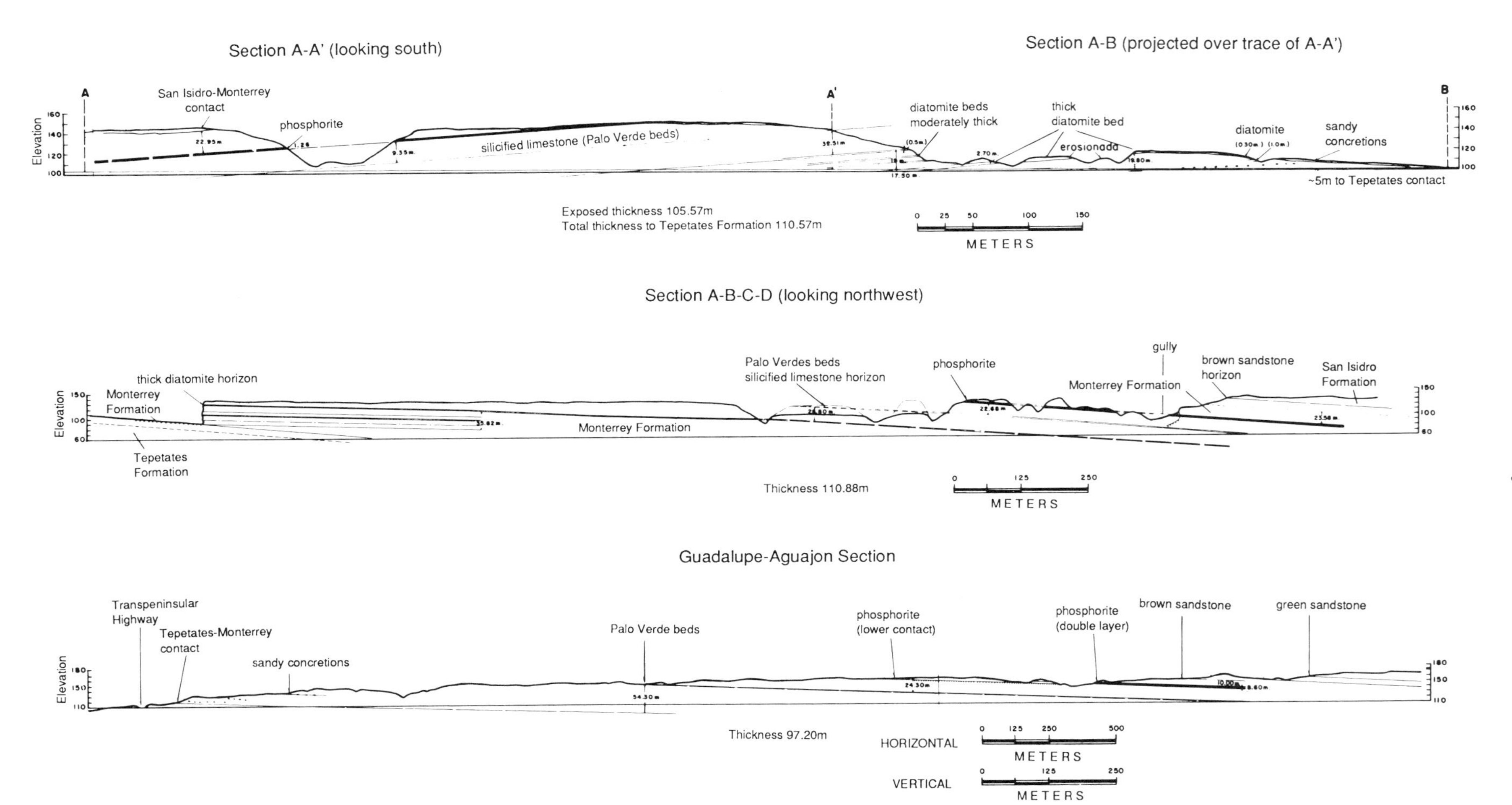

Figure 13. Representative cross sections across the strike of the phosphorite seams in the San Hilario area.

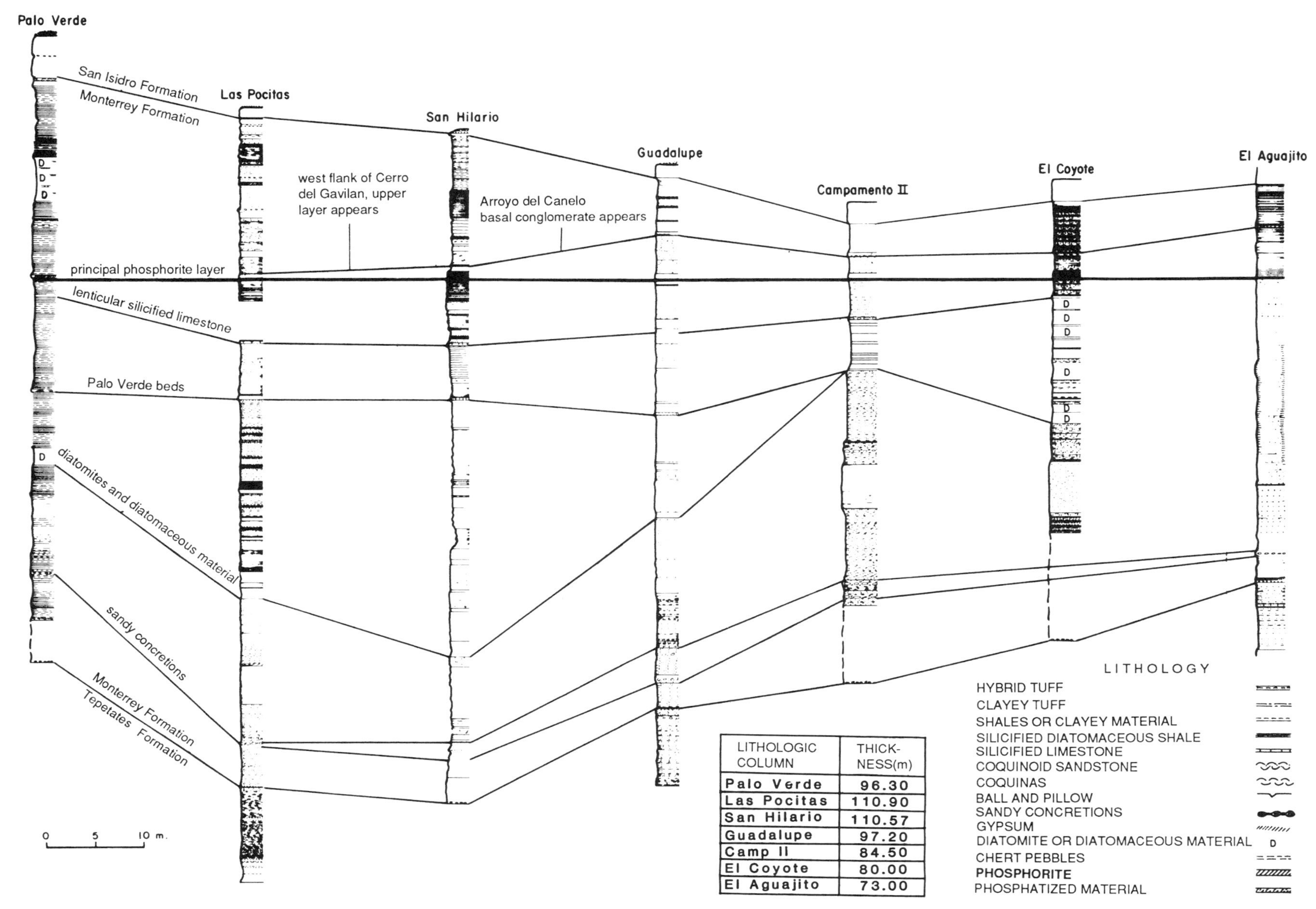

LITHOLOGIC COLUMN	THICK-NESS(m)
Palo Verde	96.30
Las Pocitas	110.90
San Hilario	110.57
Guadalupe	97.20
Camp II	84.50
El Coyote	80.00
El Aguajito	73.00

Figure 14. Representative stratigraphic columns of the Monterrey Formation measured along strike from north to south in the San Hilario district. See Figure 9 for approximate locations.

and/or 20 km from the phosphorite outcrop, to supplement the surface stratigraphic information and define phosphorite distribution at depth.

c. Based on the above, extension of regional exploration to define (at the surface and/or at depth) the limits of the original sedimentary basin.

d. Analysis of the diatomite or diatomaceous material distribution by means of specific research on quantities, quality, origin, and possibilities of its commercial development, either separately or jointly with the phosphorite.

e. Detection of rare minerals, principally uranium and vanadium, to foster integral development of the Monterrey outcrop.

REFERENCE CITED

Ojeda, R., J., 1979, Resumen de datos estratigraficos y estructurales de la Formación Monterrey que aflora en el area de San Hilario, Baja California Sur: Mexico City, Consejo de Recursos Minerales, 25 p.

Printed in U.S.A.

The Geology of North America
Vol. P-3, Economic Geology, Mexico
The Geological Society of America, 1991

Chapter 17

Geology and mineral deposits of the El Boleo copper mine, Santa Rosalía district, Baja California Sur

Ivan F. Wilson and Victor S. Rocha
Consejo de Recursos Minerales, Centro Minero Nacional, Carretera México-Pachuca, Pachuca, Hgo., México, D.F.

The El Boleo mine, second largest copper producer in Mexico during the 1920s, lies near the town of Santa Rosalía, about halfway down the east coast of Baja California Peninsula. The copper deposits occur within a recently uplifted coastal belt of Pliocene sediments that form stream-cut mesas. Mineworks are spread over an area 11 km long by 0.5 to 3 km wide.

The climate is hot and dry; rainfall is generally torrential at intervals in the course of several years. The xerophyte vegetation consists mostly of cacti and other thorny desert plants.

The region's main natural resources are copper, manganese, and nonmetallic deposits, together with prosperous fisheries in the Gulf of California and the Pacific Ocean.

The oldest rock, exposed only at small erosional windows, is a Cretaceous or pre-Cretaceous quartz-monzonite. This is followed unconformably by the mid(?)- and upper Miocene Comondú Volcanic Formation, probably at least 500 m thick, of nonmarine andesite and basalt flows, tuffs, breccias, agglomerates, conglomerates, and tuffaceous sandstones(?). The facies changes indicate that the formation was derived from the present emplacement of the Gulf of California.

When the area was invaded by the Pliocene sea the topographic relief on the Comondú Formation was at a young or early mature stage, locally attaining 450 m in relief. The Comondú is overlain with strong unconformity by the lower Pliocene Boleo Formation, which hosts the copper deposits and contains mainly interbedded tuff and tuffaceous conglomerates—ranging in composition from latites to andesites—that locally overlie a nonmarine basal conglomerate. These are followed by a quite persistent basal marine limestone, thick gypsum beds, and a good many fossiliferous sandstone lenses. Thickness of the unit ranges 50 to 250 m, averaging 140 m. It is considered to be a coastal deltaic deposit.

Unconformably overlying the Boleo is the middle Pliocene Gloria Formation of sandstone and conglomerate averaging 60 m (maximum 185 m) in thickness: a local basal conglomerate underlies marine fossiliferous sandstone, followed in turn by an overlying conglomerate that becomes nonmarine inland. These overlying units belong to the post-mineralization stage.

The 5- to 140-m-thick (average 55 m) upper Pliocene Infierno Formation, overlying the Gloria with local unconformity, consists of marine fossiliferous sandstone covered by conglomerate. This is unconformably followed by the Pleistocene Santa Rosalía Formation, a thin marine fossiliferous sandstone bed and 5- to 15-m-thick conglomerate covering some of the mesa tops.

The mesas northwest of the El Boleo mine are covered by the Pleistocene to Recent Tres Vírgenes group of volcanic rocks composed of latite, pumice, and tuff flows, locally covered by subsequent olivine basalt and breccia flows.

The mesas and terraces along the streambeds are covered by Quaternary 1- to 20-m-thick, primarily nonmarine gravel beds. The streambeds, beaches, and dry lakes hold alluvium. There is total structural unconformity between the Comondú Volcanic Formation and the more recent units. The Comondú rocks were faulted before the deposition of the Pliocene beds; in some cases, west-dipping faults with displacements of 50 to 100 m produce a series of east-tilted blocks. These rocks commonly dip 5 to 25° northeast but may reach 55°. The pre-Pliocene faults were the pathways for the mineralizing solutions.

A good proportion of the Pliocene and Pleistocene structure represents original or compactional dips developed on top of the irregular Comondú topography, ranging from 30° in the basal marine limestone of the Boleo Formation to between 3 and 20° in the clastic beds, and decreasing toward the upper part of the section. The effects of subsequent tilting and faulting are superimposed on these dips. The beds dip 3 to 10° toward the sea and are cut by numerous N10 to 45°W–trending and 65 to 70° southwest- or northeast-dipping normal faults. The faults with maximum displacement dip southwest; maximum known vertical displacement is 250 m at the Santa Agueda fault; the rest have displacements under 80 m. Many of the faults form parts of juxtaposed staggered radial systems. The strongest faulting probably took place toward the end of the Pliocene and continued with diminished intensity during the Pleistocene and Recent.

The El Boleo copper deposits are contained in the dark, bentonitic, impermeable, and soft clayey tuffs of the Boleo Formation. Five main levels of mineral in clayey groundmass have

Wilson, I. F., and Rocha, V. S., 1991, Geology and mineral deposits of the El Boleo copper mine, Santa Rosalía district, Baja California Sur, *in* Salas, G. P., ed., Economic Geology, Mexico: Boulder, Colorado, Geological Society of America, The Geology of North America, v. P-3.

been identified, each overlying conglomerates or their tuffaceous sandstone equivalents toward the sea.

Thickness of the mineral zone under development in recent years has averaged 80 cm, although in older mines it reached 1 to 2 m and exceptionally 5 m. Average grade of the mineral extracted from 1866 to 1947 was 4.81 percent copper, which, however, surpassed 8 percent initially and decreased to 3.5 percent in the latter years. Part of the mineral occurs in subparallel elongated riblike pipes. Veins and pockets are found in the Comondú and other irregular substitution bodies in different rock types.

The principal ore minerals in the sulfide zone are very fine granular chalcocite accompanied by chalcopyrite, bornite, covellite, and native copper; pyrite and galena are less common. The oxidized zone contains a large variety of oxides, carbonates, silicates, and copper oxychlorides, including rare minerals: boleite, pseudoboleite, and cumengite. The copper minerals are commonly associated with manganese oxides: cryptomelane and pyrolusite, also found concentrated in separate seams. Gangue is mainly montmorillonite together with the common rock-forming minerals, gypsum veinlets, calcite, chalcedony, and jasper.

Elements in notably high percentages within the El Boleo ores include copper, manganese, zinc, cobalt, lead, sulfur, and chlorine. In addition, nickel and silver are slightly concentrated in the El Boleo smelting products. Zinc distribution, averaging 0.8 percent (maximum 6 percent) was the subject of special study; at many localities the zinc/copper ratio increases markedly toward the hanging wall of the mineral lodes.

The origin of the El Boleo copper deposits is controversial, but most authors favor a hypothetical substitution of favorable beds by hydrothermal solutions rising along preexisting faults and fractures in the Comondú Formation. They are thought to have ascended jutting peaks and islands of the volcanic basement and spread through the Boleo beds, being trapped by the relatively impermeable clayey tuffs that cover the porous conglomerates and the sandstones. Mineralization took place in the post-Boleo–pre-Gloria interval (beginning of the Pliocene) and is believed related to the volcanic activity.

The El Boleo deposits were discovered in 1868 and subjected to exploitation since 1886 by the French-owned El Boleo Company. Total production from 1886 to June 30, 1947, was 13,622,327 metric tons of mineral, from which 540,342 metric tons of copper were extracted.

Underground workings at the El Boleo mine cover more than 588 km; mining methods similar to those applied in thin coal seams were used.

The El Boleo ores were smelted directly in oil-fueled air furnaces, producing a high copper content matte, which was then treated in converters. The end product was blister copper with an average 99.3 percent Cu content; average recovery is 87.4 percent. Oxidized and sulfide ores were smelted together, sulfur and coal being added to the furnace load. Mineral preparation tests were carried out on the El Boleo ores, which are very difficult to refine owing to the extremely fine sizing of their minerals and the expansive nature of the bentonitic clayey matrix.

Known reserves of easily exploitable copper ore are low. Exploratory drilling might reveal a few new relatively minor bodies, but probably no large unknown deposits. Several million tons of marginal reserves, averaging approximately 2 percent copper, still remain in thin, deep-seated seams.

In addition to copper, the mine contains low-grade manganese seams, development of which must await that of some commercial recovery process. Lesser quantities of zinc, lead, cobalt, nickel, and silver are present in the copper ores or in the smelting products; only very small amounts of silver have been commercially recovered from the blister copper.

Other mineral resources in the area include gypsum, limestone, calcareous sandstone, pumite, perlite, and sulfur.

Editors Note: The origin of the presently noncommercial mines of the Santa Rosalía copper district is controversial, because the copper occurs as sulfides in marine bentonitic sediments that at the same time are near to rhyolite and granite-rhyolite intrusives—a rare but universally known occurrence. Considering the degree of controversy, this chapter is only a summary, albeit without illustrations. For further details, see Wilson and Rocha (1957).

REFERENCE CITED

Wilson, I. F., and Rocha, S., V., 1957, Geology and mineral deposits of the Boleo copper district, Baja California, Mexico: U.S. Geological Survey Professional Paper 273, 134 p.

Printed in U.S.A.

The Geology of North America
Vol. P-3, Economic Geology, Mexico
The Geological Society of America, 1991

Chapter 18

Sierra Madre Occidental Metallogenic Province

Guillermo P. Salas
Geologo Consultor, Asesor del Consejo de Recursos Minerales, Centro Minero Nacional, Carretera México-Pachuca, Pachuca, Hgo., México, D.F.

INTRODUCTION

This province extends from the Mexico/United States border between Sásabe in the east, through Nogales and Agua Prieta (Sonora), to the vicinity of Estación Palomas (Chihuahua). Geographically, the metallogenic province coincides with the Sierra Madre Occidental Physiographic Province of Raisz (1959); on the west it is limited by the contact of the Gulf of California/Pacific Coastal Plains and on the east with the westernmost central Chihuahua valleys (Fig. 1). Over an area of almost 275,280 km^2, its axis runs northwest-southeast from the American border to the city of Guadalajara (Jalisco).

The Sierra Madre Occidental exhibits a variable geologic character. The physiographically high altitudes above the coastal plain—1,500 to 2,000 m on the average, some higher than 3,000 m—evolved by effects of the differential erosion of thick extrusive volcanic sequences: intercalated andesites, rhyolites, tuffs, and ignimbrites making up a volcanic assemblage.

This assemblage overlies an older basement comprising Precambrian or Paleozoic metamorphic rocks and Mesozoic sedimentary or igneous rocks.

To date no exhaustive study has been done of the province's metallogenetic processes that would allow correlation of geologic events and metallogenetic epochs for purposes of mineral prospecting. A few isolated studies have dealt with the geochronology of the basement rocks, the volcanic assemblage, and the plutons intruding them. Wisser (1966) published an excellent geological compilation of the province in connection with his research on the precious metals of northwestern Mexico.

As mentioned above, the petrological composition of the Sierra Madre Occidental is varied, as are its commercial mineral products, which however, may be related to some degree of zoning. In view of this, it is possible and desirable to undertake comprehensive scientific research in the future for purposes of subdividing this vast region, which is of such present and eventual significance in the country's mining industry.

Geographically the province comprises mountain chains, not all of which are structural, and northwest-trending valleys carved out by the Yaqui, Babispe, Tomochi, Mulatos, Moctezuma, and other streams; farther south the Fuerte, Guyapan, Piaxtla, and other river valleys flow northeast.

To the south at Nayarit (Jalisco) in the Acaponeta, Río Grande de Santiago, Bolaños, Ameca, Tomatlán, and other rivers, the course of the streams is controlled markedly by structure. Salas (1980) discusses the metallogenesis of this province in detail, as well as its relation to the plate-tectonics hypothesis, which he sees as doubtful.

The Sierra Madre Occidental contains an Eocene to Quaternary assemblage of extrusive volcanic rocks. Andesites predominate at the base of the assemblage, which is 2,000+ m thick. The upper part of the assemblage consists of successive dacitic, rhyolitic ignimbrites with thick rhyolite intervals. These volcanic flows apparently emerged along extensive northwest-trending fractures parallel to the Pacific and Gulf of California coastlines.

The northern part of the province comprises shorter isolated mountain chains occurring extensively north of Sonora and west of Chihuahua. Granitic batholiths and stocks predominate, intruding the volcanic assemblage. This is the typical zone of disseminated copper deposits including Cananea, La Caridad, Florida-Barrigón, and the Pilares de Nacozari Chimney.

According to Damon and Mauger (1966), the lava emissions were emplaced in the assemblage intermittently from the Eocene to the end of the Oligocene and the beginning of the Miocene. Further details on the Sierra Madre Occidental Province appear in Salas (1980).

REFERENCES CITED

Damon, P. E., and Mauger, R. L., 1966, Epeirogeny-orogeny viewed from the Basin and Range Province: Society of Mining Engineers Transactions, v. 235, p. 99–112.

Salas, G. P., 1980, Carta Provincias Metalogeneticas de la Republica Mexicana: Consejo de Recursos Minerales, Bulletin 21-E, 2nd edition.

Wisser, E., 1966, The epithermal precious-metal province of northwest Mexico: Nevada Bureau of Mines Report 13, pt. C, p. 63–92.

Salas, G. P., 1991, Sierra Madre Occidental Metallogenic Province, *in* Salas, G. P., ed., Economic Geology, Mexico: Boulder, Colorado, Geological Society of America, The Geology of North America, v. P-3.

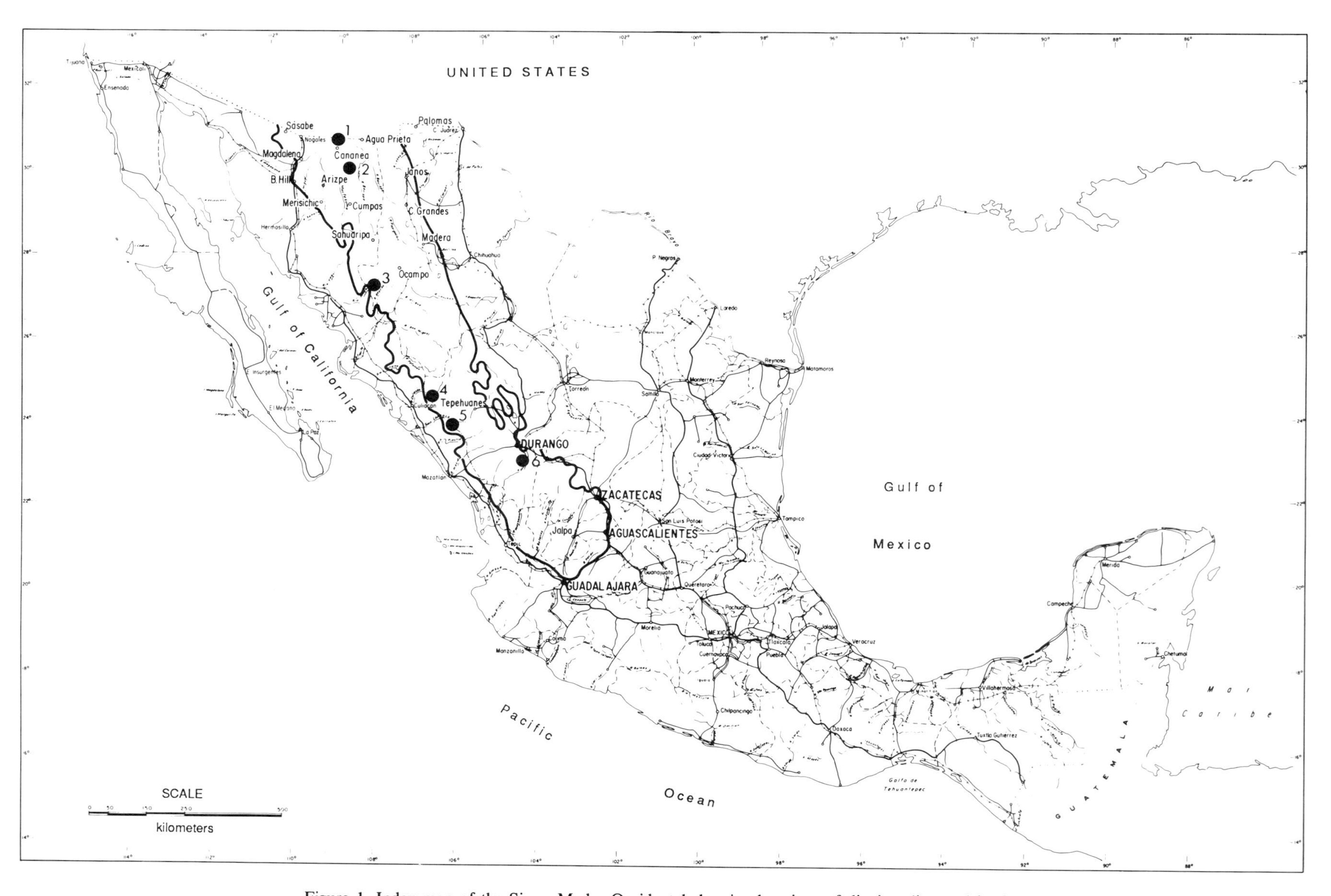

Figure 1. Index map of the Sierra Madre Occidental showing locations of districts discussed in the following chapters: 1, Cananea, Sonora; 2, La Caridad, Sonora; 3, Gochicc, Sonora; 4, Topia, Durango; 5, San Dimas–Tayoltita, Durango; 6, San Martín, Zacatecas.

The Geology of North America
Vol. P-3, Economic Geology, Mexico
The Geological Society of America, 1991

Chapter 19

Cananea copper deposit, Sonora

Guillermo P. Salas
Geologo Consultor, Asesor del Consejo de Recursos Minerales, Centro Minero Nacional, Carretera México-Pachuca, Pachuca, Hgo., México, D.F.

INTRODUCTION

Details on the Cananea copper deposits appear in several previous papers cited below. The following is taken from Richard H. Sillitoe's unpublished work dated June 1975, on file in the Council of Mineral Resources (Consejo de Recursos Minerales). The geology was described by Valentine (1936); mineralization and mineral deposit data are by Perry (1935), Velasco (1966), and Cía. Minera de Cananea, S. A. (1973). The breccias were described by Perry (1961). Sillitoe's comments are summarized below.

GEOLOGICAL ENVIRONMENT

The Cananea host rock consists mainly of early Tertiary volcanic rocks of rhyolite and andesite composition. Valentine (1936) estimated an approximate thickness of 5.5 km in the mine area for this volcanic sequence, which overlies Paleozoic quartzites and limestones unconformably. Underlying the quartzite is a coarse-grained granite dated at 1,440 ± 15 Ma (U-Pb in zircon; Anderson and Silver, 1973). Anderson and Silver (1973) believe that the granite-quartzite contact represents an unconformity and the granite is not intrusive, as did Velasco (1966).

The host rocks are intruded by a series of quartz-monzonite porphyry stocks and bodies probably overlying an extensive deep-seated plutonic intrusive. The observed breccias and alteration are probably related to these intrusives. A phlogopite sample from the old La Colorada mineral body was dated at 58.5 ± 2.1 Ma (K-Ar; Damon and Mauger, 1966). Andesite porphyritic dikes cut and intrude the mineralized bodies.

Hydrothermal breccias

The best known Cananea breccias, of collapse type, were extensively discussed by Perry (1961). Clear examples include the El Capote and Cananea-Duluth chimneys, unfortunately no longer accessible since mineworks at those localities have been suspended for a long time. However, laminated contacts and numerous plate-shaped fragments at the surface of the Cananea-Duluth chimney that dip toward the center of the chimney are evidence that this was a collapse mechanism. The apparent presence of reworked rock between the pieces may be the result of a subsequent fluidization stage.

Perry (1961) again described the La Colorada mineralized body as a collapse breccia in the initial stage of formation. However, the massive characteristics of the pegmatitic quartz, phlogopite, bornite, chalcopyrite, and molybdenite could be of pegmatitic origin, similar to the pegmatites in the neighborhood of the La Caridad deposit. Samples of a postmineral intrusion breccia containing sulfide fragments are still preserved in the geology office of the La Colorada deposit at Cananea, Sonora.

The most common breccias seem to be intrusive and occur dispersed in the open cut area where the highest-grade mineral is found; they are cemented by the reworked rock and by the sulfides themselves.

A pyritic tourmaline breccia occurring as a halo around the porphyritic stock numbered 8-110 exhibits a variety of materials cemented by very dense black tourmaline, most of which may be due to replacement.

Alteration-mineralization

The dominant alteration in the breccias, stocks, and host rock of the Cananea deposit is sericitic, comprising quartz, sericite, tourmaline, pyrite, chalcopyrite, and limited amounts of molybdenite. The hypogenic mineral attains grades of up to 0.6 percent Cu, considered to be very high for sericitic alteration. Of special interest is the fact that allunite is widespread, particularly in the breccias.

Potassic alteration at the surface has not been described because it appears to be very weak down to a depth of approximately 500 m. This suggests a change in the alteration at depth, as discussed by Lowell and Guilbert (1970) in their disseminated copper model.

Marginal alteration is propylitic (chlorite-epidote-pyrite). Paleozoic-altered limestone horizons present in the so-called

Salas, G. P., 1991, Cananea copper deposit, Sonora, *in* Salas, G. P., ed., Economic Geology, Mexico: Boulder, Colorado, Geological Society of America, The Geology of North America, v. P-3.

"Capote Basin," containing calc-silicates with abundant garnet, sphalerite, chalcopyrite, and pyrite, may be commercially prospective for zinc.

Supergene alteration

Most of the mineral reserves—1,600 million tons with 0.7 percent Cu—consist of enriched mineral in 150-m-thick seams on the average. Residual hematite at the surface indicates the presence of a multicyclic chalcosite seam at depth. Minor amounts of jarosite and goethite are also present, mainly at the edges of the mineralized body.

CONCLUSIONS

At the present erosion level, Cananea is comparable to La Caridad, except that in the former the host rocks are volcanic and not intrusive; in both cases, a weak potassic alteration begins to appear at depth. At Cananea, however, some hypogene minerals are present in the sericitic facies, whereas hypogene values are low at La Caridad.

To date some 100 million tons with 2 to 3 percent Cu have been mined at Cananea, consisting of hypogene and supergene mineral mixtures, most of the latter derived from the breccia chimneys. Cananea is thus one of the five largest known copper porphyry deposits in the world.

REFERENCES CITED

Anderson, T. H., and Silver, L. T., 1973, The Cananea granite; Implications of its Precambrian age: Geological Society of America Abstracts with Programs, v. 5, p. 534.

Cía. Minera de Cananea, S. A., 1973, Distrito Minero de Cananea, Cananea, Sonora, México, *in* Algunos Yacimientos de la República Mexicana Geología, Metodos de Extracción y de Beneficio: American Institute of Mining, Metallurgical and Petroleum Engineers, Seccion México, Comite de Tecnologia, p. 232–251.

Damon, P. E., and Mauger, R. L., 1966, Epeirogeny-orogeny viewed from the Basin-Range Province: Society of Mineral Engineers Transactions, v. 235, p. 99–112.

Perry, V. D., 1935, Copper deposits of the Cananea district, Sonora, Mexico, *in* Copper Resources of the World, v. 1: Washington, D.C., 16th International Geological Congress, p. 413–418.

——, 1961, The significance of mineralized breccia pipes: Mineral Engineering, v. 13, p. 367–376.

Valentine, W. G., 1936, Geology of the Cananea Mountains, Sonora, Mexico: Geological Society of America Bulletin, v. 47, p. 53–86.

Velasco, J. R., 1966, Geology of the Cananea district, *in* Titley, S. R., and Hicks, C. L., eds., Geology of the porphyry copper deposits, southwestern North America: Tucson, University of Arizona Press, p. 245–249.

Printed in U.S.A.

The Geology of North America
Vol. P-3, Economic Geology, Mexico
The Geological Society of America, 1991

Chapter 20

La Caridad disseminated copper deposits, Sonora

Guillermo P. Salas
Geologo Consultor, Asesor del Consejo de Recursos Minerales, Centro Minero Nacional, Carretera México-Pachuca, Pachuca, Hgo., México, D.F.

INTRODUCTION

The importance of the La Caridad disseminated copper deposits is not only the history of its exploration and discovery by modern methods and its location as the southernmost occurrence known to date of the copper belt running from Utah through Arizona in the United States to Mexico, but also its example of Mexico's copper potential. References on the detailed geology and geologic history of the deposit are included here. However, for purposes of a knowledgeable general abstract, this chapter is summarized from unpublished work in the files of the Consejo de Recursos Minerales by Richard H. Sillitoe, who was invited in 1975 to research Mexico's copper ores.

The La Caridad exploration program was described by Coolbaugh (1971) and the general geology by Saegart and others (1974); Echávarri (1971, 1975) carried out detailed petrographic and alteration studies. Consequently, Sillitoe restricted his work to certain outstanding geologic features.

GEOLOGIC ENVIRONMENT

The porphyritic quartz-monzonite stock genetically associated with the La Caridad mineralization was emplaced within intrusive rocks related to a major pluton extending over a considerable distance south of the deposit. A fine-grained rock of andesitic aspect, identified by Saegert and others (1974) as diorite and by Echávarri (1971) as porphyritic quartz-diorite, borders the intrusive stock on the west; granodiorite surrounds it on the east. The pluton intrudes Early Cretaceous or Tertiary volcanic rocks.

Postmineralization rocks are mainly rhyolite tuffs and the La Caridad fanglomerate cemented by hematite and containing rock fragments that form the caprock on the leached deposit. Postmineralization andesite dikes are also observed.

Radiometric age

Livingston (1974) determined K-Ar ages of 53.0 ± 0.4 Ma for the quartz-monzonite porphyry and sericitic alteration in the La Caridad deposit. A younger date—48.9 ± 1.2 Ma—is assigned to a second period of molybdenum mineralization in the eastern part of the deposit. The La Guadalupe pegmatite deposit was dated at 54.0 ± 1.6 Ma by Damon and others (1968).

Alteration and mineralization

The main alteration at La Caridad is sericitic (Saegert and others, 1974; Echávarri, 1971), consisting of quartz, sericite, tourmaline, pyrite, chalcopyrite, and molybdenite. This mineralization is very extensive in the central portion of the deposit, but weakens toward its edges. Sulfides are reported to make up 2 to 3 percent of the rock, and the pyrite-chalcopyrite ratio is approximately 3:1. The grade of the hypogene copper is approximately 0.25 percent. A poorly developed propylitic halo containing chlorite, epidote, calcite, and pyrite surrounds the deposit but lacks continuity.

An interesting feature, seemingly unobserved by previous workers, is the presence of potassic alteration at La Caridad, appearing in a narrow zone outward from the sericitic area and as lenses within the sericite itself. It is represented by biotite, K-feldspar, quartz, chalcopyrite, and molybdenite veinlets, especially visible in the granodiorite at Arroyo Puerto Chino. Additionally, in the vicinity of the Canutillo mine are several K-feldspar veins, several centimeters thick, with quartz, biotite, abundant molybdenite, and subordinate chalcopyrite.

The quartz-diorite porphyry in the vicinity of the Guadalupe mine contains abundant hydrothermal biotite accompanied by chalcopyrite; part of this biotite is coarse grained or in large crystals and is associated with large quartz crystals, K-feldspar, molybdenite, chalcopyrite, and pyrite.

Lenses of potassic alteration in the porphyritic rock appear within the sericitic zone and were also identified in quartz-diorite porphyry hangers of a minecut at the south flank of the deposit and at the Santa Rosa galley. These areas contain more than 1 percent copper as chalcopyrite, and a low pyrite percentage. At the Santa Rosa works, a 2-m-long block was found of quartz-diorite porphyry containing hydrothermal quartz, biotite, and

Salas, G. P., 1991, La Caridad disseminated copper deposits, Sonora, *in* Salas, G. P., ed., Economic Geology, Mexico: Boulder, Colorado, Geological Society of America, The Geology of North America, v. P-3.

chalcopyrite, together with a breccia surrounded by fragments of sericitic alteration. The potassic veinlets end abruptly at the contact of the potassic fragment, which together with the breccia is cut by later quartz-sericite and pyrite veinlets.

All of these potassic alterations are intercepted by subsequent sericitic veinlets. Even the La Caridad pegmatite, consisting of quartz and biotite, shows the almost complete later replacement of the biotite by sericite. This leads to the conclusion that the potassic alteration everywhere was prior to the main phase of sericitic alteration, and is now visible where sericitization did not take place.

Supergene alteration

The La Caridad mineral deposit is an enriched layer approximately 90 m thick. 800 million tons of 0.7 percent copper with 0.012 percent molybdenum have been estimated by Cía. Minera Asarco Mexicana, S. A. The chalcocite occurs as very fine films covering the pyrite and chalcopyrite. This deposit underlies a bed of red and black hematitic leached rock serving as indigenous caprock to the deposit, particularly in the central part. According to Saegert and others (1974), the marginal portions underlie jarosite-goethite beds, suggesting lateral movement of copper-laden solutions during the oxidation process. The presence of this hematitic escape proves the enrichment's multicyclic origin. As in the La Caridad fanglomerate neighboring the deposit, especially on the northeast, fragments of the leached caprock are found. The conclusion is that the supergene alteration was active in mid-Tertiary time, presumably ceasing immediately after, when the deposit was buried beneath volcanic rhyolite beds. A substantial amount of enrichment is thought to have taken place during the early or middle Tertiary. Erosion of the volcanics overlying the deposit bared the deposit at the end of the Tertiary. This more recent erosion has exposed the mineralized layer in the cliffs surrounding the deposit in several places. Supergene kaolinization took place both in the oxidized and supergenically enriched zones; transported kaolin can be observed all over the enriched area.

The potassic alteration halo, with its very restricted pyrite content, underlies small goethite patches, in some places accompanied by malachite, chrysocolla, neotocite, limonite, and ferromolybdite.

GEOCHEMISTRY

Osoria (1973) demonstrated that a substantial part of the La Caridad deposit underlies soils containing less than 200 ppm Cu but with 31 to 400 ppm Mo, a proportion to be expected in an acid-capped, well-leached deposit. Copper values increase to over 500 ppm at the marginal portions of the deposit, where the chalcocite layer crops out locally.

CONCLUSIONS

The La Caridad deposit has been interpreted as the upper part of a typical porphyry copper deposit, with possible potassic alteration below 500 m depth (Saegart and others, 1974). However, according to Sillitoe (unpublished work), the deposit does not fit as perfect and simple a model as proposed by Lowell and Guilbert (1970). This first alteration was potassic and more widespread than shown today by the known deposit. Where preserved, it contains low copper and high molybdenum percentages. This zone was probably related to the first pulses of the stock that generated La Caridad; evidence for this pulse has been found on drilling of the granodiorite toward the east, within the zone of strongest potassic alteration. The subsequent phase of alteration involved a more restricted zone of sericite alteration, within which the supergenically enriched mineral deposit developed later. The sericite alteration probably diminished the hypogene copper content, in view of the fact that the residual potassic patches have higher copper contents in chalcopyrite. This sericite alteration may be related to a subsequent pulse of the La Caridad stock or to the suction effect of the hydrothermal system's cooling during its last stages.

The La Caridad, La Guadalupe, and other pegmatite developments are thought to be contemporaneous with the potassic alteration. What is here termed potassic alteration, however, could be of either deuteric or hydrothermal origin. The pegmatites are considered as such only in the sense of the crystal textures, since at places they cement brecciated rock, which proves they are not true pegmatites. Of interest is that radiometric dating suggests the emplacement of the La Guadalupe pegmatite 1 m.y. prior to the sericite alteration, which supports the interpretation suggested by Sillitoe (unpublished work).

Saegert and others (1974) suggest that a late pulse of molybdenum-rich mineralization improved that mineral's grade in the eastern part of the deposit, based on the younger K-Ar date of 48.9 ± 1.2 Ma, which seems to support that idea. However, it was tentatively proposed by Sillitoe (unpublished work) that the higher molybdenum content in the eastern part of the deposit is the result of molybdenite associated with an early—not later—potassic alteration event, in which case the quartz-molybdenite veinlets crossing the sericitic alteration would form part of the process and not of a later episode. The early date will have to be explained if this hypothesis is accepted, but might be attributed to argon leakage from the fine-grained sericite used for the radiometric age determination.

REFERENCES CITED

Coolbaugh, D. F., 1971, La Caridad, Mexico's newest and largest porphyry copper deposit; An exploration case history: Society of Mineral Engineers Transactions, v. 250, p. 133–138.

Damon, P. E., and others, 1968, Correlation and chronology of ore deposits and volcanic rocks: U.S. Atomic Energy Commission Annual Report COO-689-100, 48 p.

Echávarri, P. E., 1971, Petrografía y alteración en el depósito de La Caridad, Nacozari, Sonora, México: México City, IX Convention Nacional de la Asociación de Ingenieros de Minas, Metallurgistas y Geólogos de México, Memoir, p. 49–72.

——, 1975, Estudio de la petrografía de la alteración y de la mineralización hidrotermales en el depósito de El Arco, Baja California: Industrial Minera Mexico, S. A., 45 p. [unpublished report].

Livingston, D. E., 1974, K-Ar ages and Sr isotopy of La Caridad, Sonora, compared to other porphyry copper deposits of the southern Basin and Range Province: Geological Society of America Abstracts with Programs, v. 6, p. 208.

Lowell, J. D., and Guilbert, J. M., 1970, Lateral and vertical alteration-mineralization zoning in porphyry ore deposits: Economic Geology, v. 65, p. 373–408.

Osoria, A., and de la Campa, G., 1973, Investigacion geologico-minera sobre cobre en el prospecto "La Caridad," *in* Proyecto cobre en Sonora, México: Consejo Recursos Naturales no Renovables Boletín 79, p. 23–47.

Saegart, W. E., Sell, J. D., and Kilpatrick, B. E., 1974, Geology and mineralization of La Caridad Porphyry copper deposits, Sonora, Mexico: Economic Geology, v. 6x, p. 1060–1077.

Printed in U.S.A.

The Geology of North America
Vol. P-3, Economic Geology, Mexico
The Geological Society of America, 1991

Chapter 21

Gochico mineral deposit; Geology, environment of formation, and tectonics

S. A. Rosas
Cía. Minera Peñoles, S.A. de C.V., Río de la Plata 48 esq. Lerma, Col. Cuauhtemoc, 06500 México, D.F.

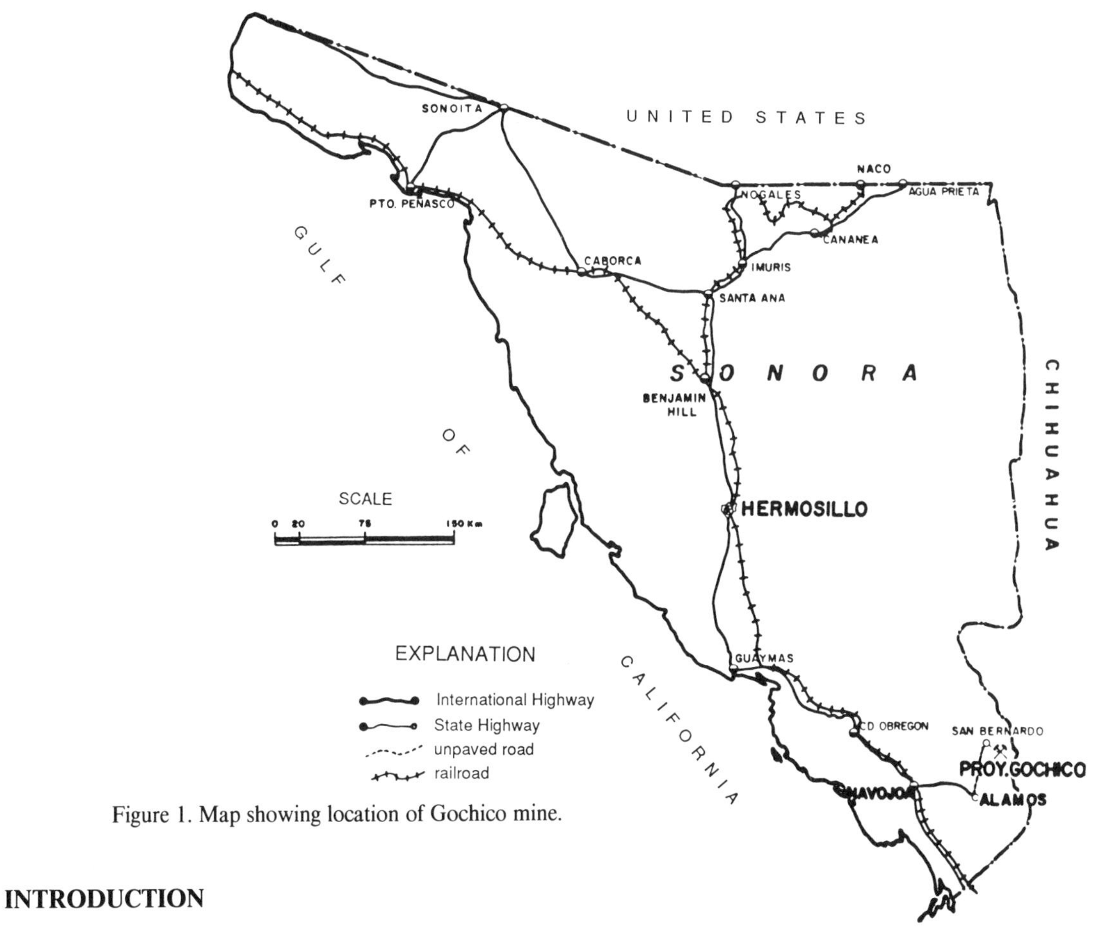

Figure 1. Map showing location of Gochico mine.

INTRODUCTION

Location

The Gochico Mine in southern Sonora (Fig. 1) is reached from Hermosillo by traveling south along a scenic highway for 350 km to the town of Navajoa, then east 50 km along the federal highway to Alamos, northeast along 63 km of earthworks to the village of San Bernardo, and from there, a 10-km dirt road to the mine. The area has a small-craft landing field; both Alamos and Navajoa have paved runways. The nearest railroad station is Navajoa.

History and production

Gochico was a small (2 m) test pit before the Peñoles Company began operations. Geophysical studies were done in 1970 by Wolf Ridge Minerals. Peñoles acquired an option on the properties in 1977, and exploration activities, including regional and local geology, structural geomorphology, geochemistry, and geophysics, supplemented by drilling and direct works, have been active since 1978. Construction of the mine unit began in January 1980, and plant operations in September 1981. 11,500 m of

Rosas, S. A., 1991, Gochico mineral deposit; Geology, environment of formation, and tectonics, *in* Salas, G. P., ed., Economic Geology, Mexico: Boulder, Colorado, Geological Society of America, The Geology of North America, v. P-3.

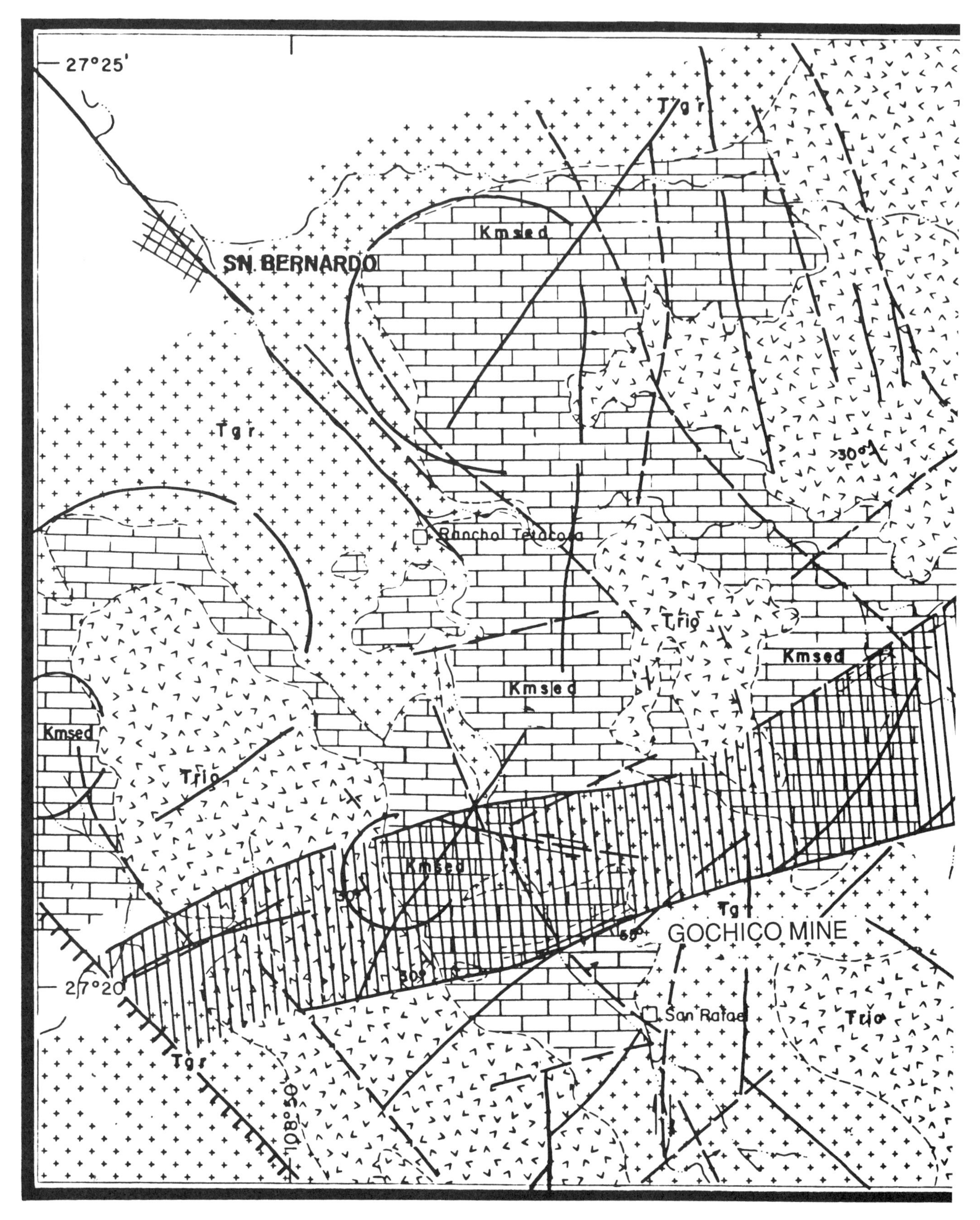

Figure 2. Geologic-tectonic map of the Gochico mine area.

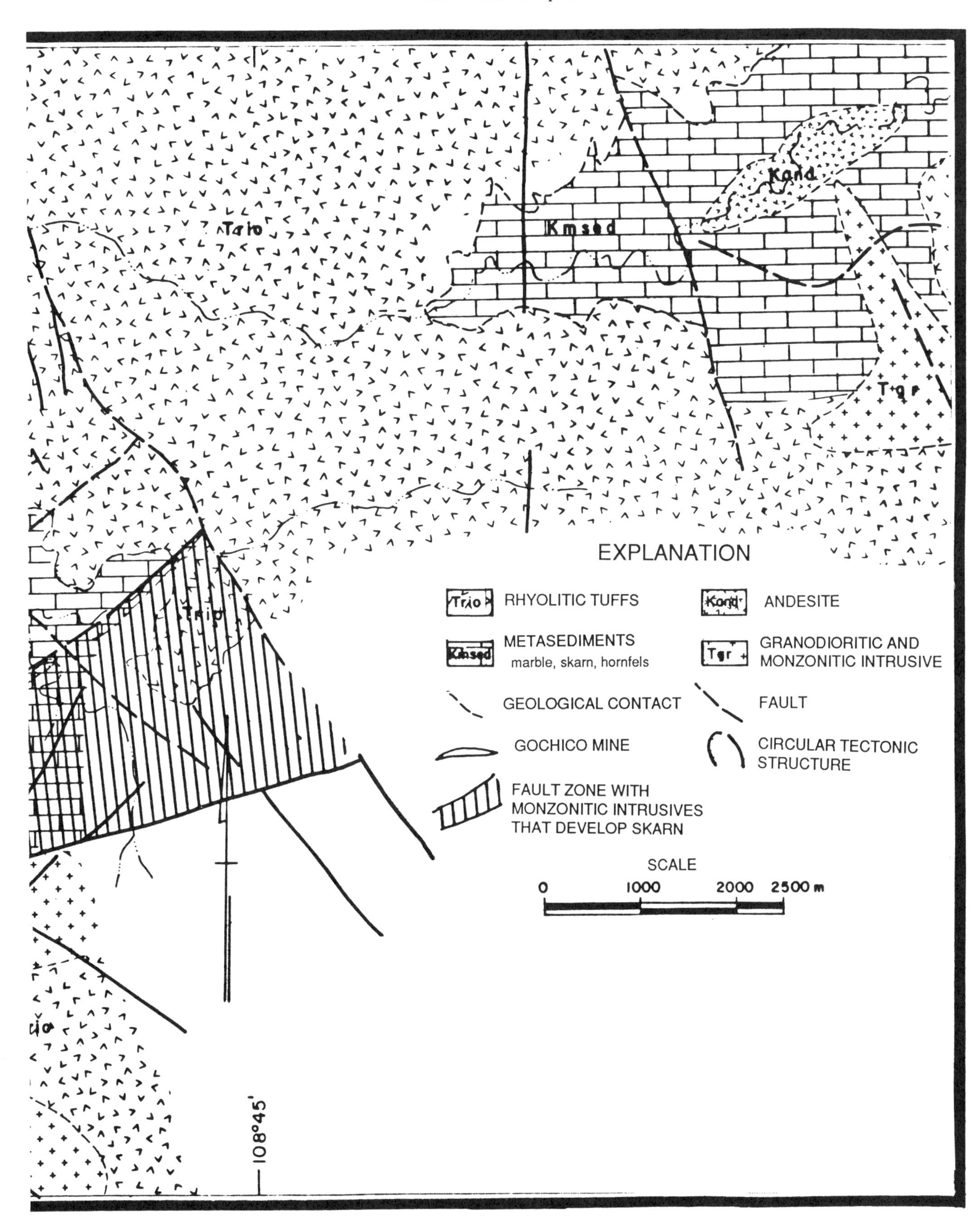
Trio
Kmsed
Kand
Tgr
EXPLANATION
Trio RHYOLITIC TUFFS
Kand ANDESITE
Kmsed METASEDIMENTS
marble, skarn, hornfels
Tgr GRANODIORITIC AND
MONZONITIC INTRUSIVE
GEOLOGICAL CONTACT
FAULT
GOCHICO MINE
CIRCULAR TECTONIC
STRUCTURE
FAULT ZONE WITH
MONZONITIC INTRUSIVES
THAT DEVELOP SKARN
SCALE
0 1000 2000 2500 m
108°45'

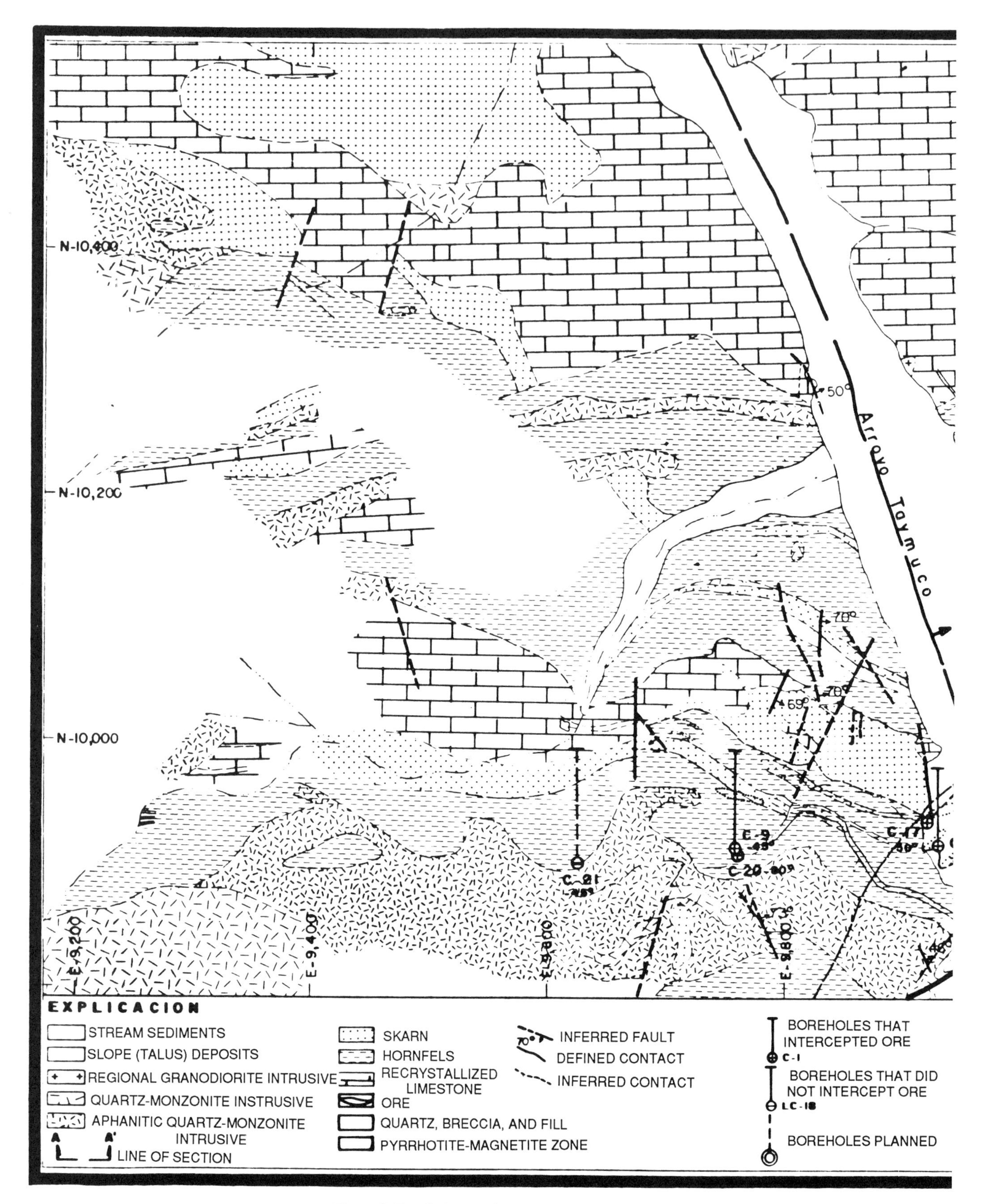

Figure 3. Detailed map of the Gochico mine area.

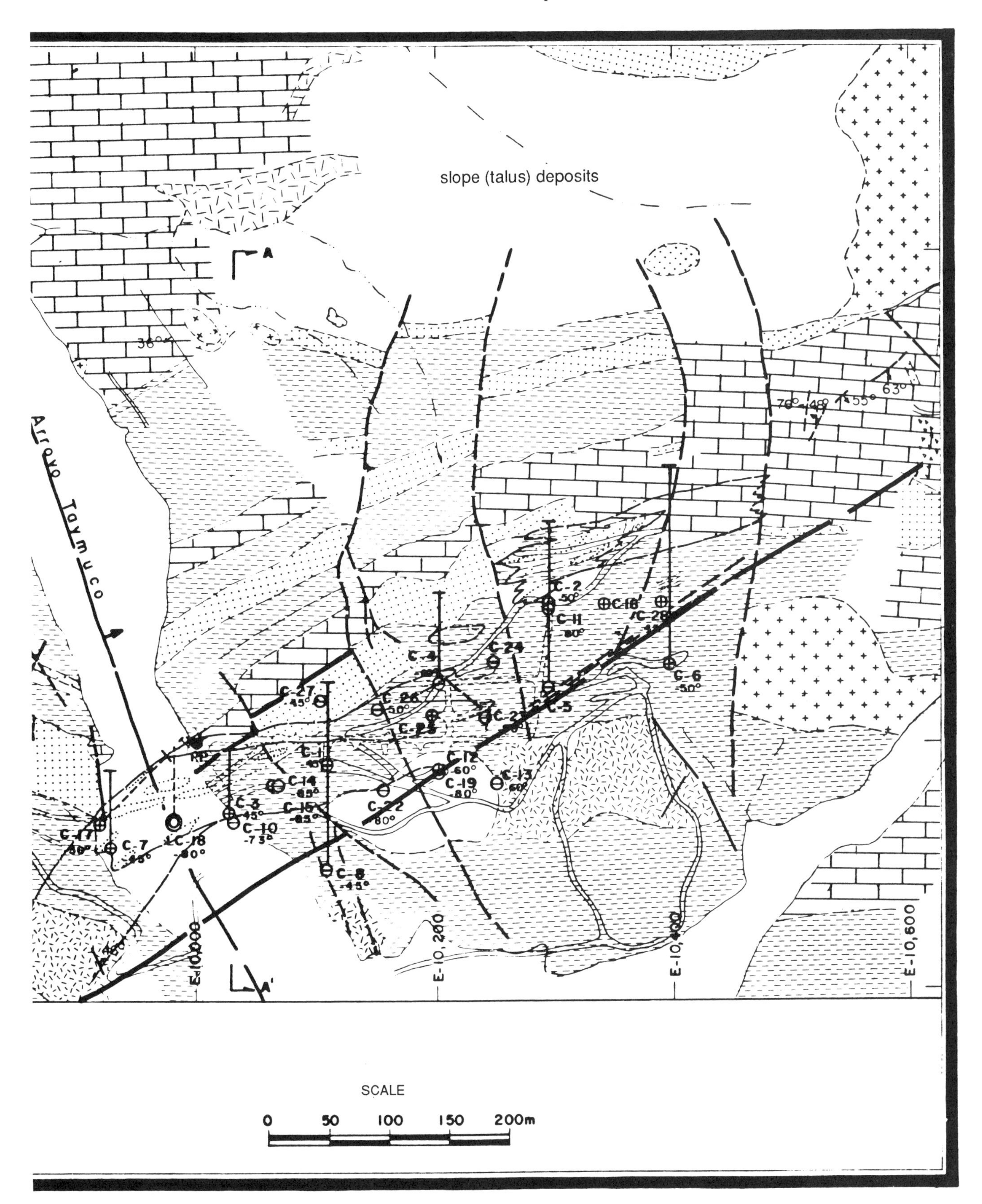
slope (talus) deposits
Arroyo Taymuco
A
A'
E-10,000
E-10,200
E-10,400
E-10,600
SCALE
0 50 100 150 200m

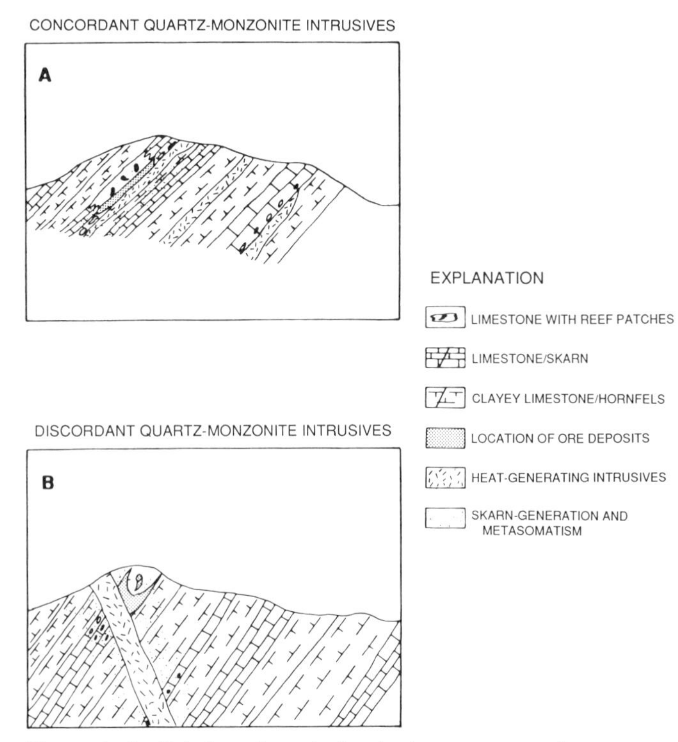

Figure 4. A. Relation of producing horizons to concordant quartz-monzonite intrusives. B. Relations of producing horizons to discordant intrusives where more continuous and prospective zones are generated; however, in areas of limestone with black chert, skarn-formation and metasomatism are more widespread.

drilling indicated 500,000 tons of ore with an average body thickness of 4.2 m and grades of 161 gr/t Ag, 9.2 percent Zn, 0.45 percent Pb, and 0.59 percent Cu. Twelve thousand tons of ore are processed at the mine's flotation plant each month.

REGIONAL GEOLOGY

Triassic-Jurassic

The oldest rocks in the mine area belong to the Barranca Formation, a metasedimentary sequence of sandstones, shales, limestones, and andesites that forms the regional basement. It is represented here only by a small andesite outcrop northeast of the mine, corresponding to the upper part of the formation.

Cretaceous

Overlying the Barranca unconformably is a Cretaceous sequence of shales, calcareous shales, and fossiliferous limestones regionally affected by metamorphism and contact metasomatism produced by quartz-monzonite intrusions along a northeast structural trend. This sequence hosts the Gochico mine skarn-type mineralization and other mineral occurrences in the area—Mexiquillo (W), Sara Alica (Au-Co), Santa Elena (Ag-Pb-Zn), and Otates (Ag-Pb-Zn)—which indicate proximity to heat sources.

Cenozoic

A thick sequence of Tertiary andesite and rhyolite tuffs overlies the Cretaceous. The basal section is breccioid and andesitic in composition, changing to rhyolitic toward the top.

The younger Tertiary is represented by conglomerates of the Baucarit Formation.

INTRUSIVES

Two intrusive periods are identified: (a) large, northwest-aligned calc-alkaline intrusions of granodiorite composition outcropping extensively mostly in the valleys and low-lying areas and related to subduction during the Cretaceous; and (b) more alkaline and less extensive, east- and northeast-oriented intrusions at the end of the Cretaceous or beginning of the Tertiary, which caused significant metamorphism and metasomatism in the area. The different evidences of Gochico-type mineralization (skarns associated with regional faults and quartz-monzonite intrusives) show similar orientations.

TECTONICS

During the Cretaceous the region's geologic environment was that of a transitional island arc to miogeosyncline in which Cu-Zn-Fe-Co skarns with lesser Pb-Ag-Au contents and veins of Pb-Zn-Ag and Au-Ag-Pb occur typically at the flanks of the large island-arc stratovolcanoes. Three fault trends are detected: (a) the oldest and best defined is northwest trending and associated with subduction and geologic activity at the similarly northwest-trending continental margin; (b) a later north trend becomes stronger east of the area; and (c) a perceptibly east to northeast trend caused important alignments in the area and paved the way for the quartz-monzonite intrusions associated with the skarn mineralization, defined regionally by extensive metamorphosed and contact-metasomatized areas. Both the skarn-type bodies and associated quartz-monzonite dikes show the same trends.

TYPE AND CHARACTERISTICS OF DEPOSIT

The Gochico mine is a distal-type skarn deposit associated with a northeast- to slightly east-trending regional fault (Figs. 2, 3), along which the numerous quartz-monzonite intrusions may be conformable or unconformable with the sedimentary stratification. The conformable occurrences exhibit better conditions for the formation of orebodies (Fig. 4).

The main body is oriented east-west, dips 60° South (Fig. 5), and is closely associated with the regional fault and the quartz-monzonite intrusions, generating what is locally termed "quartz breccia," a fault breccia filled and cemented by multiple stages of silica and mineralization. Small, lenticular orebodies

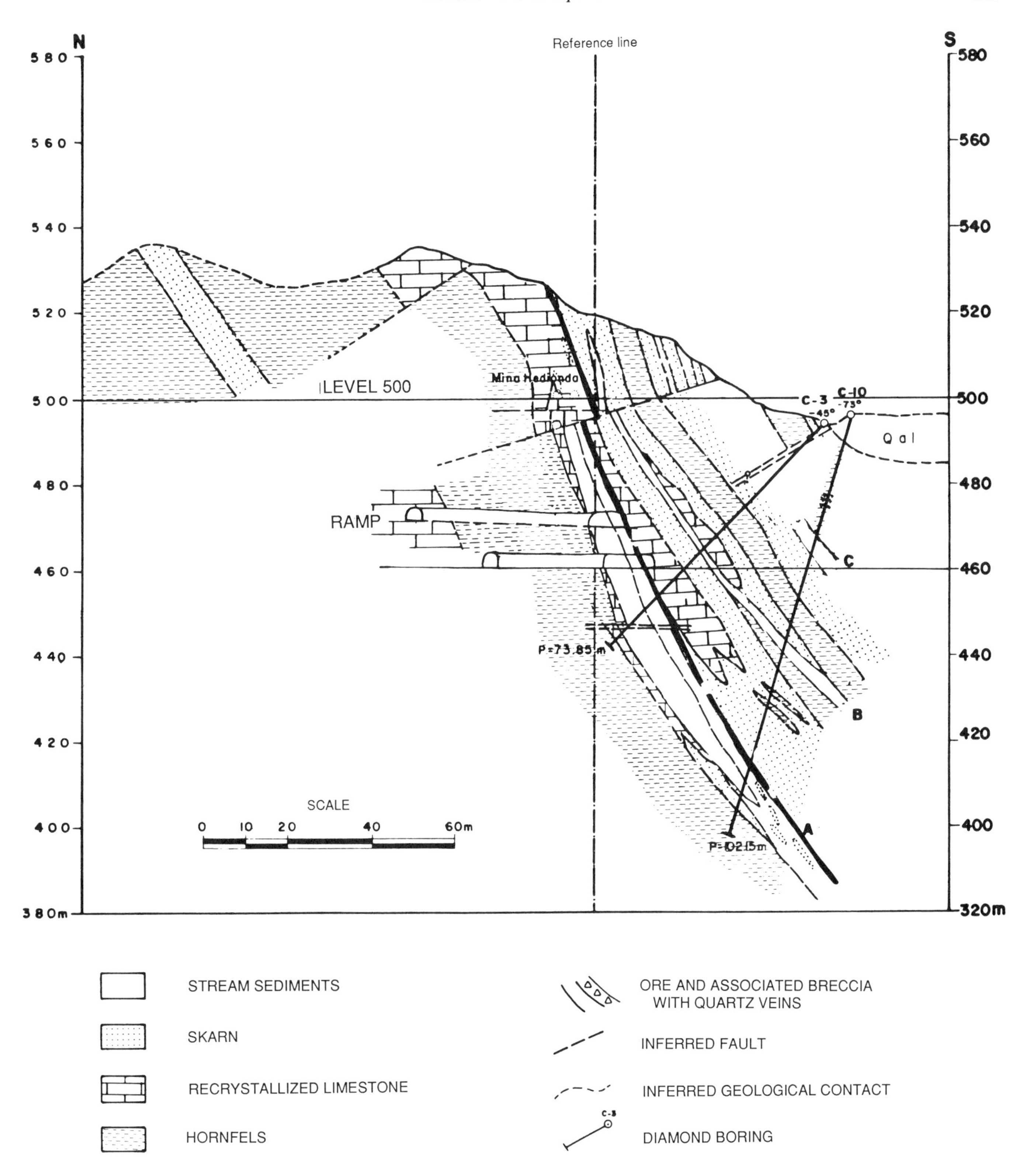

Figure 5. Geologic cross section A-A′, Gochico mine. For location, see Figure 3.

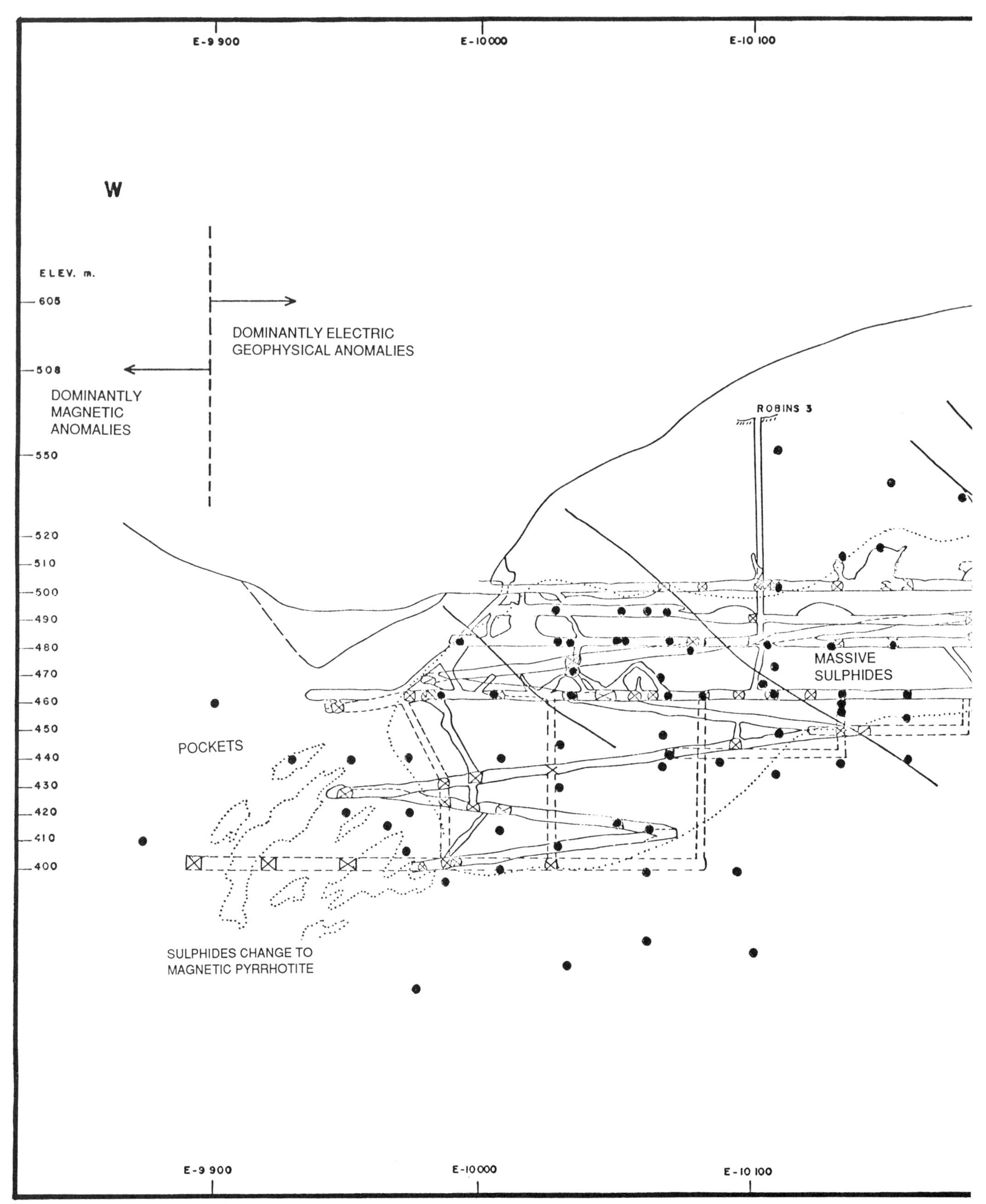

Figure 6. Longitudinal section showing the mineralogical and geophysical characteristics of the orebody.

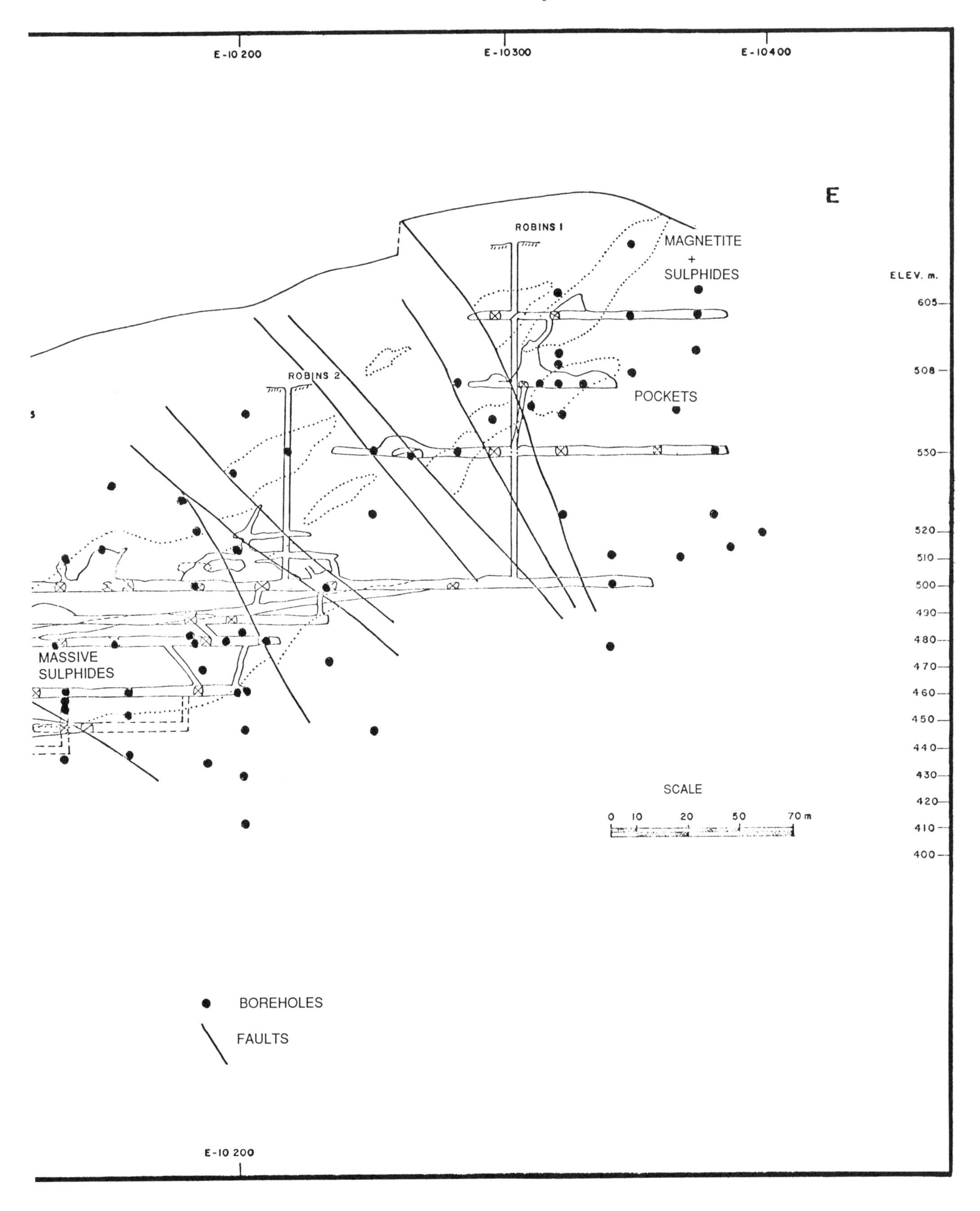
E-10 200
E-10300
E-10400
E
ROBINS 1
MAGNETITE
+
SULPHIDES
ELEV. m.
605
508
ROBINS 2
POCKETS
550
520
510
500
490
480
MASSIVE
SULPHIDES
470
460
450
440
430
SCALE
420
0 10 20 50 70 m
410
400
BOREHOLES
FAULTS
E-10 200

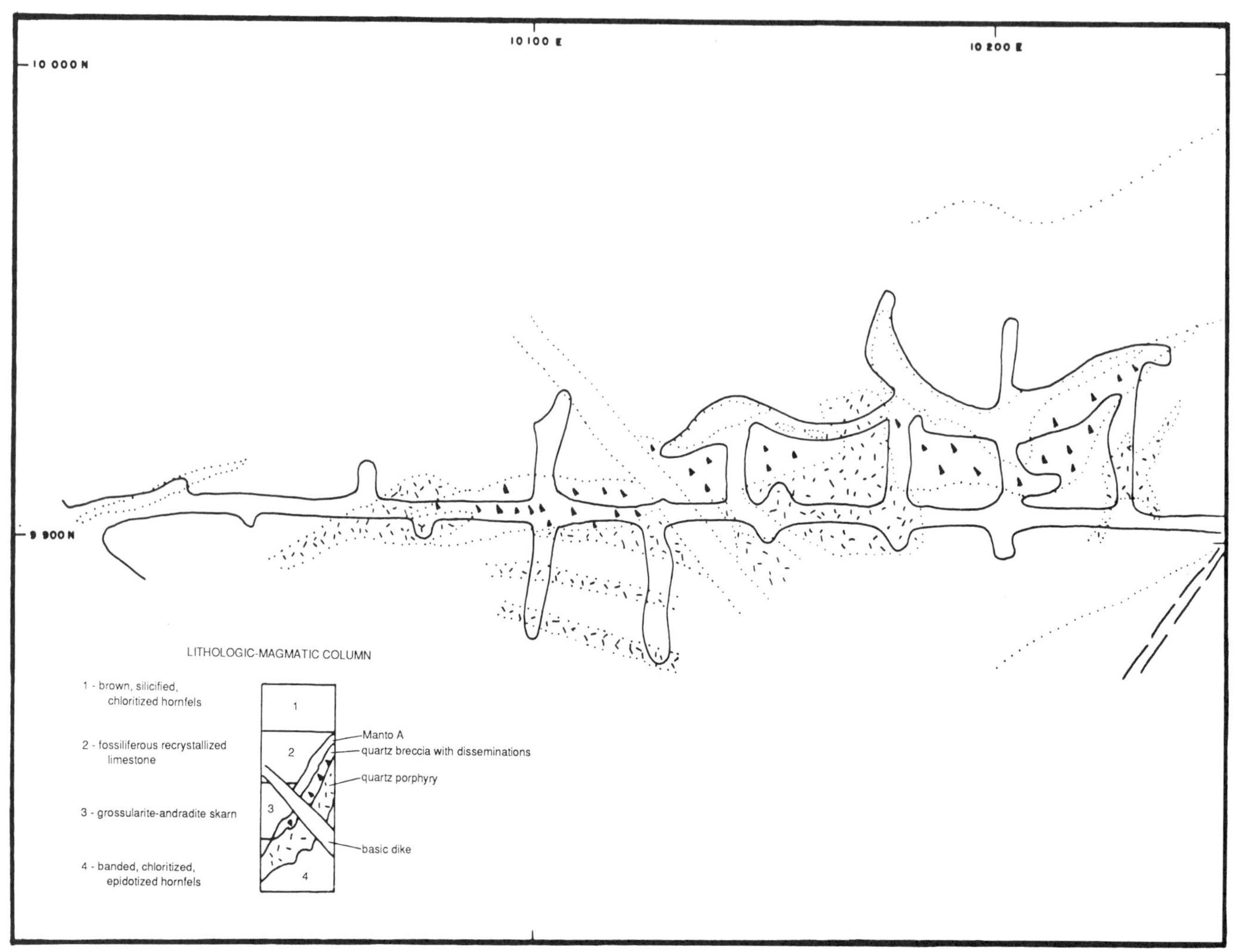

Figure 7. Simplified geology of the 500 level showing the relation between quartz porphyry, quartz breccia, and mineralization.

within the main body occur at the contact of marbleized limestone with andradite-augite and tremolite-actinolite ferroan skarns.

The dimensions of the main body are 350 m along strike and 100 m on the vertical, with thicknesses of 1 to 14 m. Mineralogy in the massive sulfides includes abundant sphalerite and pyrrhotite, scarce galena, chalcopyrite and pyrite, minor argentite, arsenopyrite, and magnetite. The massive sulfide bodies are surrounded by haloes with small pockets and dissemination of massive sulfides. The so-called quartz breccia invariably covers the massive sulfides, although it is not always associated with them.

Principal mineralization controls are structural, lithological, and thermodynamic: (a) the structural control is the east-west fault zone of weakness along which the quartz-monzonites were emplaced, generating temperature and porosity by causing brecciation; (b) lithological controls are indicated because the more favorable zones are found at the intersections of the east-west regional fault with calcareous levels; and (c) the quartz-monzonite intrusions increase in size with depth, and consequently the thermal gradient increased also, defining a marked zonation from Pb-Zn sulfides to zones with pyrrhotite and/or Co-Au-W (Figs. 6, 7).

Printed in U.S.A.

The Geology of North America
Vol. P-3, Economic Geology, Mexico
The Geological Society of America, 1991

Chapter 22

Geology and mineralization of the Topia Mining District

H. Monje H.
Cía. Minera Peñoles, S.A. de C.V., Río de la Plata 48 esq. Lerma, Col. Cuauhtemoc, 06500 México, D.F.

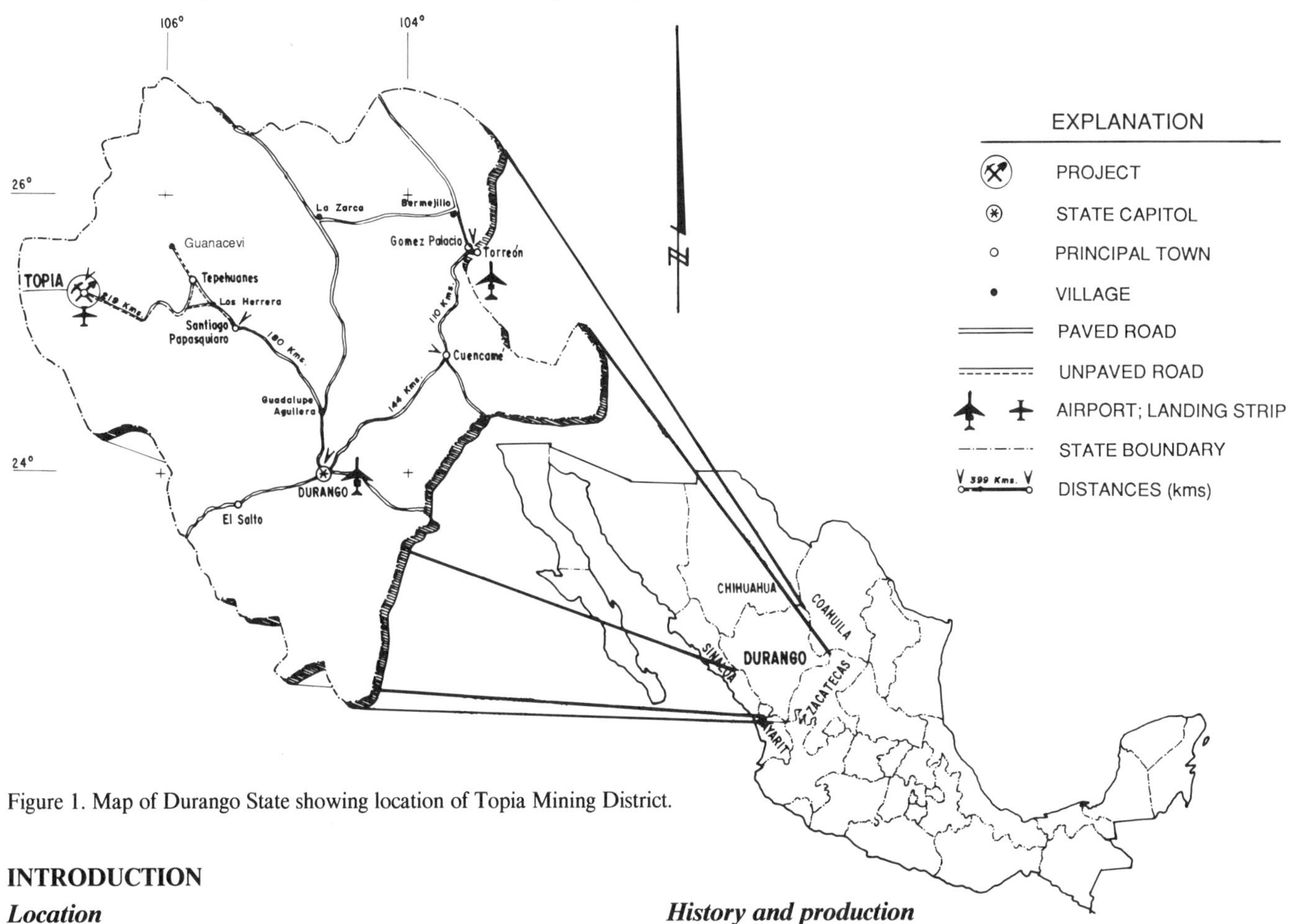

Figure 1. Map of Durango State showing location of Topia Mining District.

INTRODUCTION

Location

The Topia Mining District lies in northwestern Durango State at coordinates 25°11′N and 106°33′W, 50 km from the Sinaloa-Durango state line and 100 km east of Culiacán, Sinaloa (Fig. 1). It occupies the physiographic transition between the Volcanic Highlands and the Barrancas Zone subprovinces of the Sierra Madre Occidental Province. Access to the district is by land and air. Commercial airlines fly to Topia from Durango and from Culiacán (Sinaloa). The Durango-Topia road runs for 390 km; the first 180 km, to Santiago Papasquiaro, are paved. Topia is equipped with municipal public services and schools to high-school level.

History and production

The deposits were discovered in 1538 when small-scale mining began; the first mining concession was reportedly granted in 1602. Formal development of the deposits by five different companies began in 1870 and continued until the 1910 revolution, which brought them to an end. Thereafter, no serious effort to investigate the possible lateral or vertical extension of the mineralization was made until 1944 when the Cía. Minera de Peñoles initiated small-scale exploration and development. The company subsequently (1953) installed facilities for the processing of Pb, Ag, and Zn concentrates and continues to develop the deposits under the title Minera Mexicana Peñoles, S. A. Present extraction is 180 tons of ore per day.

Monje, H., H., 1991, Geology and mineralization of the Topia Mining District, *in* Salas, G. P., ed., Economic Geology, Mexico: Boulder, Colorado, Geological Society of America, The Geology of North America, v. P-3.

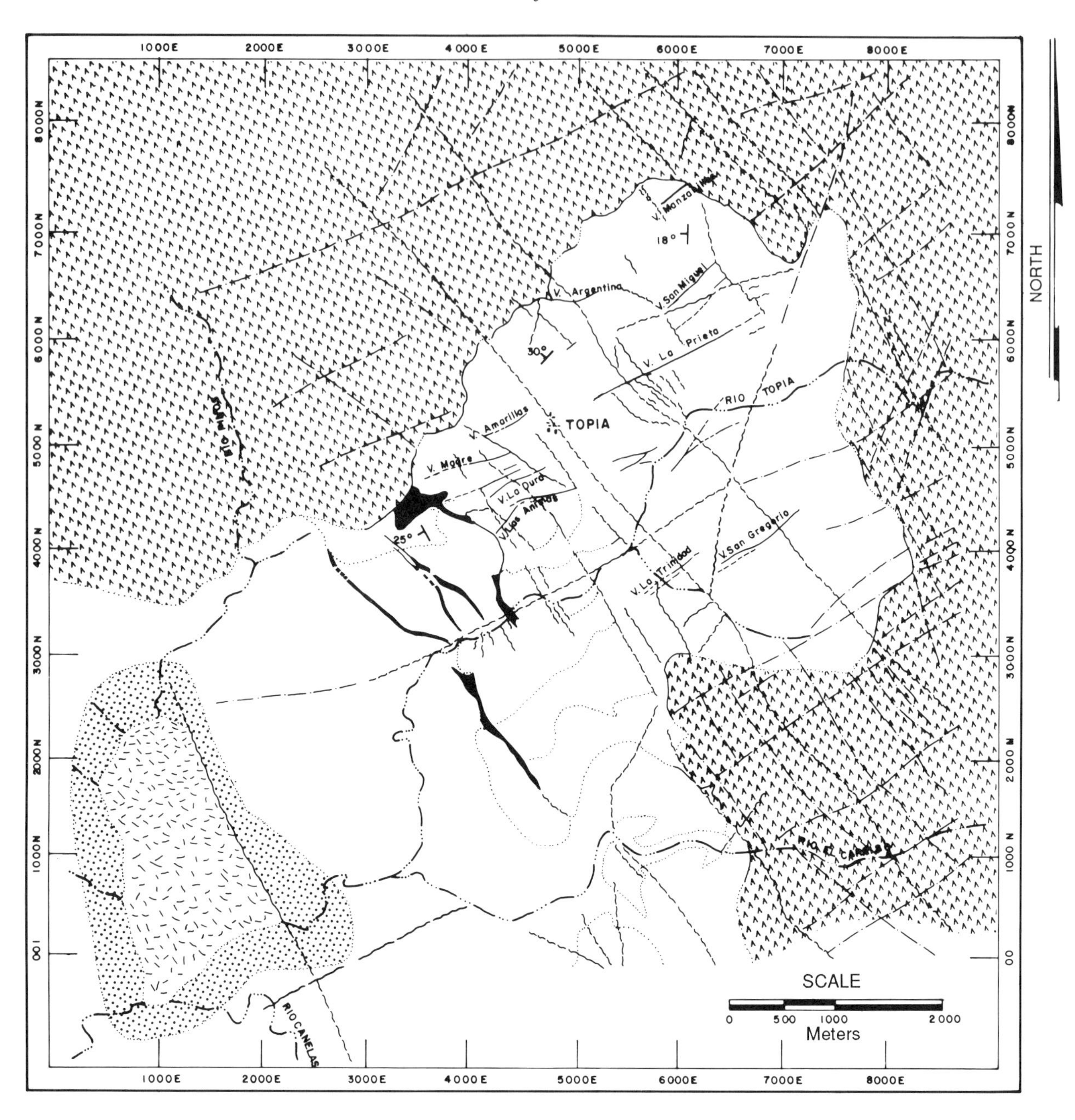

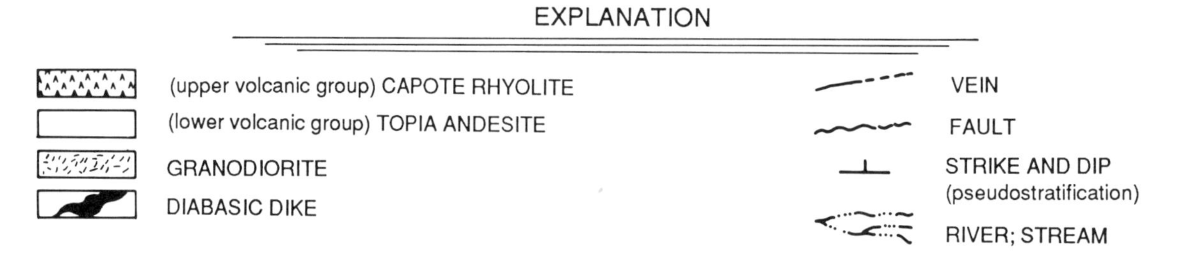

Figure 2. General geological map of the Topia District.

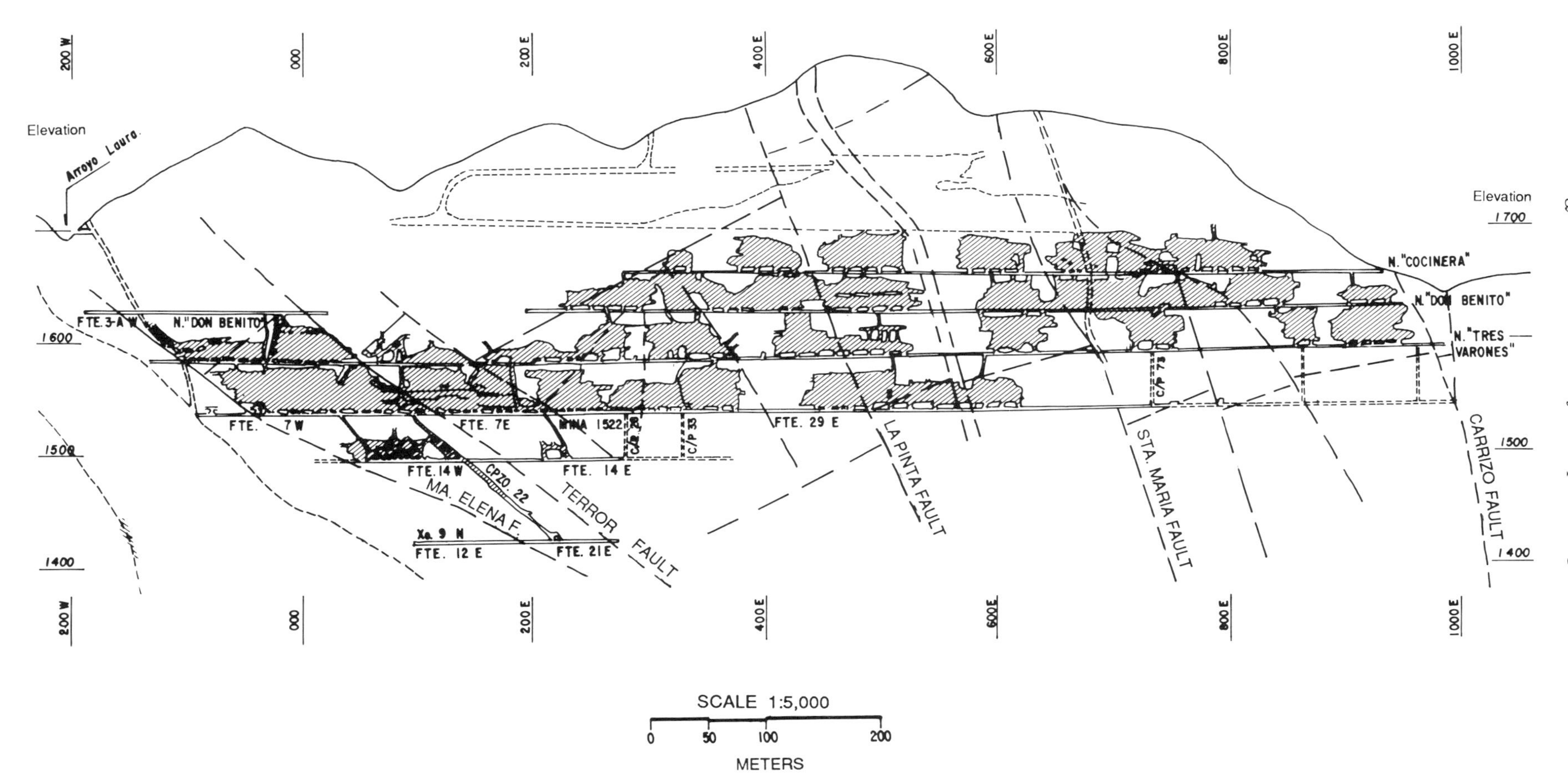

Figure 3. Longitudinal section of the Madre vein, Topia District.

GEOLOGY

Regional geology

The regional geology is dominated by Tertiary igneous extrusive rocks and, to a lesser degree, by igneous intrusions. The characteristic Sierra Madre Occidental volcanic sequence occurs at Topia, comprising two major lithologic groups: (1) the Late Cretaceous to Oligocene Lower Volcanic Group, a 1,400+ m thickness of mainly andesitic rocks, which are important mineralization host rocks throughout the Sierra; and (2) the middle Oligocene to Miocene Upper Volcanic Group of eruptive rocks—rhyolite and rhyodacite tuffs and ignimbrites interbedded with flows and tuffs of intermediate composition. Erosion has carved out extensive northwest-trending valleys, steep canyons, and high cliffs, exposing the andesitic Lower Volcanic Group.

Local geology

The pre-mineralization rocks of the Topia District make up the Topia Andesite: tuffs and lavas of the Lower Volcanic Group exposed by a large erosional window (Fig. 2). The age of the sequence in the Barrancas Subprovince is poorly known due to hydrothermal alteration and metamorphism. However, an age of 38.7 ± 0.8 Ma, determined at Guanacevi, contrasts with 52 Ma mentioned by McDowell and Clabaugh (1976) from near the Durango-Mazatlán highway. The Lower Volcanic Group ranges in age from 59 Ma(?) to 38 Ma in the Barrancas Subprovince. The andesitic sequence at Topia is intruded by granodiorite and by diabase and aplite dikes, discussed briefly below.

The Topia Andesite has been subdivided locally into three members on the basis of their position and lithology:

Santa Ana Member. Subdivided in turn into three lithologic units, the Santa Ana Member occurs extensively in the area. This estimated +545-m-thick series of tuffs and lavas forms the base of the andesite sequence over the entire district.

Carmen Member. Overlying the Santa Ana, the Carmen Member occurs most visibly in the northeastern part of the district and includes four different lithologic units: two types of lavas and two tuffaceous units with an overall thickness of approximately 400 m.

Hornos Member. The uppermost unit of the Topia Andesite, the Hornos comprises lavas toward the base and tuffs at the top, with an estimated thickness of 450 m.

Granodiorite. This rock crops out at the 800-m-high canyon of the Ríos Pinos and Topia, 5 km southwest of Topia (Fig. 2). Component minerals are quartz, orthoclase, plagioclase, biotite, and hornblende, with a compositional facies change to quartz-monzonite. K-Ar dating has indicated an age of 46.1 ± 1.0 Ma for the Topia granodiorite.

Aplite dikes trending northwestward and dipping 60° to 80° northeast outcrop continuously in the southwestern part of the area over approximately 1 km, from the Río Topia to the contact with the rhyolites, and range 2 to 5 m in thickness.

Diabase dikes are widespread within the district and trend irregularly northwestward. These intrusive bodies, ranging in width from a few centimeters to over 60 m, invade the entire Topia Andesite, except for the Upper Volcanic Group rhyolite tuffs. Their emplacement was controlled by northwest-trending faults, northeast-trending fissures, and lithologic contacts.

The post-mineralization volcanic rocks in the district, of rhyolitic character, belong to the Upper Volcanic Group. They unconformably overlie the Topia Andesite, shaping the rim of the erosional window and crowning the high portions of the Sierra (Fig. 2).

MINERALIZATION

Geology and structure of the veins

The Topia Mining District mineral deposits are a series of dominantly northeast-trending parallel veins that have been mined for silver and, to a lesser degree, for lead and zinc. These structures, restricted to the Topia Andesite, define a pre-mineral fracture and fault pattern trending N60°E and dipping 70 to 80° southeast on the average, probably caused by the early Tertiary regional tectonism. Vein mineralization is simple: essentially silver, lead, and zinc sulfides and silver sulfosalts in a quartz-rich gangue. Nine major and some 40 lesser veins have been identified within the district. The mineral lodes contain quartz, sphalerite, pyrite, galena, chalcopyrite, tetrahedrite, polybasite, and calcite.

The most noticeable vein texture is a marked alternate banding with a 3- to 10-cm thickness range. Three types of banding are present in the veins: (a) the most common is a dark band with abundant galena, sphalerite, and some quartz; (b) the second type is a dark gray quartz and pyrite aggregate; and (c) the third is similar to the first but light-colored, containing abundant crinkled-banded white quartz with patch-shaped inclusions of galena, sphalerite, pyrite, barite, or calcite.

Shape and dimensions of mineral bodies

The commercial mineralization occurs in a subhorizontal band over 1,000 m long and 250 m deep and exhibits parallelism with the present-day topography (Fig. 3). This mineralization is typical of epithermal deposits and is known as the "favorable zone." Veins are usually narrow: 0.30 to 1.0 m wide, and up to 4 km long. Extractable mineral lodes are 20 to 100 m in length along strike and are located at deviations toward the left of the vein strikes. The placement of lodes at strike deviations regardless of dip (southeast or northwest) indicates there is no significant vertical displacement.

Work by Loucks (1982) on chemical zoning patterns based on metallic ratios and fluid inclusions, indicates that most of the Topia District potential is located laterally along the veins. Such zoning studies are highly important, helping to identify vertical and horizontal intervals of commercial mineralization.

REFERENCES CITED

Loucks, R. R., 1982, Metal zoning in fissure veins, Topia, Durango, Mexico [abs.]: Florida Scientist, v. 45 (supplement 1), p. 45.

McDowell, F. W., and Clabaugh, S. E., 1976, Relation of ignimbrites in the Sierra Madre Occidental to the tectonic history of western Mexico: Geological Society of America Abstracts with Programs, v. 8, p. 609–610.

Printed in U.S.A.

The Geology of North America
Vol. P-3, Economic Geology, Mexico
The Geological Society of America, 1991

Chapter 23

Geology of the Tayoltita Mine, San Dimas District, Durango

H. M. Clark
Cia. Minera San Luis, S.A. de C.V., Campos Eliseos 400-8o. Piso, Lomas de Chapultepec, 11000 México, D.F.

INTRODUCTION

The Tayoltita Mining Unit in Durango, a mining-metallurgical industry that develops silver and gold deposits owned by Compañía Minera MSL, S. A. de C. V., is located 125 km northwest of the harbor town and prominent tourist center of Mazatlán, Sinaloa, and 150 km west-northwest of Durango, capital of Durango State (Fig. 1). A commercial airline based at Tayoltita offers year-round daily flights, stopping at Mazatlán and Durango on alternate days.

The bulk of the raw material for the mining-metallurgical operation, as well as heavy machinery and equipment, are land-transported on heavy trucks along the annually reconditioned Río Piaxtla bed between San Ignacio, Sinaloa, and Tayoltita; traffic usually ceases in June at the start of the rainy season.

The climate at Tayoltita is subtropical; temperatures range from a minimum of 10°C on winter nights to a maximum mid-day 40°C in summer. Regional annual precipitation is 69 cm, with afternoon torrential showers during the summer and lighter rains in late November and early February.

The topography of the mine area is extremely rugged. Elevations fluctuate from 450 m at Tayoltita to 1,985 m at the highest mineworks in Sierra Soledad. Tablelands surrounding the Río Piaxtla Canyon on the north, south, and east are 2,400 m high on the average, with a maximum elevation of 3,150 m at Cerro Huehuente, 19 km east of Tayoltita.

The first recorded production at Tayoltita was in 1757 when a group of Spanish adventurers melted silver- and gold-rich mineral, which was sold in Durango. Throughout the ensuing bonanza, several rich lodes in Sierra Soledad were mined until 1810 when Mexico became independent. After hostilities ceased, the near-surface mineral appeared insufficient to warrant reopening the mines, which were then worked only sporadically. William Randolph Hearst's agents acquired the principal mines in 1905 under title of the San Luis Mining Company and carried out successful exploration at depth; the presently interconnected mines have been operative since then. In 1962, Mexican investors, under title of Minas de San Luis, S. A., purchased 51 percent of "A" series (Mexican) shares in the business, which thus became Mexican property by virtue of the related 1959 Law. In 1979 a young group of Mexican shareholders headed by Antonio Madero acquired 49 percent of "B" type or foreign shares, and the company became a wholly owned Mexican enterprise: the Minas de San Luis, S. A. de C. V., and later Compañía Minera MS, S. A. de C. V., the present trade name.

The Tayoltita Unit mine has developed over 300 km of tunnels, 150 of which are accessible at 36 levels with vertical spacing of 24, 36, and 60 m. The north-trending developed block is 3 km long, 1.5 km wide, and 1.5 km deep. The present lowermost level is number 25, from which an incline is being sunk 60 m down to level 26. Total production of the district throughout the years exceeds 12 million metric tons, averaging 820 gr of silver and 16 gr of gold per metric ton.

GENERAL GEOLOGY

The Tayoltita District lies on the axis of the Sierra Madre Occidental, 60 km west of the Mexican Interior Tableland and 50 km east of the Pacific Coastal Plain. The Sierra is a cordillera-type mountain chain made up of an extensively arched complex of Late Cretaceous and Tertiary rhyolite, latite, and andesite flows commonly exceeding 2,000 m in thickness. Granodiorite batholiths and granite, diorite, and andesite stocks intrude these rocks and the underlying basal Mesozoic marine sediments. Faults (Fig. 2) with vertical displacements of hundreds to over 1,000 m divide the Sierra into structural blocks trending 10° to 30°NW and about 2 to 6 km wide.

LOCAL PETROGRAPHIC ASPECT

The rocks in and around the existing mines indicate two stages of volcanic activity separated by an erosional hiatus, during which andesitic detritus accumulated in shallow lakes. The intrusives are of different types and ages. Tables 1 and 2 are detailed descriptions of these rocks in chronological order.

STRUCTURAL GEOLOGY

The Tayoltita mineral deposit is located in an uplifted block limited by the Arana and Guamuchil high-angle normal faults

Clark, H. M., 1991, Geology of the Tayoltita Mine, San Dimas District, Durango, *in* Salas, G. P., ed., Economic Geology, Mexico: Boulder, Colorado, Geological Society of America, The Geology of North America, v. P-3.

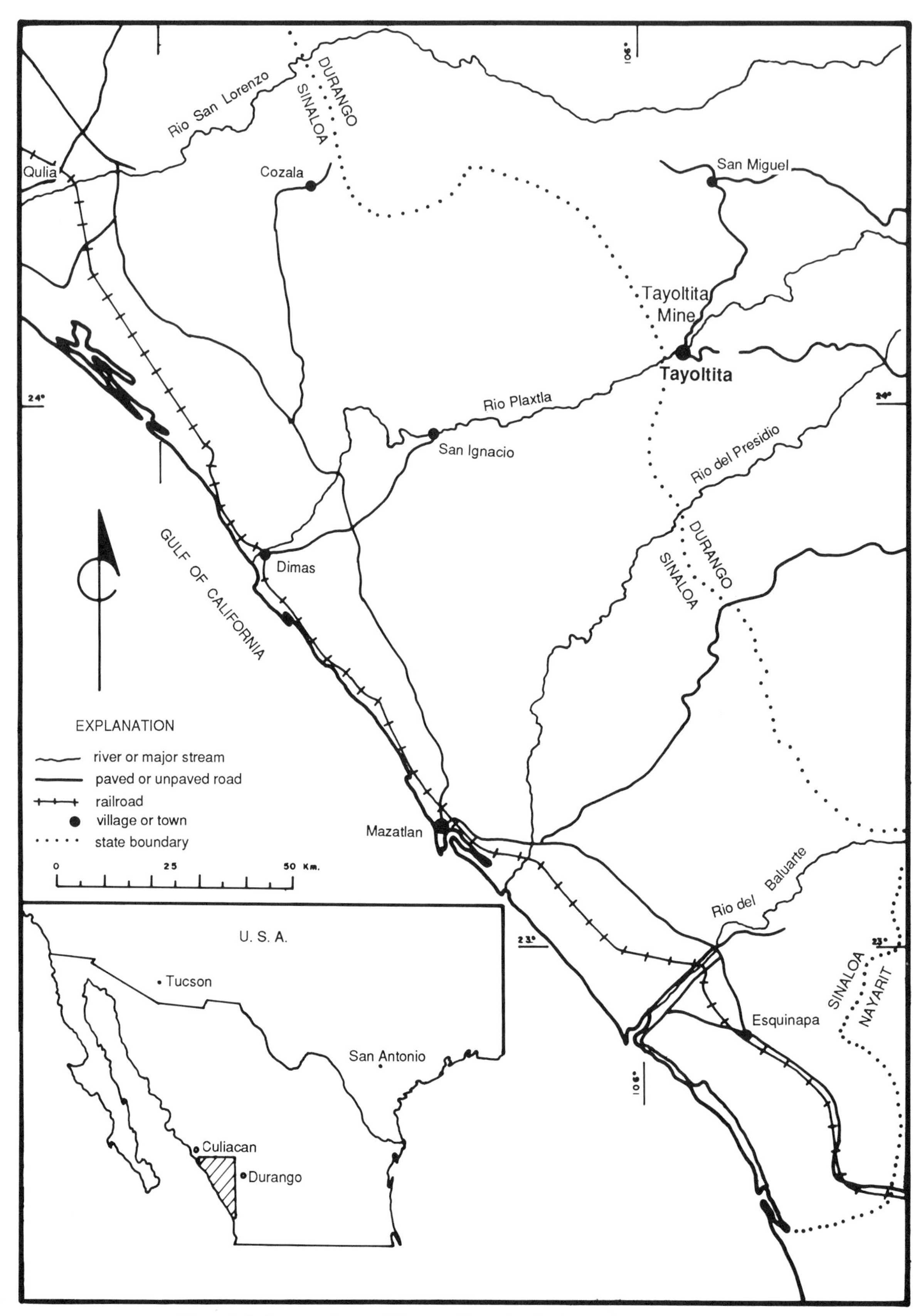

Figure 1. Index map of part of Durango State showing location of Tayoltita mine.

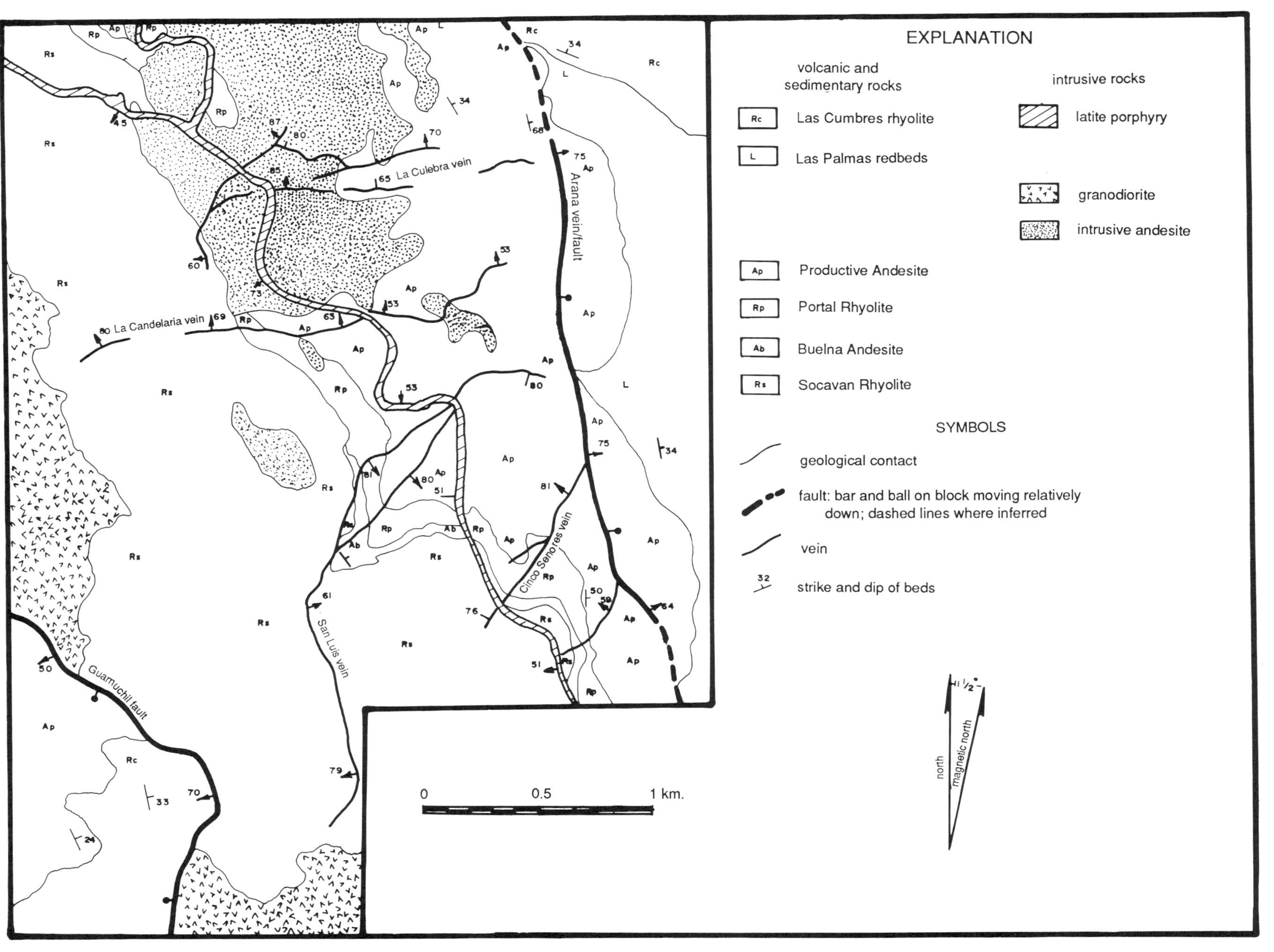

Figure 2. General surface geology of Tayoltita mine.

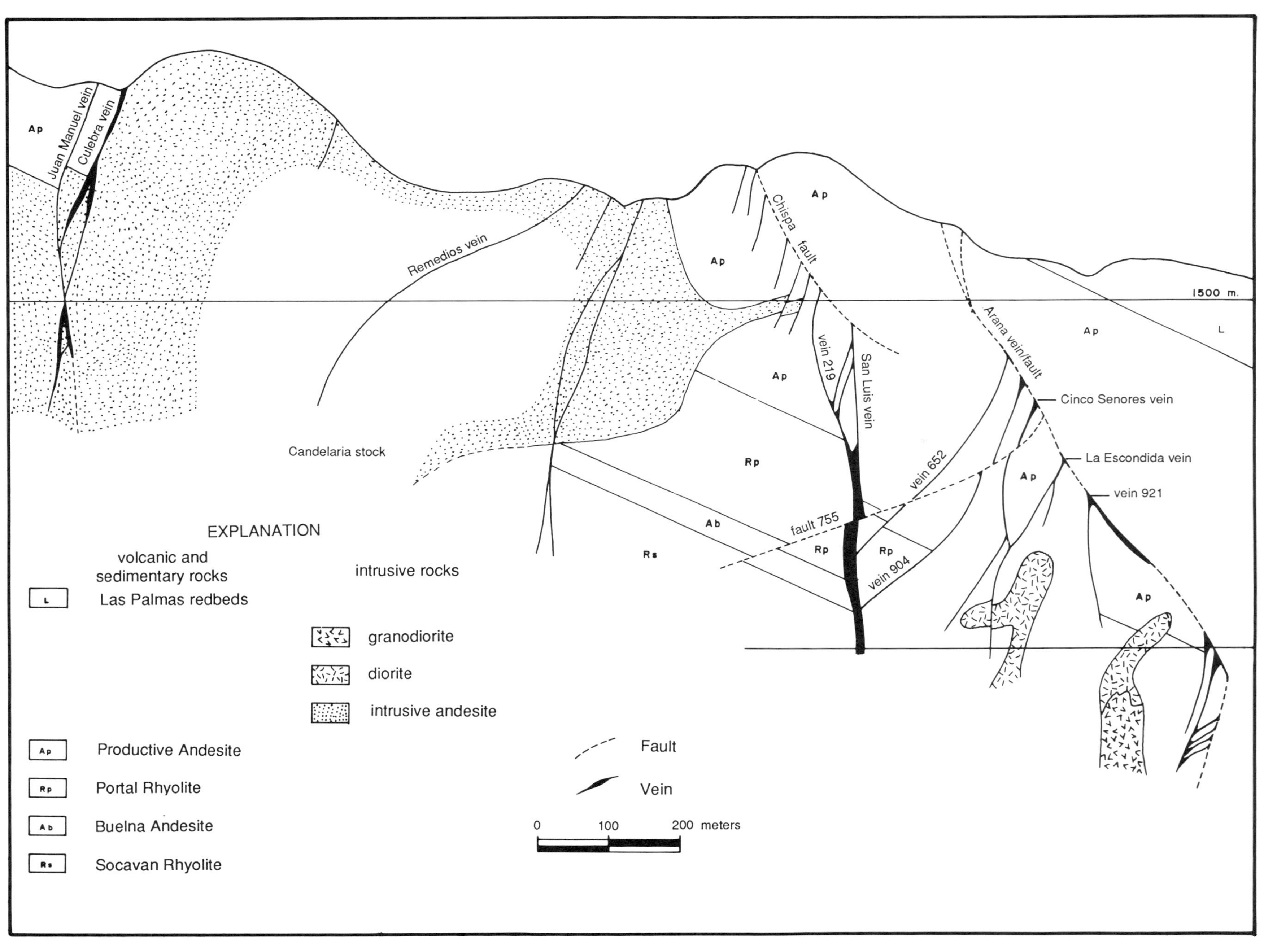

Figure 3. Geologic cross section of Tayoltita mine.

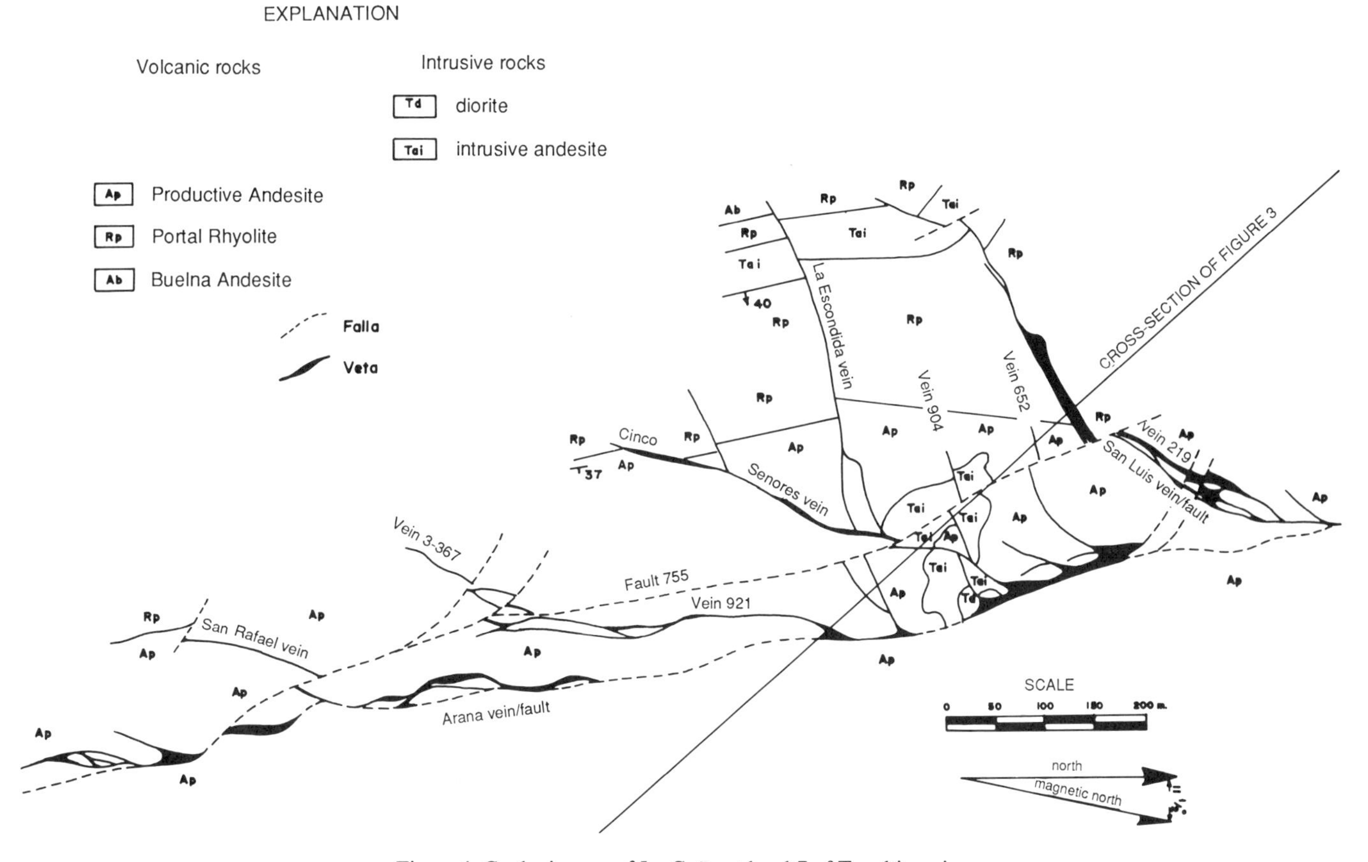

Figure 4. Geologic map of La Cuña at level 7 of Tayoltita mine.

(Fig. 3). Gold- and silver-mineralized bodies are hosted in veins 0.30 to 2.50 m thick along north-south, northeast, and east-west tension faults within the block. The dominant structural feature, a 50° wedge formed by the intersecting San Luis and Arana faults (Fig. 4), is thought to be the northern boundary of a sigmoidal fault extending horizontally 1.2 to 2.5 km (maximum) and possibly more than 1,600 m high before erosion. The closure of the northern tip of this wedge dips 45°N, and becomes vertical at 500 m depth.

Predominantly east and northeast-trending and generally north-dipping cross-veins occur within this same wedge. Drag and "drooper" veins appear exclusively below the Arana fault and trend northeast, parallel to the Arana vein itself. Staggered cross-vein–forming structures developed between intersections of several of these veins; the closely spaced cross-veins are perpendicular to the walls of the intersecting faults. All these types of veins, which range in magnitude from a few centimeters to several hundred meters, are clearly illustrated on Figure 3.

The only large mineral-producing structures to date in the northern part of the wedge are sigmoidal veins identified as 219–San Luis, Candelaria, and Culebra; the largest (1 km of vertical section) and most complex is the 219–San Luis system, with more than ten significant cross-veins and hundreds of other apparently unattractive veins (Fig. 5).

Postmineralization 35°E tilting in the Tayoltita fault block, and later staggered sliding in the northern part, produced dips of up to 55° northeast (Fig. 6), resulting in a vertical length of attractive mineralization in excess of 1,300 m in a section less than 600 m thick prior to these events.

HYDROTHERMAL ALTERATION

Andesite vein walls have been replaced to a slight degree by epidote, albite, chlorite, and calcite through intense propylitic alteration, also reflected in the breccia filling the accessory fractures. The strongest alteration affects an area extending up to 10 m away from the major fractures. Sericite replaces orthoclase and plagioclase zonally in the groundmass and phenocrysts, although this is not megascopically visible. Quartz veinlets often penetrate the vein borders, and silicification replaces the host rock along zones less than 3 m wide. Disseminated pyrite crystals usually less than 1 mm in diameter are irregularly distributed in the deposit, having developed by stages during the hydrothermal alteration and mineralization processes and possibly even after the veins solidified.

MINERALOGY

The major group of 100 mined veins in the Tayoltita area may be classified by basic mineralization patterns as follows:

TABLE 1. VOLCANIC AND SEDIMENTARY ROCKS OF THE TAYOLTITA AREA, DURANGO

Age	Unit	Description	Thickness (m)
Eocene?	Upper rhyolite	Light-colored stratified rhyolite, latites, and dacite flows forming cliffs; poorly consolidated tuffaceous ashes build up slope.	+1,500
	Unconformity		
Late Oligocene?	Las Palmas red beds	Andesites and thickly bedded to laminar sandstones overlie medium-grained andesitic pyroclastics forming mottled thin beds in slabs of +1 meter in diameter.	0–500
	Unconformity		
Cretaceous-Tertiary transition	Productive andesite	Comprises two microscopic rock types: (1) Fine-grained. Aphanitic to fine-grained red to gray andesite; occasional fractured zones with plagioclase phenocrysts or very fine fragments. Flow structure visible only in one thin section. (2) Fragmental. Compact red to purple ash, lapilli tuffs, and agglomerates with mottled fragments.	+500
Cretaceous-Tertiary transition	Portal rhyolite	Light gray to brick-red aphanitic rhyolite, often with mottled ash or lapilli-sized rhyolite fragments. Locally contains quartz "eyes." May be parallel to perpendicular to banding.	250
Cretaceous-Tertiary	Buelna banded andesite	Extremely well-consolidated bluish to gray thin- to thick-bedded lapilli and ash. Bedding is clear except where destroyed by alteration.	0–75
Cretaceous-Tertiary transition	Socavón rhyolite	Aphanitic, massive, gray quartz rhyolite porphyry, commonly with abundant ash, lapilli, and variably sized, dark gray rhyolitic fragments.	+700 Lower contact destroyed by intrusion

TABLE 2. INTRUSIVE ROCKS AT TAYOLTITA MINE, DURANGO

Age	Intrusive	Size and shape	Ag/Au mineralization ratio
Oligocene	Andesite. Very fine-grained gray to dark purple rock with oligoclase and/or augite phenocrysts more than 0.5 cm long. Weakly magnetic.	Lenticular tubes, short dikes, sills, and irregular masses with maximum Ø 1,000 m.	Premineralization
Oligocene	Diorite. Brown to green-gray rock consisting of fine- to medium-grained holocrystalline feldspars with dispersed ferromagnesian minerals. Commonly forms the center of the larger plutonic andesite bodies.	Tubes, short dikes, and irregular masses.	Premineralization
Oligocene 36.9 ± 1.8 Ma (K/Ar) in biotite*	Granodiorite. Two facies: holocrystalline pink to gray, fine- to coarse-grained plagioclase; orthoclase, quartz, and biotite. Associated with aplitic dikes.	Fine-grained facies forms dikes and dome-shaped irregular masses in a 500-m-thick zone around a coarse-grained batholitic rock.	Premineralization. Believed to be pathway for mineralizing solutions.
Oligocene	Aplite. White, light gray, and pink aplite with dispersed fine-grained biotite.	Dikes	Premineralization
Oligocene	Feldspathic porphyry. Tabular phenocrysts of Ab_{90} An_{10} in fine-grained mass in plagioclase bands.	+25-m-wide dikes. Commonly in groups.	Premineralization
Oligocene	Porphyritic granite. Holocrystalline coarse-grained quartz, orthoclase, and phenocrystalline plagioclase. May be the same differentiated magma that formed the feldspathic porphyry.	Dikes	Unknown
Eocene?	Basic andesite. Fine-grained gray to dark green rocks consisting of subophitic plagioclase with augite with abundant chlorite and magnetite; often containing calcite amygdules.	+10-m-wide dikes.	Post-mineralization
Late Eocene?	Latite porphyry, quartz, feldspar, microcrystalline quartz, and orthoclase, with partially dissolved quartz phenocrysts, albite, and orthoclase. Probably fed part of the upper rhyolites.	Dikes; only one at the mine; the area is several km long and 30 m wide.	Post-mineralization

*Paul Damon, University of Arizona, 1968.

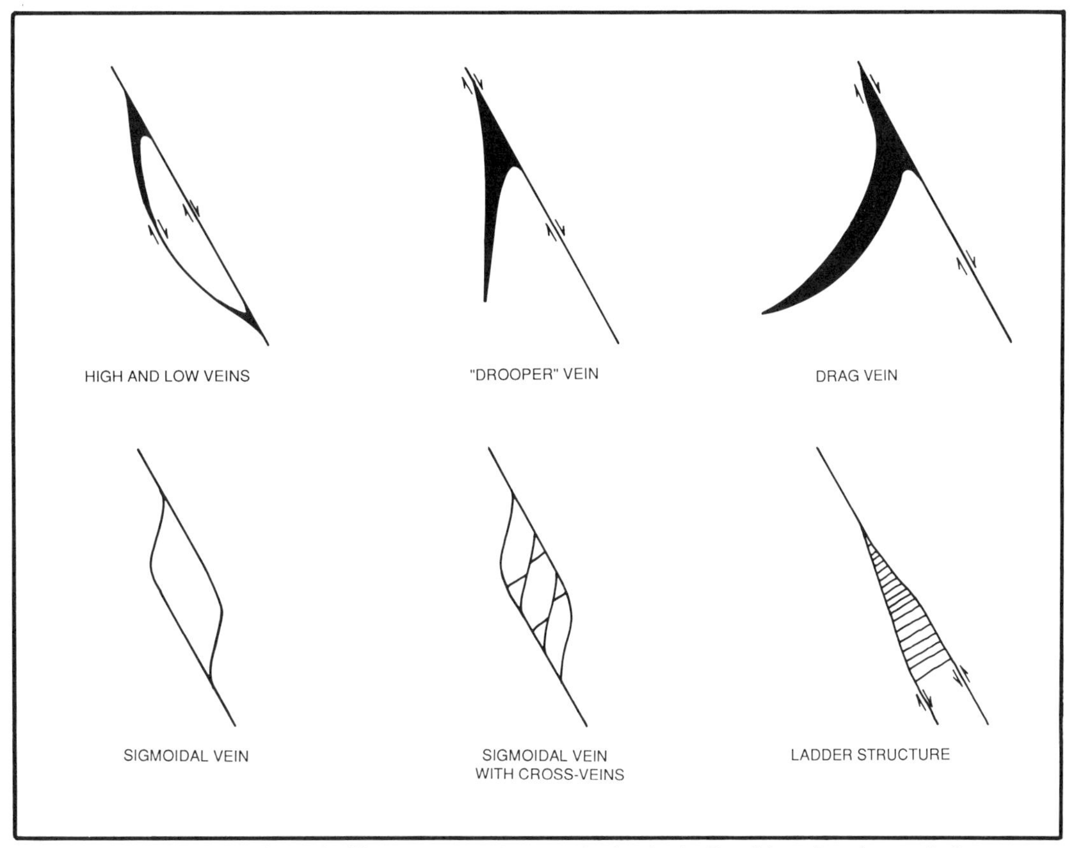

Figure 5. Cross sections of different structural types of veins in the Tayoltita mine. Arrows indicate relative direction of movement of walls.

(1) Quartz-calcite and sulfide (Culebra, Candelaria, and Cedral veins), (2) Quartz-bustamite-albite-sulfides with scarce calcite (219, 652, 904, Cinco Señores and San Luis veins), and (3) Quartz-sulfide with scarce adularia and calcite (921, San Rafael, and Arana veins). With few exceptions, these mineral series rarely intermix, and when they do, most of the vein shows one of the types.

In general, vein textures are similar; gangue and its component sulfides developed cockscomb, banded, and spiral structural types. Some veins exhibit host-rock breccia in the groundmass, in which case the rock fragments commonly served as cores around which the spiral structures developed.

GANGUE, NONCOMMERCIAL MINERALS

The dominant gangue in all of the Tayoltita veins is white, light gray, or amethyst quartz, which is always crystalline, although small pre-mineral flint patches are found occasionally. The quartz also shows abundant cavities, in particular a post-mineral variety that contains crystal-lined cavities of up to 1 m^3. In some cases this later-stage quartz was violently injected along the mineralized veins in such amounts as to dilute the original vein volume by several orders of magnitude, causing the present-day low (i.e., noncommercial) mineral content.

Calcite evolved as gangue in the veins prior to, during, and after mineral precipitation; in general the calcite that formed before and during the process is roughly crystalline and milky white, pink, or black in color, whereas post-mineralization quartz is light colored and commonly develops anhedral crystals.

Bustamite forms salmon-colored anhedral masses and often intermixes intimately with albite in white, slab-shaped crystals and irregular masses.

Adularia is typically pale pink and occurs as subhedral crystals in a mosaic pattern with quartz.

Chlorite that is microscopically disseminated in the quartz veins colors them in greenish tones; however, in type 2 and 3 veins the chlorite is laminated, dark green to black, contains interstitial argentite and electrum, and is locally termed "black chlorite mineral" when abundant.

SULFIDE MINERALIZATION

Sulfides add up to less than 5 percent of the volume in the three vein types. Galena, sphalerite, and chalcopyrite predominate though with reduced elemental content (approximately 0.1 percent lead, 0.1 percent zinc, and 0.05 percent copper). Argentite (acanthite), jalpaite, stromeyerite, native silver, and electrum replace these sulfides. Silver sulfosalts and native gold

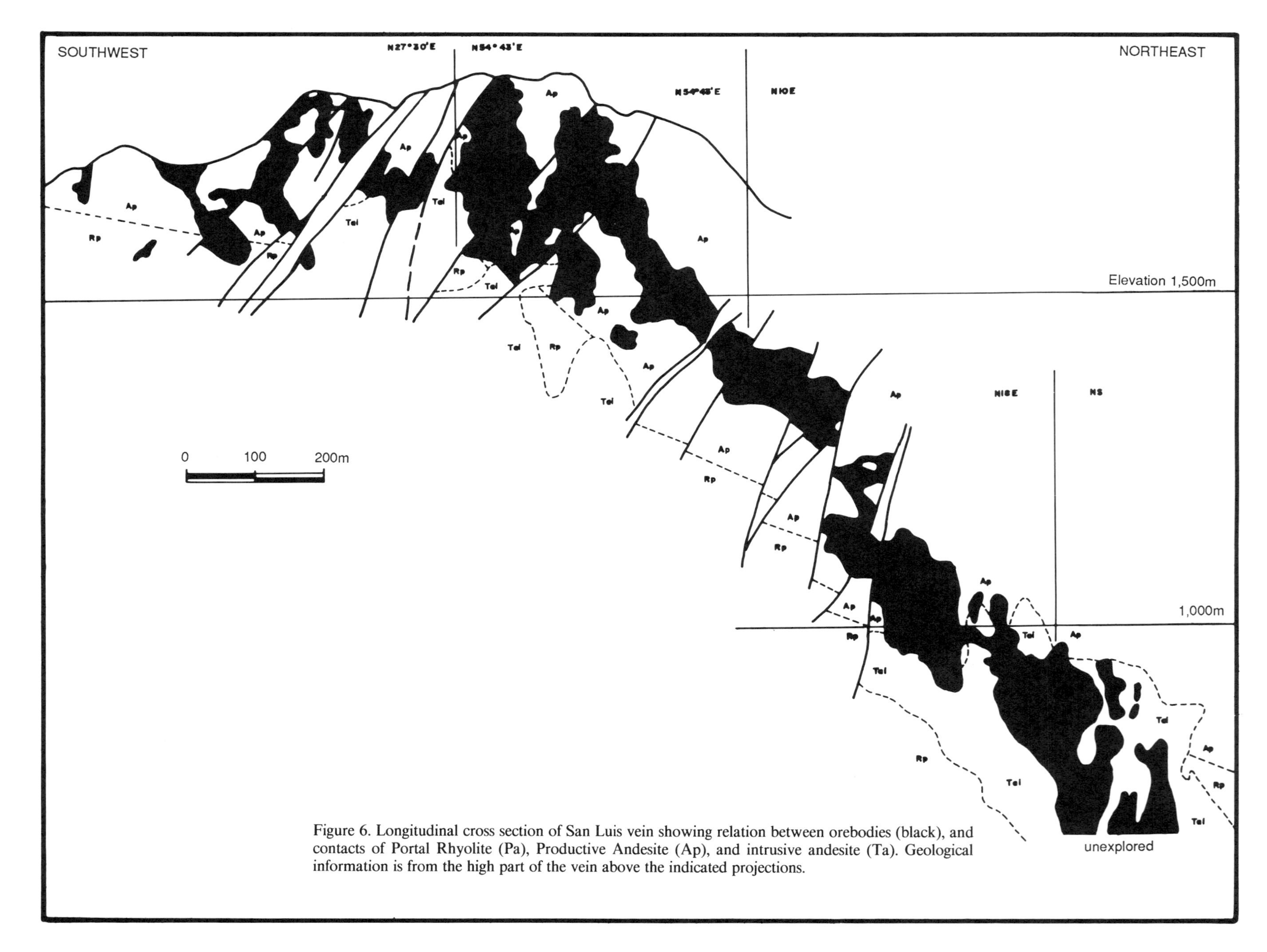

Figure 6. Longitudinal cross section of San Luis vein showing relation between orebodies (black), and contacts of Portal Rhyolite (Pa), Productive Andesite (Ap), and intrusive andesite (Ta). Geological information is from the high part of the vein above the indicated projections.

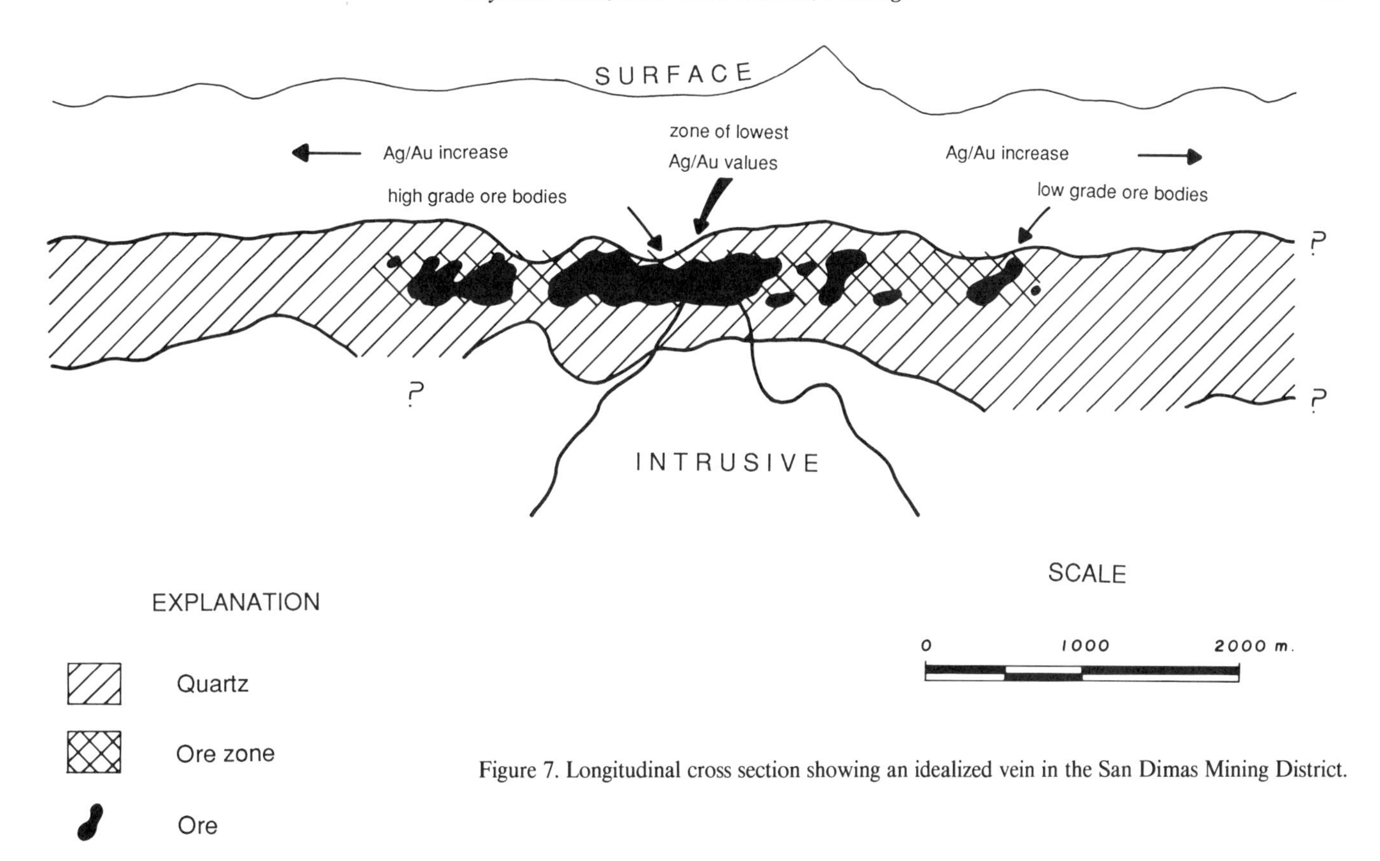

Figure 7. Longitudinal cross section showing an idealized vein in the San Dimas Mining District.

are noticeably absent or appear only as traces. Gold and silver contents in the minerals extracted to date show a 1:51 ratio (i.e., 1 kg of gold per 51 kg of silver); the gold is derived mainly from electrum.

SUPERGENIC ENRICHMENT

Strong and persistent erosion of the scarps along the stream bed where the veins crop out has made supergenic enrichment a minor factor in the formation of mineralized lodes. Near-surface parts of some veins at the southernmost boundary of the mine are slightly enriched in native silver, electrum, stromeyerite, and gold- and silver-bearing manganese oxides; however, leaching and the consequent dispersion of the precious metals predominated.

ADVANCES IN GENESIS RESEARCH

The distribution of the metal bodies, silver, gold, and vein gangue exhibits marked zoning (Fig. 7).

Metal bodies are distributed along a horizon that is normally approximately 250 m in the vertical dimension and extends horizontally for up to 5 km. Fluid inclusion studies define this level as centered approximately 500 m below the pre-mineral surface.

The Ag/Au metallic ratios exhibit zoning with relatively high Au concentrations in the central parts of the veins and relatively high Ag tending to appear at the boundaries of the level bearing the main ore body. Figure 7 shows one of these cases. In some cases the Ag/Au zoning is centered over an intrusive.

As shown on Figure 7, the quartz in a Tayoltita vein normally attains maximum thickness within the metal horizon, disappearing upward and thinning downward to a fine thread that occasionally disappears also. This observation fits the interpretation of the Tayoltita veins as a product of fossil geothermal systems in which zoning tends to extend more in the horizontal than in the vertical direction; this coincides with modern studies of geothermal system flows. On this basis, the distribution pattern of metal bodies at Tayoltita reflects the lateral flow of the mineralizing fluids. Application of this model to exploration in the San Dimas District (Durango) and Contraestaca (Sinaloa) has proved that the concept is a useful exploratory tool.

Printed in U.S.A.

The Geology of North America
Vol. P-3, Economic Geology, Mexico
The Geological Society of America, 1991

Chapter 24

Economic geology of the San Martín Mining District

P. Olivares R.
Industrial Minera México, S.A., Ave. Baja California 200, Col. Roma Sur, 06760 México, D.F.

INTRODUCTION

The San Martín Mining District lies in the Sombrerete municipality, western Zacatecas State (Fig. 1). Geographic coordinates of the Industrial Minera México, S. A., San Martín Unit are 23°40′N and 103°45′W. A paved road leading east from the installations for 6 km joins Federal Highway 45 (Panamericana) at the Zacatecas-Durango leg; the town of Sombrerete with its railroad station is about 10 km south of the junction. There is a small-aircraft landing field 15 km west of the mine. The area of the installations is at an elevation of 2,600 m; climate is temperate-humid with relatively severe winters; temperatures fluctuate between a low of −6 °C in winter and a high of 30 °C in summer.

REGIONAL GEOLOGY

The San Martín Mining District lies near the eastern boundary of the Basin and Range Subprovince in the Sierra Madre Occidental Physiographic Province; metallogenetically it belongs to the Central Mesa Province because of its location at the Sierra Madre Occidental foothills, and because of its mineral deposits, which are similar to those of that province (replacements mainly in limestones).

The sedimentary rocks typically develop low, gently rounded hills and intermittent subparallel and radial drainage, whereas cliffs and numerous dendritic streams evolve in the igneous rocks. Locally the high areas rise 300 m above the surrounding plains.

Sedimentary, igneous, and metamorphic rocks are present in the area. The Cretaceous Cuesta del Cura and Indidura Formations crop out over most of it, together with several granodiorite intrusions (Figs. 2, 3). Also, rhyolite lava flows occur abundantly throughout the district; conglomerates and alluvial material fill the valleys.

The oldest outcrops are the limestones of the Cuesta del Cura Formation (Fig. 4). In Albian–early Cenomanian time, marine waters covered the area and calcareous sediments (Cuesta del Cura Formation) accumulated at depth; interbedded shales suggest fluctuating shorelines. Depositional conditions changed during the latter part of the early Cenomanian; the clayey limestones and shales of the Indidura Formation laid down then indicate persistent regressions and transgressions until the end of the Turonian. Subsequently the cyclic sandstone/shale deposition of the Caracol Formation began and continued within the basin throughout the Late Cretaceous; the unit occurs in the vicinity, though not in the area itself.

The area's strong west to east or southwest to northeast folding is clear evidence of the Paleocene-Eocene Laramide orogeny (Fig. 3). Intrusive stocks of granodiorite composition—the Cerro de La Gloria, for example—emplaced at the end of, and perhaps simultaneously with, the orogenic phase further altered and deformed the folded rocks, giving rise additionally to a system of fractures that subsequently became conduits for the mineralizing fluids. Felsic flows invaded the area and its surroundings throughout the late Oligocene and Miocene. Strong erosion and the accumulation of large volumes of igneous and sedimentary rock fragments within the valleys, later cemented by caliche, characterize the rest of the Tertiary. Finally, the alluvial cover was laid down in Recent time.

LOCAL GEOLOGY

Sedimentary rocks

Cuesta del Cura Formation. The most widespread unit in the district, the Cuesta del Cura Formation is a dark gray limestone in undulating layers with intercalated black chert. The 10- to 40-cm-thick, well-defined, uniform calcareous beds are frequently cut by abundant small fractures filled with calcite and siderite. Late Albian–early Cenomanian ammonite molds have been collected from the rocks. Locally the unit reaches 770 m in thickness measured from the uppermost outcrops at Cerro de La Gloria to the lowermost intersections drilled at the mine site. The lithology indicates deposition in a basin environment during the Albian transgression, although the increasing early Cenomanian terrigenous sediments suggest regressive conditions for the uppermost part of the section.

Olivares R., P., 1991, Economic geology of the San Martín Mining District, *in* Salas, G. P., ed., Economic Geology, Mexico: Boulder, Colorado, Geological Society of America, The Geology of North America, v. P-3.

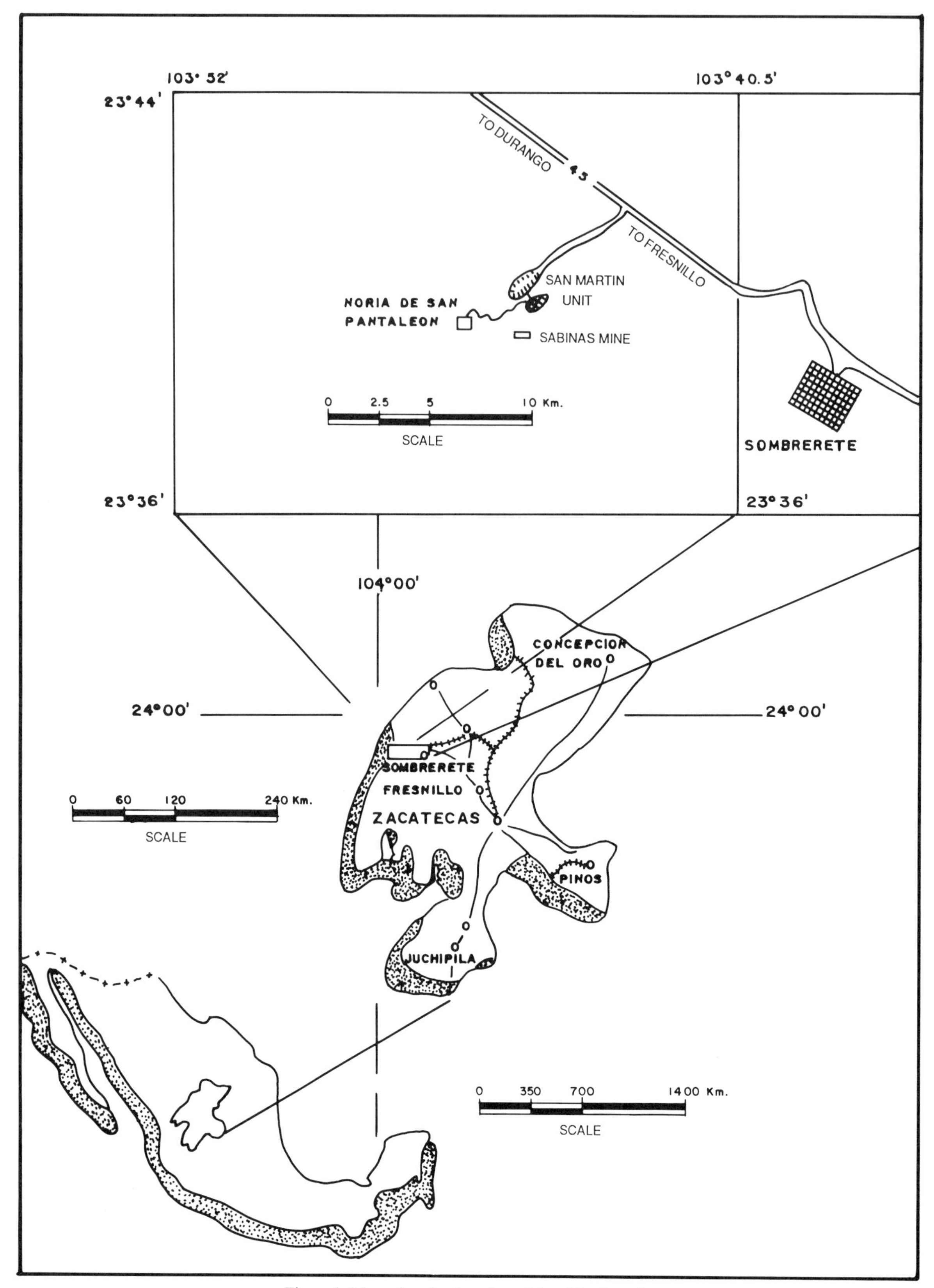

Figure 1. Maps showing location of San Martín unit.

Indidura Formation. This formation outcrops at the foot of the topographic heights of the Cuesta del Cura Formation. It consists mainly of alternating shales and fine-grained clayey limestones in 10- to 30-cm-thick layers. The shales are dark green to black, weathering in brown and yellowish tones, and strongly fissile. The limestones are very fine grained grading to clayey and dark gray to black; they weather light gray with yellowish tones. Isolated narrow black chert bands wedge out laterally. A 575-m thickness was measured at Arroyo La Herradura 2 km north of the San Martín Unit, where a Turonian index fossil was also found. The Indidura Formation was apparently deposited in shallow seas with fluctuating shorelines.

Igneous rocks

Intrusives. The Cerro de La Gloria granodiorite stock irregularly intrudes the Cuesta del Cura Limestone, appearing at the surface as a roughly shaped, north-south–elongated, 1.5- by 2-km ellipse. The rock is holocrystalline with hypidiomorphic phaneritic texture, with quartz and K-feldspar phenocrysts and lesser amounts of plagioclase and accessory biotite. Two outcrops of slightly porphyritic rock with an aphanitic groundmass north of the San Martín Mine may be apophyses of the Cerro de La Gloria intrusive.

Diamond-drilling at 25-m intervals between levels 8 and 12 (elevations 2,365 and 2,245 m, respectively) within the San Martín Mine encountered a 125-m-long by 120-m-wide apophysis of the intrusive with disseminated chalcopyrite, which acted as a structural trap for replacement bodies (Fig. 5).

Radiometric dating determined the age of the intrusive as Oligocene, which agrees with the field relations, emplacement having occurred after the Laramide orogeny and prior to the Miocene sialic volcanism. The Cerro de La Gloria stock is closely associated with the mineralization.

Extrusives. A sequence of variably thick, strongly NE-SW–fractured rhyolite flows often overlies the Cuesta del Cura and Indidura Formations, developing steep scarps at the very irregular contacts with other rocks.

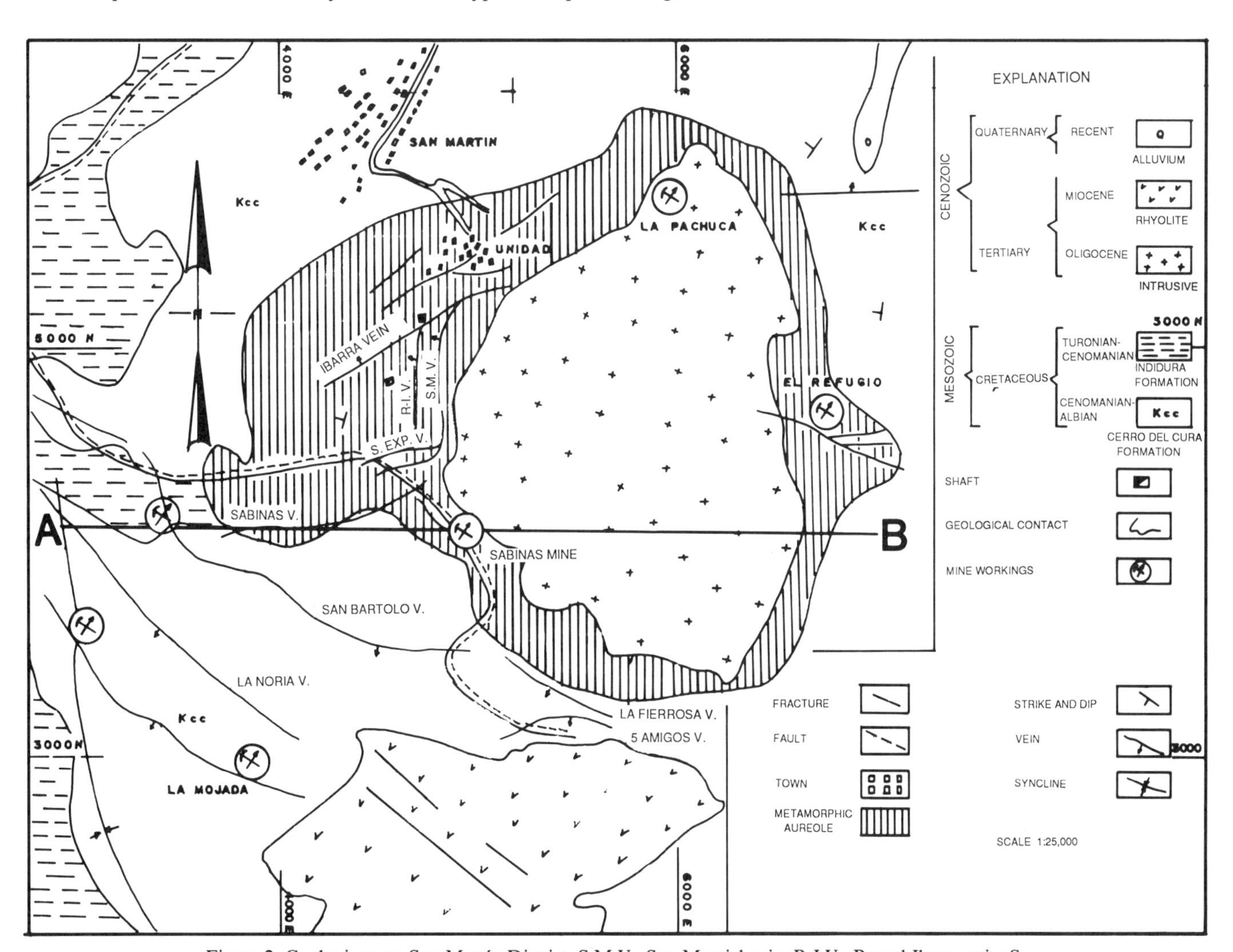

Figure 2. Geologic map, San Martín District. S.M.V., San Marcial vein; R-I.V., Ramal-Ibarra vein; S. Exp. V.

The texture of the rock is hypocrystalline and aphanitic; phenocrysts are scarce; quartz and K-feldspars predominate with small amounts of mafic minerals; color varies from pink to brown.

Metamorphic rocks

The Cerro de La Gloria granodiorite stock intrusion developed a metamorphic halo at the contacts with the Cuesta del Cura Formation (Fig. 2). The halo is asymmetrical; the northwest flank of the intrusion dips 60° northwest, and the halo extends over a much longer distance in that area than on the east side where dips may reach 80°. Other factors—such as the position of the calcareous beds relative to the stock—also contribute to the fact that at times metamorphism on the east flank extends over only 30 m, compared to 1 km on the west.

On the basis of mineralogical studies, Aranda-Gomez (1978) divided the metamorphic halo into four zones, from the intrusive outward:

Saccharoid quartz zone, restricted to the sedimentary/intrusive contact proper; quartz crystals average 0.8 mm in size; the rock exhibits light brown coloring due to its magnetite content, partially altered to hematite.

Ferrosalite-garnet zone, the commercially important interval, contains the known mineral bodies. The most abundant minerals are garnets of the grossularite-andradite series (2:1 relation); ferrosalite and vesuvianite are common; scapolite, epidote, and quartz are also frequent.

Tremolite-garnet zone, occurring at the outermost (northwest) boundary of the halo in the San Martín area, is characterized by small radial groups of very fine tremolite, and by garnets in limestone.

Tremolite-calcite zone, surrounding the intrusive on the north, south, and east, is characterized by tremolite and recrystallized calcite.

Structures

The folds, faults, and fractures in the district are of tectonic origin, a product of both orogenic movement and the Cerro de La Gloria intrusion. The sedimentary sequence was affected by

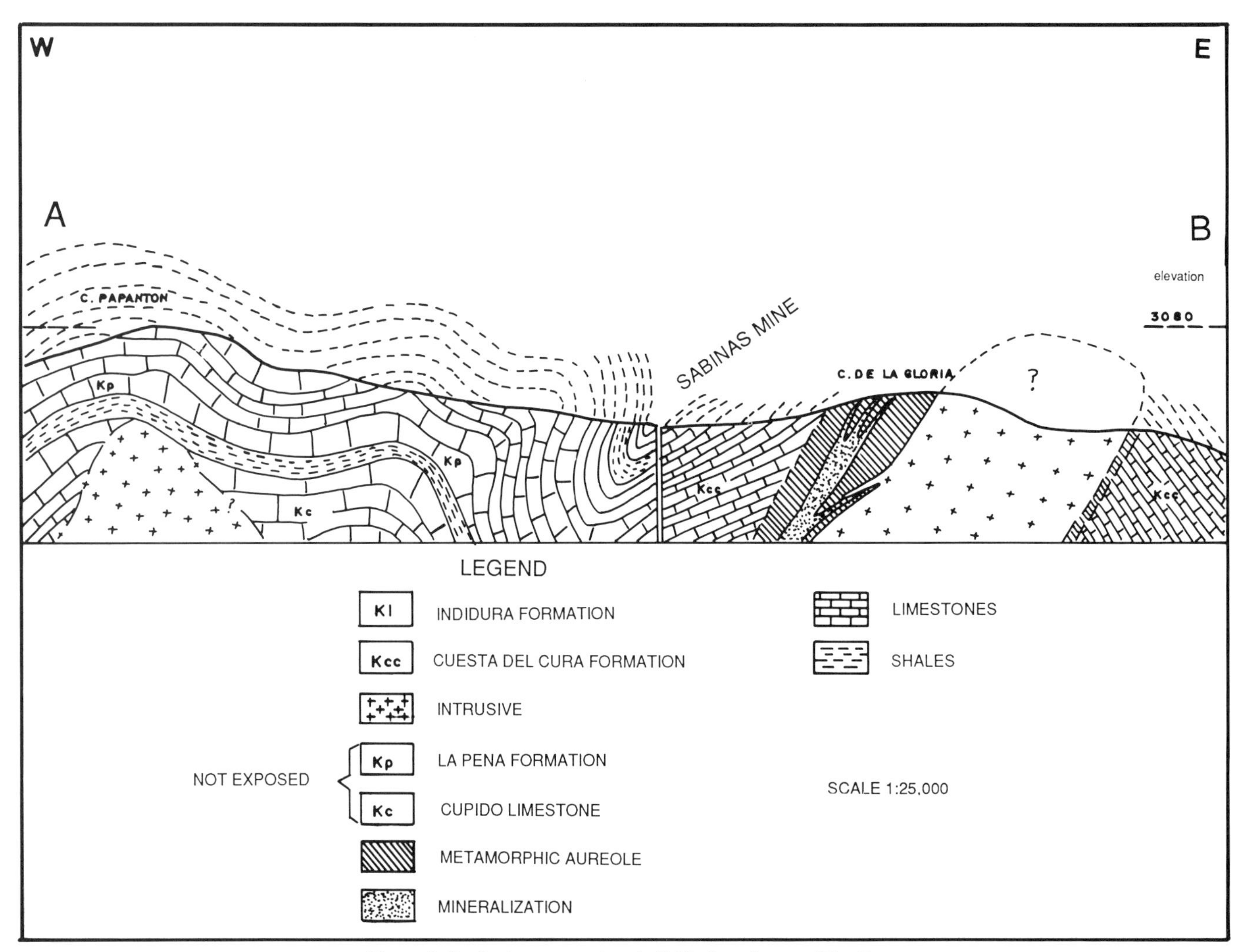

Figure 3. Geologic cross section A–B (for location, see Fig. 2).

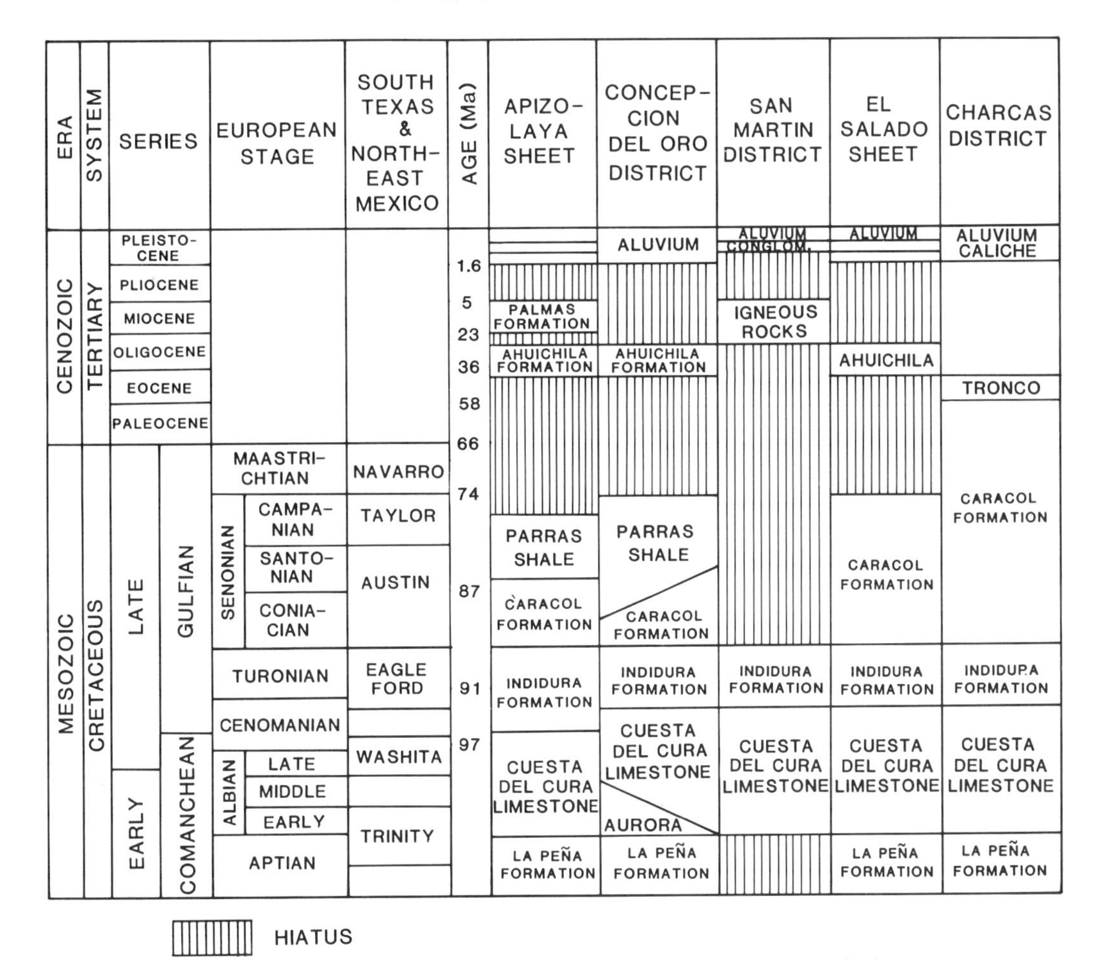

Figure 4. Correlation chart for stratigraphic units of the San Martín District.

major folding caused by forces acting from the northwest; fold axes strike northeastward and beds dip 10° to 50° to the northeast or northwest.

The Cuesta del Cura and Indira Formations exhibit a preferred northeast- and north-trending fracture system, although another major northwest-trending system is also present. All of these possibly developed during the tension and easing that followed Laramide orogenic compression.

Quick cooling caused the irregular fracturing of the rhyolite flows at the center of the district; it is of no major structural significance.

One of the most important fault and fracture systems is a product of the granodiorite intrusion: reverse faults, tangential or approximately parallel to the tactite/intrusive contact, form a generally N30°E–trending and 60°NW–dipping system that developed fault veins with clayey gouge at the hanging and footwalls of the mines. These veins appear at the surface for distances of up to 900 m; at present they are known to persist vertically down to level 16 of the San Martín Mine (elevation 2,125 m; i.e., 750 m from the surface at Cerro de La Gloria; elevation 1,875 m).

A series of staggered tensional normal faults (named 1, 2, 3, 4, and 5), transverse to the fault-vein system, developed after mineralization and possibly during emplacement of the rhyolites (Fig. 6). Within the mine the staggered faults displace the mineralized structures at the upper levels horizontally up to 20 m; at depth the displacements decrease to a few cm as the faults lessen in magnitude.

Fault 3, the most persistent, crops out for 500 m; it is known to a depth of 650 m and displaces veins by about 20 m at levels 0, 2, 4, 6, and 8; displacement becomes minimal at level 12.

MINERAL DEPOSITS

The district's most important mineral deposits are replacement veins and bodies generated in the tactites by the Cerro de La Gloria granodiorite intrusion. An extensive zone of tactites west of the intrusive hosts the San Marcial, Ibarra, and Remal-Ibarra veins, which appear at the surface for distances of up to 900 m with thicknesses of 40 cm to 3 m, paralleling the intrusive contact. Replacement bodies between them, up to 200 m in width, owe their presence to the veins, and these in turn to the intrusive.

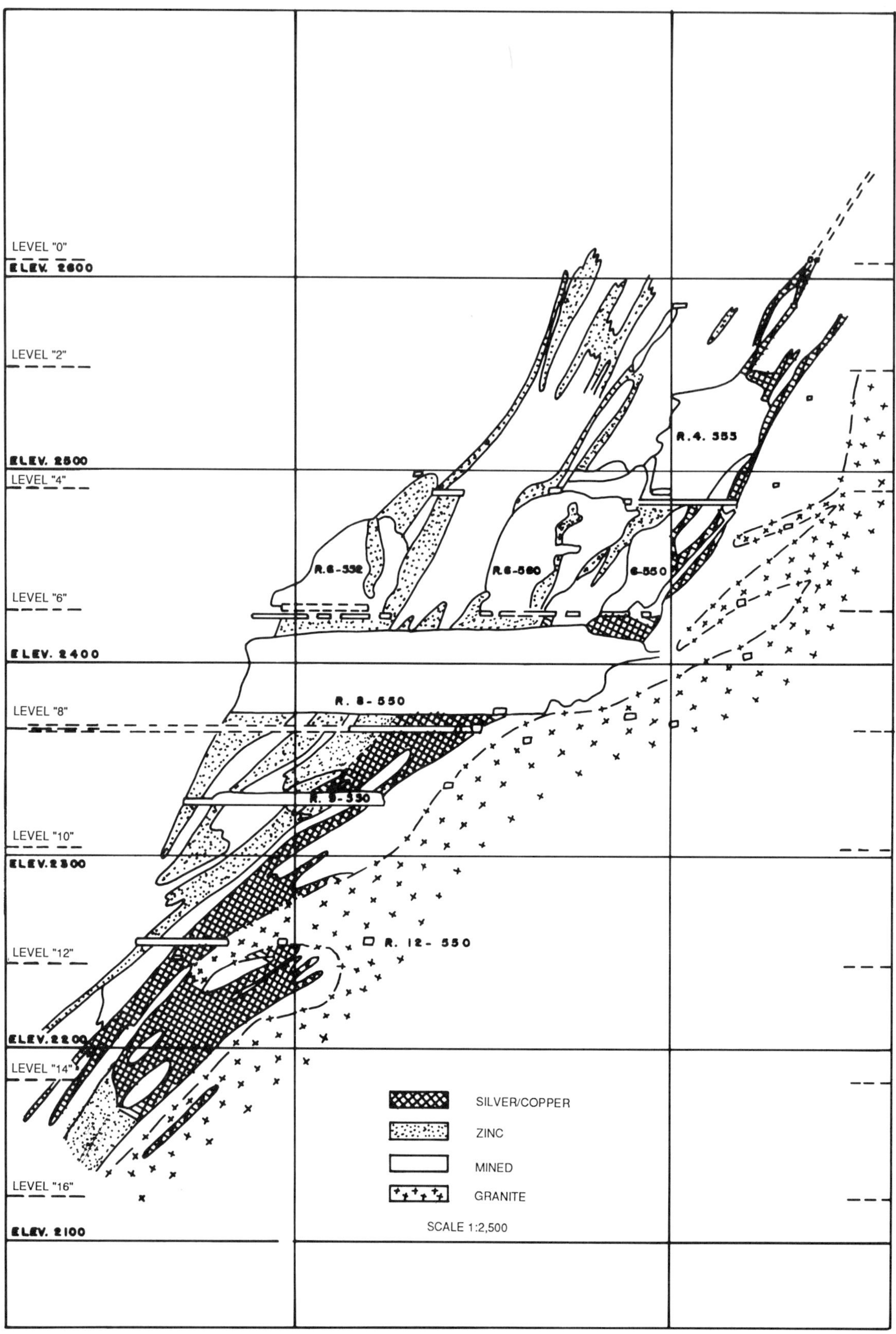

Figure 5. Cross-sectional geologic map of the San Martín Mine.

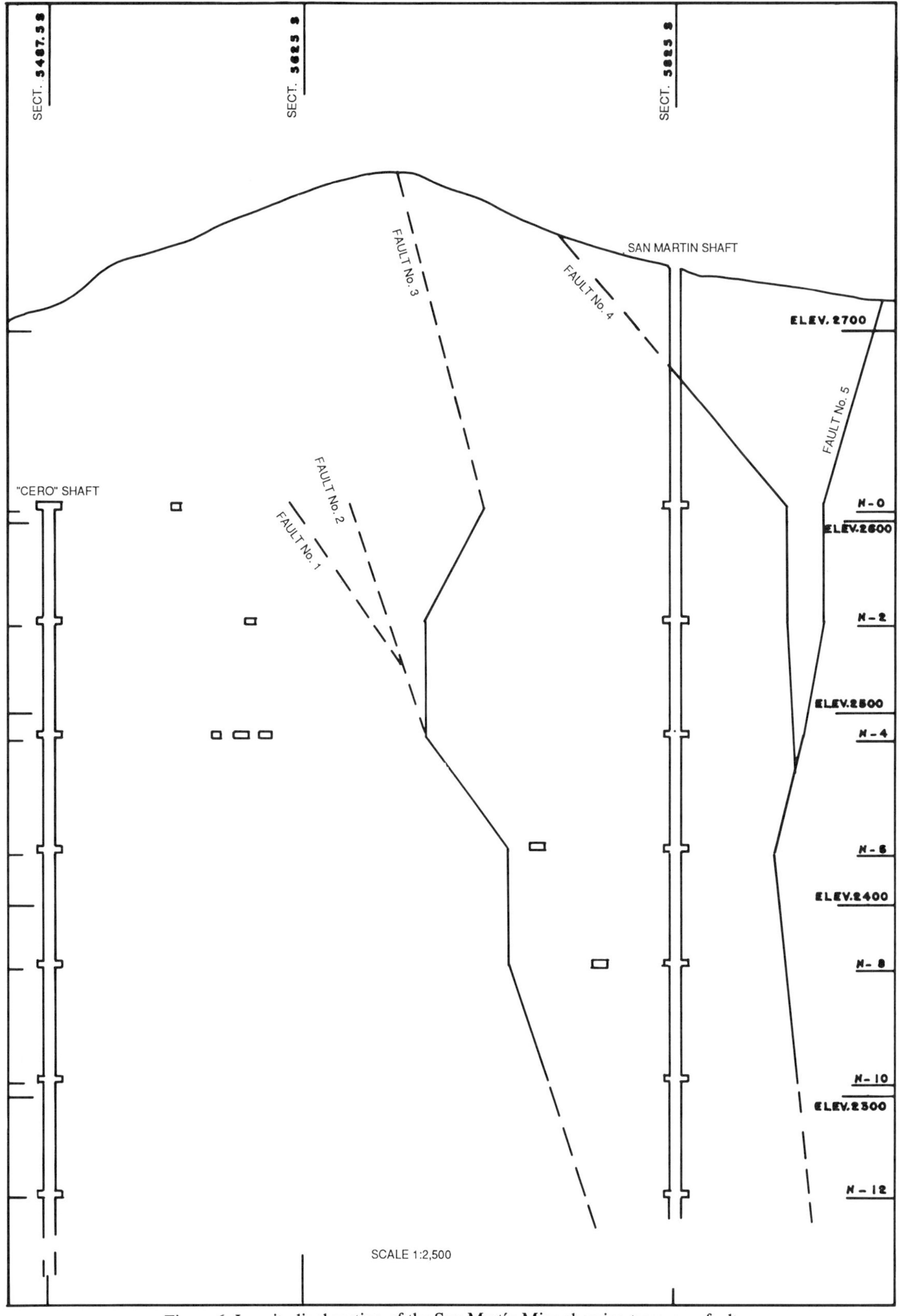

Figure 6. Longitudinal section of the San Martín Mine showing transverse faults.

The replacement bodies consist mostly of chalcopyrite, sphalerite, bornite, tetrahedrite, native silver, arsenopyrite, pyrrhotite, stibnite, and galena; tactite mineralization is disseminated within certain layers that alternate with generally barren chert beds. Vein mineralogy is similar, but the mineral is massive.

In Sabinas–Noria de San Pantaleón, another vein system located toward the south of the district at a distance of about 500 m from the intrusive, structures trend generally northwest and dip southwest. The system affects recrystallized limestones of the Cuesta del Cura Formation and is mineralized in lead and zinc with some subordinate silver.

Mineralogy and paragenesis

On the basis of evidence discussed at the end of this chapter, possibly three stages of mineralization occurred in the district (Aranda-Gomez, 1978). The minerals in the paragenetic sequence are described individually below:

Arsenopyrite, the first sulfide deposited during the principal mineralization period (middle stage), is found in the silver/copper–rich zone next to the intrusive, occurring usually as +2 mm euhedral to subhedral crystals.

Pyrite is quite rare, occurring as slightly anisotropic euhedral crystals with small pyrrhotite inclusions. One of the observed crystals partially engulfs an arsenopyrite idiomorphic crystal, suggesting the somewhat later entry of the pyrite, which owes its anisotropism to the arsenopyrite mixture.

Pyrrhotite, appearing rarely at the higher levels of the San Martín Mine but abundantly at depth, is associated with zones of high copper concentration near the intrusive in the Ibarra vein and at the intersection of the three principal veins. It occurs as medium- to coarse-grained aggregates of subhedral and anhedral crystals that replace the arsenopyrite and are replaced in turn by chalcopyrite and sphalerite. In diamond-drill deep cores it appears in association with sphalerite and galena; textures suggest replacement of the galena by pyrrhotite, which further suggests a first (early stage) mineralization period. The pyrrhotite may be magnetic or not; in larger proportions it is usually magnetic.

Bornite is abundant from the higher levels to level 8, below which chalcopyrite tends to predominate. Bornite replaces arsenopyrite and galena and is invariably massive, occurring either alone or associated with other sulfides.

Chalcopyrite is the most abundant copper sulfide in the deposit. Allotriomorphic chalcopyrite masses replaced the bornite and were in turn substituted by galena. Its temporal relation with sphalerite is unclear: at times the chalcopyrite seems to replace it, at other times the opposite appears to be the case. It seems more reasonable to assume that the chalcopyrite is somewhat prior to the sphalerite. Chalcopyrite commonly occurs massively; some crystals show striation.

Sphalerite is very abundant and generally massive, although occasional crystals are also present. The mineral is light to dark brown in color with glassy to metallic luster and appears to have been deposited after the bornite; the iron content is quite variable (Burton, 1975) suggesting that the mineral developed within a wide temperature range and under diverse chemical conditions.

Galena occurs principally in the veins and in very minor amounts in the tactites. On a textural basis it was probably one of the first minerals to be deposited, occurring as cubic crystals associated with chalcopyrite and occasionally with sphalerite. Pyrrhotite is seen to replace it at the deepest levels of the deposit.

Native silver is usually associated with bornite in thin veinlets cross-cutting the sulfides and at times also associated with pyrargirite. It is most abundant at level 8, at the footwall of the intersection of the three veins, becoming much rarer deeper down.

Tetrahedrite is abundant at the higher levels of the mine where it occurs massively; at depth it appears as crystals in calcite-filled fractures; it is commonly associated with chalcopyrite and galena.

Stibnite appears associated with calcite as radial aggregates of prismatic crystals in cavities; it is generally found in the mineralized zone farthest from the intrusive.

Fluorite is the common gangue mineral associated with zinc, occurring in anhedral masses with a well-defined octahedral cleavage; it is also found in the barren tactites and cornalites.

Calcite is abundant throughout the deposit. Its relations with the sulfides in postmineral cavities and fractures indicate discontinuous deposition throughout the mineralization process. Burton (1975) identified two different forms: (1) transparent to white, massively crystallized calcite I with well-developed rhombohedral cleavage, associated with the massive sulfides; and (2) calcite II, appearing in post-mineral cavities as prisms with scalenohedral and rhombohedral terminations.

Quartz occurs rarely as occasional small crystals on cavity walls.

In summary, the interpreted textures indicate that the tactite was replaced by massive sulfides, and arsenopyrite was the first deposited mineral (from early galena), followed by pyrrhotite, bornite, chalcopyrite, and sphalerite (Fig. 7).

Field observations and microscope studies reveal three typical associations in the San Martín sulfides—chalcopyrite-bornite, chalcopyrite-galena-tetrahedrite, and sphalerite-chalcopyrite—which remain unchanged from one place to another; however, their variable relative abundances result in marked zoning, both horizontal and vertical, within the deposit.

Bornite and chalcopyrite are abundant toward the hanging wall of the main replacement body near the intrusive between the San Marcial (S.M.V.) and Ramal-Ibarra (R-I.V.) veins (Fig. 2). Bornite diminishes and the sphalerite-chalcopyrite association increases away from the intrusive; at the Ibarra vein sphalerite becomes considerably more frequent. No textural evidence indicating that one mineral developed before the other in the bornite-chalcopyrite association has been found; presumably both were deposited simultaneously under identical physicochemical conditions.

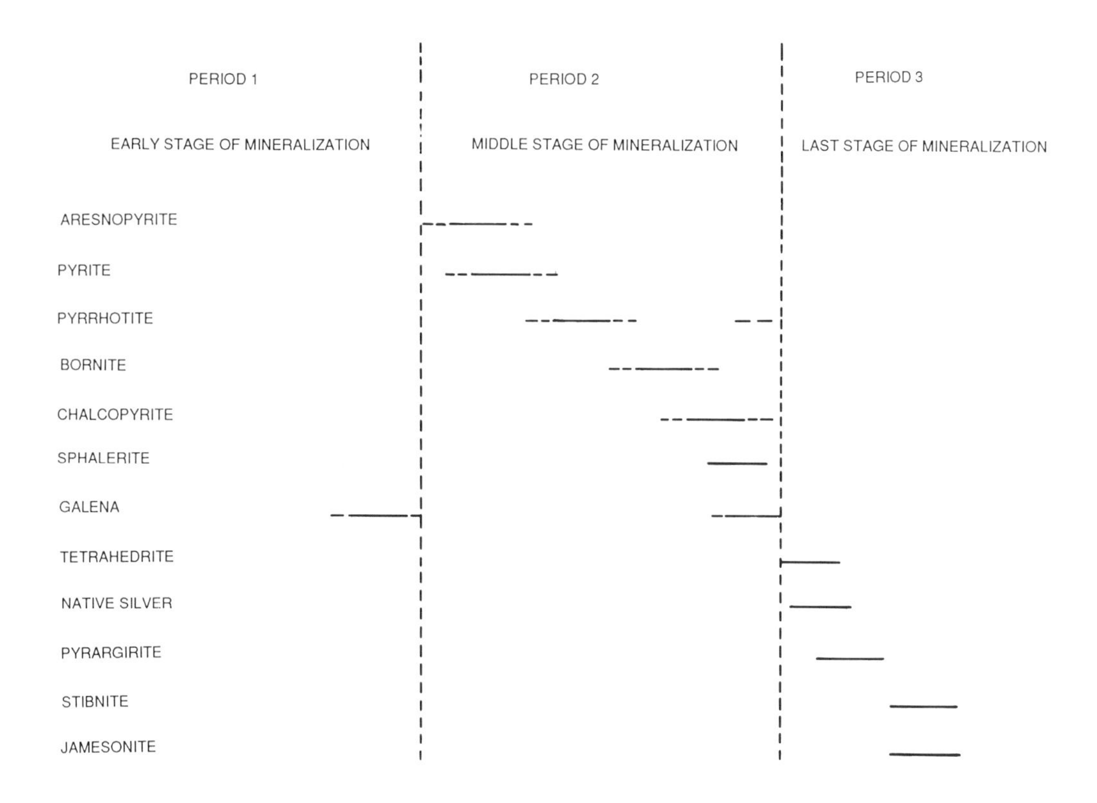

Figure 7. Paragenetic sequence of mineralization, San Martín Mine (from Aranda-Gomez, 1978).

The characteristic chalcopyrite-galena-tetrahedrite association occurs in both the silver/copper- and zinc-rich zones.

In the sphalerite-chalcopyrite association—where the former is dominant—both minerals replace garnets (grossularite and andradite), preserving the habits of the latter, which indicates the intensive pre-mineralization process undergone by the rock.

The garnet-idocrase-diopside-wollastonite tactite developed prior to the sulfide deposition; for this reason the barren tactite zones within the body of massive sulfides represent inadequate or unfavorable material for replacement.

ECONOMIC GEOLOGY

History and production

The San Martín Mining District is one of the oldest in Mexico; the first vein was discovered by the Spaniards in 1555. Estimated production from 1555 to 1821 was on the order of 2,500,000 tons of ore, with an average 450 g/ton of silver and 0.5 g/ton of gold. No reliable production information is available for the remaining nineteenth century and first decades of the twentieth, but mine workings were sparse. Small-scale development of the primary sulfides began in the mid-1930s, producing about 450 g/ton of silver, 1 to 3 percent lead, 1 to 4 percent copper, and 6 percent zinc.

The present-day San Martín District producing mines (Fig. 1) are the Industria Minera México, S. A., San Martín unit (6,800 tons/day of 120 g/ton silver, 1 percent copper, and 5 percent zinc), and the Sabinas Mine (400 tons/day of 160 g/ton silver, 3 percent lead, 0.5 percent copper, and 4.5 percent zinc).

Noria de San Pantaleón, the principal mine from 1930 to 1940, remains closed down. Operations at El Refugio and Pachuca (Fig. 2) have produced minor volumes of high silver content in recent years.

Principal mines

San Martín Mine, the largest and most important in the district, is located in the western part of the Cerro de La Gloria intrusive granodiorite; veins crop out for distances of up to 900 m and are presently known down to level 20 (750 m underground). Mining is by the slim pillars and hydraulic fill method; a 5-km-long service shaft at the footwall of the main mineralized body allows access to the levels; the lower levels are vertically spaced

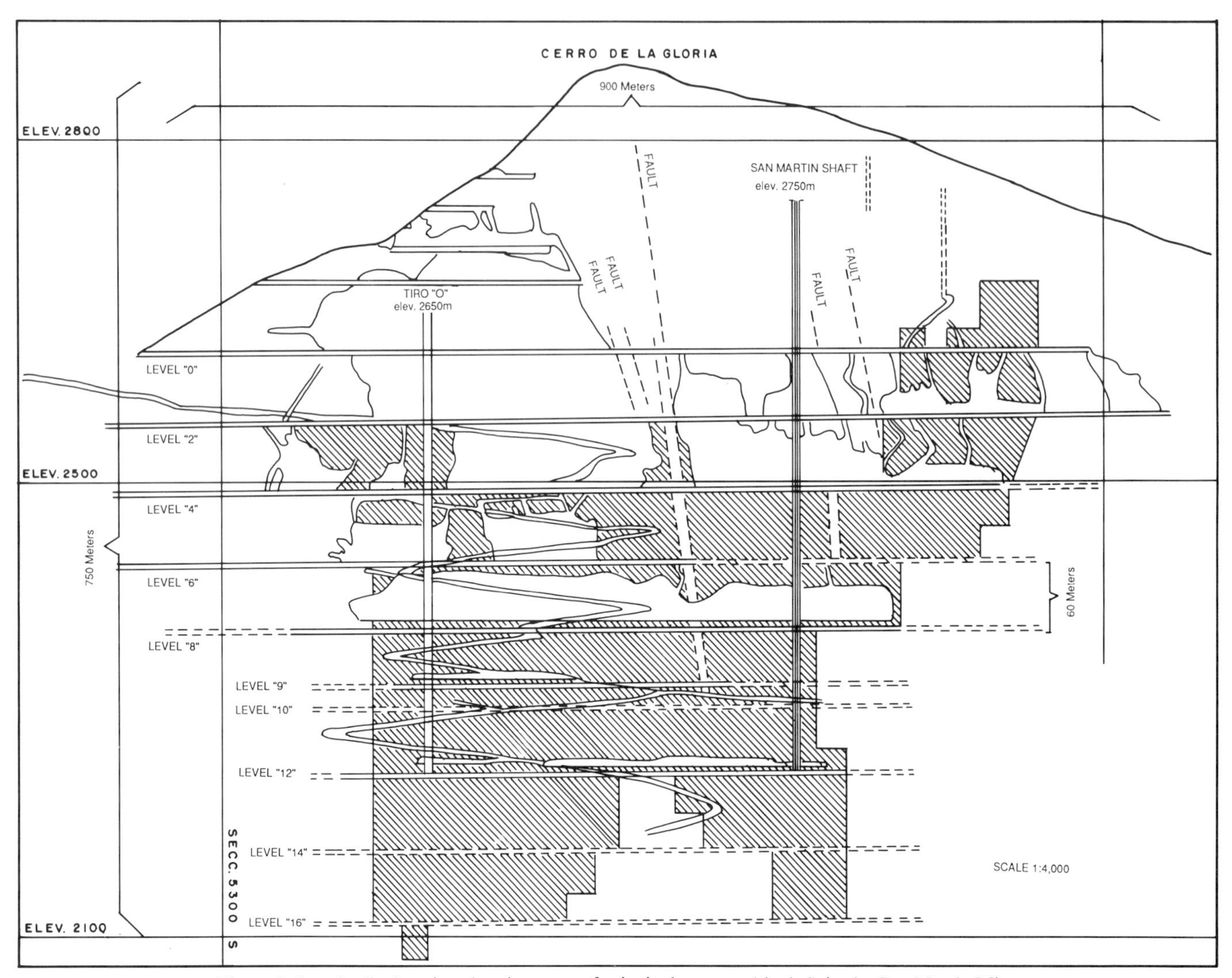

Figure 8. Longitudinal section showing areas of principal reserves (shaded) in the San Martín Mine.

every 60 m. There are two raises for mineral extraction "Zero," from 0 level (elevation 2,650 m) to level 12; and "San Martín" from the surface (elevation 2,750 m) also to level 12, at an elevation of 2,240 m (Fig. 8).

Sabinas Mine, located in the southern portion of the district, contains part of the metamorphic halo surrounding the granodiorite intrusive. Replacement bodies in east-west veins of irregular thickness contain sphalerite, galena, and chalcopyrite as economic sulfides. Mining is by conventional hydraulic fill.

REFERENCES CITED

Aranda-Gomez, J. J., 1978, Metamorphism, mineral zoning, and paragenesis in the San Martín Mine, Zacatecas, Mexico [M.Sc. thesis]: Golden, Colorado School of Mines, 83 p.

Burton, B. H., 1975, Paragenetic study of the San Martín Mine, near Sombrerete, Mexico [M.Sc. thesis]: Minneapolis, University of Minnesota, 86 p.

Monziváis Hernández, A., 1983, Metalogenia del Distrito San Martín, Sombrerete, Zac., *in* 15th Convención Nacional de la Asociación de Ingenieras de Minas, Metalurgístas y Geólogos de México, Memoir, p. 3–27.

Printed in U.S.A.

The Geology of North America
Vol. P-3, Economic Geology, Mexico
The Geological Society of America, 1991

Chapter 25

Sierra Madre Oriental Province

Guillermo P. Salas
Geologo Consultor, Asesor del Consejo de Recursos Minerales, Centro Minero Nacional, Carretera México-Pachuca, Pachuca, Hgo., México, D.F.

INTRODUCTION

This metallogenic province covers 420,880 km^2 (Salas, 190, p. 69–77) and coincides with the huge Mexican Geosyncline, a miogeosyncline extending from Alaska through the western United States and crossing the Mexican border in the vicinity of Ciudad Juarez (Chihuahua) to form the Sierra Madre Oriental.

The Sierra is a chain of elongated mountains formed by large synclines in Jurassic and middle Cretaceous limestones and intercalated clastic sedimentary rocks several thousand meters thick; in some areas (Cañon del Novillo, west of Ciudad Victoria,

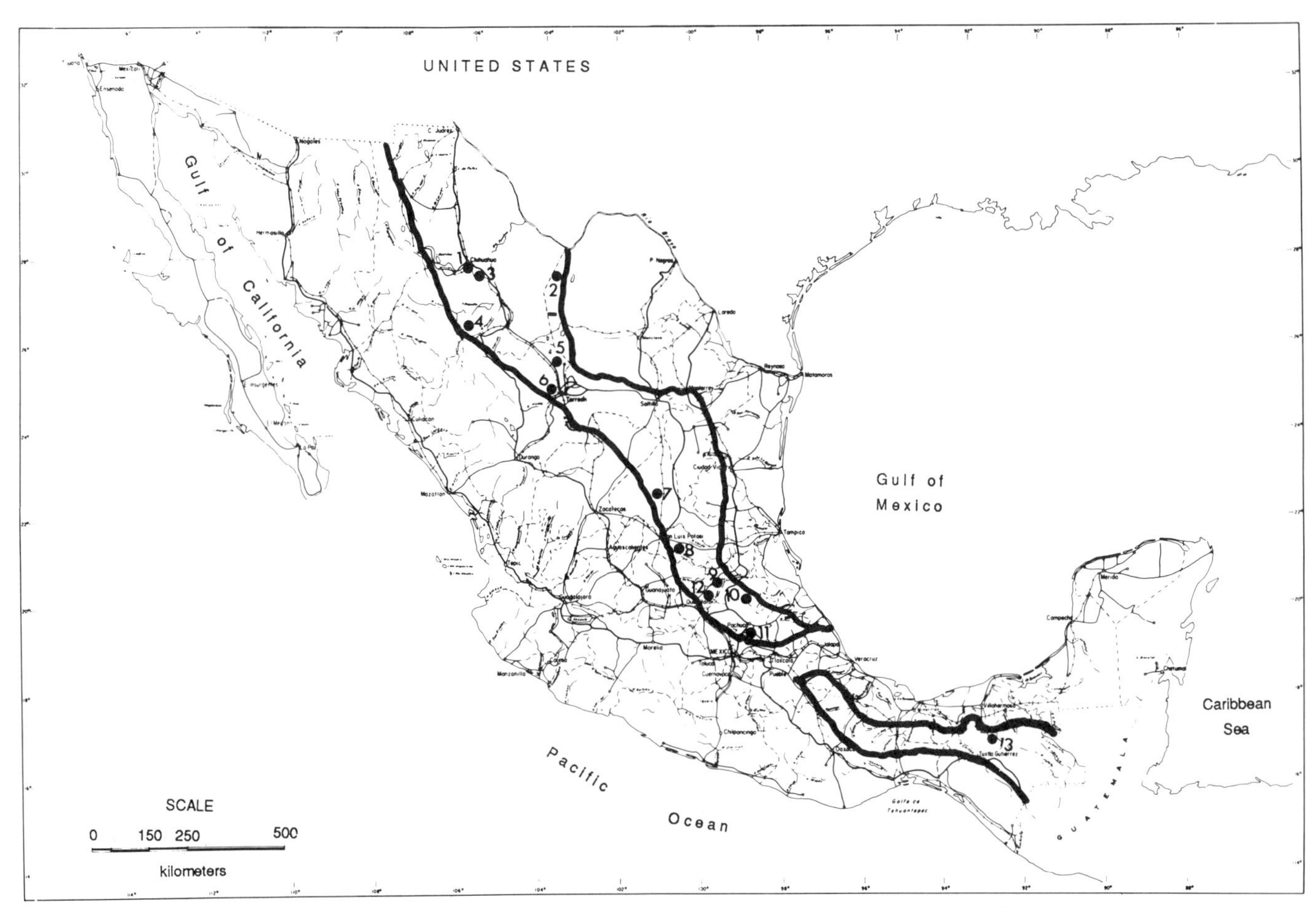

Figure 1. Map showing the area of the Sierra Madre Oriental Province, and distribution of the mining areas discussed in the following chapters. 1, Santa Eualia, Chihuahua; 2, La Encantada, Coahuila; 3, Naica, Chihuahua; 4, San Francisco del Oro, Chihuahua; 5, Antares, Durango; 6, Velardeña, Durango; 7, Charcas, San Luis Potosí; 8, El Realito, Guanajuato; 9, La Negra, Queretaro; 10, Zimapan, Hidalgo; 11, Molango, Hidalgo; 12, Pachuca–Real del Monte, Hidalgo; 13, Santa Fe, Chiapas.

Salas, G. P., 1991, Sierra Madre Oriental Province, *in* Salas, G. P., ed., Economic Geology, Mexico: Boulder, Colorado, Geological Society of America, The Geology of North America, v. P-3.

Tamaulipas) the Mesozoic section is unconformable over Precambrian metamorphic rocks, schists, and gneisses. Figure 1 shows the extent of the province, interrupted at its southern end by the Neo-Volcanic Axis, as seen on the ERTS-1 Satellite image mosaic.

The large folds and faults in Mesozoic limestones trend generally northwest except between Torreon (Coahuila) and Monterrey (Nuevo León) where the Great Mexican Geosyncline deviates at the southern end of the Coahuila paleopeninsula and structures are almost east-west. The hinge of this deviation, trending east-west to northwest, and the Tertiary intrusives, coincide with the Concepcion del Oro Mining District (Zacatecas).

Worth noting is the fact that the emplacement of granodiorite intrusives (San Carlos, Tamaulipas, del Oro, Santa Eulalia, Parral, Chihuahua Ranges, and others farther south) follows the orientation of the geosynclinal alignments, proving the tectonic relation with the generation of acid magma. The geology of this province is discussed in detail by Salas (1980, p. 69).

For as yet unexplained metallogenic reasons, the metals in this province are very different from those in the Sierra Madre Occidental. The deposits contain both metallic and nonmetallic minerals; polymetallic deposits with variable silver, lead, zinc, and copper contents are abundant, occasionally with very subordinate gold. However, their main value is their silver content. Mineralization is of epithermal origin and is only occasionally metasomatic. Vein mineralization predominates in both cases. On occasion, polymetallic minerals occur in mantos, such as Fresnillo (Zacatecas), Santa Eulalia, Naica, and some mines in Parral (Chihuahua).

Excellent nonmetallic deposits are also present in the province: fluorite in San Luis Potosí and northeastern Guanajuato; sulfur in Guascama, and San Luis Potosí (abandoned); barite, asbestos, and talc in Tamaulipas; and bentonite in Durango.

Typical districts are described in the following chapters (Fig. 1): 1, Santa Eulalia, Chihuahua; 2, La Encantada, Coahuila; 3, Naica, Chihuahua; 4, San Francisco del Oro, Chihuahua; 5, Antares, Durango; 6, Velardeña, Durango; 7, Charcas, San Luis Potosí; 8, El Realito, Guanajuato; 9, La Negra, Queretaro; 10, Zimapan, Hidalgo; 11, Molango, Hidalgo; 12, Pachuca–Real del Monte; Hidalgo; 13, Santa Fe, Chiapas.

REFERENCES CITED

Salas, G. P., 1980, Carta Provincias Metalogeneticas de la Republica Mexicana: Consejo de Recurson Minerales, Bulletin 21-E, 2nd Edition.

Printed in U.S.A.

The Geology of North America
Vol. P-3, Economic Geology, Mexico
The Geological Society of America, 1991

Chapter 26

Economic geology of the Santa Eulalia Mining District, Chihuahua

D. Maldonado E.
Industrial Minera México, S.A., Ave. Baja California 200, Col. Roma Sur, 06760 México, D.F.

INTRODUCTION

The Santa Eulalia Mining District, covering approximately 30 km^2 is located 25 km east of the town of Chihuahua in the Sierra Santo Domingo (Fig. 1). It is divided into two mineralized zones: the East Field (San Antonio Mine), and the West Field (Buena Tierra and Potosí Mines) (see Fig. 5). The orebodies are principally chimneys and mantos, consisting of minor filling and replacement along fractures and disseminated minerals in tactites.

Access to the district is by paved road from Chihuahua; a continually reconditioned dirt road runs for 11 km between the two fields. Temperatures in the dry steppe climate range from –4°C (winter low) to 38°C (summer high); the normal torrential rain period is from May to September with lighter rains in November and December. Mining is the region's main economic activity, employing about 90 percent of the working-age males; cattle-raising and agriculture are of domestic scale.

REGIONAL GEOLOGY

The Santa Eulalia Mining District lies within the Basin and Range Province in a folded and faulted NNW-trending belt of pre-Tertiary rocks belonging to the Sierra Madre Oriental metallogenic province. The Sierra Santo Domingo, approximately 20 by 11 km, rises about 600 m above the surrounding valleys to an average elevation of 1,900 m. A NNW-trending and west-dipping fault with a minimum displacement of 1 km truncates the range's western end; on the east it slopes gently eastward to the valley.

The regional stratigraphic section consists of thick Cretaceous lacustrine and marine carbonates and evaporites unconformably overlain by Tertiary volcanic rocks and volcaniclastic sediments.

The Sierra Santa Eulalia (or Santo Domingo) is a wide, doubly displaced, NNW-trending, low-dipping anticline; the mineralized areas occur on its east and west flanks in the southern displaced block.

The calcareous rocks that outcrop in the north of the district are covered by thick Tertiary volcanics in the south, where only limestone windows are to be found (Fig. 2).

LOCAL GEOLOGY

Stratigraphy

Cretaceous sediments. The oldest known formation in the district—a sequence of pure white anhydrites with thin black shale and limestone intercalations—has been correlated with the upper Aptian Cuchillo Formation (Figs. 3, 4). The limestones and shales are very bituminous, fetid when freshly cut, and locally contain abundant rudist fragments and framboidal pyrite; no commercial mineralization is known in this unit.

The carbonate sequence overlying the Cuchillo Formation has been divided locally into five members on the basis of lithology and commercial importance and correlated with the lower to mid-Albian Glen Rose and Lágrima Formations, and the Albian to Cenomanian Aurora Formation. The Black Limestone Member appears to be identical to the Cuchillo dark sediments but lacks evaporites; however, thin medium- to fine-grained gray limestone horizons have been found in the member. The entire section contains scattered pyrite crystals and pyrite-filled veinlets in lesser abundance, particularly at the top; no economic ores are found in these rocks.

The Blue Limestone Member is the most important unit from a commercial viewpoint because it hosts most of the known mineral bodies; this is a dark bluish gray, homogeneous, microgranular limestone bearing no fossils except in the upper part of the section.

The Lower Fossiliferous Limestone Member is mostly a massive, light gray to white, medium- to coarse-grained recrystallized limestone with saccaroidal texture at intervals, containing a few chert nodules and abundant pelecypod fossils. Sulfide and oxide orebodies in mantos (stratiform bodies) and chimneys have been mined from this rock.

The Intermediate Fossiliferous Limestone Member is a compact medium-bedded argillaceous rock containing considerable manganese at some levels and dispersed chert nodules; orebodies in this member occur as chimneys but not mantos.

The Upper Fossiliferous Limestone Member is a series of light gray to gray, strongly recrystallized calcareous rocks with abundant and extensive chert bands parallel to the bedding; the

Maldonado E., D., 1991, Economic geology of the Santa Eulalia Mining District, Chihuahua, *in* Salas, G. P., ed., Economic Geology, Mexico: Boulder, Colorado, Geological Society of America, The Geology of North America, v. P-3.

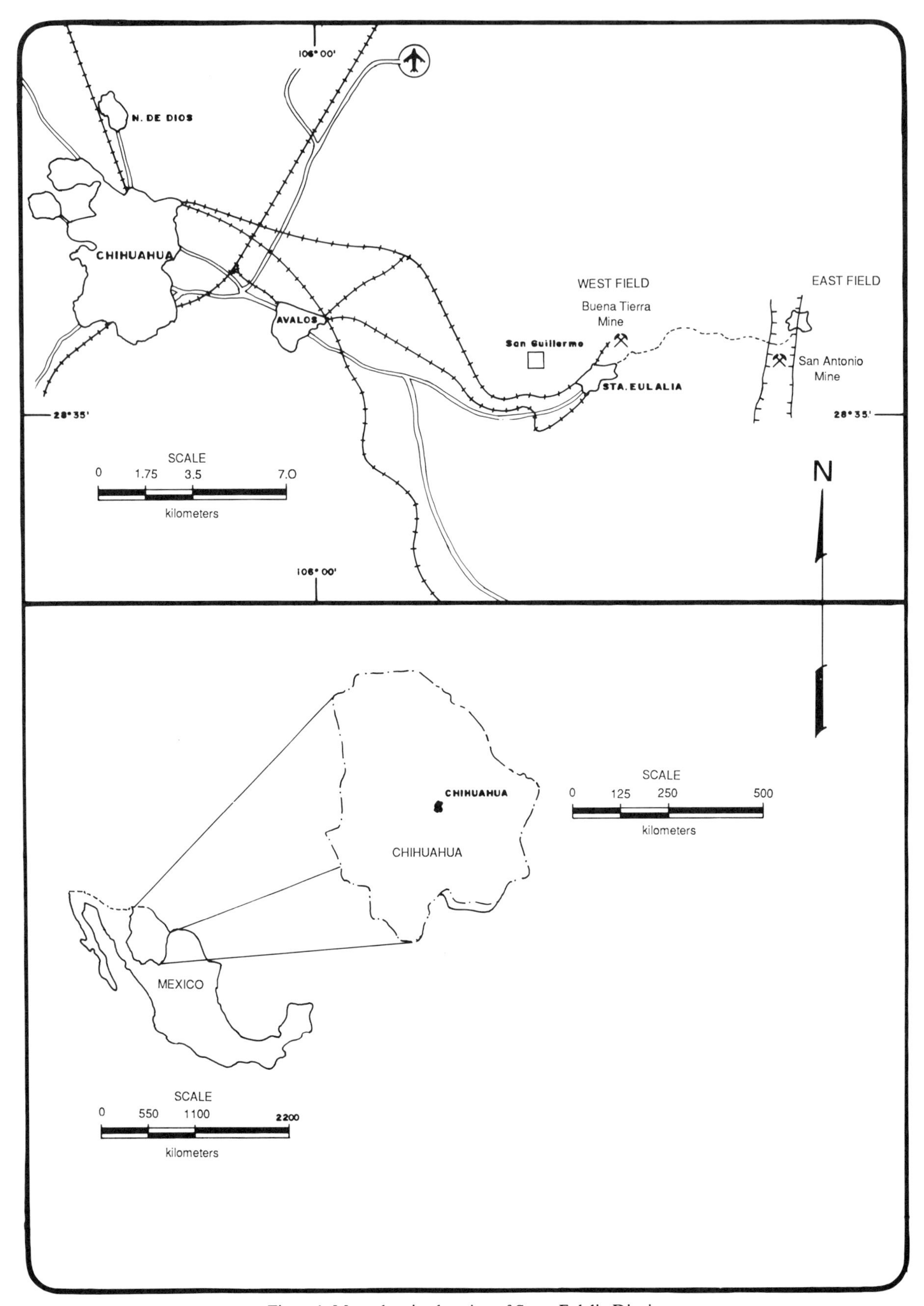

Figure 1. Maps showing location of Santa Eulalia District.

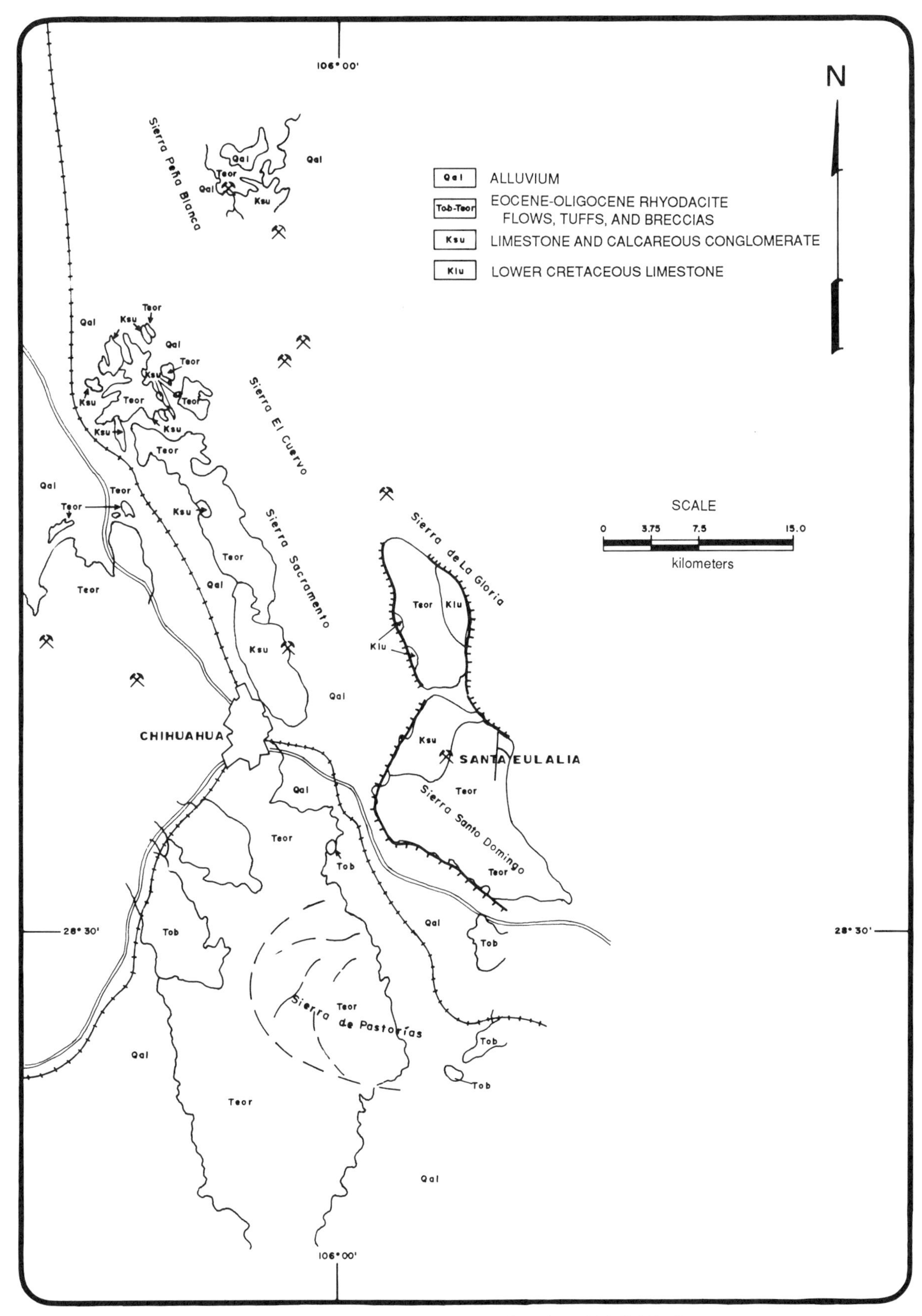

Figure 2. Geologic map of the region of the Santa Eulalia District.

largest commercially mineralized mantos in the district occur at the base of this sequence.

Tertiary volcanic rocks. A volcanic-sedimentary unit unconformably overlying the limestones consists mainly of basal conglomerate, andesite and rhyolite tuffs, rhyolite flows, sandstones, and volcaniclastic sediments. The basal conglomerate is the most important, consisting of subrounded limestone fragments in a calcareous-argillaceous groundmass with some andesitic material that increases toward the top. The unit contains narrow oxidized orebodies at the San Antonio Mine (East Field), and rather small and irregular mantos-type orebodies are found in the West Field at the contact between the carbonate sedimentary sequence and the volcanic-sedimentary overburden.

The upper units (tuffs, flows, sandstones) contain only very narrow veinlets filled with quartz, fluorite, barite, galena, and pyrite of no commercial significance. Rhyolite in the upper part of the sequence dated 31.7 ± 0.7 Ma (Megaw and McDowell, 1983) has been correlated with the volcanic rocks of the Sierra de Pastorías southwest of Santa Eulalia (Fig. 2).

Tertiary intrusive rocks. Most of the large intrusives known in the area are related to the ore deposits; the principal bodies are described below.

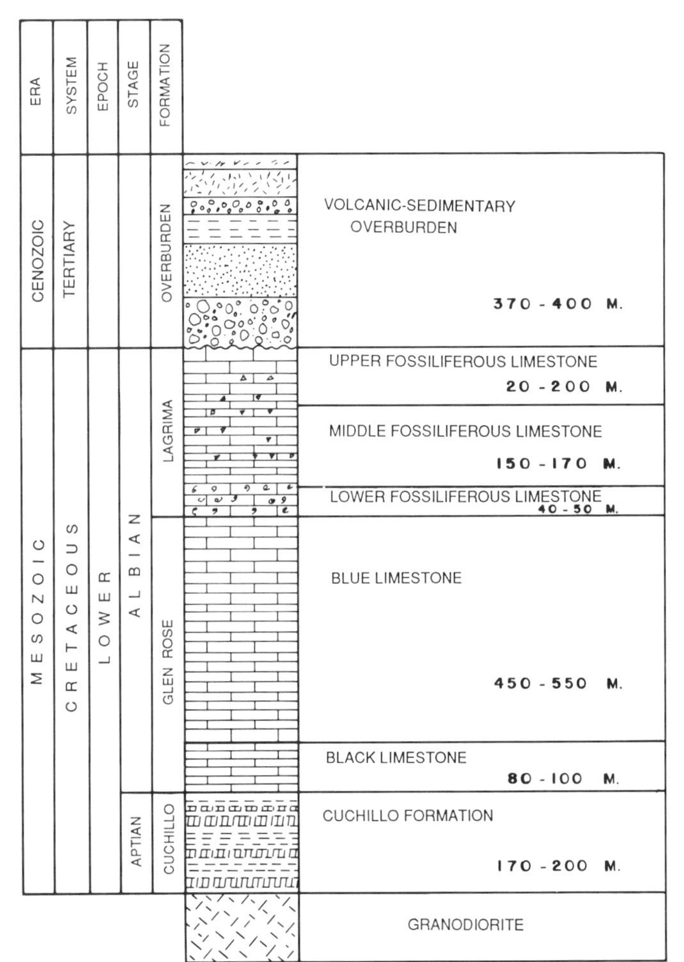

Figure 3. Stratigraphic column for the Santa Eulalia District.

Granodiorite. This unit underlies and intrudes the Cuchillo Formation and the Black Limestone Member in West Field deep drillholes (Buena Tierra Mine) (Fig. 5). It contains plagioclases, K-feldspar, quartz, and hornblende phenocrysts, is uniformly altered to chlorite, calcite, and pyrite, and shows sparsely disseminated sphalerite, galena, and chalcopyrite. Where the rock intrudes the Cuchillo Formation, an alteration rim (calc-silicates) about 0.5 m thick is developed. Present information does not indicate any direct association of the granodiorite with the known orebodies.

Diorite-diabase. These intrusive rocks occur in both fields as sills of variable thickness, texture, composition, and alteration according to their location; 1- to 70-m-thick sills of aphanitic to phaneritic textures and mineralized in the upper, lower, and middle parts are known. At times these are intruded by rhyolites, developing traps for mineral deposition at the intersections. Clark and others (1979) report their age of 37.0 ± 0.8 Ma. At their margins these rocks served as passageways for mineralizing fluids, with the aid of pre-mineral fracturing.

Rhyolites (felsites). These are most closely related to the mineralization, especially in the East Field. They occur as white, aphano-porphyritic to aphanitic dikes and sills composed mainly of quartz, orthoclase, and albite, often containing disseminated pyrite and minor amounts of sphalerite, galena, and arsenopyrite euhedral crystals in banded textures. The mineralization appears at the dike walls where it exhibits spatial zoning, and both above and below in the sills, which contain most of the ore. These intrusives are often altered and brecciated at the edges, indicating that they served as channels for mineralizing fluids. A West Field felsite date of 26.8 ± 0.6 Ma (Clark and others, 1979) is somewhat doubtful owing to the alteration of the rock sample.

Rhyolite porphyry. This occurs in the district as a persistent east-west dike outcropping for about 5 km across the entire calcareous and volcanic-sedimentary sequence. Silver mineralization in its upper part may represent remobilization of earlier mineralization (De la Fuente L., 1969).

Other dikes—such as the andesitic Mina Vieja and Potosí—apparently exerted some control over the oxide, carbonate, sulfide, and sulfosalt orebodies in the Upper and Lower Fossiliferous Limestone Members (Fig. 3).

Structural geology

Two very gentle fold systems, north-south and east-west, predominate locally; limestones dip 5 to 10°, steeper dips occurring only at the fault zones. The fault and fracture systems briefly described below played an important role in mineralization control.

North-south system. This is considered to be the main control for mineralization in the West Field. It consists of normal, mostly east-dipping, almost vertical, closed faults of small vertical displacement and variable continuity, both along strike and at depth. The largest orebodies are associated with faults of this system (Chorro, Potosí, J-Norte, Peñoles, Tiro Alto, and Bustillos; Fig. 3), which favored the development of bodies in the sulfide zone and considerably extensive mantos in the oxide zone where they acted in combination with the N50°E system.

N50°E system. Faults and fractures in this system show regular strikes and more variable dips than the north-south system. West Field mantos in the sulfide and oxide zones are strongly controlled by faults and fractures in conjunction with intrusive bodies and favorable limestones, resulting in considerably extensive bodies.

N20°E system. This includes faults and fractures affecting Cretaceous and Tertiary rocks and is considered to be the most important system in the East Field, generating the San Antonio tectonic graben bounded by persistent (both along strike and at depth) normal faults of strong displacement. These faults controlled both the emplacement of premineral intrusives and the mineralization itself.

Other minor systems (east-west, N70°W, N30°E) acted locally within restricted areas.

MINERAL DEPOSITS

The two mineralized areas in the district, at the east and west flanks of the anticline (Fig. 5), exhibit different mineralizations, controls, shapes, and dimensions of the bodies.

East Field (San Antonio Mine)

The orebodies occur mostly as elongated chimneys and, to a lesser extent, as mantos and fissure fillings (Fig. 4). The chimneys are controlled by N20°E–trending rhyolite dikes. Variable amounts of Zn, Pb, Cu, and Fe sulfides appear mainly in bodies within the tactites at cleavage planes or in fractures. Massive sulfide mineralization is not rare but occurs only in small bodies such as the central manto at level 13.

The minerals bearing the metallic elements are listed in order of abundance: pyrite-marcasite, sphalerite, galena, pyrrhotite, chalcopyrite, arsenopyrite, pyrargirite, and magnetite in the sulfide zone; cassiterite, hematite, Cu and Fe carbonates, cerusite, and smithsonite in the oxidation zone. The most abundant silicates are grossularite and andradite associated with epidote, tourmaline, tremolite, actinolite, and hedenbergite; quartz, calcite, and fluorite are also present.

On the basis of radiometric ages and field data, the diabase intrusions preceded emplacement of the felsite dikes and sills whose margins served as passageways for subsequent mineralization in Ca, Mg, Fe, Al, and Na silicates derived from alteration of the limestone. Figure 6 shows mineralization paragenesis at the

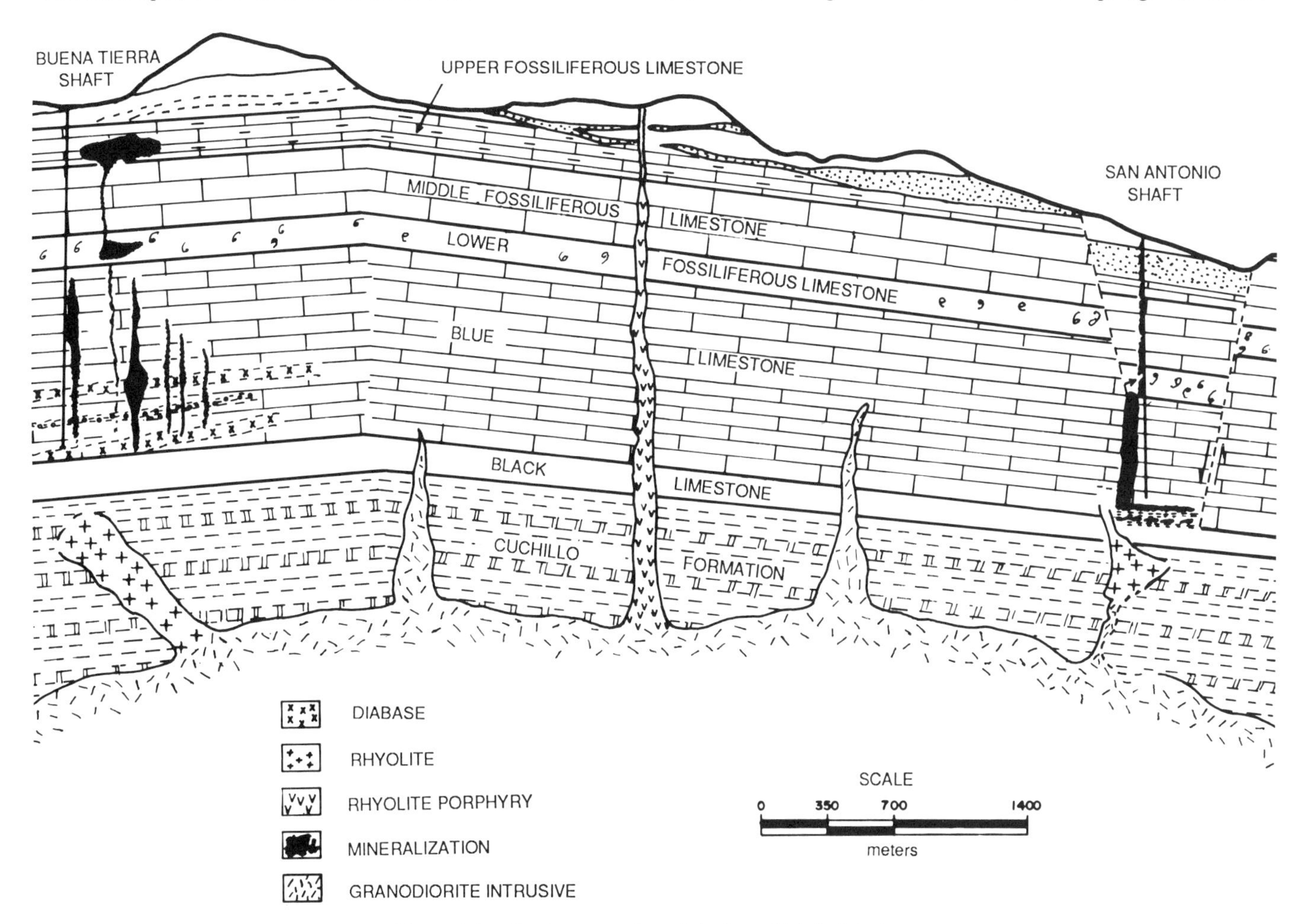

Figure 4. Idealized cross section of the Santa Eulalia District.

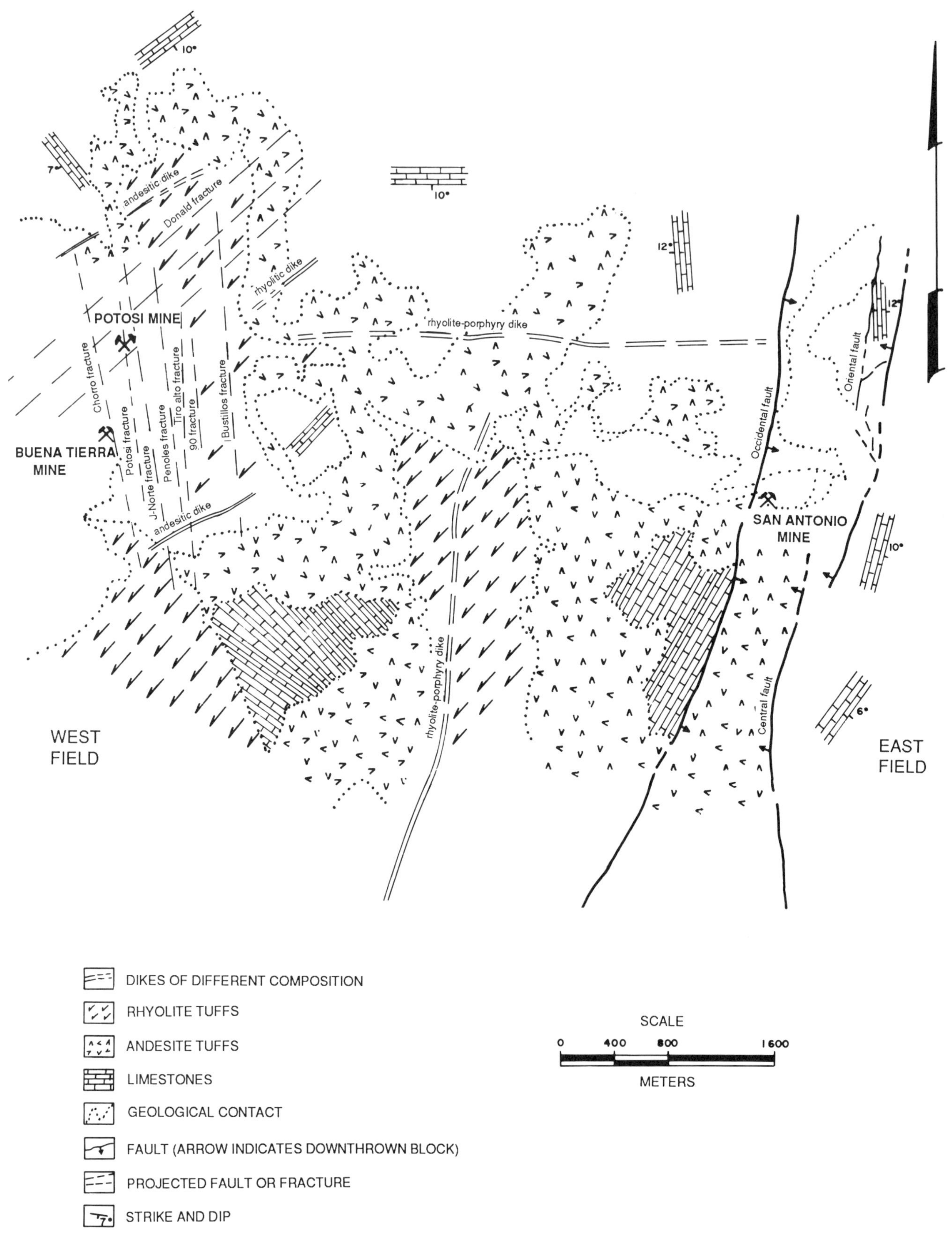

Figure 5. Detailed geologic map of the Santa Eulalia District.

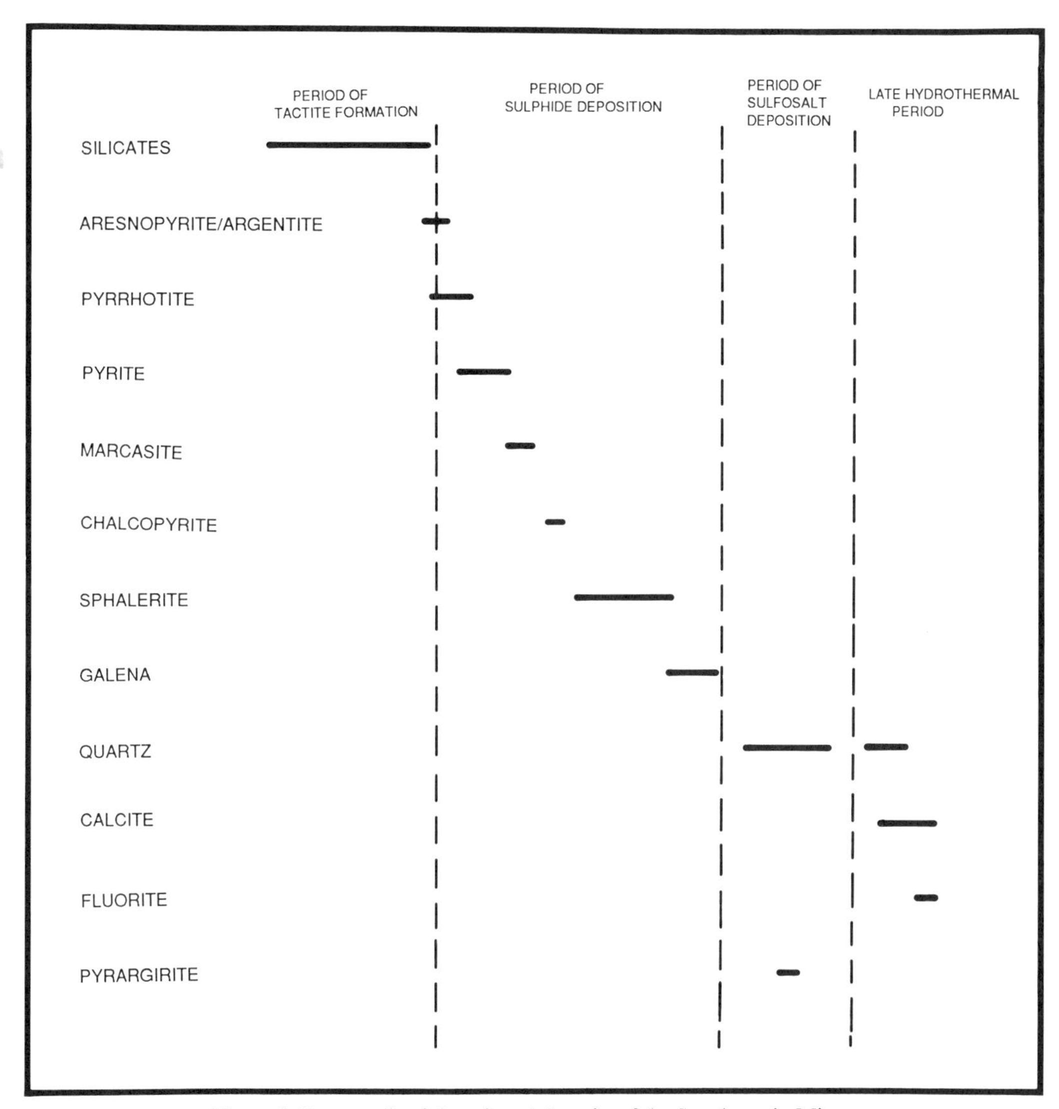

Figure 6. Paragenesis of the mineral deposits of the San Antonio Mine.

San Antonio Mine. The chalcopyrite-sphalerite, galena-arsenopyrite, and galena-argentite associations are conspicuous.

Adjacent to the dike the silicates are zoned as follows: epidote-chlorite-fluorite-grossularite-actinolite followed by massive sulfides and limestone. Chalcopyrite is common in the epidote-chlorite-grossularite zones, whereas sphalerite occurs mainly in the grossularite-actinolite zone. Massive bodies of sphalerite, pyrite, pyrrhotite, or galena-sphalerite in polycrystalline agglomerates with scarce silicates may occur at the contact with the limestone (Fig. 7). Pyrrhotite tends to increase at depth in the chimneys, whereas galena decreases; calcite exsolutions in sphalerite also increase.

Stanniferous veins and chimneys mined in the East Field contain cassiterite as the principal mineral associated with specularite, quartz, topaz, fluorite, magnetite, tourmaline, and wolframite. The largest tin-bearing body is an approximately 300-m-high chimney next to the San Antonio dike. Hewitt (1943) suggests a late volatile stage related to the deposition of the sulfides as a possible mechanism for the tin mineralization. Burt and others (1982) speculate on the possible association of topaz rhyolites and replacement bodies containing this element. The mineralogy of the San Antonio tin-bearing bodies is very similar to the topaz rhyolites they describe, and the Santa Eulalia District lies within a known topaz rhyolite belt, which suggests that it may be an example of replacement mineralization related to such rocks (Fig. 8).

West Field

Four types of mineralization are found in this field. The so-called normal sulfides consist of pyrite, pyrrhotite, sphalerite, and galena with small amounts of chalcopyrite and arsenopyrite. The silicates—fayalite, ilvaite, and knebelite—are not very abundant. This type of mineralization usually appears as small, irregular bodies with commercial sphalerite and galena; the sphalerite predominates in chimneys and the galena in mantos,

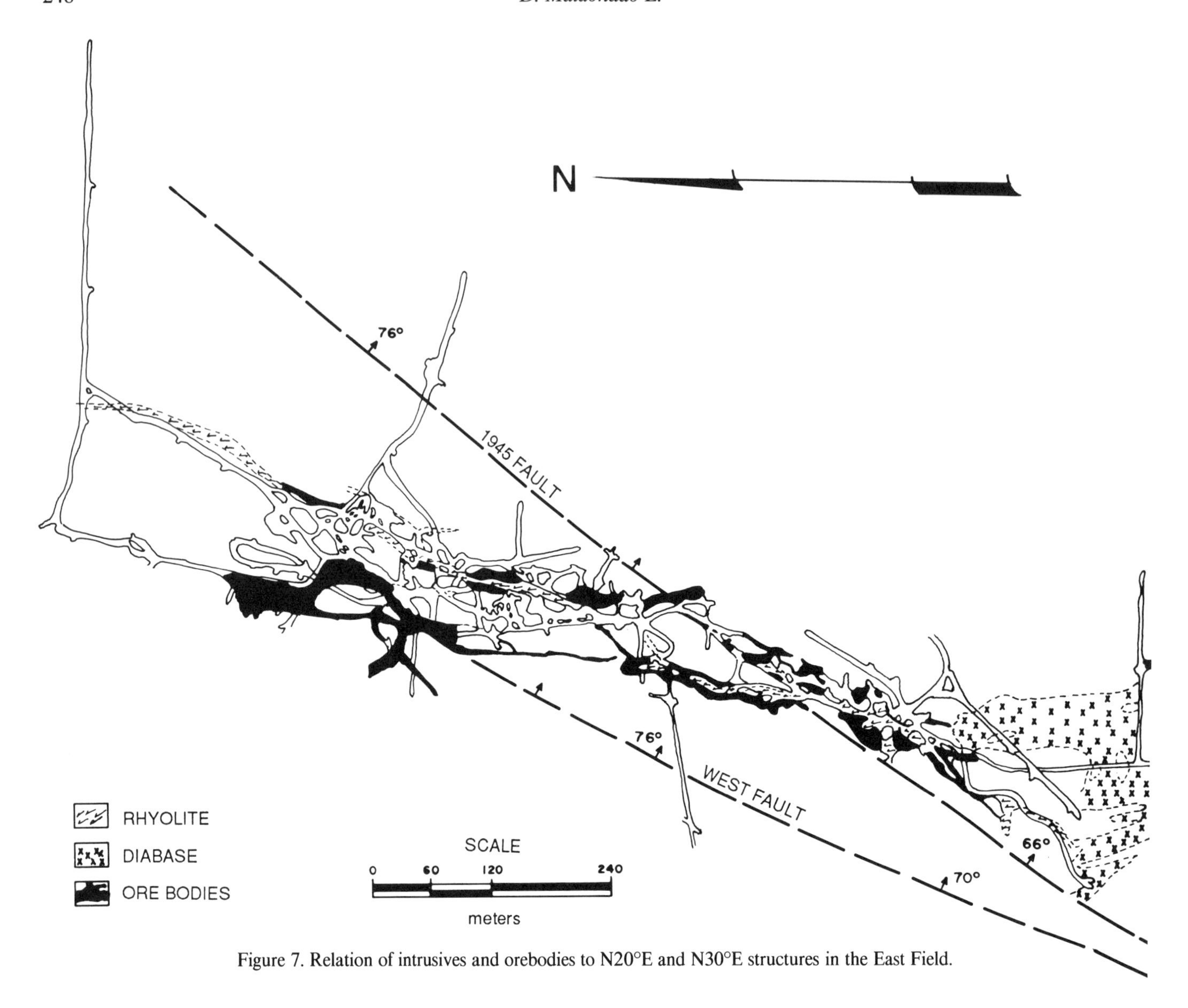

Figure 7. Relation of intrusives and orebodies to N20°E and N30°E structures in the East Field.

with silver values. Their distribution is irregular, there being no well-defined zoning (Fig. 9).

Oxidized bodies containing Pb, Zn, and Fe carbonates with noticeable amounts of manganese and calcite as gangue were heavily exploited in the past.

A fourth type occurs in the upper part of the limestone sequence where the rocks contain abundant manganese associated with small particles (4 to 15 microns) of native silver and pyrargirite (Torres A. and Muñoz C., 1981). Silver content in these rocks ranges from 30 g/ton in small very isolated zones, to several hundred grams per ton.

In this field sulfide mineralization predominates in the Blue Limestone Member; the Upper and Lower Fossiliferous Limestone Members contain most of the oxidized bodies at the upper levels (Fig. 10).

Genesis

Disagreements as to the origin of the Santa Eulalia mineral deposits are long-standing; different arguments favor open space filling, epigenetic hydrothermal, and syngenetic-diagenetic replacements.

Open caverns are frequent; however, typical cavity-fill textures are not. The evidence suggests that fluids ascended from below and advanced laterally along restricted zones, replacement thus becoming the selectively active process. Replacements were presumably guided along narrow tabular zones, which would explain why the mantos show such consistent mineralogies and mineral proportions over distances of 4 km along strike.

Recent sulfur isotope studies of Cuchillo Formation sulfates, Black Limestone syngenetic pyrite, and mined ore sulfides furnish

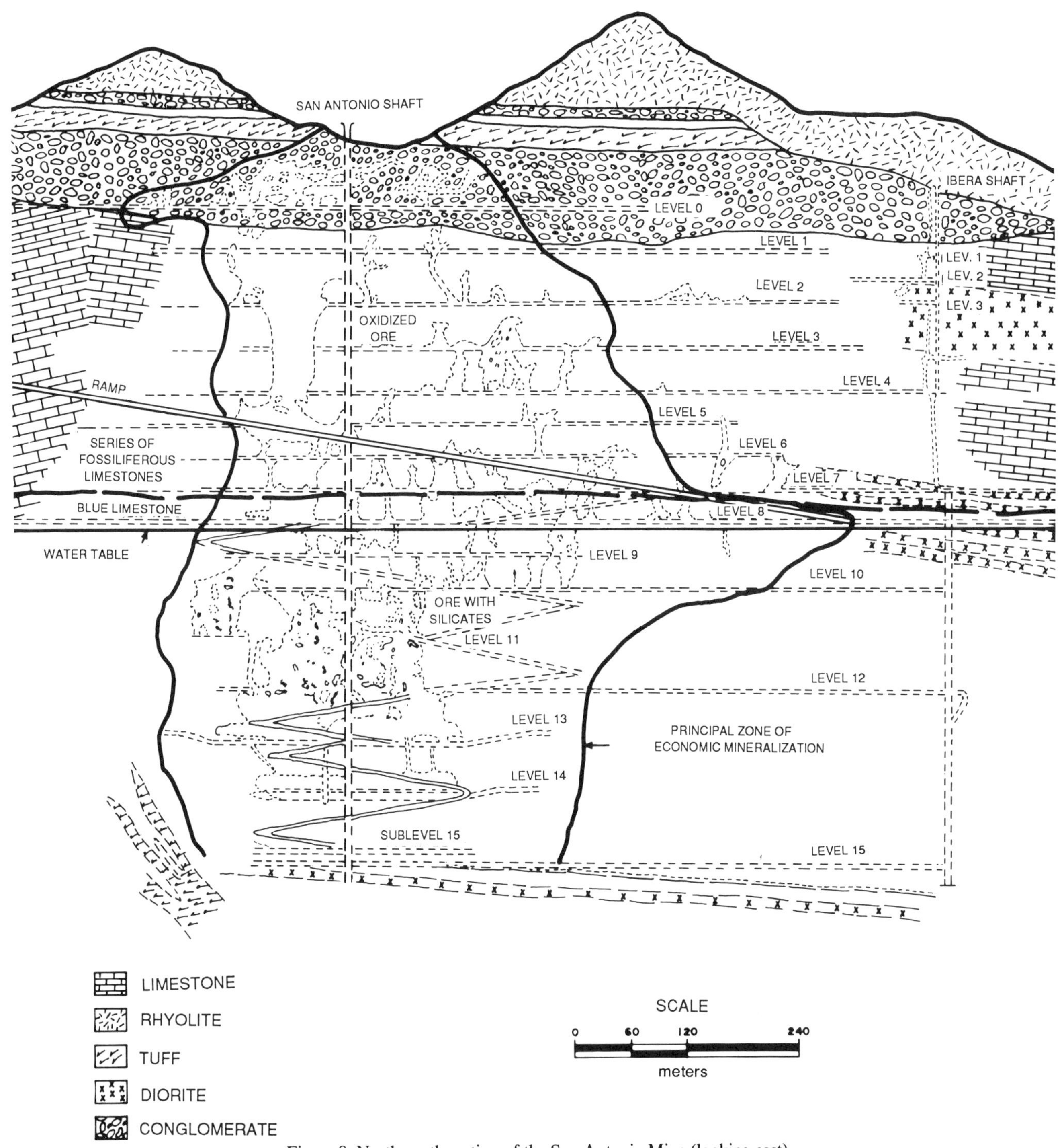

Figure 8. North-south section of the San Antonio Mine (looking east).

the following isotope values: +17 in Cuchillo Formation sulfides, –27 in Black Limestone pyrite, +4 in diabase pyrite, –7 to –12 in West Field sulfides, –4 to +2 in San Antonio chimney tactites, and –7 to –9 in San Antonio central manto sulfides.

These figures suggest that the sulfur in the Cuchillo Formation sulfates was not remobilized to form the sulfides and other minerals in the Santa Eulalia District tactites. It may also be assumed that the sulfides in the different West Field mineralized structures belong to the same mineralization stage and source. Likewise the San Antonio central manto massive sulfides were presumably derived from a different source and belong to a mineralization stage other than that of the tactite-hosted bodies (chimneys) in the same area.

Temperature studies of fluid inclusions in fluorite, quartz,

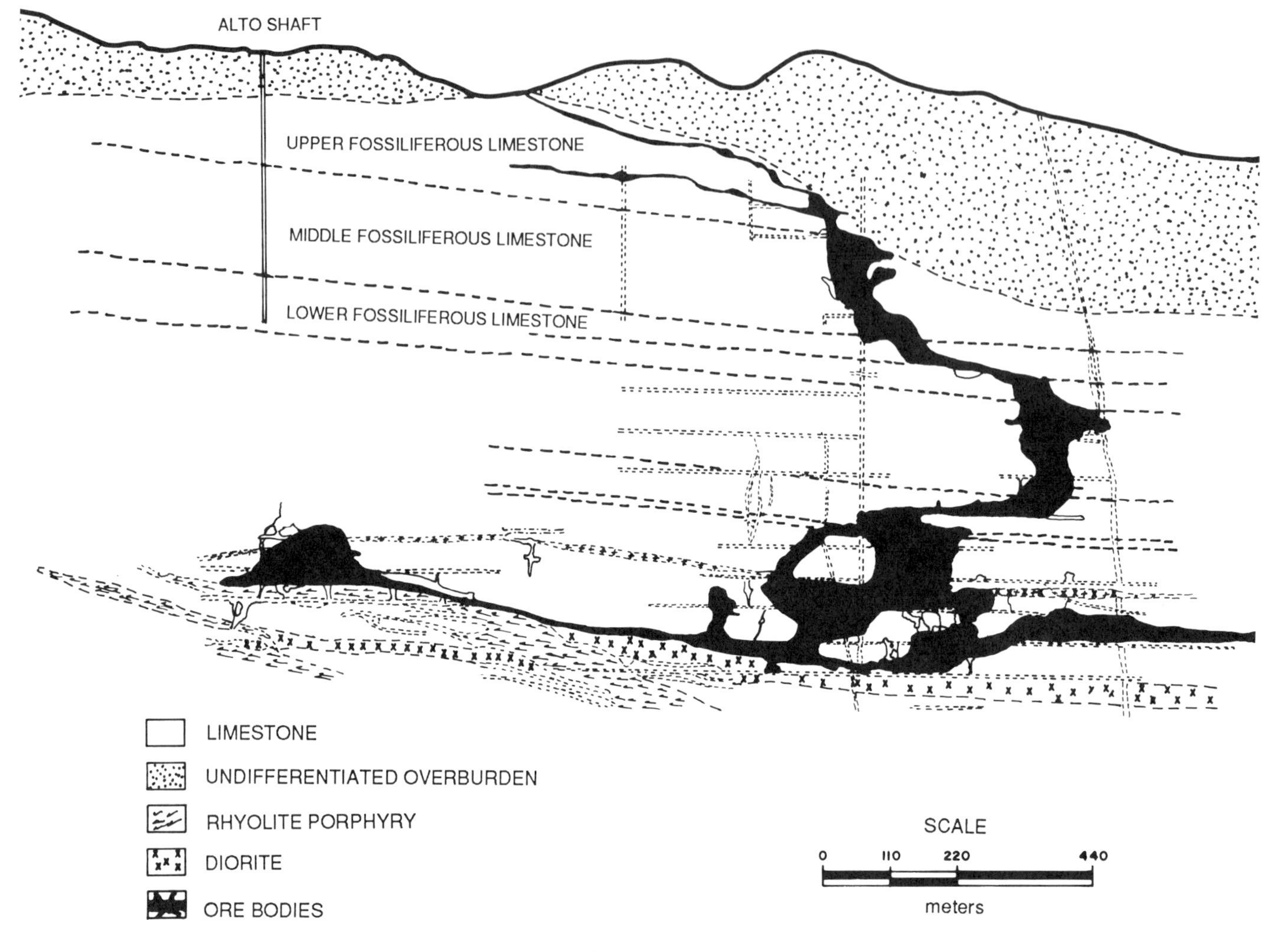

Figure 9. Generalized longitudinal section of the Q-Denis–Tiro Alto structure in the West Field.

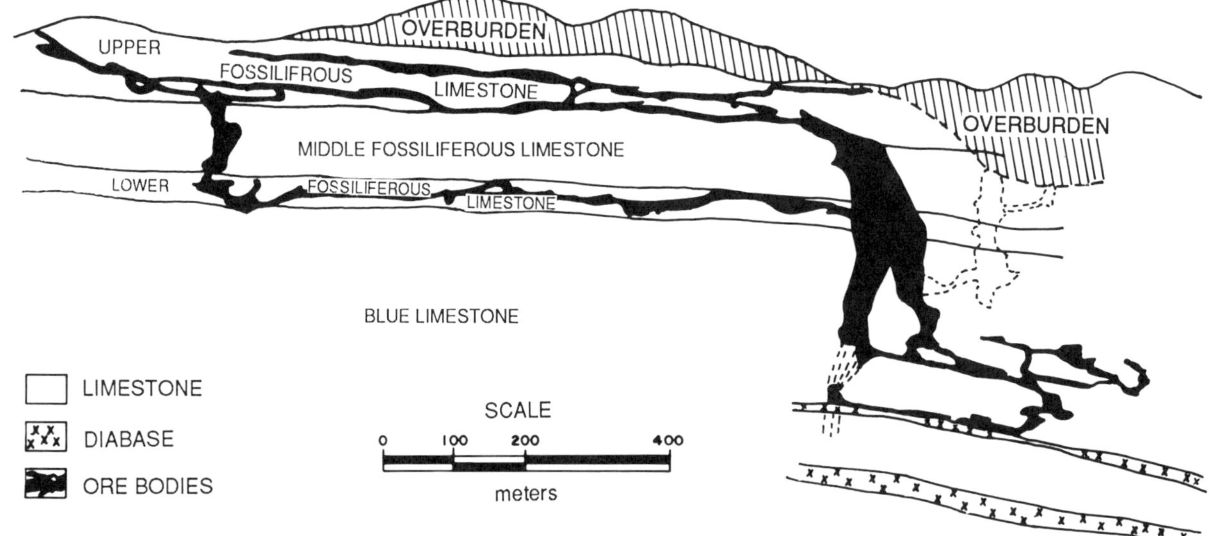

Figure 10. Longitudinal section of chimneys and mantos in the West Field.

TABLE 1. HISTORIC PRODUCTION OF THE SANTA EULALIA DISTRICT, 1912–1984

Mine	Type of mineral	Period	Tons	Au (g/ton)	Ag (g/ton)	Pb (%)	Zn (%)	Sn (%)	V (%)
Mina Vieja	Oxides and carbonates	1912–1961	1,800,000	....	400	2.5	1.5	....	....
Velardeña	Oxides	1931–1964	500,000	....	300	4.0	2.0	....	....
Reina de Plata	Oxides and carbonates	1935–1945	100,000	1.0	500	1.8	2.5	....	....
San Antonio	Smelting zinc	1915–1917	5,000	....	....	....	33.0	....	....
San Antonio	Smelting lead	1912–1971	300,000	....	434	18.2	7.0	....	....
San Antonio	Tin	1930–1940	440,535	....	389	15.6	4.4	1.22	....
San Antonio	Vanadium	1935–1940	70,748	....	102	11.0	6.4	0.30	1.0
San Antonio	Pb and Zn	1912–1983	4,000,000	....	122	2.9	9.5	....	....
Buena Tierra	Pb and Zn	1935–1984	3,000,000	....	200	7.0	6.0	....	....
Potosí	Pb and Zn	1927–1940	5,000,000	....	250	10.0	10.0	....	....
Potosí	Pb and Zn	1941–1984	7,000,000	....	150	6.0	5.0	....	....

and massive sulfides of the San Antonio mantos and chimneys establish a 200 to 400°C temperature range of formation, which indicates a magmatic-hydrothermal origin, although the presence of tactites demands the additional intervention of metasomatic processes.

Rare-earth studies of the district intrusives show no genetic relation between the rhyolites and diabases of both fields nor between the latter and the granodiorite. This suggests that different magmatic chambers generated different intrusive rocks, spatially related with mineralizing fluid emissions throughout the district.

Complementary studies are under way to define temperature distributions, index-element patterns, ages, and stable-isotope relations, which could help explain the characteristics of Santa Eulalia–type deposits.

ECONOMIC GEOLOGY

The East and West Fields contain orebodies whose characteristics make them different. A series of chimneys controlled by rhyolite dikes emplaced along staggered faults occurs in the East Field. These tabular bodies, including oxides and sulfides, range in thickness from a few to 45 m over a distance of 1 km and extend vertically down to 600 m. Present diamond-drill exploration confirms the continuity of these bodies at depth both north and south of the San Antonio Mine.

Mineworks to develop chimneys are typical in the West Field. The Chorro chimney was one of the largest deposits in the district (700 m high, 200 m long, and as much as 80 m wide). Other replacement bodies associated with the J-Norte, Peñoles, Tiro Alto, Q-Denis, and 90 faults containing large commercal ore volumes have made this the most important field of the region in the past. The West Field also contains the largest mantos—horizontally continuous over 4 km—in the Upper and Lower Fossiliferous Limestones as shown on Figure 10, which also illustrates the close relation between mantos and chimneys.

Replacement veins are less common, though still present, in both fields; examples are the East Field Potosí vein and the West Field M vein. The latter is important because of its different mineralization from that of the nearby chimneys and represents an attractive objective for future exploration, not being well known at depth.

The district has been mined since its discovery in 1591; production declined comparatively between 1790 and 1880 owing to the minework depths, the relatively scarcity of oxides and the country's constant social, political, and economic problems.

Several foreign-capital companies arriving in the district in 1880 launched another period of prosperity; the Potosí Mining Company (present-day Minerales Nacionales de México, S.A.) and A.S.E.R. Co. (today's Industrial Minera México, S.A.), began operations at that time, and at present control practically all the important activity in the area.

Silver-bearing oxidized ore was mined in both fields during the 1707–1790 period. The Chihuahua Royal Exchange recorded 11,902,626 silver marks (~8 oz), although according to several authors a large part of the ore was smuggled out to evade the heavy taxes of the time. Considering the average grade of the 6 to 8 ounces per load or 1,500 g/ton (1 oz/load = 225 g/ton) the value would represent 2,000,000 tons of mineral. It should be noted, however, that since the silver content was improved by adding cobbing within and outside the mines, the extracted mineral is estimated to have been twice that amount. Taking into account that the dollar's original silver content equaled that of the silver "duro" or "peso," the production of the period represented roundly 100,000,000 dollars of that time.

Silver production during the decline of mining activity (1790–1880) is estimated in about 20,000,000 dollars.

Production figures from 1881 to 1920 are insufficient in spite of the heavy activity during that period. The known partial records and the history of contemporary mines suggest an extraction of about 7.5 million tons of raw mineral averaging 500 g/ton of silver and 12 percent lead.

Production for the 1912–1984 period, from the largest producing mines only, is shown on Table 1.

According to figures published by De la Fuente L. (1984), the district's estimated production throughout the years has totalled 37,000,000 metric tons of raw mineral from which 396,000,000 oz of silver, 2,760,000 metric tons of lead, and 1,862,000 metric tons of zinc have been extracted, averaging 320 g/ton of silver, 8.0 percent lead, and 7.1 percent zinc, in addition to 5,000 metric tons of copper, 4,000 metric tons of tin, 700 metric tons of vanadium, and 1 metric ton of gold.

Mines presently operating in the primary sulfides zone are San Antonio and Buena Tierra (Industrial Minera México, S.A.) and Potosí (Minerales Nacionales de México, S.A.).

The most important is San Antonio (East Field) with a daily production of 750 tons averaging 112 g/ton of silver, 3.0 percent lead, 6.8 percent zinc, and 0.15 percent copper; iron content is 11 percent. The mine's main shaft collar is at an elevation of 1,600 m; 16 levels (0 to 15) comprise a total depth of 674 m; there are two auxiliary shafts and two Robbins raises for ventilation. A 1,600-m-long access ramp runs from the surface to the lower drifts, in addition to a 1,000-m-long interior ramp connecting drifts 12 and 15. Other access ramps are being continued to connect the important areas.

Exploitation is by the cut and hydraulic fill method and poses serious problems because of the large amounts of water under high pressure in the rocks and even in the orebodies; most of the mining operations are carried out beneath the local groundwater level (1,121.94 m). Average pumping is 1,400 gpm, which must be considerably increased by additional workings—presently under way—in order to continue operating.

The San Antonio Mine's reserves of exploitable and prospective ore are 9,500,000 tons containing 107 g/ton of silver, 2.2 percent lead, 7.1 percent zinc, and 0.18 percent copper.

The West Field Buena Tierra Mine is practically paid out. Elevation of the main shaft collar is at 1,859 m; 21 drifts—some of which connect with the Potosí Mine—comprise a total depth of 830 m. Present exploitation is by the room-and-pillar method in small mantos related to the already mined large bodies and by shrinkage stoping in narrow almost vertical mineralized structures. Remaining reserves are about 60,000 tons in small bodies.

The Potosí Mine—also in the West Field—is the second most important mine in the district. Elevation of the main shaft is at 1,859.7 m; 22 drifts comprise 970 m of total depth; there are four auxiliary shafts and the combined horizontal mine-works cover more than 300 km. Daily production since 1941 is 550 tons from mantos, a small tonnage as compared to the previously mined large bodies. No exact reserve figures are available, but there seems to be no problem in continuing at the present production rate.

In the present century the state of Chihuahua has been the foremost domestic producer of silver, lead, and zinc, to which Santa Eulalia has contributed considerably. In addition the district mines have been the region's main source of employment for many generations.

REFERENCES CITED

Burt, D. M., and others, 1982, Topaz rhyolites; Distribution, origin, and significance for exploration: Economic Geology, v. 77, p. 1818–1836.

Clark, K. F., and others, 1979, Magmatismo en el Norte de México en Relacion a los Yacimientos Metaliferos: 13th Convencion Nacional Asociación de Mineros, Metalurgistas y Geólogos de México Memoir, p. 8–57.

De la Fuente L., F. E., 1969, Geologia de la Mina El Potosí, Distrito Minero de Santa Eulalia, Chihuahua [Professional thesis]: Universidad Nacional Autonoma de México, 138 p.

Hewitt, W. P., 1943, Geology and mineralization of the San Antonio Mine, Santa Eulalia District, Chihuahua: Geological Society of America Bulletin, v. 64, p. 173–204.

Megaw, P.K.M., and McDowell, F. W., 1983, Geology and geochronology of volcanic rocks of the Sierra de Pastorias area, Chihuahua, Mexico, *in* Geology and mineral resources of north-central Chihuahua: Guidebook of the 1983 Field Conference, p. 195–203.

Torres A., J. M., and Muñoz C., F., 1981, Mineragrafía de las Calizas de Mina Vieja de Santa Eulalia, Chihuahua: IMMSA Internal Report, 7 p.

Printed in U.S.A.

The Geology of North America
Vol. P-3, Economic Geology, Mexico
The Geological Society of America, 1991

Chapter 27

Geology and mineralization of the La Encantada Mining District, Coahuila

B. Solano R.
Cía. Mineras Peñoles, S.A. de C.V., Río de la Plata 48 esq. Lerma, Col. Cuauhtemoc, 06500 México, D.F.

INTRODUCTION

The La Encantada District is located in the Ocampo municipality, Coahuila, at coordinates 28°22′N and 102°35′W, at an elevation of 1,800 m. The nearest important town is Múzquiz, Coahuila, 120 km directly southeast on the Múzquiz–Boquilla del Carmen road, which is paved for 100 km and earthworks for another 120 km; bypaths connect La Encantada with Ocampo, Coahuila (160 km), and Ciudad Camargo, Chihuahua (Fig. 1); a light-aircraft runway is 20 km away.

La Encantada is one of the few mining districts in Mexico that was not discovered during the Spanish colonization. The first indications were found by local people in 1956, and Cía. Minera Los Angeles immediately began to mine the San José, Guadalupe, La Escondida, and San Francisco orebodies; La Prieta, the best producer to date, was discovered in 1963.

La Encantada, S.A., began operations in the area in 1967, installing a magnetic separation plant in 1973 that was later substituted by a flotation plant (1978). Two important discoveries are the "660" contact orebody and the high-grade mantos zone between levels 635 and 710.

Total tonnage in the known orebodies is about 4.0 MMT, averaging 450 g/ton Ag and 12 percent Pb (Rico, Aguilera, 1983). Mineral extracted by La Encantada, S.A., exceeds 1.8 MMT, averaging 350 g/ton Ag and 5.0 percent Pb (Trejo, de la Cruz, 1983).

GENERAL GEOLOGY

The area is characterized by a sequence of Cretaceous limestones and sedimentary detrital rocks intruded by bodies of diorite to granodiorite composition locally emplaced along late Oligocene faults and fractures. Less extensive metamorphic rocks include marble, skarn, and hornfels derived from calcareous and pelitic sediments, respectively, which occur in association with the granodiorite intrusion.

La Encantada lies structurally on the east flank of a regional anticline affected by mostly northwest-trending faults (Fig. 2).

Stratigraphy

Regionally the stratigraphic sequence present in the area consists of upper Lower Cretaceous to lower Upper Cretaceous calcareous and argillaceous rocks of the Cupido, La Peña, Aurora, Cuesta del Cura(?), Del Río(?), and Buda(?) Formations, briefly described below from older to younger.

Cupido Formation (upper Hauterivian–Barremian). The Sierra La Vasca, 30 km northwest of La Encantada, is a series of thin, massive limestones, dolomitic limestones, and dolomites, whose upper contact (Cupido–La Peña) is transitional and thus difficult to place. The transgressive Cupido Formation was probably deposited on the irregular surface of an extensive paleoplatform.

La Peña Formation (Aptian-lower Albian). At the Sierra La Vasca, this formation consists of a lower member of interbedded shales, laminar argillaceous limestones, and thinly bedded fossiliferous limestones; a middle member of thickly bedded black dolomites alternating with dolomitic limestones at the base and top; and an upper sequence of alternating shales, siltstones, laminar argillaceous dolomites, and black dolomitic limestones.

At the mine area the unit is found in drillholes below level 635 and contains thin black shales interbedded with bituminous and carbonaceous black limestones of reducing depositional environment at the water/sediment interface.

Aurora Formation (lower and middle Albian). This is the most widespread unit of the area and contains most of the orebodies. Light gray, thickly bedded to massive cryptocrystalline to microcrystalline limestones show a change in facies at the base, grading to a more near-coast pelitic-calcareous sequence to the north and northeast, corresponding to the Glen Rose Formation.

Cuesta del Cura Formation(?) (middle Albian–lower Cenomanian). A sequence of thin- to medium-bedded oolitic limestones containing abundant chert nodules, especially toward the top, is tentatively correlated with the Cuesta del Cura Formation. The upper part of the unit is either eroded or covered by talus; its thickness is therefore estimated to be 275 to 355 m,

Solano R., B., 1991, Geology and mineralization of the La Encantada Mining District, Coahuila, *in* Salas, G. P., ed., Economic Geology, Mexico: Boulder, Colorado, Geological Society of America, The Geology of North America, v. P-3.

increasing southeast of the area. The lower contact with the Aurora Formation is hard to locate owing to the lithologic similarities of the formations and has been placed here at the point where the chert nodules disappear. According to Smith (1970) this sequence may well be included in the Aurora Formation.

Del Río Formation(?) (upper Albian-lower Cenomanian). Very restricted in the area—it occurs only as small erosional valleys at the flanks of the anticlines—this sequence of microcrystalline limestones, argillaceous limestones, shales, and calcareous sandstones in thin and laminar beds is mostly covered by talus and is estimated to be about 45 m thick.

Buda Formation(?) (lower Cenomanian). Also very restricted, the unit is a light gray limestone, medium bedded at the base and laminar toward the top, with scarce chert nodules. It is 62.0 m thick in the Sierra Chilicote, and 105.5 m in the Sierra La Encantada.

Igneous rocks

Several intrusive rocks of different compositions and textures are present in the area. A green holocrystalline porphyritic rock with a fine-grained phaneritic groundmass consisting of feldspars, ferromagnesian minerals, pyrite, and small epidote-filled fractures occurring at depth beneath the tactite zone is classified as altered granodiorite porphyry of as yet undefined distribution and geometry. However, magnetometric and drill-hole data suggest its considerable size and more or less horizontal attitude. A second type of porphyry—a light gray rock with an aphanitic groundmass composed of feldspars, quartz, scarce mafics (hornblende), and disseminated pyrite—is classified as rhyolite porphyry and considered to be the probable hypabyssal equivalent of the granodiorite porphyry.

A third type of porphyry is a green rock with Na-plagioclases as main components, secondary biotite-chlorite, and accessory opaque minerals; it is classified as andesite porphyry.

The fourth intrusive type is greenish gray to dark green phaneritic holocrystalline rock composed of plagioclases and mafics (mostly hornblende) and classified as diorite.

The two last occur mainly as dikes in association with the intrusive bodies at depth and are presumably contemporaneous.

Metamorphic rocks

The granodiorite/Cretaceous sedimentary contact is characterized by a diopside-garnet tactite usually consisting of a granoblastic diopside and garnet mosaic with occasional subordinate quartz, plagioclases, tremolite, actinolite, calcite, magnetite, and commonly 5 to 15 percent of fluorite. The rock was presumably

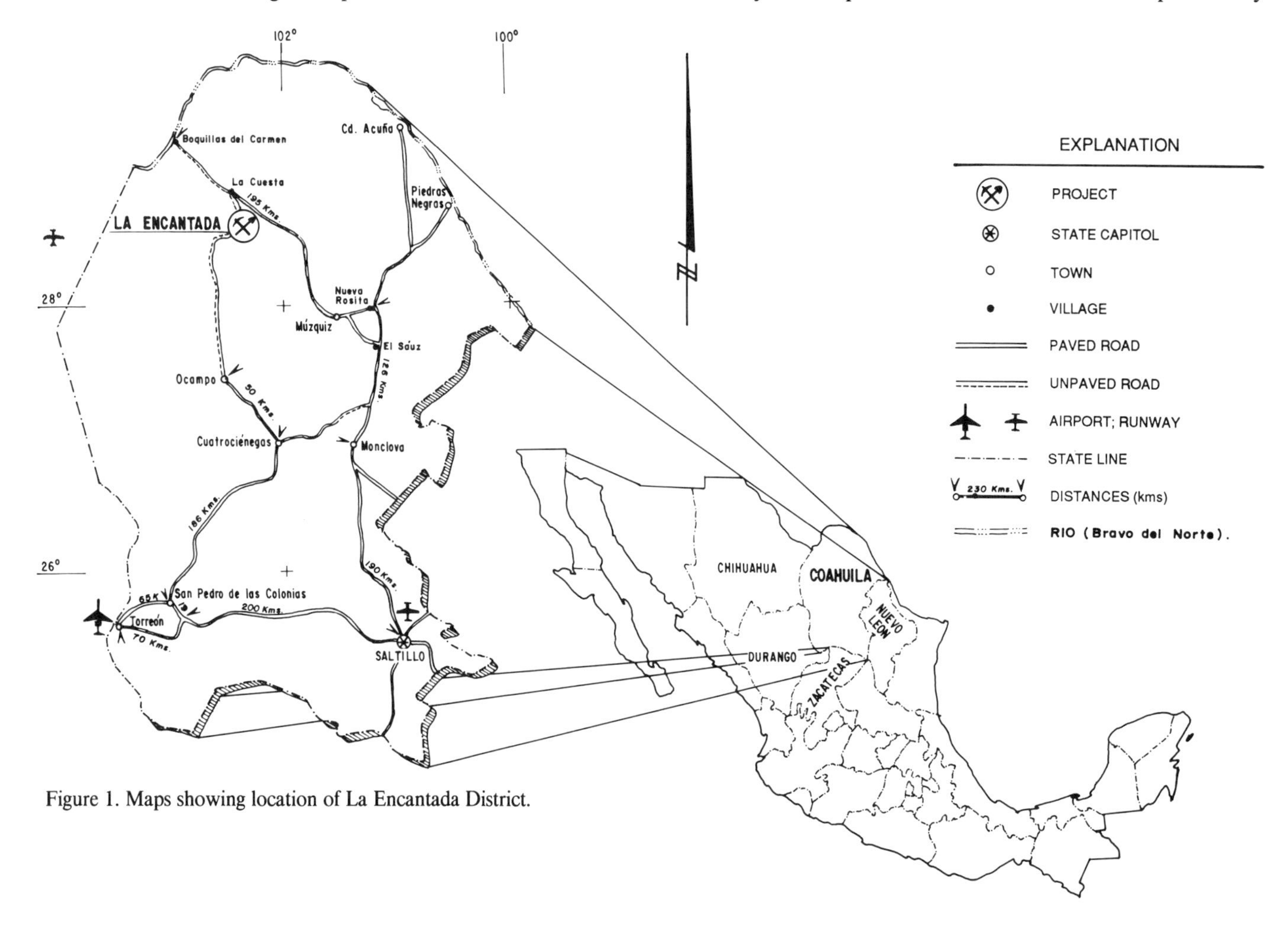

Figure 1. Maps showing location of La Encantada District.

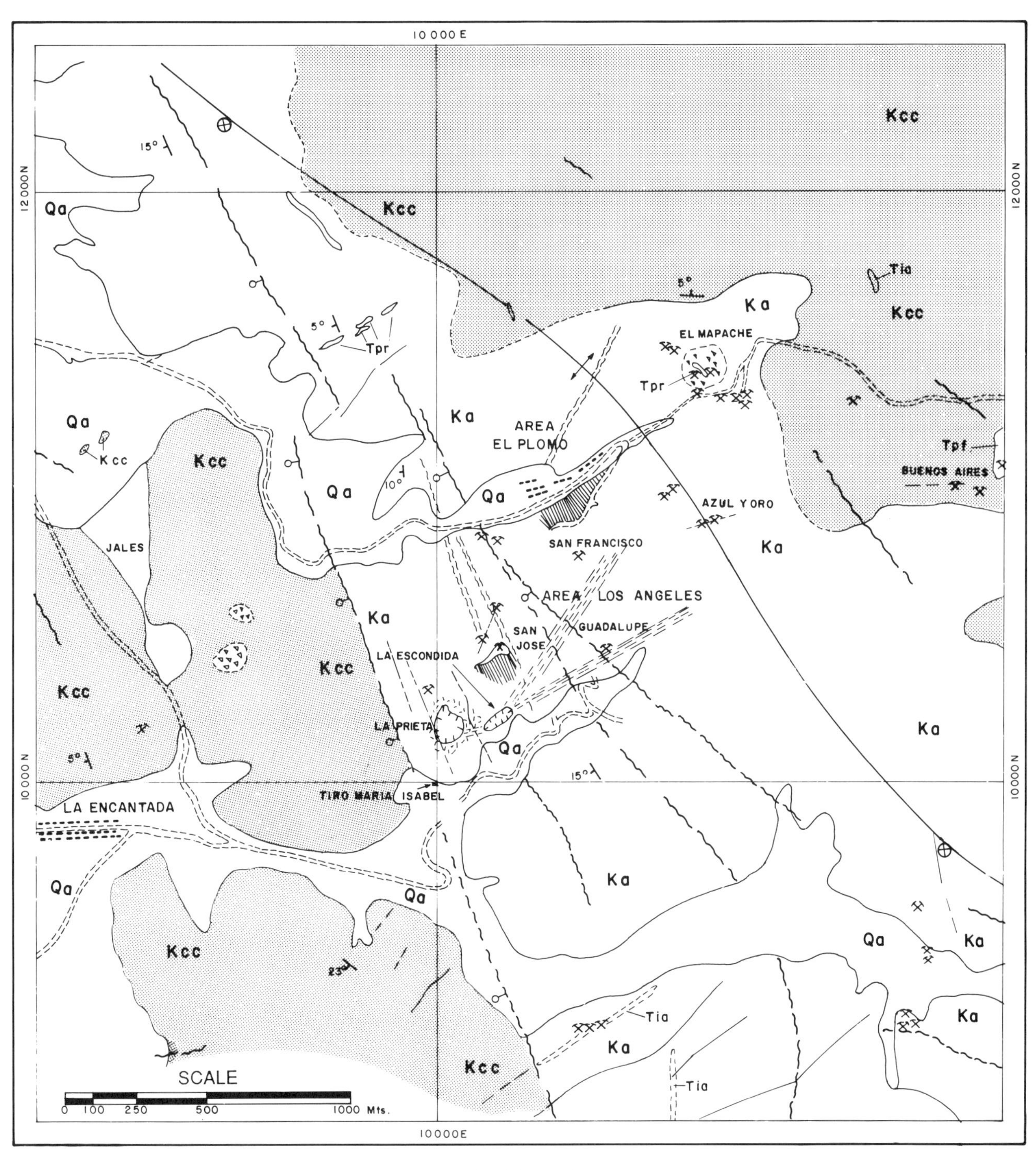

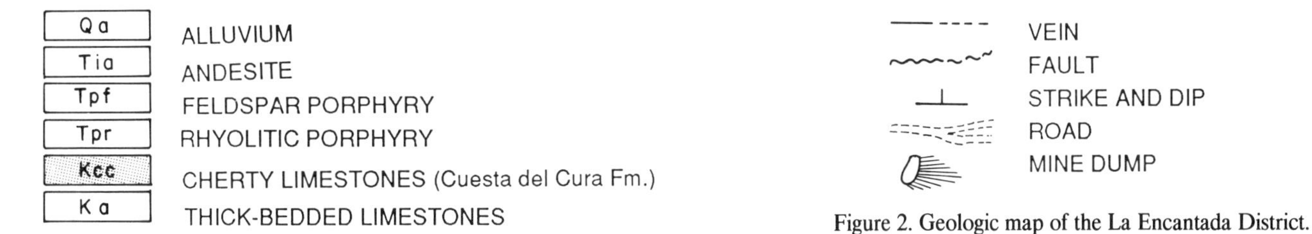

Figure 2. Geologic map of the La Encantada District.

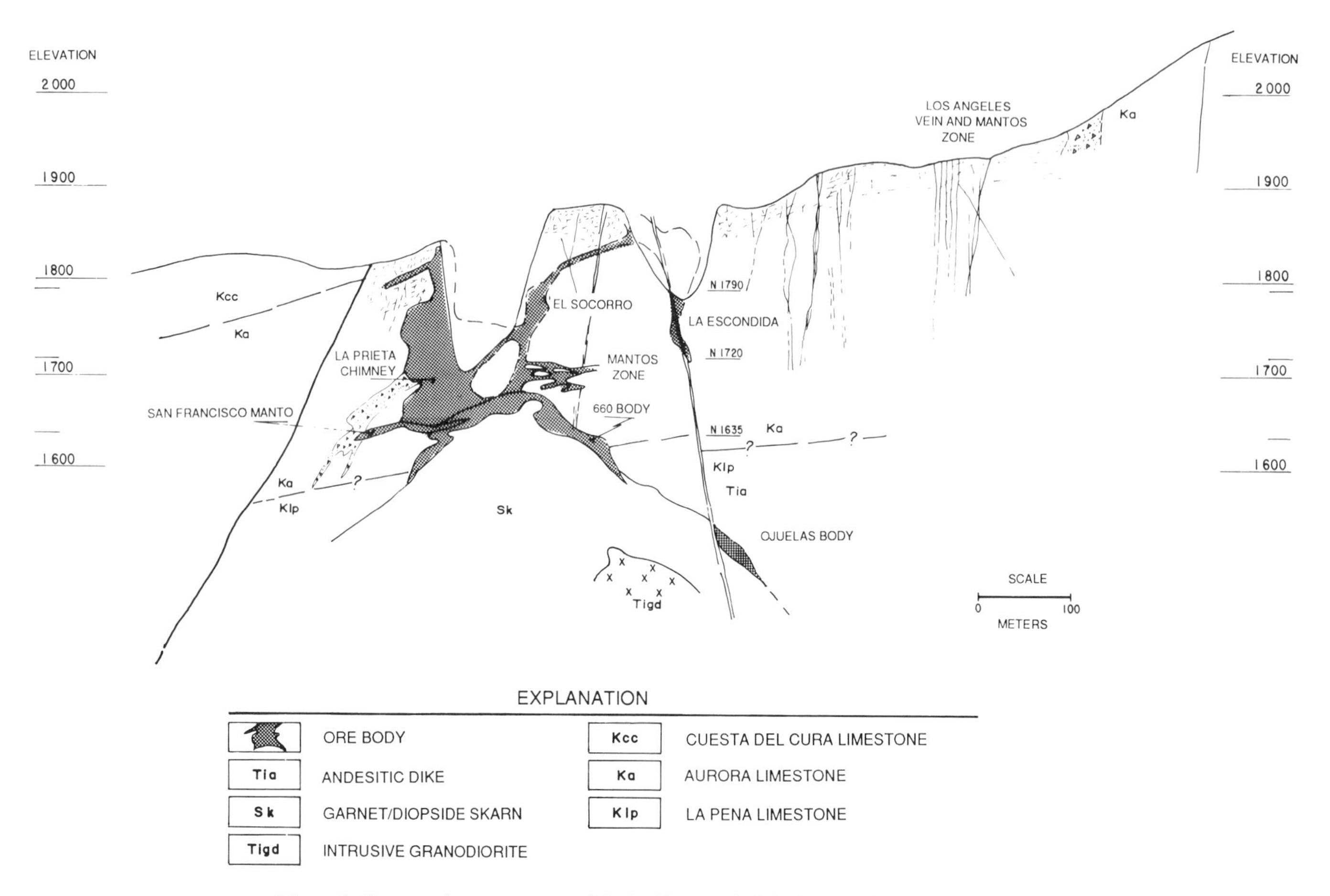

Figure 3. Cross section across part of the La Encantada District, looking to the northwest.

derived by contact metamorphism at less than 800°C from an older calcareous-dolomitic, locally argillaceous sequence. According to fluid inclusion determinations, the fluorite mineralization developed at temperatures of 360 to 370°C (Kesler, 1974, *in* Luján-Garcia, 1975).

Structures

The La Encantada Mine is on the southeastern flank of a wide, northwest-trending anticlinorium truncated by a regional parallel fault (La Encantada fault). Beds trend generally northwest with slight dips, which locally may be up to 40° southwest. Most of the faults in the district are normal and appear to be of two main types: either longitudinal, paralleling the structural axes developed by post-folding tension, or perpendicularly crosscutting the axes. The mineralized structures are mostly aligned N60° to 80°E (known chimneys) and N30°W (some of the veins); the veins appear to be younger, although both were premineral. Contrasting with the weak regional deformation caused by the intrusives, the granodiorite intrusion developed a dome-shaped skarn structure in the lower part of the La Prieta chimney. Deep drilling and mine development reveal the tactite dome as a northeast-oriented semi-elliptical structure almost 200 m in the vertical dimension at its widest part and with steep flanks; it adopts a subhorizontal attitude at approximately 1,400 to 1,450 m above sea level. Also worth noting is the fact that the limestones are more highly fractured in the areas of the main orebodies; a concentric or radial structural pattern is developed symmetrically with respect to the tactite dome's approximate center.

MINERALIZATION

Description of the deposits

The various types of ore bodies are chimneys, irregular contact bodies, mantos, and veins (Fig. 3), in order of importance.

Chimneys are bodies of almost elliptical cross section and considerable vertical development: examples are La Prieta, El Socorro, and La Escondida. The La Prieta chimney tops at level 790 and widens at depth to a maximum northeast-oriented 130 by 75 m (level 720); the section decreases to the bottom at level 630 over a total 160 m of vertical development. The structure is characterized by a high-grade central core, an Fe-rich surrounding halo with lesser Ag values, and is positioned above the tactite dome described above. Original tonnage was an estimated 1.7 MMT, averaging 646 g/ton Ag and 23.5 percent Pb.

Contact bodies are located at the tactite dome's outer contact with limestone and/or marble: 660 and Ojuelas bodies are examples. 660 is a tabular body flexing above the contact, with 290 m of maximum longitudinal development between elevations 1680 and 1580; it is connected locally with the La Prieta chimney root at level 635. 1.3 MMT of 418 g/ton Ag and 5.8 percent Pb were originally estimated. Sulfide mineralization—mainly marmatite—occurs at the base of 660 within the tactite.

Mantos are tabular bodies that follow defined horizons or the general trend of the beds. Orebodies of this type were mined initially in the Los Angeles, San José, Guadalupe, San Juan, El Refugio, Esperanza, and Fortuna areas. Since 1981, however, the so-called "mantos zone" adjacent to the La Prieta chimney between level 635 and level 710 has been under development. The Los Angeles mantos are found above 1,900 m elevation, whereas the mantos zone occurs in the interval between 1635 and 1760 m. The Los Angeles mantos show strong brecciation and filling features whereas replacement was possibly more important in the case of the mantos zone. In both zones, irregular tabular bodies are commonly found interconnecting mantos at different elevations to develop a complex grid around limestone centers. Selective exploitation of the mantos zone produces oxidized ores averaging 1,250 kg/ton Ag and 20 percent Pb.

Veins occur mostly in the area known as El Plomo in the northeast part of the district. Outstanding among these is the San Francisco vein, lodged along a northeast to southwest fracture extending to the La Escondida chimney.

Mineralization

The La Encantada Ag-Pb deposits are characterized by a complex mixture of Fe, Ag, Pb, and Zn oxides of secondary origin denoting the strong supergene alteration of the original primary minerals. The most common oxides are hematite, goethite, limonite, and cerussite with lesser amounts of anglesite, hemimorphite, smithsonite, and rare native silver and cerargirite. Relict pyrite, galena, marmatite, and proustite-pyrargirite are found locally.

Geological Model

The various types of La Encantada ore deposits possibly form part of a tectonic and thermodynamic model, in which the intrusion caused mineralogical, textural, and deposit-type zoning to develop as follows:

Deep zone. Contact bodies with skarn and hornfels development; contact metasomatism (660 and Ojuelas orebodies).

Intermediate zone. Partial replacement and fill bodies: chimneys and mantos of La Prieta—and mantos zone—types, respectively, probably interconnected and serving as "connecting vessels."

Upper zone. Characterized by mantos, veins, and fracture-fill zones probably reflecting deep and/or laterally mineralized structures.

REFERENCES CITED

Hunter, N. J., 1972, Regional geology and environment of ore deposition at La Encantada mine, Coahuila, Mexico: Tormex, S. A., Internal Report, April 1971.

Lozano Chavez, G., 1981, Reconocimiento Estratigráfico del Area "La Encantada" Mpio, de Ocampo, Coah: SIPSA de CV, Internal Report, June 1981.

Lozej Gian, P., and Beales, F., 1977, Stratigraphy and structure of La Encantada Area, Coahuila, Mexico: Geological Society of America Bulletin, v. 88, p. 1793–1803.

Luján-Garcia, M., 1975, Geología e Interpretación de los cuerpos de plomo-plata de La Encantada, Coahuila: 11th Convencion Nacional Asociación de Mineros, Metalurgistas y Geologos de México Memoir.

Mining Magazine, 1980, La Encantada: Mining Magazine, p. 12–23.

Rico Aguilera, J. J., 1983, Explotación de Minerales de Alta Ley en La Encantada: 15th Convencion Nacional Asociación de Mineros, Metalurgistas y Geologos de México Memoir.

Smith, C. I., 1970, Lower Cretaceous stratigraphy, northern Coahuila, Mexico: Texas Bureau of Economic Geology Report of Investigations 65, 101 p.

Solano-Rico, B., and Flores-Saucedo, G., 1982, Reporte Geológico Superficial de La Encantado, Coahuila: SIPSA de CV Unidad David Contreras C, Internal Report, June 1982, 19 p.

Trejo de La Cruz, P., 1982, El Modelo de Mineralización en La Encantada, Coahuila: La Encantada, S. A., Internal Report.

Printed in U.S.A.

The Geology of North America
Vol. P-3, Economic Geology, Mexico
The Geological Society of America, 1991

Chapter 28

Geology and genesis of the Naica mineral deposits, Chihuahua

H. A. Palacios M., F. Querol S., and G. K. Lowther
Cía. Fresnillo, S. A., de C. V., Río de la Plata 48 esq. Lerma, Col. Cuauhtemoc, 06500 México, D.F.

INTRODUCTION

The Naica mining district is located in the municipality of Saucillo in south-central Chihuahua State, 110 km directly southeast of Chihuahua, the state capital, at an altitude of 1,382 m and geographic coordinates 27°52′00″N, 105°26′15″W (Fig. 1). Access to the district is by a 26-km-long paved road joining the Panamerican Highway at Conchos Station and another road, also paved, that runs for 40 km to Ciudad Delicias; the México–Ciudad Juárez Central Railroad stops at Conchos Station.

The first mining concession in the district was granted in 1794 (Aldama, 1945). The deposits were discovered by prospectors, and small-scale mining began in 1828, during the rainy seasons only owing to the scarcity of water (Lambert, 1892). In the last years of the nineteenth century the Compañía Minera de Naica, S. A., began industrial operations, which were later (1911) suspended when the mine reached ground-water level, and also because of the country's political instability at the time. In 1924 several companies renewed the extraction of oxidized ores. This work continued until 1951, when the Fresnillo Company acquired the Naica Mine and installed electric pumps and a 400 ton/day sulfide-processing flotation plant (in 1952), which doubled production capacity the following year. In 1956 the Fresnillo Company acquired all the deposits in the district and is presently the only producing company under title of Compañía Fresnillo, S. A. de C. V.; since that date the plant capacity gradually increased to 3,000 tons/day in 1985.

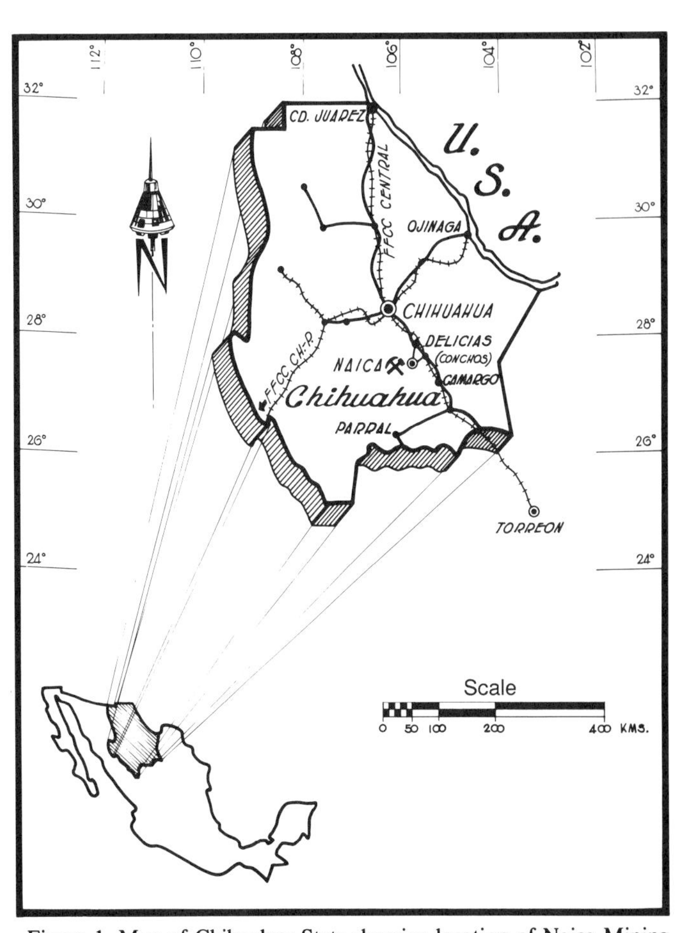

Figure 1. Map of Chihuahua State showing location of Naica Mining District.

REGIONAL GEOLOGY

Physiography

In the Raisz (1959) classification the area lies in the Basin and Range Province where small, northwest-oriented, isolated ranges stand out in extensive plains or bolsons serving as catchment basins.

The mine is located on the northeast flank of a 12-km-long by 7-km-wide dome, the Sierra de Naica, elongated northwest-southeast and affected by small secondary folding, faulting, and erosion. This actually comprises the three smaller ranges, the Sierra de La Mina, Sierra de Enmedio, and Sierra del Monarca (Fig. 2). The physiographically mature system, as a whole, has an average elevation of 1,700 m; parallel drainage develops alluvial fans at the mouths of several streams.

Stratigraphy

The Sierra de Naica consists almost entirely of Albian sedimentary rocks overlying the Aptian Cuchillo Formation evapo-

Palacios M., H. A., Querol S., F., and Lowther, G. K., 1991, Geology and genesis of the Naica mineral deposits, Chihuahua, *in* Salas, G. P., ed., Economic Geology, Mexico: Boulder, Colorado, Geological Society of America, The Geology of North America, v. P-3.

ritic sequence, which has not been detected in the district, although it has been extensively described in neighboring areas (De Ford and Haenggi, 1970; Hewitt, 1943). The following units are present in the area:

Aurora Formation. This formation was described by Franco (1978) and appears at the surface as a 370-m-thick, medium- to mostly thick-bedded limestone. The lower contact is not exposed, but more than 1,000 m have been drilled. In outcrop the unit consists of yellowish gray to dark gray oolitic calcarenites with fine- to coarse-grained bioclasts and intraclasts and occasional chert lenses and nodules, intercalated with bioclastic micrites and micrites having light to dark gray bioclasts and abundant chert. The lower part of the formation is highly recrystallized and within the mine is totally marbleized.

The unit is correlated with the Edwards Formation of south Texas (Franco, 1978) on the basis of its upper Albian microfau-

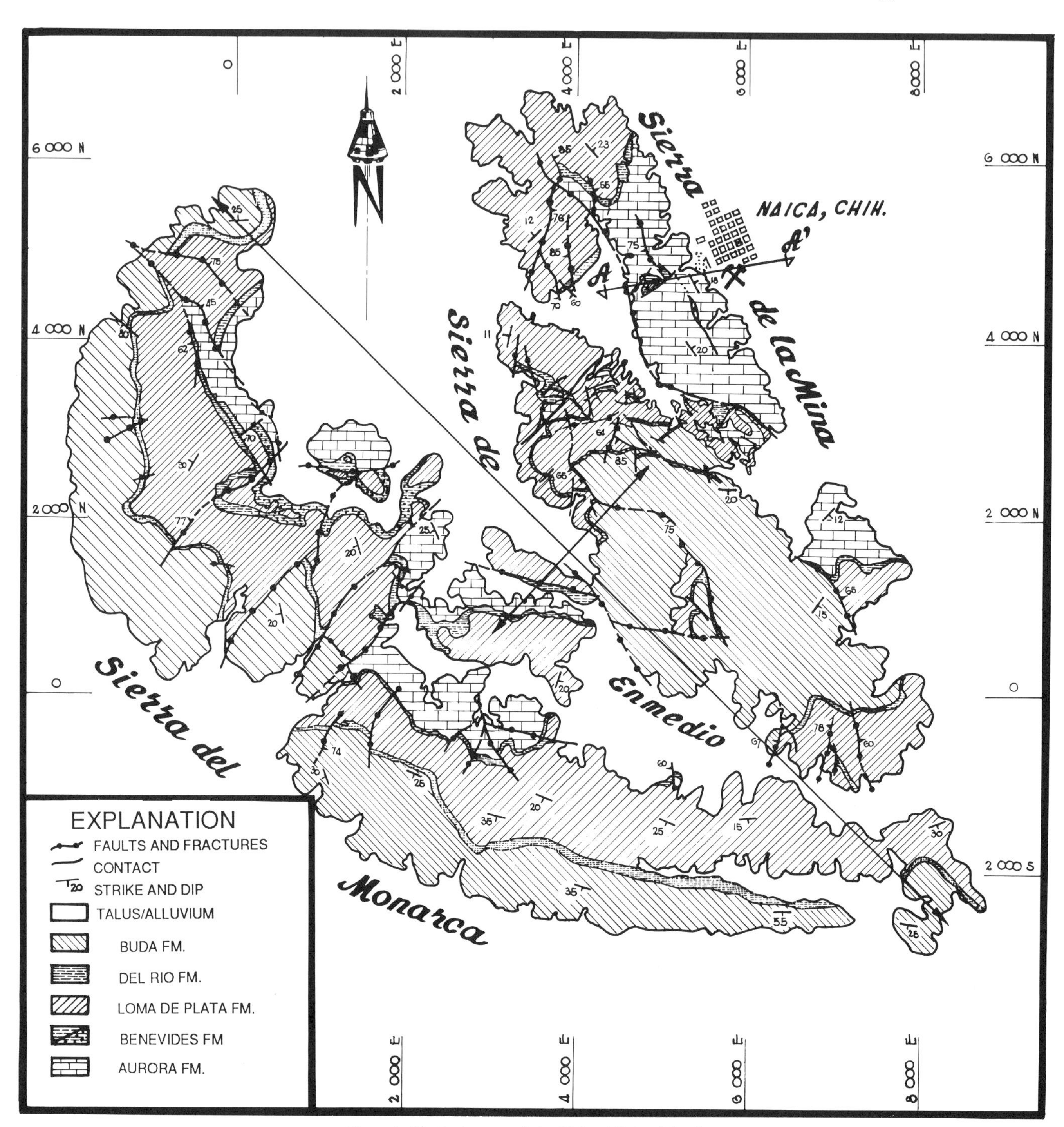

Figure 2. Geologic map of the Naica Mining District.

nal content. Wilson (1956) divided the Aurora Formation at the mine into three members—white to light gray marbleized limestone, medium to dark gray marbleized limestone, and nonmarbleized limestone—according to the rock's response to metamorphism and metasomatism caused by mineralization, a division that relates directly to the mineralization process rather than to well-defined lithological differences.

Benevides Formation. This formation conformably overlies the Aurora and consists of dark gray, ochre-weathering calcareous shales ranging 30 m (Sierra de La Mina) to 60 m (Sierra del Monarca) in thickness, with disseminated pyrite and abundant macrofossils (*Mortoniceras leonense, Prohystoceras, Gryphaea, Neithea texana,* and *Homomya;* Wilson, 1956) of late middle Albian age, correlative with the Kiamichi Formation in south Texas.

Loma de Plata Formation. This approximately 450-m-thick unit occurs throughout the Sierra de Naica where two members may be identified: a lower 30- to 40-m-thick medium-bedded (0.6 to 0.8 m) light gray micrite with scarce microfossils, and an upper thick-bedded to massive, cliff-forming limestone. Both contacts are conformable and well marked with the overlying Del Río and the underlying Benevides Formations. The unit is of late Albian age, correlative with the Georgetown Formation in central Texas.

Del Río Formation. This formation is mainly composed of gray, thinly bedded shales 20 to 30 m thick, and argillaceous sandstones exposed only in Sierra de Enmedio and Sierra del Monarca, being absent by erosion at the Sierra de La Mina. Lower and upper contacts are conformable with the underlying Loma de Plata and overlying Buda Formations. The unit is late Albian to Cenomanian in age and correlative with the Eagle Mountain Formation in southwest Texas.

Buda Formation. This formation has a distribution similar to that of the Del Río. It includes gray-buff thick-bedded to massive limestones of unknown thickness, the top of the formation having been eroded in the area. It is of early Cenomanian in age and correlates with the upper Washita Group.

Igneous rocks

The only igneous rocks in the Sierra de Naica are felsite dikes and sills. Under the microscope the rock shows holocrystalline and alliotromorphic fine-grained texture of K-feldspar and quartz, described as mostly albite by Wilson (1956). On a mineralogical basis it has been termed albitite or alaskite. Pyrite, calcite, fluorite, and chalcopyrite traces are the accessory minerals. Within the mine the rock is light gray with conchoidal fracture and marked banding parallel to the host rock; in the mineralized areas it is usually replaced by skarn. At the surface it appears as discontinuous reddish lenses, which frequently show Liesegang rings. Clark and others (1979) report an isotopic (K/Ar) date of 26 Ma for these rocks.

Structural geology

The Sierra de Naica as a whole forms a roughly dome-shaped structure approximately 12 km long by 7 km wide. Deep magnetometry indicates the presence of a more strongly magnetic basement dome approximately 2,000 m below the surface (D. F. Coolbaugh, personal communication), later interpreted as an igneous dome.

The mineralized zone lies on the dome's eastern flank, toward the center of a subsidiary northeast-dipping structural nose; strong fracturing, pre-, post-, and contemporaneous with the mineralization, is present. The premineralization fracturing developed a northwest-trending fault and fracture system that appears to have guided the emplacement of dikes and mantos. A second system of generally N40°E-trending, approximately 60°-northwest- and southeast-dipping mineralized faults and fractures—some of which show displacements of tens of meters—is assumed to be contemporaneous with the mineralization, because it occasionally displaces the mantos and controls the sulfide mineralization in chimneys at the upper levels of the mine (level 290 to surface). The most important faults in the system are the Torino, Tehuacán, Descubridora, and Ramón Corona (Fig. 3). Following these, and also the mineralization, a third system of northwest-trending faults comprises the southwest-dipping Gibraltar fault with 50 m of vertical displacement, the Naica fault, also southwest-dipping with about 200 m of displacement, and the northeast-dipping Montaña fault, which merits special attention because it hosts the famous Naica selenite crystal caves in the oxidation zone; among these is the conspicuous Cueva de Las Espadas, where crystals up to 2 m in size completely line the cavity walls.

ORE DEPOSITS

Introduction

The Naica district belongs to the Sierra Madre Oriental Metallogenic Province (Salas, 1975), which is characterized by large deposits of basic sulfides in calcareous sediments folded by the Laramide orogeny and intruded in the Tertiary by silica- and alkaline-rich igneous bodies, intimately associated with the hydrothermal processes that generated the deposits. Besides Naica, the most important districts of similar nature in the province are Santa Eulalia (Chihuahua), Concepción del Oro (Zacatecas), Velardeña (Durango), Santa María de La Paz and Charcas (San Luis Potosí), and Zimapán (Hidalgo). The Naica ore deposit is characterized by large replacement bodies in limestones and by skarn with massive Pb and Zn sulfides, some Cu, and lesser amounts of Au, Ag, tungsten, and molybdenum.

Mineralogy

To date, the mineralogy of the Naica ore deposits has not been fully defined at the lower levels; the only known descrip-

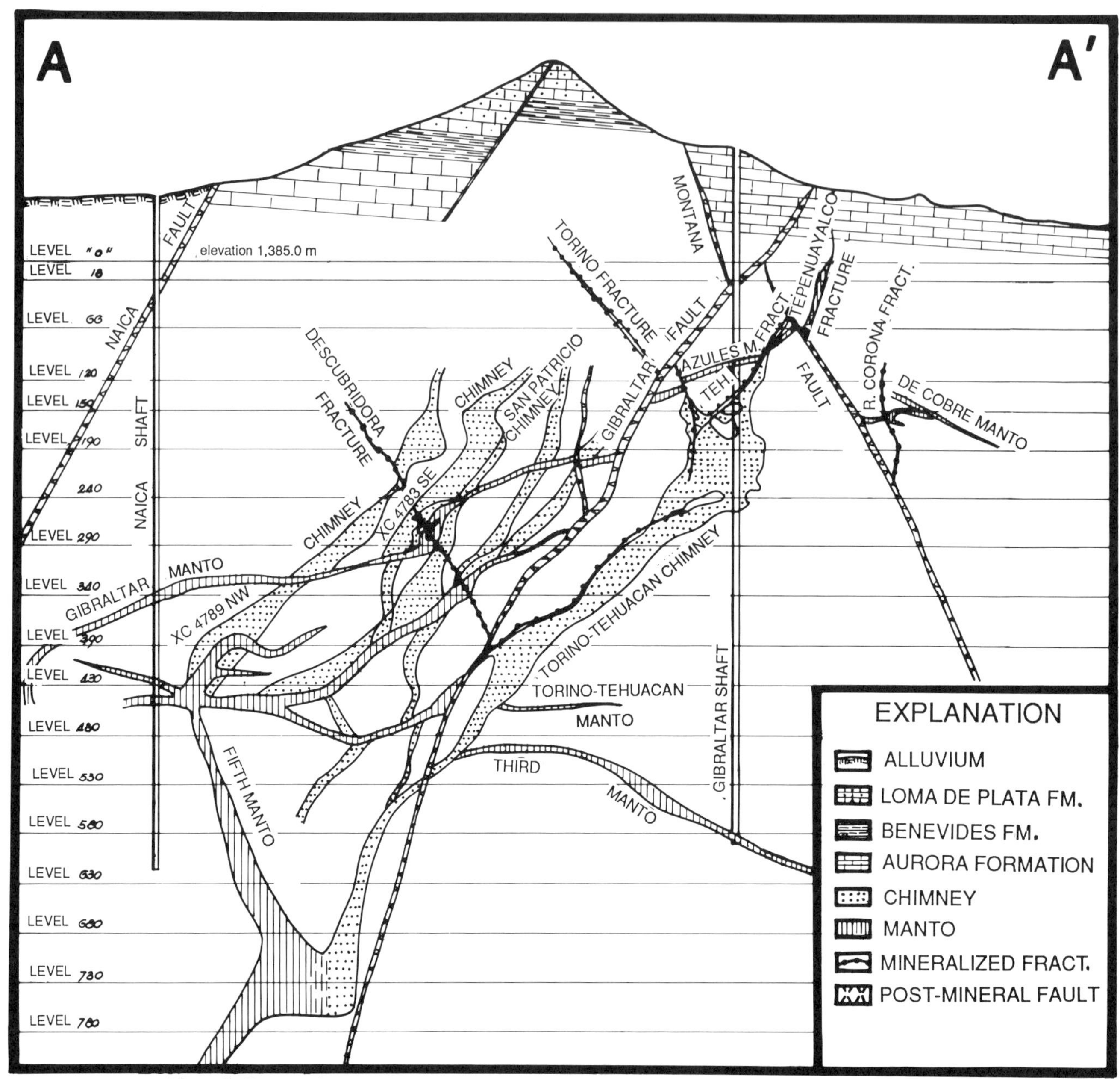

Figure 3. Cross section showing the main orebodies of the Naica Mine. Location of section shown on Figure 2. (Modified from Bravo, 1968.)

tions are from the higher drifts (Stone, 1959). Ruiz (1985) indicated that—at least in the available samples—the garnets in the mineralized skarn do not show the same composition throughout the mine: Gr_{75}–Sp_2 in the mantos and Ad_{75}–Sp_2 in the chimneys. The difference relates to the respective exoskarn and endoskarn host rocks. However, hedenbergite appears to show the same composition: Jo_{40}–Hd_{40}, classified as manganese hedenbergite, in both types of orebodies.

Characteristics of the mineralization

Chimneys and mantos are the two geometric types of deposits. The orebodies of massive sulfides with occasional silicates that are roughly cylindrical, tending to the vertical (dips of more than 45°), and as much as 4,000 m^2 in cross section are described as chimneys. The term "manto" has lost its original meaning, which referred to tabular bodies almost parallel to the bedding; as development advanced, the dips of these bodies were seen to steepen, at times becoming almost vertical, and the term is presently applied to all of the usually unconformable tabular orebodies of silicates with lesser sulfide content than the chimneys.

Mantos. These are generally northwest-trending tabular skarn bodies with dips ranging from the horizontal, when they appear to follow bedding planes, to vertical along preexisting erratically shaped fractures. To date, more than 17 mostly interconnected mantos, ranging from centimeters to 25 m in thickness, have been discovered. Some mantos stretch for over 500 m and occur at all levels, from near-surface to the lowermost drifts, having been drilled down to more than 800 m beneath the zero level. Ore volumes are variable, exceeding 3 MMT in both the Gibraltar and Quinto mantos (Fig. 3).

In his study of the Azules Navacoyán manto, an extension of the Gibraltar, Stone (1959) defined two zones: a central high-temperature zone characterized by garnet (grossularite), vesuvianite, and wollastonite, which changes laterally to a hedenbergite, quartz, and calcite zone. We do not agree, however. Skarn mineralization in the mantos is garnet (grossularite-andradite), vesuvianite, wollastonite, hedenbergite, quartz, calcite, fluorite, molybdenite, pyrite, pyrrhotite, and marcasite. The ore is galena, sphalerite, chalcopyrite, and scheelite. Silver appears as microscopic crystals of bismuth-rich sulfosalts included in the galena. On the basis of their spatial relations with the felsite sills and dikes, the mantos have been divided into exoskarn and endoskarn.

Endoskarn mantos are characterized by a larger grain size than exoskarn mantos, significant scheelite and molybdenite concentrations, and a higher Ag/Pb ratio than the exoskarns (Table 1), and are found replacing the felsite either occasionally at the flanks or erratically to the point of complete replacement. Examples are the Gibraltar, Segundo, Tercero, and Quinto mantos.

Exoskarn mantos are fine-grained carbonate and silicate orebodies with rare scheelite and molybdenite and an Ag/Pb ratio similar to that of the chimneys. They are usually associated at one end with an endoskarn from which they break off, eventually disappearing in silicified limestone. Generally smaller than the endoskarns, most are conformable or almost conformable with the calcareous beds. The most typical example is the Torino-Tehuacán manto (Fig. 3).

Mineralization. Sulfide mineralization in the mantos is generally lodged at the skarn contacts and usually replaces marble, although it is also in the skarn proper. It is worth noting that the skarn and occasionally the felsite contain disseminated sulfides with higher, often commercially prospective, silver contents (Fig. 4). Scheelite and molybdenite are never associated with the massive sulfide zones, but are disseminated in the skarn. As a rule the exoskarns show more abundant Pb-Zn mineralization.

Chimneys are tabular-shaped orebodies of massive sulfides with variable silicate content that dip more than 45°, with horizontal elliptical or irregular cross sections, in some cases of as much as 4,000 m^2. About 60 chimneys, ranging from 70 m (Torino-Tehuacán) to 3 m in diameter, are known in Naica. Most appear to end in the oxidation zone and do not surface, except Torino-Tehuacán, which crops out over an area 4 m in diameter. Almost all the chimneys join the mantos at depth; others end in small veins. The deepest chimney to date is Torino-Tehuacán with over 800 m of vertical extension and 3,300,000 tons of ore in the primary sulfide zone. The upper-level chimneys are controlled by the northeast-trending fracture system. Stone (1959) recognizes two kinds of chimneys: sulfide/silicate chimneys, and massive sulfide chimneys.

Sulfide/silicate chimneys. These are the most important with regard to volume. Silicates are present mostly toward the central part of the chimney; their percentages increase with depth. These chimneys rise from the mantos at depth, usually at points of change in strikes and dips, or at intersections with premineral fractures. The most abundant silicates are garnet, hedenbergite, quartz, and wollastonite, at times pseudomorphically replaced by pyrite, galena, sphalerite, arsenopyrite, chalcopyrite, and pyrrhotite, accompanied by calcite, fluorite, and anhydrite (Fig. 5). The increase in silicates with depth causes zoning in the chimneys and also increasing Ag/Pb ratios.

Massive sulfide chimneys. As originally described by Stone (1959), these included some chimneys that turned out to be the upper parts of silicate chimneys as mining operations advanced. However, a series of smaller chimneys (less than 100,000 tons) were also discovered, characterized by the total absence of silicates, and by abundant pyrite, galena, and sphalerite (marmatite), which in contrast with other chimneys, end both above and below in small veinlets that are spatially unrelated to any manto. The possible source channels for the mineralization of these chimneys is as yet unknown. The most conspicuous of these is the San Patricio chimney between levels 129 and 390, with 260 m^2 of average cross section. This type of chimney is found mostly at the edges of the mine's mineralized zone. For purposes of comparison, Ag/Pb metallic ratios in the several types of Naica orebodies are shown on Table 1.

The similar values in exoskarn mantos and sulfide/silicate chimneys indicate analogous formation temperatures. Because all the silver is included in the galena, the higher ratios may be related to the higher formation temperature of the galena.

Origin of the chimney deposits. Fluid inclusions in fluorite from samples of the Torino-Tehuacán chimney and Gibraltar manto (Erwood and others, 1979) indicate that the hydrothermal solutions that formed the deposit belong to one of three different types according to their composition. Type "A" inclusions consist of liquid and gas with homogenization temperatures of 119° to 379°C and salinities less than 20 percent NaCl. Type "B" inclusions consist of liquid + gas + halite with homogenization temperatures of 237° to 369°C, 31 to 43 percent NaCl, and less than 12 percent KCl. Type "C" inclusions consist of liquid + gas + halite + silvite with homogenization temperatures of 277° to 490°C, 52 to 63 percent NaCl, 22 to 31 percent KCl, and some high homogenization temperature (565 to 684°C) inclusions. Because all three types occur throughout the deposit and are completely miscible, Erwood and others (1979) conclude that the differences are due to the period of formation and therefore reflect the changes undergone by the mineralizing solutions with time. Most probably type "C" evolved to type "B" and finally to "A," which suggests a considerable decrease in salinity as the system evolved and a barely perceptible decrease in temperature. The high salinity of the original solutions and the persistent anhydrite in the

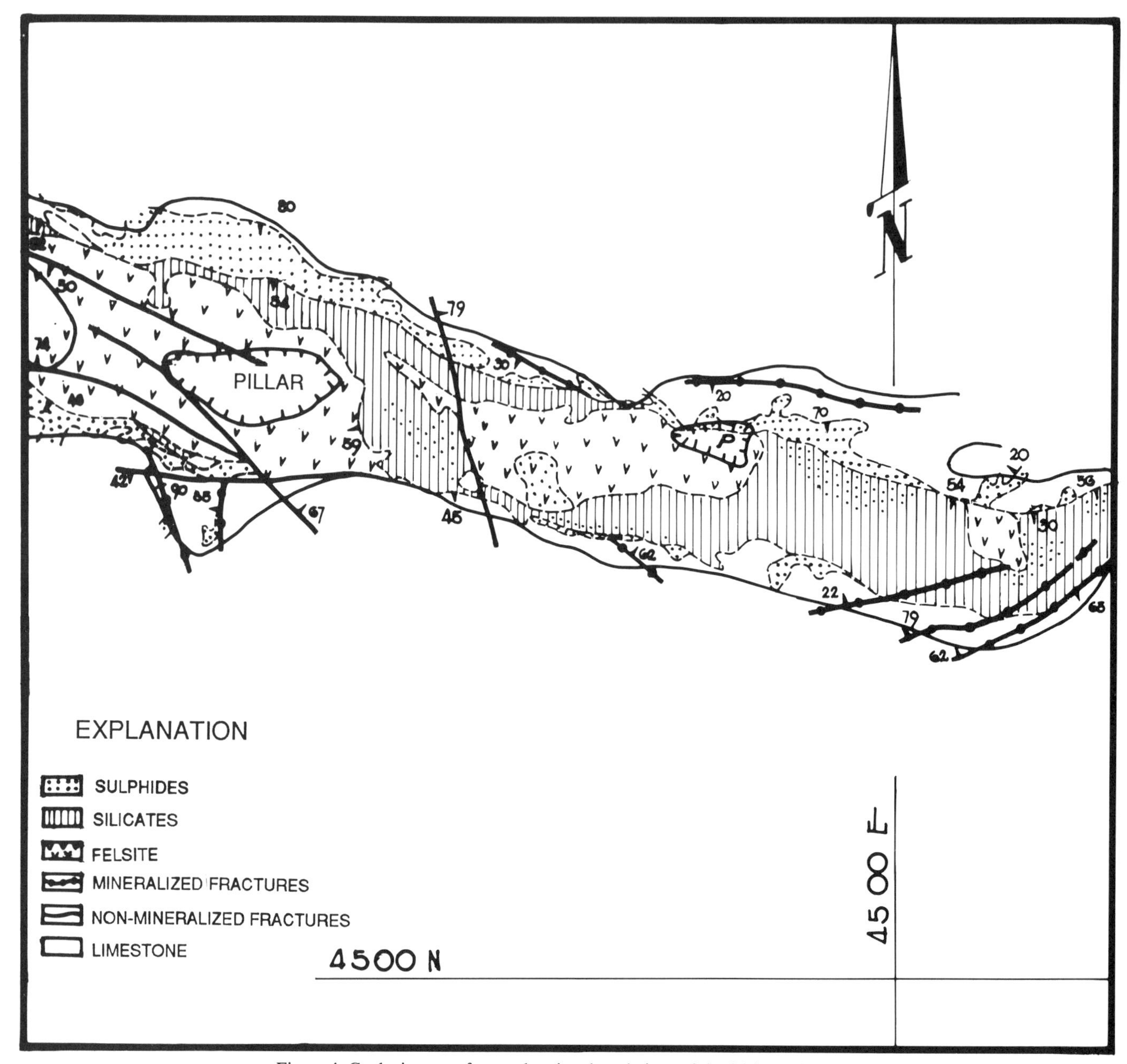

Figure 4. Geologic map of a cut showing the relations of the felsite and skarn.

deposit could be explained assuming that the solutions originated in the connate waters of the underlying evaporite sequence.

All the evidence points to the hydrothermal origin of the Naica ore deposits. The geothermal system was generated by intrusion below the 2,000 m depth—as suggested by the magnetic anomaly—that interacted with the connate waters of the sedimentary sequence, giving rise to a briny hydrothermal system of high metal-transport capacity. The orebodies evolved from brines rich in silica, aluminum, manganese, iron, lead, zinc, silver, fluorine, and sulfur that followed zones of weakness, such as the contact between the almost simultaneously emplaced felsite dikes and the limestone host rock, replacing both of them and incorporating part of their component elements (Ca, Mg) to form new minerals. Fluid inclusions, textures, and mineralogies indicate that the hydrothermal solutions varied in their composition as the system evolved. The most indicative change is the decrease in chlorine fugacity at the same rate as the increase in sulfur fugacity, causing precipitation of the sulfides, in some cases at the expense of previously formed silicates. The formation temperatures of mantos and chimneys also varied locally, as shown by Ag/Pb ratios, higher in the high-temperature (endoskarns and deep parts of chimneys) than in the lower temperature bodies (massive sulfides at the edges of mantos and in chimneys).

On the basis of structural position and geometry, the mantos or silicate bodies developed prior to the chimneys (i.e., during the phase of maximum energy in the geothermal process). Isotope dating places the hydrothermal process in late Oligocene time, probably coinciding with the final stages of Sierra Madre Occidental magmatism.

Average tonnage and grades

To December 31, 1984 the Compañía Fresnillo, S. A. de C. V., had produced 15,486,138 tons of sulfide ores in Naica, averaging 0.34 ppm Au, 177 ppm Ag, 5.5 percent Pb, 4.3 percent Zn, and 0.34 percent Cu. Estimated production prior to 1951 (González Reyna, 1956) was a little under 1 MM metric tons (m.t.) of oxide and sulfide ores, from the upper levels to less than 10 m below the ground-water table, with about 16 percent Pb and 400 ppm Ag. Total production to 1984 would thus be in the range of 16.5 MM m.t. of ore. According to estimated reserves as of September 1984, proved and probable ores total 4.6 MM m.t. of 0.2 ppm Au, 152 ppm Ag, 3.6 percent Zn, and 0.42 percent Cu. These figures represent only the known ores above the last explored drift level (530); mineralization down to level 800 has not been evaluated.

Mining and metallurgical methods

At present, 62,000 tons per month are mined by cut-and-fill; the ores are treated by selective Pb-Zn flotation. The principal commercial metals are silver, gold, lead, zinc, copper, and a small percentage of tungsten and cadmium. Gold, silver, and copper are obtained from the lead concentrates; cadmium is a by-product of the zinc concentrates; the final tailings in the flotation are treated separately to recover tungsten. Average pretreated ore grades and recovery percentages are shown in Table 2.

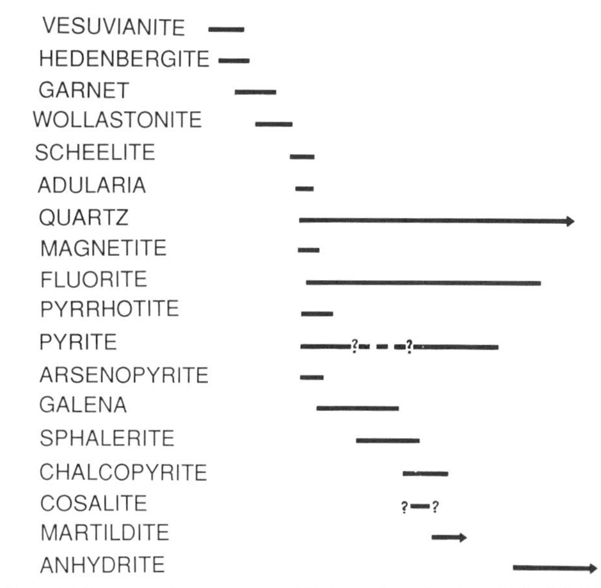

Figure 5. Paragenetic sequence of Naica mineralization. (Modified from Stone, 1959.)

TABLE 1. Ag/Pb RATIOS IN NAICA ORE BODIES

Type of body	Ag/Pb ratio
Endoskarn manto	62.5
Exoskarn manto	31.4
Sulfides-silicate chimneys	30.7
Sulfides chimneys	25.5

TABLE 2. AVERAGE PRE-TREATED ORE GRADES AND RECOVERY PERCENTAGES FOR NAICA ORE

Au	Ag	Pb	Zn	Cu	WO_3
ppm		%			
0.32	153	4.36	3.46	0.40	0.17
with the following recoveries:					
26.9	90	94.5	84	70	32

REFERENCES CITED

Almada R., F., 1945, Geografía de Estado de Chihuahua: Chihuahua, Mexico, Ruiz Sandoval, S. A., 628 p.

Clark, E. K., Damon, P. E., Schutter, S. R., and Shafiqullah, M., 1979, Magmatismo en el norte de México en relación con los yacimientos metalíferos: 13th Convencion Nacional Asociación de Mineros, Metalurgistas y Geologos de México Memoir, p. 8–57.

De Ford, R. K., and Haenggi, W., 1971, Stratigraphic nomenclature of Cretaceous rocks in northeastern Chihuahua, *in* The geologic framework of the Chihuahua Tectonic Belt: West Texas Geological Society Publication 71-59, p. 175–196.

Erwood, R. J., Kesler, S. E., and Cloke, P. L., 1979, Compositionally distinct saline hydrothermal solutions, Naica Mine, Chihuahua, Mexico: Economic Geology, v. 74, p. 95–108.

Franco R., M., 1978, Estratigrafía del Albiano-Cenomaniano en la region de Naica, Chihuahua, y su relacion con los yacimientos de plomo y zinc [M.S. thesis]: Universidad Nacional Autonoma de Mexico.

González Reyna, J., 1956, Memoria geológico-minera del Estado de Chihuahua: 20th International Geological Congress, Mexico.

Hewitt, W. P., 1943, Geology and mineralization of the San Antonio Mine, Santa Eulalia District, Chihuahua, Mexico: Geological Society of America Bulletin, v. 54, p. 173–204.

Lambert, A. D., 1892, Zona Minera de Naica: Boletin Agricultura y Minera, p. 143–183.

Raisz, E., 1959, Land forms of Mexico: Cambridge, Massachusetts, Map with text, scale 1:3,000,000.

Ruiz, J., and Mark, D. B., 1985, Geology and geochemistry of Naica, Chihuahua, Mexico: Society of Mining Engineers, American Institute of Mining and Metallurgical Engineers, preprint 85–43.

Salas, G. P., 1975, Carta y provincias metalogeneticas de la Republica Mexicana: Consejo Recursos Minerales Publicacion 21-E, 242 p.

Stone, J. G., 1959, Ore genesis in the Naica District, Chihuahua, Mexico: Economic Geology, v. 54, p. 1002–1034.

Wilson, I. F., 1956, El Distrito Minero de Naica, Chihuahua, México: 20th International Geological Congress, Excursions A-2 and A-5, p. 63–75.

Printed in U.S.A.

The Geology of North America
Vol. P-3, Economic Geology, Mexico
The Geological Society of America, 1991

Chapter 29

San Francisco del Oro Mining District, Chihuahua

Guillermo P. Salas
Geologo Consultor, Asesor del Consejo de Recursos Minerales, Centro Minero Nacional, Carretera México-Pachuca, Pachuca, Hgo., México, D.F.

INTRODUCTION

This is one of several important silver mines in Chihuahua. It is located south of central Chihuahua and forms part of the Parral Super district, which also includes the Santa Bárbara, Palmillas, Parral, and La Esmeralda Districts. The San Francisco del Oro District is approximately 100 km south of the city of Chihuahua, the state capital, and 20 km southwest of Hidalgo del Parral (Fig. 1). It is accessible by an excellent road system.

Numerous mineralized bodies have been located to date. The famous Mina de Agua, discovered in 1567, gathered a horde of prospectors, and mining at La Prieta started in 1629. The San Diego de Minas Nuevas bonanza in 1645 renewed prospecting activity that developed dozens of silver and gold mines in the area, many of which are still in production. By 1659, more than 115 mines were operating.

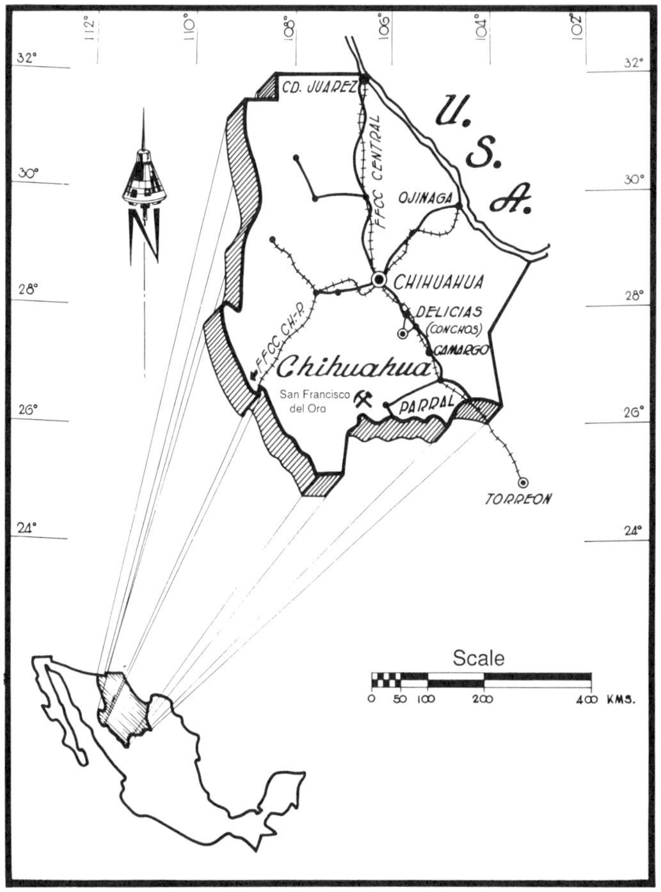

Figure 1. Map of Chihuahua State showing location of San Francisco del Oro Mining District.

GEOLOGY

The district extends over 300 km^2 at the transition between the Sierra Madre Occidental and Central Mesa Metallogenic Provinces and shows geologic features of both.

The gently west-dipping upper Aptian–lower Albian Parral Formation, containing shales with some marl and sandstone toward the base, correlates with the Trinity Formation of Texas. A thin conglomerate at the base marks the Cretaceous marine transgression. Andesites and rhyolites typical of the Sierra Madre Occidental overlie the marine section, and the whole sequence was intruded by quartz and rhyolite dikes (Fig. 2).

During the upper Tertiary the great northwest-southeast–trending Santiaguillo fault depressed an extensive area east of San Francisco del Oro, known today as the Casa Colorada Valley.

STRUCTURE

The original structure from San Francisco del Oro to Santa Bárbara is an anticline trending N28°E at San Francisco del Oro. The veins, ranging from 0.7 to 3.0 in width, were generated by fault and fracture filling (Fig. 3). In the San Francisco del Oro mine the veins are short when compared to the Santa Bárbara District, where they may extend for as much as 5 km.

Faulting took place following the mineralization. Vertical displacement of the veins reaches 100 m in the Los Clarines Mines; however, displacements are usually about 10 m.

Salas, G. P., 1991, San Francisco del Oro Mining District, Chihuahua, *in* Salas, G. P., ed., Economic Geology, Mexico: Boulder, Colorado, Geological Society of America, The Geology of North America, v. P-3.

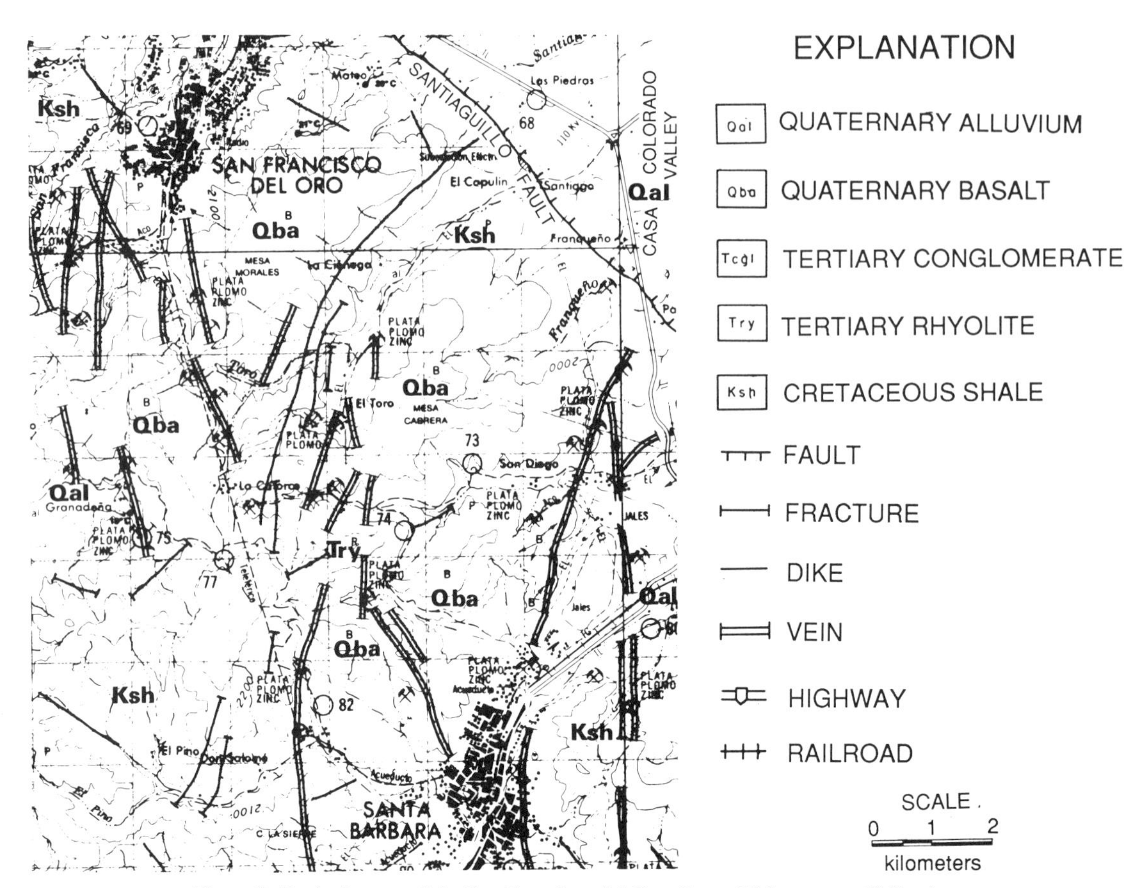

Figure 2. Geologic map of the San Francisco del Oro–Santa Bárbara area, Chihuahua.

MINERALOGY

San Francisco del Oro orebodies produce mainly gold, silver, lead, zinc, and copper. Assay averages are: 14 percent zinc, 6 percent galena, and 2 percent chalcopyrite; associated minerals are quartz (as much as 33 percent) and fluorite (11 percent) in the sulfide zone. Fluorite is currently produced commercially by treating the tailings, and in some areas may reach 50 percent of the tailings (Santa Cruz, Frisco, etc.). Cerussite, anglesite, malachite, azurite, jarosite, and limonite are the oxidation zone minerals.

This district appears to be in the intermediate zone of epithermal precious metals emplaced in volcanic strata. The breccia chimney and manto deposits are metasomatic replacements in the Cretaceous limestone sequence. Normal metal contents are 0.38 g/t Au, 151 g/t Ag, 4.28 percent Pb, 6.45 percent Zn, 0.48 percent Cu, and 14.6 percent CaF_2. Diamond-drilling during recent years (1984) in search of additional reserves averaged 7,000 m/year. Production of the San Francisco del Oro District is 840,000 t/year with the above grades.

REFERENCE CITED

Salas, G. P., 1980, Carta y Provincias Metalogeneticas de la Republica Mexicana: Consejo de Recursos Minerales, Bulletin 21-E, 2nd edition, 242 p.

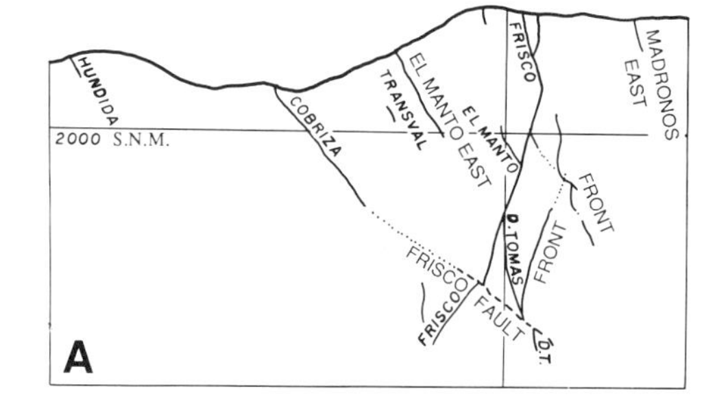

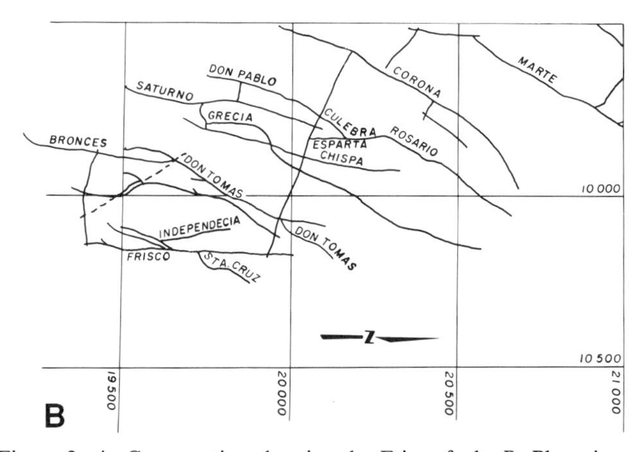

Figure 3. A, Cross section showing the Frisco fault. B, Plan view of Frisco fault system.

Printed in U.S.A.

The Geology of North America
Vol. P-3, Economic Geology, Mexico
The Geological Society of America, 1991

Chapter 30

Economic geology of the Velardeña Mining District, Durango

I. Hernández C.
Unidad Minera de Velardeña, Industrial Minera México, S.A., Ave. Baja California 200, Col. Roma Sur, 06760 México, D.F.

INTRODUCTION

The Velardeña Mining Unit, owned by Minerales Metálicos del Norte, S.A., a member of Grupo Industrial Minera México, S.A. de C.V., is located in northeastern Durango State within the Cuencamé municipality, 150 km northeast of the city of Durango and 85 km south-southwest of Torreón. Geographic coordinates of Velardeña Village, in the vicinity of the mineworks, are 25°04′N and 103°44′W. Altitude above sea level ranges from 1,400 m in the valley separating Sierra de Santa María and Sierra de San Lorenzo to about 2,000 m at the San Lorenzo peaks. Climate and vegetation are typical of northern Mexico's semidesert zone. Federal Highway 49 runs approximately 6 km northwest of Velardeña, crossing the Durango-Torreón railroad track 2.5 km farther on at Pedriceña, near which is a small light-aircraft runway. Velardeña has telephone and telegraph facilities, schools, and other services for a population of 5,000, most of whom depend on the mines for a living (Fig. 1).

REGIONAL GEOLOGY

The Velardeña Mining District lies in the Sierra Madre Oriental Metallogenic Province on the western flank of the Sierras Atravesadas, which consists of a folded belt of clastic sediments approximately 3,500 m thick. The oldest outcrops are Lower Cretaceous limestones of the Aurora (Kia) and Cuesta del Cura (Kicc) Formations, often intruded by quartz-latite dikes. The two main structures within the regional tectonic framework are the San Lorenzo anticlinorium and the Santa María dome (Fig. 2).

The Aurora and Cuesta del Cura Formations were laid down in marine waters at the beginning of the Cretaceous. The thick Aurora Limestone section indicates stable depositional conditions; thinner Cuesta del Cura calcareous layers interbedded with parallel chert bands testify to slow uplift (Levich, 1973), which escalated at the end of the Cretaceous.

Thin limestone and shale beds of the Indidura Formation, overlain by siltstones, sandstones, and conglomerates at the south flank of the Santa María dome suggest the onset of rapid tectonism and uplift, coinciding with the first rumblings of the Laramide orogeny.

Orogenic compression in this area gave rise to the Santa María dome and the San Lorenzo anticlinorium; the folding was accompanied by N50°W and N60°W faults and fractures that later guided intrusive emplacements. The present-day ranges and valleys resulted from movements along these fracture trends, which generated a series of horsts and grabens, the latter being subsequently filled by Quaternary alluvium, Tertiary volcanic fragments, and calcareous conglomerates of the Tertiary Ahuichila Formation.

The San Lorenzo anticlinorium, northeast of the Santa María dome, is separated from it by an alluvium-covered structural low; on the north it joins the Rosario and on the southeast the San José and Jimulco anticlinoria. Its east flank is depressed by faulting into the Sangrita basin.

The Aurora Limestone crops out over most of the northern half of this tectonic unit; metamorphosed Cuesta del Cura and Indidura Formation rocks are exposed at the northwestern flank. Tertiary volcanic acid rocks and the Ahuichila Formation unconformably overlie the Cretaceous folded sequence at several localities on both the west flank and north ends.

Post-depositional tectonic deformation is indicated by the slightly arched structural attitude of the Tertiary rocks. Also, Tertiary volcanic rocks occur very near the crest of the anticlinorium.

The structure contains numerous symmetrical to overturned closed folds of 50 to 200 m in amplitude; metamorphism has destroyed the evidence of folding in the northwest flank. The central part of the San Lorenzo anticlinorium is cross-cut by a canyon that developed along a fault zone; the faults crossing this feature show very slight vertical displacements. Southwest of these faults the folding generated a series of anticlinal ridges and synclinal valleys.

Igneous intrusions such as the very large San Diego and El Cobre bodies mostly affected the central part. Smaller intrusions, generally of diabase and diorite, are present at the western flank.

The Santa María dome is an isolated structure 5 km southwest of the San Lorenzo anticlinorium, extending approximately 7 by 5 km, with its major axis pointing N45°W (i.e., parallel to the San Lorenzo fold axis). An alaskite intrusion ex-

Hernández C., I., 1991, Economic geology of the Velardeña Mining District, Durango, *in* Salas, G. P., ed., Economic Geology, Mexico: Boulder, Colorado, Geological Society of America, The Geology of North America, v. P-3.

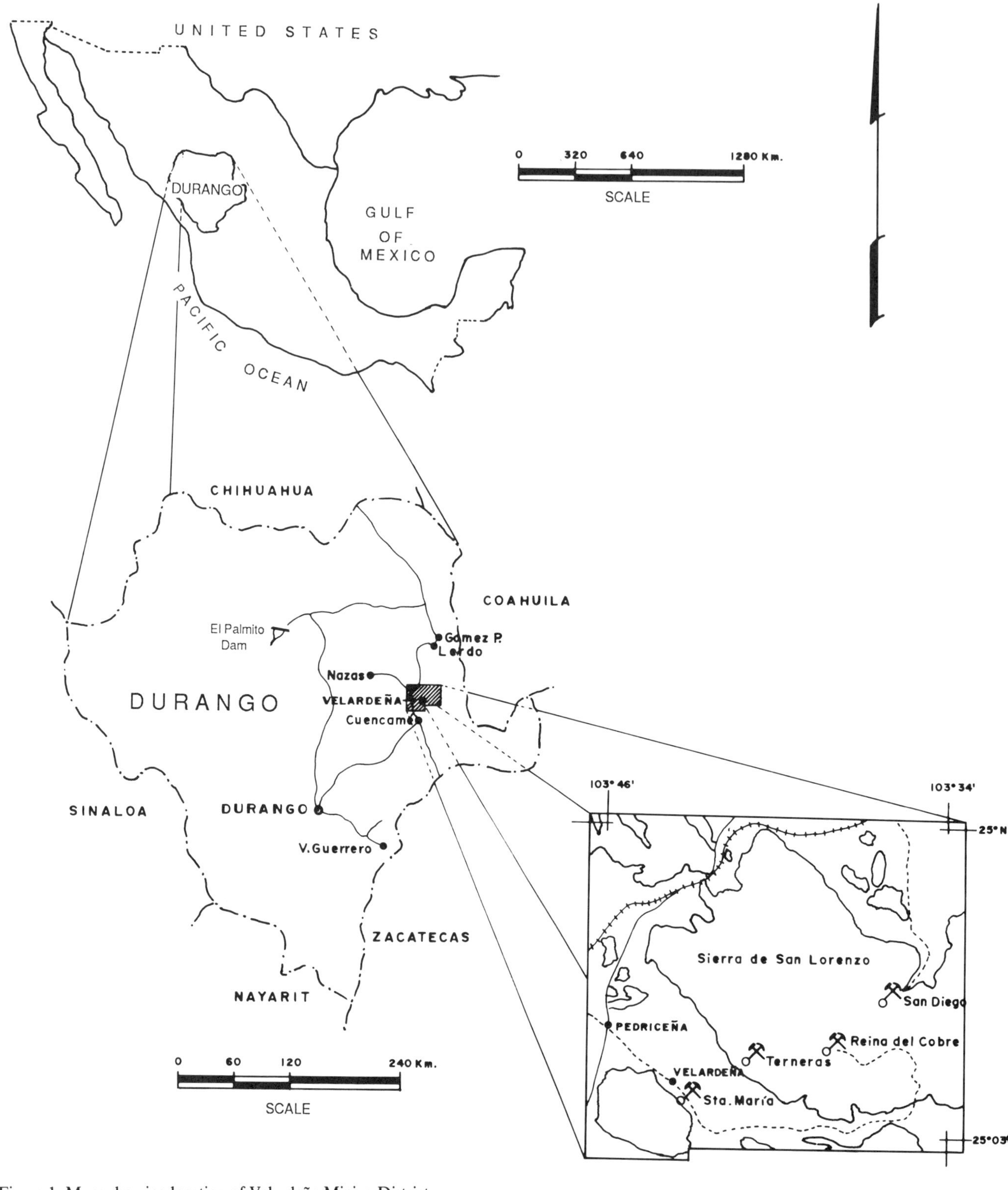

Figure 1. Maps showing location of Velardeña Mining District.

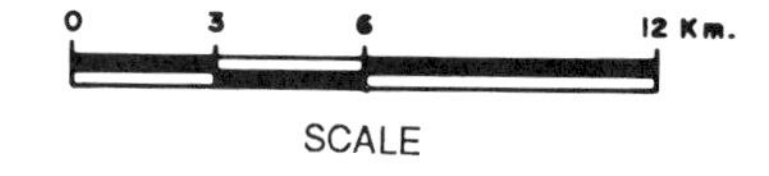

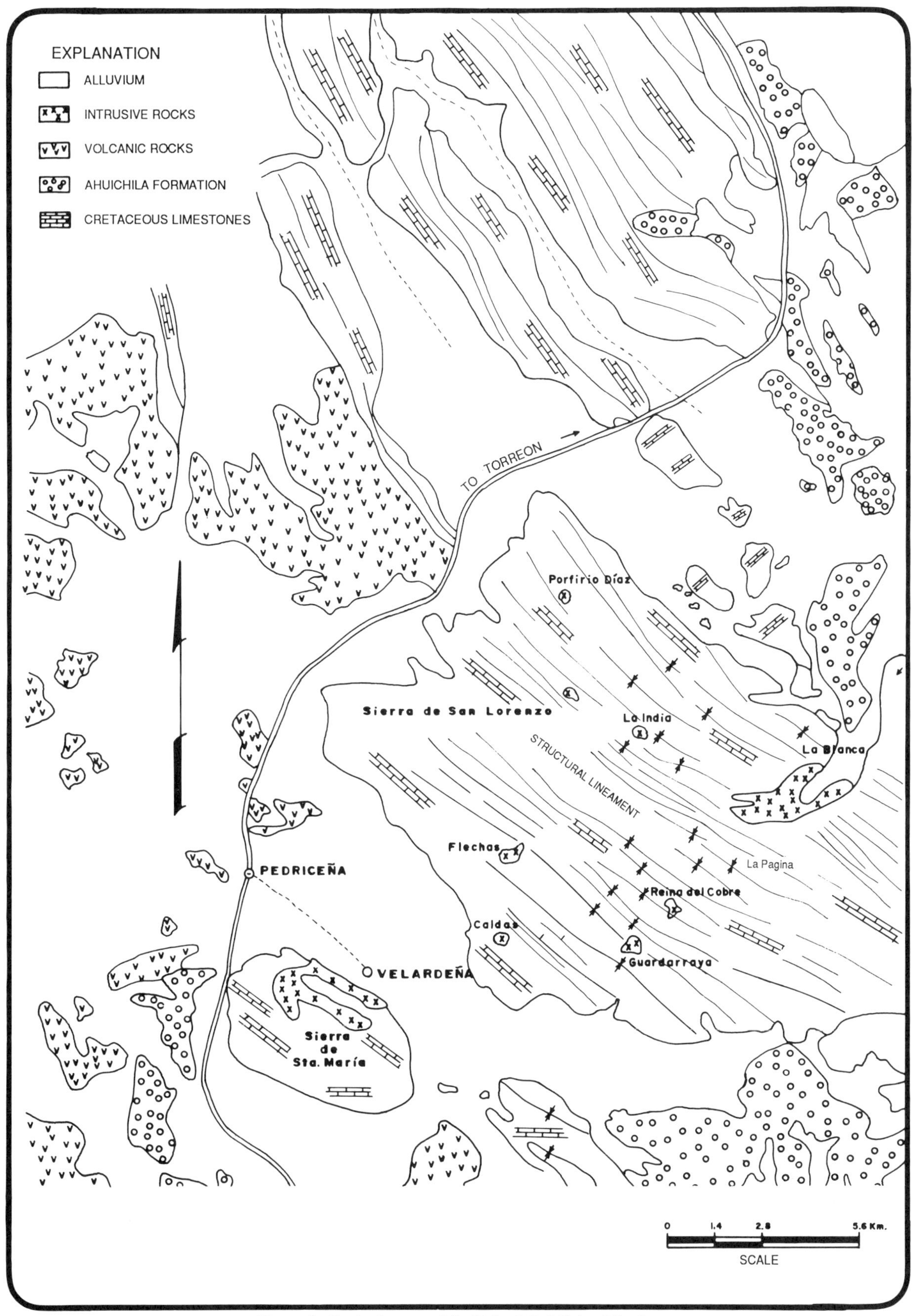

Figure 2. Geologic map of the Velardeña Mining District.

posed at the center of the dome caused steep tilting of the Aurora and Cuesta del Cura rocks except on the northeastern side; the Indidura Formation crops out at the southwest flank. The intrusive body exhibits laccolith geometry and features.

LOCAL GEOLOGY

Sedimentary, igneous, and metamorphic rocks are present in the Velardeña Mining District. The first make up the thick Cretaceous limestone sequence of the Aurora and Cuesta del Cura Formations.

The Aurora Limestone, underlying the Cuesta del Cura Formation and forming the central part of the Santa María dome, is the oldest sedimentary unit in the Sierra de Santa María. It is distinguished from the Cuesta del Cura by its thicker (1.3 to 3 m) beds, more abundant ammonite fossils, and lesser allochemical content. Because of its relative purity, a metamorphic aureole of garnet (grossularite) and wollastonite tactites develops at the contacts with the intrusive bodies.

The Cuesta del Cura Formation outcrops in both the Sierras de Santa María and San Lorenzo as a series of light to dark gray calcareous beds, crossed by calcite veinlets with no preferred trend, and abundant black chert lenses. The generally N55°E–trending, 60°-northeast–dipping, 30- to 50-cm-thick, yellowish-weathering beds are considerably argillaceous and therefore unfavorable for replacement processes. The unit hosts silver-lead-zinc deposits at the contacts with the intrusives, particularly the dike at the Santa María Mine, but contains no substitution bodies because of its high allochemical content.

Igneous rocks in the area are products either of magmatic injection or flows. The former occur in the Sierras de Santa María and San Lorenzo as stocks and dikes intruding the older limestones; in the Sierra de Santa María these rocks are mainly stocks of alaskitic composition and trachyte dikes.

The alaskite is a gray to pink porphyritic rock with conspicuous quartz and Na-Ca feldspar "augen" or phenocrysts in a phaneritic crystalline groundmass. The main body, approximately 3 km long by an average 400 m wide, trends N50°W and dips 55°SW (Fig. 3). Mafics are noticeably absent. A smaller dike-like alaskite body exposed at the Los Azules Mine (Fig. 4) is closely related to the mineralization.

The most conspicuous trachyte outcrop is the 2.5-km-long and (average) 25-m-wide Santa María dike, a compact, light gray aphanitic volcanic rock trending parallel to the axis of the Sierra de Santa María and dipping southwest, which is an important mineralization control in the area. No skarns develop at the contact with the Cretaceous rocks, owing probably to the low silica content in the intrusive fluids and the high clay content of the sedimentary host rocks.

Andesite flows and tuffs crop out at the northwest flank of the Sierra de Santa María. The rocks are greenish gray and porphyritic with conspicuous andesite phenocrysts; the greenish hue is due to propylitic alteration. The absence of economic ores in this lithologic unit and the extensive propylitization lead to the conclusion that the alteration was deuteric and post-magmatic.

The Velardeña Mining District shows evidence of both metamorphic and contact metasomatic processes. Metamorphism caused widespread limestone marbleization. Contact metasomatism generated irregularly distributed tactite zones—with grossularite, wollastonite, and olivine, principally—next to the contacts of certain intrusives with the calcareous rocks. The most conspicuous appears at the Reina del Cobre Mine (Fig. 2), where a quartz-monzonite dike intrudes the Aurora limestones and commercial mineralization is restricted to the tactite.

The local structural features are fracture and fault systems, folds, and the two major structures (Santa María dome and San Lorenzo anticlinorium). Two types of faults and fractures are present. The Laramide event generated a system of staggered northwest-trending faults and fractures controlling the emplacement of intrusive bodies, which enhanced the fracturing and at times also the limestone brecciation. Subsequent tectonism caused another fracture system affecting the intrusive rocks, which probably relates directly to the mineralization, as discussed below.

Small solution caves occur in the area, but only where the limestone is sufficiently pure (i.e., Santa María Mine Level 9). These are probably post-mineralization features, there being no paleokarst-type ore deposits in the district.

Silicification, propylitization, and kaolinization are the existing types of alteration. Widespread silicification in the vicinity of the intrusives indurates the calcareous rocks. Propylitization—either hydrothermal or deuteric—appears next to the tactite zones in hydrothermal situations and affects only the andesites in deuteric situations. Kaolinization occurs mostly at the surface, presumably because of chemical weathering. Oxidation appears down to the 30-m depth; the Sierra de Santa María gossans of locally high pyrite content are conspicuous.

ORE DEPOSITS

Santa María and Los Azules are the two principal mines in the Sierra de Santa María.

The ore deposit in the Santa María Mine is exogenous epigenetic; mineralization was subsequent to the host rocks (trachyte and limestone) and of different origin. The orebodies are located at both contacts of the large trachyte dike with the Cuesta del Cura limestones where no tactites developed, which eliminates the possible pyrometasomatic provenance of the deposit.

Los Azules is a fissure-filling and replacement endogenous epigenetic deposit; mineralization followed the host rock, but both have a common origin. In contrast with the Santa María Mine, mineralization at Los Azules occurs as fracture fills in the alaskite and trachyte dikes, or disseminated deposits replacing the host rock. The alaskite dike trends N17°W and dips west, whereas the trachyte strikes N60°W, cross-cutting the dike. An important ore shoot is lodged at the intersection. Mineralization is

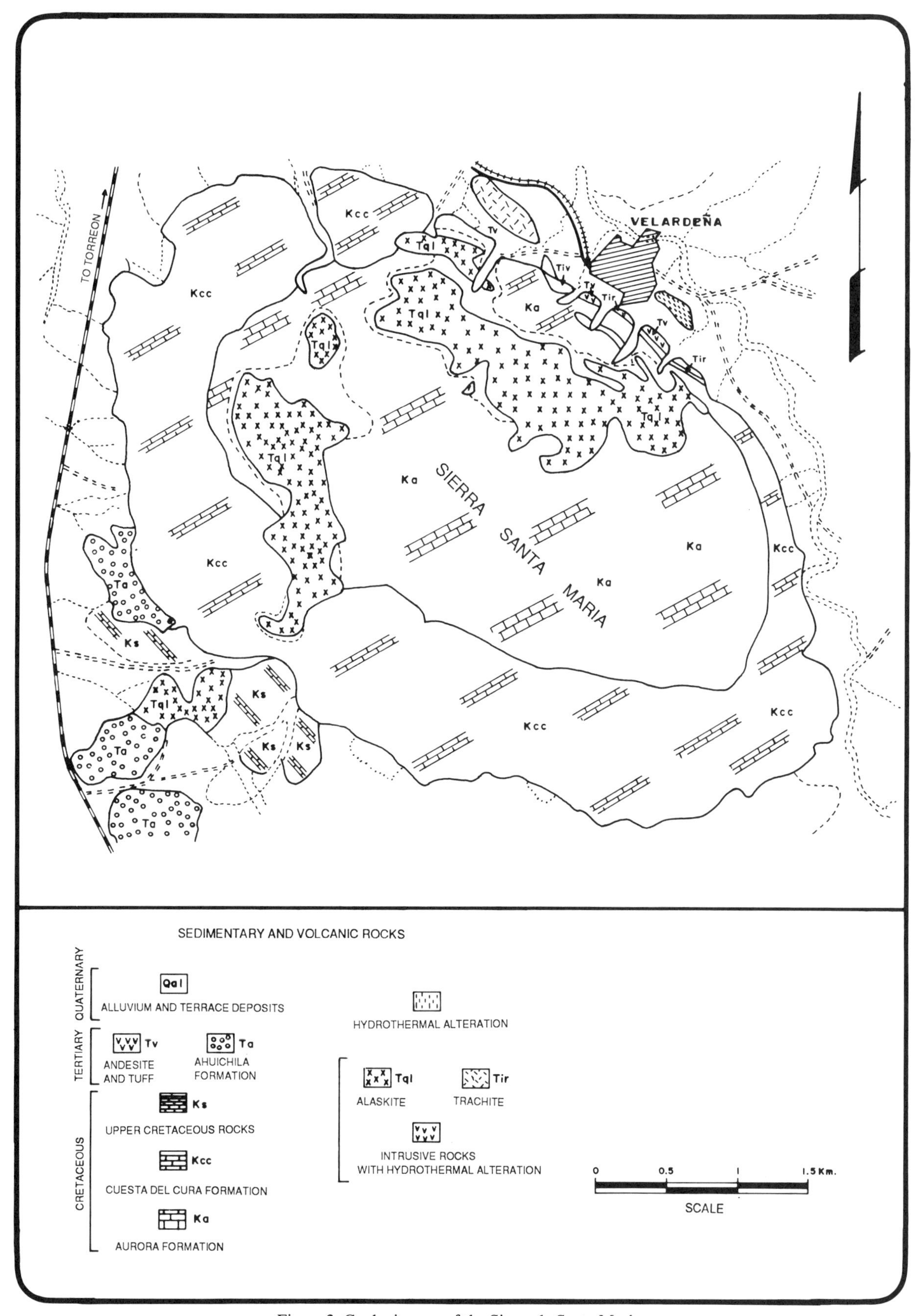

Figure 3. Geologic map of the Sierra de Santa María.

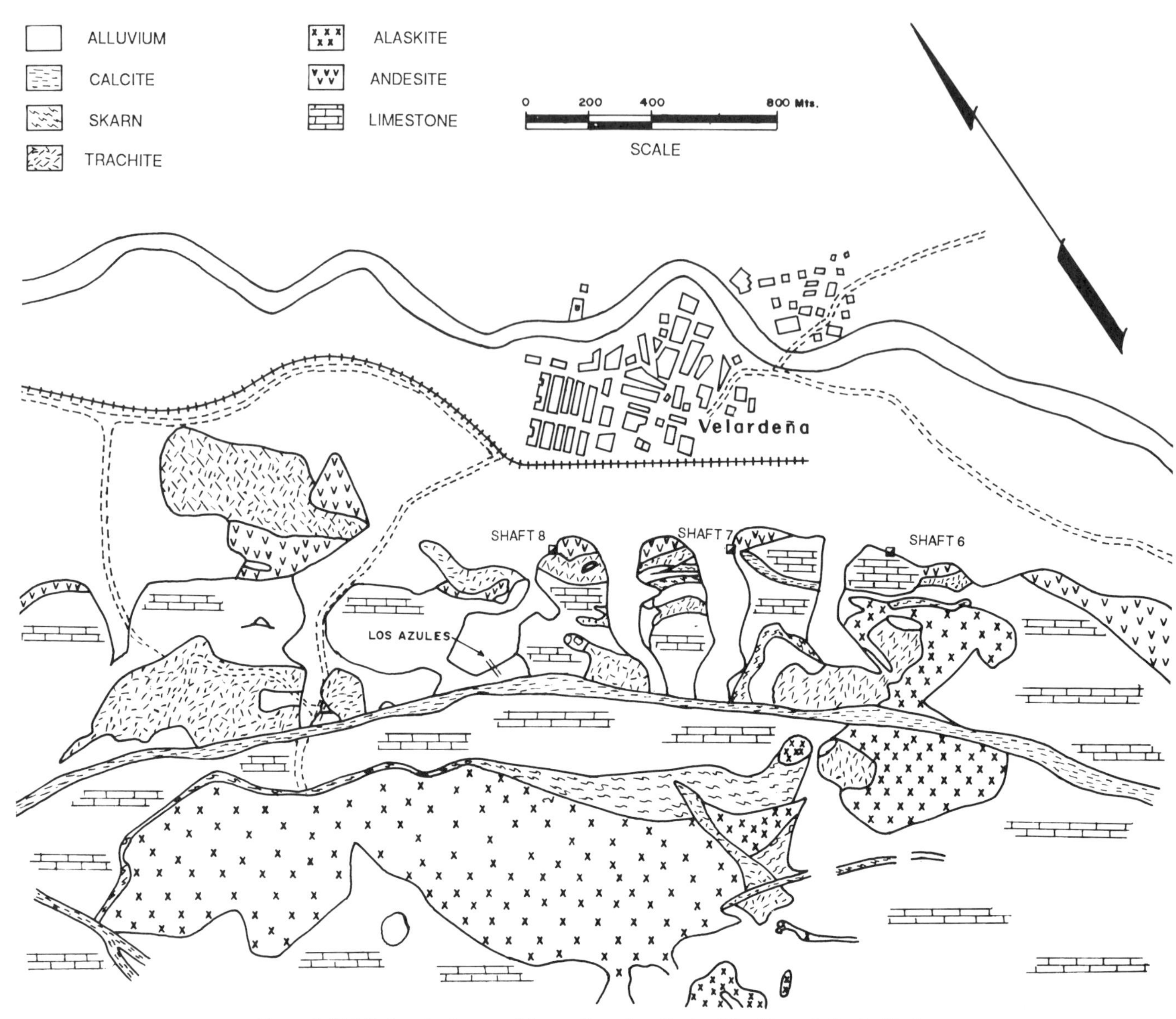

Figure 4. Detailed geologic map of the northwestern flank of the Sierra de Santa María.

generally more abundant in the alaskite than in the trachyte, probably because, being more homogeneous, it is more easily fractured. The metallic sulfides, in order of decreasing abundance, are arsenopyrite, pyrite, marcasite, silver-bearing galena, and sphalerite.

Levich (1973) suggests the following paragenetic sequence (Fig. 5): quartz deposited first, reappearing at the final stage of mineralization after an interval of nondeposition, followed by pyrite during the initial and intermediate stages; third, calcite, also appearing at the beginning and the end and continuing to deposit after all the remaining minerals, followed by widely time-extensive arsenopyrite deposition and the subsequent simultaneous deposition of sphalerite, pyrrhotite, and chalcopyrite, after which galena appears and later fluorite, ending the sequence.

The mineralogy of the various ore bodies indicates horizontal zoning with highest temperature minerals at the center of the dome and lowest temperature minerals at the contacts of the trachyte dike in the Santa María Mine orebody, representing the silver-lead-zinc zone. Away from this zone a large zinc orebody shows sharply decreased silver-lead and increased zinc values, sphalerite becomes marmatite, and pyrrhotite appears; lead-silver values are practically absent at some places.

The principal mine in the Sierra de San Lorenzo is Reina del Cobre, with significant copper values in the ample tactite zone surrounding the quartz-monzonite intrusion.

ECONOMIC GEOLOGY

The Velardeña Mining District contains several types of epigenetic ore deposits: replacements (some of the Sierra de Santa María orebodies), cavity fillings (Terneras vein system), and stockworks (Los Azules) and contact metasomatism (Reina del

Cobre). In all of these the intrusives (mainly dikes) of trachyte, diorite, quartz-latite, and quartz-monzonite composition are closely related to the origin of the mineralization.

Mineralized structures persist both along strike and vertically. The Terneras veins and the Sierra de Santa María contact-mineralized dikes extend over distances of 1.5 km and to depths of 300 m. The Reina del Cobre orebodies, with geometries approaching that of chimneys, extend for 250 m (maximum) horizontally and 250 to 500 m vertically.

Commercially mineralized zones are irregularly distributed in the orebody or structure containing them (i.e., although many intervals or areas may be subcommercial, they are not completely barren). Favorable mineralization control conditions fostered development of the large ore shoots mined in the past and at present. The ores are generally silver, lead, and zinc; subordinate gold appears on occasion, with significant values at Reina del Cobre. The gangue is commonly quartz, calcite, pyrite, marcasite, pyrrhotite, and garnets.

Two vein systems, east-west and northeast-southwest, occur in the district, the former being the most important. With the exception of the Terneras system, vein deposits are small and lack the potential for mechanized exploitation.

Dikes associated with mineralization in the Sierra de Santa María trend generally northwest. The Santa María Mine on the northeast flank of that range is the largest producer in the district. The longest mineworks extend for 2 km and up to 230 m in depth. The present main shaft (Shaft No. 6) is located east of the orebody, which has been almost completely mined down to the 150-m depth level (Level 6); most of the reserves are below this depth (Figs. 6, 7, and 8).

The mine produces 600 t/day of 156 g/t Ag, 3.8 percent Pb, and 5.2 percent Zn; iron content is about 14 percent. This production is combined with that of Los Azules (150 t/day of 120 g/t Ag, 2.1 percent Pb, 0.8 percent Zn, and 7.3 percent Fe) and Reina del Cobre (150 t/day of 101 g/t Ag, 3.4 percent Pb, 2.5 percent Zn, 0.4 percent Cu, and 5.7 percent Fe).

Other less important, mostly abandoned mines are briefly described below.

Northwest-trending veins were mined down to 50 m at the La Página Mine. Mineworks are up to 5 m wide, and pillar samples show values of up to 4 g/t Au, 156 g/t Ag, and 1.6 percent Pb; no intrusives are exposed in the area, but the strong metamorphic halo indicates their probable presence nearby.

The San Isidro Mine is also abandoned; 50-m-deep mineworks were developed in northwest-trending, 1-m-wide veins in tactite and recrystallized limestone host rocks; recent assays show 1 g/t Au, 97 g/t Ag, 1.9 percent Pb, and 1.4 percent Zn.

An intrusive diorite outcrop at the area of the Porfirio Díaz Mine developed a narrow metamorphic halo; numerous old small-scale mineworks followed short N60°W–trending, 40-cm-wide veins assaying 1.2 g/t Au, 34 g/t Ag, 0.8 percent Pb, and 0.3 percent Zn.

A diorite intrusion exposed at Las Cocinas caused strong and extensive metamorphism of the surrounding limestones, generating large garnetized and epidotized areas. The (abandoned) mineworks were developed in veins and manto-like replacement orebodies up to 2 m wide. Reconnaissance sampling at accessible sites assayed 1.7 g/t Au, 160 g/t Ag, 4.7 percent Pb, and 4.0 percent Zn. The behavior of these bodies at depth is unknown.

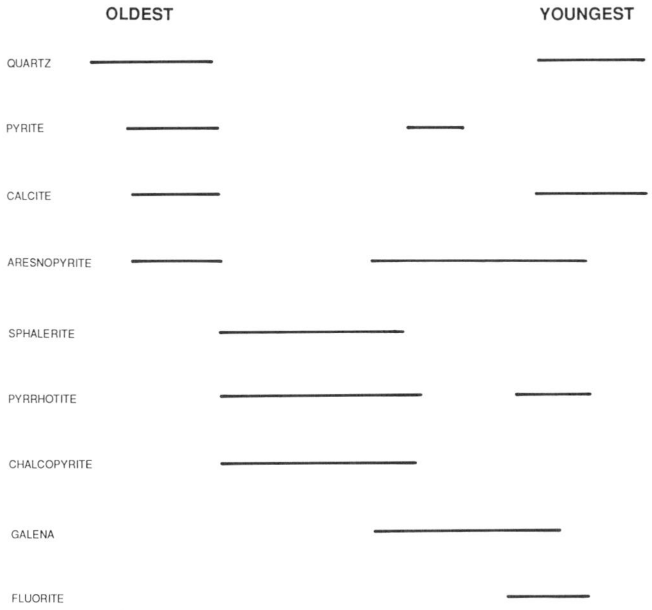

Los Libres is another diorite intrusive outcrop area presently mined at the contact by prospectors who extract ores with 5.1 g/t Au, 232 g/t Ag, 3.0 percent Pb, and 1.7 percent Zn. These are longitudinally small bodies presently mined down to a maximum 50 m. Narrow northwest-trending veins of scarce potential remain unworked.

A small tonalite intrusive cropping out in the immediate vicinity of La India Mine produced an ample metamorphic zone varying in intensity from garnet tactites at the contact to simple recrystallization at the most distant localities. The orebodies are mantos developed to date by small-scale mining. Assay averages are 0.2 g/t Au, 255 g/t Ag, 0.8 percent Pb, and 0.6 percent Zn.

Another small diorite intrusive exposed at the San Diego mine developed an abundantly garnetized and silicified metamorphic zone. Conspicuous northwest-trending veins up to 1.5 m thick, with commercial ores down to 300 m of maximum depth, may be followed for up to 2 km along strike. Assays average 1.3 g/t Au, 175 g/t Ag, 4.1 percent Pb, and 1.1 percent Zn. Most of these veins are payed out and are presently being worked selectively by prospectors.

Highly oxidized and silicified volcanic rhyolites cropping out at the La Blanca Mine area lodge several 1.5-m-wide veins with maximum values of 1.3 g/t Au, 242 g/t Ag, and 1.9 percent Pb. Although not very persistent, their simultaneous development would probably yield acceptable ore tonnages.

OLDEST YOUNGEST

QUARTZ

PYRITE

CALCITE

ARESNOPYRITE

SPHALERITE

PYRRHOTITE

CHALCOPYRITE

GALENA

FLUORITE

Figure 5. Paragenetic sequence of mineralization in the ore deposits of the Sierra de Santa María. Age gets increasingly younger to the right.

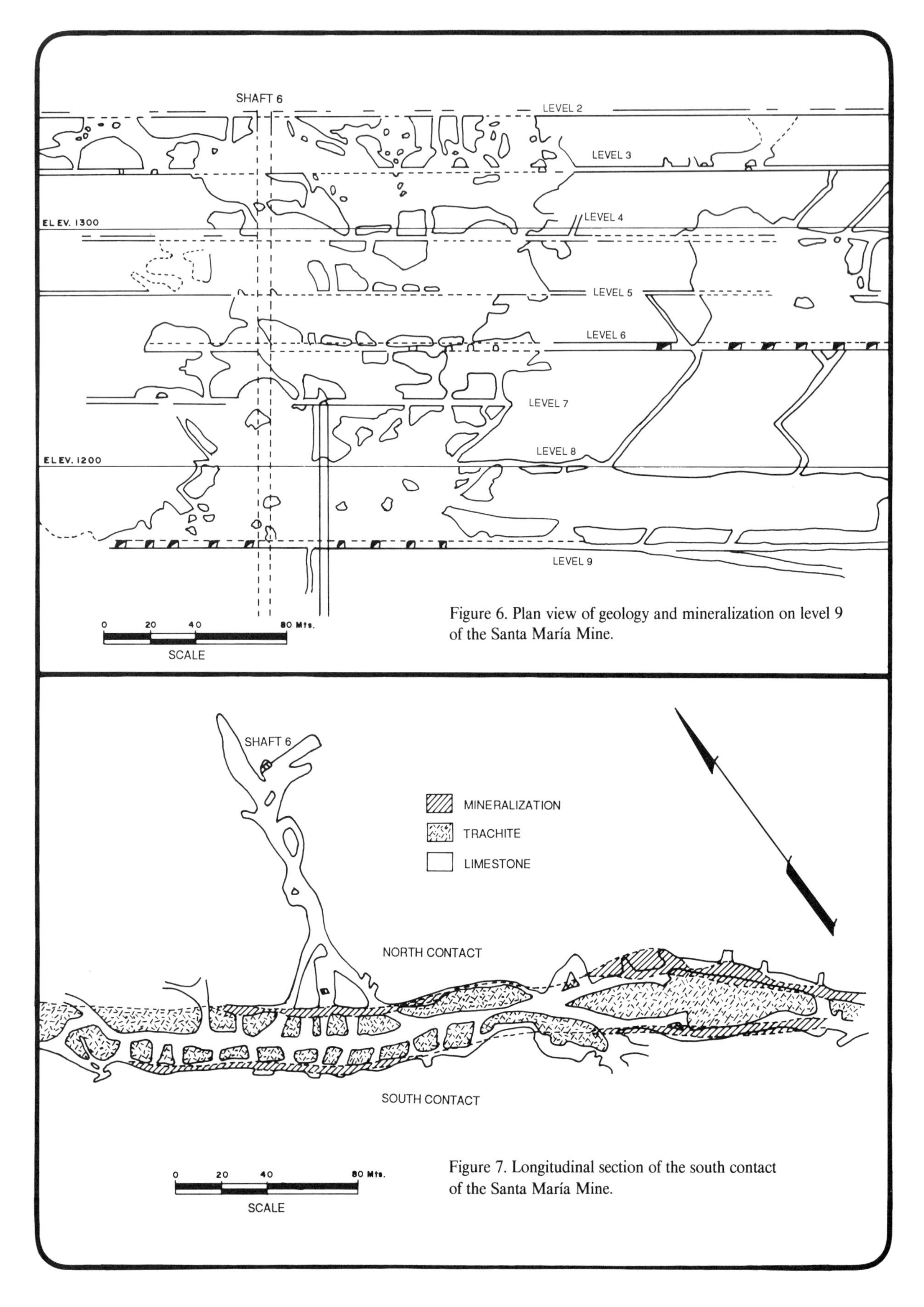

Figure 6. Plan view of geology and mineralization on level 9 of the Santa María Mine.

Figure 7. Longitudinal section of the south contact of the Santa María Mine.

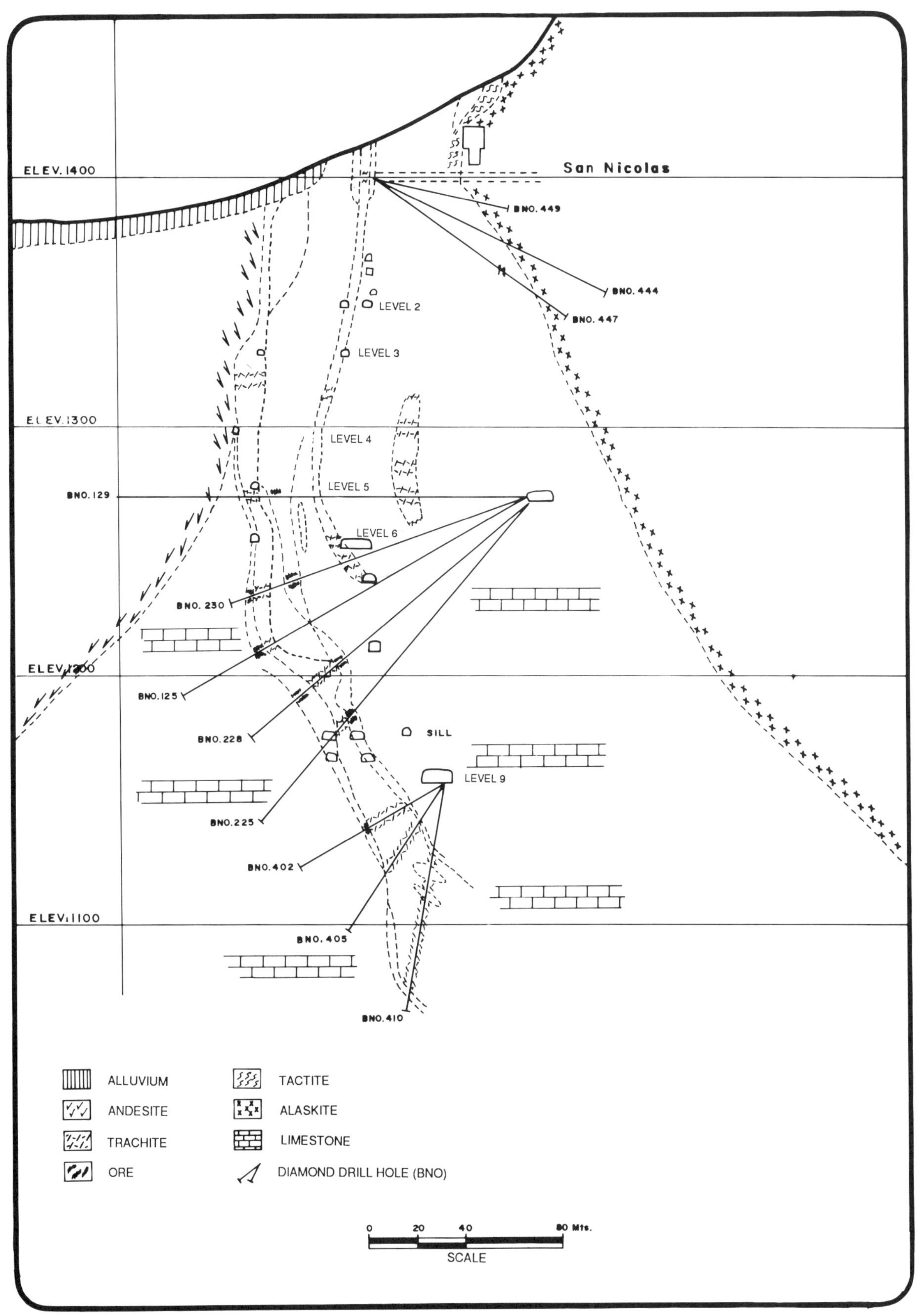

Figure 8. Cross section 31-50, N36°E, Santa María Mine, looking to the southeast.

The Velardeña Mining District has undergone several stages of development. In the first half of the sixteenth century the Spaniards worked shallow silver-rich oxidation zones; exploitation was continuous during colonial times and intermittent for most of the nineteenth century owing to the country's political instability. A smelter installed by the Velardeña Mining and Smelting Company in 1888 launched modern mining in the district. The company was acquired in 1902 by American Smelting and Refining Company, who increased production and built a more modern 2,500-ton/day smelter for ores from the main producing mines of the period: Santa María, Ternaras, and Reina del Cobre. The first quarter of the present century was the heyday for silver and lead production at Velardeña.

In 1926, technical and socioeconomic problems forced the American Smelting and Refining Company to close down and dismantle the smelter, one of the largest in the country at the time. Thereafter, small-scale mining was carried on by private enterprise and minor companies until 1968 when Asarco Mexicana, S.A. (today's Grupo Industrial Minera México, S.A. de C.V.), launched a development and mineworks reconstruction program. As a result of redefined reserves, a plant with a sulfide processing capacity of 300,000 tons/year was installed.

REFERENCES CITED

Ambriz M., D., 1979, Geología y yacimientos minerales de la Mina Santa María en Velardeña, Durango: 13th Convencion Nacional de la Asociación de Ingenieros de Minas, Metalurgístas y Geológos de México, Memorias, p. 225–260.

Clemons, R. E., and McElroy, D. F., 1965, Hoja Pedriceña 13-R-1 (4) y Hoja Torreón 13-R-1 (1): Instituto de Geología, Universidad Nacional Autonoma México, maps.

Castillo, R. J., and others, 1979, Estudio fotogeológico del Domo Santa María y Sierra de San Lorenzo, Velardeña, Durango: Private Report, Industrial Minera Mexicana, Sociedad Anonima, 61 p.

Imlay, R. W., 1937, Geology of the middle part of the Sierra de Parras, Coahuila, Mexico: Geological Society of America Bulletin, v. 48, p. 587–630.

Levich, R. A., 1973, Geology of ore deposits of Sierra de Santa María dome, Velardeña, Durango, Mexico [M.S. Thesis]: Austin, University of Texas, 84 p.

Spurr, J. E., and others, 1907, Ore deposits of Velardeña District, Mexico: Economic Geology, v. 3, p. 688–725.

Printed in U.S.A.

The Geology of North America
Vol. P-3, Economic Geology, Mexico
The Geological Society of America, 1991

Chapter 31

Economic geology of the Charcas Mining District, San Luis Potosí

F. Castañeda A.
Industrial Minera México, S.A., Ave. Baja California 200, Col. Roma Sur, 06760 México, D.F.

INTRODUCTION

The Charcas Mining District is located about 110 km north of the city of San Luis Potosí at geographic coordinates 23°07′N and 101°06′W, with an average altitude of 2,200 m. The climate is temperate and semiarid; annual precipitation averages 450 mm.

Access is by a paved road that extends north for 80 km from Highway 57, and by 130 km of paved road leading south to San Luis Potosí; the México-Laredo Railroad runs 13 km east of the district (Fig. 1), which also has a light-aircraft runway. Local schools reach technical and preparatory levels.

Mining is the main economic activity followed by cattle raising, trade, and agriculture. Present population is 25,000, of which 1,300 family groups depend directly on the mining activity for a livelihood.

REGIONAL GEOLOGY

The Charcas District occupies the east-central part of the Central Mesa Physiographic Province, which includes tectonic ranges and intermontane valleys. It belongs to the Sierra Madre Oriental Metallogenic Province. The lithology controls topographic relief.

Easily eroded sandstones and shales of the Late Triassic Zacatecas Formation (Fig. 2) exposed in low-lying areas of the San Rafael Valley form the core of an anticlinorium flanked by gently sloping rounded ridges of almost north-trending—and both east- and west-dipping—Jurassic and Cretaceous limestones.

The erosional cycle is at the stage of advanced maturity, having bared the Triassic rocks and developed extensive dendritic, south-directed drainage.

Late Triassic transgressive seas deposited the sandy-argillaceous sediments of the Zacatecas Formation. Emergence at the end of the Triassic and beginning of the Jurassic gave rise to strong erosion of the marine sediments. Volcanic activity took place simultaneously with the deposition of the Middle Jurassic continental La Joya Formation, as indicated by interbedded lavas and igneous conglomerates at some localities.

Renewed transgression in the Oxfordian covered the Mexican Geosyncline almost entirely and deposited carbonate sequences (Zuloaga Limestone) containing shallow-water faunas representing a low-energy environment. Subsidence and movements in the final stages of the Late Jurassic introduced terrigenous material from the Kimmeridgian onward, best represented by the Kimmeridgian/Tithonian La Caja Formation.

Marine deposition continued in somewhat deeper waters from the Neocomian to the beginning of the lower Aptian, with a slight contribution of fine clastics during the deposition of the Taraíses Formation; these fine clastics are absent in the Cupido Formation, which has rare benthonic fossils. The erosion of tectonic units (Coahuila Island and Valles–San Luis Potosí Platform) presumably uplifted during late lower Aptian to upper Aptian time provided the clastic sediments of the La Peña Formation.

The Cuesta del Cura Limestone, containing scarce fossils and abundant chert bands, was laid down in deeper waters during Albian to Cenomanian time.

Finally, the general area of the geosyncline was subjected to compression during the Hidalgoan (Laramide) orogeny; the Valles–San Luis Potosí platform, reacting to forces from the west-southwest, developed the region's north-trending elongated structures typified by the San Rafael anticlinorium.

The simultaneous emergence and consequent erosion of the area filled the taphrogenic tectonic troughs. The orogenic effects persisted from Paleocene to upper Eocene time, during which large batholiths—clearly evident at Charcas (Fig. 3)—were emplaced. Continuing erosion exposed the Triassic rocks in the San Rafael anticlinorium.

LOCAL GEOLOGY

Stratigraphy

Zacatecas Formation. This lithologically varied formation—principally shales, sandstones, and phyllites—forms the basement in the area. Shales or sandstones predominate at some localities, and flysch in others; thick, sometimes massive metaconglomerates associated with metatuffs are also present. The formation crops out in the San Rafael valley, with an estimated

Castañeda A., F., 1991, Economic geology of the Charcas Mining District, San Luis Potosí, *in* Salas, G. P., ed., Economic Geology, Mexico: Boulder, Colorado, Geological Society of America, The Geology of North America, v. P-3.

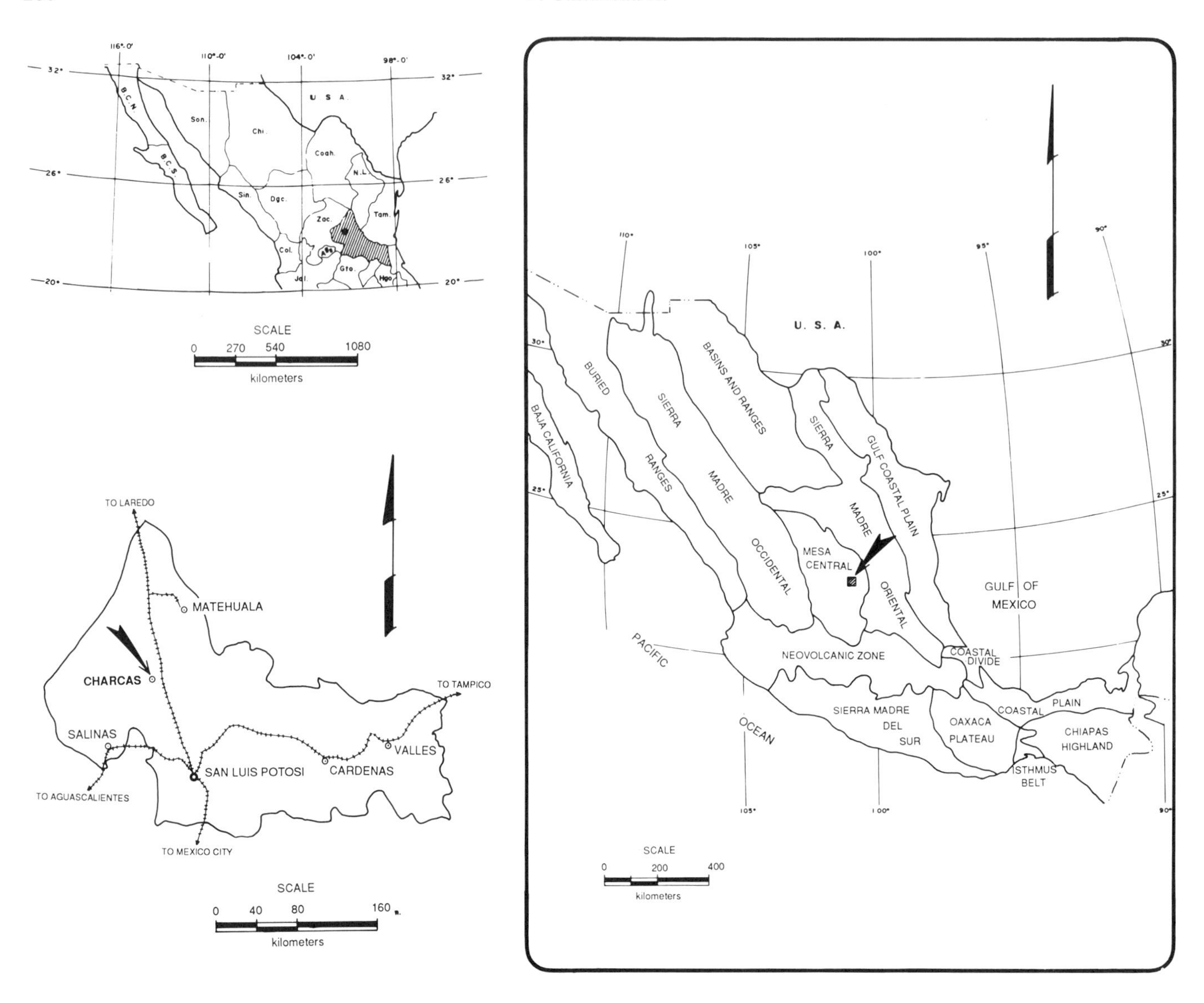

Figure 1. Maps showing location of the Charcas Mining District.

thickness of 1,500 m. Fossils collected from the Zacatecas northwest of Charcas indicate an Upper Triassic age. The unit hosts fissure-fill silver, lead, copper, and zinc ores.

La Joya Formation. This unit has an average thickness of 70 m in the area and is well exposed in west-directed streambeds near the Morelos Mine. The base comprises reddish shales and fine-grained pseudostratified tuffs followed by sandy conglomerates containing sandstone, shale, and metamorphic fragments of the Zacatecas Formation. The contact with the latter is an angular unconformity. In the absence of fossils, it is assigned to the Late Triassic and mostly Lower Jurassic on the basis of its stratigraphic position. The formation shows erratic narrow mineralized veinlets of no commercial significance.

Zuloaga Limestone. This 600-m-thick unit is divided into a lower member, approximately 450 m thick, and composed of 1-m-thick dense limestones that alternate with 3- to 4-m-thick finely crystalline limestones; the upper (approximately 160-m-thick) member is thickly bedded dense microcrystalline limestone with black chert lenses. The unit unconformably overlies the La Joya Formation.

The fossil content (gastropods and coral remains) of the Zuloaga Limestone suggests deposition in Late Jurassic shallow warm waters. The limestone contains replacement orebodies (Morelos Mine) at the base and very irregular mantos (Santa Eulalia Mine) in the upper part of the section.

La Caja Formation. Of Oxfordian-Kimmeridgian age by its fossil content (pelecypods and ammonites), this formation has four lithologically distinct members: a lower gray argillaceous limestone with black calcareous concretions overlies finely crystalline limestone that grades upward to gray-buff, very argillaceous laminar limestone, topped by bluish gray limestone with black chert bands. The formation is conformable above the Zuloaga Limestone. Irregular mantos occur at the base in the Las Lupes area.

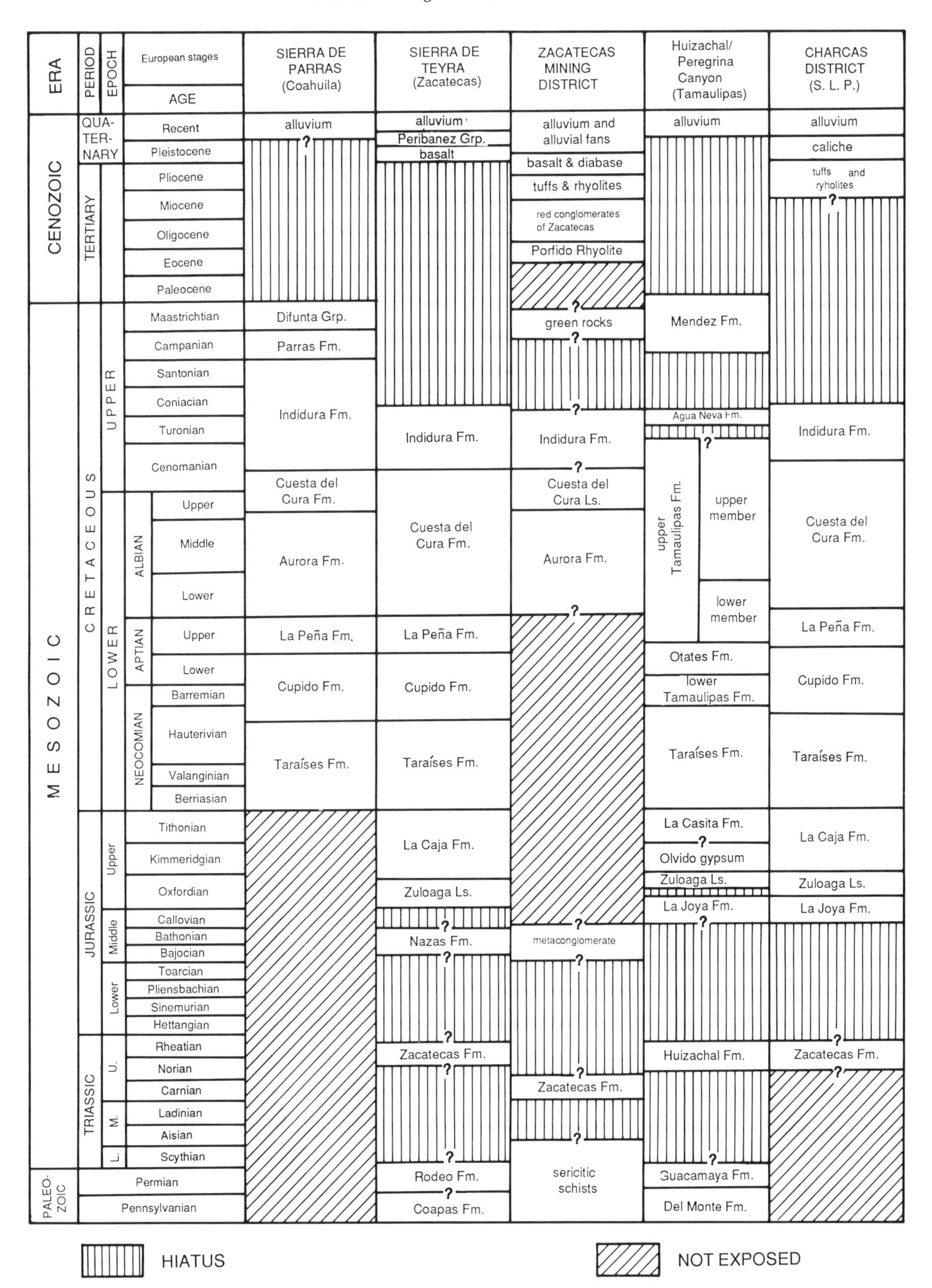

Figure 2. Correlation chart of stratigraphic units found in the Charcas Mining District.

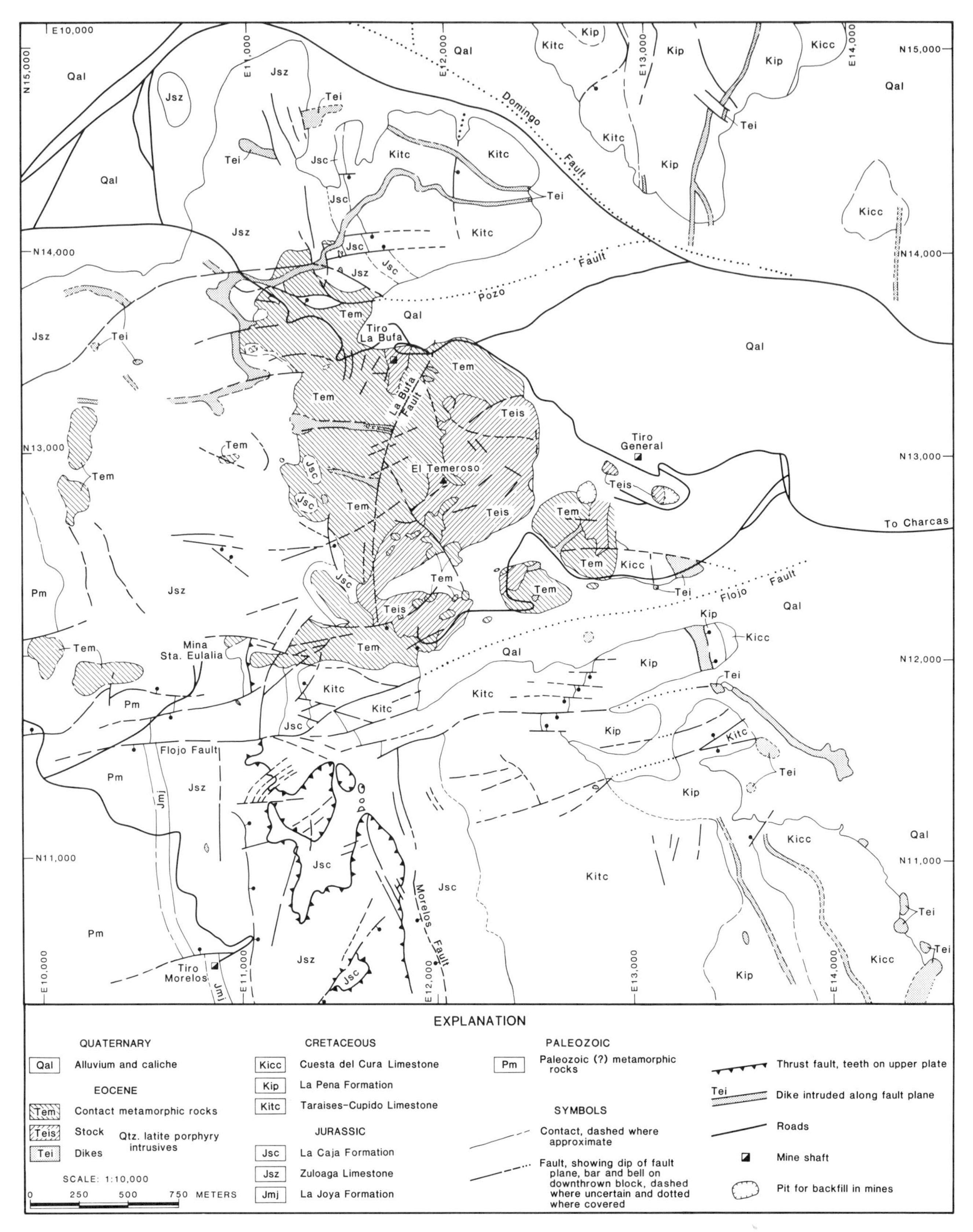

Figure 3. Simplified geologic map of the Charcas Mining District. Lines indicate relative time of formation and relative duration with respect to other minerals. (From Butler, 1972.)

Taraíses and Cupido Formations. These formations cannot be differentiated in Charcas where they consist of very dense medium- to thick-bedded argillaceous limestone with iron nodules that cause its light yellowish gray color. Ammonites in the rock are of Valanginian-Hauterivian age. These units contain most of the orebodies in the district.

La Peña Formation. In the area this is usually a thin sequence of pink and gray calcareous shales and argillaceous limestones with black chert bands assigned to the Aptian on the basis of calcispherules, ammonites, and foraminifera found north of Charcas.

Cuesta del Cura Formation. This thin- to medium-bedded limestone with argillaceous intercalations and abundant black chert bands is best exposed at hillslopes north and south of Charcas; Albian and Cenomanian small semievolute ammonites have been identified. Veins and replacement orebodies occur in the formation.

Igneous rocks

The main intrusive body in the area is the El Temeroso quartz-latite porphyry stock (Fig. 3), dated by K/Ar at 46.6 ± 1.6 Ma. The outcrop's longer axis (1,300 m) trends approximately north-south, and the shorter (400 m) east-west. This intrusion determined mineralization control in the district, having caused radial and peripheral fractures that were subsequently mineralized. The numerous rhyolite dikes emplaced at the regional east-west and north-south fractures had little effect on the adjacent carbonate rocks; they are pre-mineral because they host some fissure-fills.

The isolated and not very extensive extrusive rocks, which outcrop as tablelands occasionally with scarped rims, are tuffs and lithic tuffs with quartz and K-feldspar crystals in a pumicite groundmass. Reddish brown rhyolites, in which quartz predominates over K-feldspar, and Na-plagioclase with occasional biotite are also present. These rocks are exposed about 18 km north of the principal mines, and at Charcas toward the south.

Structural geology

The dominant structures in the region are folds associated with faults in Mesozoic carbonate rocks. The folds, either asymmetric or recumbent toward the east or north, developed by gravity sliding at the La Joya–Zuloaga and La Caja–Taraíses contacts during the early Tertiary Hidalgoan (Laramide) orogeny. The basement fault blocks that developed during deposition of the La Joya Formation were reactivated in the Eocene and Pleistocene; their regional east-west trend cross-cuts the north-trending Mesozoic folds.

The main local structure is the north-trending San Rafael asymmetric overturned anticline, approximately 20 km long; the Charcas District is located on the east flank of the structure. Jurassic block-faulting reactivated in the Tertiary divided the district into four structurally different areas. The most important, bounded by the Pozo fault in the north and by the Flojo fault in the south, contains the El Temeroso intrusive, the metamorphic halo, and the main source of ores. The stratigraphic sequence is normal and dips 20 to 60° east.

ORE DEPOSITS

Description of the mineralization

There are two commercially important types of ore deposits in the district. Triassic and the mid-Cretaceous carbonate rocks host fissure-fill orebodies (veins). In the Cretaceous rocks, replacement occurred at both vein walls when the limestone host rocks were favorable to substitution processes. At present, the large ore volumes of the irregular and mantos-type replacements in calcareous rocks make them the most commercially important. The main deposits—under development by the Charcas Unit of Industrial Minera México, S.A.—are associated with local and regional fracture systems trending generally N65°W to N80°E and dipping 60 to 70° northeastward toward the intrusive at the Leones and Santa Isabel veins, or else 30 to 60° southwestward toward the limestone at the Nueva, Santa Rosa, Tigres, and other smaller veins.

The fissure-fill mineralization parallel to the intrusive/limestone contact near the Tiro General Mine includes the major Leones and Santa Isabel veins, which converge toward the east to form the Combinación vein. Toward the west the Leones vein is hosted in limestone, whereas the Santa Isabel penetrates the intrusive with considerably reduced width. Fissure development is attributed to a normal fault that developed when the intrusive contracted and subsided during the cooling process.

Each of these veins is cut by the Principal fault, which is roughly parallel to the El Temeroso stock boundaries and younger than the veins but represents an old weakness zone, as indicated by the numerous replacement orebodies occurring along the fault.

The Leones–Santa Isabel trend continues farther west of the Principal fault, generating the El Rey and La Reina replacement orebodies, which are the largest producers in the district. Toward the southwest, the La Campaña and San Sebastián veins are subparallel to the intrusive contact. The La Bufa fault, also almost parallel to the El Temeroso contact, controls replacement mineralization on both sides.

Commercial replacement mineralization extends west of the intrusive to the Triassic-Jurassic contact that contains the La Aurora, Las Lupes, and Las Bibianas deposits, whereas the fissure-fills extend in the same direction to a distance of about 7 km from the intrusive outcrop, at the Potosí, La Trinidad, San Agustín, and San Diego mines. Ten km southwest are the N60°W–trending and 45- to 60°-dipping El Membrillo and San Francisco fault veins. Closer to the intrusive in the same direction is the Mina Morelos replacement ore deposit in a N45°W–trending, 65°NE–dipping fault at the La Joya/Zuloaga Limestone contact.

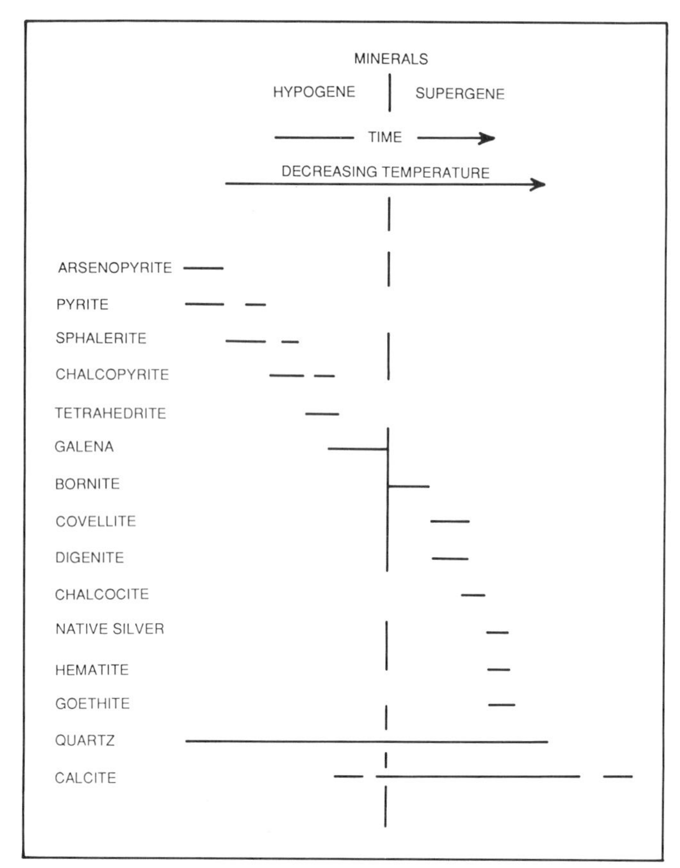

Figure 4. Paragenesis of the minerals in the principal orebodies. (From Butler, 1972.)

With some exceptions the mineralogy of the replacement orebodies is similar to that of the veins, to which they are occasionally connected. The replacements are rather irregularly shaped though sometimes tabular and conformable with the stratification, indicating selective replacement of some beds.

Mineralogy and paragenesis

The paragenetic sequence reported by Butler (1972) is shown on Figure 4. In addition, the following minerals are also present: diaphorite, jalpaite, stromeyerite, argentite, schirmerite, matildite, aguilarite, stibnite, cuprodescloizite, hemimorphite, smithsonite, hematite, and native gold and silver.

Geochemistry of the orebodies

Orientation studies in the district indicate distributions and genetic associations that may be important aids in exploration. For instance, boron metasomatism in the vicinity of zinc deposits within the metamorphic halo suggests boron enrichment in areas of high zinc concentration. Figure 5 compares boron and zinc contents in a drillhole through the Santa Rosa replacement orebody; although the Zn/B ratio is not a simple one, it might be a guide in geochemical exploration.

Another example suggests that vanadium and strontium were dispersed in the limestones by mineralizing fluids and developed haloes at the outer boundaries of the orebodies.

Geochemical orientation studies indicate Zn, Cd, Cu, Ag, Pb, and B as the most useful elements in exploration, given the consistency of the halos they generate. Table 1 compares the migration distances and concentrations of the index minerals in the El Rey and La Reina orebodies.

Zoning

Field observations indicate marked vertical and horizontal zoning in the district. Based on the Leones vein, Pb values remain more or less constant (about 3 percent) from the surface to the 250 m depth, decreasing thereafter to less than 1 percent at 600 m. Cu (approximately 0.3 percent near the surface) increases to 2.5 percent with depth. Pb and Ag values decrease longitudinally toward the east in the La Bufa Mine, whereas Zn and Cu increase considerably. The only two known small gold deposits occur north of La Trinidad Mine and south of Las Hormigas in the vicinity of El Temeroso. Antimony appears in a stibnite vein outside the metamorphic halo, although also in veinlets near the intrusive stock at levels 14-00 and 18-00 of the Tiro General Mine, which possibly indicates another mineralization phase at greater depth.

Origin of the deposit

The fissures caused by regional faulting and later by the cooling of the intrusive magma affected both the limestone and the stock itself, favoring the emplacement of orebodies in both rocks. This suggests that the intrusion had solidified when mineralization took place and consequently was not the source. The coexistence, near highly permeable zones, of veins and replacements containing the same mineralogical species suggests their common origin.

The paragenesis and mineralogy of the orebodies are typical of Pb-Zn-Cu-Ag deposits in carbonate rocks. The general mineral depositional sequence followed the order of stability of complex sulfides.

The high metasomatic boron content and silicification of the orebodies suggest that the material was transported to the surface from a deep area of high fluid pressure. Mineralization took place

TABLE 1. GEOCHEMICAL INFORMATION

Mineral	Migration distance (m)	Range of values (ppm)
Ag	80	8–12
Pb	80	200–350
Zn	230	500–700
Cd	155	30–50

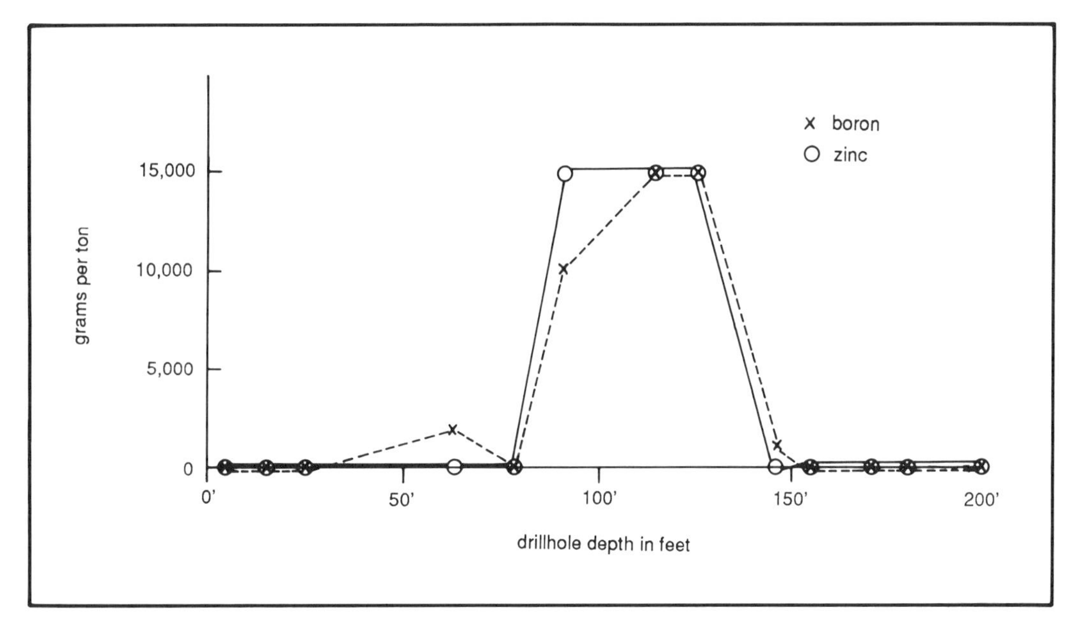

Figure 5. Comparative contents of boron and zinc in drillholes in the Santa Rosa emplacement orebody.

in two different stages: the first (Leones vein type) is silver/lead/zinc–rich with calcite and borosilicates as the main gangue minerals, whereas copper and silver predominate in the second (Santa Isabel type) with a high quartz content in the gangue.

In conclusion, the Charcas ore deposits were probably generated by complex sulfide-carrying fluids derived from deep-seated magma. The mineral elements were transported upward in solution through highly permeable regional fault zones and along the boundaries of the El Temeroso intrusive stock. Precipitation followed the relative order of stability of complex sulfides and caused the zoning present in the district. The age of the mineralization has been determined as 38.6 Ma.

ECONOMIC GEOLOGY

Ore deposits

Industrial Minera México, S.A., is the main producer in the Charcas Mining District. A new processing plant, finished in 1985, produces 3.450 ton/day of lead and zinc concentrates at the Charcas Unit (i.e., an annual total of 1,069,500 tons).

The group of mines run by Industrial Minera México, S.A., includes San Sebastián, Morelos, La Bufa, and Tiro General; at present the El Rey and La Reina orebodies between the two latter are under development; the La Aurora and Las Lupes bodies east of La Bufa are being explored.

The Consejo Minerales is presently exploring a fissure-fill deposit that includes the conspicuous El Membrillo and San Francisco veins 10 km southwest of the Tiro General Mine. Lesser antimony deposits are under development west and north of the Tiro General Mine.

Production

Modern production figures of the principal mines are shown in Table 2.

Small mineworks at Hacienda de Charcas Viejas in 1570 were the first reported in the district. The present mine area was discovered in 1583. The most productive Leones and Santa Isabel veins (today inactive though not paid out) were opencut mined jointly near the San Bartolo Shaft, yielding up to 1.5 kg/t Ag near the surface. Estimated value of production from 1865 to 1895 was 13,924,900 Mexican pesos of the period, with a profit of 5,113,437 pesos. 250,000 tons were extracted from 1906 to 1910 of 485 g/t Ag, 4.4 percent Pb, 2.14 percent Cu, and 11.4 percent Zn.

Precise historical data on numerous smaller mines operated by successive owners and prospectors at different times is unavailable due to the absence of records.

TABLE 2. PRODUCTION FIGURES FOR THE PRINCIPAL MINES OF THE CHARCAS DISTRICT

Mine	Period	Extracted tonnage	Grades			
			Ag (g/ton)	Pb (%)	Cu (%)	Zn (%)
T. General	1926–1983	13,352,960	164	1.71	0.50	7.63
S. Sebastián	1956–1974	662,926	147	2.29	0.49	6.27
La Bufa	1942–1957	682,065	72	0.35	0.34	9.50
Morelos	1944–1953	71,854	116	2.29	0.39	10.30
S. Diego	1933–1934	7,433	414	0.11	0.06	0.40
L. Eulalias	1968	830	119	5.00	0.56	14.40
Potosí	1952	217	127	0.11	1.60	1.20

Ore reserves

Reserves of the Charcas Unit as of January 1, 1984 are shown on Table 3. The Consejo de Recursor Minerales has calculated the El Membrillo–San Francisco reserves as of October 1983 to be 833,500 tons of 0.92 g/t Au, 115 g/t Ag, 0.86 percent Pb, and 2.92 percent Zn.

TABLE 3. RESERVES

Classification	Tons	Grades			
		Ag (g/ton)	Pb (%)	Cu (%)	Zn (%)
Exploitable	9,961,190	103	0.71	0.28	4.66
of interest	6,488,200	74	0.27	0.43	6.07

REFERENCES CITED

Butler J., H., 1972, Geology of the Charcas Mineral District, San Luis Potosí, Mexico [Ph.D. thesis]: Golden, Colorado School of Mines, 170 p.

Castañeda A., F., 1977, Geología, controles de la mineralization y exploración indirecta en el Distrito de Charcas, S.L.P.: 1st Symposium on Mining, Asociación de Ingenieros de Minas, Metalurgístas y Geólogos de México.

Humara G., G., 1971, Geología y mineralización de la Mina Tiro General en Charcas, S.L.P.: 9th Convencion Nacional de la Asociación de Ingenieros de Minas, Metalurgístas y Geólogos de México, 15 p.

Madrigal L. J., 1979, Controles e interpretación mineralógica estructural de los depositos de reemplazamiento en el Distrito de Charcas, S.L.P.: 13th Convencion Nacional de la Asociación de Ingenieros de Minas, Metalurgístas y Geólogos de México, 39 p.

Printed in U.S.A.

The Geology of North America
Vol. P-3, Economic Geology, Mexico
The Geological Society of America, 1991

Chapter 32

Geology and mineralization of the El Realito Mine, Río Santa María District, Guanajuato, México

P. Fraga M.
Cía. Minera Peñoles, S.A. de C.V., Río de la Plata 48 esq. Lerma, Col. Cuauhtemoc, 06500 México, D.F.

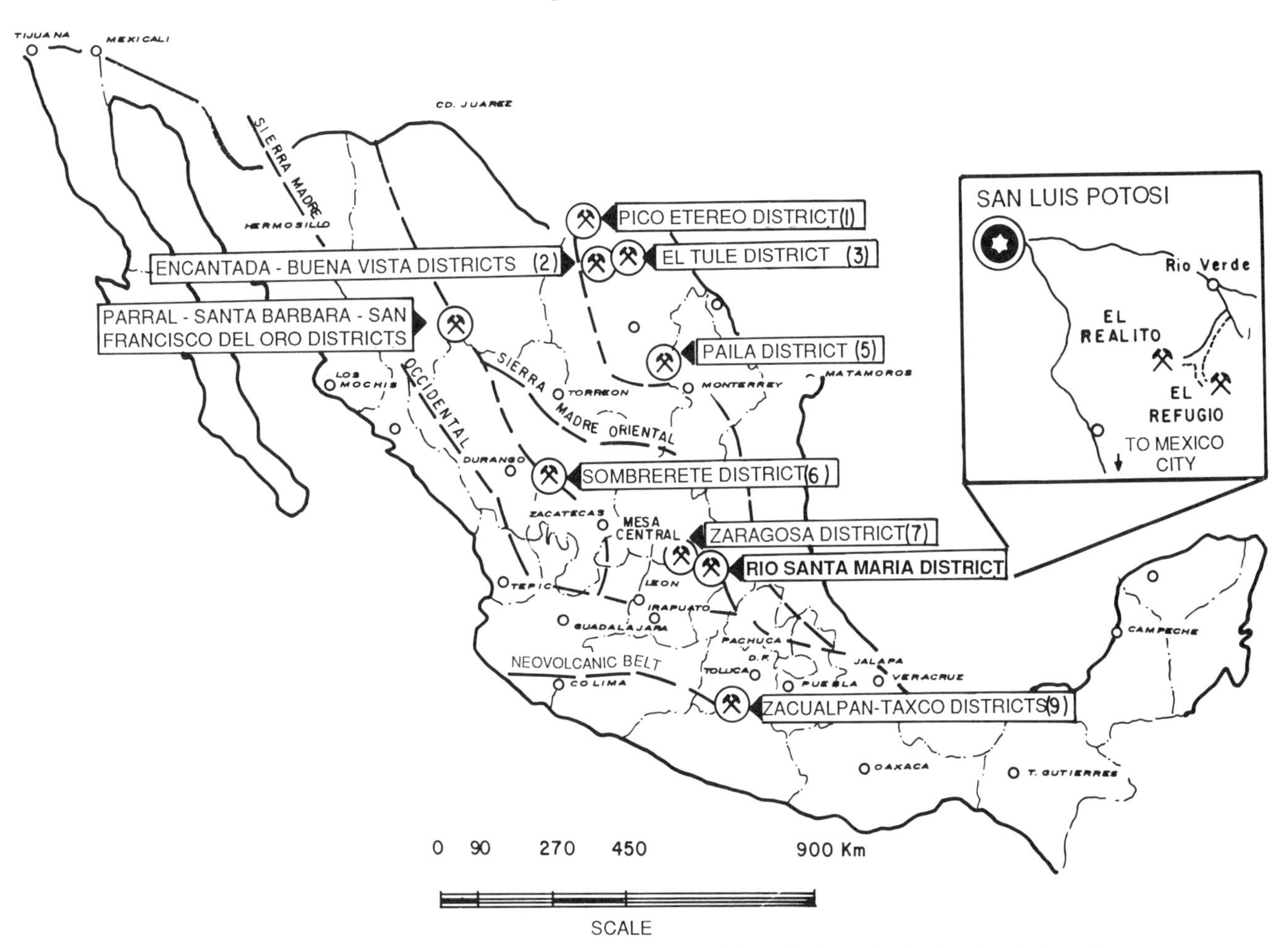

Figure 1. Maps showing location of the Río Santa María Mining District, and other fluorite districts of Mexico.

INTRODUCTION

The Río Santa María Fluorite District lies at the Guanajuato–San Luis Potosí state line 100 km directly S50°E of the city of San Luis Potosí. The nearest important town is Río Verde 45 km N40°E away (Fig. 1). Access to the El Realito and El Refugio Mines is from Río Verde by a paved road for 16 km that turns southwest for 43 km to Alamos de Martinez at the center of the district. The nearest railroad station is Ciudad Fernandez, west of Río Verde. The nearest seaport is Tampico on the Gulf of Mexico, 250 km east of Río Verde.

The district covers 240 km^2 and lies at the transition between the Sierra Madre Oriental and the Central Mesa (Fig. 1). The stratigraphy is presented on Table 1.

Fraga M., P., 1991, Geology and mineralization of the El Realito Mine, Río Santa María District, Guanajuato, México, *in* Salas, G. P., ed., Economic Geology, Mexico: Boulder, Colorado, Geological Society of America, The Geology of North America, v. P-3.

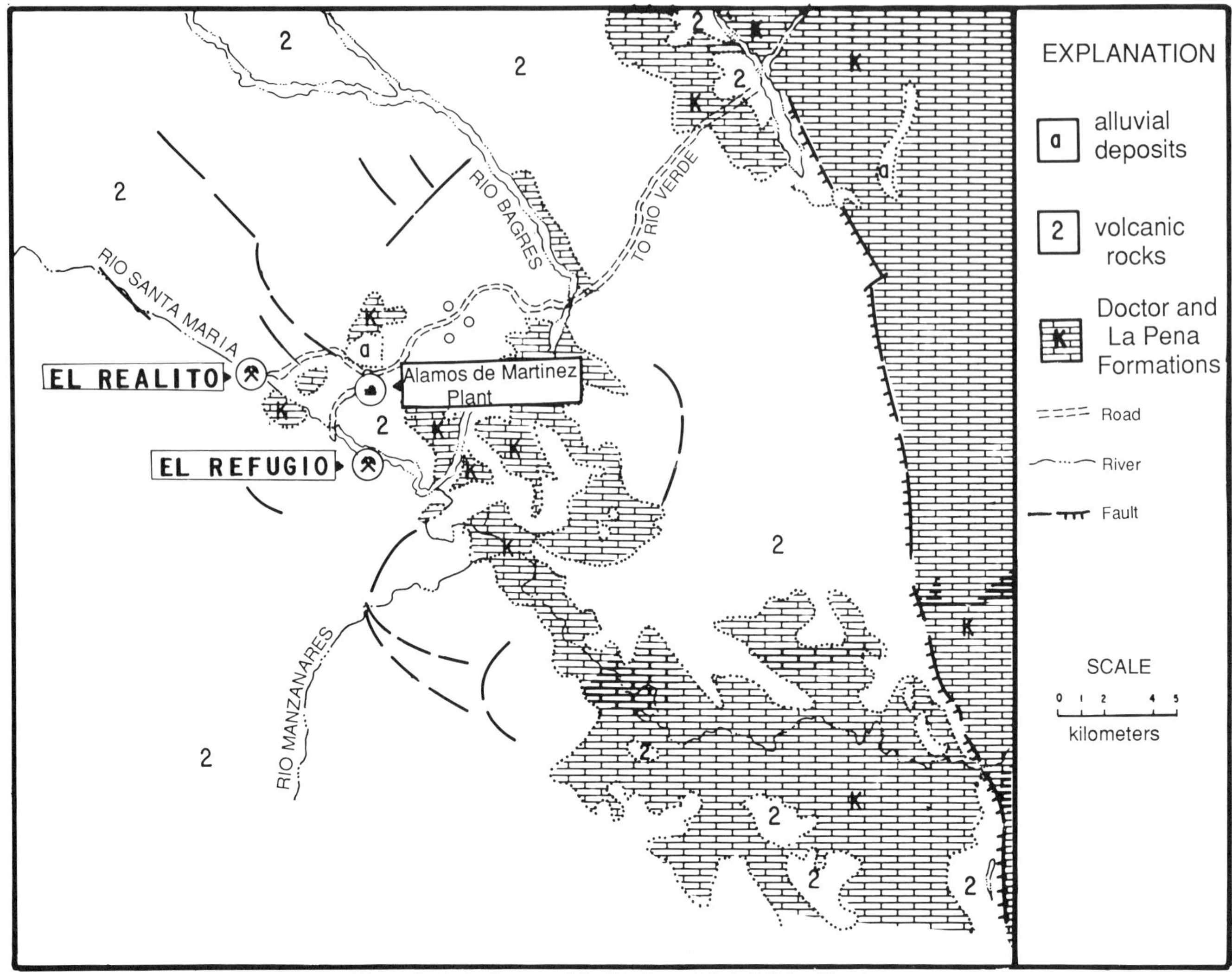

Figure 2. Geologic map of the Río Santa María Mining District.

TABLE 1. REGIONAL STRATIGRAPHY, SANTA MARÍA DISTRICT

Quaternary		Alluvium deposits
	Unconformity	
Tertiary	Andesite and rhyolite volcanic and pyroclastic rocks.	
	Intrusive rocks; rhyodacitic subvolcanic porphyries.	
	Unconformity	
Lower Cretaceous	Doctor Formation:	Cerro Ladrón limestone facies.
		Socavón limestone facies.
		San Joaquín limestone facies.
		La Negra limestone facies.
	La Peña Formation:	Black calcareous shales.
		Argillaceous limestones.

History

Prospectors seeking small mercury concentrations in the upper portion of some of the fluorite orebodies began to mine the district in 1950. Several companies later carried out exploration until 1970 when Grupo Industrias Peñoles, S.A., took control over most of the district and started operating the El Realito and El Refugio Mines, the two largest fluorite producers (Fig. 2).

REGIONAL TECTONICS

The fluorite deposits are regionally associated with caldera-type structures following a northwest fault trend along the course of the Río Santa María.

Various volcanic and pyroclastic units of andesite and rhyolite composition, together with rhyodacite subvolcanic intrusives genetically and spatially associated with the fluorite mineralization, are present in these Tertiary calderas, whose development

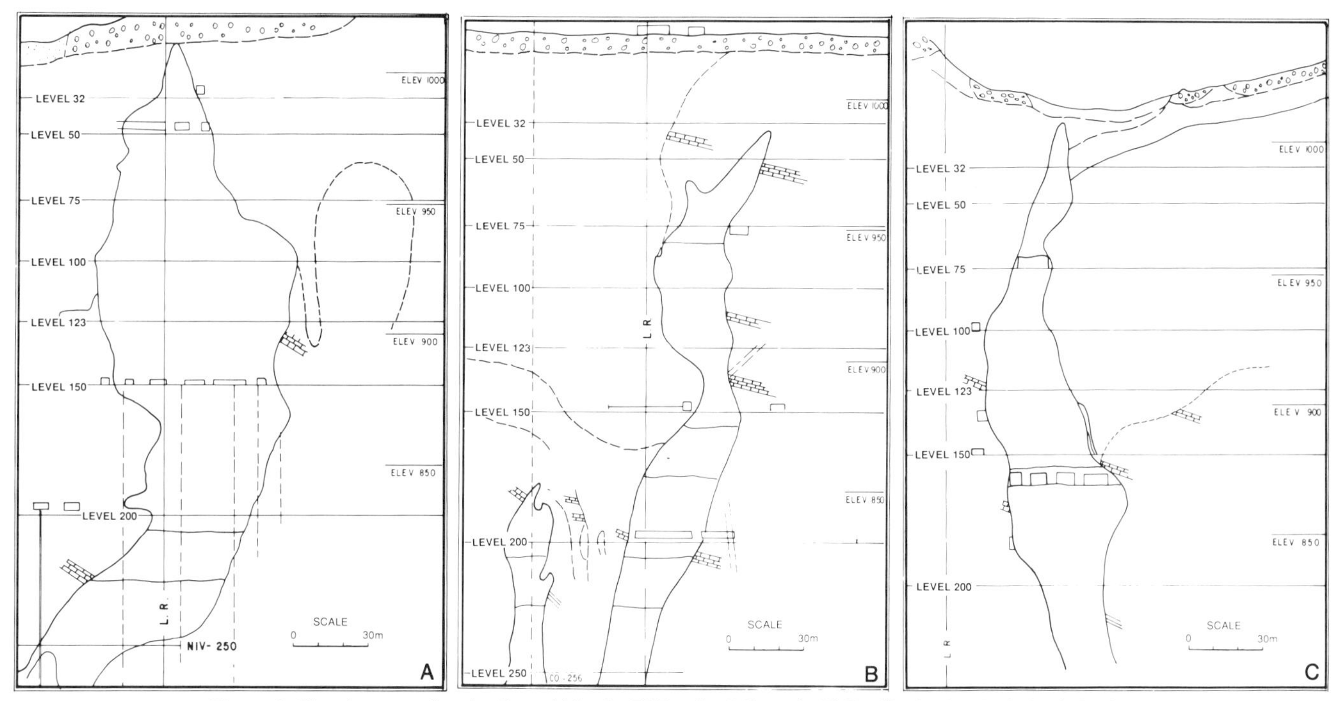

Figure 3. Sketch maps of orebodies within the El Realito Mine. A, El Realito body; B, Colonia body; C, El Quatro body.

affected the Santa María District area by generating the lithologic and structural conditions for subsequent fluorite deposition.

Pre-mineral structures—faulting, brecciation, folding—which developed in the calcareous rocks of the Doctor Formation during the formation of the calderas, determined the placement of the orebodies. The predominant pre-mineral structures (caldera annular margins, radial or central faults) trend northwest and host the principal known orebodies.

Post-mineral, northeast-trending faults affect some of the known bodies by displacing the principal northwest-trending structures.

GEOLOGY OF THE EL REALITO DEPOSIT

Surface geology in the area surrounding the El Realito Mine is represented by scattered outcrops of limestones of the Doctor Formation on the right bank of the Río Santa María. These rocks have been gently folded in northwest-trending anticlines, and they are unconformably overlain by Tertiary volcanic rocks. Small, intrusive, strongly kaolinized, and pyrite-oxidized rhyolite outcrops are visible among the limestones.

Within the El Realito Mine, three separate chimney-type orebodies—Cuerpo Colonia, El Cuatro, and El Realito—have been located in the limestone aligned along a regional northwest-trending fault zone from northwest to southeast. Exploration and mining have revealed the depth of the orebodies to exceed 300 m with varying diameters, and the long axes to be respectively 150, 40, and 17 m at level 200 (Fig. 3).

Mineralization

Fluorite appears in varied colors (brown, honey, reddish, or gray), generally massive, banded, or brecciated, filling open spaces, and replacing the calcareous sediments.

Printed in U.S.A.

The Geology of North America
Vol. P-3, Economic Geology, Mexico
The Geological Society of America, 1991

Chapter 33

Geology and mineralization of the La Negra mining unit, Querétaro

P. Fraga M.
Cía. Minera Peñoles, S.A. de C.V., Río de la Plata 48 esq. Lerma, Col. Cuauhtemoc, 06500 México, D.F.

INTRODUCTION

La Negra Mine is located in the Maconí Mining District, Querétaro State, in the eastern part of the Sierra Guanajuato Gorda (Fig. 1). The nearest important town is Querétaro, 215 km southwest of the mine unit; the nearest village is Maconí, 5 km away from the mine. Access to La Negra Mine is from Querétaro on the Cadereyta-Maconí road, which takes about 2 hours. The nearest railroad station is San Nicolás, 120 km away on the México-Querétaro railroad. The mine is at an elevation of 2,000 m.

This summary is based mostly on work by José E. Gaytán-Rueda (1975), presently the Director of Peñoles Group Exploration-Mines; the Special Studies group of Industrias Peñoles, S. A., Exploration Division; and unpublished reports by the unit's Geological Department.

History

The district was mined for oxidized Pb, Zn, and Ag ores in colonial times; prospectors did small-scale mining in the last century. The district has had different owners in this century; in the 1950s a mining company carried on diamond-drill exploration with negative results. Grupo Industrias Peñoles, S.A., acquired the main properties in the district in 1960; an exhaustive exploration program, which included diamond drilling, resulted in the discovery of the chimney-type La Negra and El Alacran orebodies, and the mining unit began operations in 1971.

REGIONAL GEOLOGY

The La Negra Mining District lies in the Sierra Madre Oriental physiographic province. Regional stratigraphy is represented by Late Jurassic Las Trancas Formation and the Lower and Upper Cretaceous Doctor, Soyatal, and Mezcala Formations, unconformably overlain by Tertiary volcanic rocks and intruded by granodiorites or diorites, also of Tertiary age (Table 1).

The Doctor Formation, containing the principal orebodies in the La Negra facies, is the commercially most important unit.

REGIONAL TECTONICS

The area is characterized regionally by a circular geomorphologic anomaly caused by intrusive dikes, apophyses, and irregular bodies typical of a tectonomagmatic activation zone. These intrusives metamorphosed the limestones of the La Negra facies of the Doctor Formation, developing a well-defined grossularite-andradite skarn zone, together with structural channels favorable to subsequent mineralization.

LOCAL GEOLOGY

The principal outcropping unit in the La Negra Mine area is the La Negra facies of the Doctor Formation (Fig. 2) comprising usually thinly bedded black argillaceous limestones, trending N30 to 40°W, and dipping 20 to 80° southwest, with occasional black chert nodules. The facies is partially metamorphosed by northeast- and northwest-trending granodiorite-diorite apophyses.

TABLE 1. REGIONAL STRATIGRAPHY, LA NEGRA DISTRICT

Quaternary		Alluvium deposits
---Unconformity---		
Tertiary		Volcanic rhyolites, andesites, and basalts. Intrusive granodiorites, diorites.
---Unconformity---		
		El Morro Conglomerate.
---Unconformity---		
Upper Cretaceous	Mezcala Formation	Limestone, argillite limestones
	Soyatal Formation	Shales, sandstones
Lower Cretaceous	Doctor Formation	Cerro Ladrón limestone facies, Socavón limestone facies, San Joaquín limestone facies, La Negra limestone facies.
---Unconformity---		
Upper Jurassic	Las Trancas Formation	Sandstones, graywackes, and shales.

Fraga M., P., 1991, Geology and mineralization of the La Negra mining unit, Querétaro, *in* Salas, G. P., ed., Economic Geology, Mexico: Boulder, Colorado, Geological Society of America, The Geology of North America, v. P-3.

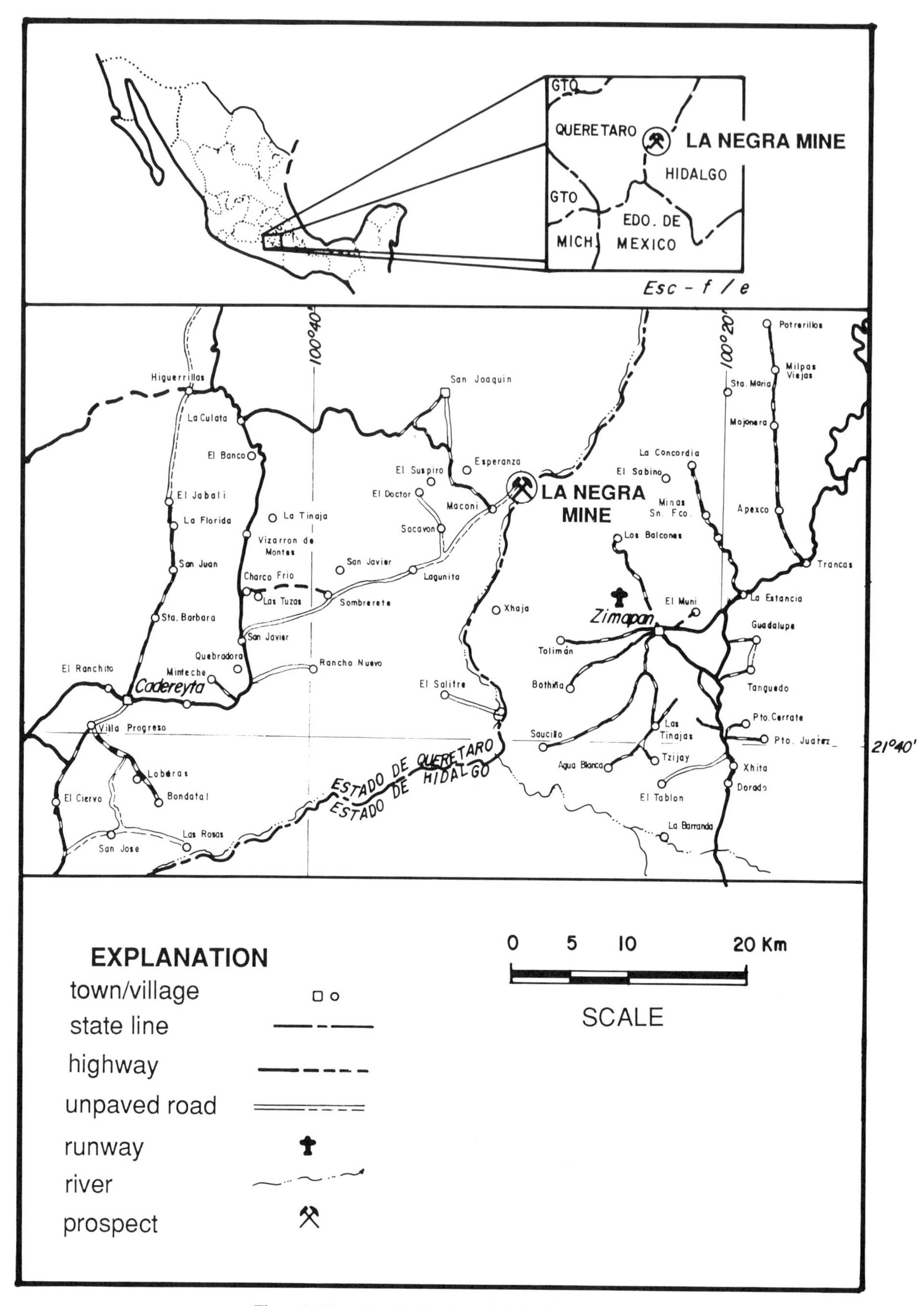

Figure 1. Maps showing location of the La Negra mine.

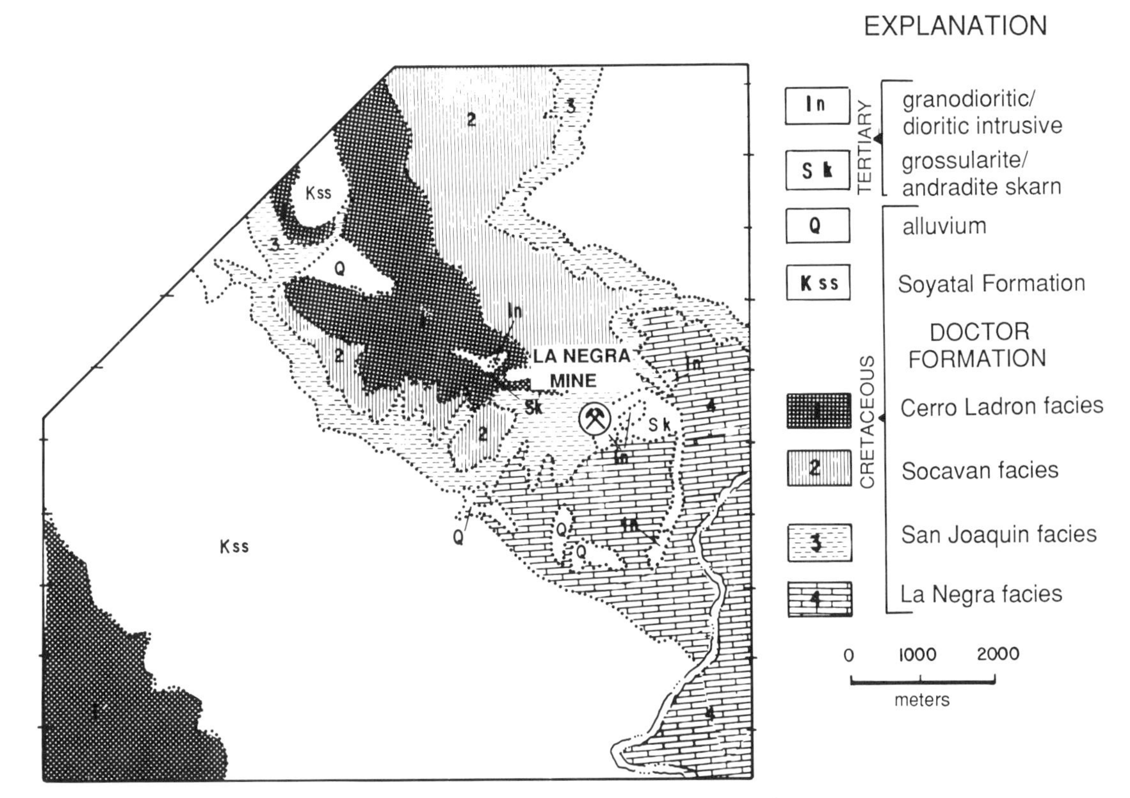

Figure 2. Regional geology at the La Negra mine.

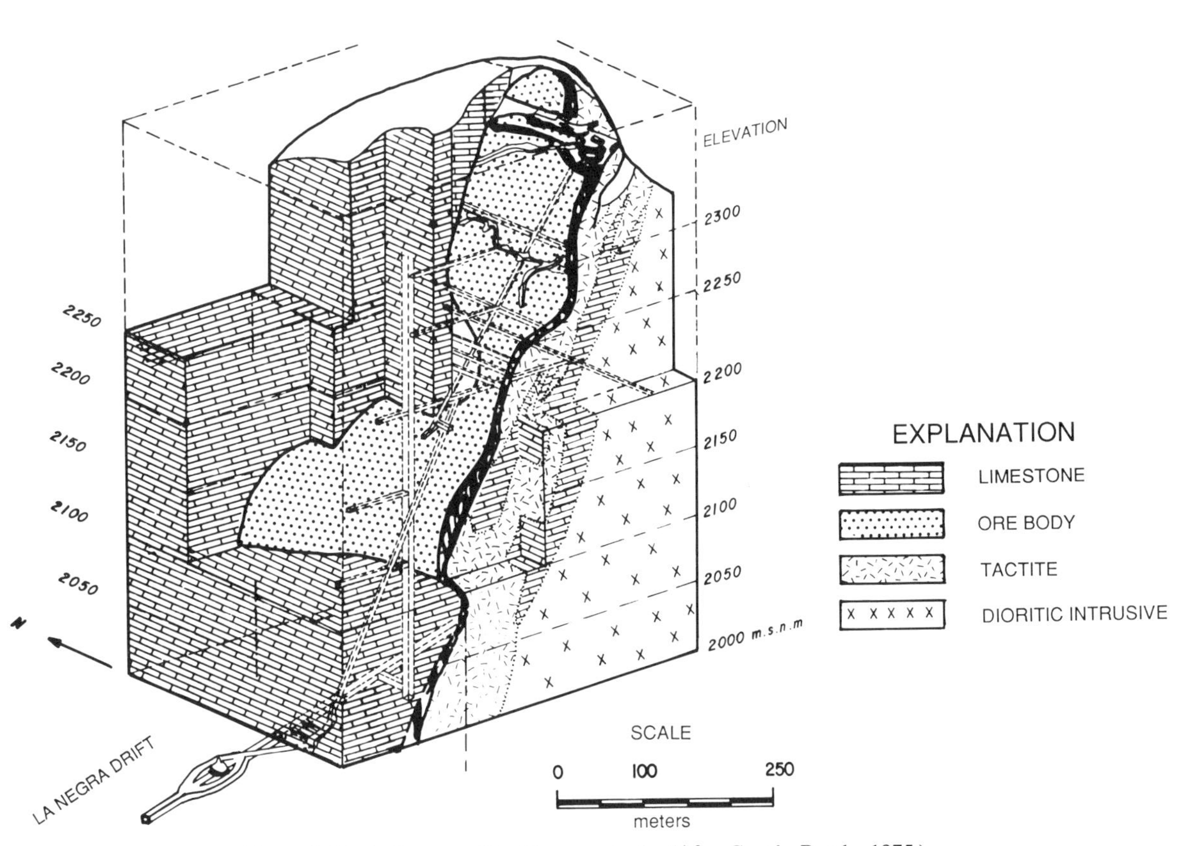

Figure 3. Block diagram of La Negra orebody. (After Gaytán-Rueda, 1975.)

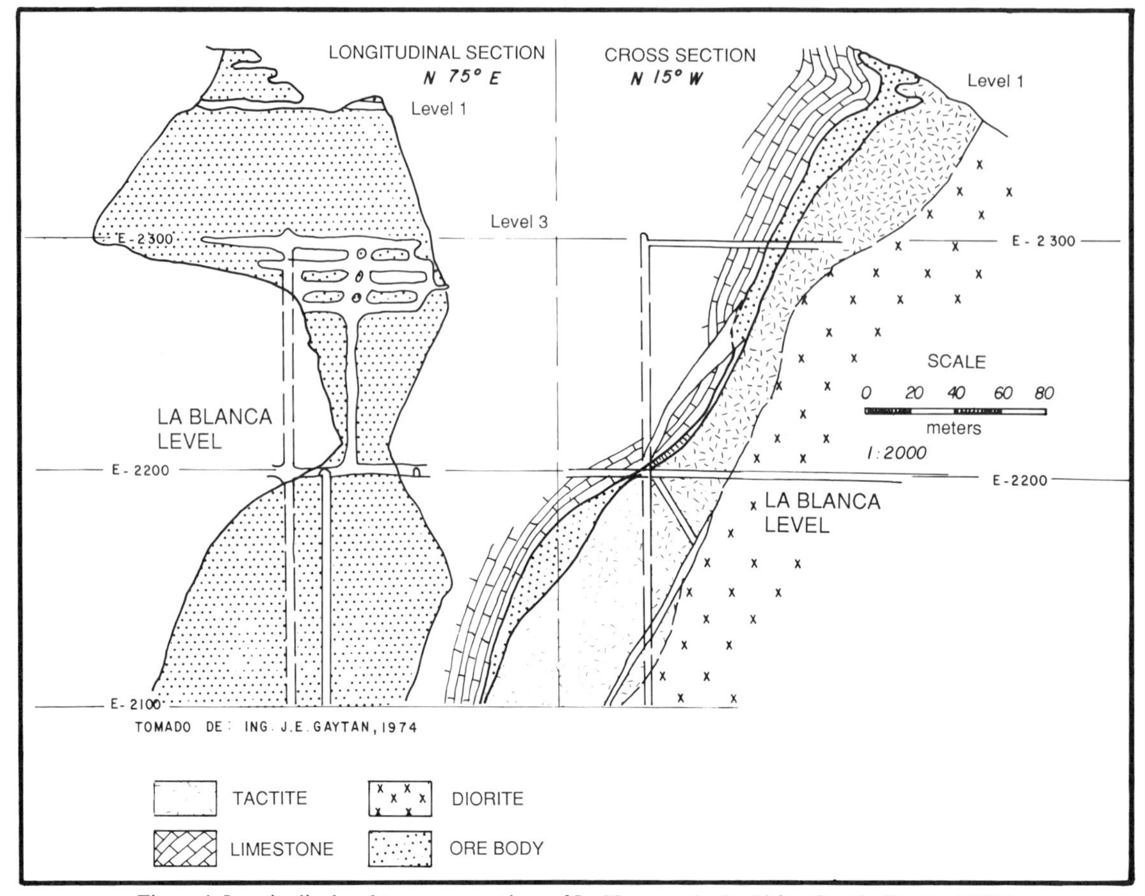

Figure 4. Longitudinal and transverse sections of La Negra orebody. (After Gaytán-Rueda, 1975.)

The orebodies are located mostly at the transition from garnet to limestone; orebody mineralogy is predominantly hessite, galena, marmatite, and chalcopyrite, with pyrite, arsenopyrite, and pyrrhotite as gangue.

MINERALIZATION

The La Negra Mine orebodies (Figs. 3, 4) exhibit well-defined structural and mineralogical vertical zoning. Disseminated ores with noncommercial metal content at topographically and stratigraphically higher parts, become manto-type bodies at depth with commercially prospective metal contents and volumes. Deeper down, between levels 2,000 and 2,400, the mantos become the major silver-producing chimney-type orebodies (La Negra, Alacrán). At greater depth the Ag-Pb contents decrease and Cu increases.

The Negra-Alacrán-Silvia-Esperanza–type chimneys are variably sized. Each of the principal and best-known (Negra, Alacrán, Silvia) contain minimum volumes of 500,000 tons; similarly sized orebodies are presently under exploration and only partially known.

From 1971 to 1983 the La Negra Mine has produced 2,600,000 tons of 157 g/t Ag, 1.04 percent Pb, 2.80 percent Zn, and 0.20 percent Cu at an average rate of 300,000 m.t./year (i.e. 25,000 m.t./month); proven reserves to 1983 are close to 3,000,000 tons.

REFERENCE CITED

Gaytán-Rueda, J. E., 1975, Exploration and development of the La Negra Mine, Maconí, Querétaro, México [M.S. thesis]: Tucson, University of Arizona, 98 p.

Printed in U.S.A.

The Geology of North America
Vol. P-3, Economic Geology, Mexico
The Geological Society of America, 1991

Chapter 34

Description of some deposits in the Zimapán District, Hidalgo

G. García G. and F. Querol S.
Cía. Fresnillo, S.A. de C.V., Río de la Plata 48 esq. Lerma, Col. Cuauhtemoc, 06500 México, D.F.

GEOGRAPHIC LOCATION

The Zimapán Mining District, in the Zimapán municipality, western Hidalgo State, is located at geographic coordinates 20°47′N and 99°24′W. Access is by Federal Highway 85 (México-Laredo) at m 207 (Fig. 1).

Unpaved roadways lead from the village of Zimapán to the district's main mine areas at Carrizal, 7 km northwest, and El Monte, 8 km north (Fig. 2); Zimapán has all the basic services as well as a light-aircraft runway.

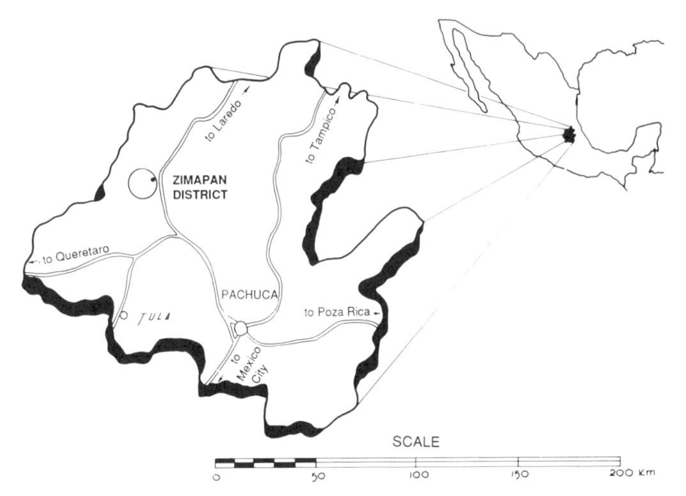

Figure 1. Map showing location of the Zimapán Mining District.

BRIEF HISTORY OF THE DISTRICT

The district was discovered at the beginning of the seventeenth century; the very sparse information on its development, published by Simons and Mapes (1957), is summarized below.

The Lomo de Toro Mine, discovered by Lorenzo del Sabra in 1632, was worked continuously for silver and lead until 1810 when the Spaniards were ousted from the area during the War of Independence. No production figures are available, but at the time, possibly 40 smelters operated around Zimapán, and another 12 around El Monte. Production from 1785 to 1788 was valued at approximately 1,985,720 dollars at the time. From 1791 to 1793, Don Angel Bustamante, Earl of Batopilas, extracted ores valued at about $100,000 from the Santa Rita mine and 5,700 metric tons of lead from Lomo de Toro. The value of the district's production from 1830 to 1840 is estimated at $2 million; at the time, Lomo de Toro was under management by Compañía Minera Real del Monte y Pachuca, and the La Cruz, San Fernando, and Guadalupe Mines were run by the Compañía Anglomexicana. In 1840, a Mr. Juarez extracted lead and silver values at more than $50,000 from El Monte. Production was practically suspended from 1840 to 1872 due to the country's political and economic instability; in 1878, only four smelters and four small furnaces processed ores values at 35 to 64 dollars per ton. Lomo de Toro was revived from 1890 to 1901, continuing with smaller-scale production until 1910 when it again closed down until 1929. The Hidalgo Copper Mining and Smelting Company began working the Concordia and perhaps other mines in the El Monte area in 1913, smelting 1,741 metric tons of ores with 21 percent Pb and 806 ppm Ag. Mining activity increased during the 1920s, and by 1929, six companies operated 18 mines throughout the district. Smelting continued at Zimapán until 1939 when the American Smelting and Refining Company launched the San Luis Potosí foundry.

The present large oxidized orebodies of the Lomo de Toro Mine were discovered in 1945; by 1947 the roadway to the Carrizal area allowed increased production, mostly from Lomo de Toro and Los Balcones. In the 1929 to 1945 period, Lomo de Toro produced a total of 35,000 metric tons of high-grade lead-silver ores.

The Compañía Fresnillo began exploring the El Monte area in 1948, and exploration of the Carrizal area began in 1950. At the time the ore was hauled by burros over 10.5 km to Zimapán. The combined ore production was processed at the company's Fresnillo plant until a selective flotation plant with a capacity of 900 metric tons/day of sulfides was installed at El Monte in 1972.

García G., G., and Querol S., F. 1991, Description of some deposits in the Zimapán District, Hidalgo, *in* Salas, G. P., ed., Economic Geology, Mexico: Boulder, Colorado, Geological Society of America, The Geology of North America, v. P-3.

GEOLOGIC LOCATION

Mesozoic marine rocks and Cenozoic continental and volcanic deposits crop out in the Zimapán District area.

Physiography

The district lies in the Mexican Plateau at the boundary zone of the Central Mesa and Sierra Madre Oriental Physiographic Provinces. The western foothills of the Sierra Madre Oriental are represented north of Zimapán by the Sierra El Monte (Fig. 2).

Major physiographic features are the Sierra El Monte, the Zimapán alluvial fan, and the Río Tolimán. The Río Tolimán drains into the Moctezuma, which marks the Hidalgo-Querétaro state line.

The east-west–trending Sierra El Monte (highest peak: San Nicolás at 2,720 m asl) forms the north flank of Zimapán valley, developing low volcanic hills toward the east as far as the port of La Estancia on the Panamerican Highway. On the west the range is cut by the imposing Tolimán canyon, which is several kilometers long, about 1,500 m deep, and rimmed by steeply dipping (more than 40°) scarps of highly deformed Cretaceous sedimentary rocks. The Carrizal area lies at the bottom of the canyon, which becomes wider and shows more gently sloping walls after crossing the contact with Jurassic argillaceous rocks. Local relief at Carrizal is 800 m. The drainage, strongly controlled by the lithology, develops a rectangular pattern of streams running parallel to the calcareous strata.

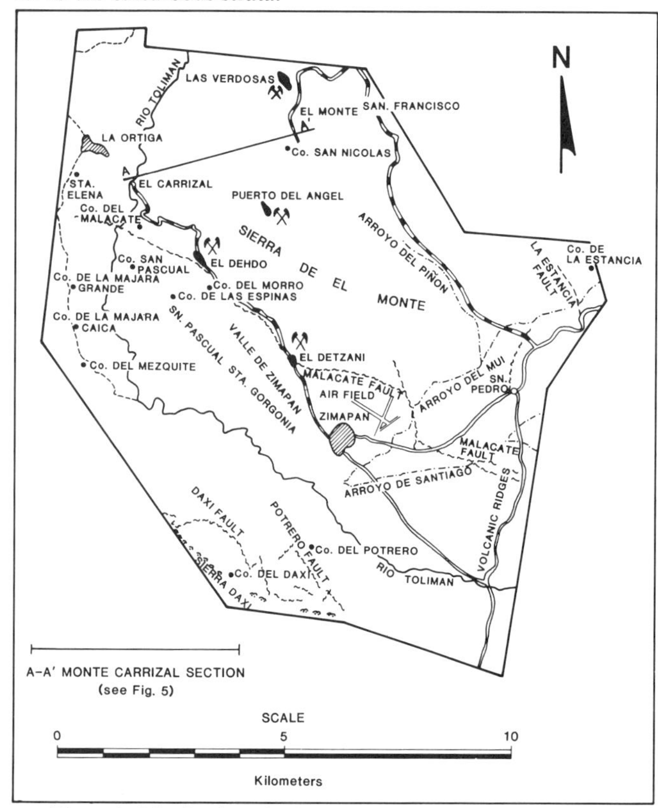

Figure 2. Distribution of mines in the Zimapán Mining District.

Regional stratigraphy

Upper Jurassic to Late Cretaceous outcrops form the mountains over most of the district; Tertiary continental and volcanic rocks, igneous plutonic and hypabyssal monzonites associated with the mineralization, and small andesite dikes are also present (Figs. 3, 4). Late Tertiary and Quaternary semiconsolidated alluvial fans cover the low-lying areas.

Jurassic: Las Trancas Formation. Composed of thin-bedded, gray, shiny, phyllitic, calcareous shales, the Las Truncas Formation is interbedded with dark gray limestones exposed in the northern part of the district with an estimated thickness of 800 m, the base not being exposed. The Kimmeridgian-Tithonian age indicated by macrofossils is extended by some to the Aptian. The unit unconformably underlies the El Doctor Formation and forms the basement of the mineralization (Simons and Mapes, 1957).

Cretaceous: El Doctor Formation. Encompassing most of the district's mountain areas, the El Doctor Formation is the host rock for the mineralization. Some 750 to 900 m have been measured in complex folds in the Ixmiquilpán-Actopán area (Hidalgo). The formation is made up of gray and dark gray limestones in 10- to 30-cm-thick beds, with intercalated black chert layers that become white near the mineralized zones, and massive, lenticular, occasionally very fossiliferous, possibly reefal limestones up to 30 m thick and hundreds of meters long, assigned to the middle Albian–lower Cenomanian on the basis of their fossil content (López Ramos, 1979). The formation is conformable over the Las Trancas and unconformably overlain at times by the El Morro fanglomerate, volcanic rocks, or Recent alluvium.

Cretaceous: Soyatal Formation. This formation, an estimated 1,000 m thick, is a sequence of yellow shales alternating with black micritic limestones and marls; the calcareous phase increases or decreases laterally. The underlying contact with the El Doctor Formation is gradual and conformable. Macro- and microfossils indicate its age as Cenomanian-Maastrichtian, and it is unconformable beneath the Tertiary El Morro Fanglomerate, Las Espinas volcanics, and Quaternary deposits.

Tertiary: El Morro Fanglomerate. A poorly sorted, red to purple-gray fluviolacustrine conglomerate of subangular limestone and volcanic fragments with some sandy beds, the El Morro Fanglomerate interfingers with volcanic agglomerates and flows and attains a maximum thickness of 360 m west of the district, although it is widely variable laterally. Its age is unknown, but by correlation with the Guanajuato Conglomerate, it may be Eocene-Oligocene (Simons and Mapes, 1957).

Tertiary: Las Espinas Formation. This formation is a series of mostly andesite and dacite flows and agglomerates occasionally including basalts and quartz-latites. The Las Espinas is exposed south of the district with thicknesses of up to 150 m; it interfingers at the base with the El Morro Fanglomerate, which makes them contemporaneous. Absolute ages have not been determined.

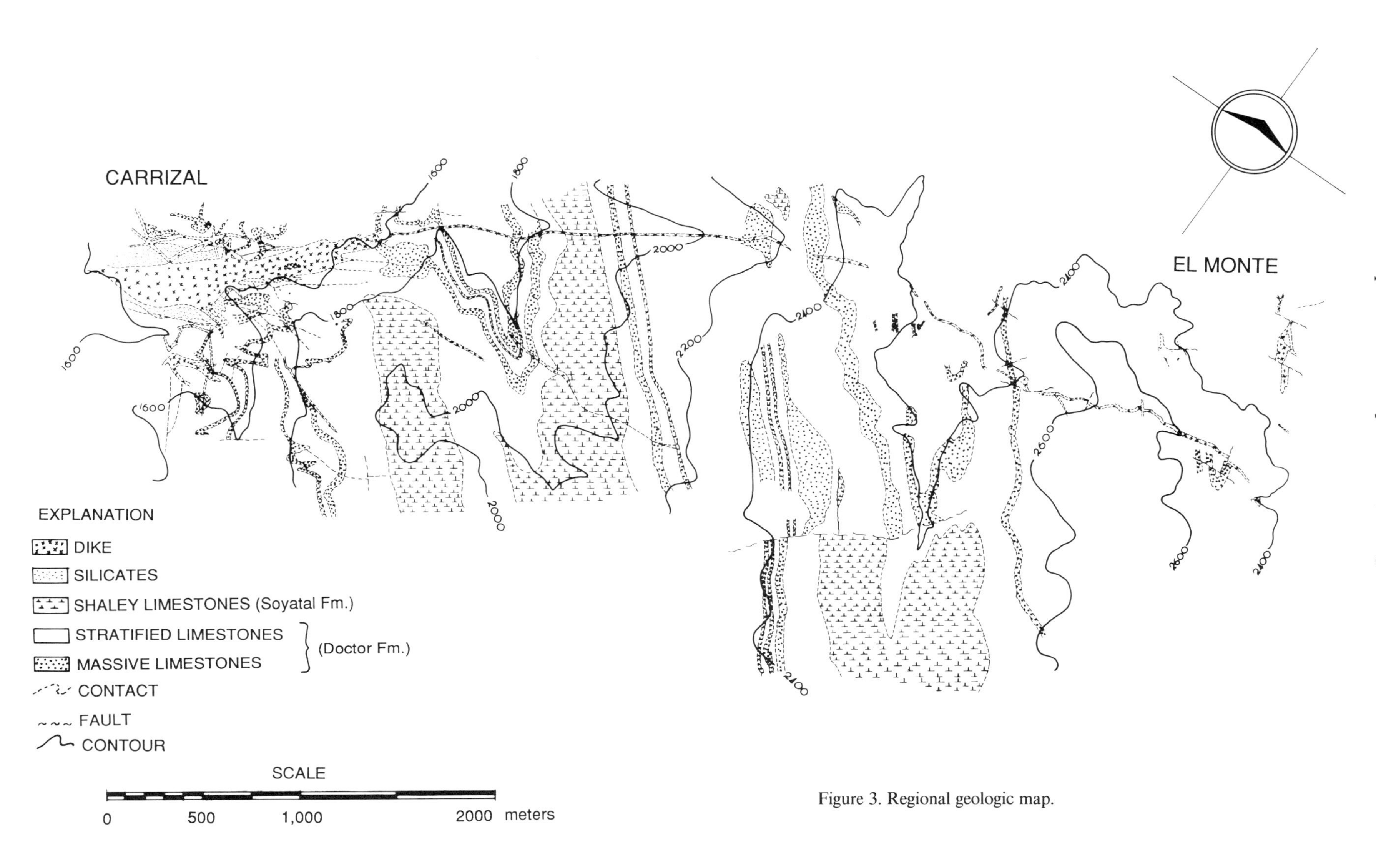

Figure 3. Regional geologic map.

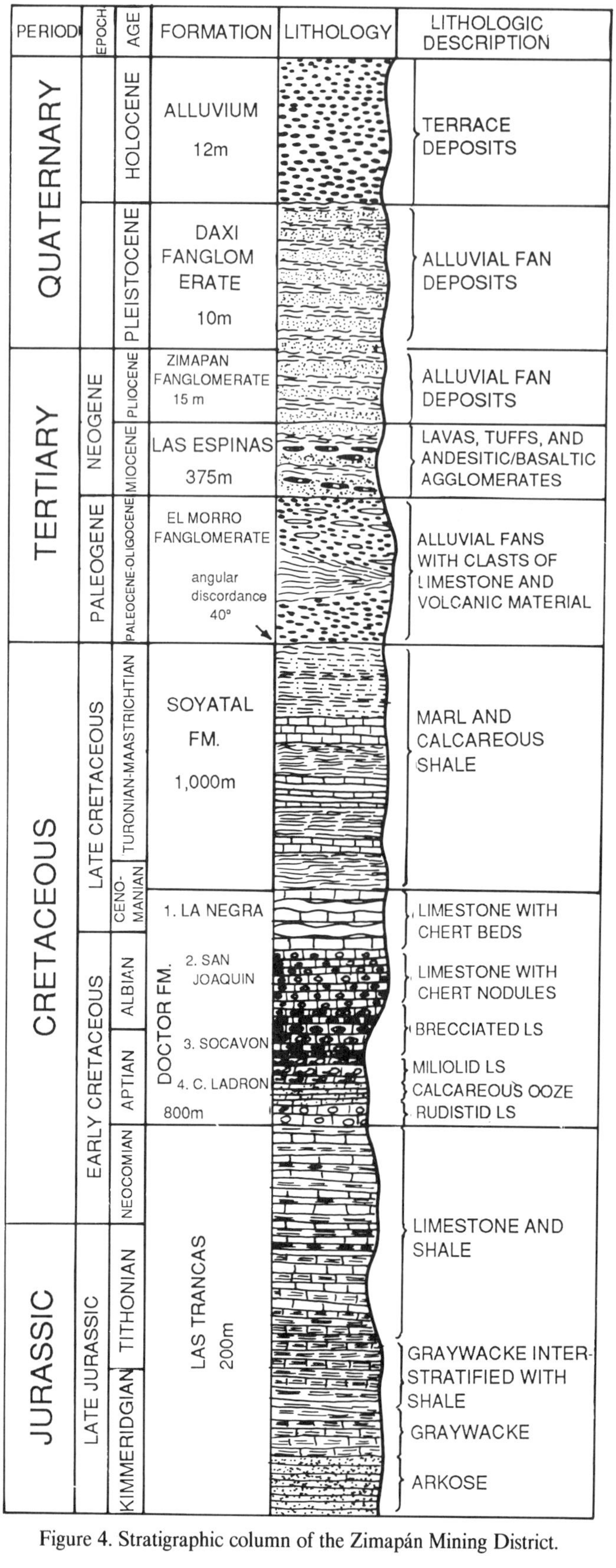

PERIOD	EPOCH	AGE	FORMATION	LITHOLOGY	LITHOLOGIC DESCRIPTION
QUATERNARY		HOLOCENE	ALLUVIUM 12m		TERRACE DEPOSITS
QUATERNARY		PLEISTOCENE	DAXI FANGLOM ERATE 10m		ALLUVIAL FAN DEPOSITS
TERTIARY	NEOGENE	PLIOCENE	ZIMAPAN FANGLOMERATE 15 m		ALLUVIAL FAN DEPOSITS
TERTIARY	NEOGENE	MIOCENE	LAS ESPINAS 375m		LAVAS, TUFFS, AND ANDESITIC/BASALTIC AGGLOMERATES
TERTIARY	PALEOGENE	PALEOCENE-OLIGOCENE	EL MORRO FANGLOMERATE; angular discordance 40°		ALLUVIAL FANS WITH CLASTS OF LIMESTONE AND VOLCANIC MATERIAL
CRETACEOUS	LATE CRETACEOUS	TURONIAN-MAASTRICHTIAN	SOYATAL FM. 1,000m		MARL AND CALCAREOUS SHALE
CRETACEOUS	LATE CRETACEOUS	CENO-MANIAN	DOCTOR FM. 1. LA NEGRA		LIMESTONE WITH CHERT BEDS
CRETACEOUS	EARLY CRETACEOUS	ALBIAN	2. SAN JOAQUIN		LIMESTONE WITH CHERT NODULES
CRETACEOUS	EARLY CRETACEOUS	ALBIAN / APTIAN	3. SOCAVON		BRECCIATED LS
CRETACEOUS	EARLY CRETACEOUS	APTIAN	4. C. LADRON 800m		MILIOLID LS; CALCAREOUS OOZE; RUDISTID LS
CRETACEOUS	EARLY CRETACEOUS	NEOCOMIAN	LAS TRANCAS 200m		LIMESTONE AND SHALE
JURASSIC	LATE JURASSIC	TITHONIAN	LAS TRANCAS 200m		GRAYWACKE INTERSTRATIFIED WITH SHALE
JURASSIC	LATE JURASSIC	KIMMERIDGIAN	LAS TRANCAS 200m		GRAYWACKE; ARKOSE

Figure 4. Stratigraphic column of the Zimapán Mining District.

Regional structures

The district is located in the northwest-trending El Piñón anticlinorium, formed by part of the Cretaceous sediments of the Sierra Madre Oriental, which was folded during the Laramide orogeny. The anticlinorium, 11 km wide along the Río Tolimán canyon (Fig. 5), is bounded on the northeast by the Aguacate syncline and on the southwest by the Maconí syncline. Its southwestern flank is a series of recumbent folds with subhorizontal axial planes in the El Doctor and Soyatal Formations, clearly identifiable along the Tolimán Canyon. In the Carrizal area the folds affect massive calcareous strata and are known as the San José Volador anticline, La Carolina syncline, and La Paz anticline.

Several internal anticlines and synclines (San Felipe and Puerto del Angel), with closely spaced subvertical axial planes, appear at both the top and flank of the El Piñón anticlinorium and are best defined in the area by the upper contact of the Soyatal Formation. The anticlinorium attains over 1,800 m of structural relief, with fold amplitudes of 100 to 180 m.

Regional magmatism

The igneous activity in the district was mostly Early Oligocene magmatism (38.7 ± 0.8 Ma; Gaytán, 1957) represented by several mainly monzonite intrusives and predominantly intermediate volcanics.

An extensive volcanic series south and east of the district covering almost the entire southern portion of Hidalgo State consists of intercalated lava flows, tuffs, and agglomerates interbedded with lacustrine deposits of the Pachuca Group, which includes the Las Espinas Formation.

Zimapán Mining District

Most of the mines in the very extensive Zimapán District are closely associated with intrusive monzonite and quartz-monzonite bodies (Simons and Mapes, 1957) distributed over four areas: Santa Gorgonia–San Pascual, El Monte–San Francisco, El Carrizal (Fig. 2) and La Luz–La Cruz, just west of the map area.

The El Monte–San Francisco and El Carrizal areas are discussed in detail in this chapter; general features of the other two are summarized below.

Santa Gorgonia–San Pascual area lies 5 km northwest of Zimapán within easy reach. Host rocks are the El Morro Fanglomerate and Las Espinas volcanics overlying Soyatal calcareous shales; latite and monzonite dikes and irregular intrusive bodies cross the entire area. Orebodies are associated with igneous dikes and hornfels areas caused by the metasomatism of the shales and fanglomerates. The ores replace and fill both Las Espinas and fanglomerate host rocks and occasionally Upper Cretaceous calcareous shales. The mineralized structures are sometimes irregular veins trending NW30 to 70°SE and dipping

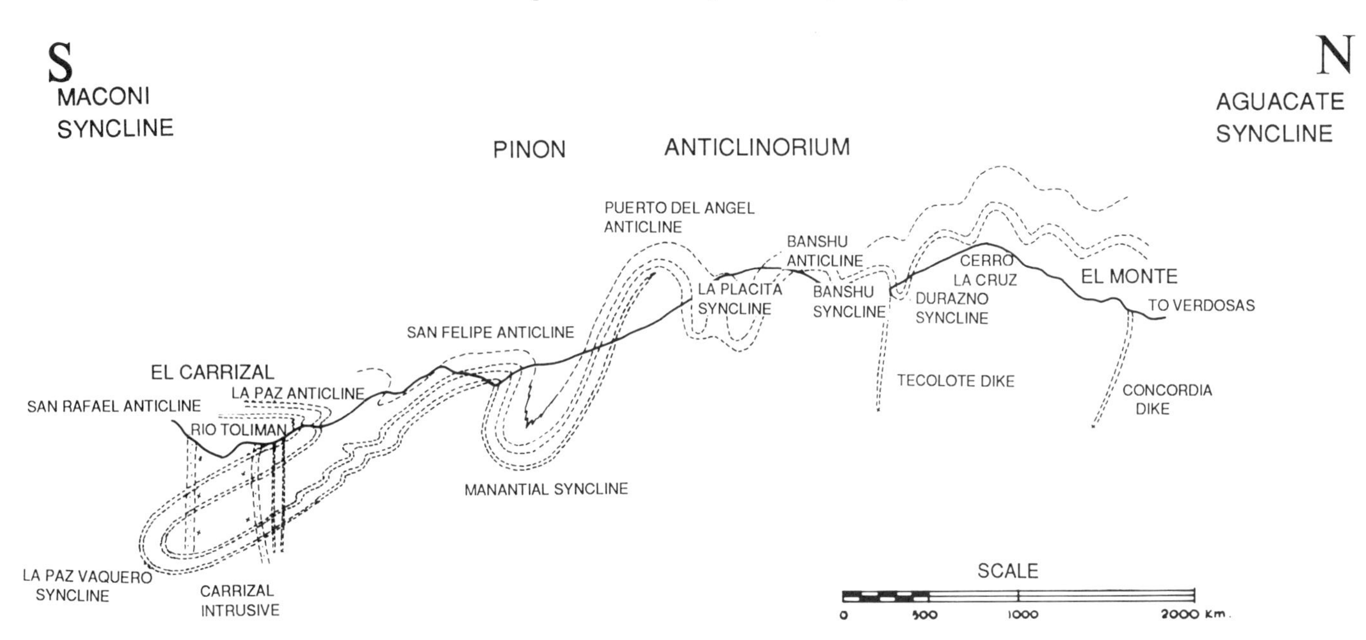

Figure 5. Cross section of the El Piñón anticlinorium.

steeply southwest (and occasionally northeast). Mineralization appears within the veins as small isolated and discontinuous oreshoots. Production from this area, mainly in the oxides zone, has been sparse and sporadic (Simons and Mapes, 1957).

La Luz–La Cruz area. 6.3 km west-northwest of Zimapán, apparently had considerable production and installations at the beginning of this century but is not presently under development. The host rock is mostly monzonite and to a lesser degree Soyatal shales. In contrast with the high Pb-Zn content in other district areas, La Luz–La Cruz is a copper-silver orebody attaining 10,000 ppm Ag in the oxidation zone; structures are veins lodged at fault contacts or faults in the quartz-monzonite. No production figures are available (Simons and Mapes, 1957).

MINERAL DEPOSITS

El Monte area

Lithology—sedimentary rocks. Only calcareous rocks of the El Doctor Formation's Socavón, San Joaquín, and La Negra Members are present in the El Monte Mine (Fig. 3); the lithology from older to younger is described below.

Socavón Member is a series of 15- to 80-cm-thick beds of sedimentary breccia with calcareous intraclasts of varying size and degree of roundness in a coarse-grained groundmass, accumulated in an erosional and subaqueous transport environment. Ground instability caused by the fragile character of these limestones poses problems within the mine.

San Joaquín Member contains 0.5- to 1-m beds of gray, compact cryptocrystalline limestone, with black chert nodules 2 to 30 cm in diameter.

La Negra Member consists of 10- to 30-cm-thick beds of gray, fine-grained limestone with bands and rare nodules of black chert that change in color to white in the proximity of igneous bodies due to hydrothermal metasomatism.

The sequence of members extends in age from Neocomian to lower Turonian.

Igneous rocks are represented in the El Monte area by monzonite-latite dikes, including the Concordia and Tecolote dikes, which are conspicuous for their size and commercial importance and are metasomatized both within (endoskarn) and at the edges and contacts with the limestones (exoskarn).

Structures. The El Monte orebodies are located on the northwestern flank of the NW45°SE–trending Verdosas anticline (part of the Piñón anticlinorium), which is cross-cut by a generally NE65°SW–trending regional system of normal faults with left-lateral horizontal displacements of up to 40 m and unknown horizontal extension; these faults include the prominent Concordia, Miguel Hidalgo, and Guadalupe faults (Fig. 6). The igneous dikes are randomly emplaced through strongly influenced by the orientation of the anticlinal axis, as shown by the NW35°SE–trending Concordia dike. It should be noted that the dikes exhibit maximum skarn and mineralization development along these trends.

Description of the deposit. The metasomatic El Monte skarn-type ore deposits are products of hydrothermal solutions that interacted with the igneous and limestone host rocks of the San Joaquín and La Negra Members. Hydrothermal alterations associated with the orebodies, in order of decreasing importance, are: silicification, pyritization, propylitization, and argillization; sericitization is local. The mineralization appears in exoskarn, endoskarn, and igneous rock. Skarn mineralogy is wollastonite, garnets (andradite and grossularite), and quartz, with subordinate calcite, chlorite, dolomite, and sericite. In addition, disseminated phases of the sulfide group are also found, usually replacing the wollastonite, garnet, and quartz: pyrite, sphalerite, arsenopyrite,

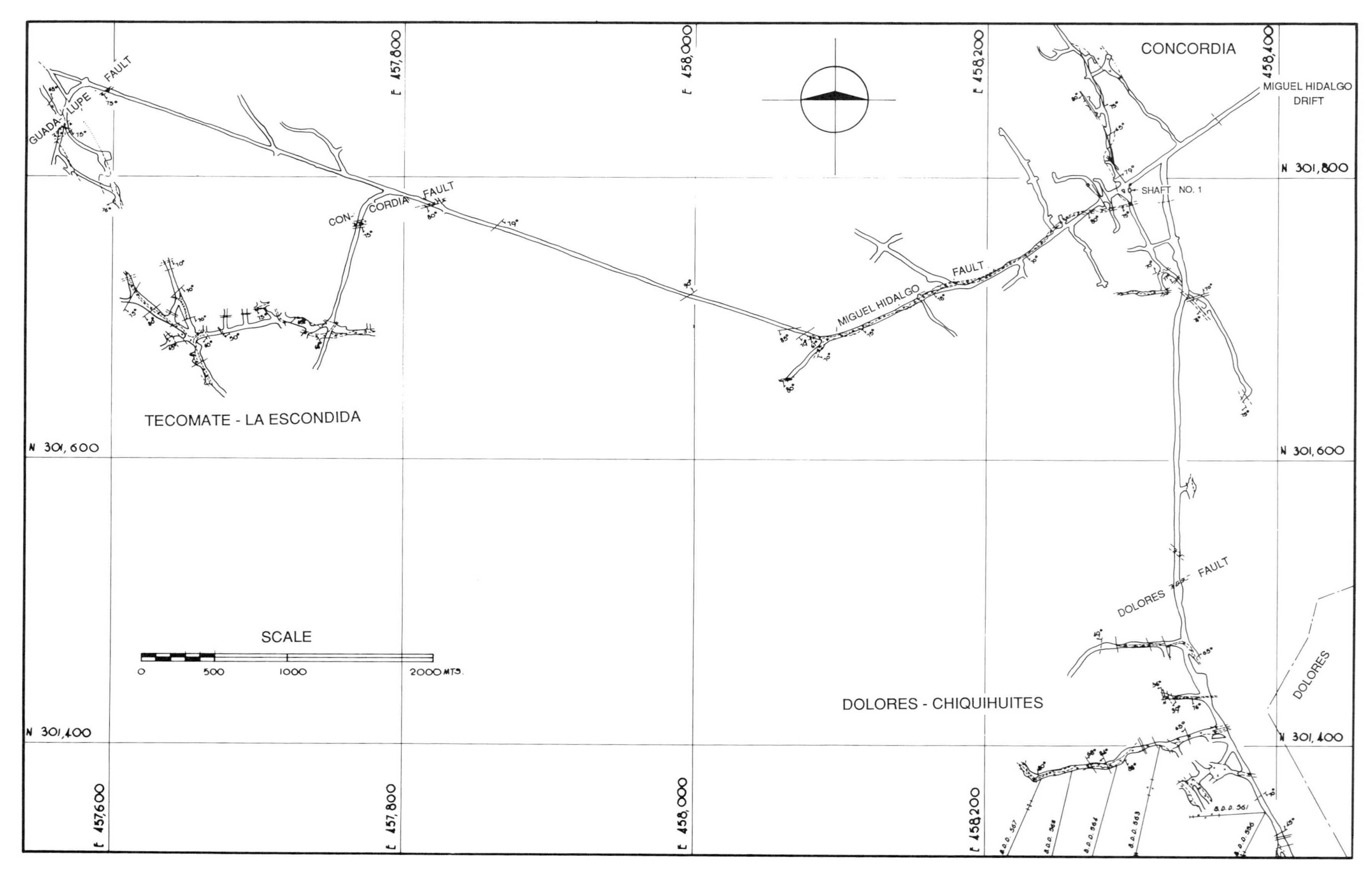

Figure 6. Geologic map of the 0 level, El Monte area.

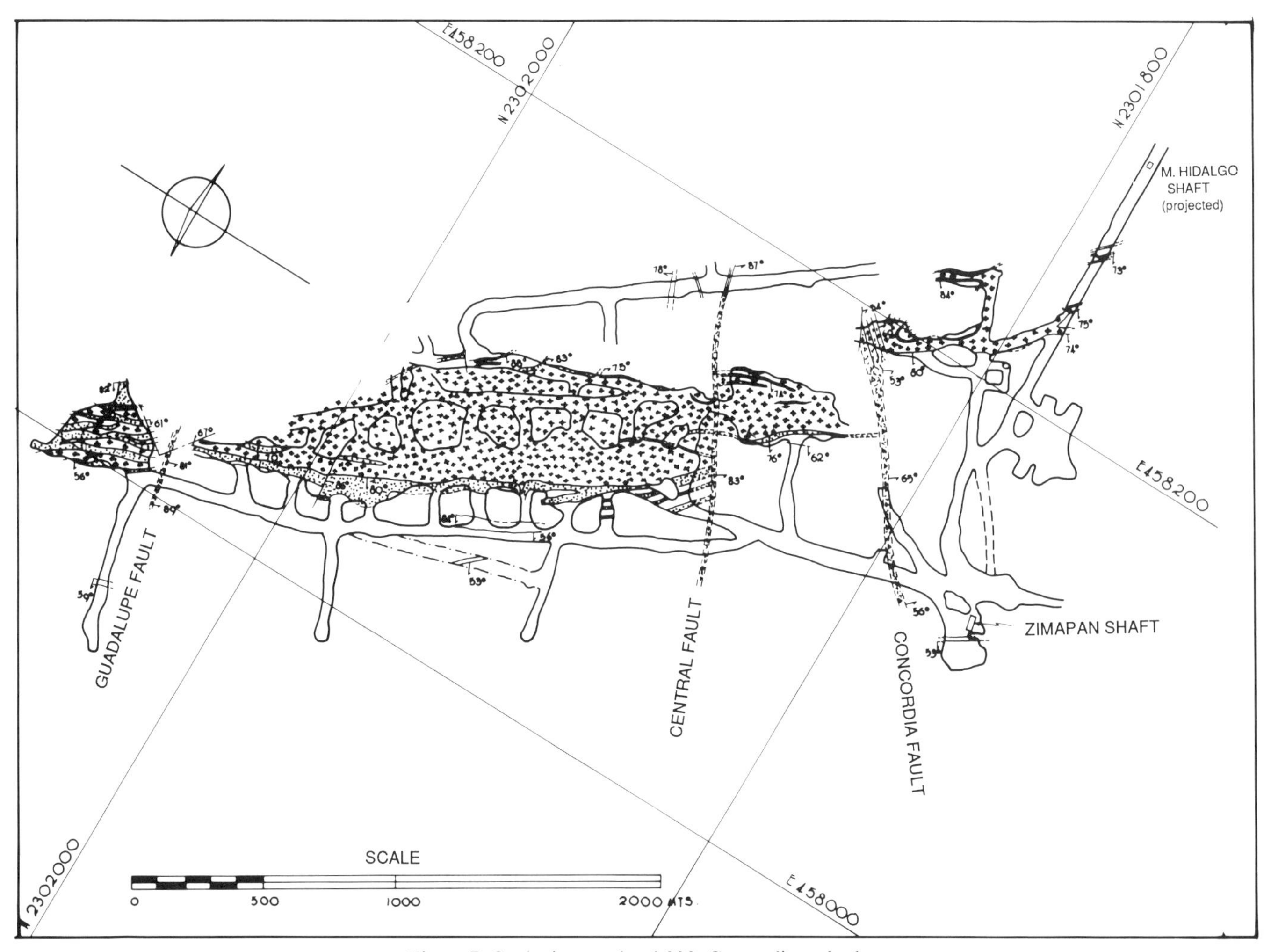

Figure 7. Geologic map, level 333, Concordia orebody.

galena, chalcopyrite, and lesser amounts of bornite, argentite, tetrahedrite, and boulangerite; scheelite is another commercially important phase present.

As mentioned above, the igneous dike rocks belong to the monzonite-latite family. The unaltered rock contains 50 percent An_{38} plagioclase, 40 percent K-feldspar, and 10 percent accessory minerals (augite, biotite, and sphene). In general, the endoskarn mineralogy produced by alteration of the dikes is the same as in the exoskarn. In addition to the commercially prospective orebodies associated with the dikes, which are described below, the area includes a series of small, geometrically discontinuous orebodies, mantos, veins, and minor skarn chimneys with sulfides and massive sulfides of scarce economic value at present.

Detailed description. For purposes of description the El Monte ore deposit may be divided into three major areas: the Dolores-Chiquihuites, Tecomate–La Escondida, and Concordia areas (Fig. 6).

Dolores-Chiquihuites includes 80- to 140-cm-thick, olive-green aphanitic latite dikes whose boundaries are masked at the contacts by exoskarns with an exoskarn-latite volumetric ratio of 2:3. The area is characterized by high Zn and Cu, and low Ag, values. Mineralization occurs disseminated in exoskarn zones around the igneous dikes, and as skarn and/or sulfide chimneys at points where the dikes interact, bifurcate, or bend. The area also contains the generally N75°E–trending, but abundantly flexed, 5- to 12-m-thick Tecolote dike of monzonite composition and phaneritic texture. Exoskarn distribution along this dike is unequal, increasing in thickness along NW35 to 50°SE trends (i.e., parallel to the Concordia dike); the disseminated metallic sulfides occur in these areas of abundant skarn.

Although widely separated from the above, the Tecomate–La Escondida area exhibits the same features and requires no further description.

The Concordia area is characterized by a northwest-trending, 70°-dipping, 390-m-long phaneritic monzonite dike averaging 33 m in width and cross-cut by generally NE65°SW–trending parallel faults that dip 65° southeast (except the 65°-northwest-dipping Central fault), called Guadalupe, Central, and Concordia faults (Fig. 7). As previously discussed, these are normal faults with oblique displacements of up to 40 m toward the northwest. For development purposes the Concordia area is divided into the Guadalupe, Northwest, Central, and Southeast blocks. The Concordia dike shows apophyses at its ends, bifurcating to the southeast and splitting into five dikes to the northwest. A small exoskarn zone averaging 1.5 m in thickness appears at its ceiling contact. Exoskarn areas within the intrusive body in the Northwest and Central blocks represent totally replaced limestone xenoliths, which at the southeastern end are only marbleized and lack silicates. The exoskarn-monzonite volumetric ratio is 1:20. The barely altered contact at the floor shows a maximum 20 cm of silicate development. Figure 11 is a skarn thickness distribution plot in the northwest Concordia block.

Five fracture systems, of which only two are mineralized, appear in the Concordia:

NW35°SE	Mineralized in primary sulfides and parallel to the body.
N65°SW	Mineralized in primary sulfides and perpendicular to the igneous body; parallel to the main fault system.
N–S	Very continuous fractures with Fe oxides and dips similar to the igneous body. Mining problems are caused by infiltrating water and the continuity of the fractures.
NW60°SW	Barren or mineralized in primary sulfides.
Horizontal	Dips may vary to 15°. Continuous, barren fractures that pose serious mining problems; fractures were presumably caused by decompression of the rock after partial mining, producing major groundfalls.

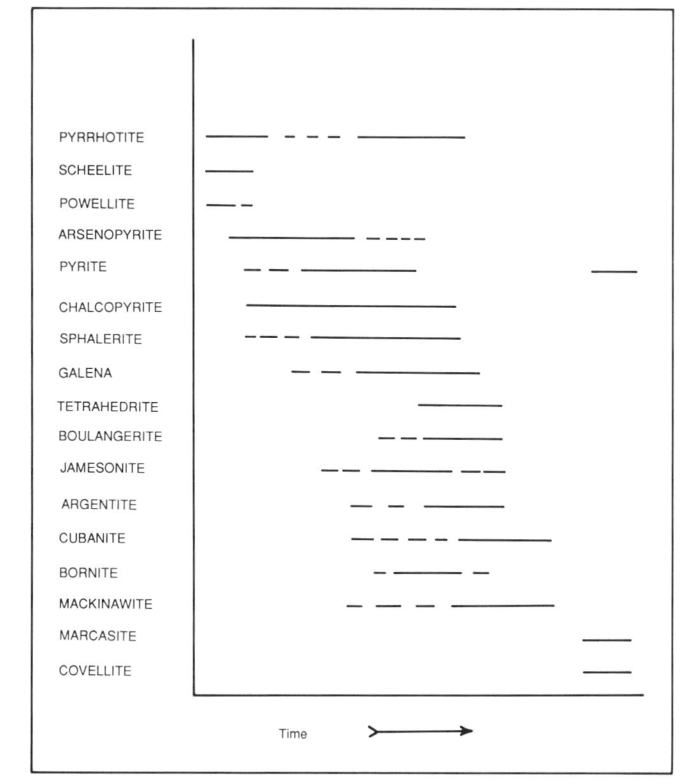

Figure 8. Paragenetic sequence of mineralization of the Concordia orebody.

TABLE 1. RANGES OF VALUES AMONG ELEMENTS IN THE CONCORDIA AREA

Elements	Very high	High	Medium	Low	Very low
Pb	>1.13	0.95–1.15	0.76–0.94	0.54–0.75	<0.54
Zn	>2.06	1.73–2.06	1.39–1.72	0.97–1.38	<0.97
Cu	>0.39	0.32–0.39	0.26–0.31	0.19–0.25	<0.19
Ag	>150	126–150	100–115	70–99	<70

Mineralization varies in the Concordia area depending on the block, which indicates that the fault blocks developed prior to mineralization. This is confirmed by the fact that the three principal faults separating the blocks contain veins of varying thickness in Guadalupe (35 cm, basic sulfides), Central (40 cm, basic sulfides), and Concordia (up to 5 m, mineralized breccia). The mineralization appears in small veinlets, 1 mm to 2 cm thick, of differing mineralogy according to their vertical position (i.e., the NW35°SE–trending veinlets at level 229 contain galena, sphalerite, and some pyrite, whereas the same system at level 333 has mostly pyrite and arsenopyrite (Fig. 9). Lateral variations from one block to another are due to the degree of oxidation in the fractures, the type of hydrothermal alteration (argillization is associated with high values), and the proximity to mineralized faults.

The paragenetic mineral sequence of the Concordia orebody described by P. Gemmell (written communication, 1983) is shown graphically on Figure 8. Chalcopyrite, arsenopyrite, and high-temperature pyrrhotite appear as early sulfides, followed by other high-temperature basic sulfides such as galena, sphalerite, and chalcopyrite, followed in turn by subsequent exsolution during the cooling phase of the ore when sulfosalts and other sulfides were exsolved.

Metallic ratios. For purposes of defining the orebody's geochemical behavior the distribution of Ag, Pb, Zn, and Cu grades (Table 1) were plotted on graphs (Fig. 9), analysis of which led to the conclusion that the highest concentrations appear close to major fault or fracture zones cross-cutting the igneous body.

Metallic ratios were also studied with the object of correlating mineralization with (1) igneous dike thickness and (2) skarn volume, represented by its average thickness in mine blocks (Figs. 10, 11). Ratios were obtained from grade averages in approximately 10- by 10-m-wide blocks of the orebody, determined by systematic sampling every 2 m along lines, and representing each block by a point at its center (Table 2). The corresponding plots of these values show areas of specific ratio ranges. The graphs (Fig. 12) present the spatial distribution of the Ag/Cu, Ag/Pb, Ag/Zn, Pb/Zn, Pb/Cu, and Zn/Cu quotients compared with Figures 10 and 11. The following conclusions may be drawn:

1. The greatest thickness of the igneous body is located at sections 4 to 15 and tends to grow downward toward the southeast (Fig. 10).

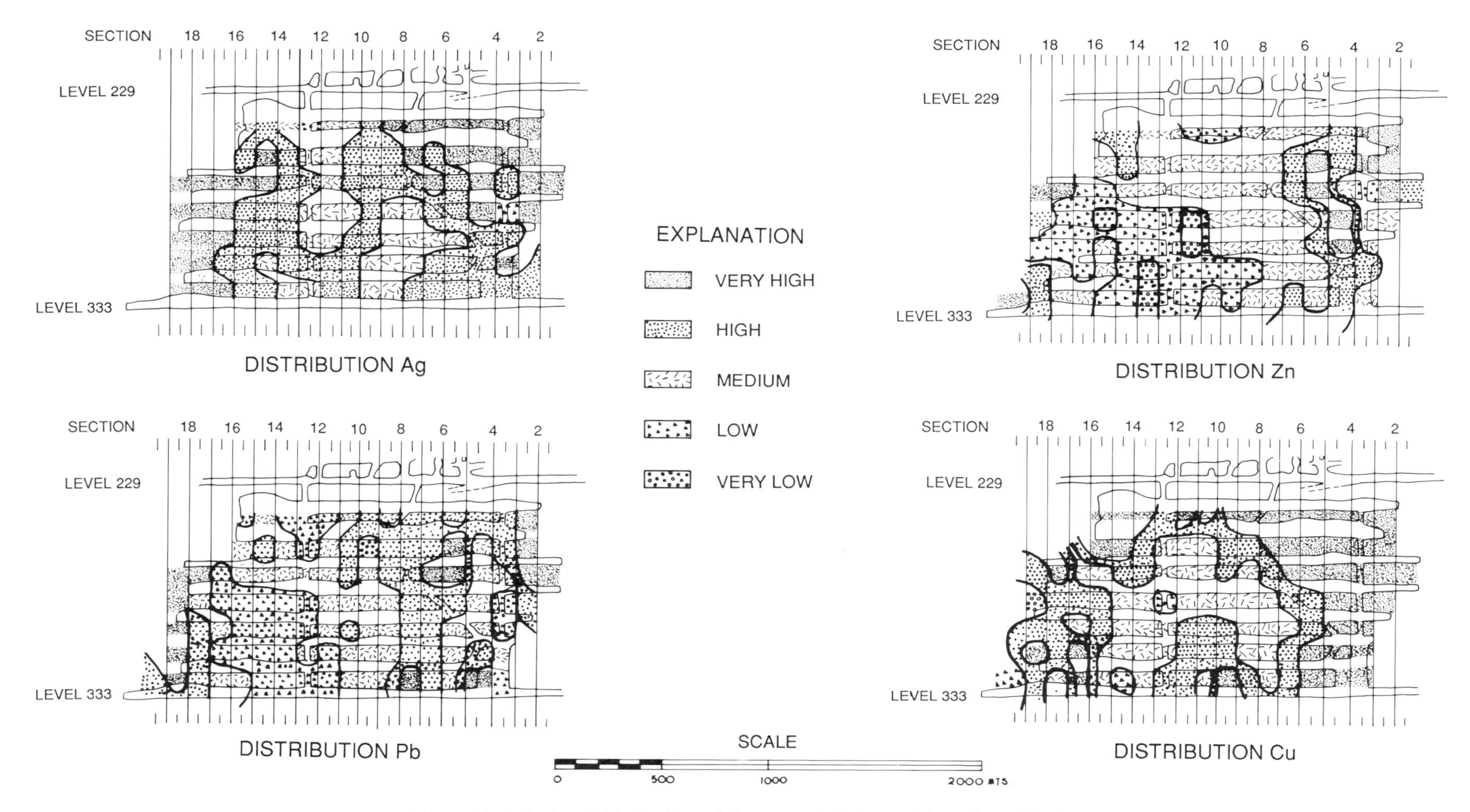

Figure 9. Distribution of Pb, Ag, Zn, and Cu grades in the Concordia northwest block.

2. Skarn thickness distribution is apparently random, unrelated to dike thickness and certainly controlled by local secondary permeability of the limestone. Likewise the distribution of metallic ratios, which does not follow a skarn-defined pattern, leads to the assumption that the processes were not contemporaneous (Figs. 9, 11, and 12).

3. The distribution of Ag/Zn and Pb/Zn ratios shows no defined pattern, and ratios are small, proving the almost contemporaneous deposition of galena and sphalerite and the intimate association of silver and lead ores.

4. The Ag/Pb and Ag/Zn distributions scarcely show variation along the orebody, which means that Ag, Pb, and Zn values are genetically related, and their ratios suffer no great change along the extension of the deposit. The anomalies present can be explained: (1) the zone of oxidation and enrichment at Section 7 was caused by very open fractures; (2) a pattern of high silver values relates to late hydrothermal silver enrichment along fractures of the NE65°SW system coinciding with areas of strong argillization.

5. The distribution of Ag/Cu, Pb/Cu, and Zn/Cu ratios shows high values at the thickest portion of the igneous dike. The superimposed anomaly in the northwestern zone is due to skewed ratio values at that location, which represent only the ceiling of the orebody and the exoskarn, the remainder belonging to properties owned by others.

Tonnage and average grades. El Monte production figures prior to 1975, when the processing plant was installed, are completely unknown. From 1975 to December 1984, total ore production was 1,419,974 metric tons of 0.04 ppm Au, 153 ppm Ag, 1.04 percent Pb, 2.23 percent Zn, and 0.53 percent Cu. Proven reserves at El Monte are 1,047,750 metric tons of 132 ppm Ag, 0.91 percent Pb, 1.52 percent Zn, and 0.42 percent Cu.

Mining method and metallurgy. Owing to the 70° dip and roughly tabular shape of the orebody, the Concordia body is mined by longhole drill at sublevels using the open-stope level variant (Fig. 13).

The Concordia body has been divided into seven sublevels with 10-m blocks; each sublevel is widened by compressed-air and cleared by face-loaders. Preparation works consist of: (1) service raises in highly altered and indurated rock at the hanging wall, (2) widening of sublevels in the ore, and (3) ventilation and runoff raises, with end raises for slotting.

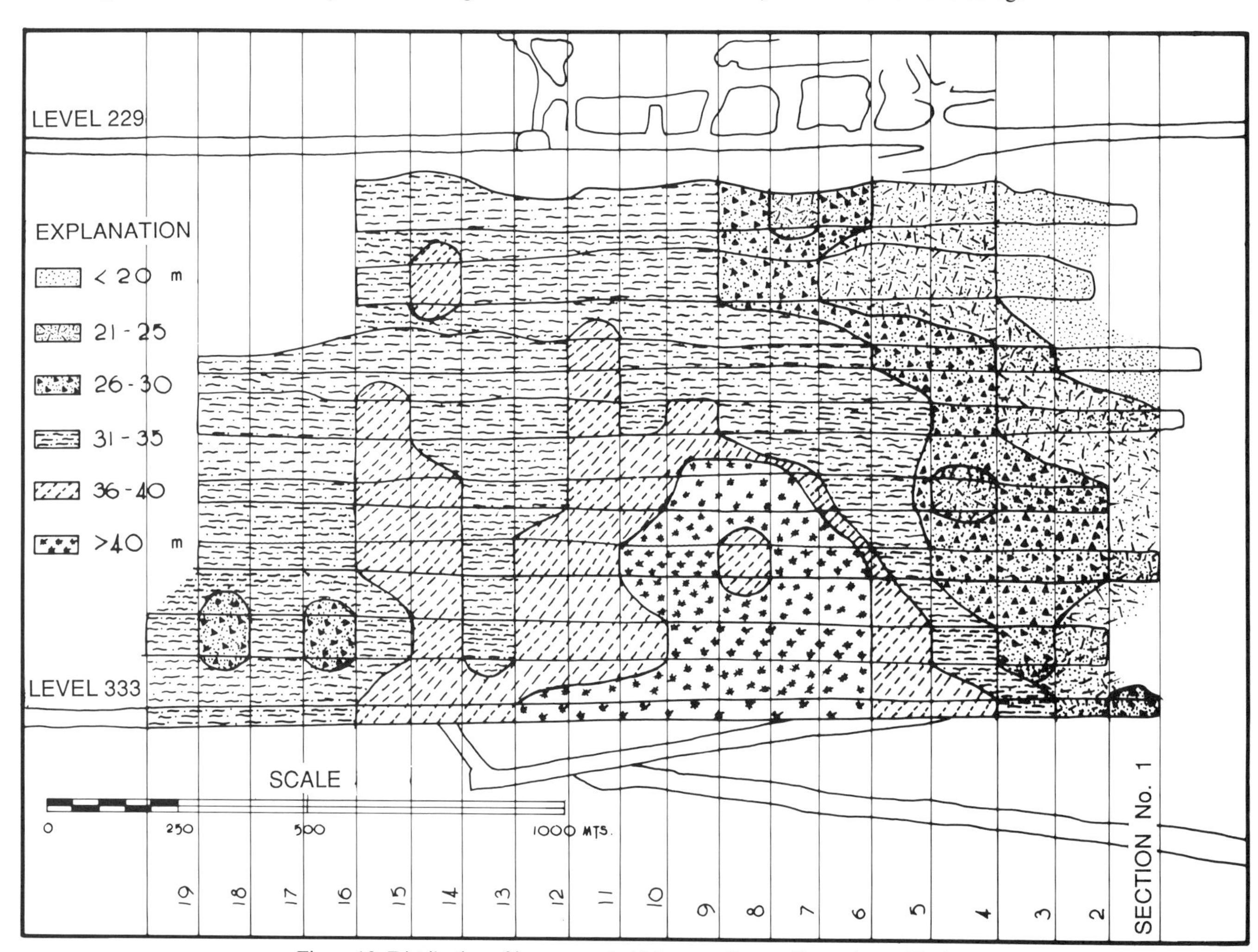

Figure 10. Distribution of igneous rock thicknesses, Concordia northwest block.

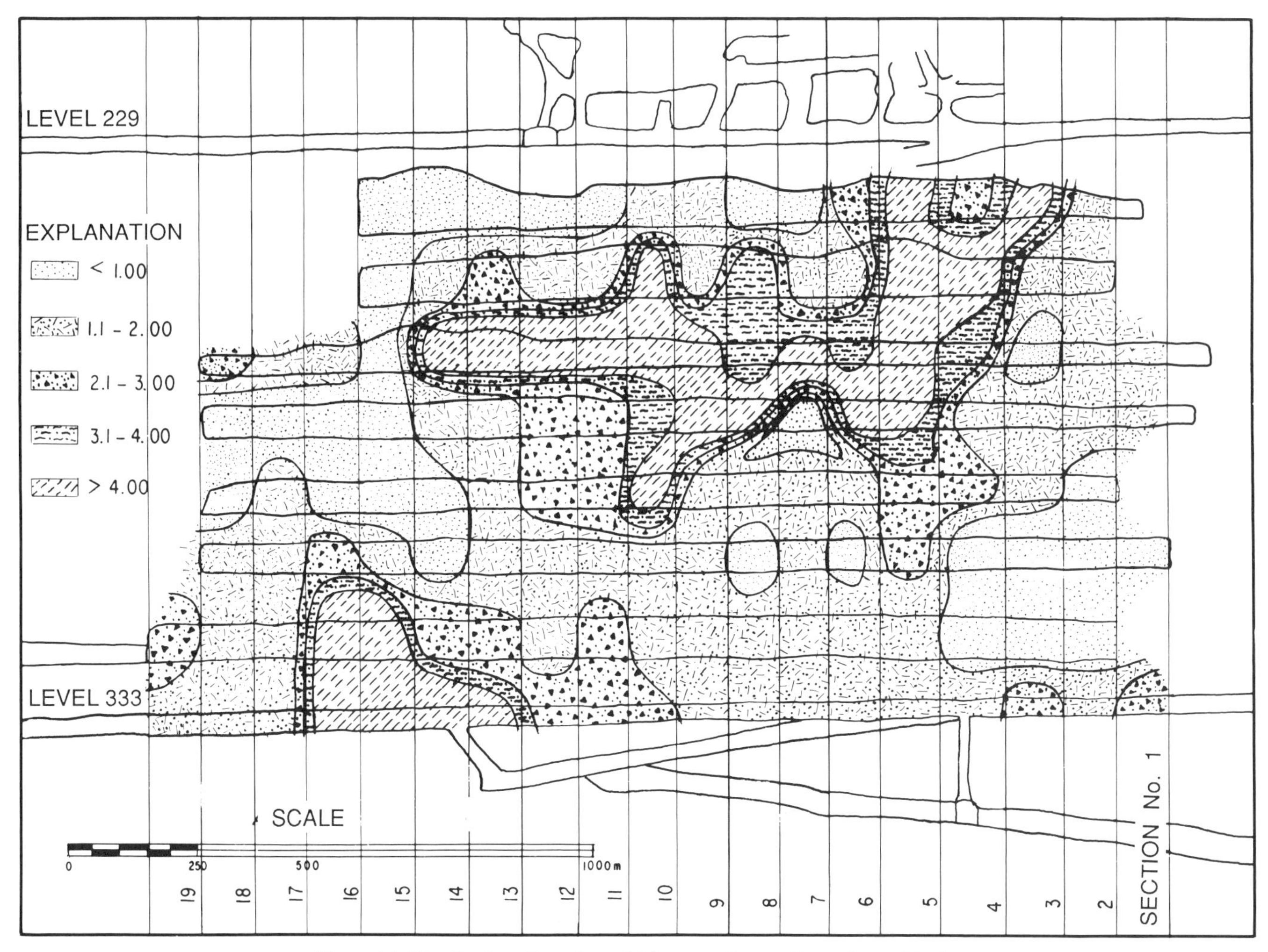

Figure 11. Distribution of exoskarn thicknesses, Concordia northwest block.

Preparation works start with an incline from the upper level down and crossholes toward the body ceiling at the sublevels. The ramps serve as access for face-loaders, equipment, and services. These works are simultaneous with the widening of the lower level (extraction) and developing access and extraction faces, as well as an uprising slot. Exploitation is by parallel floor drills using Wagon-Drill machines.

The ores are processed outside the mine at the company plant by selective flotation of the sulfides, producing lead and zinc concentrates. Total plant recovery is 73 percent Ag, 83 percent Pb, 59 percent Zn, and 59 percent Cu.

Carrizal area

Introduction. The Carrizal ore deposits are contained in the La Negra and San Joaquín Members of the Doctor Formation (Fig. 14). All the orebodies relate closely to an intrusive quartz-monzonite stock with apophyses in the form of phacoliths. By their geometry and location in space the orebodies may be grouped in veins, mantos, and chimneys.

TABLE 2. RANGES OF VALUES AMONG ELEMENTAL RATIOS IN THE CONCORDIA AREA

Ratios	Very high	High	Medium	Low	Very low
Ag/Cu	>600	500–599	400–499	300–399	<300
Ag/Pb	>250	200–249	150–199	100–149	<100
Ag/Zn	>150	100–149	<100		
Pb/Zn	>0.9	0.70–0.89	0.50–0.69	0.30–0.49	<0.30
Pb/Cu	>5.0	4.0–4.9	3.0–3.9	2.0–2.9	<2.0
Zn/Cu	>6.0	5.0–5.9	4.0–4.9	3.0–3.9	<3.0

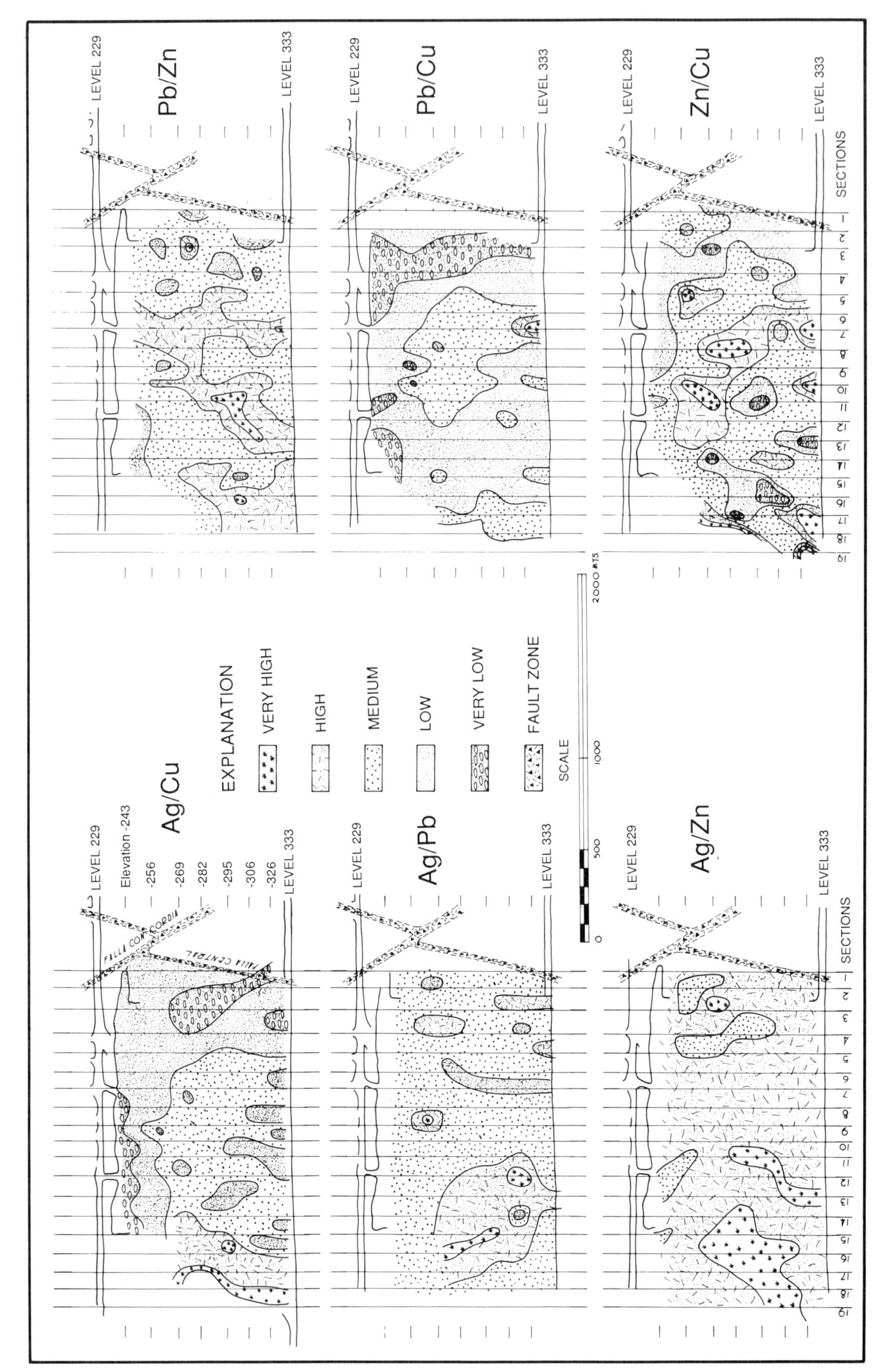

Figure 12. Distribution of metallic ratios, northwest Concordia block, El Monte area.

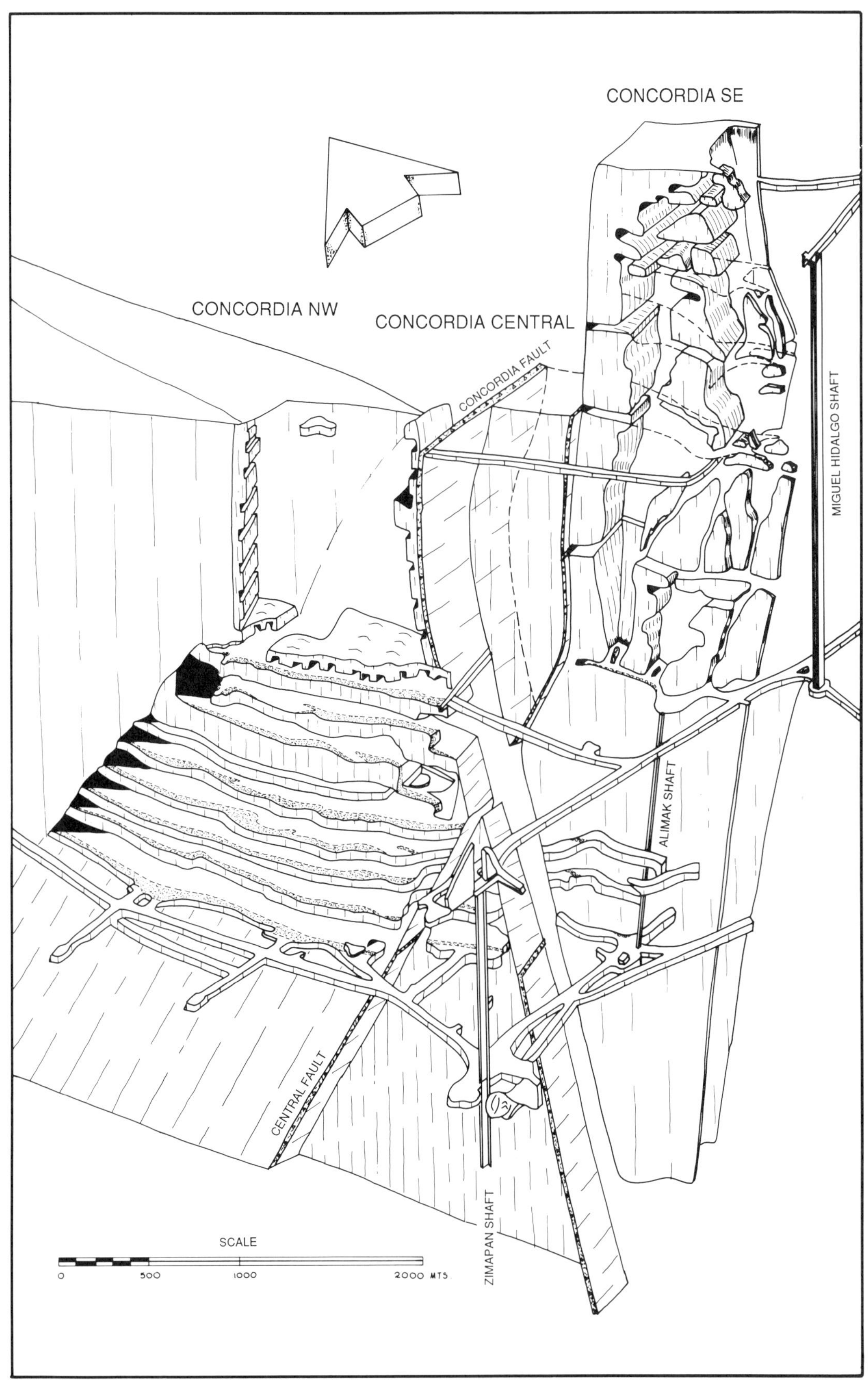

Figure 13. Isometric projection of the Concordia orebody.

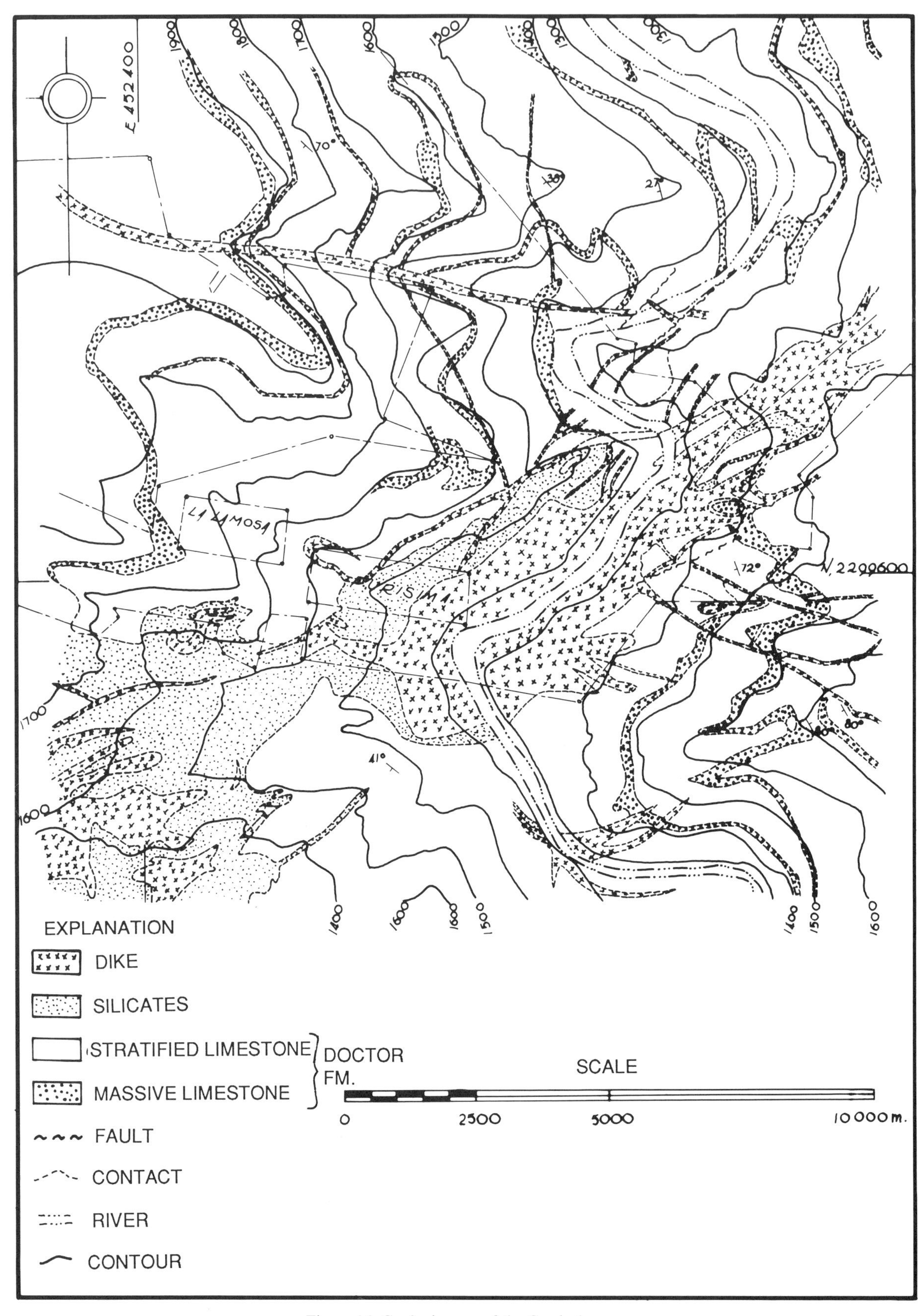

Figure 14. Geologic map of the Carrizal area.

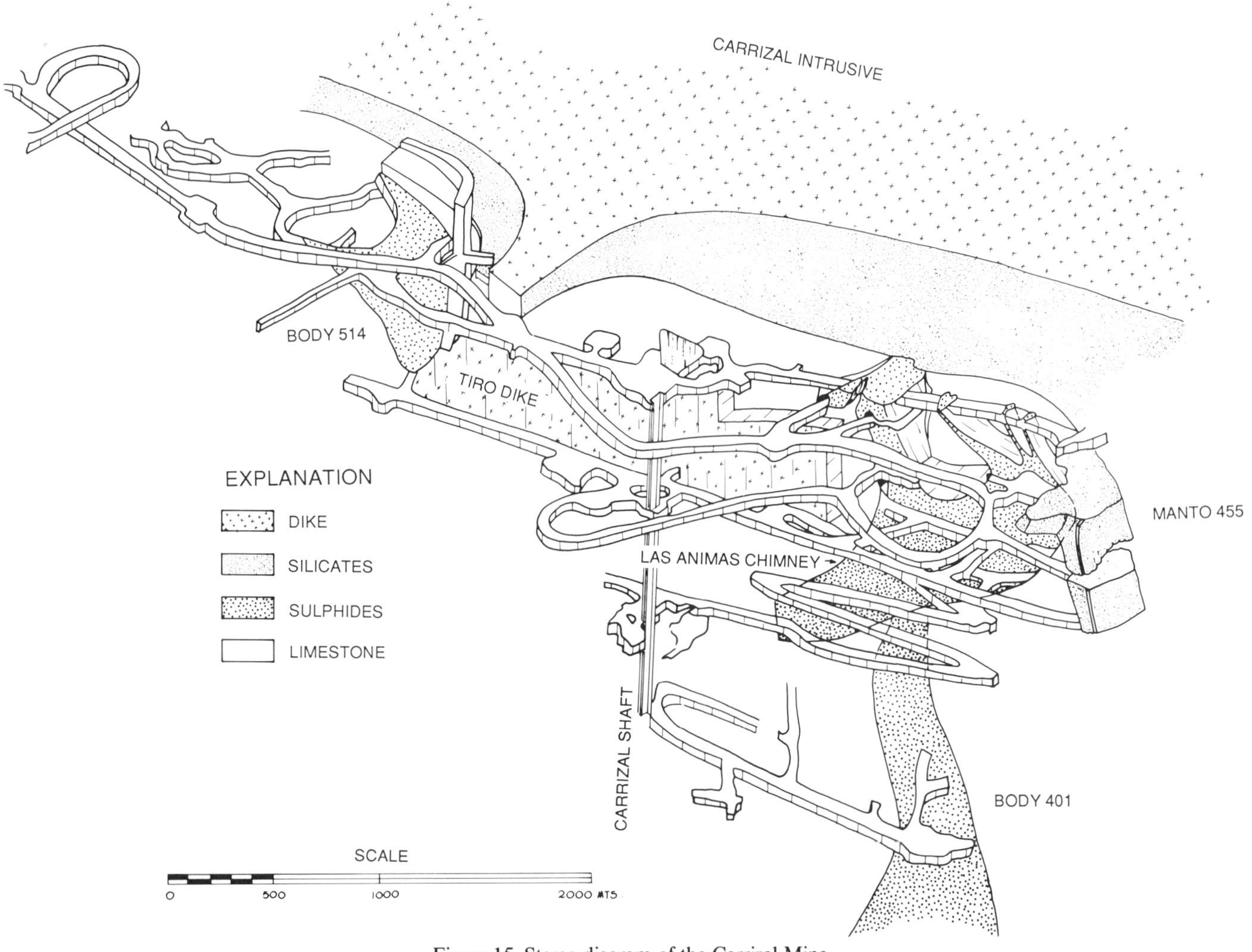

Figure 15. Stereo diagram of the Carrizal Mine.

Structures. The ore deposits occur on the recumbent flank of the San Rafael overturned anticline (part of the El Piñón anticlinorium; Fig. 5), whose 30°NW–dipping axial plane trends NW50°SE, the same as the anticlinorium. Abundant fractures following the regional system—parallel to the folding—appear continuously for over 3 km at the surface and control the emplacement of the mineralized veins.

The northeast-trending Carrizal intrusive in the central part of the area is characterized by a highly eccentric elliptical shape, 1.5 km by 0.25 km in cross section, and numerous dikes, trending mostly northwest (i.e., parallel to the regional structure) and northeast (parallel to the extension of the principal stock). The latter is a downward-spreading truncated cone whose walls generally dip 82° in opposite directions. The major, almost vertical dikes, such as El Tiro and 514 (Fig. 15), show thicknesses of up to 15 m.

Lithology. Limestones in the mineralized area are slightly recrystallized and occasionally totally replaced by silicates (exoskarn) with the following mineralogy, in order of decreasing abundance: garnet, calcite, wollastonite, diopside, epidote, chlorite, quartz, and opaque minerals. Exoskarns surrounding the main stock become thicker and more consistent when developed in igneous dikes and on occasion include the mineralization.

Types of deposits (Fig. 15). *Mantos.* All the mantos occur in stratiform exoskarn or igneous phacoliths(?) that follow the calcareous bedding. The best known is the cylindrical (20 by 40 m) southwest-dipping 455 manto, which is parallel to other stratiform exoskarn bodies and branches out from the main igneous stock; sphalerite, chalcopyrite, and pyrite are its main sulfides. Ore tonnage is approximately 20,000 metric tons of 145 ppm Ag, 0.58 percent Pb, 1.55 percent Zn, and 0.46 percent Cu.

Veins. The vein system follows the northwest regional orientation, dips 45°NE, extends an average 55 m, and is emplaced in both massive and stratified limestone. The veins are always "born" in the igneous stock and wedge out as they move away; mineral distribution in all directions is extremely irregular, making development difficult (Fig. 11).

In general, the veins are the most recently mineralized structures, postdating mantos, chimneys, and igneous dikes, which become enriched by them at the intersections.

Vein mineralogy is pyrite, fine-grained galena, scarce sphalerite, and very fine-grained calcite, all of which appear as crustification along the structure. Veins 401, 412, and 413 are the chief ones in the Las Animas Mine (Fig. 16), averaging 179 ppm Ag, 2.21 percent Pb, 4.17 percent Zn, and 0.45 percent Cu.

Chimneys. The most commercial orebodies in the area, chimneys are irregular but mostly cylindrical in shape, unconformable with the limestone host rock and appreciably parallel to the igneous bodies, and dipping mostly 60 to 80° northeast. They usually appear at intersecting structures, varying widely in size from 24 to 2,000 m^2 in horizontal cross section and up to 250 m in vertical extension. They contain fine- to coarse-grained massive sulfides. Pyrrhotite, pyrite, and some magnetite (indicating a high Fe content), sphalerite, galena, chalcopyrite, and lead and antimony sulfosalts make up the mineralogy. Las Animas, 401, and 514 are the most important chimneys in the area (Fig. 15). Las Animas, with a maximum diameter of 100 m and trending northeast, extends down to 343 m, branching upward at the 243-m level into several mantos and veins. At 433 m it begins to wedge out and ends at the contact with the Carrizal intrusive exoskarn. Ore thickness distribution in the Las Animas chimney clearly demonstrates its cylindrical shape. Some 70 percent of the orebody has been mined; remaining reserves are 130,000 metric tons; average grades are 225 ppm Ag, 3.54 percent Pb, 4.84 percent Zn, and 0.21 percent Cu.

Analysis of values distribution. For purposes of defining the large-scale mineralization process, isopleths of Pb, Zn, Cu, and Ag grades were plotted for the Las Animas and 401 chimneys, which have the maximum tonnage (Fig. 17; Table 3). The distribution pattern of complex values in the Las Animas chimney clearly shows high values in two well-defined areas: at the chimney footwalls, and extending vertically upward from level −122, which suggests two sources of mineralizing fluids: at veins 412 and 401.

Metallic ratios analysis in Las Animas and 401 chimneys. The Pb/Zn, Ag/Zn, and Cu/Zn ratios display similar behavior (Fig. 18; Table 4). The corresponding surfaces of equal metal content indicate two anomalies: beneath level −122, and between levels −72 and −89, both at the 412 vein footwall. The deeper anomaly, confirmed by the Ag/Pb and Pb/Cu ratios, is

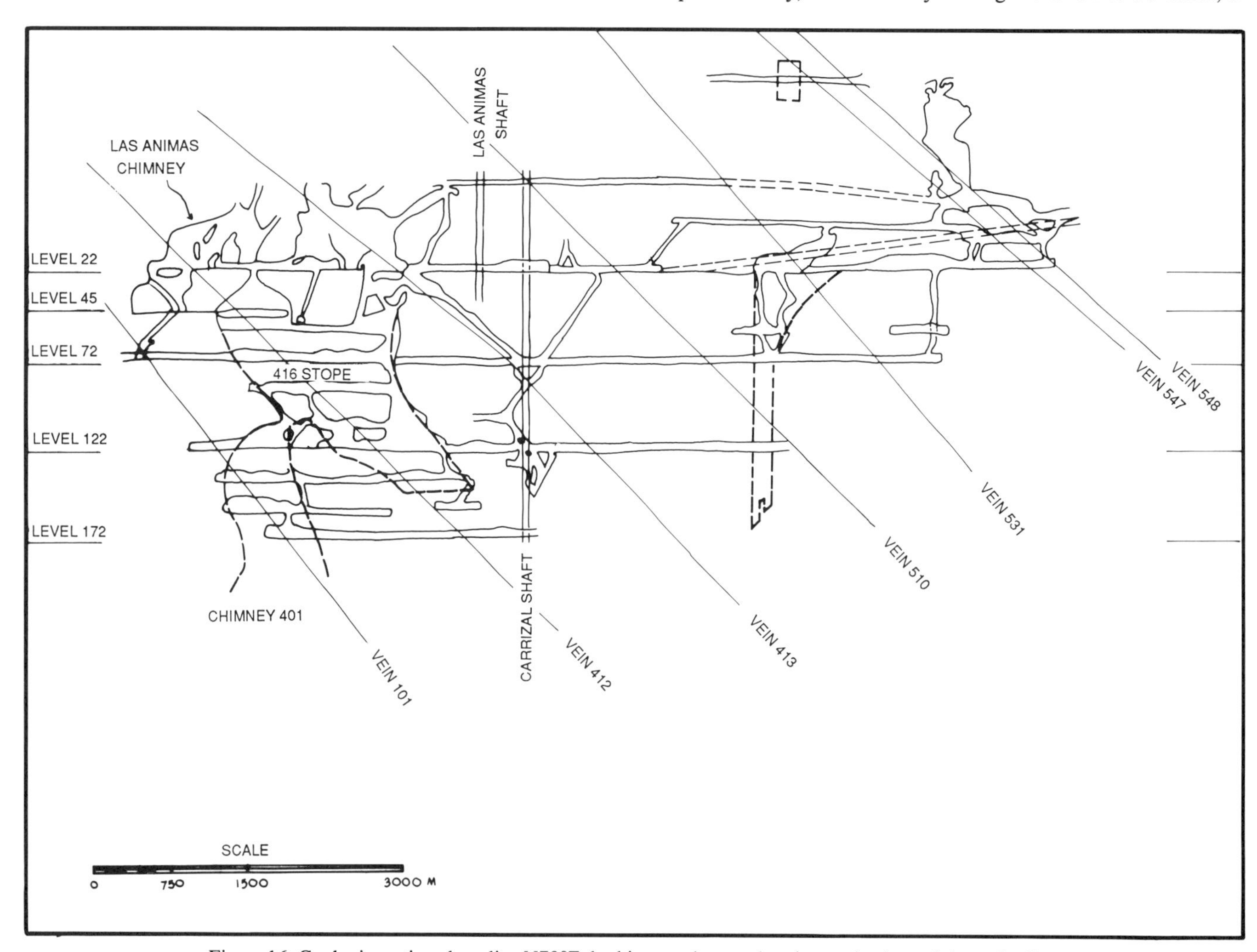

Figure 16. Geologic section along line N70°E, looking northwest, showing projections of the orebodies in the Carrizal Mine.

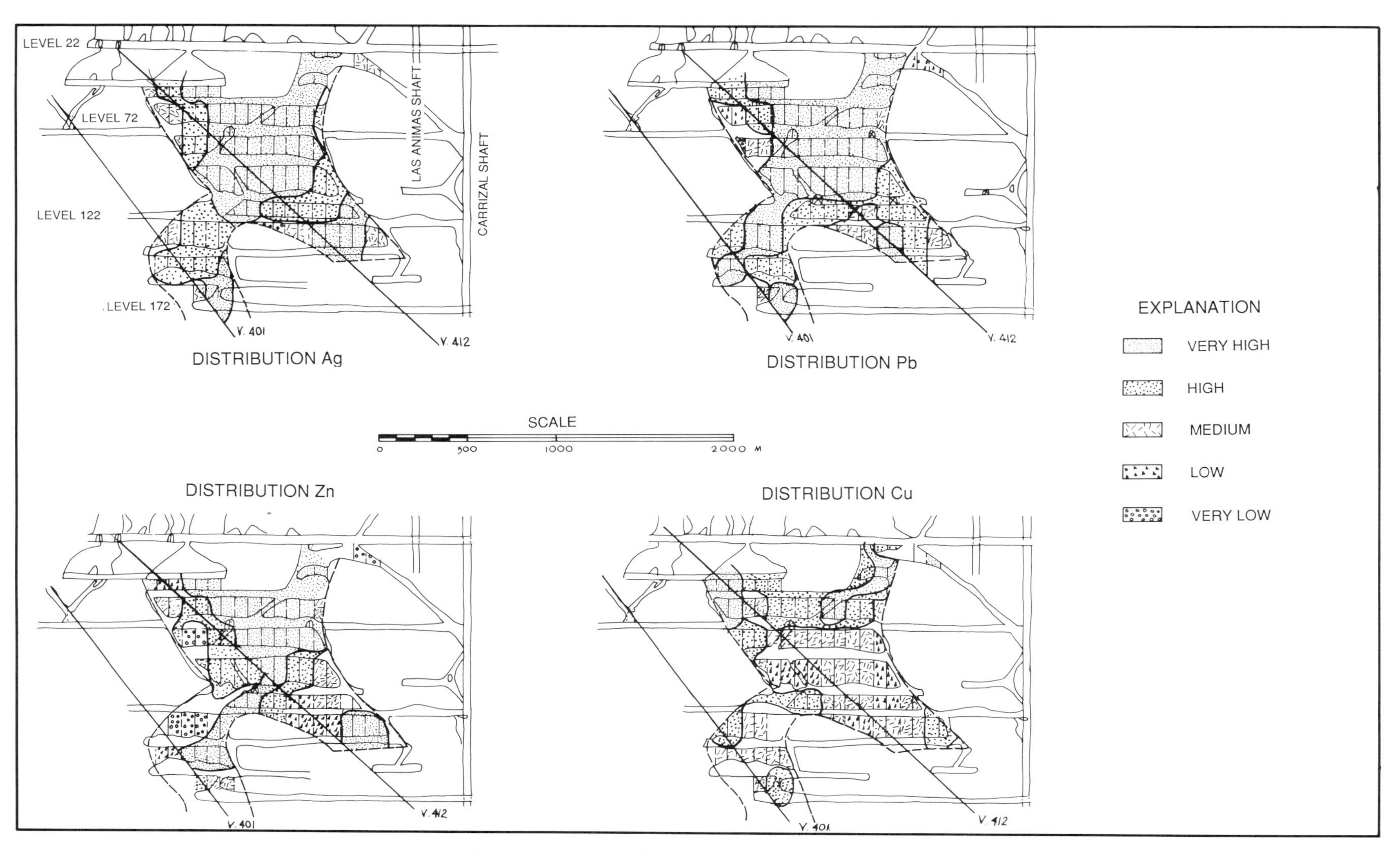

Figure 17. Distribution of Pb, Zn, Cu, and Ag values in the Las Animas and 401 chimneys, Carrizal Mine.

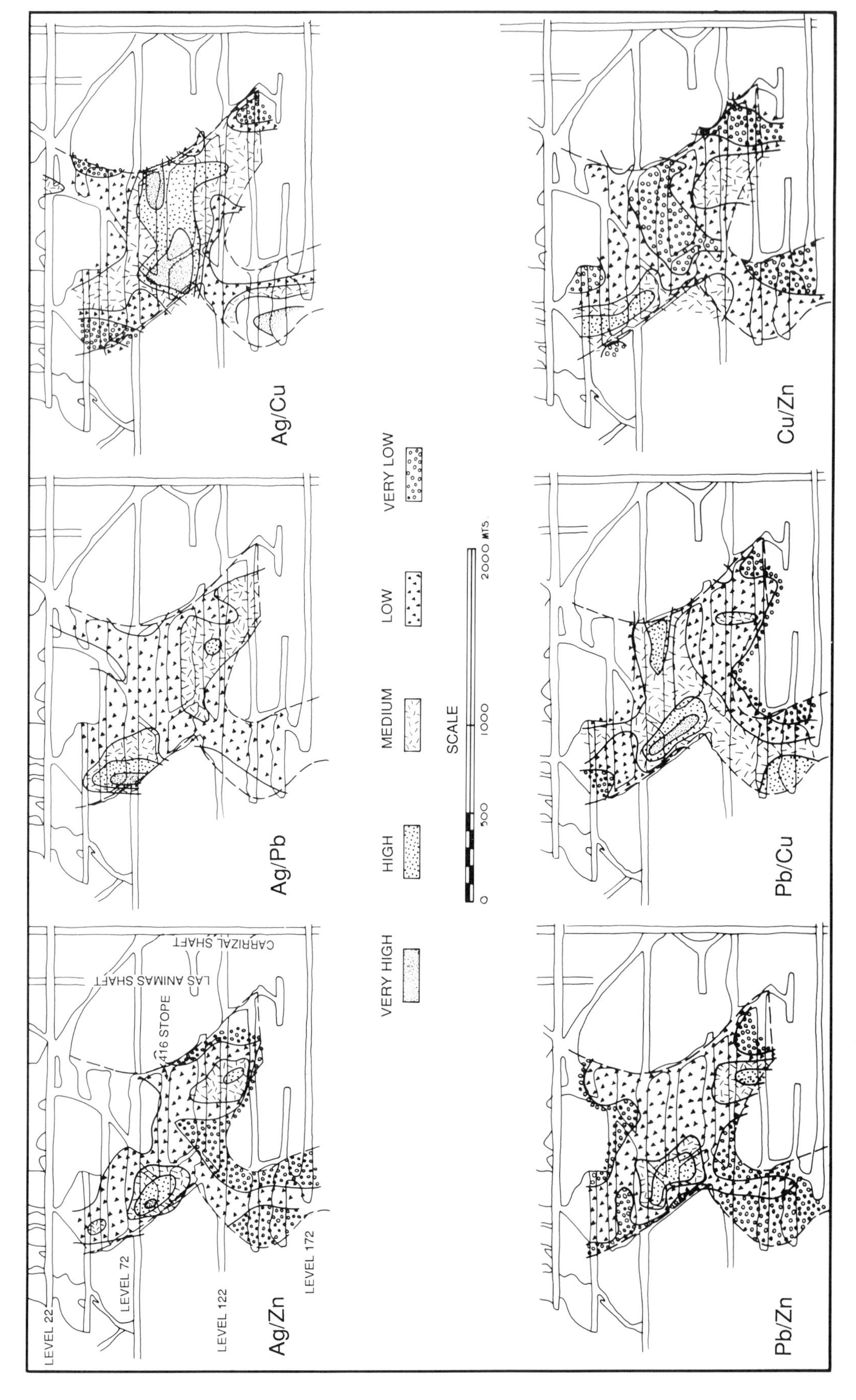

Figure 18. Distribution of metallic ratios in the Las Animas and 401 chimneys of the Carrizal Mine.

explained by the intersection of vein 412 with the Las Animas chimney, apparently enriching the latter in lead-silver sulfosalts after the original mineralization. The shallower, much higher anomaly shows absolutely no correlation with this vein and is probably associated with the underlying 401 chimney, which connects directly through the Tiro dike hanging wall. The hypothesis proposed here—that later solutions formed chimney 401, permeated the Las Animas chimney, and dissolved part of the sphalerite, replacing it with lead-silver-iron sulfosalts—will be tested by petrological study of the ore at strategic points in both chimneys.

Tonnage, average grade, and mining method. To 1984, the total tonnage extracted from the Carrizal area Las Animas Mine was 389,107 metric tons of 196 ppm Ag, 2.91 percent Pb, 4.47 percent Zn, and 0.36 percent Cu. Total reserves, including all the orebodies in the mine, are 612,700 metric tons of 205 ppm Ag, 2.97 percent Pb, 4.57 percent Zn, and 0.31 percent Cu. The Las Animas chimney is thus the best producer in the mine at present. Extraction is by the sublevels method; mantos and veins are mined by overhead stoping. All the mineral is hauled out from level –22 by 10-mt dump trucks to the El Monte area plant.

GENESIS OF THE DEPOSITS

As mentioned previously, the genesis of the Zimapán District ore deposits is intimately linked to the emplacement of igneous intrusive monzonite dikes, which together with the metasomatic skarn developed by the consequent hydrothermal process, laid the ground for the emplacement of large and small orebodies during several mineralization episodes, some almost contemporaneous with the skarn but mostly postdating it. The typically epigenetic mineralization is much younger than the host rocks. Fractures in the igneous bodies and the skarn, and the contacts of these bodies with the host rocks were the main pathways. On occasion the recumbent or vertical portions of massive limestones, given their primary fracturing and permeability, also served as pathways and deposit sites for the ores. Mineralization appears both as fracture fills and replacement, depending on the host rock and depth of emplacement. The deposits may be classified in general as mesothermal, occasionally epithermal. Estimated mineralization depths range from –1,300 m to the early Pliocene paleosurface, when mineralization presumably took place. However, the major replacement orebodies were emplaced between 700 and 1,300 m below the paleosurface (Simons and Mapes, 1957).

TABLE 3. RANGES OF VALUES AMONG ELEMENTS IN THE CARRIZAL AREA

Elements	Very high	High	Medium	Low	Very low
Pb	>3.50	3.50–2.63	2.61–1.76	1.75–1.23	<1.23
Zn	>5.25	5.25–3.95	3.94–2.65	2.64–1.85	<1.85
Cu	>0.30	0.30–0.22	0.21–0.16	0.15–0.11	<0.11
Ag	>200	200–151	150–101	100–71	<71

TABLE 4. RANGES OF VALUES AMONG ELEMENTAL RATIOS IN THE CARRIZAL AREA

Quotients	Very high	High	Medium	Low	Very low
Cu/Zn*	>100	100–92	91–48	47–32	<32
Pb/Cu	>40	40–31	30–21	20–10	<10
Ag/Cu	>1500	1500–1201	1200–901	900–600	<600
Ag/Pb	>100	100–80	79–60	59–40	<40
Pb/Zn	>1.50	1.50–1.21	1.20–0.91	0.90–0.60	<0.60
Ag/Zn	>100	100–81	80–61	60–40	<40

*Cu/Zn values are x10^{-3}.

REFERENCES CITED

Gaytán-Rueda, J. E., 1975, Exploration and development at the La Negra Mine, Maconí, Querétaro, Mexico [M.S. thesis]: Tucson, University of Arizona, 98 p.

López Ramos, E., 1979, Geología de Mexico, Vol. II: privately published, p. 447–448.

Simons, F., and Mapes, E., 1957, Geología y yacimientos minerales del Distrito Minero de Zimapán, Hidalgo: Instituto Nacional Investigones Recurses Minerales, Boletin 40, 282 p.

Printed in U.S.A.

The Geology of North America
Vol. P-3, Economic Geology, Mexico
The Geological Society of America, 1991

Chapter 35

Geology of the Molango Manganese District, Hidalgo

R. Alexandri R. and A. Martínez V.
Cía. Minera Autlan, S.A. de C.V., M. Escobedo 510-5o. piso, Lomas de Chapultepec, 11000 Mexico, D.F.

INTRODUCTION

The Molango Manganese District covers an approximate area of 1,250 km^2 in northeasternmost Hidalgo State (Fig. 1). In spite of the extensive erosion of the orebody, vast manganese ore volumes are spread over approximately 180 km^2 (Frakes and Bolton, 1984).

The center of the area, which includes the villages of Molango, Xochicoatlan, Lolotla, Tepehuacan, and Tlanchinol, is at 98°45′W and 20°55′N. The Tetzintla area, presently under development, is located in the northern part of the Manganese District, about 20 km north of Molango (Hidalgo), at 90°41′15″W, and 20°57′30″N. The paved Mexico-Tampico short route crosses the district from south to north; at "Las Casetas" on km 157, a 16-km paved roadway connects the Tetzintla Mine with the Compañía Minera Autlán's industrial and housing (Otongo) areas.

GEOLOGY

In the Molango Manganese District (Alexandri and others, 1985), Precambrian metamorphic rocks unconformably underlie upper Paleozoic to Lower Cretaceous sedimentary sequences and Tertiary intrusive and extrusive rocks (Fig. 1).

Precambrian

Gneissoid metamorphic rocks and metaconglomerates crop out at Cañon de Tlaltepingo, east of the village of Otongo toward the core of the Huayacocotla anticlinorium (Fig. 1).

Megascopically identified gneisses consist mainly of quartz, feldspars, and biotite, with accessory apatite, zircon, and occasionally abundant garnet. The metaconglomerates are reddish brown and consist of quartz and gneiss fragments cemented by siliceous material.

Fries and Rincón-Orta (1965) determined the age of 1.210 ± 140 Ma in gneiss outcrops near Huiznopala, 7 km southwest of Tezintla.

Permian

Guacamaya Formation. The lower and middle Paleozoic stratigraphic column is missing (Fig. 2). Flysch-type argillaceous-arenaceous sediments of the Guacamaya Formation, represented by a 200-m-thick rhythmic sequence of black to dark gray shales, sandstones, and a few conglomerates, crop out 3 km northeast of Tetzintla Mine.

The Guacamaya Formation unconformably underlies the Triassic Huizachal Formation at Cañon de Chipoco (Fig. 1). It contains abundant fusulinids, brachiopods, and pelecypods, and is lithologically correlatable with the Cañon Peregrina Permian sedimentary outcrops northwest of Ciudad Victoria (Tamaulipas).

Triassic

Huizachal Formation. This formation is a series of continental, reddish gray, sandy shales, sandstones, and conglomerates with abundant milky quartz clasts in a quartzitic groundmass (Imlay, 1948). *Otozamites,* an important Late Triassic index fossil plant, is found in the upper part of the formation. Lower and Middle Triassic sediments are absent. At Cañon de Chipoco the Huizachal Formation is unconformable on the Guacamaya and beneath the Huaycocotla Formations.

Jurassic

Huaycocotla Formation. At Cañon de Tetzintla, this formation comprises a 100-m-thick sequence of conglomerates, with quartz and sandstone fragments in an argillaceous-arenaceous groundmass, with sandstones, and dark gray carbonaceous shales (Imlay, 1948). This is the most widespread Mesozoic formation in the district. At Cañon de Tetzintla, it overlies the Huizachal Formation with slight angular unconformity and unconformably underlies the Cahuasas Formation. Ammonites in the shales identified as *Arnioceras* by Imlay, as well as worm specimens, some pelecypods, and fossil plants in the middle part, place the formation in the Sinemurian.

Alexandri R., R., and Martínez V., A., 1991, Geology of the Molango Manganese District, Hidalgo, *in* Salas, G. P., ed., Economic Geology, Mexico: Boulder, Colorado, Geological Society of America, The Geology of North America, v. P-3.

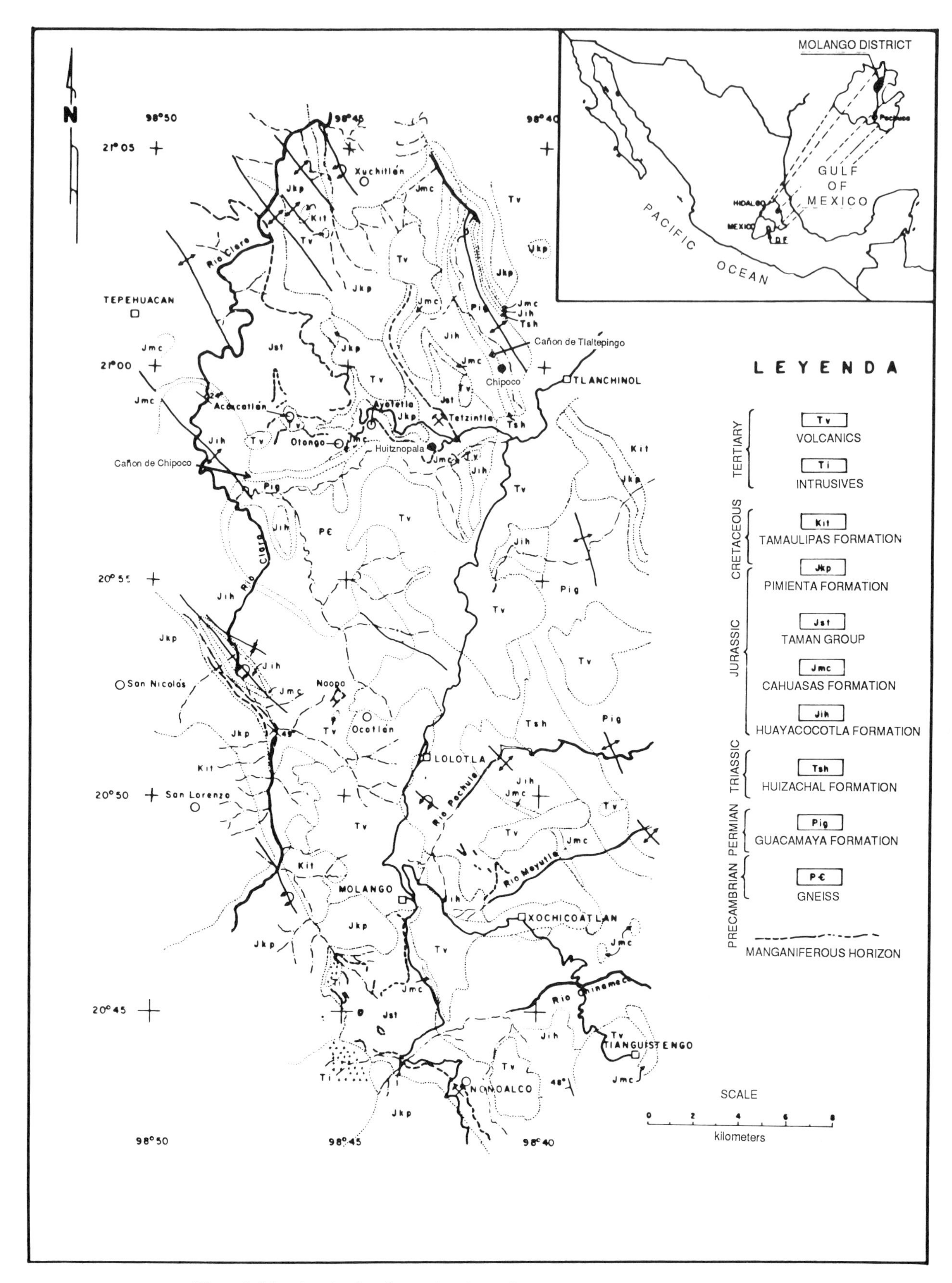

Figure 1. Map showing location and geology of the Molango Mining District.

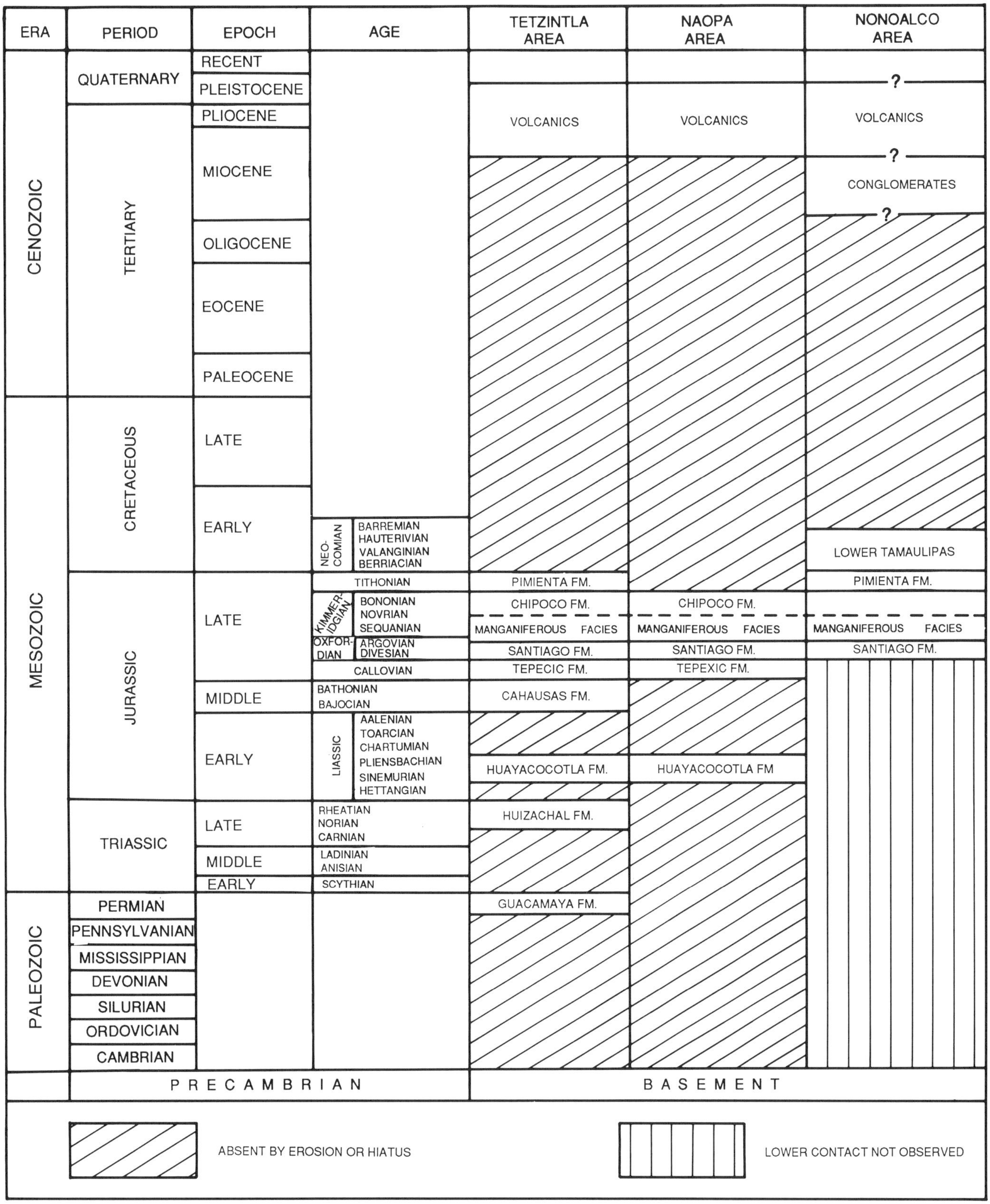

ERA	PERIOD	EPOCH	AGE	TETZINTLA AREA	NAOPA AREA	NONOALCO AREA
CENOZOIC	QUATERNARY	RECENT				
		PLEISTOCENE		VOLCANICS	VOLCANICS	? VOLCANICS
	TERTIARY	PLIOCENE				
		MIOCENE				? CONGLOMERATES ?
		OLIGOCENE				
		EOCENE				
		PALEOCENE				
MESOZOIC	CRETACEOUS	LATE				
		EARLY	NEOCOMIAN: BARREMIAN, HAUTERIVIAN, VALANGINIAN, BERRIACIAN			LOWER TAMAULIPAS
	JURASSIC	LATE	TITHONIAN	PIMIENTA FM.		PIMIENTA FM.
			KIMMERIDGIAN: BONONIAN, NOVRIAN, SEQUANIAN	CHIPOCO FM. / MANGANIFEROUS FACIES	CHIPOCO FM. / MANGANIFEROUS FACIES	MANGANIFEROUS FACIES
			OXFORDIAN: ARGOVIAN, DIVESIAN	SANTIAGO FM.	SANTIAGO FM.	SANTIAGO FM.
			CALLOVIAN	TEPECIC FM.	TEPEXIC FM.	
		MIDDLE	BATHONIAN, BAJOCIAN	CAHAUSAS FM.		
		EARLY	LIASSIC: AALENIAN, TOARCIAN, CHARTUMIAN, PLIENSBACHIAN, SINEMURIAN, HETTANGIAN	HUAYACOCOTLA FM.	HUAYACOCOTLA FM	
	TRIASSIC	LATE	RHEATIAN, NORIAN, CARNIAN	HUIZACHAL FM.		
		MIDDLE	LADINIAN, ANISIAN			
		EARLY	SCYTHIAN			
PALEOZOIC	PERMIAN			GUACAMAYA FM.		
	PENNSYLVANIAN					
	MISSISSIPPIAN					
	DEVONIAN					
	SILURIAN					
	ORDOVICIAN					
	CAMBRIAN					
	PRECAMBRIAN			BASEMENT		

ABSENT BY EROSION OR HIATUS

LOWER CONTACT NOT OBSERVED

Figure 2. Stratigraphic correlation chart for the rocks of the Molango Mining District.

Cahuasas Formation. This formation consists of siltstones and argillaceous sandstones, with a 4-m-thick, poorly sorted basal conglomerate that includes subangular red to dark gray fragments of quartz sandstone (Carrillo Bravo, 1958). At Chipoco Village, the unit has an average thickness of 260 m. At the Tlaltepingo Dam, along the road to Otonga, about 100 m west of the Otongo Social Center, the formation appears unconformably above the Huaycocotla Formation and also beneath the Tepexic Formation. No fossils have been found in the unit; by its stratigraphic position it may be younger than Pliensbachian and older than Callovian.

Tepexic Formation. This formation is represented by 20 m of dark gray, wavy-bedded calcarenites with abundant fossil oysters, which at times become coquinas. The contact with the unconformably underlying Cahuasas Formation is visible 200 m east of the Otongo Clinic-Hospital; the formation is conformably overlain by the Santiago Formation. The unit is assigned to the middle Callovian on the basis of its relation with the Santiago Formation.

Santiago Formation. Originally described as a member of the Taman Formation, this unit was recently raised to formation status by Cantu Chapa (1969). The formation is a sequence of dark gray, brown, or ochre-weathering, carbonaceous, calcareous siltstones containing small disseminated pyrite crystals, intercalated with argillaceous limestone beds up to 40 cm thick; the rocks display cleavage diagonal to the bedding planes and conchoidal fracture. Calcareous nodules appear intercalated toward the top of the formation.

Manganese mineworks at Cañon de Tetzintla, particularly the opencut mine, have exposed the full 360-m thickness of this formation; the lower Santiago/Tepexic contact and the upper contact with the manganese-bearing member of the Chipoco Formation are both conformable. Specimens of *Ostrea* sp., *Gryphaea,* and *Reineckia* sp. indicate its age as middle Callovian to late Oxfordian.

Chipoco Formation. Originally termed Taman Mixto within the group of that name, this formation is the most important unit from the economic viewpoint, enclosing at the base what is considered the largest manganese deposit in North America, and one of the largest in the world.

The base of the formation includes a 40-cm-thick dark gray manganese-bearing limestone of breccoid structure produced by strong folding of the entire sedimentary sequence, and of low manganese content. The overlying strata include 4 to 9 m of dark gray, thinly bedded, fine-grained manganese-bearing limestone with banded structure containing interbedded laminar bands of pyrite. The predominant manganese carbonate is rhodochrosite; the average chemical composition of the manganese carbonates presently under development is Mn, 27.2 percent; Fe 5.7 percent; SiO_2, 9.21 percent; Al_2O_3, 2.11 percent; CaO, 5.9 percent; Mg, 6.74 percent.

Manganese content decreases sharply at the top of this sequence and disappears at about 50 m above the Santiago-Chipoco contact. Pyrite content also decreases substantially toward the upper part of the formation. On the other hand, CaO content increases upward; as bedding becomes thicker (40 cm to 1 m) and grain size coarsens, the rock becomes a calcarenite, occasionally with oolitic texture. A persistent, 50-cm to 2-m-thick sponge-spicule siliceous horizon appearing 18 m above the contact with the Santiago Formation, occurs extensively throughout the manganese district.

Crystalline nonmanganiferous limestones (grainstones) 60 to 225 m in size alternate with greenish gray calcareous siltstones. Macrofossils are restricted to certain limestone horizons where they may become very abundant, as in the case of the *Aulacomyella* concentrations toward the top of the formation.

The Chipoco Formation is conformable over the Santiago and transitional to the overlying Pimienta Formation. It is assigned an early to middle Kimmeridgian age on the basis of paleontological determinations by Imlay (1948).

Pimienta Formation. The lower part of this unit comprises thinly bedded, very argillaceous limestones intercalated with siltstones and abundant dark gray to black chert lenses (Heim, 1926). These grade upward to thin to medium-bedded light gray limestones with chert bands that decrease toward the top of the formation.

The transitional Chipoco/Pimienta (Aguayo, 1977) contact is exposed along the road connecting the industrial zone and the Tetzintla pit. Toward the Chiconcoac farm settlement, the Pimienta Formation is covered by volcanic (basalt) flows. At both the north and south ends of the district, the lower Tamaulipas Formation of buff, fine-grained, medium to thickly bedded limestones with well-developed stylolites and irregular chert nodules overlies the Pimienta Formation transitionally.

The Pimienta Formation is Tithonian in age and may be correlated both chronologically and lithologically with the subsurface Pimienta Formation of the Tampico (Tamps.) region. The lower Tamaulipas Formation is of early Cretaceous age.

Cenozoic igneous rocks

The entire late Paleozoic and Mesozoic continental and marine sedimentary sequence is partly covered by basaltic fissure-flows corresponding to the northernmost manifestations of Miocene volcanism in the Trans-Mexican Axis. Intrusive granodiorites, tentatively assigned to the late Tertiary, affect the whole sedimentary section within the district.

GEOLOGIC HISTORY

Major tectonism took place in late Precambrian or perhaps Paleozoic time; the Grenville orogeny gave rise to the gneisses and metaconglomerates exposed in the central part of the Huaycocotla anticlinorium.

Erosion of these uplifted areas led to the accumulation of thick, flysch-type conglomeratic and sandy-argillaceous sediments (Guacamaya Formation). The onset of regional uplift and normal faulting in the Mesozoic, possibly related to the first pulses of the Gulf of Mexico opening, developed topographically

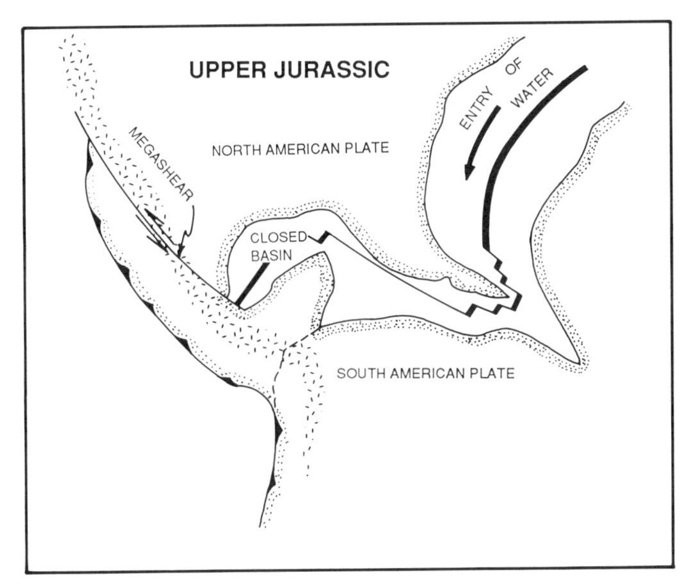

Figure 3. Sketch showing a possible early stage in the opening of the Gulf of Mexico following relative southeastward movement of Mexico along the postulated Mojave-Sonora megashear. (After Silver and Anderson, 1974.)

irregular and abrupt tectonic troughs that received the thick continental deposits of the Huizachal Formation.

The Lower Jurassic (Sinemurian) marine transgression laid down the Huayacocotla Formation. Geological information is as yet insufficient for a complete tectonic reconstruction of the Gulf of Mexico; consequently the aulacogen structure in the Huayacocotla region suggested by Schmidt-Effing (1980) is still to be confirmed.

At the end of the Lower Jurassic the area was uplifted and strongly folded; marine regression led to deposition of the continental Cahuasas Formation. The tectonism that resulted in the opening of the Gulf of Mexico evolved during this period.

The Atlantic rift probably reached the Gulf region, separating the North and South American Plates and allowing the gradual invasion of seas from the east that initiated the major Late Jurassic–Early Cretaceous transgression. Because the western part of the proto-Gulf remained closed and isolated from the Pacific, stagnant basins developed. On the other hand the Yucatán Peninsula remained linked to South America and, together with western Mexico, drifted to the southeast along the regional Mojave-Sonora megashear (Silver and Anderson, 1974) that crosses the whole of Mexico from northwest to southeast, emerging near the western coast in the Mojave Desert region (Fig. 3). The Tepexic, Santiago, and Chipoco Formations were deposited in the course of these events.

The megashear ceased to move at the end of the Jurassic (Tithonian-Neocomian). The South American Plate separated from western Mexico and Yucatán and continued drifting toward the southwest. The western end of the proto-Gulf thus opened up, allowing Pacific waters to circulate and the Pimienta and lower Tamaulipas Formations to be laid down (Dickinson and Coney, 1980; Fig. 4).

The entire sedimentary sequence was strongly folded, though not metamorphosed, during the Laramide orogeny in the Late Cretaceous (Campanian-Maastrichtian), forming the Huayacocotla anticlinorium. Intensive igneous intrusive and extrusive activity took place during the Tertiary.

GENETIC HYPOTHESIS

Normal manganese concentration in present-day seas is 0.00013 ppm; however, the solubility of this element has been proved to increase 500 times in anoxic marine waters.

Manganese studies in the Black Sea reveal massive manganese precipitation at the interface zone where deep anoxic waters intermingle with surface waters taking in oxygen from the atmosphere (Cannon and Force, 1983; Brewer and Spencer, 1974; Fig. 5).

The marine transgression initiated in the Callovian laid the ground for the formation of the Molango deposit. Oxygen demand for the decomposition of organic matter increased as the seas invaded lands with high biologic activity.

It should be kept in mind that, at the time, the proto-Gulf was not connected with the Pacific; the invading Atlantic seas covered a topographically irregular surface developing restricted stagnant basins. The large proportion of disseminated pyrite, absence of benthonic faunas, content of organic matter, and argillaceous character of the rock reflect the reducing environment of the Santiago Formation, which was probably deposited at depths of 200 to 300 m in very still waters unaffected by wave action. Under these conditions, assuming the basin filled with calcareous muds that had a high manganese content in solution, possibly

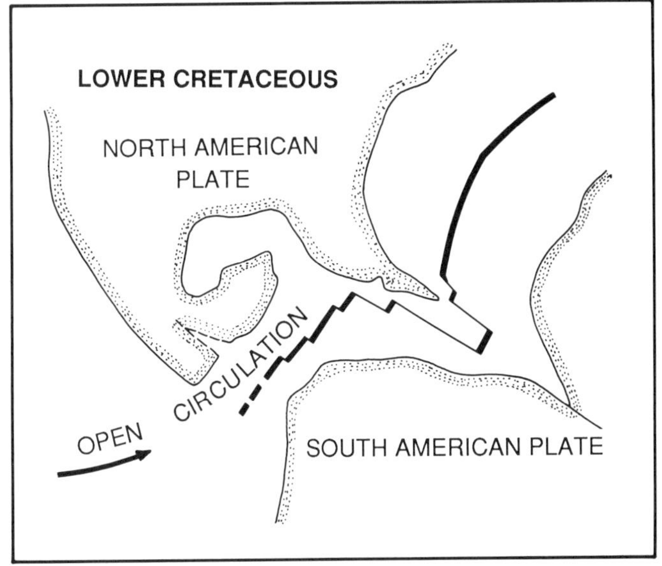

Figure 4. Sketch showing the postulated state of opening of the Gulf of Mexico in the Late Cretaceous, when the present Gulf area was connected to the eastern Pacific Ocean. (After Dickinson and Coney, 1980.)

derived from the oceanic spreading center that generated the proto-Gulf, the model emerges of a stratified sea with manganese-rich calcareous muds at the bottom overlain by anoxic waters, with well-oxygenated waters toward the surface.

Finally, the more thickly bedded calcarenite horizons of the Chipoco Formation that occur 30 m above the contact with the Santiago could represent a momentary reversal of the transgressive process. This would force oxygenated waters to descend and intermingle with anoxic waters trapped in the interstices of the calcareous muds, and cause the precipitation of manganese and the replacement of calcareous sediments by rhodochrosite ($MnCO_3$) in an early diagenetic phase.

Renewed transgression in the middle Kimmeridgian initiated the proto-Gulf–Pacific connection and allowed increased water circulation that did away with the reducing conditions; the upper Chipoco and Pimienta Formations were laid down, the latter in a pelagic, well-oxygenated environment.

CHEMICAL COMPOSITION OF THE DEPOSIT

Mineralogy

Calcite ($CaCO_3$) forms part of the groundmass and is also present in veinlets normal to the bedding. Calcite in the groundmass contains some manganese and is classified as manganocalcite: $CaCO_3$ (CaMn).

Kutnahorite (CaMn) $(CO_3)_2$ is analogous to dolomite, with manganese occupying the place of magnesium.

Rhodochrosite ($MnCO_3$) is an important part of the crystalline groundmass, occurring in fractures and as rhombohedral crystals in druses, and at times in botryoidal and globular form.

The predominance of one or another mineral component depends on the rock's manganese content. In general, rhodochrosite, containing 61 percent Mn, contributes most to the commercial value of the deposit. Accessory minerals are clays, quartz, magnetite, pyrite, and interbedded serpentine.

Manganese dioxide deposits

Manganese oxides directly derived from the carbonates and classified as recently formed epigenetic deposits occur in the Tezintla ara. All are products of secondary alteration of the manganese-bearing sediments caused by oxidation and leaching of the outcrops under the appropriate conditions.

Chemical and lithologic composition. These epigenetic deposits contain black manganese dioxides: predominantly nsutite, pyrolusite, cryptomelane, hausmannite, and birnessite, with interbedded ochre-yellow clay, hematite, and limonite in fractures and cavities.

Average analyses of this ore are 37.6 percent Mn; 8.0 percent Fe; 2.9 percent SiO_2; 2.0 percent Al_2O_3; 0.4 percent CaO; 1.0 percent MgO; and 0.1 percent C.

Origin. Carbonate rocks were leached by acid meteoric waters that dissolved Mn^{++}, Fe^{++}, and Mg^{++} ions. When these

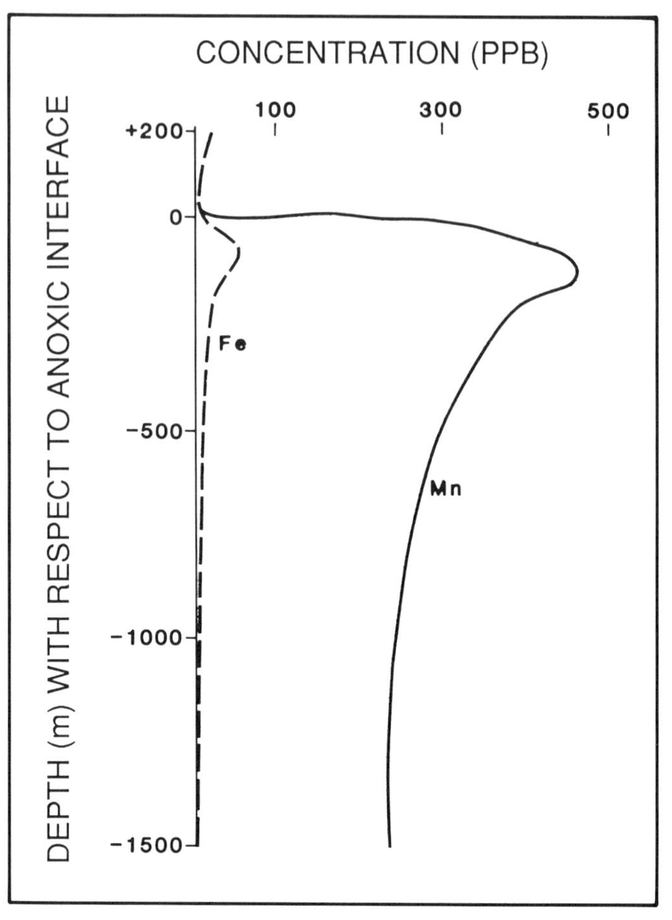

Figure 5. Distribution of dissolved manganese and iron in the Black Sea (Brewer and Spencer, 1974).

acid solutions are neutralized in the presence of oxygen, manganese and some iron hydroxide may precipitate and finally crystallize as oxides.

All the local conditions: exuberant vegetation, confined circulation of the solutions, and high permeability of the upper beds (in contrast with the impermeable rocks of the Santiago Formation) favor this process, which can be chemically explained by the following reactions.

The formation of carbonic acid is

$$H_2O + CO \rightarrow H_2CO_3. \quad (1)$$

When carbonic acid (H_2CO_3) attacks the carbonates, the following reaction ensues:

$$2MnCO_3 + 2H_2CO_3 \rightarrow 2MnO + 4CO_2 + 2H_2O. \quad (2)$$

MnO is hydrated in the presence of water producing Mn hydroxide:

$$MnO + H_2O \rightarrow Mn\,(OH)_2. \quad (3)$$

This hydroxide produces Mn bioxide when oxydized:

$$2Mn(OH)_2 + O_2 \rightarrow 2MnO_2 + H_2O. \quad (4)$$

The distribution of the manganese dioxide deposits is very local, being restricted to areas of manganese carbonate outcrops and to the depth and intensity of the weathering process.

CONCLUSION

The Molango Manganese District at present is the eighth largest in the world, based on its reserves. It is one of the lithologically best preserved deposits of its type, which makes it an ideal subject for future research on the physicochemical conditions and tectonic environment of its formation.

REFERENCES CITED

Aguayo, C.J.E., 1977, Sedimentación y diagénesis de la Formación Chipoco (Jurásico Superior) en afloramientos, Estados de Hidalgo y San Luis Potosí: Revista de Instituto de Mexicana Petroleos, v. 9, no. 2, p. 11–37.

Alexandri, R., Jr., Force, E. R., Cannon, W. R., Spiker, E. C., and Zantop, H., 1985, The sedimentary manganese carbonate deposits of the Molango District, Mexico: The Geological Society of America Abstracts with Programs, v. 17, p. 511.

Brewer, P. G., and Spencer, D. W., 1974, Distribution of some trace elements in Black Sea and their flux between dissolved and particulate phases, *in* The Black Sea; Geology, chemistry, and biology: American Association of Petroleum Geologists Memoir 20, p. 137–143.

Cannon, W. F., and Force, E. R., 1983, Potential of high-grade shallow-marine manganese deposits in North America, *in* Shanks, W. C., III, ed., Cameron volume on unconventional mineral deposits: New York, Society of Mining Engineers, p. 175–189.

Cantu C., A., 1969–1976, El contacto Jurásico-Cretácico, La estratigrafia del Neocomiano y amonitas del Pozo Bejucol (Centro Este de Mexico): Sociedad Geológica de México Boletin, v. 37, no. 2.

Carrillo B., J., 1958–1961, Geologia del Anticlinorio Huizachal-Peregrina-al NW de Ciudad Victoria, Tamaulipas: Asociación Méxicana de Geológos Petroleros Boletin, v. 13, no. 1–2, 98 p.

Dickinson, W. R., and Coney, P. J., 1980, Plate tectonic constraints on the origin of the Gulf of Mexico, *in* Pilger, R. H., Jr., ed., The origin of the Gulf of Mexico and the early opening of the Central North Atlantic: Baton Rouge, Louisiana State University, p. 27–36.

Frakes, L. A., and Bolton, B. R., 1984, Origin of manganese giants; Sea level change and anoxic-oxic history: Geology, v. 12, p. 83–86.

Fries, C., Jr., and Rincón-Orta, C., 1965, Contribuciones del Laboratorio de Geocornometria, Pt. 2., Neuves aportaciones geocronológicas y técnicas ampleadas en al Laboratorio de Geochronometria: Instituto de Geología, Universidad Nacional Autonoma de México Boletin 73, p. 57–133.

Heim, A., 1926–1940, The front ranges of the Sierra Madre Oriental México, from C. Victoria to Tamoxunchale: Eclogae Geologae Helvetiae, v. 33, p. 313–362.

Imlay, R. W., 1948, Stratigraphic relations of certain Jurassic Formations in eastern Mexico: American Association of Petroleum Geologists Bulletin, v. 32, p. 1750–1761.

Schmidt-Effing, R., 1980, The Huaycocotla aulocogen in Mexico (Lower Jurassic) and the origin of the Gulf of Mexico, *in* Pilger, R. H., Jr., ed., The origin of the Gulf of Mexico and the early opening of the Central North Atlantic: Baton Rouge, Louisiana State University, p. 79–86.

Silver, L. T., and Anderson, T. H., 1974, Possible left lateral early to middle Mesozoic disruption of the southwestern North American craton margin: Geological Society of America Abstracts with Programs, v. 6, p. 955–956.

Printed in U.S.A.

The Geology of North America
Vol. P-3, Economic Geology, Mexico
The Geological Society of America, 1991

Chapter 36

Pachuca–Real del Monte Mining District, Hidalgo

C. Fries
Instituto de Geología, U.N.A.M., Cd. Universitaria, Deleg. Coyoacan, 04510, México, D.F.

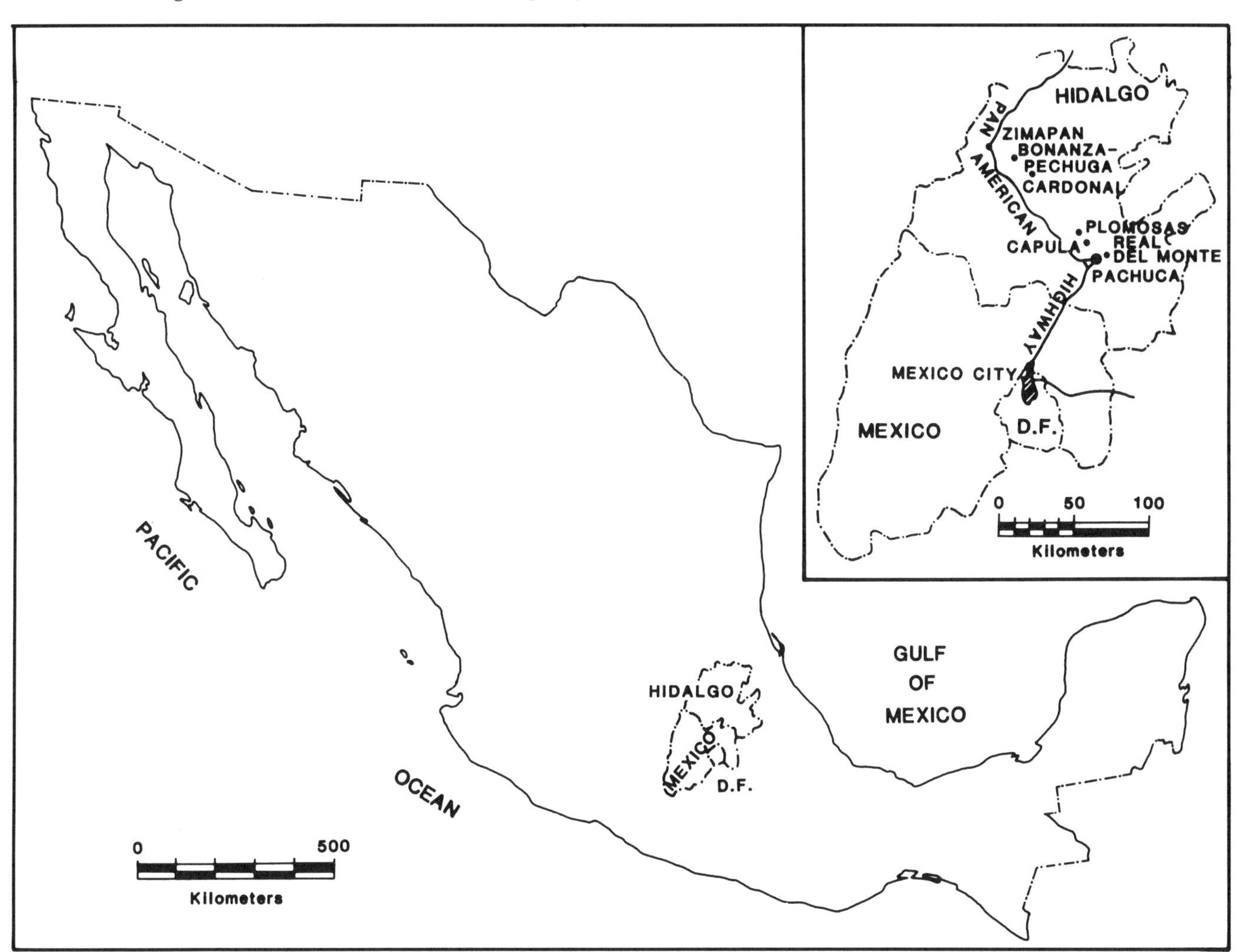

Figure 1. Index map showing location of Pachuca–Real del Monte Mining District.

INTRODUCTION

This district, one of the most prolific silver producers in the world, is located in central Hidalgo State, about 100 km north-northeast of Mexico City (Fig. 1), and has remained in production almost continuously since its discovery in 1522. Almost 25 million ounces of silver and 134,000 oz of gold were produced from 1973 to 1981 from more than 2,000 m of underground mineworks. Past and present diamond-drillholes total more than 100 km. Accumulated production from discovery to 1982 is estimated at 1,311,418,000 oz Ag and 6,643,000 oz Au.

The geology and development of the Pachuca District, set forth in detail by Fries and Geyne (1963), are briefly summarized below.

Fries, C., 1991, Pachuca–Real del Monte Mining District, Hidalgo, *in* Salas, G. P., ed., Economic Geology, Mexico: Boulder, Colorado, Geological Society of America, The Geology of North America, v. P-3.

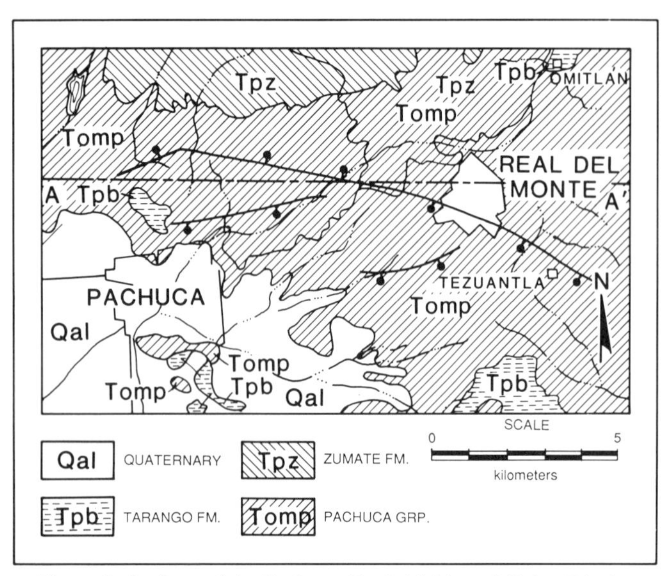

Figure 2. Geology of the Pachuca–Real del Monte Mining District.

GEOLOGY

The district lies approximately at the center of the Neovolcanic Axis Province (Salas, 1975). A Cenozoic series of andesites, rhyolites, and pyroclastics unconformably overlies at depth the tightly folded and faulted Mesozoic Alpine-type structures forming the Sierra Madre Oriental (Fig. 2). Deepest production is from 600 m; consequently, the Mesozoic limestones have not been reached, and their possible mineralization remains unproved. At present this sedimentary section is being sought by deep diamond drilling.

The stratigraphy of the Cenozoic mineralized section, more than 2,600 m thick, comprises lower Oligocene and upper Pliocene volcanic rocks of ten different formations.

The mineral accumulation is controlled by probably Miocene concurrent fracturing and faulting. The principal faults trend northwestward and dip from vertical to 65°. North-trending faults occur in the northeastern part of the district. Mineralization probably took place in the Pliocene when intrusive dacite and rhyolite dikes were emplaced along the faults. Intensive propylitization, albitization, seritization, and carbonatization occur together with abundant injected quartz in veins formed by hydrothermal metallogenic processes.

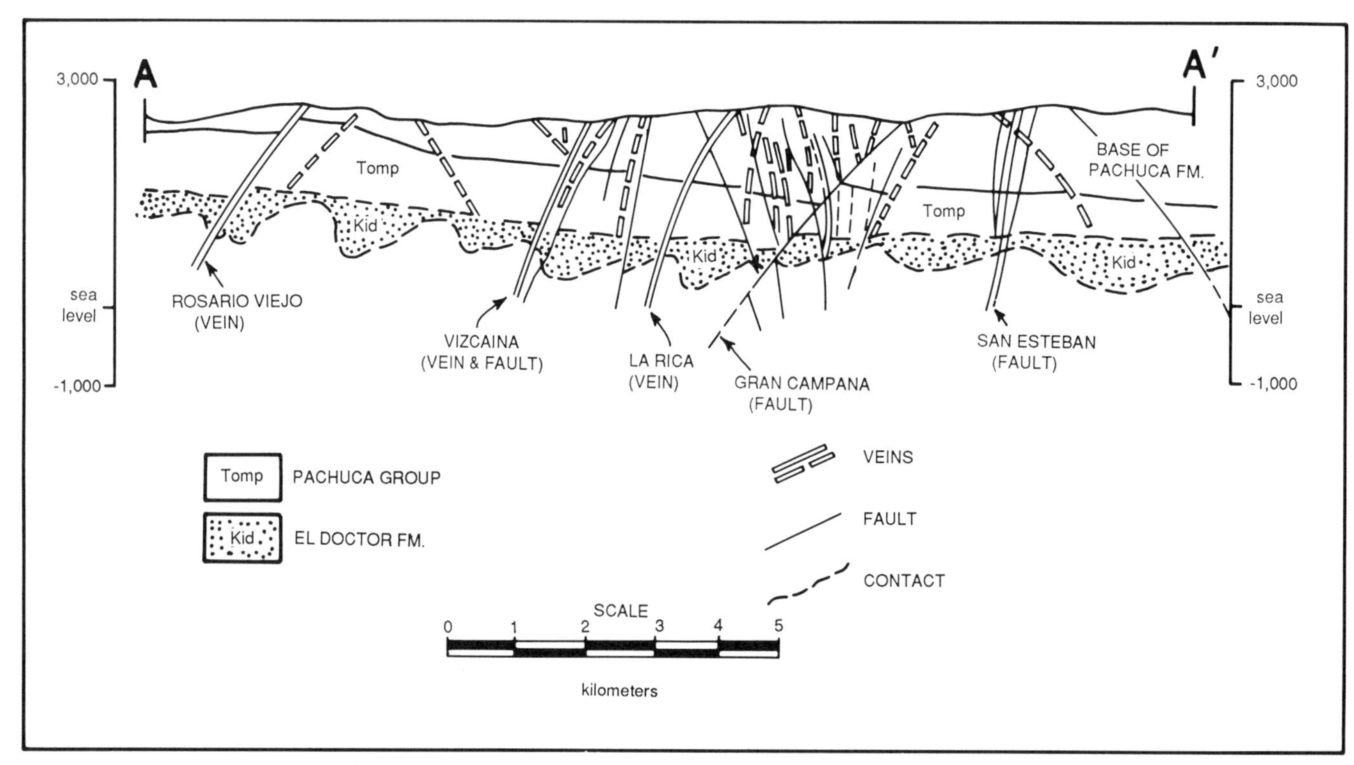

Figure 3. Geologic cross section of Pachuca–Real del Monte Mining District.

ECONOMIC GEOLOGY

Northwest-trending veins along major faults predominate, but prolific veins along east-west and north-south faults are also abundant. In some areas, vein thicknesses range from 1 to more than 40 m. Silver values range from several grams/ton to several kg/ton in mineral bonanzas of thin high-value veinlets (Fig. 3).

Acanthite and argentite are the principal silver ores together with lesser percentages of myrargirite, pyrargirite, proustite, and native silver. Present average assays are 180 g/t Ag and 1.2 g/t Au.

Main producing areas are the Pachuca area in the west of the district, beneath the town, and the Real del Monte area to the east (Figs. 1, 4, and 5). As mentioned above, the deepest levels are above the 700 m depth, and polymetallic production may be present deeper down in the Sierra Madre Oriental limestone section underlying the entire district.

I believe that these silver deposits developed originally in vast fractured areas, possibly during the lower Pliocene, and were

TABLE 1. OFFICIAL PRODUCTION FIGURES (TROY OZ.), COMPANIA DE REAL DEL MONTE Y PACHUCA, S.A.

	Silver		Gold	
Year	Annual	Accumulated	Annual	Accumulated
1973	2,735,675.99	2,735,675.99	14,148.16	14,148.16
1974	2,616,450.53	5,352,126.52	13,729.30	27,877.46
1975	2,801,675.86	8,153,802.38	9,280.68	37,158.14
1976	3,544,550.41	11,698,352.79	14,948.70	52,106.84
1977	3,265,959.62	14,964,312.41	19,396.39	71,503.23
1978	2,955,550.62	17,919,863.03	25,029.83	96,533.06
1979	2,607,938.24	20,527,801.27	14,555.91	111,084.97
1980	1,982,647.37	22,510,440.64	11,156.23	122,241.20
1981	2,252,218.56	24,762,667.20	11,709.13	133,950.33

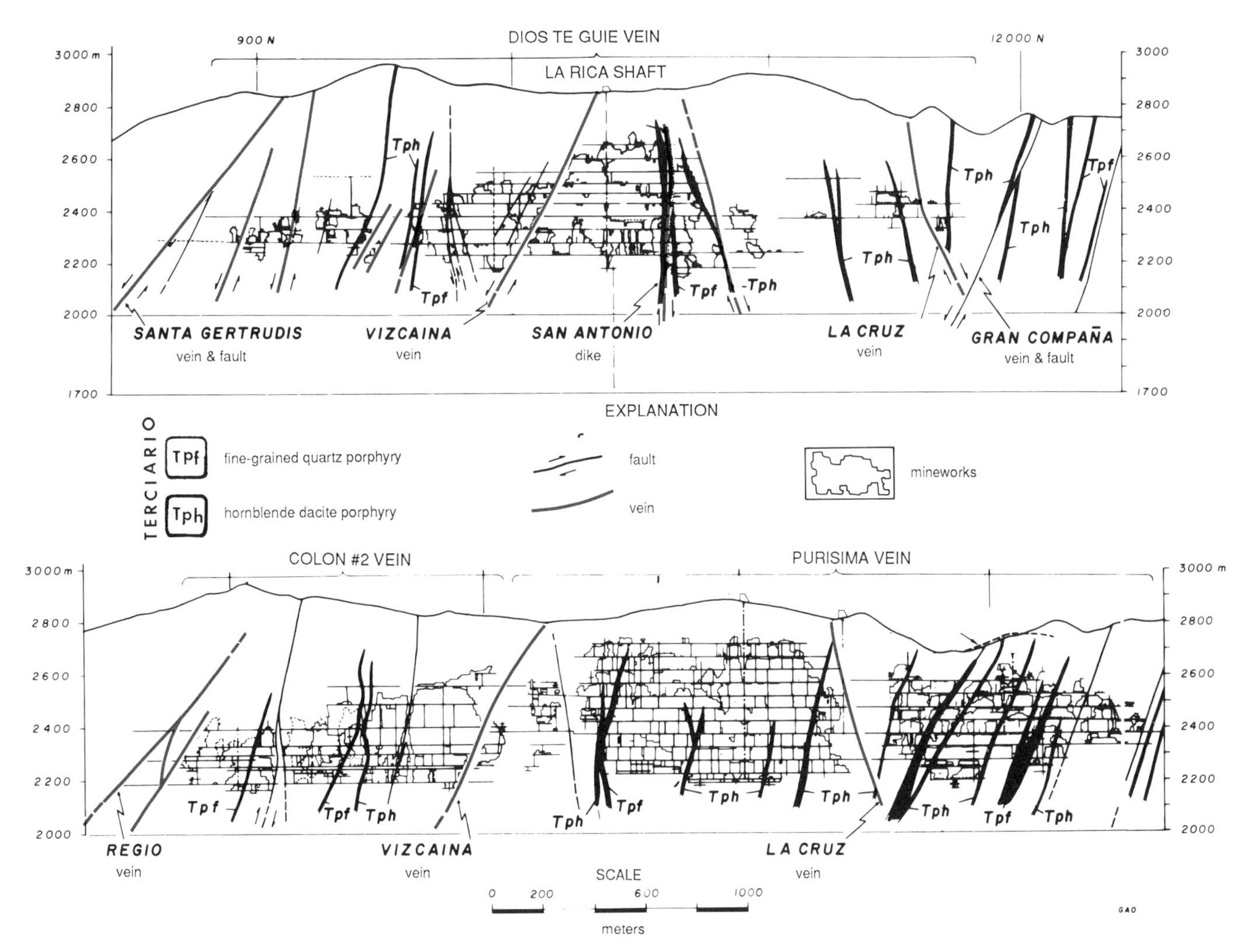

Figure 4. Sections showing the principal veins in the Pachuca–Real del Monte Mining District.

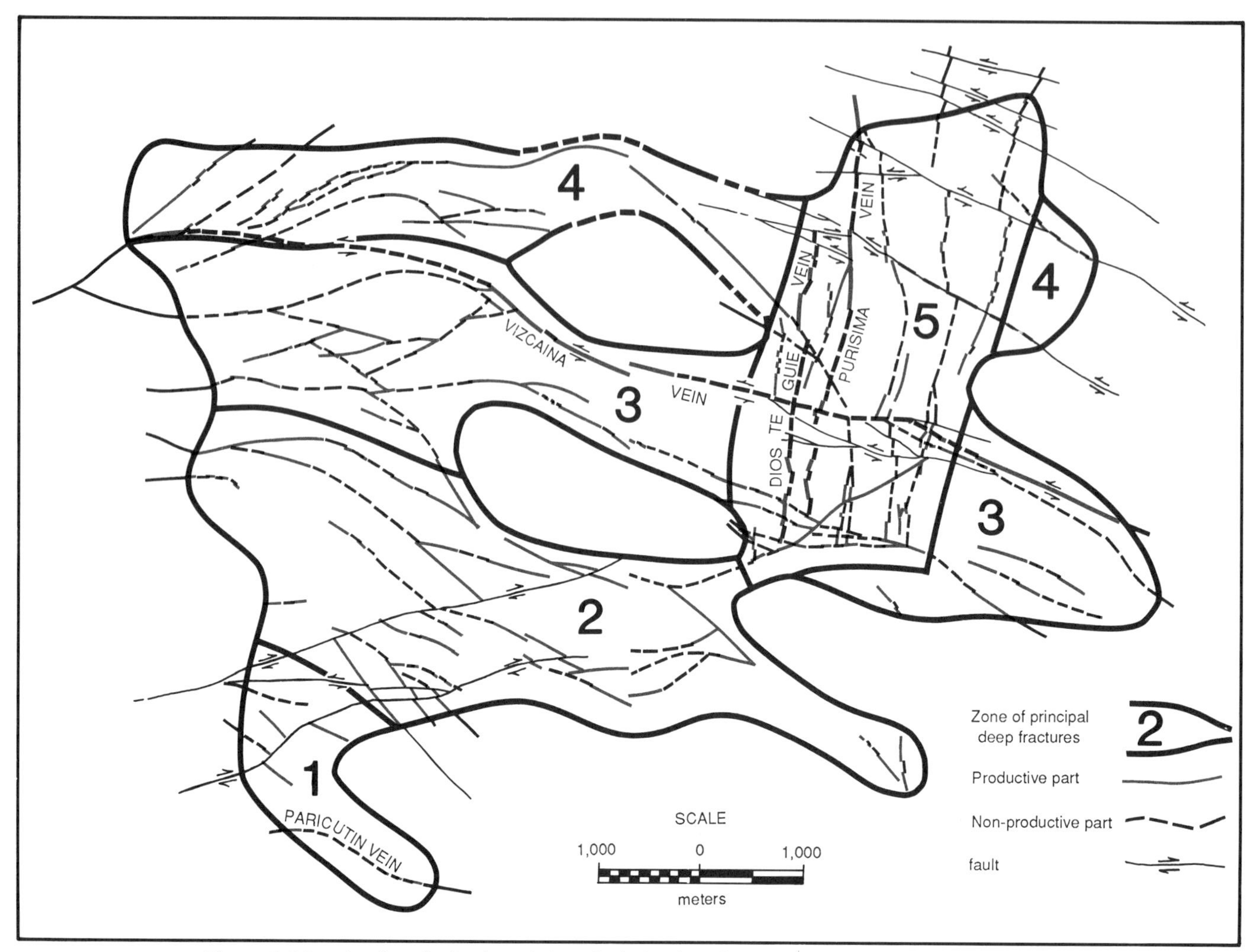

Figure 5. Map showing distribution of the principal veins of the Pachuca–Real del Monte Mining District.

redeposited by remobilization resulting from subsequent (upper Pliocene or lower Pleistocene) volcanic activity in the region.

Tables 1 and 2 show silver and gold production from 1973 through 1981.

Reserves are 8,000,000 tons of 208 g/t Ag and 0.99 g/t Au, of which 4,000,000 tons are ready for immediate production.

TABLE 2. DOMESTIC PRODUCTION (TROY OZ)

Year	Silver		Gold	
	Annual	Accumulated	Annual	Accumulated
1973	38,773,790	38,773,790	132,558	132,558
1974	37,552,064	76,325,854	132,454	267,012
1975	38,034,325	114,360,179	144,714	411,722
1976	42,631,881	156,992,060	162,811	574,533
1977	47,004,382	203,996,442	212,709	787,242
1978	50,798,169	254,794,611	202,003	989,245
1979	49,287,084	304,081,695	188,000	1,177,245
1980	52,084,199	356,165,894	195,991	1,373,236
1981	53,203,977	409,369,871	199,000	1,572,236
1982	49,840,752	459,210,623	196,248	1,768,484

REFERENCES CITED

Fries, C., and Geyne, A. R., 1963, Geología y yacimientos minerales del Dto de Pachuca–Real del Monte, Hidalgo, México: CRNNR, Publicación 5-E, 215 p.

Salas, G. P., 1975, Carta y provincias metalogenéticas de la República Mexicana: México City, Consejo de Recursos Minerales, Publicación 21-E, 199 p.

Printed in U.S.A.

The Geology of North America
Vol. P-3, Economic Geology, Mexico
The Geological Society of America, 1991

Chapter 37

The Santa Fe Mine, Chiapas

J. Pantoja A.
Instituto de Geología, U.N.A.M., Cd. Universitaria, Deleg. Coyoacan, 04510 México, D.F.

INTRODUCTION

The Santa Fe Mine is located in the municipality of Solosuchiapa in Chiapas State near the Chiapas/Tabasco state line. The town of Pichucalco (Chiapas) lies 40 km directly northwest, and Teapa (Tabasco) about 30 km north.

Access is by the Pichucalco–Tuxtla Gutiérrez unpaved roadway for 50 km through Ixtacomitán and Solosuchiapa to El Beneficio Ranch where a 3-km-long trail leads to the mine entrance. The roads are passable year-round. Paving of the Pichucalco–Tuxtla Gutiérrez road is scheduled in the near future (Fig. 1).

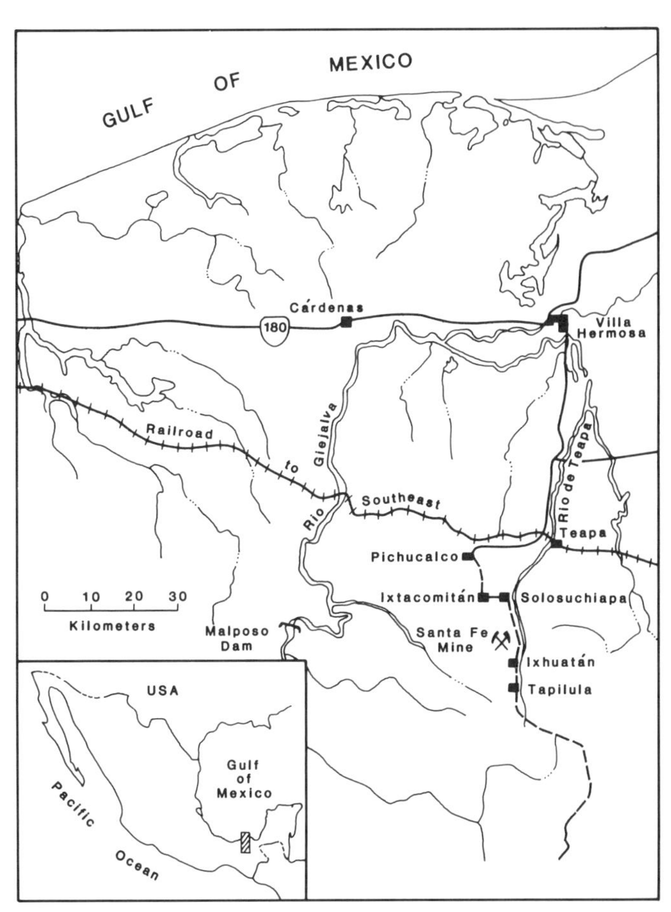

Figure 1. Map showing location of the Santa Fe Mine.

GENERAL OBSERVATIONS

This chapter discusses the results of exploration from July 1, 1967, to May 15, 1968, for purposes of calculating the Santa Fe Mine copper-silver reserves. With this and preexisting work, the Santa Fe Mine is now capable of extraction of 20 to 30 t/day of copper ores, which can be increased to 50 t/day by adding the appropriate hoisting facilities to the Santa Fe and Providencia shafts. Present mine capacity exceeds 200 t/day if the prospective open cast mining of wollastonite ores is included.

To determine water availability for an eventual mill, Arroyos La Danta and El Pino were gaged at several points; the two have a volume in excess of 4,000 L/min in the driest month, which is May.

GEOLOGY

At the beginning of the Tertiary the region was invaded by acid and intermediate intrusions, followed—possibly due to residual magmatic concentration—by basic to ultrabasic intrusions that involved the previously emplaced rocks and the overlying Cretaceous cover. Subsequent volcanism in the area is represented by largely eroded intrusive andesite dikes.

Silica-rich magmas replaced the calcareous sediments immediately after the granodiorite stock was emplaced in Santa Fe and generated the present-day extensive wollastonite orebody. Diorite and small ultrabasic dikes later intruded these, and preexisting rocks and the accompanying residual solutions originated the principal copper mineralization.

Ore studies suggest at least two periods of copper-silver mineralization with low gold grades. The first is contemporaneous with the wollastonite replacement, occurring disseminated in small pockets within the latter. The second period, immediately following the ultrabasic and diorite intrusions, filled faults, fractures, and contacts between these rocks and the wollastonite. Subsequent secondary enrichment mostly by wollastonite dissolu-

Pantoja A., J., 1991, The Santa Fe Mine, Chiapas, *in* Salas, G. P., ed., Economic Geology, Mexico: Boulder, Colorado, Geological Society of America, The Geology of North America, v. P-3.

tion generated the richest mines, developed in the past. The second period of mineralization possibly extends to the final intrusive phase of the andesite dikes marking the onset of Tertiary volcanism in the region.

Genetically the Santa Fe is a hydrothermal, medium-temperature, shallow to medium-depth deposit. Primary minerals are bornite, chalcopyrite, chalcocite, enargite, argentite, galena, and sphalerite. Oxidized minerals are chalcocite, enargite, malachite, azurite, and chrysocolla. Gold, though in minimal proportions, is intimately associated with bornite, chalcopyrite, and enargite. However, free gold occurs in wollastonite together with copper ores in the upper part of the Santa Fe Mine, known as El Portillo. The main silver ores have not been studied but are presumed to be argentite and other silver and copper complex sulfides. Besides wollastonite, the accompanying gangue in the principal veins includes quartz, chalcedony, and calcite. The Nivel del Cobre ore has a clayey gangue with abundant iron oxides and hydroxides and lesser amounts of manganese oxides. Pyrite is common in all the deposits.

RESERVES

Ten thousand tons of "probable reserves" averaging 6.0 gr Au, 189 gr Ag and 2.47 percent Cu per metric ton have been estimated for the Santa Fe mine. Tonnage distribution for individual orebodies is shown on Table 1; ore grades are the result of the total weighed average according to the sampling, including all the mineralized areas and incorporating values that appear in sampling maps of previous companies.

The disseminated mineralization and erratic occurrence of the orebodies precluded precise calculations of positive reserves, which consequently are rated as probable.

Considering possible new finds in the course of development, and that the above figures exclude the low-grade mineralized areas and several local pyrite outcrops, the potential reserves of the area are estimated at 5,000 tons of as yet unknown grade.

TABLE 1. ESTIMATED RESERVES, SANTA FE MINE

Orebodies	Reserves (tons)	Au (g/ton)	Ag (g/ton)	Cu (%)
El Cobre	5,000	4.7	191	1.60
El Jardín	1,000	3.5	167	2.47
Los Arcos 1 and 2	1,000	4.4	315	3.20
Verde Vein	600	5.0	100	2.50
Santa Maria	200	25.0	200	10.00
Goyens Vein	600	5.0	150	2.50
Taylor Vein	600	20.0	250	6.00
Santa Fe Stopes	300	4.0	100	2.50
Providencia Stopes	300	4.0	100	2.50
El Portillo	400	10.0	150	2.50

Printed in U.S.A.

The Geology of North America
Vol. P-3, Economic Geology, Mexico
The Geological Society of America, 1991

Chapter 38

Antimony deposits of the Los Tejocotes region, state of Oaxaca

R. Guíza, Jr., and D. E. White
Comite Directivo para la Investigacion de los Recursos Minerales de México, Consejo de Recursos Minerales, Centro Minero Nacional, Carretera México-Pachuca, Pachuca, Hgo., México, D.F.

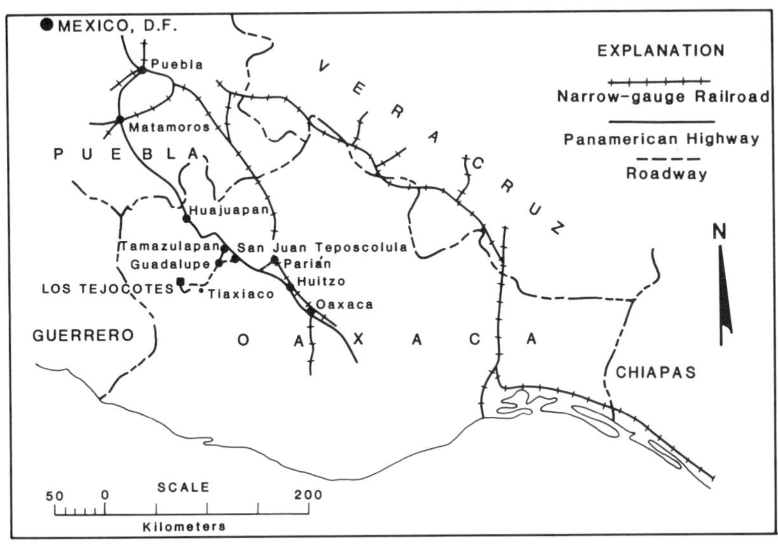

Figure 1. Map showing the Los Tejocotes region.

The Tejocotes mines in the state of Oaxaca have produced more antimony per year since 1938 than any other area in Mexico. Average annual production from 1938 to 1943 was 4,300 metric tons of shipped ores containing 56 to 58 percent of metallic antimony (Fig. 1).

The mid-Jurassic complexly folded and faulted sedimentary section comprises three limestone and three shale formations, all overlain by sandstones and affected by quartz and feldspar porphyry intrusions.

The antimony deposits occur in limestones as: (1) irregular or roughly elongate orebodies, (2) similarly shaped masses in argillaceous material that replaced the limestone, (3) fracture-fill veins, and (4) veinlets or disseminated masses. Types 1 and 2 are found near and generally more or less parallel to the quartz porphyry contacts. Each orebody averages several tons of mineral for export, and some attain 500 tons.

Types 3 and 4 are less restricted, and are usually oriented normal to the porphyry contacts. At least two type 3 veins have produced up to 300 tons of export ore.

Stibnite (Sb_2S_3) was the main antimony mineral species originally present in all the deposits. Near the surface it has oxidized in situ and is presently found almost completely altered to antimony oxide in colors ranging from yellow to reddish brown.

Geologic conditions determining the genesis of the orebodies were as follows: (1) central portions of anticlines in which the lower limestone is overlain by the lower shale, (2) east sides (usually located at the hanging walls) of contacts between the limestone and irregular mantos of intrusive quartz porphyry, and (3) irregular portions of porphyry/limestone contacts.

In some places, replacements of the limestone by silica and clay minerals serve as guides to locate antimony ores. In several deposits the antimony may have concentrated near the surface by selective dissolution of the limestone and removal of clay through erosion.

Recommendations made in White and Guíza (1947) may serve as guides in exploration for new ore deposits (Fig. 2, next page).

REFERENCES CITED

White, D. E., and Guíza, Jr., R., 1947, Antimony deposits of the Tejocotes region, state of Oaxaca, Mexico: U.S. Geological Survey Bulletin 953-A, 26 p.

Guíza, Jr., R., and White, D. E., 1991, Antimony deposits of the Los Tejocotes region, state of Oaxaca, *in* Salas, G. P., ed., Economic Geology, Mexico: Boulder, Colorado, Geological Society of America, The Geology of North America, v. P-3.

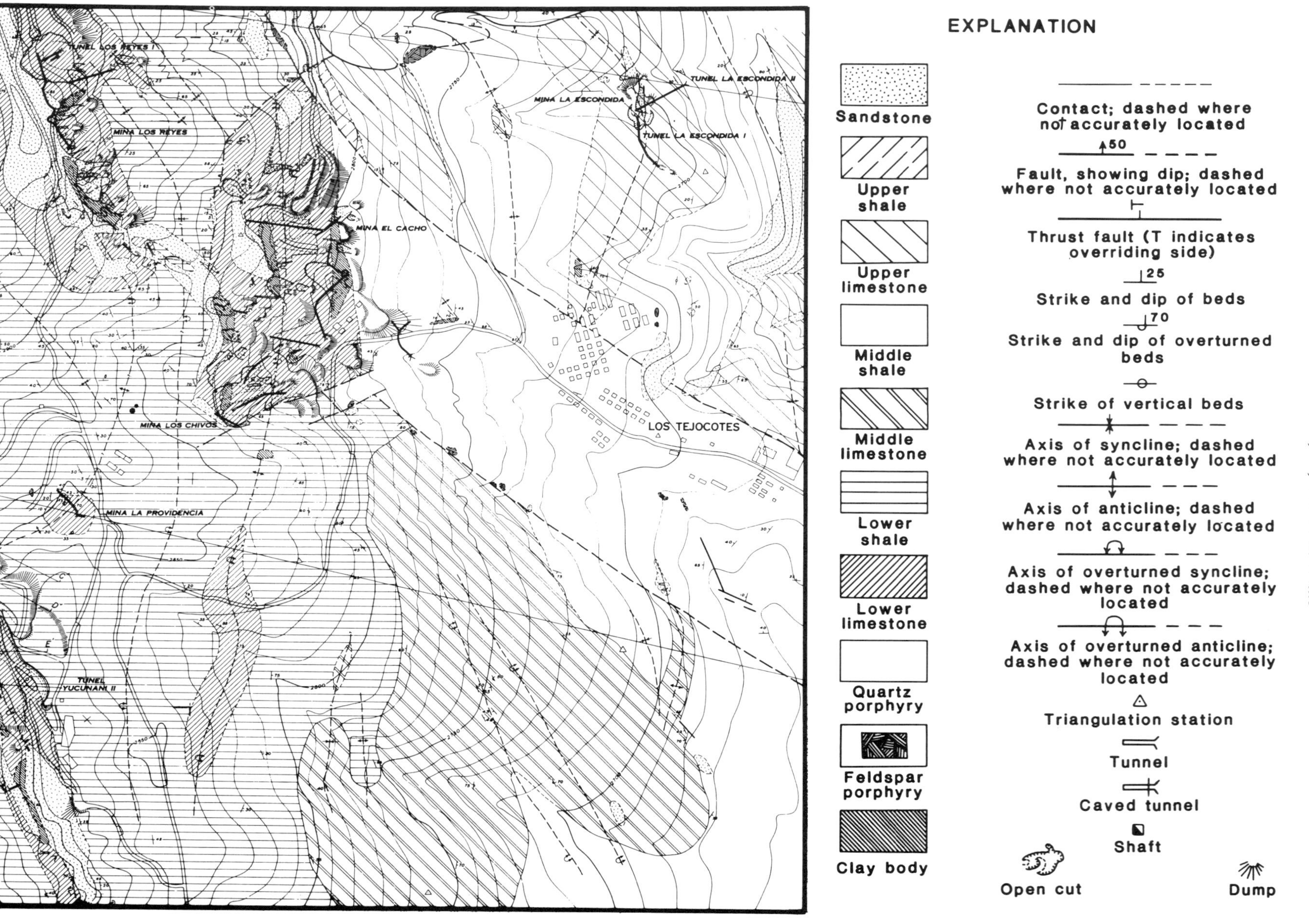

Figure 2. Geologic map of the Los Tejocotes region. (From White and Guíza, 1947.)

The Geology of North America
Vol. P-3, Economic Geology, Mexico
The Geological Society of America, 1991

Chapter 39

Sierra Madre del Sur Metallogenic Province

Guillermo P. Salas
Geologo Consultor, Asesor del Consejo de Recursos Minerales, Centro Minero Nacional, Carretera México-Pachuca, Pachuca, Hgo., México, D.F.

This province (Fig. 1) is named after the mountain chains with which it coincides physiographically. Though these are not continuous, their Mesozoic sedimentary formations, folded and faulted during the Eocene, follow the trend of the Pacific coast. The older (Precambrian and Paleozoic) units strike north-south, whereas the post-Eocene mineralizing intrusions generally follow northwesterly trends (Fig. 1).

The mountain chains developed in thick sequences of effusive and pyroclastic rocks unconformably overlying considerable sections of northwest-trending, folded and faulted Mesozoic limestones, which in turn overlie Paleozoic formations near the Tomellín Canyon in northeastern Oaxaca. Late Precambrian (750 to 850 Ma) foliated schists and gneisses crop out both at Oaxaca and in southern Chiapas.

Extensive schists and gneisses unconformably underlying the Cretaceous sedimentary section in northern Oaxaca and southwestern Guerrero are assigned to the Mesozoic on the basis of isotope geochronology; the schists are host rocks for part of the Taxco (Guerrero) ore deposits.

The mineralization in Michoacán, Guerrero, Oaxaca, and the few deposits known in Chiapas are of Oligocene–upper Miocene age.

Mineralization appears associated with felsitic intrusive stocks and dikes; it occurs at the periphery of the stocks or above them when they are not exposed.

The abundant southeastern Guerrero and southwestern Oaxaca metasomatic iron ore deposits are low-volume, which makes them uncommercial for the time being (1984). The Cemento Cruz Azul Company is presently developing a small iron-ore deposit (El Carmen) for clinker.

Small- and medium-sized polymetallic deposits abound in Oaxaca, some with high gold and silver grades (besides lead, zinc, and copper). Especially important is the Natividad and Anexas deposit (Fig. 1, location 5), for many years a prolific gold and silver producer from veins in Cretaceous limestones in contact with granodiorite.

Of particular interest in Oaxaca is the Concepción Pápalos asbestos deposit (Fig. 1, location 3) a few kilometers north-northeast of the Huitzo titanium (ilmenite with magnetite and fluorapatite) and the crystalline graphite deposits (Fig. 1, location 4), the latter being at present an important commercial producer for export. The asbestos fiber, consisting almost completely of slip chrysotile with only very small amounts of cross-fiber veinlets, has not been developed. The three deposits occur in Precambrian schists and gneisses within a zone of scarcely explored pegmatites containing allanite, thorium-rich betafite, and large quartz, mica, and other crystals.

There are no presently producing mines in the Sierra Chiapas, possibly due to lack of exploration. Small bauxite deposits are in the process of being evaluated. South of Solosuchiapa, the Santa Fe Mine, of copper and wollastonite at the contact of Cretaceous limestones with late Miocene granodiorite, has been abandoned for many years. Other mineralized localities are discussed in Salas (1975).

Examples of ore deposits studied in Guerrero, Michoacán, and Oaxaca are described in the following six chapters: (1) La Minita, Michoacán; (2) Las Truchas, Michoacán; (3) Concepción-Pápalos, Oaxaca; (4) Huitzo, Oaxaca; (5) Natividad and Anexas, Oaxaca; and (6) Telixtlahuaca, Oaxaca (see Fig. 1).

REFERENCES CITED

Salas, G. P., 1975, Carta y provincias metalogeneticas de la Republica Mexicana: Mexico City, Conzejo de Recursos Minerales, Publicación 21-E, 199 p.

Salas, G. P., 1991, Sierra Madre del Sur Metallogenic Province, *in* Salas, G. P., ed., Economic Geology, Mexico: Boulder, Colorado, Geological Society of America, The Geology of North America, v. P-3.

Figure 1. Map showing location of districts discussed in the following six chapters. 1, La Minita, Michoacán; 2, Las Truchas, Michoacán; 3, Concepción-Pápalos, Oaxaca; 4, Huitzo, Oaxaca; 5, Natividad and Anexas, Oaxaca; 6, Telixtlahuaca, Oaxaca.

The Geology of North America
Vol. P-3, Economic Geology, Mexico
The Geological Society of America, 1991

Chapter 40

Geology and genesis of the La Minita barite deposit, Michoacán

G. De La Campa J.
Cía. Minera Peñoles, S.A. de C.V., Río de la Plata 48 esq. Lerma, Col. Cuauhtemoc, 06500 México, D.F.

INTRODUCTION

Brief history

Barite has been mined in the Coalcomán area since the 1950s. Compañia Minera Autlán undertook a manganese exploration drilling program at the end of that decade with negative results; however, Zn and Ag values, of little interest to the company at the time, were significant.

What is now known as the La Minita Unit of Minera Capela, S.A. de C.V., was not discovered until 1976, when the Peñoles San Luis Potosí (S.L.P.) Regional Office, under Jose E. Gaytán Rueda, began searching for volcanogenic deposits.

Location and access

Geographic coordinates of the La Minita Mine area are 18°52′06″N and 103°17′06″W within the Coalcomán municipality, Michoacán, about 60 km directly southeast of Colima (Colima; Fig. 1).

Access to the deposit from the town of Colima on the road to Pihuamo takes 2½ hours to reach the junction with the 36-km-long unpaved roadway leading to El Encino Mine. This unpaved road is followed for 18 km, to the turnoff toward Coalcomán. From this junction it is 40 km to La Minita. From Guadalajara it takes 4½ hours along the road leading south to Ciudad Guzmán and Pihuamo, which reaches the junction with the Las Encinas roadway 12 km south of Pihuamo, for an approximate total distance of 200 km.

Guadalajara (Jalisco) and Manzanillo (Colima) have international airports; the town of Colima also has a small airport, and a small, unpaved light-aircraft runway lies 30 km away from La Minita.

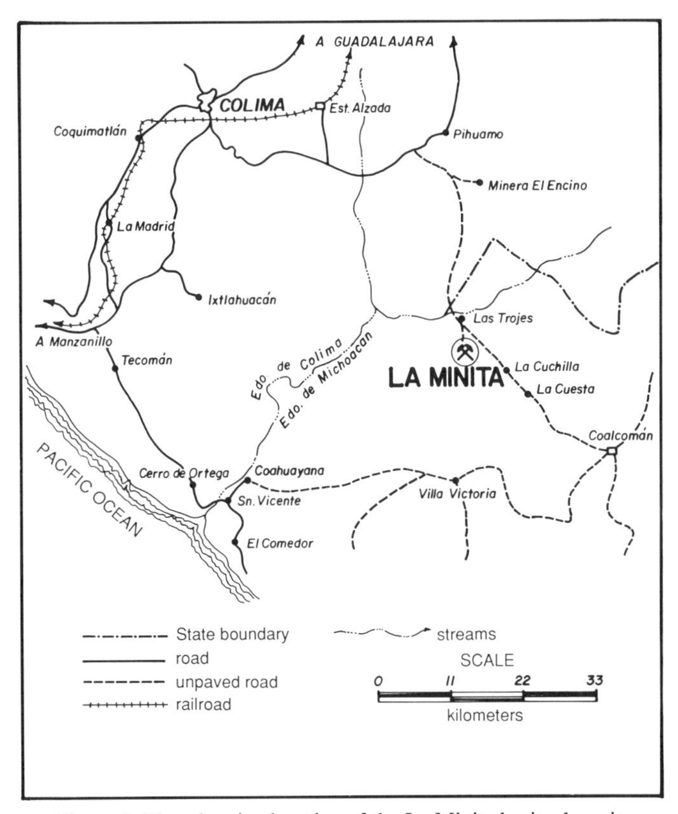

Figure 1. Map showing location of the La Minita barite deposit.

REGIONAL AND LOCAL GEOLOGY

Geomorphologically the region comprises gently sloping hills in shales, sharp karstic topography in limestones, and rounded volcanosedimentary rock crests; elevations range from 700 to 1,700 m.

Physiographically the area belongs in the Sierra Madre del Sur Balsas–Mezcala Basin Subprovince of Paleozoic metasediments overlain by Cretaceous marine sediments and continental deposits.

The local stratigraphic column enclosing the event that generated the La Minita deposit comprises three lithologic types, from older to younger, tentatively assigned to the Cretaceous (Aptian-Albian): mainly volcanic, volcanosedimentary, and predominantly sedimentary (Fig. 2).

De La Campa J., G., 1991, Geology and genesis of the La Minita barite deposit, Michoacán, *in* Salas, G. P., ed., Economic Geology, Mexico: Boulder, Colorado, Geological Society of America, The Geology of North America, v. P-3.

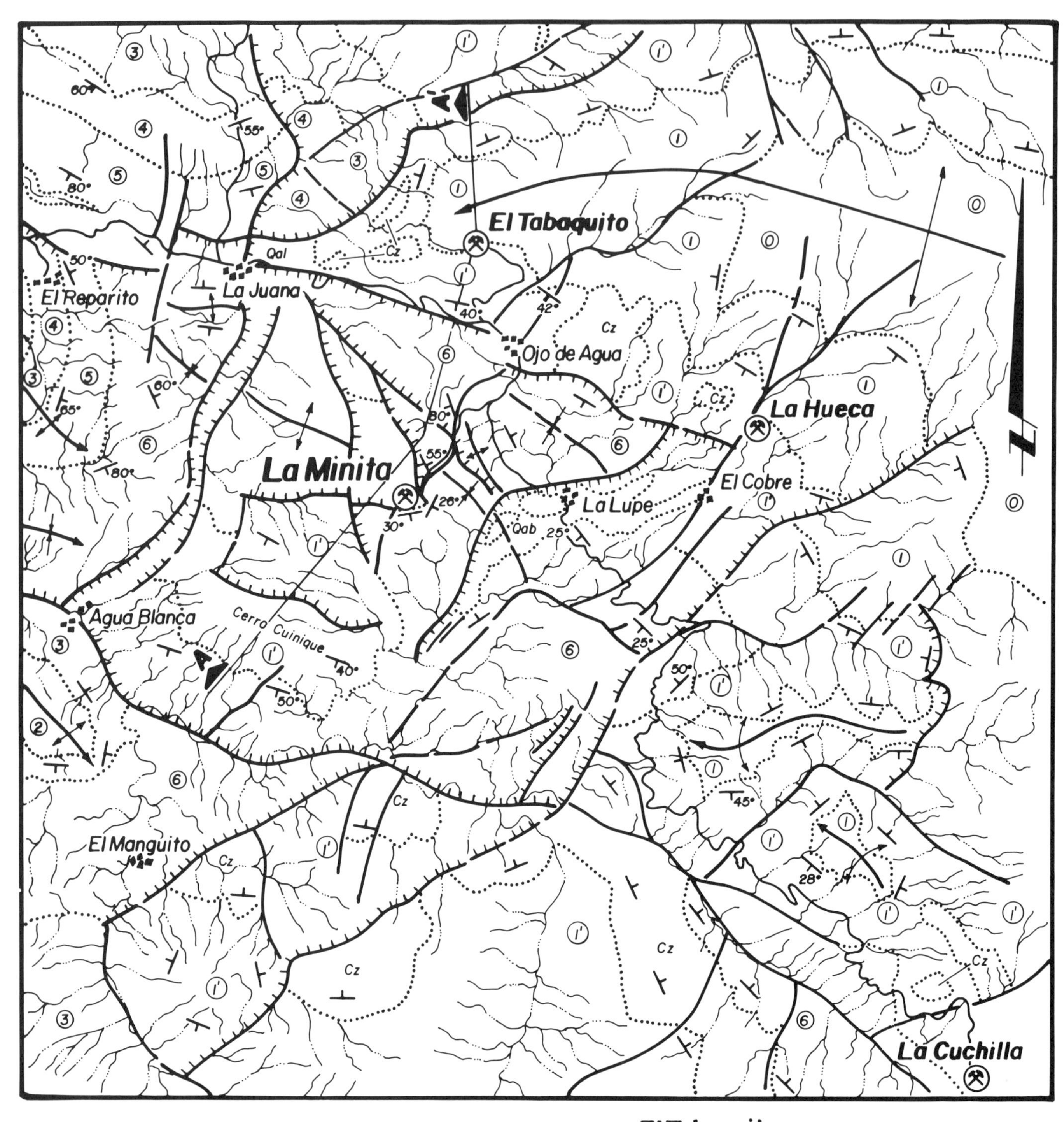

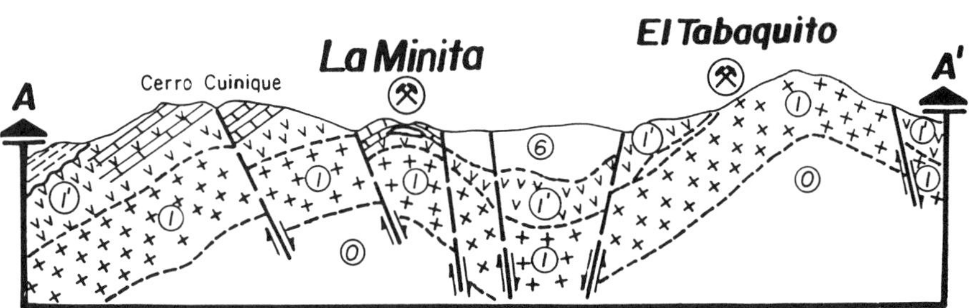

Figure 2. Geologic map and stratigraphic column for the La Minita barite deposit.

STRATIGRAPHIC COLUMN

Age	Symbol	Description
Quaternary	Qal, Qab	ALLUVIAL DEPOSITS
Tertiary	IT	INTRUSIVES
CRETACEOUS	6	LITHOSTRATIGRAPHIC UNIT No. 6 Shales, siltstones, argillaceous limestones and sandstones
	5	LITHOSTRATIGRAPHIC UNIT No. 5 Coarse-grained tuffaceous ss, quartz and/or chalcedony frags.
	4	LITHOSTRATIGRAPHIC UNIT No. 4 Fine-grained ss interbedded with black argillaceous ls beds.
	3	LITHOSTRATIGRAPHIC UNIT No. 3 Fine-grained ss, fossiliferous coarse-grained ss.
	2	LITHOSTRATIGRAPHIC UNIT No. 2 Argillaceous black ls inter-bedded with sh and ss.
	1', Cz	LITHOSTRATIGRAPHIC UNIT No. 1' Felsic tuffs, jasper and manganese horizons. Thick reefal calc. bodies Stratified Ba, Zn, Ag, Pb orebodies
	1	LITHOSTRATIGRAPHIC UNIT No. 1 Medium to coarse-grained lithic tuff
	0	LITHOSTRATIGRAPHIC UNIT No. 0 Massive rock: spherulitic structure, slightly reddish-brown due to iron oxides in microfractures

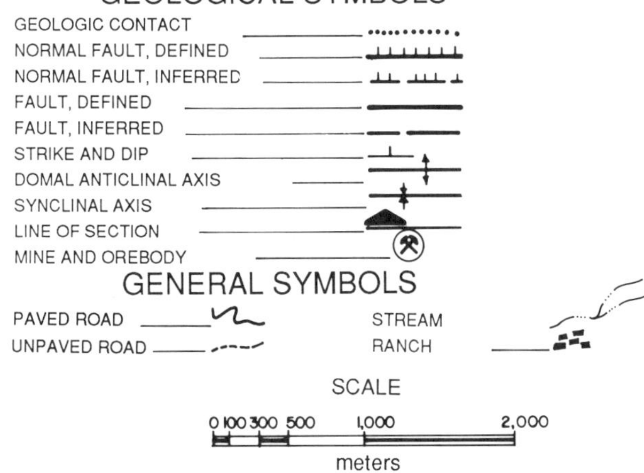

Volcanic sequence

This sequence contains three units: (1) a lower series of spherulitic andesite tuffs; (2) an intermediate group of highly silicified and chloritized pyroclastic volcanic rocks of andesitic composition, locally impregnated with manganese oxide and slightly magnetic, with copper, zinc, and magnetite mineralization in the form of mantos; and (3) an upper unit of massive, thinly bedded felsic tuffs, rhyolites, and small rhyolite-dacite intrusives, which is the most important commercially because it contains the known orebodies.

Volcanic-sedimentary sequence

This sequence contains reefal bodies with abundant rudists, argillaceous clays, sandstones, and tuffaceous shales.

Sedimentary sequence

The deposition of these rocks was structurally controlled by grabens: small basins filled with alternating shales, sandstones, and limestones in thin beds.

GEOLOGY OF THE DEPOSIT

La Minita is a stratiform polymetallic deposit of sulfides, sulfates, carbonates, and oxides, genetically related to Early Cretaceous(?) submarine intermediate and felsic volcanism.

Mineralization appears both massively and in veinlets. The massive type is the most commercial, consisting essentially of barite with Ag, Pb, and Zn minerals.

Form and dimension

Structurally the La Minita deposit occupies the southeast nose of a doubly plunging, N60°W–trending regional domal structure cut by two N30°E–trending normal faults. On the whole the orebody adopts the same domal shape, elongated parallel to the regional structural axis, with an area of 500 by 250 m and an average thickness of 14 m (Fig. 3).

Mineralogy and zoning

This deposit comprises two zones of mineralization as mentioned above; however, only the commercial massive type, vertically zoned into three subzones, will be briefly described.

Lower subzone. This subzone contains the largest concentration (6 percent) of zinc sulfides, low Ag (60 gr/ton) and Pb (0.3 percent) contents, less than 0.2 percent Cu, and lesser amounts of barite than in the two upper subzones. Minerals present are barite, sphalerite, galena, stromeyerite, tetrahedrite, chalcopyrite, pyrite, magnetite, quartz, and calcite.

Sulfides occur massively, banded, breccioid, and in veinlets; sphalerite shows colloform texture; galena is very fine-grained, micro- and cryptocrystalline, and silver occurs in solid solution with the galena, as stromeyerite and as extremely fine tetrahedrite. These textures indicate recrystallization, replacement, and cavity filling.

Intermediate subzone. This subzone shows similar features, except for the relative concentrations of the commercial minerals and their partial oxidation.

Zinc sulfides occur in lesser amounts; Pb increases as well as Ag, which represents the subzone's main commercial value; barite increases; cerussite, anglesite, massicot, hematite, smithsonite, and magnetite appear.

Upper subzone. In this subzone, barite is practically the only mineral; Ag, Pb, and Zn also appear in lesser amounts, possibly owing more to enrichment than to primary distribution.

Genesis

Research results identified five main structural types, suggesting that this is a submarine volcanosedimentary deposit.

Widespread banded and laminar structures parallel to bedding planes of the overlying tuffs point to a sedimentary origin. Gradually bedded ores indicate that the constituent particles were deposited by effects of water and gravity. Brecciated barite ores in chaotic arrangement with mostly angular fragments of jasper (ferruginous chert) indicate that the main body was deposited on a paleotopographic slope and subsequently slumped down. Brecciated jasper and limestone with disseminated sulfides in veinlets and replacing preexisting rocks are widespread in the northwestern part of the orebody, suggesting thermal activity after the sedimentary ores were deposited.

Stockworks in rhyolites clearly visible in the northwestern part of the orebody, together with the dissemination, veinlets, and replacements in the overlying reefal limestones, definitely indicate their formation in preexisting rocks, whereas the mineralization in other parts of the area is of sedimentary origin with subsequent transport and tectonism.

In general the La Minita deposit contains two types of ores: (1) massive and stratiform, generated by hydrothermal fluids unloaded onto the sea floor and subsequently brecciated by transport and tectonism; and (2) typically epigenetic veinlet and disseminated ores formed by the reaction of hydrothermal fluids with preexisting rocks (Fig. 4).

ORE RESERVES

Some 78 drillholes—ranging 50 to 170 m in depth, for a total 5,960 m (that intersected 1,224 m of mineralization)—were distributed so as to provide the most reliable information for calculation of reserves, which are rated here as proven, given the

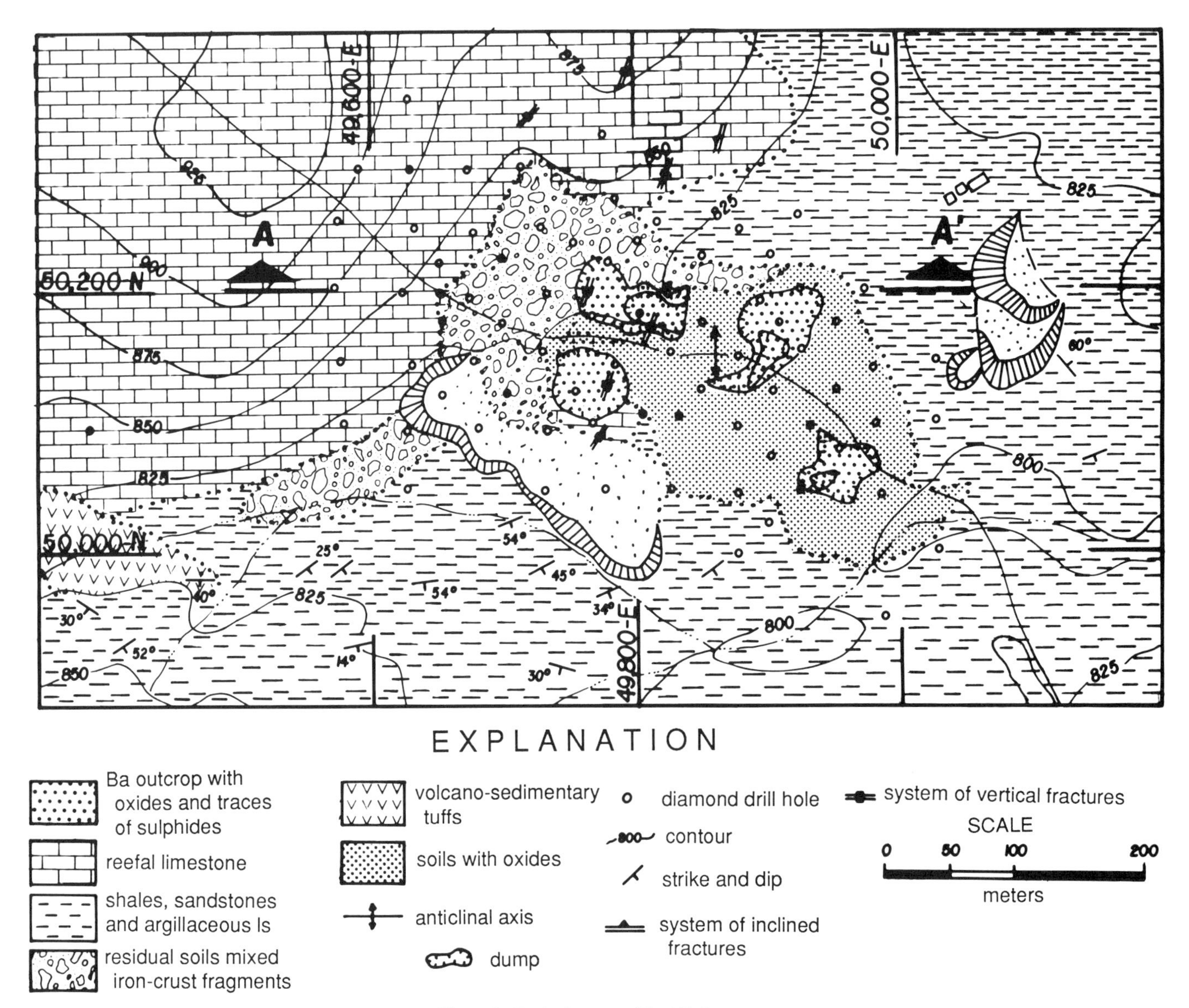

Figure 3. Geologic map of La Minita cut.

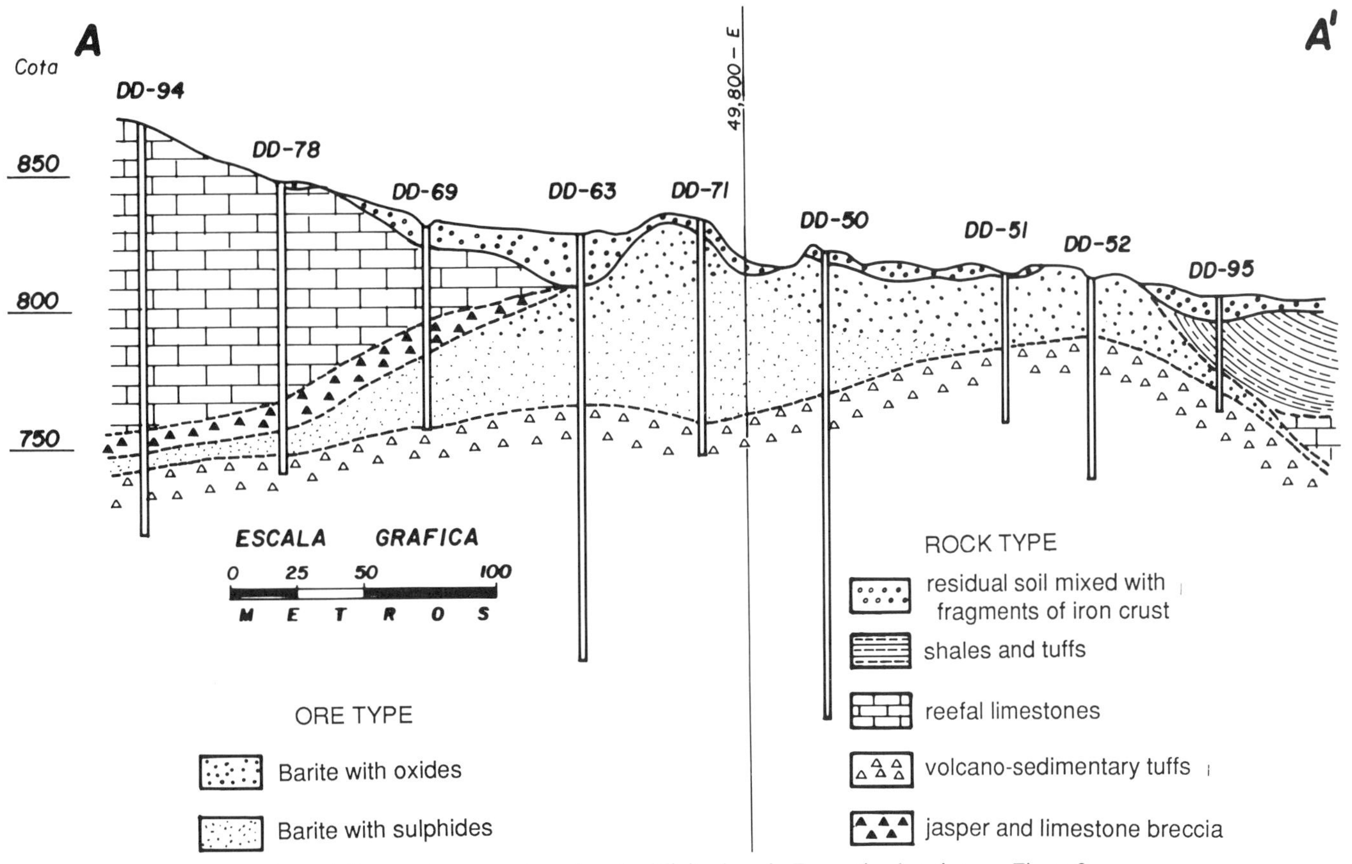

Figure 4. Structural cross section of the La Minita deposit. For section location, see Figure 3.

recovery of cores from the mineralized zone, which exceeded 90 percent.

Ore reserves were calculated by the polygonal and block method. Structural contours drawn at the top and base of the orebody defined its boundaries; 10-m-high blocks were considered. The prisms defined by each polygon contain information provided directly by diamond drilling at their centers; thus all the blocks are considered as proven.

Complementing the reserves calculations, metallurgical assays were done to define the characteristics of the ores and their responses to different processes. Results show they are readily concentrated by selective flotation with the exception of silver, recovery of which is very low in the sulfide zone. Recoveries in semioxidized ores are in the range of 70 percent Ag, 70 percent Pb, 40 percent Zn, and 95 percent $BaSO_4$, and of 55 percent Ag, 75 percent Pb, 90 percent Zn, and 95 percent $BaSO_4$ in the sulfide zone.

La Minita Mine thus shows proven reserves of 6,250,000 tons, averaging 78 gr/ton Ag, 0.33 percent Pb, 4 percent Zn, and 48 percent $BaSO_4$, including a dilution factor of 10 percent that affects grades and tonnage.

ACKNOWLEDGMENTS

Special mention must be made of the work done on this project by the Peñoles exploration technical staff: Edgardo Arévalo, Jose Bravo, Alfonso Rosas, Víctor de la Garza, Benito Noguez, Roberto Téllez, Pedro Fraga, and José Rueda. The ideas and participation of this staff in specific field activities and reserves calculations, under the supervision of José E. Gaytán Rueda, together with the valuable exploration concepts contributed by Pedro Sánchez Mejorada, led to the discovery of this volcanogenic polymetallic deposit, at present a successful mining unit.

Printed in U.S.A.

The Geology of North America
Vol. P-3, Economic Geology, Mexico
The Geological Society of America, 1991

Chapter 41

Las Truchas iron deposits, Michoacán

Eduardo Mapes
Consejo de Recursos Minerales, Centro Minero Nacional, Carretera México-Pachuca, Pachuca, Hgo, México, D.F.

INTRODUCTION

The Las Truchas deposits cover approximately 14 km^2 at the crests of Cerros Santa Clara, El Campamento, Las Truchas, El Volcán, Valverde, El Tubo, La Bandera, and Leopardo, forming part of the southwestern mountain massif known as Sierra Arteaga in the Sierra Madre del Sur, that extends along part of the Pacific coast and toward the southeastern end of the Sierra Madre Occidental (Figs. 1, 2).

The deposits are located in the Melchor Ocampo municipality of Michoacán State, 4.5 km directly northwest of the coast town of Playa Azul (Fig. 1).

GEOLOGY

General concepts

The Las Truchas iron ores occur in the Sierra Madre del Sur foothills facing the Pacific Ocean; the chain is a result of orogenic movements and strong Tertiary magmatic activity, which generated mostly granite, granodiorite, monzonite, and diorite intrusions, as well as considerable volcanic flows.

These plutonic intrusions invaded older sedimentary and metamorphic rocks, fostering ore deposition at their contacts. The Las Truchas iron ores and many other deposits at the Pacific watershed evolved during this period.

The warm, humid regional climate deeply weathered all the rocks in the district, which are mostly covered by thick rubble, making geological observation difficult. Lithologic relations can only be discerned in some streambeds, particularly Las Truchas, and where exposed by the mineworks.

Geologic formations

Outcrops in the Las Truchas District range from Lower Cretaceous (Albian?) to Recent in age. Strongly weathered and metamorphosed fossils are found only in the scarce limestone remnants, and ages are therefore mostly inferred.

The sedimentary rocks are limestones, andesitic conglomerates, fluvial fans (Brisa Mar fan), marine detritus (coastal plain), fluvial terraces (Ríos Acalpica, Las Truchas, and La Cañada), and Recent alluvium.

Igneous outcrops are diorite, quartz diorite, and granodiorite plutons; aplite, spessartite, andesite, and granophyre dikes; and small or irregular andesite porphyry bodies. The iron ores developed at the diorite/limestone contacts, together with many products of contact metamorphism such as amphibolite, skarn, hornfels, and autoclastic breccias. Marbleization is conspicuous, and silicification was minor (Mapes, 1957, Plate 1).

MINERAL DEPOSITS

Description, distribution, and occurrence of orebodies

Las Truchas iron orebodies spread out discontinuously over a general area approximately 4 km wide by 10 km long, grouped as follows: west of El Volcancito, El Volcán, Las Truchas, El Mango, El Campamento, Santa Clara, La Bandera, Potrero de Tanila, El Leopardo, Valverde, El Tubo, and El Bordón; 23 well-defined outcrops of considerable size, as well as numerous smaller exposures, extend over the whole area (Fig. 2).

The orebodies are separated from each other by distances that vary from tens of meters to several kilometers. However, more than 7 percent of the available reserves are located along a northwest-trending, 3-km-long and 1-km-wide belt oriented along the main axis of a series of hills known as Cerros Santa Clara, El Campamento, El Mango, Las Truchas, El Volcancito, and El Volcán, which contain the largest deposits in the district (Mapes, 1957, Plate 1).

Geologic and structural relations of the orebodies

The geologic conditions under which the Las Truchas iron ores evolved are extremely difficult to define due to the advanced erosion of the region. Merely isolated, relatively thin caprocks remain as vestiges of what was probably one large, continuous body. The limestone, affected by plutonic intrusion, is the oldest unit and of considerable importance in mineralization control.

Mapes, E., 1991, Las Truchas iron deposits, Michoacán, *in* Salas, G. P., ed., Economic Geology, Mexico: Boulder, Colorado, Geological Society of America, The Geology of North America, v. P-3.

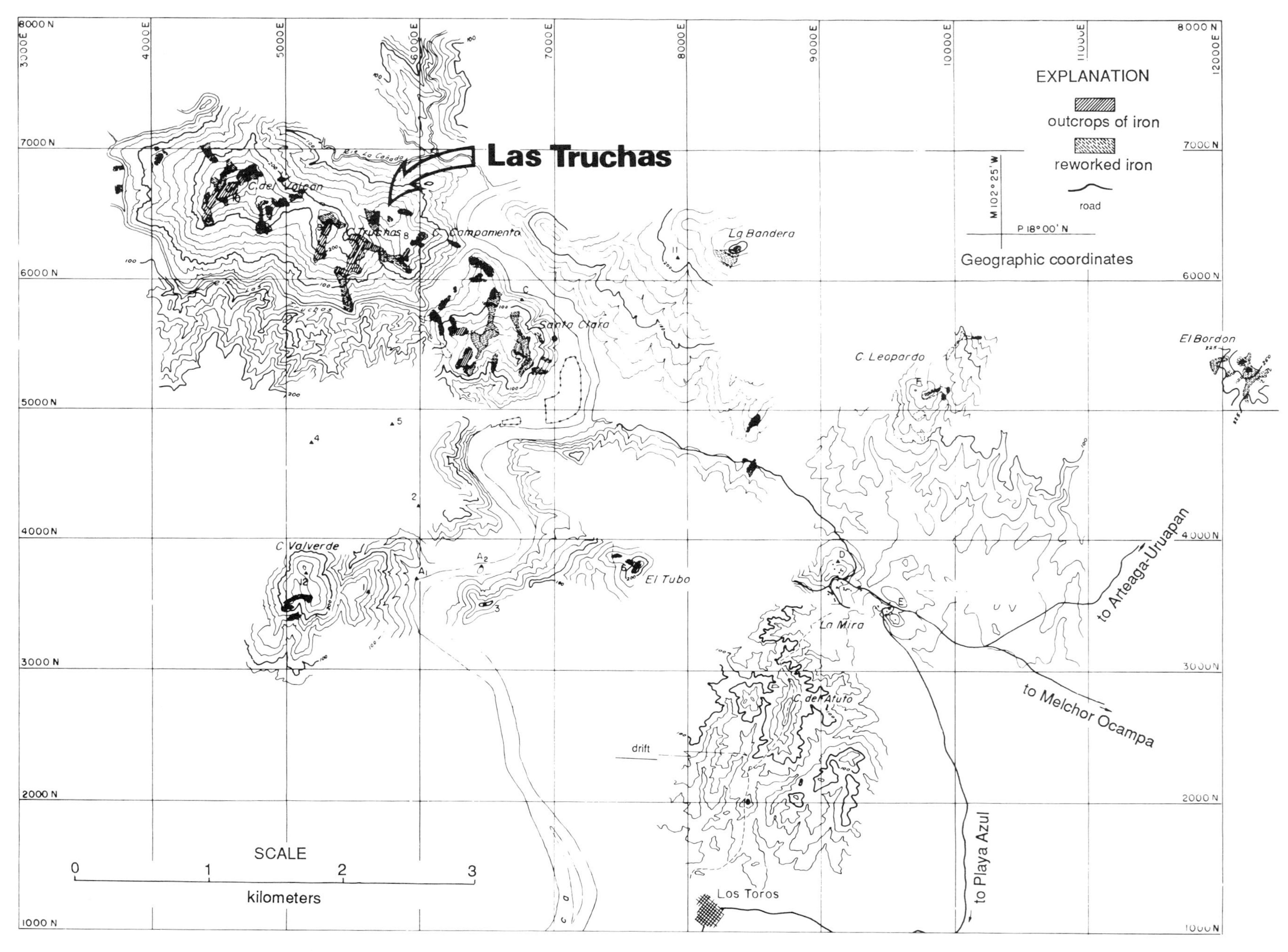

Figure 1. Map showing location of the Las Truchas iron deposits and related port facilities, Michoacán.

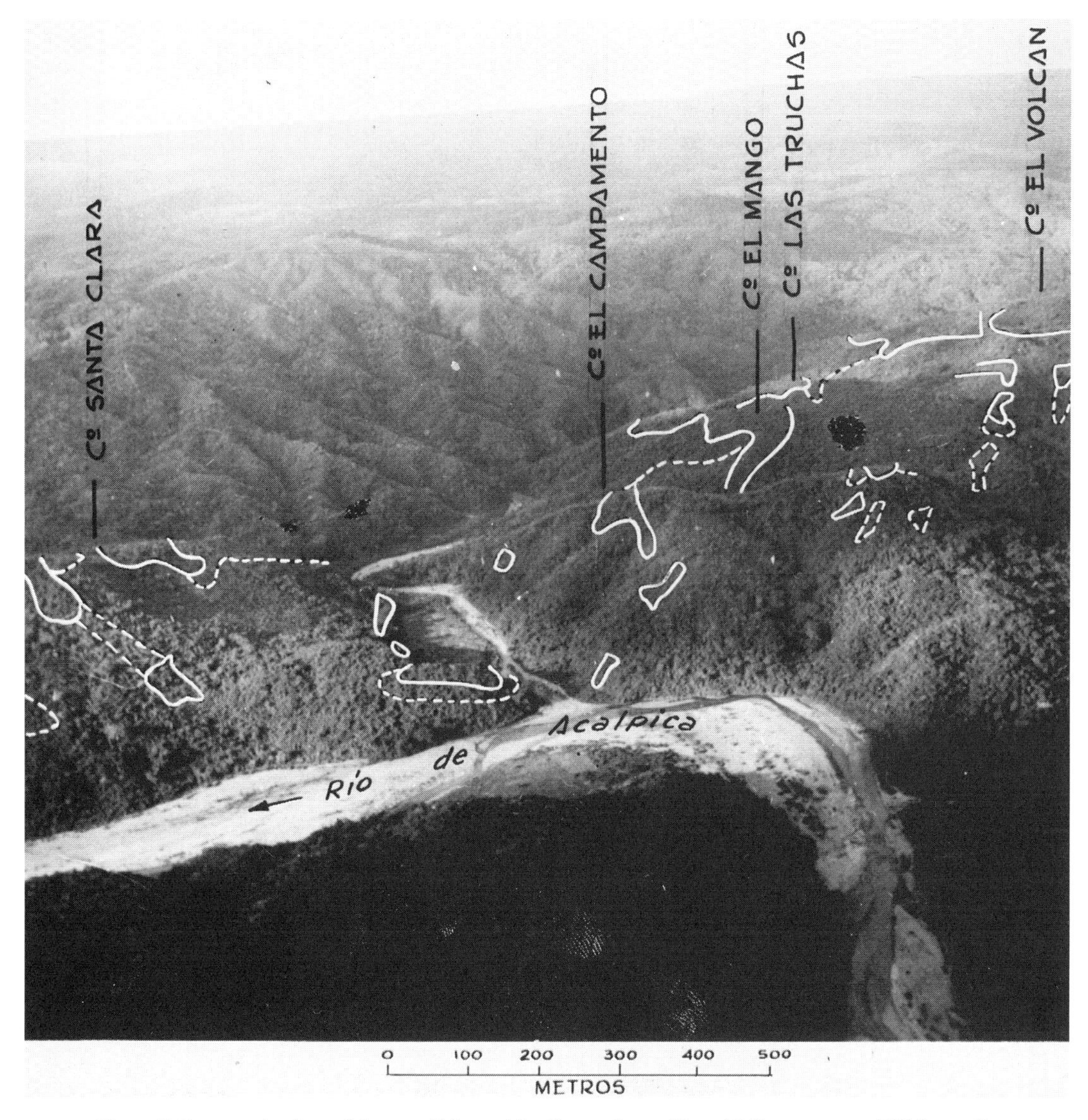

Figure 2. Panoramic view of the massif formed by Cerros Santa Clara, El Campamento, El Mango, Las Truchas, and El Volcán. White lines outline iron-ore outcrops; dashed lines show areas of reworked ore.

The iron ores, almost without exception, are lodged in diorites. The diorite/iron contact is sometimes sharp, showing no alteration in the intrusive rock; however, iron ores are often separated from the diorite by a zone of metamorphic minerals (garnets, hedenbergite, pyroxenes, abundant epidote) and by amphibolite or diorite impregnated with iron and pyrite.

The iron ore/limestone contact is found in very few places and for only tens of meters; at these localities there is much more interpenetration at the contact with the iron ore or diorite than at the iron/intrusive contact, and the iron ore/limestone contact is therefore much more irregular.

Calcite is occasionally recrystallized, with large calcite crystals often showing a thin iron oxide film at cleavage planes, or else abundant specularite, garnet, quartz, epidote, hedenbergite, magnetite, and pyrite. These skarn areas are irregularly shaped and probably penetrated the limestone for considerable stretches from the contact.

DESCRIPTION OF THE ORE

General surface characteristics

Outcrops, thin areas of reworked or fragmented mineral exposed by trenches and pits, and Fe hydroxide–impregnated soils, show that the ore is quite homogeneous at the surface, consisting exclusively of magnetite and hematite (martite), with specularite, limonite, and goethite in very minor proportions. The mineral is almost wholly massive, heavy, and homogeneous as regards texture, hardness, and Fe content; in most cases crystallization is very fine grained to aphanitic. The ore occurs mainly as thick masses at the hill crests and develops cliffs and ridges at hill tops and slopes by its resistance to erosion. The few impurities remaining at the surface—most having been dissolved and transported by atmospheric agents—are clays and chlorites derived from host-rock alteration, as well as pyrite, some epidote, quartz, calcite, and probably some unreplaced rock.

LAS TRUCHAS IRON ORE RESERVES

General

As mentioned above, ore bodies in this deposit are very irregularly shaped, both at the surface and even more so at depth. The exploration program comprised 39 diamond-drillholes for a combined length of about 3,000 m, of which about 1,350 m were in iron ores, as well as cleanout and timbering of eight collapsed drifts, and several trenches and pits in reworked or disintegrated mineral. These works were deemed adequate for calculation of reserves.

Measured reserves

For measured reserves calculations (Table 1) the mineralized area was divided into 12 zones, 11 of which appear on the geologic map and the twelfth on the individual plat of the El Bordón outcrops; each was treated individually. Tonnages were calculated by multiplying areas of each outcrop or section by their thicknesses and average specific weights.

RESULTS OF THE PRELIMINARY CALCULATION OF RESERVES

Orebodies	Reserves (metric tons)
Iron orebodies	60,804,000
Reworked ore	5,977,000
Measured ore	66,782,000
Probable ore	7,106,000
Total measured and probable ore	73,888,000

REFERENCES CITED

Cardenas V., J., Arriaga G., G., Estrada B., S., and Farias G., R., 1968, Estudio Geologico-Geofisico de los yacimientos ferriferos de Las Truchas, Michoacán.
Fried, Yacimientos ferriferos de Las Truchas, Michoacán, México: Essen.
Mapes V., E., 1957, Los yacimientos ferriferos de Las Truchas, Michoacán: INIRM Boletin, no. 43, CRNNR Boletin 46.

Printed in U.S.A.

The Geology of North America
Vol. P-3, Economic Geology, Mexico
The Geological Society of America, 1991

Chapter 42

Geology of the Pegaso asbestos deposit, Concepción Pápalo, Cuicatlán, Oaxaca

J. C. Ramírez L.
Cía. Minera Pegaso, S.A.

INTRODUCTION

The Pegaso deposit is located north of the town of Oaxaca in the Concepción Pápalo municipality, Cuicatlán District (Oaxaca), at coordinates 17°51′N and 96°52′W (Fig. 1).

Access to the mine is by the México-Tehuaca–Teotitlán del Camino–Oaxaca Federal Highway for 316 km to the village of Cuicatlán, from which a year-round passable unpaved roadway runs for 23 km to the mine area. Cuicatlán has a light-aircraft runway; also the México-Oaxaca railroad stops at Cuicatlán.

The weather is temperate to warm from February to June, with heavy rains from July to September, and otherwise is cold with sparse rainfall. Conifers in the higher mountains and cacti and bushes in the lowlands make up the regional vegetation.

The area lies in the north part of the Sierra Madre del Sur Physiographic Province. The rugged regional topography is mostly scarped, particularly at some of the streams; coalluvium and asbestos develop gentle slopes in the middle parts of the mountains; metamorphic rocks (schists) support steep scarps of extremely sharp relief at the high and low parts of the mountains; the mine is at an average altitude of 1,750 m.

The regional hydrology is a dendritic-type drainage system flowing north to the San Lorenzo and south to the Chiquito tributaries of the Río Grande.

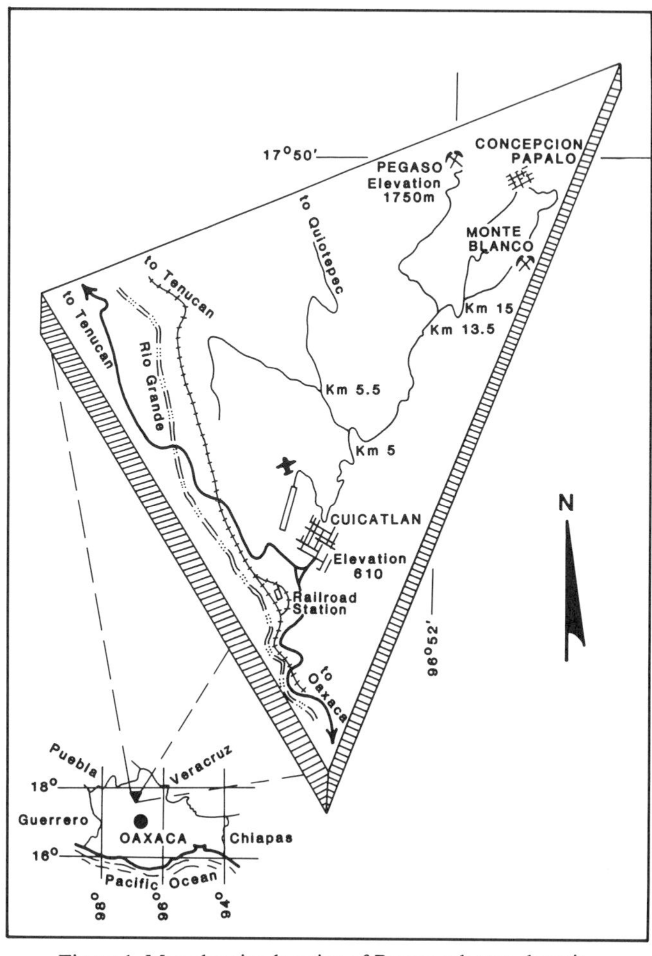

Figure 1. Map showing location of Pegaso asbestos deposit.

GEOLOGY

Geologic provinces

The area occupies the northernmost end of the Oaxaca Peninsula, bounded to the north by the Zongolica, west by the Tlaxiaco, and south by the Sierra Madre del Sur Geologic Provinces (Fig. 2).

Regional geology

The regional geology includes metamorphic and sedimentary rocks (Fig. 3).

Metamorphic rocks. Schists and some phyllite horizons of possibly sedimentary origin contain the Pegaso asbestos deposit and may be correlated with the Paleozoic Acatlán Formation.

Sedimentary rocks. A sedimentary sequence comprising three units occurs in Tomellín Canyon, toward the western part

Ramírez L., J. C., 1991, Geology of the Pegaso asbestos deposit, Concepción, Pápalo, Cuicatlán, Oaxaca, *in* Salas, G. P., ed., Economic Geology, Mexico: Boulder, Colorado, Geological Society of America, The Geology of North America, v. P-3.

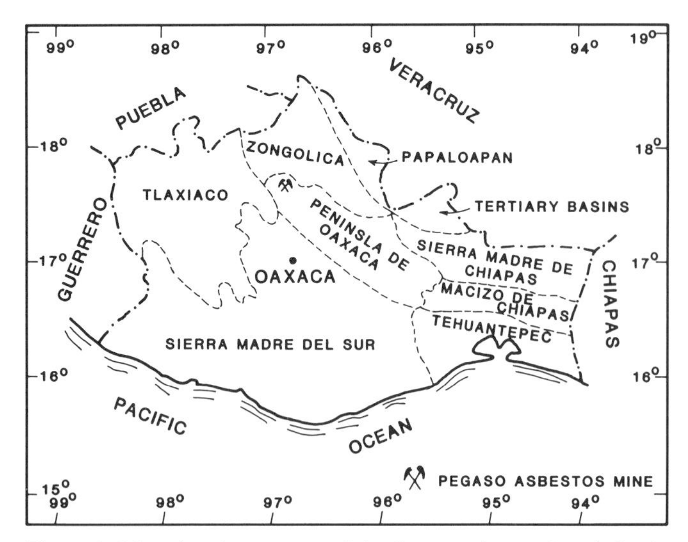

Figure 2. Map showing context of the Pegaso asbestos deposit in the geologic provinces of Oaxaca.

of the area (Fig. 4). The lowest unit includes yellowish brown, medium-bedded sandstones (graywacke) interbedded with reddish siltstones (Fig. 3), and is a flysch deposit. The sandstone beds show gradation from coarse and semirounded fragments of serpentine, schists, and mostly intrusive acid rocks at the base, to finer grained material at the top.

The middle unit contains medium- to coarse-grained reddish sandstones (arkoses) interbedded with reddish conglomerates and horizons of explosive-type intermediate volcanic rocks, and marks the onset of molasse-type deposition.

Its sharp contact with the lower unit, marked by a change in color from yellowish brown to reddish tones, is exposed at the base of Tomellín Canyon. The sandstones and conglomerates are composed of semirounded, mostly metamorphic (with some igneous) fragments cemented by carbonates and iron oxides.

The upper unit has thick layers of reddish conglomerates unconformably overlying metamorphic rocks and is formed mostly of semirounded metamorphic fragments embedded in fine-grained iron-oxide material.

The lower contact of this sequence is not exposed; however Pedrazzini (1970) defined a similar sequence 50 km to the east in the Nochixtlán area, unconformably overlying the Tertiary (Eocene-Oligocene) calcareous Tecomatlán Conglomerate.

Age. The lowest two members of the sequence may be correlated with the upper Tertiary (Miocene-Pliocene) Yanhuitlán Formation, and the upper with the upper Tertiary (Pliocene) Sosola Formation. The sequence has been informally termed the Cuicatlán group, of possibly Cenozoic (Miocene-Pliocene) age.

Local geology

The geology of the present area consists of low-grade folded and fractured metamorphic rocks, coalluvial and alluvial sediments, and some igneous rocks (Fig. 5).

Metamorphic rocks include quartz-muscovite-chlorite schists, quartzites, phyllites, and very sporadic metagranites. The thickly bedded schists increase in thickness toward the higher section of the village of Concepción Pápalo, where they are highly folded and fractured. The metamorphic sequence contains veinlets and isolated nodules of mostly amethyst-type quartz. The rocks are intruded by probably Recent (unmetamorphosed) hypabyssal diabase dikes 2 to 3 m thick.

These metamorphic rocks are correlatable with the Paleozoic Acatlán Formation of southwest Puebla and west Oaxaca States, with its type-section at Acatlán, Puebla.

Sedimentary rocks are represented by alluvial and coalluvial deposits in small, localized depressions along creeks and rivers, mostly at the middle range heights.

Coalluvial deposits in the area consist of semiconsolidated rocks composed of poorly sorted, subrounded to rounded fragments in a clayey groundmass. The apparently normal faulting in the metamorphics is represented in the field by zones of irregular foliation, and at some places by highly oxidized and fractured areas with carbonates as alteration products; such fault zones range in thickness from a few centimeters to about 3 m.

Geologic history

The geologic history of the region includes a positive landmass (craton) of metamorphic rocks, and a depositional sedimen-

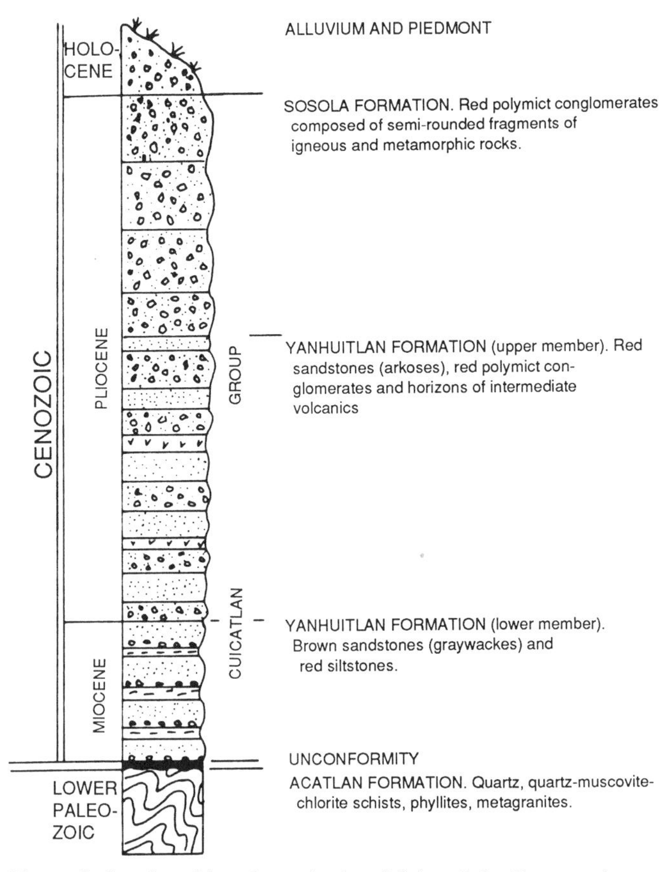

Figure 3. Stratigraphic column in the vicinity of the Pegaso asbestos deposit.

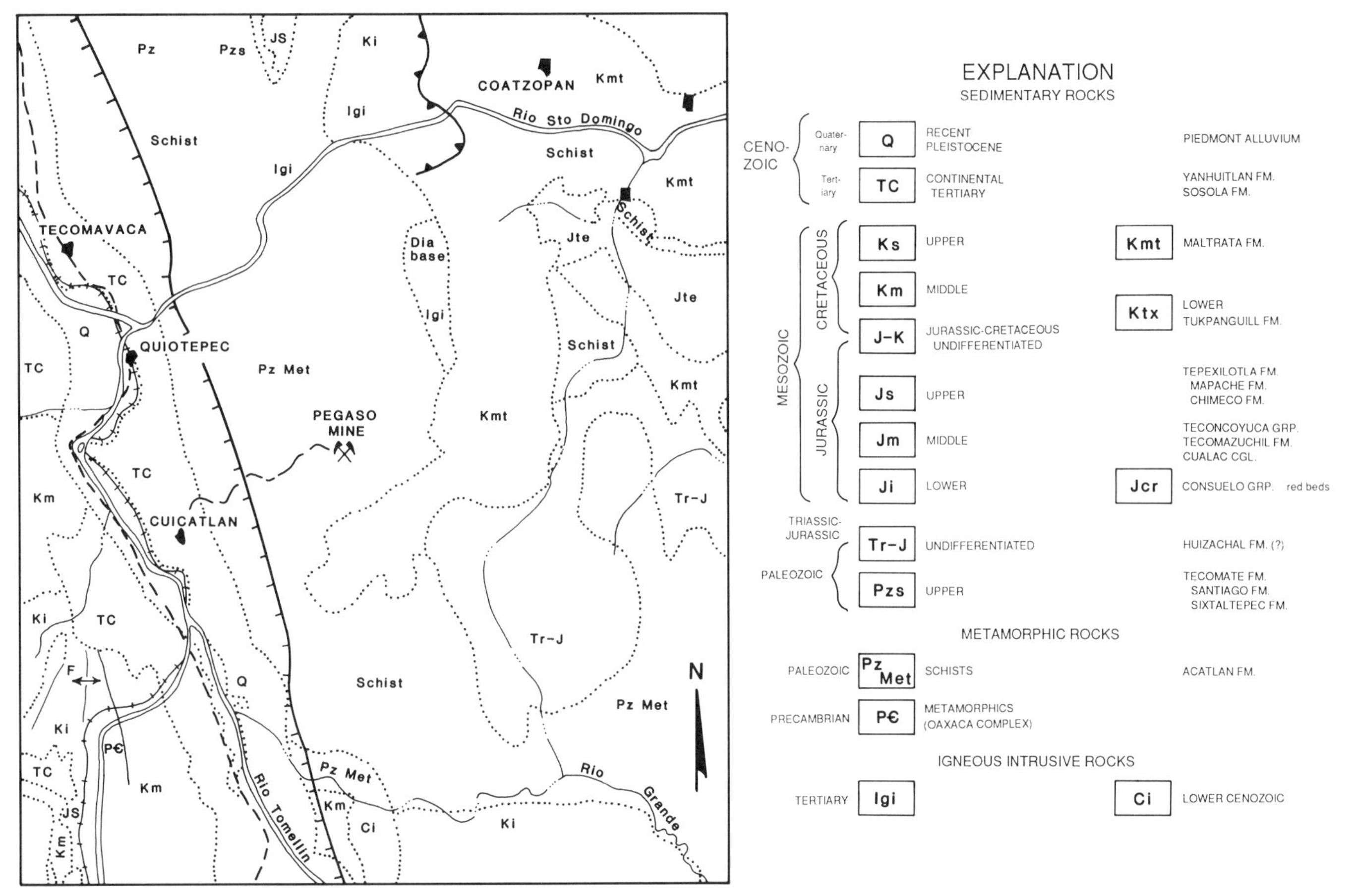

Figure 4. Regional geologic map of the Pegaso mine area.

tary basin as tectonic elements. Sedimentary deposition began in the Miocene when periodic continental uplift gave rise to the flysch sandstones (graywackes) and siltstones of the lowest unit of the local sequence. Stronger subsequent uplifts caused increasingly coarse deposition in an oxidizing environment, associated with intermediate periods of volcanic activity, producing the middle unit as the basal molasse deposition.

Finally, the area was entirely uplifted, and the deposition of thick conglomerate layers in an oxidizing environment terminated the sequence.

Tensional (taphrogenic) forces in the final stages of this tectonic cycle fractured the entire sequence; basic volcanic (diabase) activity is attested by the dikes exposed in the metamorphic sequence. The taphrogenic fracturing, together with weathering and erosion, generated major landslides that accumulated in the more stable areas of the existing topography.

MINERAL DEPOSITS

The Pegaso asbestos mine is one of the major mines in the world and the largest in the country, with large potential mineral reserves. For this reason the mineralogy and its distribution within the deposit must be clearly defined to establish the geologic framework in which the orebody developed.

Results of chemical analyses of the Pegaso deposit serpentine at the Johns Manville Research Center are shown on Table 1.

Mineralogy

The mineralogy of the deposit includes silicates, oxides, and carbonates.

Silicates. (a) Serpentine-group chrysotile is the commercial mineral, occurring in the deposit as slip fiber and very sporadically as cross-fiber veinlets within the serpentinite rock. (b) Serpentine occurs massively within the deposit, forming part of the asbestos host rock and enclosing fragments of igneous rocks. (c) Tremolite, a fibrous mineral of the amphibole group, occurs in veinlets as fracture-fill, and is disseminated at the contacts of the serpentine with some of the dikes. (d) Pectolite is a radial fibrous mineral found in veinlets both in the serpentine and in dikes. (e) Talc is widespread in the deposit, occurring disseminated in the massive asbestos (talc schists) and in laminar form. (f) Cordierite occurs massively in association with talc. (g) Gar-

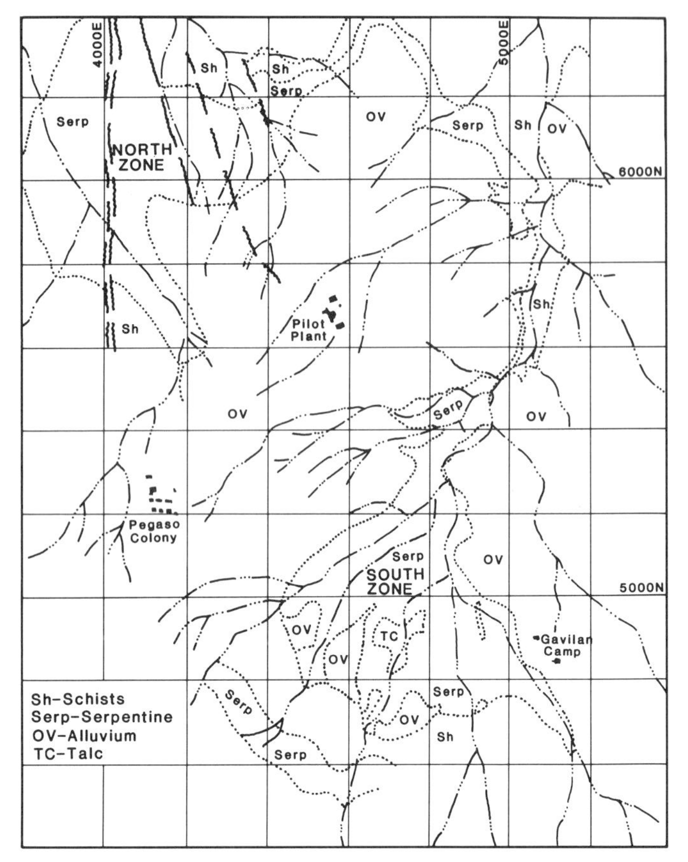

Figure 5. Detailed geologic map of the Pegaso mine area.

TABLE 1. CHEMICAL AND SPECTROGRAPHIC ANALYSIS OF PEGASO ASBESTOS

	Old Sample (1), FE No. 4374 (AX 3474 M)	New Sample (2), FE No. 1016
X-ray analysis	Chrysotile with Fe	Chrysotile, Magnetic rock, Dolomite traces, Magnesite
Mixture at 105 °C	1.0	1.0
Loss of ignition at 1,800 °F	14.0	12.3
SiO_2	40.6	39.9
Fe_2O_3	4.8	6.03
Al_2O_3	0.22	0.26
TiO_2	0.006	0.007
Cr_2O_3	0.10	0.26
Mn_3O_4	0.11	0.10
NiO	0.42	0.47
CaO	0.74	0.74
MgO	38.8	38.1
Na_2O	0.1	0.08
K_2O	0.1	0.01
SO_3	Traces	0.05
CO_2	0.9	0.6

nierite occurs very sporadically disseminated in the deposit as small granules within the fiber.

Oxides. (a) Magnetite is one of the most common minerals in the deposit, occurring disseminated as patches of small crystals within the fiber. (b) Brucite, a fibrous mineral, is found filling fractures and in veinlets around some of the dikes. (c) Pyrolusite occurs disseminated in the serpentinite and in the dikes.

Carbonates. These minerals are widespread within the deposit and include (a) aragonite, a radial and fibrous mineral, both disseminated and in veinlets within the asbestos; (b) hydromagnesite, as fracture-fill and in some veinlets within the asbestos; and (c) calcite nodules, mainly in highly fractured zones within the deposit.

Geologic characteristics of the Pegaso deposit

The deposit is made up of northern and southern zones, separated by colluvium with small outcrops of finely fractured serpentine, particularly in some of the streambeds (Fig. 5).

South zone. Approximately 16 hectares of exposed ore is surrounded both at the upper and lower contacts and on the east by metamorphic rocks of the Acatlán Formation. Both contacts are wavy and irregular, marked by finely fractured serpentine with zones of serpentinite in some localities. Near the contacts the metamorphic rock contains a thin bed of laminar talc, carbonate, silicates, and locally diabase dikes, conformable with the contacts. The eastern lateral contact is unconformable and tectonic, marked by a fault zone within the serpentine. The similar western contact places serpentine in contact with colluvium, the serpentine continuing at depth.

Talc schists in the central part of the zone are surrounded first by a thin aureole of cordierite, and then by serpentinite with cross-fiber veinlets and tremolite. A N10°W–trending diabase dike partially intrudes the asbestos body in the eastern part of the zone with no surrounding alteration or metamorphism. In the west a similar irregular diabase dike shows a serpentinite aureole together with tremolite and brucite veinlets and thin cross-fiber veinlets. Serpentinite occurs as semirounded blocks of varying sizes, some of which exhibit certain foliation. The main alterations in the deposit are aragonite veinlets and disseminated magnetite and talc. Fractures trend N50°W and dip 30 to 40°W, showing no displacement.

North zone. This zone covers approximately 12 hectares of exposed mineral.

As in the South zone, the upper and lower contacts are conformable and irregular, and the lateral contacts are unconformable and tectonic with schists in the west and with coalluvium of the intermediate zone in the east.

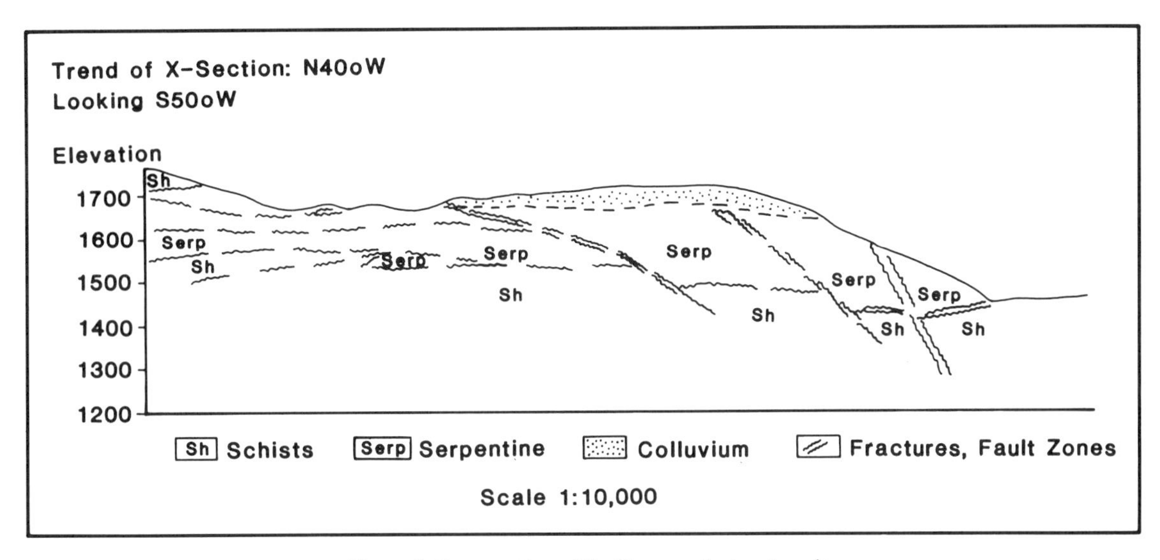

Figure 6. Cross section of the Pegaso asbestos deposit.

In the south-central part of the zone a large serpentinite area surrounds a north-trending, highly fractured granitic rock (metagranite) partially intruding the serpentine. At some localities, rodingite dikes(?) appear surrounded by serpentinite. The serpentinite appears as semirounded fragments of various sizes, as do some diabase and gabbro dike fragments up to 2 m long.

The main alterations in this deposit are disseminated magnetite and pectolite, and carbonate (magnesite) veinlets. Fracturing is usually normal, trending N15°W and dipping an average 15°W (Fig. 6).

Origin of the asbestos deposit

The origin of the deposit begins with an ophiolite complex generated by initial basic magmatism (gabbros, diabases, and peridotites) during the geosynclinal phase, emplaced within the continent and enclosed by metamorphic rocks of the Oaxaca peninsula. The complex was serpentinized by magmatic silica-rich aqueous solutions, according to the following chemical reaction:

$$\underset{\text{Peridotite}}{3\,(MgFe)_2SiO_4} + SiO_2 + 4H_2O \rightarrow \underset{\text{Serpentinite}}{2\,(MgFe)\,(OH)_4Si_2O_5}$$

Some fractures were subsequently filled with chrysotile gel that crystallized perpendicular to fracture loads producing cross-fiber veinlets.

Compression during orogenesis folded the Mesozoic sedimentary sequence at the end of the Cretaceous and beginning of the Tertiary. It also affected the continent, particularly the Oaxaca peninsula, generating a thrust fault upon which the serpentine acted as a lubricant and was sheared, giving rise to massive asbestos (slip fiber) together with second-generation serpentine.

The fracturing caused melange within the serpentine, described by Muratov (1977) as a serpentine groundmass with embedded fragments of gabbro, diabase, schists, gneiss, and rare limestones of different ages.

In the course of the thrusting, solutions rich in Ca and Al silicates circulated within the serpentine, acting upon the serpentine and some schist fragments to produce rodingites, talc schists, and pectolite veinlets.

The effects of these metasomatic processes appear in the central part of the South zone where talc schists and local pectolite veinlets occur. In the North zone, isolated rodingites included in the serpentinite occur as dense, yellow-pink to light yellow-green, compactly structured rocks composed of garnet (grossularite-andradite), diopside, brucite, calcite, hematite, and limonite. Pectolite veinlets are also present within the serpentine and some of the dikes.

In addition, ground waters circulating within the serpentine reacted with Ca and Mg and developed hydromagnesite and aragonite veinlets within the fiber and at the serpentine/schist contacts.

In the final stages of the tectonic cycle, tensional forces developed a system of normal faults that displaced and fractured the serpentine body and through which magmatic activity caused basic intrusions, such as the diabase dike in the southeastern part of the South zone and a fault-bound granitic intrusion in the northeastern part of the North zone. This last magmatic activity scarcely affected the asbestos deposit, causing only slight recrystallization.

REFERENCES CITED

Morales C., M., 1967, Explotación de asbesto en el municipio de Cuidad Victoria, Tamps.: [Professional Thesis].

Muratov, M. V., 1977, The origin of continents and ocean basins: Moscow, Mir, 186 p.

Printed in U.S.A

The Geology of North America
Vol. P-3, Economic Geology, Mexico
The Geological Society of America, 1991

Chapter 43

Titanium deposits in Huitzo and Telixtlahuaca, Oaxaca

F. J. Diaz T.
Consejo de Recursos Minerales, Centro Minero Nacional, Carretera México-Pachuca, Pachuca, Hgo., Mexico

LOCATION

Although incompletely explored, the best known of the titanium deposits in the state of Oaxaca are those in the Etla District, covering parts of the San Francisco Telixtlahuaca, San Pablo Huitzo, and Santa María Tenexpan municipalities (Fig. 1). The deposits are predominantly ilmenite, with iron, phosphorous, and occasional very subordinate rutile. They form part of 2,450 hectares of the Mineral Reserves Zone protected by the Mining Development Commission (as published in the *Diario Oficial* of Oct. 8, 1973) in the Sierra Madre de Oaxaca physiographic province. The geologic setting is almost exclusively a region of late Precambrian metamorphic rocks in an area of more or less uniform relief.

Trails lead to the area from the villages of Telixtlahuaca and Huitzo, 120 km from Oaxaca on either the Oaxaca-Mexico highway or by railroad. Year-round passable trails that formerly interconnected the ilmenite deposits during the exploration period have been abandoned.

MINERALOGY

The ilmenite occurs either disseminated or massively in mantos parallel to the gneissic foliation of the host rock. Disseminated ilmenite deposits may be as much as 20 m thick; the more consolidated mantos, 2 to 6 m thick, are affected by local tensional and compressional folding and are horizontally displaced along small faults.

The host rock is quartzo-feldspathic to hornblende gneiss of the Precambrian Oaxaca basal metamorphic complex. Most of the mineralization is in the quartzo-feldspathic gneiss, possibly as the result of magnetite replacement in intermediate intrusives that was subsequently remobilized during regional metamorphism. Another hypothesis proposes deposition of marine heavy-mineral sands (ilmenite, magnetite, pyrite, and apatite) along broadly extensive coastlines, which were subsequently metamorphosed to originate the mantos.

Locally, the gneisses are sporadically invaded by tabular or irregular pegmatites (quartz-feldspar, containing some mica and radioactive minerals [betafite and allanite]) that crosscut (65 to 80°) the gneissic foliation. Very subordinate rutile often appears in mantos and lenses in the northernmost part of the area near the Oaxaca-Mexico highway.

Ilmenite outcrops trending S19 to 25°E and dipping 40 to 50° southwest extend for 1 km or more along strike and recur vertically at intervals of 150 to 400 m.

DESCRIPTION OF THE DEPOSITS

Tracing of the mineral deposits along strike in trenches, and detailed geologic-topographic surveys of three explored mineralized sectors (Arañas, Tejón, and Tenexpan; Fig. 2) allowed demarcation of the orebodies and an estimate of approximate volumes and grades. The ilmenite mixture with iron oxides (2 to 39 percent magnetite content) is perceptibly magnetic. Sixty-nine analyzed samples showed 5 to 12 percent titanium oxide and 2 to 16 percent apatite (P_2O_5). The very limited opportunity for surface sampling both laterally and vertically led to the inclusion of a geophysical study and subsequent diamond-drill evaluation in the recommendations below. The estimations given in the following deposit descriptions are based on the figures in Tables 1 and 2.

Arañas sector (Fig. 2A)

Four parallel 2 to 5-m-thick mineralized mantos spaced 130 m apart on the average were mapped along outcrops extending for 200 to 500 m; on this basis (considered by G. P. Salas to be incomplete; personal communication, 1985) about 534,000 tons of possible and 900,000 tons of potential volume are estimated.

Tenexpan sector (Fig. 2B)

Two mineralized parallel mantos separated by a 100-m interval over 1,100 m of outcrop showed an estimated 765,200 tons of possible and 708,700 tons of potential reserves.

Tejón sector (Fig. 2C)

Estimated volumes are 3,450,000 tons of possible and 1,769,000 tons of potential reserves in three parallel mineralized mantos spaced 450 m apart, over 1,500 m of outcrop; no considerable deepening was done to establish vertical spreads.

Diaz T., F. J., 1991, Titanium deposits in Huitzo and Telixtlahuaca, Oaxaca, *in* Salas, G. P., ed., Economic Geology, Mexico: Boulder, Colorado, Geological Society of America, The Geology of North America, v. P-3.

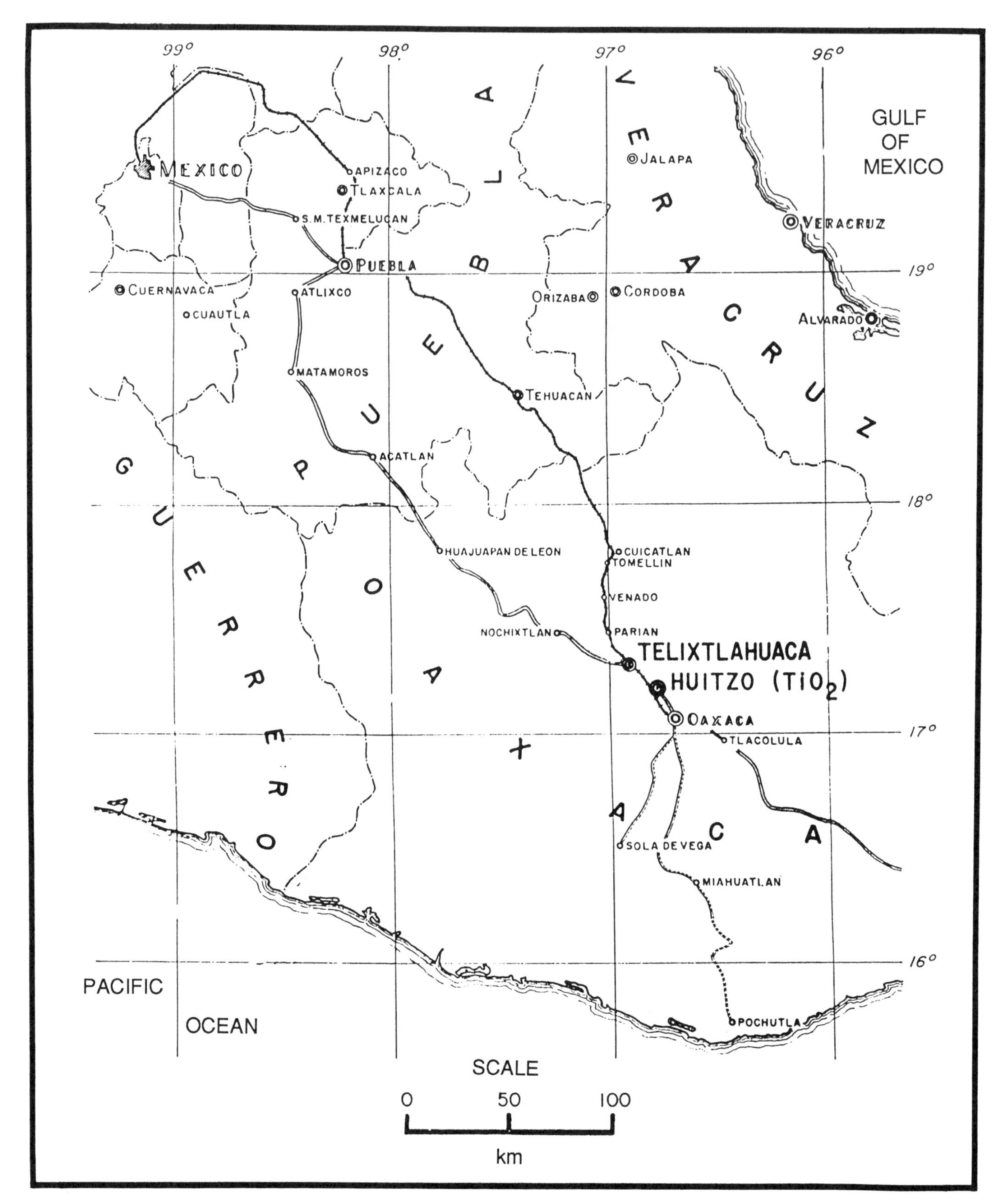

Figure 1. Location map for the Huitzo TiO_2 deposit, Oaxaca.

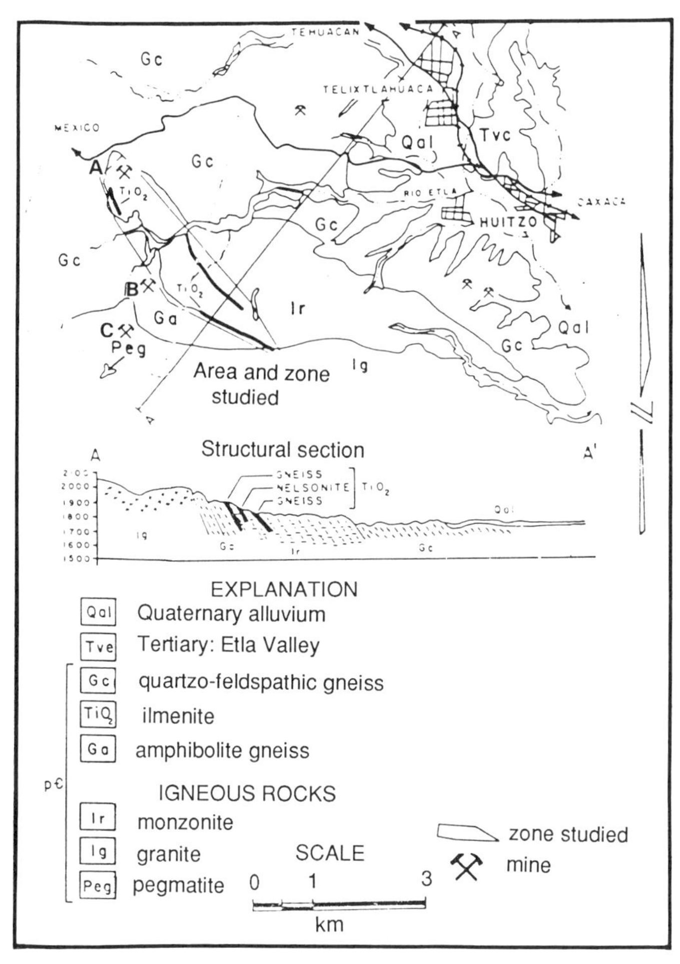

Figure 2. Geologic map and cross section, Huitzo ilmenite (TiO_2) deposit, Oaxaca. Three explored mineralized sectors are shown on map: A, Arañas sector; B, Tejón sector; C, Tenexpan sector.

TABLE 1. AVERAGE GRADES

Sector	Manto Number	TiO_2 (%)	Fe_2O_3 (%)	P_2O_5 (%)
Arañas	1	9.09	17.72	4.17
	2	8.90	24.16	14.12
	3	8.96	27.95	10.94
	4	10.63	24.54	7.17
Tejón	1	6.70	20.45	3.37
	2	9.70	22.73	2.37
	2	10.27	23.90	11.40
Tenexpan	1	10.35	21.50	11.80
	2	9.92	25.90	8.12

TABLE 2. TYPES OF RESERVES

Sector	Possible (tons)	Potential (tons)	TiO_2 (%)
Arañas	534,000		8.97
Arañas		900,000	
Tejón	3,455,000		8.85
Tejón		1,769,000	
Tenexpan	765,000		10.05
Tenexpan		709,000	10.12
Total	4,754,000	3,378,000	

CONCLUSIONS

1. The origin of this disseminated and massive ilmenite deposit is debatable. The two principal possibilities are (a) the notion of subsequently metamorphosed hydrothermal magnetite replacement, or (b) heavy-mineral–rich marine sands recycled by late Precambrian regional metamorphism (as proposed by G. P. Salas, personal communication, 1985).

2. The mineralization occurs in 2 to 20-m-thick mantos or thin layers parallel to the host-rock gneissic foliation over distances of 200 to 1,500 m; continued detailed exploration may increase these figures.

3. At least three main areas were located: the Arañas, Tejón, and Tenexpan sectors representing 4,754,000 tons of possible reserves. Average grades for these areas are presented in Table 1.

4. Based on the morphology and dimensions of the mantos, additional potential reserves of 3.4 million tons are estimated.

5. Total evaluated reserves of titanium minerals in the Huitzo area are 8,132,000 tons (see Table 2).

6. The mineralized areas rated as important have the necessary infrastructure for future development: good roads, water, power supply, railroads, and proximity to the state capital. Since no exhaustive economic study is yet available, opencast and underground development costs and ways to increase reserves remain to be established.

7. Studies in the Mineral Resources Council's "El Rutilo" sector, northwest of the Mining Development Commission's National Mine Reserve, show that the ilmenite mantos do not continue northwest; however, they may extend toward the southeast.

Recommendations

1. The main recommendation is to apply exploratory geophysics and diamond-drilling and carry out metallurgical tests on the titanium concentration of the ores for domestic industrial purposes.

2. In view of the type of deposit and the metallurgical test results, exploration for thorough evaluation should continue, since development is feasible in spite of the titanium-iron mixture.

3. For more accurate evaluations of reserves, the continuity along strike and dip of the mineralization should be defined by diamond-drilling, in view of the plasticity and incompetence of the host-rock foliation, which could make the orebodies either wedge out or thicken.

4. Diamond-drilling will establish whether production is economic and ore grades are stable and sufficiently commercial for the market, and also the feasibility of producing apatite (P_2O_5), which occurs in concentration of 3 to 14 percent in the deposit.

5. Present geological information on the Huitzo area indicates a prospective sector toward the southeast; regional exploration should continue in that direction.

REFERENCES CITED

Schmitter, V. E., 1956, Nelsonita de ilmenita y magnetita de la región do Huitzo, Estado de Oaxaca: Mexico, D.F., 20th International Geological Congress, Sociedad Geológico de México Excursión México y Oaxaca, 15 p.

Unpublished sources of information, all from the Mineral Resources Council

De Pelsmaeker, F. and Diez T., A., 1959, Conjunto minero-metalúrgico para la obtención de escoria litonífera, arrabio y cepatilo partiendo de minerales de S. Pablo-Huitzo, Oax.

Díaz Tapia, F. J., 1983a, Estudio evaluativo del yacimiento de titanio localizado en el distrito de Etla, Huitzo, Oax.

——, 1983b, Informe técnico sobre la ilmenita de Huitzo, San Pablo Huitzo, Etla, Oax.

Elweel, P. and James, P., 1968, General and economic appraisal of the ilmenite deposits of the Huitzo area, district of Etla, Oaxaca.

Franco López, M., 1962, Informe sobre los depósitos de ilmenita cercanos a Huitzo, Oax.

Pérez Larios, J., 1966, Yacimiento titanífero en las proximidades de Huitzo, Oax.: Archivo Técnico CRM.

Pesquera Velázquez, R., and Carbonell Córdoba, M., 1966, Yacimiento de ilmenita de la región de Huitzo, Oax.: Archivo Técnico CRM.

Ramírez Vázquez, C., 1977, Informe geológico evaluativo de los cuerpos de titanio Alfonso y Francisco, Huitzo, distrito de Etla, Oax.

Schulze, G., 1959, Estudio genética de los yacimientos de titanio en Huitzo y en la zona de Pluma Hidalgo-Apango, Oax.: Archivo Técnico CRM.

Ugalde V., H., 1971, Exploración Titanio Huitzo, Oaxaca, Mpio. de Huitzo Oax.

Ugalde V., H., and Eliazer, O. G., 1967, Breves notas preliminares sobre los depósitos titaníferos de Huitzo, Oax.

Printed in U.S.A.

The Geology of North America
Vol. P-3, Economic Geology, Mexico
The Geological Society of America, 1991

Chapter 44

Neovolcanic Axis Metallogenic Province

Guillermo P. Salas
Geologo Consultor, Asesor del Consejo de Recursos Minerales, Centro Minero Nacional, Carretera Mexico-Pachuca, Pachuca, Hgo., Mexico, D.F.

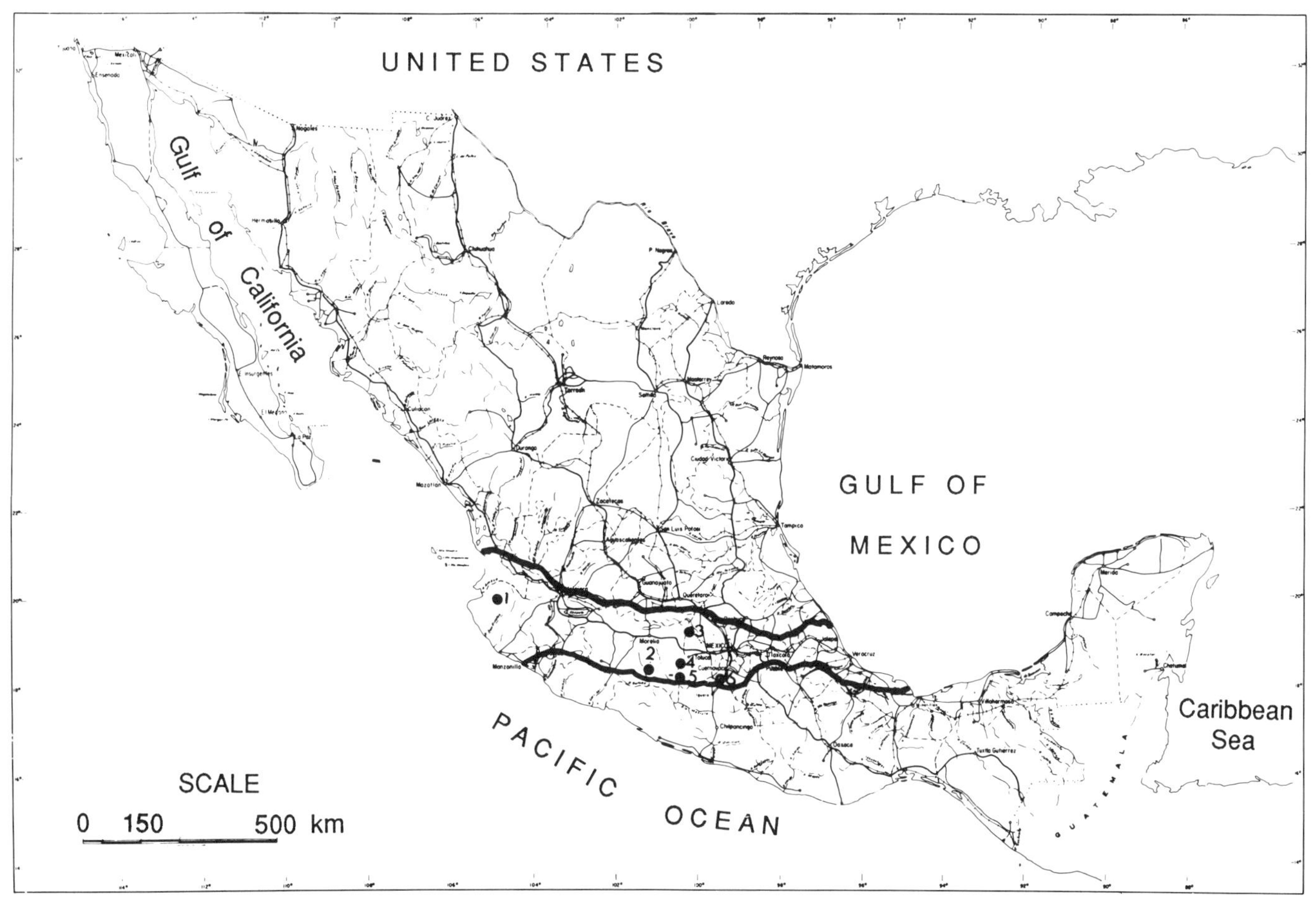

Figure 1. Map showing locations of mineral deposits of the Neovolcanic Axis Metallogenic Province discussed in the following chapters. 1, Cuale, Jalisco; 2, Inguaran and 3, El Oro–Tlalpujahua, Michoacán; 4, Zacualpan and 5, Tizapa, México; 6, Taxco, Guerrero.

This province does not fit the established concept of a metallogenic province. However, its geology and physiography seem to control certain metallogenic aspects related to tectonics and epigenetic hydrothermalism over an estimated 185,000 km^2 (Salas, 1975, p. 107), an area that almost crosses the country from east to west: from the Los Tuxtlas volcanic complex in Veracruz on the Gulf of Mexico to the Bahía de Banderas on the Pacific Coast of the state of Jalisco (Fig. 1).

Mineralization varies considerably within the province. Massive sulfide deposits (Tizapa and La Esmeralda, state of Mexico; La Dicha, Campo Morado, and Copper King in Guerrero, etc.) seem to predominate in older metamorphic rocks. The extrusive sequence consists mostly of andesites and pyroclastics intruded by granodiorites that generated polymetallic deposits of silver, lead, zinc, copper, and occasionally small amounts of gold (CRM, 1980).

Salas, G. P., 1991, Neovolcanic Axis Metallogenic Province, *in* Salas, G. P., ed., Economic Geology, Mexico: Boulder, Colorado, Geological Society of America, The Geology of North America, v. P-3.

Chapters 45 through 50 describe typical examples of these deposits. Some important mineralized districts not described in these chapters are found at the eastern end of the Neovolcanic Axis. These include the Tetela gold deposits, about 80 to 100 km northeast of Tlaxcala in northern Puebla State; copper deposits at the contacts between limestones and post-Cretaceous intrusive rocks in the Teziutlán area near the town of that name (Puebla); the presently abandoned Tatatile iron mines; the Las Minas copper and gold deposit 30 km northwest of Jalapa; and an opal deposit 20 km northwest of Las Minas in the state of Veracruz.

Mineralizations in this eastern part of the province are related to the Sierra Madre Oriental orogeny and may have been generated by intrusion of the Teziutlán Massif and neighboring intrusives; they are included in the Neovolcanic Axis province because they may have been affected by post-Laramide volcanic episodes after their original emplacement.

Volcanic processes affected this province considerably at various times, possibly from the Oligocene to the Pleistocene, and may have been a factor in the enrichment of previously formed deposits. Such volcanogenic enrichment is the possible case in the Pachuca and Real del Monte, Angangueo, El Oro, and Tlalpujahua Districts, and probably also in the Taxco District (Guerrero), although this remains to be proved.

REFERENCES CITED

CRM, 1980, Carta y provincias metalogenéticos de la república Mexicana: México, D.F., Consejo de Recursos Minerales, Boletín 21-E (2nd edition), 200 p.

Salas, G. P., 1975, Carta y provincias metalogenéticas de la república Mexicana: México, D. F., Consejo de Recursos Minerales Publication 21-E, 242 p.

Printed in U.S.A.

The Geology of North America
Vol. P-3, Economic Geology, Mexico
The Geological Society of America, 1991

Chapter 45

Geological description of the Cuale District ore deposits, Jalisco, Mexico

G. Berrocal L. and F. Querol S.
Cía. Fresnillo, S.A. de C.V., Río de la Plata 48 esq. Lerma, Col. Cuauhtemoc, 06500 Mexico, D.F.

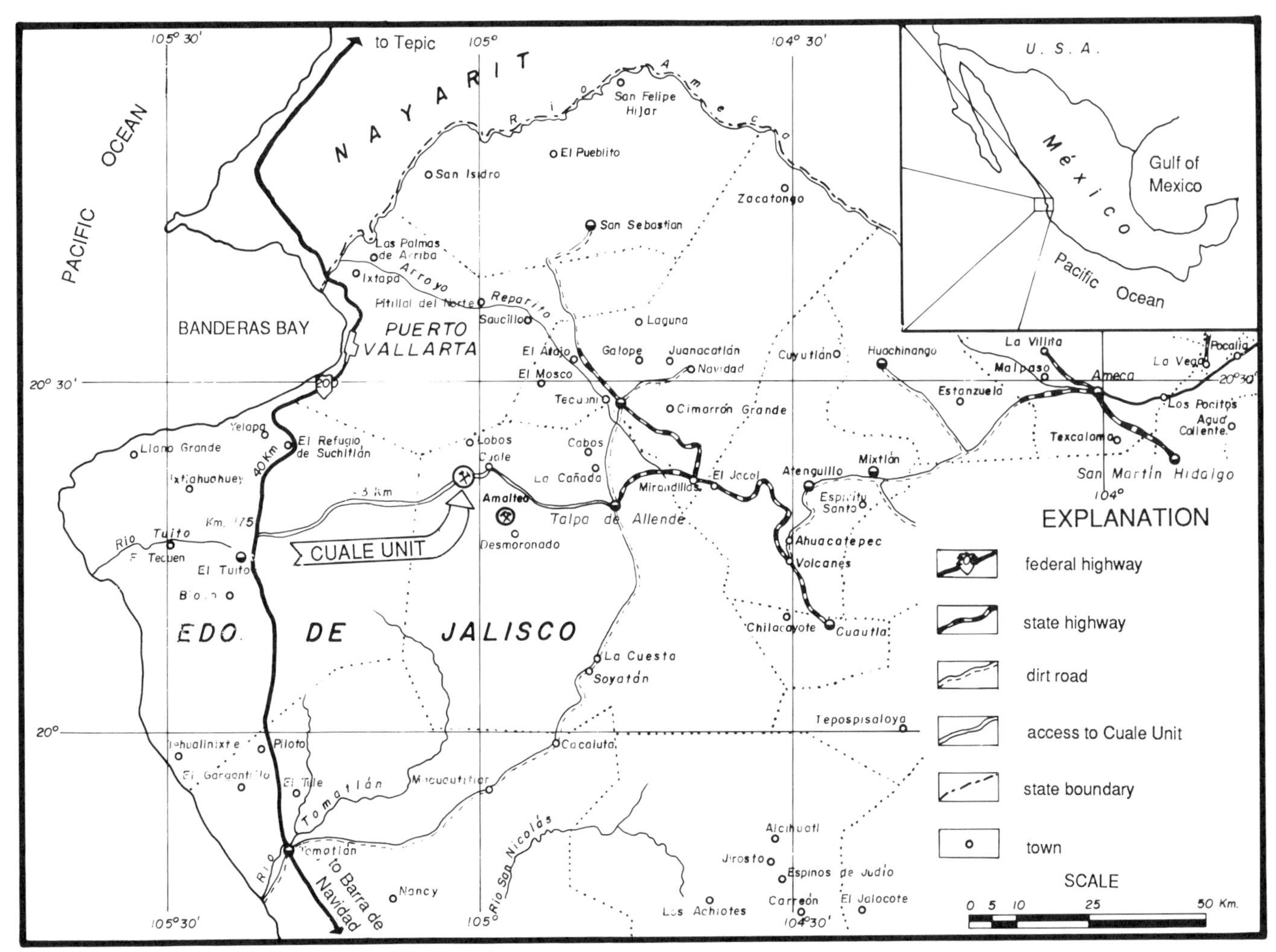

Figure 1. Map showing location of the Cuale Mining District, Jalisco, Mexico.

INTRODUCTION

The Cuale Mining District, covering approximately 30 km^2, is located in the northwestern part of Talpa de Allende Municipality (Jalisco) (20°22′N, 105°07′W), 30 km southeast of Puerto Vallarta (Fig. 1). Access to the district is by the Puerto Vallarta–Barra de Navidad road for 40 km, from which point a 43-km year-round passable dirt road runs east to the area.

The La Prieta Mine discovery in 1804 launched mining activity in the district; deposits were mined intermittently by the Hernández family until 1854 and more regularly by the Unión de Cuale Company from 1854 to 1899 (Beatty, 1899). In the past

Berrocal L., G., and Querol, S., F., 1991, Geological description of the Cuale District ore deposits, Jalisco, Mexico, *in* Salas, G. P., ed., Economic Geology, Mexico: Boulder, Colorado, Geological Society of America, The Geology of North America, v. P-3.

century an estimated 250,000 tons of selected ores with 900 to 1,500 g/ton Ag were extracted and processed by roasting, amalgamation, and leaching in numerous small haciendas around Cuale. From 1900 to 1954, several companies (the Esperanza Company, El Oro, México, Compañía Minera Peñoles, and Eagle Picher Company) attempted unsuccessfully to reactive the mines. Local prospectors mined Minas del Oro deposits in rudimentary fashion (Triplett, 1938).

Zimapán's acquisition of the area in 1965 consolidated its control over all the district. In 1967 and 1973 the company mined the neighboring Amaltea Unit, producing 266,500 tons of ore composed of 1 g/t Au, 154 g/t Ag, 2.6 percent Pb, 13.6 percent Zn, and 0.97 percent Cu.

Exploration at Cuale began in 1971 and was stepped up from 1975 to 1978 when access roads became available.

Exploration by other companies prior to 1960 had focused on the La Prieta Oreshoot. Zimapán however aimed exploration efforts toward the search for mantos that could be more readily studied, leaving La Prieta for later investigation. TURAM electromagnetic surveying in 1971 located the Coloradita, Chivos de Arriba, and Chivos de Abajo anomalies, which together with the Socorredora and Naricero indications (Fig. 2), made the area an attractive prospect.

To evaluate the anomalies and indications, and also the La Pirita and Gradenza (Minas del Oro) orebodies, 215 diamond-drillholes for a total 11,038 m were drilled, and direct underground exploration and rehabilitation of older mineworks were carried out from 1972 to 1978. Results of this work indicated 1,471,000 tons of ore with an average composition of 1.15 ppm Au, 169 ppm Ag, 1.27 percent Pb, 4.89 percent Zn, and 0.34 percent Cu.

On this basis the Zimapán S.A. smelter was built and put into operation in January 1981; it presently processes around 20,000 metric tons (mt)/month.

PHYSIOGRAPHY AND GEOMORPHOLOGY

The Cuale District is included in the Meseta del Norte physiographic province west of the Meseta Neovolcánica province (Raisz, 1964). Present-day geomorphology was shaped by endogenous processes: igneous granitic intrusions fractured and faulted the host rocks, volcanism produced lava flows and pyroclastic deposition, and simultaneous volcanic and sedimentary episodes are represented by stratified and folded rocks. Thereafter, weathering and fluvial erosion developed the Sierra de Cuale, a geomorphologically young mountainous area of high peaks, deep steep-sided (45 to 70°) gorges, and dendritic drainage.

The district is located in the higher parts of the Sierra de Cuale, a northwest-trending mountain complex deeply dissected by north-south gorges, in which Cerro de El Cantón (elevation

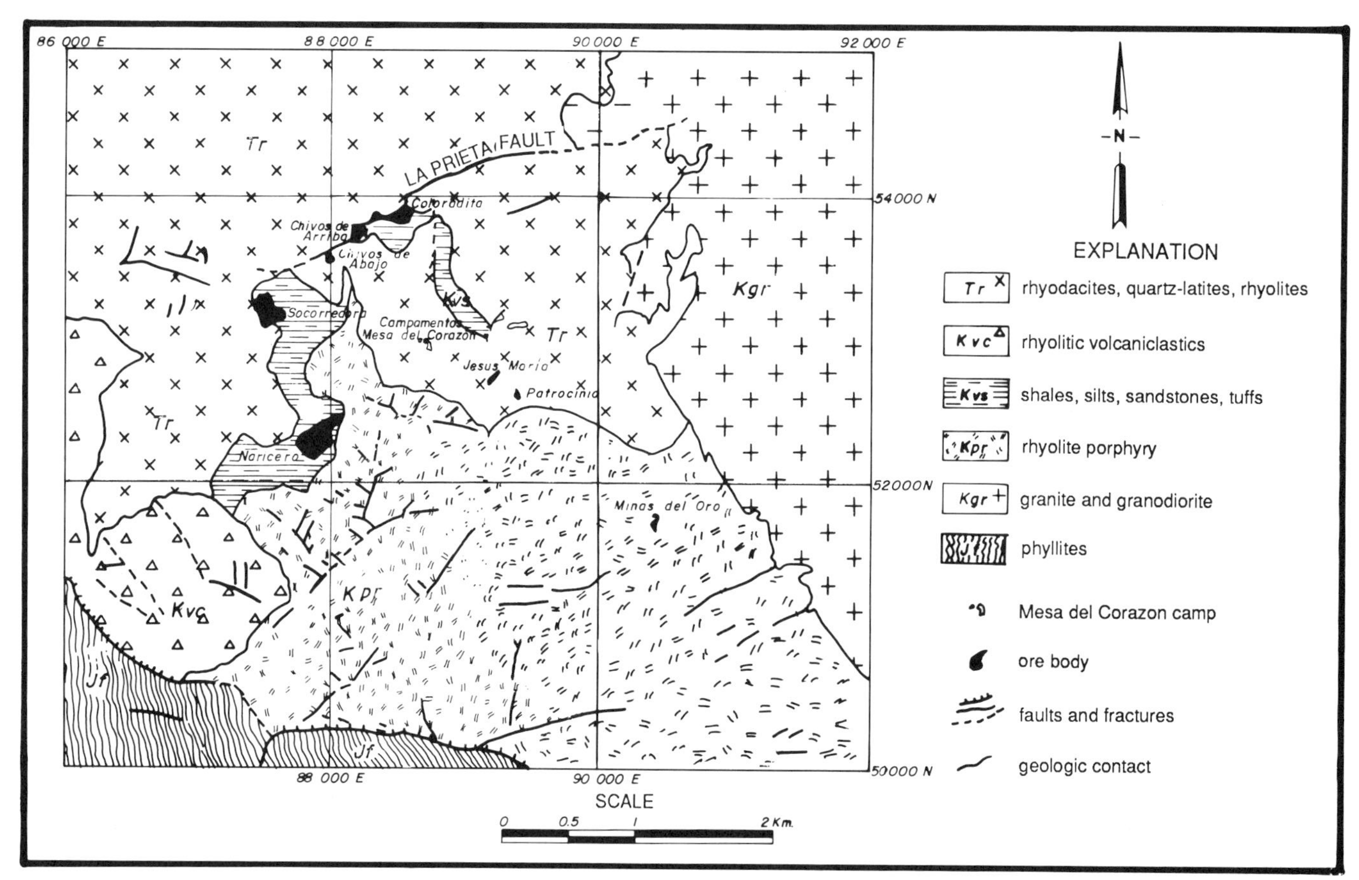

Figure 2. Photogeologic map of the Cuale Mining District (modified from Priego de Witt, 1977).

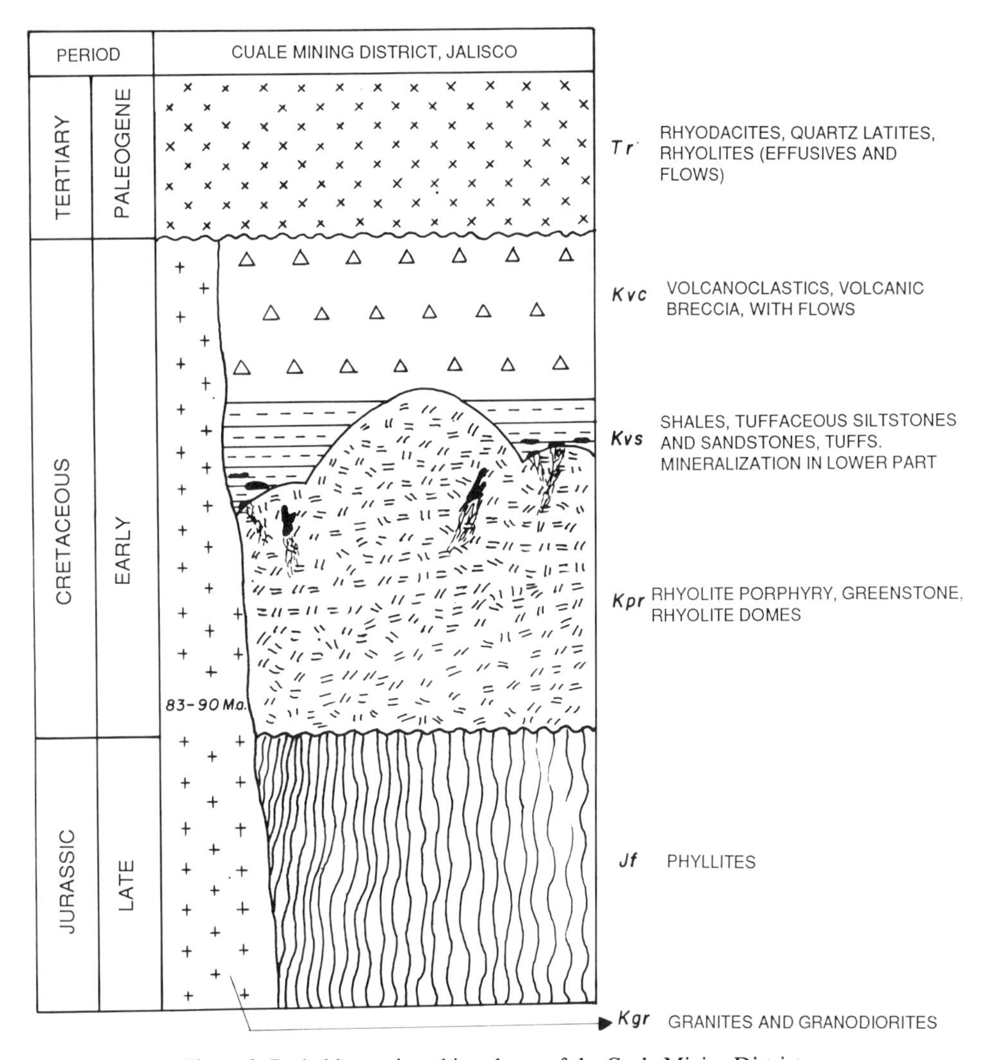

Figure 3. Probable stratigraphic column of the Cuale Mining District.

2,400 m) and Cerro del Caracol (elevation 2,200 m) are the most conspicuous. The deposits occur in a horseshoe-shaped area between 1,800 and 2,200 m above sea level; the mine installations are at Mesa del Corazón in its north-central part.

LOCAL GEOLOGY

No regional geological field mapping has been carried out to date owing to the large size and rugged relief of the district. Figure 2 is a photogeologic map suggesting regional lithology and structures. The rocks exposed in the district are described below, and their stratigraphic relations are shown on Figure 3.

Jurassic

Phyllites (Jf) outcropping in the west and southwest, and observed between kms 25 and 37 of the access trail to the mining area, consist of micaceous phyllites with abundant segregated quartz in lenses parallel to the foliation or in irregular veins; and chlorite and sericite in elongated packets of thin, large lenses or veinlets cross-cutting the foliation. Priego de Witt (1977) correlates these rocks with metasedimentary sequences at several localities in Michoacán and southern Jalisco States, which are assigned to the Jurassic-Cretaceous by the UNAM (Universidad Nacional Autónoma de México) Geological Institute. Estimated thickness is approximately 800 m. Locally the sequence exhibits slight folds with limbs dipping generally 10 to 35°; however, regional folding is inconsiderable, and the general structure is an east-striking, gently north-dipping homocline.

Cretaceous

Rhyolite porphyry (Kpr). A series of dome-like occurrences of volcanic rocks found over most of the district is composed of rhyolites (porphyritic, spherulitic, vitrophyric) and rhyolite porphyries, all of which are light gray and porphyritic, with abundant euhedral quartz phenocrysts in normally vitreous or devitrified groundmass, and in most cases showing phyllitic or argillaceous alteration. These are host rocks of disseminated sulfides in flake structures. Several authors have attempted to separate the numerous textural variations of the porphyries, but weathering and cover by the vegetation make this impractical in exploration and the practice was abandoned. Field data not yet confirmed in the laboratory suggest that all these rocks are cogenetic and penecontemporaneous. The rhyolites, locally named "lower rhyolite," enclose the Gradenza (Minas del Oro) and underly part of the Coloradita, Chivos de Abajo, and Chivos de Arriba orebodies. The upper part of the interval is interbedded with the upper volcanoclastic sequence.

Volcanic-sedimentary sequence (Kvs). This sequence, of restricted distribution in the district, encloses all the stratiform deposits and comprises alternating black shales, siltstones, tuffaceous sandstones, and rhyolite tuffs exposed in the Naricero, Las Bolas, Socorredora, Chivos de Abajo, Chivos de Arriba, Coloradita, and the vicinity of the Jesús María orebodies. These mostly stratified rocks represent a shallow marine environment with small pools, now forming black shale lenses, most of which are mineralized. In the Naricero and Socorredora areas, the sequence trends northwest and dips 5 to 30° north. Macomber (1962) measured a 200-m thickness of this sequence in the Socorredora arroyo and distinguished 26 units or members; however, facies changes are very frequent. Silicified radiolaria and sponge spicules reported in the shales may be of Late Jurassic to Early Cretaceous (Macías R. and G. Solís, 1985) and Middle Jurassic or Early Cretaceous (Ortigosa, 1983) ages. By correlation with other volcanic rocks in the area they are tentatively assigned to the Early Cretaceous, and definitely confirmed as pre-Tertiary.

Volcanoclastic breccia (Kvc). These rocks, composed of angular to subangular fragments of rhyolitic composition in an equally rhyolitic groundmass, may be seen in the area known as El Reliz on the access trail to the mining area in the southeastern part of the district. The unit is propylitized and intruded by light gray to white rhyolite dikes probably belonging to the Kpr unit.

Granites and granodiorites (Kgr). These rocks are found in the northeast and east-central parts of the district as well as along the access trail to the mining area and the Federal Highway south of Puerto Vallarta. They are strongly weathered, usually rounded in shape, and occasionally very friable, generally exhibiting phaneritic texture with euhedral quartz and feldspar crystals and locally variable biotite and hornblende contents. Isotopic dating (83 to 90 Ma; Gastil and others, 1979) indicates multiple intrusions. The rocks intrude phyllites (Jf) and porphyritic rhyolites (Kpr).

Tertiary

Rhyolites, rhyodacites, and quartz latites (Tr). This unit is the most widespread in the district and is probably correlatable with the Sierra Madre Occidental Oligocene-Miocene acid continental volcanism. It is a complex series of lithic and crystalline tuffs and flows considered as ignimbrites.

In the Naricero Mine area the unit is unconformable above the volcanic-sedimentary sequence (Kvs); locally it has been named the "upper rhyolites."

MINERAL DEPOSITS

The Cuale deposits, generally associated with the volcanic-sedimentary sequence and acid rocks, are lenticular discontinuous mantos of banded, fine-grained massive sulfides, mantos of fragmental sulfides, or massive oreshoots enclosed in rhyolite tuffs and black shales. An exception is the Grandeza (Minas del Oro) orebody, in which mineralization occurs as a flaked structure of multidirectional microfractures.

Orebodies vary in both thickness and extension; underground they also are affected by the gentle folding of the host rocks. The distribution of the various orebodies in the district, which overlie the mixed volcanic-sedimentary sequence, with the exception of Jesús María, Patrocinio, and Grandeza orebodies that are associated with acid volcanic rocks, is shown on Figure 2.

At present, Zimapán is in the process of developing or preparing the following: Naricero, Socorredora, La Prieta, Coloradita, and Grandeza (Minas del Oro). Chivos de Arriba, Chivos de Abajo, and Jesús María are completely paid out; Patrocinio has only been diamond-drilled. Presently worked orebodies or mines are briefly described below.

Coloradita Mine

This mine is in a complex deposit possibly formed over a volcanic source or exhalative center. Diamond-drilling has located the possible pathway for the mineralizing solutions, a lattice (stockwork?) of pyrite veinlets with occasional copper values in rhyolite with phyllitic alteration. Two massive sulfide lenses overlie the exhalative center. The first lens is pyrite (80 percent) with gold and copper values (yellow ore), which under the microscope shows advance edges on galena and occasional straight pyrite contacts, indicating two stages of pyrite mineralization. Sphalerite appears only at the edges of galena or as islands within anhedral pyrite. Small amounts of quartz form the gangue.

The other lens, immediately above, is composed mainly of sphalerite (85 percent), pyrite, galena, chalcopyrite, and tetrahedrite (black ore), with chlorite and sericite as gangue.

Above the massive sulfide lenses is an upper enveloping zone with lenses of chaotically distributed sphalerite, galena, and pyrite disseminated in rhyolite tuff. Being very close to the surface, many of these lenses are completely oxidized, forming silver- and gold-enriched gossan zones.

Incipiently bedded tuffaceous sandstones crown the whole sequence (At, Fig. 4).

Owing to the arrangement of massive sulfide lenses over a veinlet stockwork, the deposit is classified as proximal. The ore mantos are slightly tilted 15 to 35° southwest, tending to steepen to 45 to 50° near the flank of the volcanic structure.

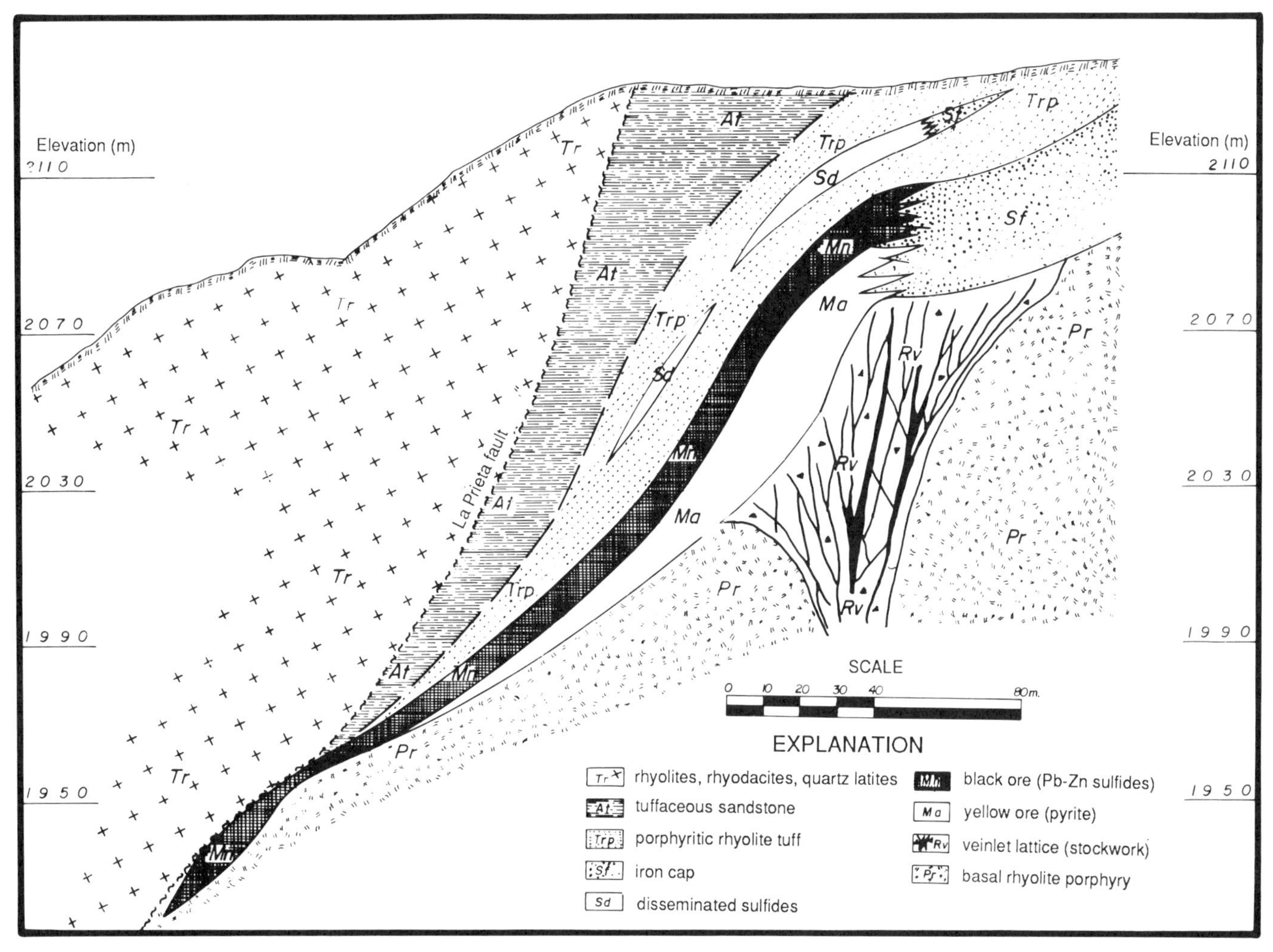

Figure 4. Idealized cross section of the Coloradito orebody (modified from Luna, 1979).

Original dimensions of this practically paid-out orebody were: length, 120 m; width, 80 m; average thickness, 11.0 m (Luna, 1979).

Socorredora Mine

This deposit is a series of interbedded individual ore lenses in almost horizontal attitude, with slight undulations of as much as 20°, that accumulated jointly with volcanic-sedimentary material in a small sea-floor pool, which makes the deposit appear as a gently sloping syncline (Fig. 5).

The deposit is roughly elliptical in shape, tapering gradually in every direction. In general, the ore lenses occur within a heterogeneous section of stratified rhyolite tuffs and discontinuous lenticular bodies of black shale, all overlain by rhyolite tuffs and porphyritic rhyolites.

Though there is no clear mineralogical zoning (as in Coloradita), two distinctive ores—black and yellow—are present. The yellow ore comprises several pyrite stages in uniformly sized grains, some with sphalerite inclusions. Polished surfaces reveal sphalerite replacing pyrite, with irregular contours in individual grains, and chalcopyrite inclusions. Galena is usually in contact with pyrite, rarely with sphalerite, and is associated with boulangerite. Quartz, cristobalite, devitrified glass, sericite, and rock fragments make up the gangue. Silicification and sericitization are the predominant alterations.

Under the microscope, the black ore consists of large, evenly sized sphalerite crystals surrounded or cross-cut by pyrite veinlets; galena appears in large masses with pyrite and sphalerite inclusions. Silver-bearing minerals include tenantite and stromeyerite associated with galena. The gangue is composed of quartz, barite, tuffaceous material, and some gypsum. Silicification, kaolinization, and carbonatization are common alterations.

The massive mineralization generally occurs as oreshoots in kaolinized and friable rhyolite tuffs; these massive ores change laterally to disseminated ores with sphalerite and galena included in an argillaceous groundmass along with quartz and barite.

Original dimensions of the orebody, presently at an advanced stage of development, were: length, 200 m; width, 100 m; average thickness, 7.5 m (Luna, 1979).

Naricero Mine

This mine is developed in undoubtedly the largest orebody in the district. The orebody consists of lenticular mantos (at least two clearly identified) composed of subangular fragments of massive and disseminated sulfides with reworked volcanic material, all in a tuffaceous groundmass devitrified to quartz and feldspars (Fig. 6). The mineralized tuffaceous interval is approximately 35 m thick at the center and tends to thin out in every direction. The general structure seems to be a N20 to 40°W–trending, 25°NE–dipping homocline.

Subsequent faulting caused block movements that displaced the orebody from a few centimeters to over 30 m.

Orebodies are commonly found interbedded with discontinuous carbonaceous shale lenses. The mineralogy exhibits "fragmental transported" texture in which the principal ore minerals are disseminated or evenly sized large sphalerite grains with chalcopyrite exsolutions and veinlets, as well as mixed pyrite-galena grains; galena crystals are observed within the sphalerite. Sphalerite is more abundant than pyrite, whereas galena proportions are very low. Pyrite is mostly disseminated in the gangue as subhedral or rounded grains with corroded edges. Lesser amounts of tetrahedrite-tenantite, geocronite, and proustite-pyrargirite are also present (J. G. Bravo, personal communication, 1985).

According to its structural (the deposit is topographically 100 m lower than all the other known orebodies), textural (presence of reworked fragments), and mineralogic characteristics, Naricero is a distal deposit generated by the destruction of a proximal body.

Original dimensions were: length, 390 m; width, 195 m; average thickness, 2.72 m (Luna, 1979).

Grandeza Mine (Minas del Oro)

This mine shows different features from the foregoing (Fig. 7; i.e., lack of massive black and yellow orebodies and associated volcanic-sediments). The Grandeza Mine probably forms the base of an exhalative center. The orebody develops a flaked structure with erratic gold and zinc values in narrow (few mm to 5 to 10 cm), irregular, multidirectional veinlets of no

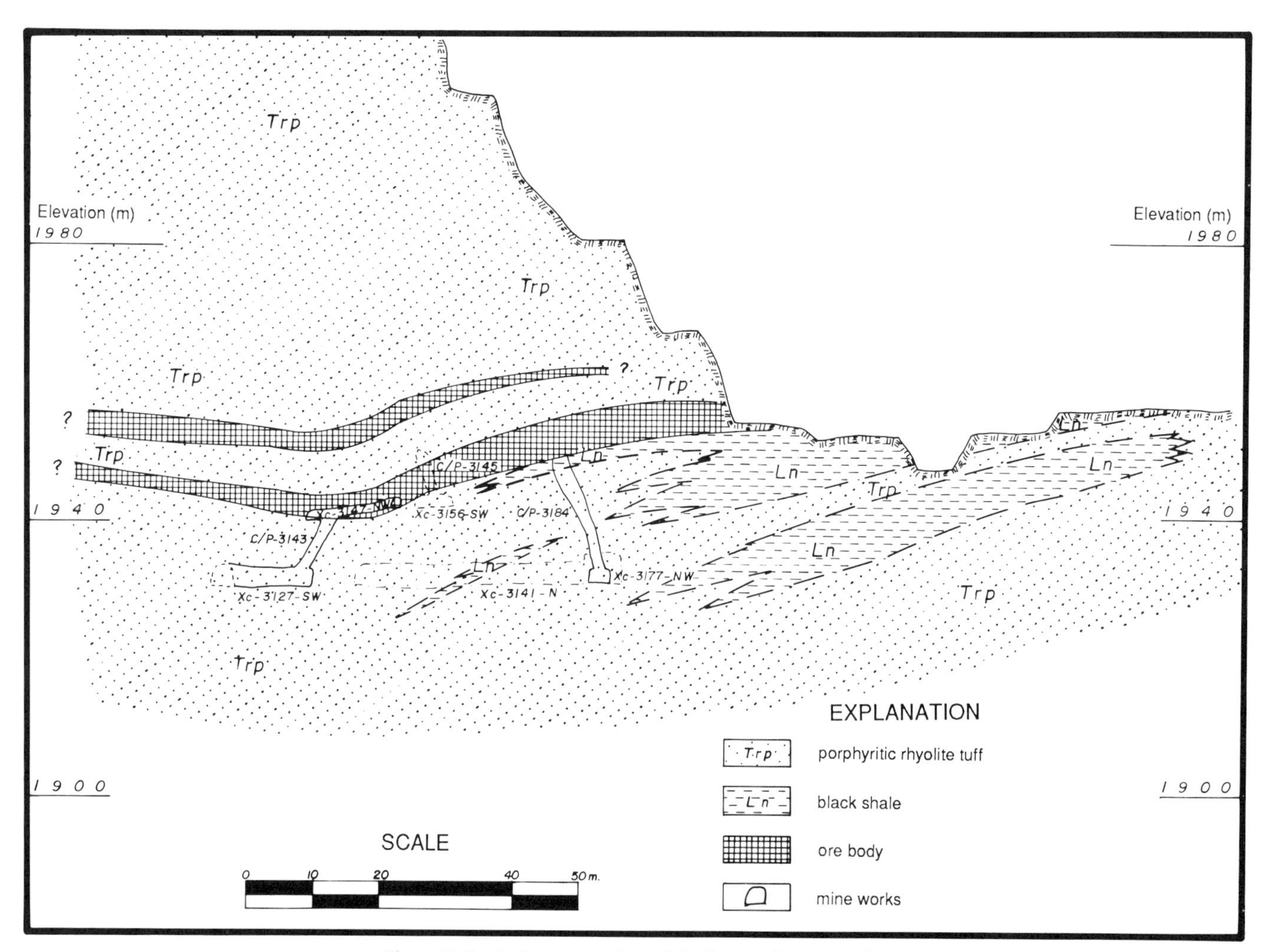

Figure 5. Geologic cross section of the Socorredora orebody.

preferred dip. The host rock is porphyritic rhyolite, with flow texture in some parts and porphyritic in others, with quartz and plagioclase phenocrysts in a silicified aphanitic groundmass. The orebody demarked to date is roughly elliptical in outline, 20 to 50 m wide, more than 125 m long in an almost north-south direction, and 70 m high.

The most abundant minerals are pyrite and sphalerite. Pyrite appears as idiomorphic crystals occasionally ending in curved surfaces; it is also very fractured with galena filling the fractures. Though less abundant, sphalerite is the main economic mineral with galena, chalcopyrite, and pyrite inclusions. Other minerals present are bornite, covellite, and gold, the latter as microscopic inclusions within pyrite. Grandeza, also known as Minas del Oro, is the only orebody showing readily identifiable gold under the microscope.

La Prieta Mine

This mine is the most important structure of the district, at the so-called La Prieta fault zone, which trends NE52° and dips 50 to 65° NW, extends for 2.5 km, and bounds all the known orebodies associated with it in the downthrown block. Along the La Prieta structure the deposit comprises several massive oreshoots that accumulated in independent stream beds along the slopes of a presumed submarine volcanic plateau, and chaotic deposits of disseminated mineral throughout the entire slope, indicating heterogeneous transport not far from the source. The La Prieta oreshoots are no more than the extension toward the northwest and downward of the Chivos de Arriba and Coloradita massive sulfide mantos.

Figure 8 shows the clear connection with the Chivos de Arriba massive sulfide mantos, which having been deposited very close to the edge of the plateau, partially slid down and accumulated on the slope, generating the oreshoot at stope 3785 (La Prieta Mine).

The following features characterize the oreshoots. At the bottom of the orebody, a zone of primary sedimentary clays was deposited at the time of mineralization (because it contains abundant galena and sphalerite disseminations); the clays exhibit compaction slide structures that resemble faults. Striae and sub-

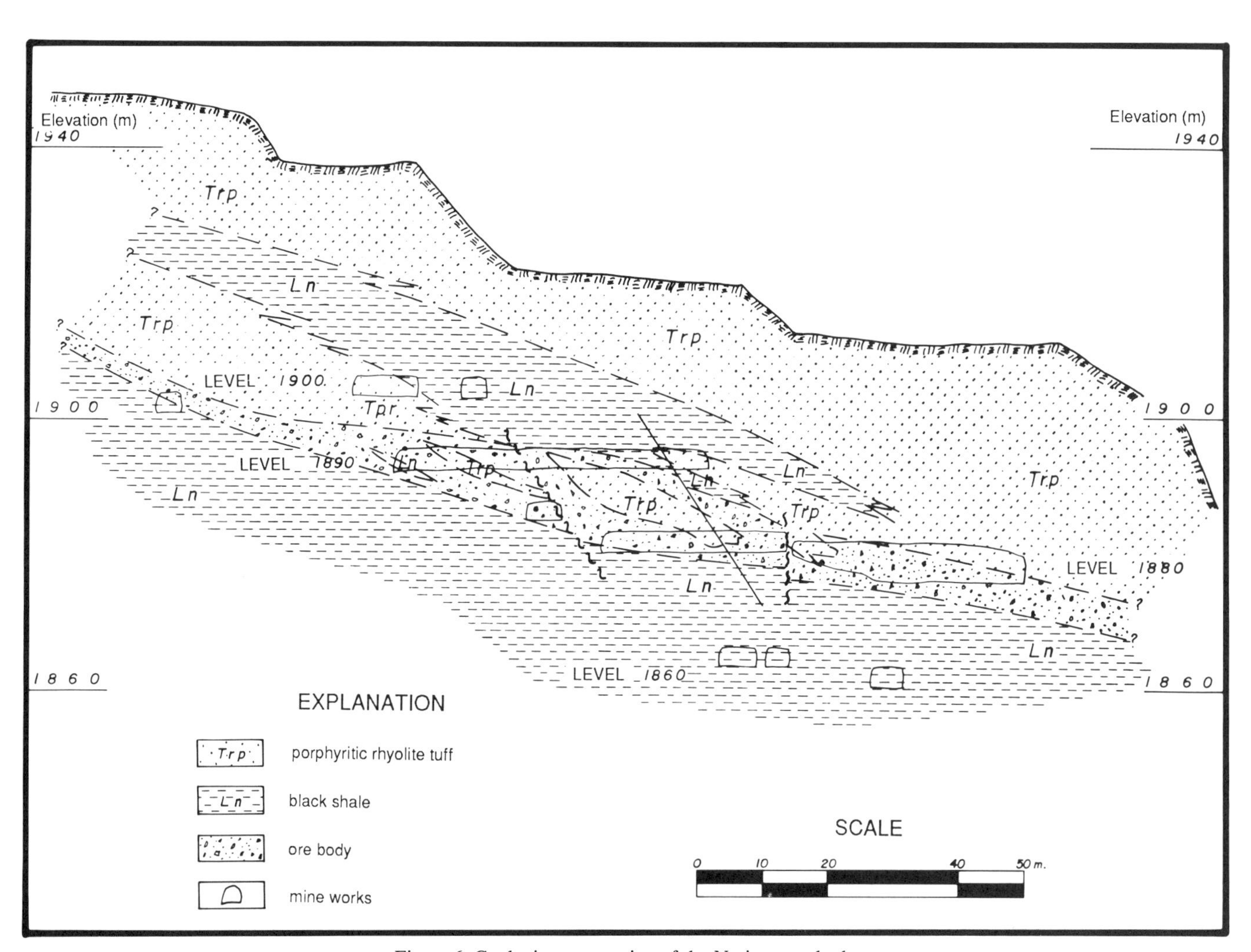

Figure 6. Geologic cross section of the Naricero orebody.

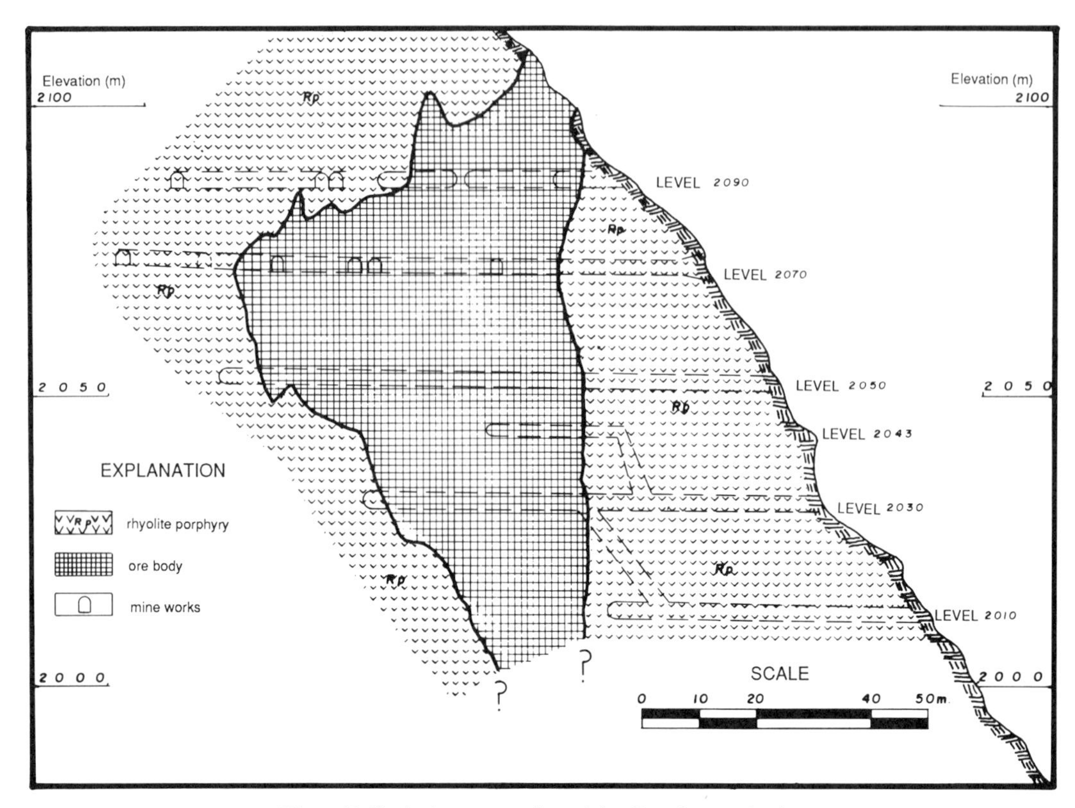

Figure 7. Geologic cross section of the Grandeza orebody.

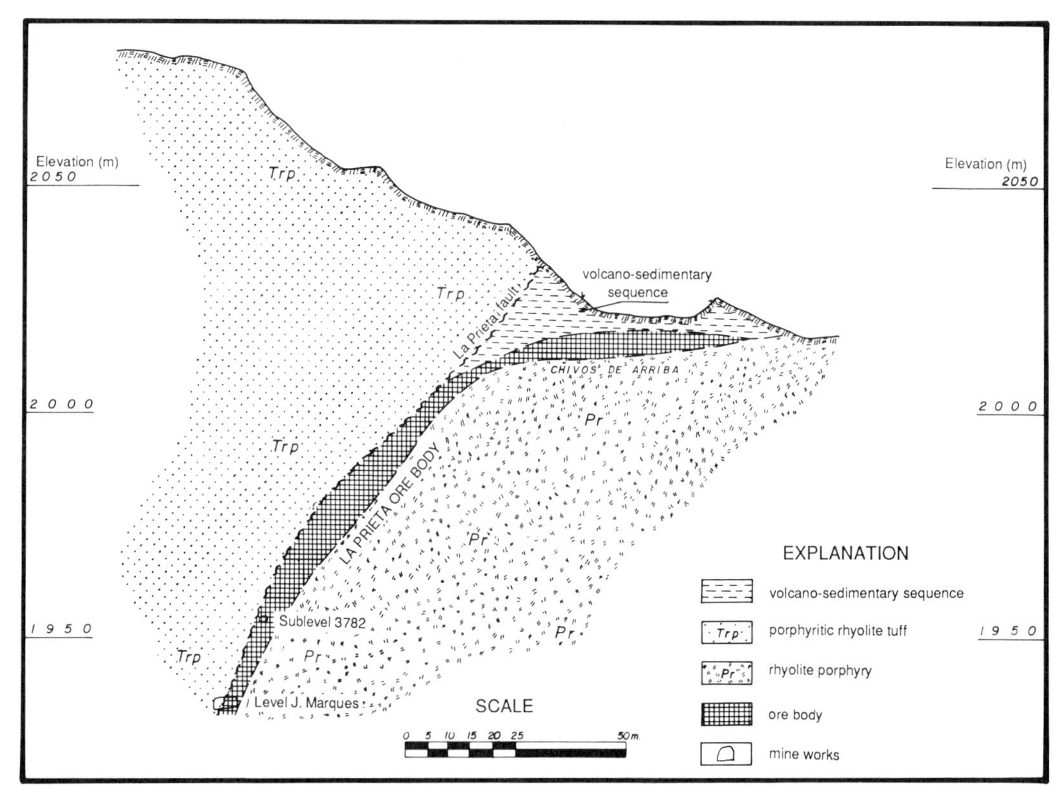

Figure 8. Geologic cross section of the La Prieta orebody.

angular clasts within the clay groundmass are used as guides to define fault zones, in contrast to the sedimentary clays associated with the mineralization. Massive sulfides of Pb-Zn (below) and Fe-Cu with Au values (above) frequently showing sedimentary banding, overlie the mineralized clays. Immediately above the massive ores is a zone of chloritized clays with occasional elongated siliceous nodules and a clearly brecciated zone representing the La Prieta fault. Propylitized rhyolite tuffs and rhyolites compose the hanging wall.

Under the microscope the massive mineralization shows fine-grained sphalerite, tetrahedrite, chalcopyrite, stannite, and pyrite traces in a barite-rich groundmass; when sphalerite is massive, fine grains of euhedral pyrite, and galena and chalcopyrite traces are observed.

The mixed, already diluted, La Prieta ore has relatively high metallic sulfide content (approximately 32 percent), which may reach 65 percent in the massive "oreshoots." These may be from a few centimeters to 10 m thick, 30 to 70 m long, and 70 to 80 m high.

Average volumes and grades

The estimated tonnage at the start of Cuale Unit operations in January 1981 was 1.47 million metric tons (mt) of 1.15 ppm Au, 169 pm Ag, 1.27 percent Pb, 4.89 percent Zn, and 0.34 percent Cu. Total production to December 1984 was 771,200 mt of 0.58 ppm Au, 156 ppm Ag, 1.74 percent Pb, 4.89 percent Zn and 0.39 percent Cu (J. Bravo N., personal communication, 1985).

Metallurgical operation

Metallurgical recovery is by selective flotation producing lead and zinc concentrates; average recovery is 63 percent Au, 69 percent Ag, 74 percent Pb, 78 percent Zn, and 72 percent Cu. Total production of metal in concentrates in 1980 to 1984 was 270 kg Au and 77.4 mt Ag.

REFERENCES CITED

Gastil, G. D., Krummenbacher, M. A., and Jensky, H., 1979, Reconnaissance geology of west-central Nayarít, México; Summary: Geological Society of America Bulletin, v. 90, p. 15–18.

Raisz, E., 1964, Landforms of Mexico, 2nd ed.: Cambridge, Massachusetts, Harvard University Press.

Unpublished sources of information

Beatty, E. C., 1899, Report on the property of the "Unión in Cuale Company," State of Jalisco, México: Compañía Minera Peñoles, private report.

Luna, B. R., 1979, Yacimientos minerales y exploración del cuerpo "Naricero" en el Distrio Cuale, Municipio de Talpa de Allende, Jalisco [professional thesis]: I. P. N.

Macías R., C., and G. Solís, P., 1985, Mineragrafía, microtermometría e isotopía de algunos yacimientos del Distrito Minero de Cuale, Jalisco [professional thesis]: Universidad Nacional Autonoma de Mexico, Facultad de Ingenieria.

Macomber, B. E., 1962, Geology of the Cuales Mining District, Jalisco, Mexico [Ph.D. thesis]: New Brunswick, Rutgers University, 387 p.

Ortigosa, F., 1983, Geología de Cuale, Jalisco [professional thesis]: Universidad Nacional Autonoma de México, Escala 1:5,000.

Priego de Witt, M., 1977, Estudio fotogeológico del area Cuale-Cuatro Minas-Amaltea, Mpio. de Talpa de Allende, Jalisco, Mexico: Compañía Fresnillo, S.A. de C.V., private report.

Triplett, W. H., 1938, Report of the Cuale Mining District, Jalisco, Mexico: Compañía Minera Peñoles, S. A., private report.

Printed in U.S.A.

The Geology of North America
Vol. P-3, Economic Geology, Mexico
The Geological Society of America, 1991

Chapter 46

Economic geology of the Inguarán Mining District, Michoacán

A. Osoria H.
Industrial Minera México, S.A., Ave. Baja California 200, Col. Roma Sur, 06760 Mexico, D.F.
N. Leija V.
Comisión Federal de Electricidad, Río Rodano 14, 06500 Mexico, D.F.
R. Esquivel
Industrial Minera México, S.A., Ave. Baja California 200, Col. Roma Sur, 06760 Mexico, D.F.

INTRODUCTION

The Inguarán Mine unit–owned by Zinc de México, S.A. (of Grupo Industrial Minera México, S.A. de C.V.)–is located in La Huacana Municipality, 100+ km north-northeast of the town of Morelia in east-central Michoacán State; geographic coordinates are 19°48′ to 19°58′N and 101°35′ to 101°45′W (Fig. 1).

From Morelia a four-hour drive on federal highways reaches La Huacana and from there a wide, 28-km-long dirt road leads to Inguarán. The mine area is at an elevation of 700 m; climate is subtropical, with abundant rainfall in summer and dry winters. Vegetation is typically subtropical.

The regional economy is based on seasonal agriculture and extensive cattle-raising; mining is practically restricted to the Inguarán Unit. La Huacana has telephone and telegraph facilities; a light-craft runway near the unit was kept in excellent condition during the period of its activity.

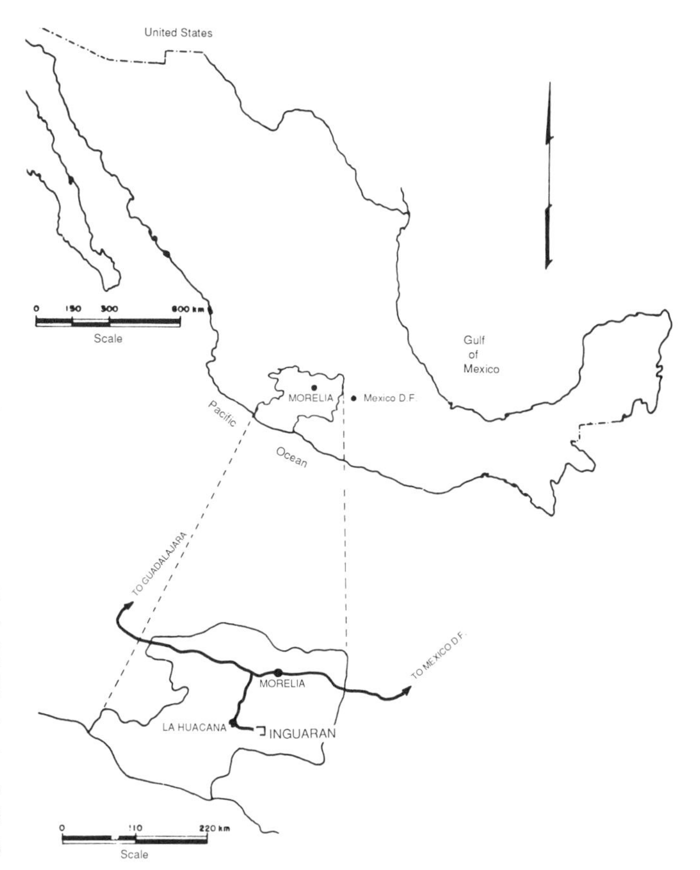

Figure 1. Map showing location of the Inguarán Mining District.

REGIONAL GEOLOGY

The district lies in the south-central part of the Mexican Neovolcanic Axis Metallogenetic Province. Granite and granodiorite igneous intrusive rocks are predominant, partially covered by andesite and basalt flows. The oldest and most widespread unit is a medium- to coarse-grained, pink-hued, reddish-weathering granite intruded by porphyritic light-gray granodiorite. A series of 10-m-wide (maximum width at the surface), light-green, andesitic porphyry dikes with very conspicuous plagioclase phenocrysts is also present. Mainly andesitic tuffs overlie the above units toward the east in the area; extensive vesicular basalt flows occur in the north and west-central parts (Fig. 2).

Three conspicuous fault systems trend predominantly north (N45°E, N60°E, and N75°W); the last two are the most important.

Osoria H., A., Leija V., N., and Esquivel, R., 1991, Economic geology of the Inguarán Mining District, Michoacán, *in* Salas, G. P., ed., Economic Geology, Mexico: Boulder, Colorado, Geological Society of America, The Geology of North America, v. P-3.

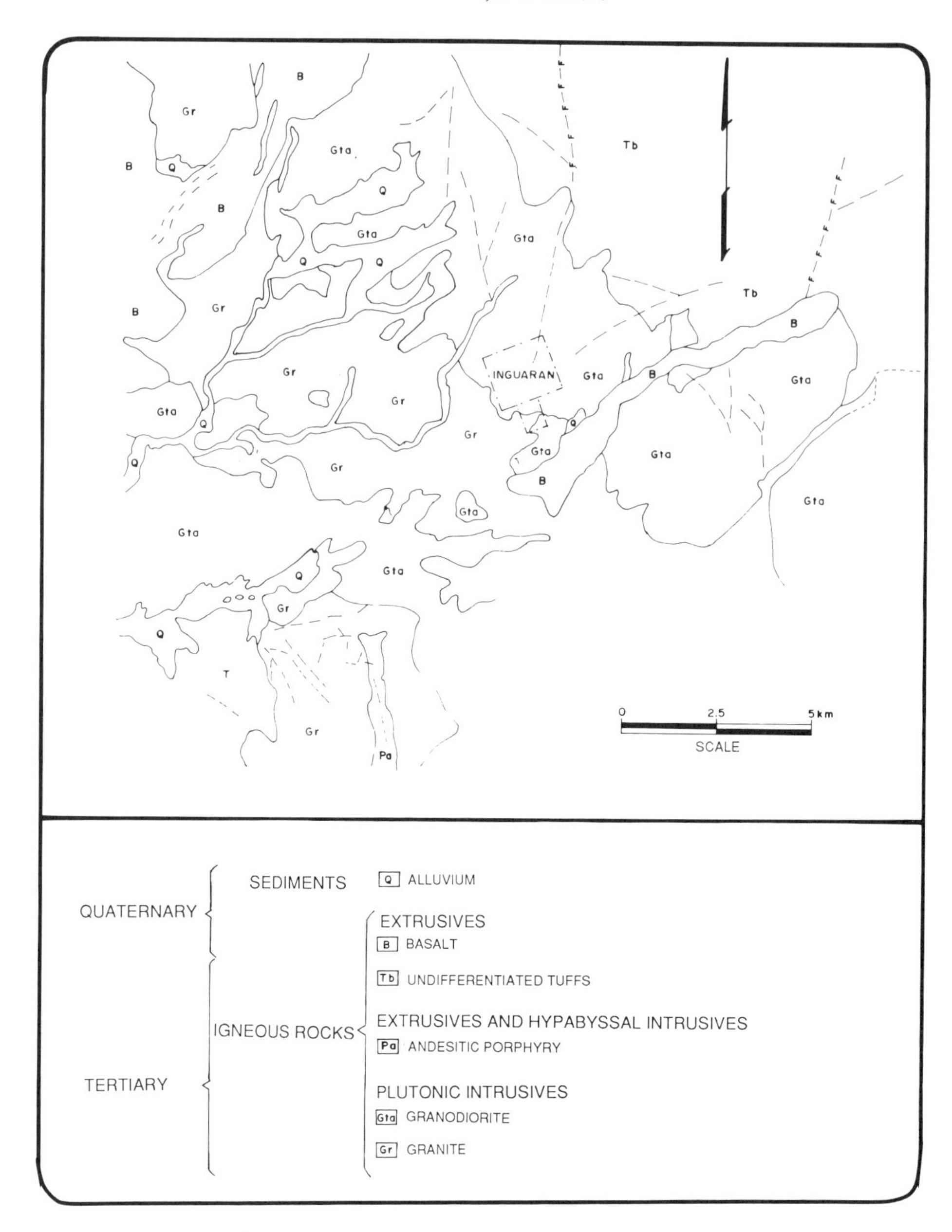

Figure 2. Geologic map of the Inguarán Mining District.

LOCAL GEOLOGY

The predominantly granodioritic Inguarán Mine area proper also has a quartz-monzonite differentiation facies, with transitional contacts between both rock types.

The regional fault systems apparently acted as emplacement controls, in both types of intrusive rocks and sometimes at the contacts, of various brecciated chimneys that range from virtually sterile to commercially mineralized (copper) (Fig. 3).

Outcrops exhibit considerable widespread quartz-sericite alteration with occasional tourmaline in irregular concentrations, whereas breccias show strong propylitic alteration haloes extending only a few meters from the host rock.

MINERAL DEPOSITS

Three types of ore deposits have been identified in the district: veins, injection breccias, and brecciated chimneys.

Veins

Mineralized veins consist of 1- to 15-m-thick tabular bodies of copper or lead sulfides or oxides in milky quartz–tourmaline–epidote gangue, which are not persistent along strike and therefore are of insufficient volume for large-scale commercial development.

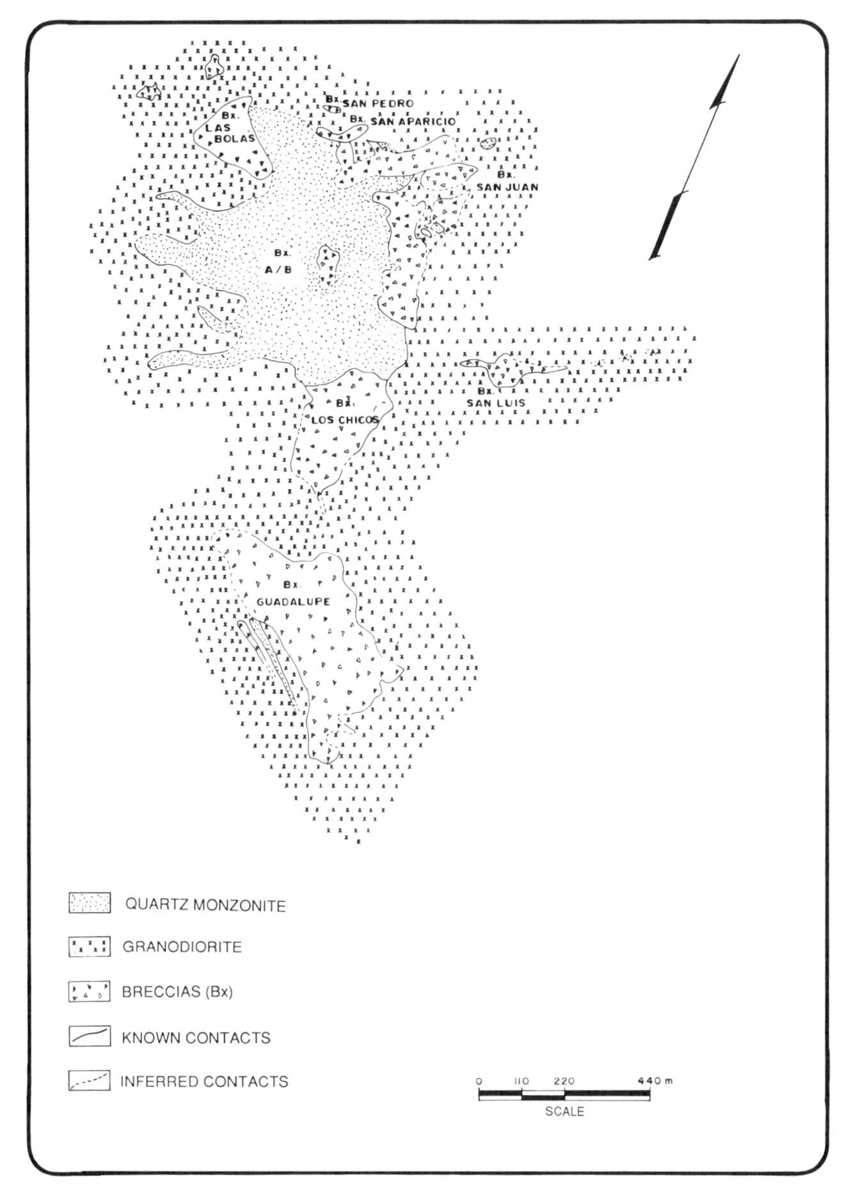

Figure 3. Geologic map of the principal producing part of the Inguarán Mining District.

Injection breccias

The mineralized injection breccias are tabular brecciated bodies composed of subrounded igneous fragments in quartz cement and showing flow features; some are strongly chloritized. These are also low-volume ores, in spite of a few rich shallow zones.

Brecciated chimneys

Brecciated chimneys are the most important orebodies. These have been developed by large-scale mechanization. They are of roughly elliptical shape, with major axes of several hundred meters and shorter axes one-third to one-fifth of this figure; some bodies extend 350 m downward, although they usually do not extend below 200 m.

Most authors agree that these are typical gas-fluxing breccias. At times the surfaces of the rounded and subrounded breccia fragments suggest mechanical abrasion by small particles suspended in the gases when the breccia was formed. These small particles are now rock dust cementing the fragments; the quantity of small particles varies from one chimney to another and is an important element in the degree of mineralization. Bodies containing little rock dust are usually poorly mineralized, or else their mineralization is related to post-breccia fracturing.

Most breccia fragments are relatively unaltered; the dust is usually strongly propylitized. Hydrothermal mineralization in the principal commercial orebodies produced chalcopyrite-quartz-calcite-actinolite-epidote deposition, with lesser pyrite and scheelite, mostly in open spaces between the fragments; some material filling small fractures across the fragments formed veinlets.

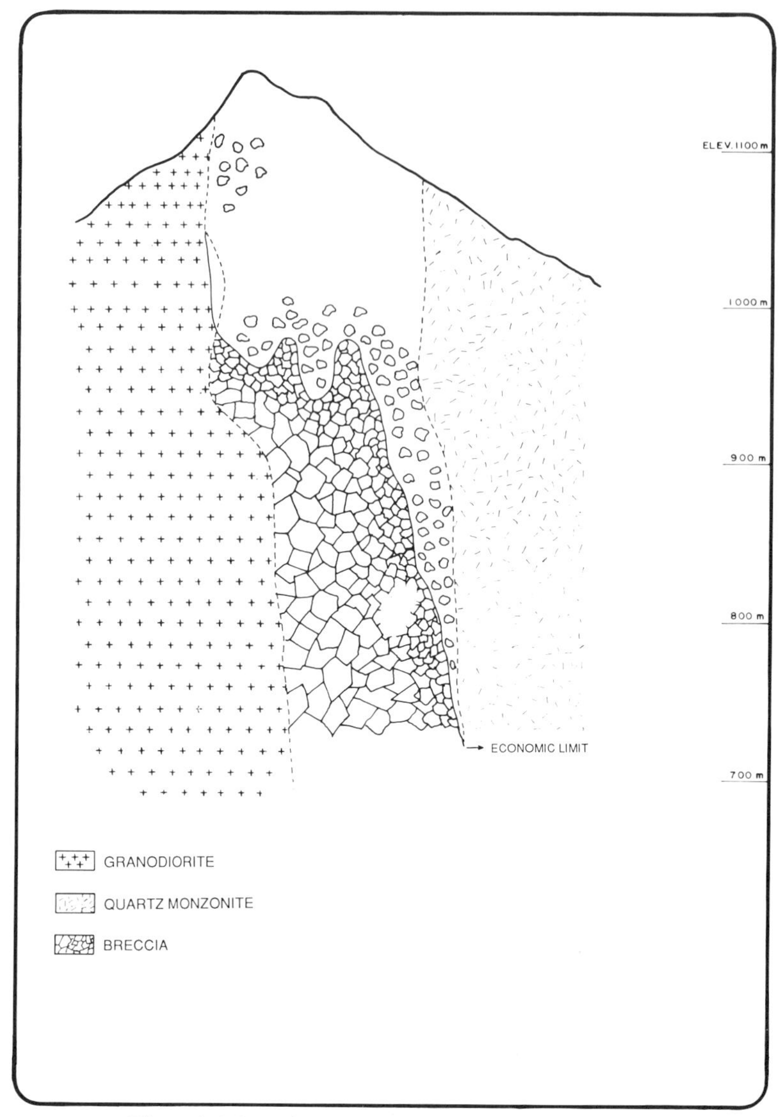

Figure 4. Schematic cross section of the San Juan breccia.

ECONOMIC GEOLOGY

The mining history of the district is poorly documented; however, the rich surface zones, mostly at the A/B Breccia (Fig. 3), are known to have been mined intermittently since the end of the past century to the 1930s. These older small and shallow works soon disappeared when the modern Inguarán mineworks were initiated. In any case, extracted volumes (of copper only) would have been small, because the area lacked adequate access routes and remained completely isolated.

In the early 1960s, Asarco Mexicana (now Industrial Minera México, S.A. de C.V.), a long-time concessionaire of properties that included the A/B Breccia outcrops, began exploration activities, including diamond-drilling, which led to an estimation of sufficient volumes of commercial ores to justify the installation of a 2,200 t/day plant. Postmineralization faulting of the A/B Breccia separated and horizontally displaced (by approximately 100 m) the two segments (A and B), which were mined individually.

Subsequent exploration revealed other brecciated chimneys with varying degrees of mineralization, but most production was from the A/B Breccia, with lesser volumes from the San Luis and San Juan (Fig. 4) breccias. Mining costs of the remaining orebodies were too high for their commercial development.

From 1971 to 1982 a little over 7 million tons of ores were extracted, with grades ranging from 1.38 percent Cu at the start to 1.0 percent at the end of the operations, when milling capacity had been increased to 3,200 t/day; average grade of the entire production was 1.2 percent Cu.

In the final years of the Inguarán Unit activity, commercially acceptable tungsten concentrates were obtained by hydraulic mining and gravimetric concentration from the 0.02 and 0.04 percent of tungsten contained in the tailings.

Printed in U.S.A.

The Geology of North America
Vol. P-3, Economic Geology, Mexico
The Geological Society of America, 1991

Chapter 47

Zacualpan Mining District, State of Mexico

B. Noguez A., J. Flores M., and A. Toscano F.
Cía. Campana de Plata, S.A., Rio de la Plata 48 esq. Lerma, Col. Cuauhtemoc, 06500 Mexico, D.F.

INTRODUCTION

The Zacualpan District is located in the Zacualpan municipality near the northern Guerrero stateline, southwestern State of México, 94 km directly southwest of Mexico City and 164 km by land through Toluca; geographic coordinates are 18°44′00″N and 99°48′30″W. Average elevation is 2,060 m (Fig. 1). It lies at the northern border of the Cuenca del Balsas–Mezcala subprovince of the Sierra Madre del Sur physiographic province. The rugged mountain relief is in the mature stage of the erosional cycle; drainage follows a rectangular pattern. The climate is predominantly semitropical to subhumid; annual rainfall averages 800 to 1,600 mm and temperatures range 14° to 22°C. Vegetation is characterized by white oak (*Quercus candinas*), pine (*Pinus ocote*), arbutus (*Arbustus variant*), and cedar (*Juniperus* sp.).

Access to the area from Mexico City is by Highways 130 (México–La Marquesa–Ixtapa de la Sal–Zacualpan) and 55 (México–Toluca–Almoloya de Alquiciras–Zacualpan).

The name Zacualpan comes from the Náhuatl words *za-*

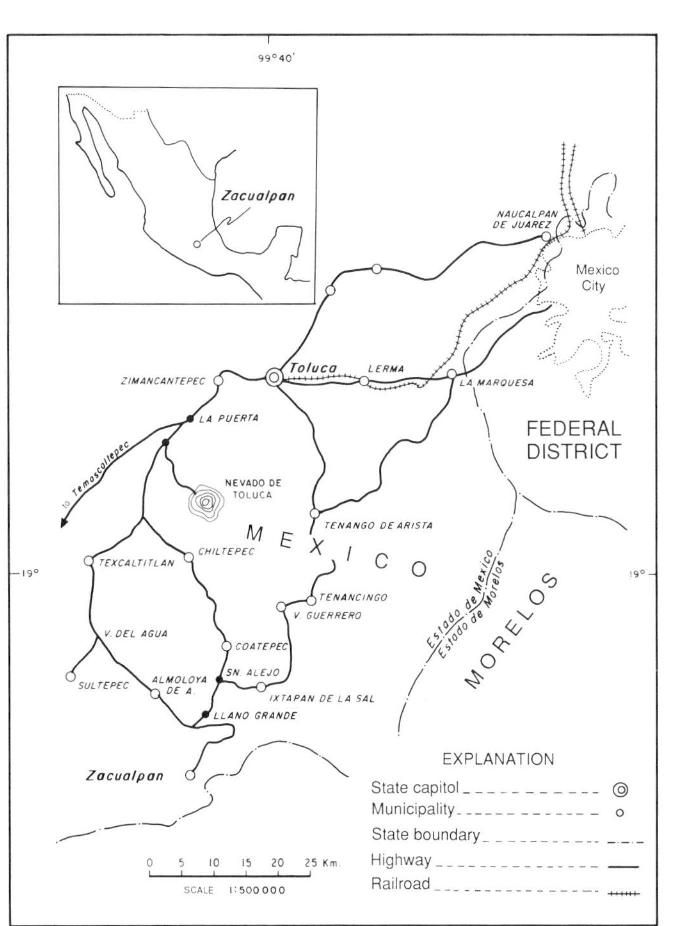

Figure 1. Map showing location of the Zacualpan Mining District.

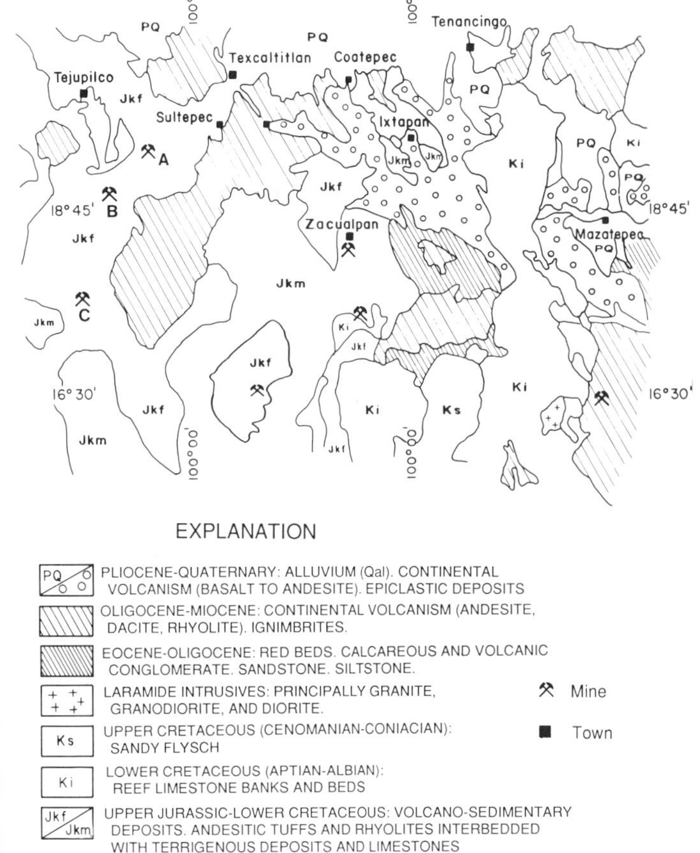

Figure 2. Regional geologic map of the Zacualpan Mining District. Currently operating mines: A, Guadalupe; B, Pachuqueno; and C, Capsa-Regenerador.

Noguez A., B., Flores M., J., and Toscano F., A., 1991, Zacualpan Mining District, State of México, *in* Salas, G. P., ed., Economic Geology, Mexico: Boulder, Colorado, Geological Society of America, The Geology of North America, v. P-3.

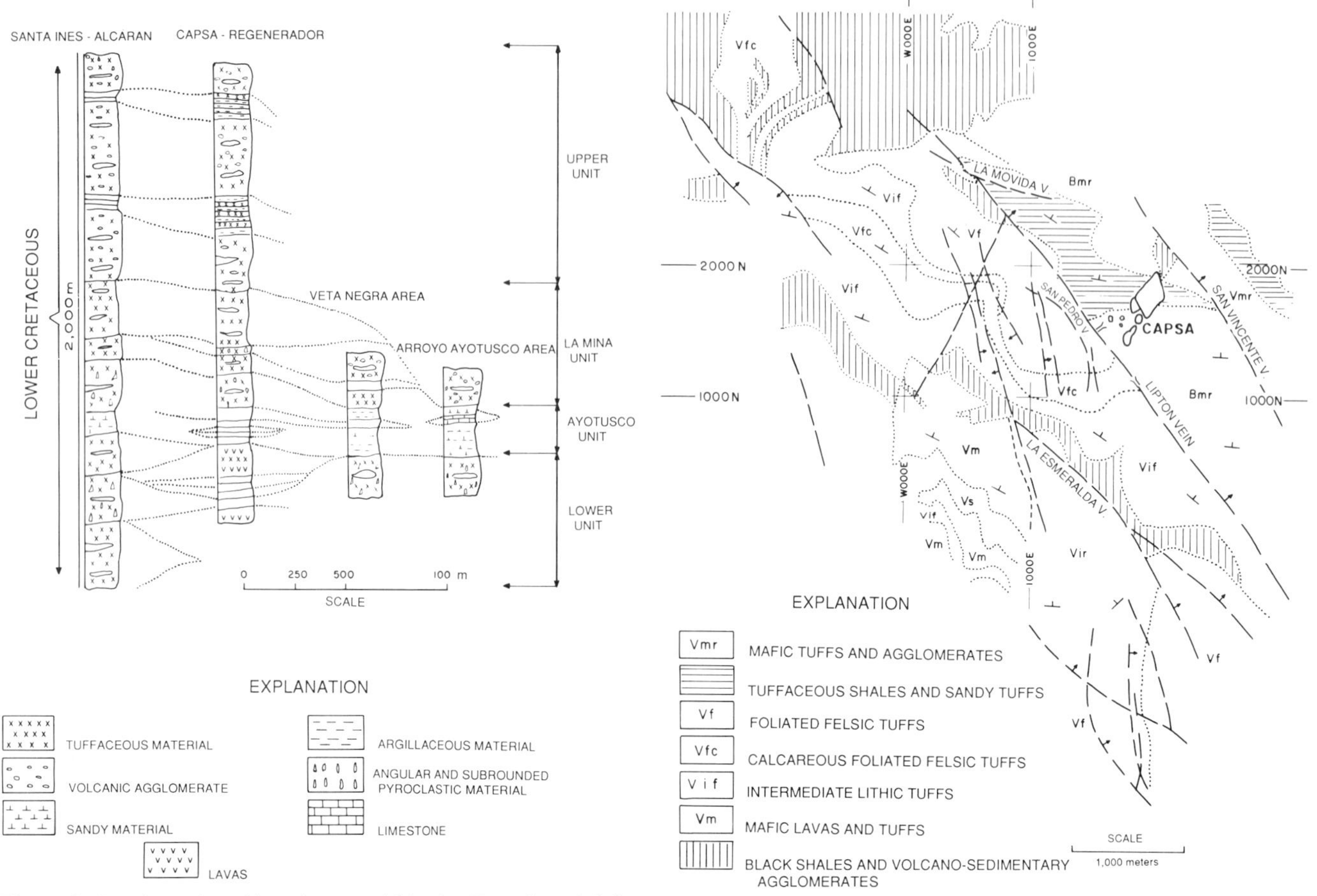

Figure 3. Local stratigraphic columns within the Zacualpan Mining District.

Figure 4. Map of the principal vein systems in the Zacualpan Mining District.

cualli (pyramid) and *pan* (above): "place above the pyramid." Very small-scale mining of ore veins began in pre-Hispanic times; the mining district as such was founded by the Spaniards in 1528; in September or November 1532 it became the first "Real de Minas" of Mexico and consequently of America.

No production figures are available for the first four centuries, but innumerable bonanzas were reported at different localities over the years. Production figures for one of the district's ten or eleven mine zones suggest its economic importance: from 1967 to 1982 the Guadalupe, Pachuqueño, and Capsa-Regenerador Mines (Fig. 2; the only ones in operation at present) produced approximately one million tons of ores from which 9,000,000 oz of silver were recovered. Production prior to this period is estimated to have been similar, though with somewhat lower volumes.

REGIONAL GEOLOGY

Regionally outcropping metavolcanic and metasedimentary rocks consist of tuffs and mainly andesite-dacite-rhyolite lavas and agglomerates, and black shales, volcanic-sedimentary agglomerates, sandstones, and scarce limestones, respectively (Fig. 2). Transitions from one lithology to another, facies changes, and interdigitation are common.

These rocks have been locally grouped into four lithologic units: Lower Unit, Ayotusco Unit, La Mina Unit, and Upper Unit. The Lower Unit is composed of mafic lavas and tuffs with intermediate lithic tuffs, "foliated" felsic tuffs, and intercalated tuffaceous shales and sandy tuffs. It is conformably overlain by the Ayotusco Unit: thinly bedded black carbonaceous shales with disseminated pyrite, volcanic-sedimentary agglomerate lenses, and calcareous, sandy, and tuffaceous lenses. The La Mina Unit, conformable above the Ayotusco Unit, includes "foliated" felsic tuffs, calcareous felsic tuffs, and intermediate lithic tuffs. The La Mina Unit is topped conformably by the Upper Unit, which consists of mafic agglomerates and tuffs, mafic crystalline tuffs and intercalated tuffaceous shales, and sandy tuffs. Argillaceous lenses occur commonly throughout the sequence, which totals >1,600 m of measured thickness (Fig. 3).

The whole sequence is regionally metamorphosed to the near-greenschist facies (prehnite-pumpellyite). The rocks are genetically associated with island-arc submarine volcanism that was

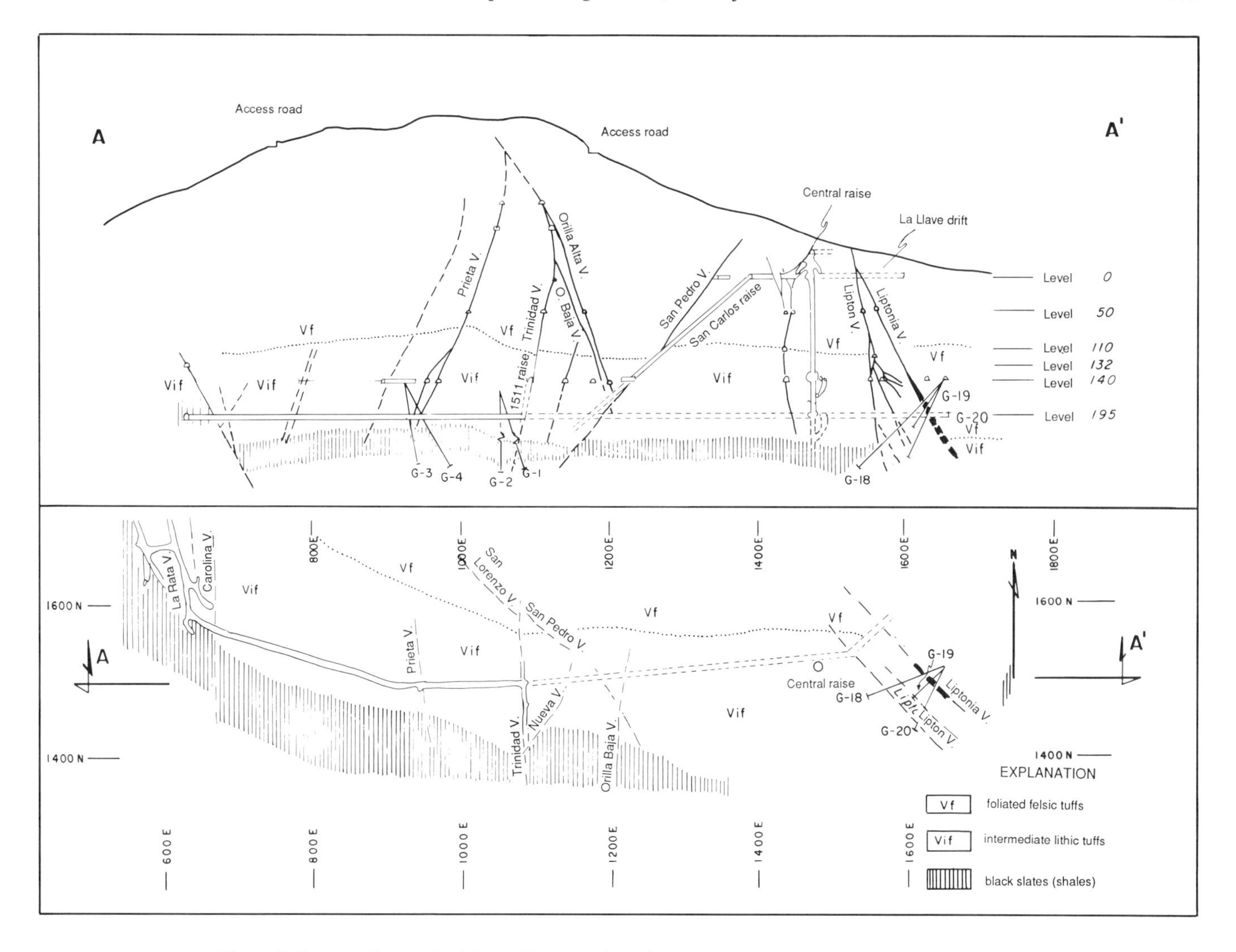

Figure 5. Cross section (top) of Capsa-Regenerador Mine area, and plan view (bottom) of Level 195.

active at the Pacific coast during the Late Jurassic and Early Cretaceous.

GEOLOGY OF THE CAPSA-REGENERADOR DEPOSIT

The Zacualpan District ore deposits are veins, bonanzas, or epithermal deposits (i.e., generated at low temperatures and at shallow to intermediate depths); ore shoots are separated by barren parts within the same structure. Their association with young (Tertiary) volcanic rocks is not evident, the latter having disappeared by erosion in the vicinity of the deposits (Figs. 4, 5).

Main production is silver and some lead, with accessory gold and zinc. Vein ore minerals in the oxidation zone are native silver, kerargyrite, cerussite, and hemimorphite in gangue composed of hematite, goethite, jarosite, and various limonites with Fe and Mn, besides secondary calcite and aragonite. In the hypogenic zone, the minerals are argentite, proustite-pyrargirite, polybasite, native silver, galena, sphalerite, and scarce chalcopyrite in gangue composed of calcite, quartz, dolomite, rhodocrosite, fluorite, and chalcedony. Gold appears in native form and as electrum.

Main alterations are propylitization, sericitization, silicification, and carbonatization, the first two on a regional scale and the others much more locally.

The mineralization appears in veins or lodes filling preexisting cavities of extensional fault and/or fracture zones or longitudinal conjugated systems. These tabular-shaped bodies, dipping 55° to almost vertical (average 70°), occur in the La Mina Unit host rocks described above.

Mineralized structures group into three generally well-defined systems trending northwest, north-south, and northeast (Table 1).

TABLE 1. DESCRIPTIONS OF MINERALIZED STRUCTURES

NORTHWEST TREND (Lipton Vein, La Movida Vein, etc.)	
a. Orientation	N45°W, with variation to N60°E.
b. Dip	60°NE; variation dips SW.
c. Thickness and grades	1 to 6+ m; low-medium, occasionally high-grade.
d. Lateral extension of structures	Regional, more than 2 km in length.
e. Dimensions of known ore shoots	Vertical: 80 to 220 m; lateral: 135 to 300 m; oval in longitudinal cross-section, with major axes usually laterally oriented.
f. Potential	High tonnage; low to medium grade.
NORTH-SOUTH TREND (Trinidad Vein, La Rata, etc)	
a. Orientation	North-south, with variation to N20°W.
b. Dip	Average 70°W or E (conjugated system).
c. Thickness and grades	20 cm to 1.5 m; generally high to very high grade.
d. Lateral extension of structures	Locally 500 m average length.
e. Dimensions of known ore shoots	Vertical: 50 to 335 m; lateral: 50 to 455 m; semicircular and oval in longitudinal cross-section; major axis laterally oriented.
f. Potential	Low to medium tonnage; high grade.
NORTHEAST TREND (San Diego Vein, Milton Vein, etc.)	
a. Orientation	N25°E.
b. Dip	65°NE.
c. Thickness and grades	40 cm to 2.5+ m; medium to high grade (200 to 380 g/ton Ag).
d. Lateral extension of structures	400 to 700 m in length.
e. Dimensions of known ore shoots	Vertical: 100 to 180 m; lateral: 150 to 500 m; oval cross-section; major axis usually laterally oriented.

REFERENCES CITED

Campa M., F., and others, 1978, La evolución tectónica de Tierra Caliente: Geological Society of Mexico Bulletin, v. 39, p. 33–42.

Ortega y Larsen, C., 1933, Generalidades sobre las zonas de Zacualpan, Estado de México y Tetipac, Gro.: Revista Industrial, v. 1, p. 101–119.

Robles R., R., 1935, Generalidades sobre Zacualpan y paragénesis de la veta "La Esmeralda": Mexico, D.F., Universidad Nacional Autonoma de México, Boletín Minero, 132 p.

Villafana, J., 1909, Las minas de Coronas y Anexas, pertenecientes a la Seguranza Mining Compañía: Memoir Sociedad Cientifica Antonio Alzate, v. 28, p. 23–51.

Villarello J., D., 1906, Descripción de algunas minas de Zacualpan, Estado de México: Memoir Sociedad Cientifica Antonio Alzate, v. 23, p. 251–266.

Unpublished sources of information

Diaz García V., M., 1977, El contacto esquisto Taxco–Roca Verde, Taxco Viejo, en la región de Zacualpan, Estado de México [professional thesis]: México, D.F., Universidad Nacional Autonoma de Mexico, p. 60–79.

Noguez A., B., Oropeza O., C., and Herrera G., B., 1982, Estudio geológico superficial del Distrito Minero de Zacualpan, Edt. de México: Industrias Peñoles, internal report.

Novelo, L. F., 1979, Estudio geológico-estructural del Distrito Minero de Zacualpan, Estado de México: Industrias Peñoles, internal report.

Oropeza O., C., 1983, Prospección geológica en el Distrito Minero de Zacualpan, Estado de México [professional thesis]: México, D.F., Universidad Nacional Autonoma de México, 126 p.

Printed in U.S.A.

The Geology of North America
Vol. P-3, Economic Geology, Mexico
The Geological Society of America, 1991

Chapter 48

Geology of the Tizapa Ag, Zn, Pb, Cu, Cd, and Au massive polymetallic sulfides, Zacazonapan, Mexico

J. de J. Parga P. and J. de J. Rodríguez S.
Consejo de Recursos Minerales, Centro Minero Nacional, Carretera México-Pachuca, Pachuca, Hgo., Mexico

INTRODUCTION

In 1977, Guillermo P. Salas, general director of the Mineral Resources Council, commissioned special studies manager José Luis Lee Moreno to evaluate the geologic-mining potential of the Neovolcanic Axis Metallogenetic Province of Salas (1975). Detailed studies of ERTS-1 satellite images over approximately 10,000 km^2 of the province had revealed important tectonic features suggesting as yet undiscovered ore deposits. The evaluation project began in 1977 under Raúl Cruz Ríos.

Regional geology (Nieto and others, 1977) indicated extensive volcanic-sedimentary outcrops at the boundary with the Sierra Madre del Sur province, and exploration for volcanogenic massive sulfides was recommended. By mid-1978 the first Tizapa samples had tested successfully, and the Metamorphic Rocks Project was set up in January 1979. Since that time, exploration and preliminary evaluation of the deposit have continued without interruption.

Tizapa lies 67 km directly 60° southwest of Toluca and 4 km southeast of Zacazonapan in southwestern Mexico State (Fig. 1). Highway 130 from Toluca leads to Temascaltepec, from where unpaved roadways reach Zacazonapan.

Fieldwork included a semidetailed regional geologic survey using part of the "Valle de Bravo" Sheet E-14-A-46 (Direccion General de Estudios del Territorio Nacional, scale 1:50,000) as topographic base; subsequent 1:1,000 topographic mapping over approximately 1 km^2 as a basis for detailed geological and aerial geophysical electric surveys; direct diamond-drilling with core recovery; and a 30-m prospecting tunnel.

REGIONAL GEOLOGY

The Tizapa metamorphic sequence crops out from northern Guerrero State throughout southwest Mexico State and eastern Michoacán State (De Cserna, 1978). In Tizapa the sequence is unconformable over granitic basement (augen gneiss); an average estimated thickness of 2,000 m includes quartzo-feldspathic schists, graphitic phyllites, chlorite schists, muscovite schists, tremolite/actinolite schists, biotite schists, and metarhyolites, overlain with apparent conformity by Cretaceous volcanic and sedimentary rocks, and on occasion unconformably by Plio-Quaternary basaltic andesite and olivine basalt flows (Fig. 2). The metamorphic sequence is intruded by a hornblende diorite stock, rhyolite bodies, and dikes of acid to intermediate composition.

The sequence underwent three compressional deformations (Parga, 1981): the first (D_1) is characterized by penetrative axial foliation (S_1) associated with isoclinal folding; the second (D_2) folds the S_1 foliation into isometric angular folds and locally produces fold cleavage (S_2); the third (D_3) is defined by a 10 to 42° northwest-dipping regional anticline. Two subsequent extensional tectonic phases occurred, probably from the Miocene to the Quaternary.

The sequence was involved in three metamorphic episodes. The first and strongest (M_1, syntectonic with D_1) reached the lower amphibolite facies and is estimated to have occurred at temperatures of 500 to 550° C, pressures of about 2 kb, under approximately 7,500 m of rock cover, and in an area of high geothermal gradient (approximately 70° C/km), typical of low pressure/high temperature areas. The second metamorphic event (M_2, syntectonic with D_2) remained in the greenschist facies at estimated temperatures of 350 to 400°C and pressures below 2 kb. The third event was an extensive, strong and irregular retrograde metamorphism affecting the whole sequence, possibly prior to the D_3 deformation.

The Tizapa metamorphic rocks are tentatively assigned a late Paleozoic age on the basis of their stratigraphic and structural position and degree of deformation and metamorphism.

The Tizapa deposit contains lenticular massive sulfide bodies lodged in graphitic phyllites, muscovite schists, and chlorite-muscovite schists. Based on correlation of several bodies detected by the drill, orebodies may be centimeters to 20 m thick (4 m on the average) and extend laterally for tens to 400 m. Underground these bodies are affected by what appears to be a gentle secondary folding in the host rocks, which the 100 by 100 m drill grid was not able to define.

Parga P., J. de J., and Rodríguez S., J. de J., 1991, Geology of the Tizapa Ag, Zn, Pb, Cu, Cd, and Au massive polymetallic sulfides, Zacazonapan, Mexico, *in* Salas, G. P., ed., Economic Geology, Mexico: Boulder, Colorado, Geological Society of America, The Geology of North America, v. P-3.

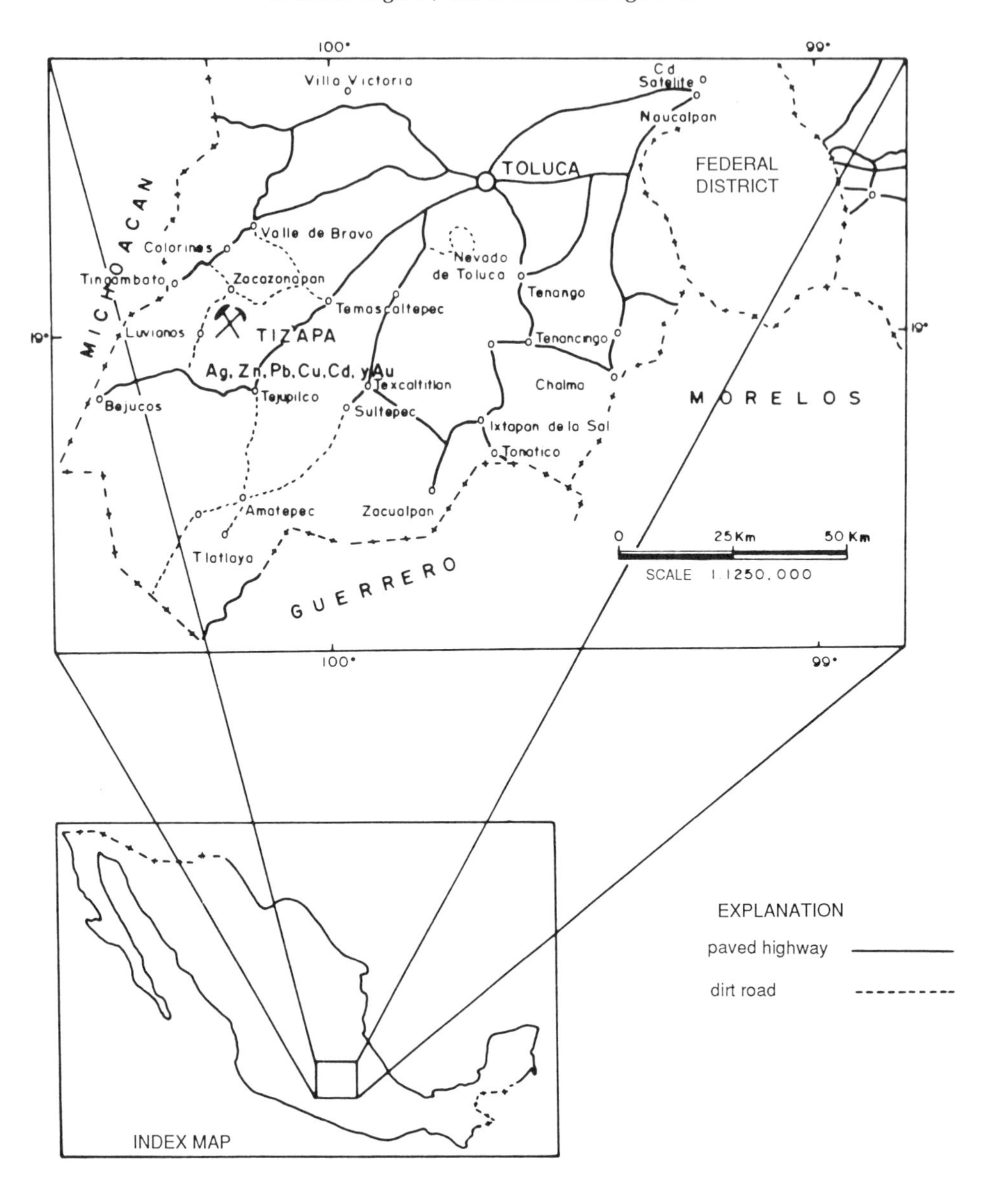

Figure 1. Map showing location of the Tizapa area, Mexico.

Distribution of orebodies

Sulfide bodies have been encountered in drillholes over an area approximately 400 by 400 m, mostly below the basalts that form the cap of the Mesa de Tizapa (Fig. 2). The true lateral and vertical range of the deposit is still unknown, but its extension to the north and northwest seems most probable; other sulfide bodies may be present under the Mesa de Tenayac basalts about 2 km to the southeast.

Most of the drilling was done in Mesa de Tizapa, a 30- to 85-m-thick volcanic cover that together with an underlying 5- to 40-m thickness of poorly consolidated lacustrine sediments forms an overburden 50 to more than 100 m thick covering the metamorphic sequence. The deepst orebody is 231 m below the Mesa de Tizapa surface, and the shallowest is 75.65 m; however, massive sulfide lenses crop out at the Arroyo Tizapa.

Toward the south a normal fault zone displacing the orebodies at depth, and small rhyolite intrusives, combine to prevent clear definition of their distribution in that direction (Fig. 2).

Relations with the host rocks

Graphitic phyllites, muscovite schists, and chlorite-muscovite schists are the sulfide-bearing rocks (Fig. 3); the orebodies are conformable with the prevailing foliation of the host rocks and are often lodged at the contacts.

In general, orebodies are found in both phyllites and chlorite- and/or muscovite schists. The fact that most of the drilled bodies occur in the schists may indicate associated sericitization in many cases, but this is difficult to prove due to the regional metamorphic alteration. Drilling detected barren intervals (of a few to over 50 m) separating the orebodies.

Mineralogy and texture of the ores

Tizapa polymetallic sulfides consist of pyrite, sphalerite, galena, chalcopyrite, and silver-bearing tetrahedrite (freibergite) in order of decreasing abundance, all of which usually show fine-grained massive textures in which 1-mm anhedral pyrite and sphalerite crystals are sometimes clearly visible. Megascopically, small dark gray and yellowish bands representing sphalerite and pyrite-chalcopyrite zones, respectively, are common and also conformable with the predominant host-rock foliation. Sulfides in the orebodies do not exhibit zoning; the silver, lead, zinc, and copper values reported in chemical analyses show no defined trend. In general, the mineralization occurs as black ore; yellow ore containing commercial copper only was encountered in some drill holes.

Under the microscope, the ores show granular texture defined mostly by anhedral pyrite crystals embedded in anhedral sphalerite and some chalcopyrite, tetrahedrite (var. freibergite) occurs as minute anhedral grains included in sphalerite; the latter also appears as euhedral crystals with galena intergrowths and inclusions. Tin, yittrium, and arsenic have been reported quantitatively with no mineral identification.

Alteration

Zoned alteration patterns surround kuroko-type massive sulfides and Archaean volcanogenic massive sulfide deposits in Canada. Sericite is always present at the core (siliceous ore) of the kuroko alteration pattern, surrounded by a zone of chlorite and sericite. In contrast, the strongest alteration in the Archaean chimneys consists almost exclusively of chlorite and quartz; sericite and chlorite occur as alteration at the marginal zone.

Unfortunately it has not been possible to define alteration zones in the Tizapa deposit; sericitization, chloritization, kaolinization, and silicification are all present but cannot be differentiated from the metamorphic alterations, owing to lack of the necessary equipment for oxygen and hydrogen isotope quantification and fluid inclusion studies.

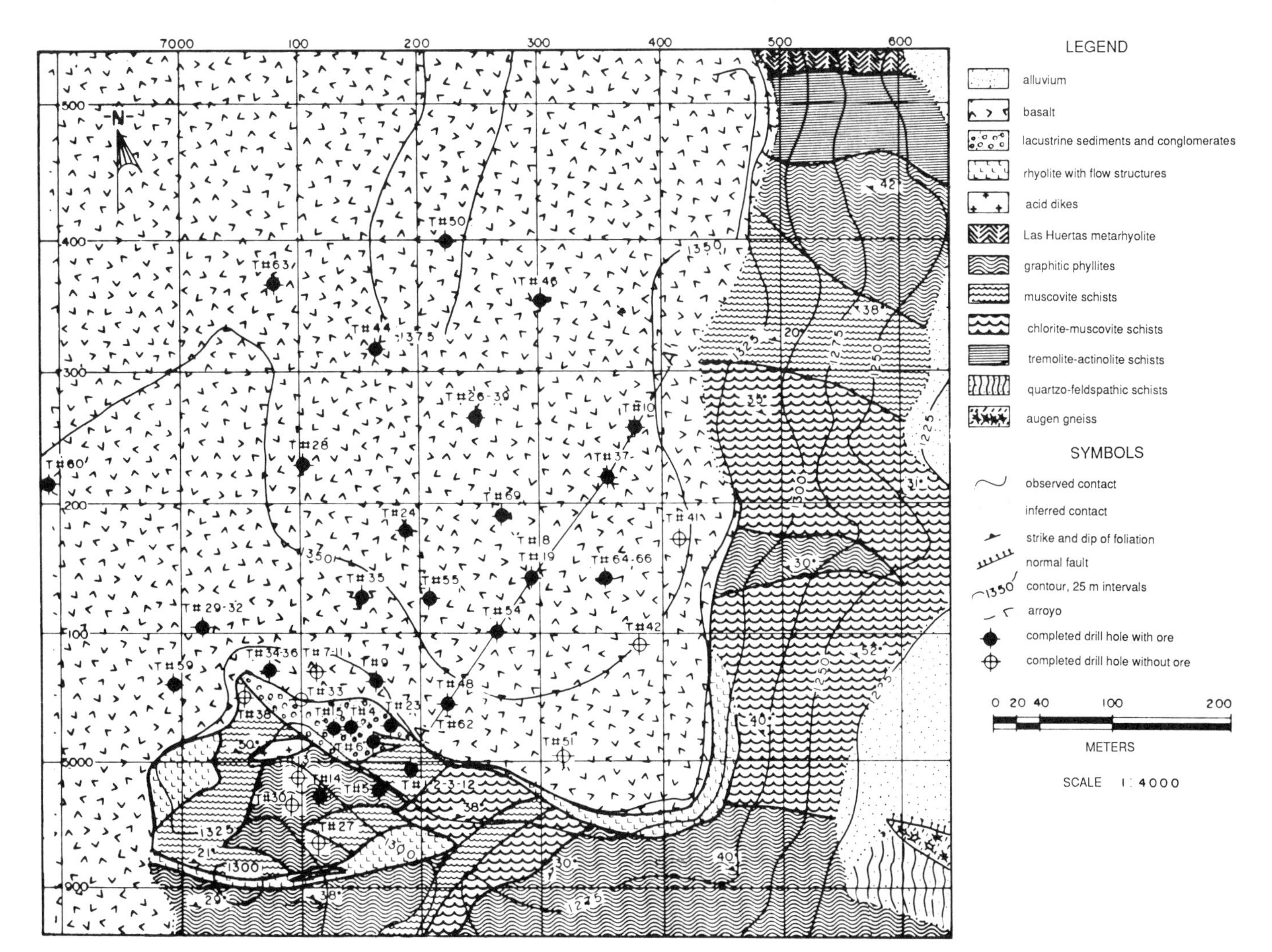

Figure 2. Geologic map of the Tizapa area.

GENESIS OF THE DEPOSIT

The origin of the Tizapa mineralization is difficult to prove. The intrusion of sulfide bodies by rhyolites and aplitic dikes demonstrates that mineralization was pre-Tertiary; however, its relation with the host-rock tectonic evolution remains unclear. Some drill cores show ores affected at least by the final folding of the metamorphic rocks (Fig. 3), probably in Late Cretaceous to Early Tertiary time.

Stratiform massive sulfide deposits are generated on or immediately beneath the sea floor by metal-rich exhalations of high-temperature hydrothermal systems. Their common characteristic is that they are all more or less stratified, lenticular, fine-grained equigranular masses of mostly iron, zinc, lead, and/or copper sulfides, frequently associated with gold, silver, and lesser proportions of cadmium, tin, indium, arsenic, antimony, and other minerals, which are occasionally recoverable as subproducts.

On the basis of host-rock lithology, these stratiform deposits are classified as "volcanogenic" or "sedimentogenic" massive sulfides (Franklin and others, 1981). Recent investigation shows them to be genetically more closely related to submarine hydrothermal activity than to either volcanic or sedimentary events, the conclusion being that the host-rock composition has little direct influence on their origin.

The Tizapa deposits are here classified tentatively as "sedimentogenic massive sulfides" of Zn, Pb, and Cu with associated Ag, Au, and Cd on the following basis: (a) the typically lenticular shape and conformity of the orebodies with respect to the metasedimentary host rocks; (b) the rhythmic stratification (banding) of the ores; (c) contemporaneous metamorphism and folding affecting both host rocks and orebodies at the same time; (d) the presence of pressure shadows surrounding some sulfide crystals, indicating pretectonic deposition; and (e) brecciation structures in both orebodies and metasedimentary host rocks are evidence of phreatic explosions, which reflect violent boiling that brecciated the orebodies and/or caused gravity slides of ores and sediments before their consolidation.

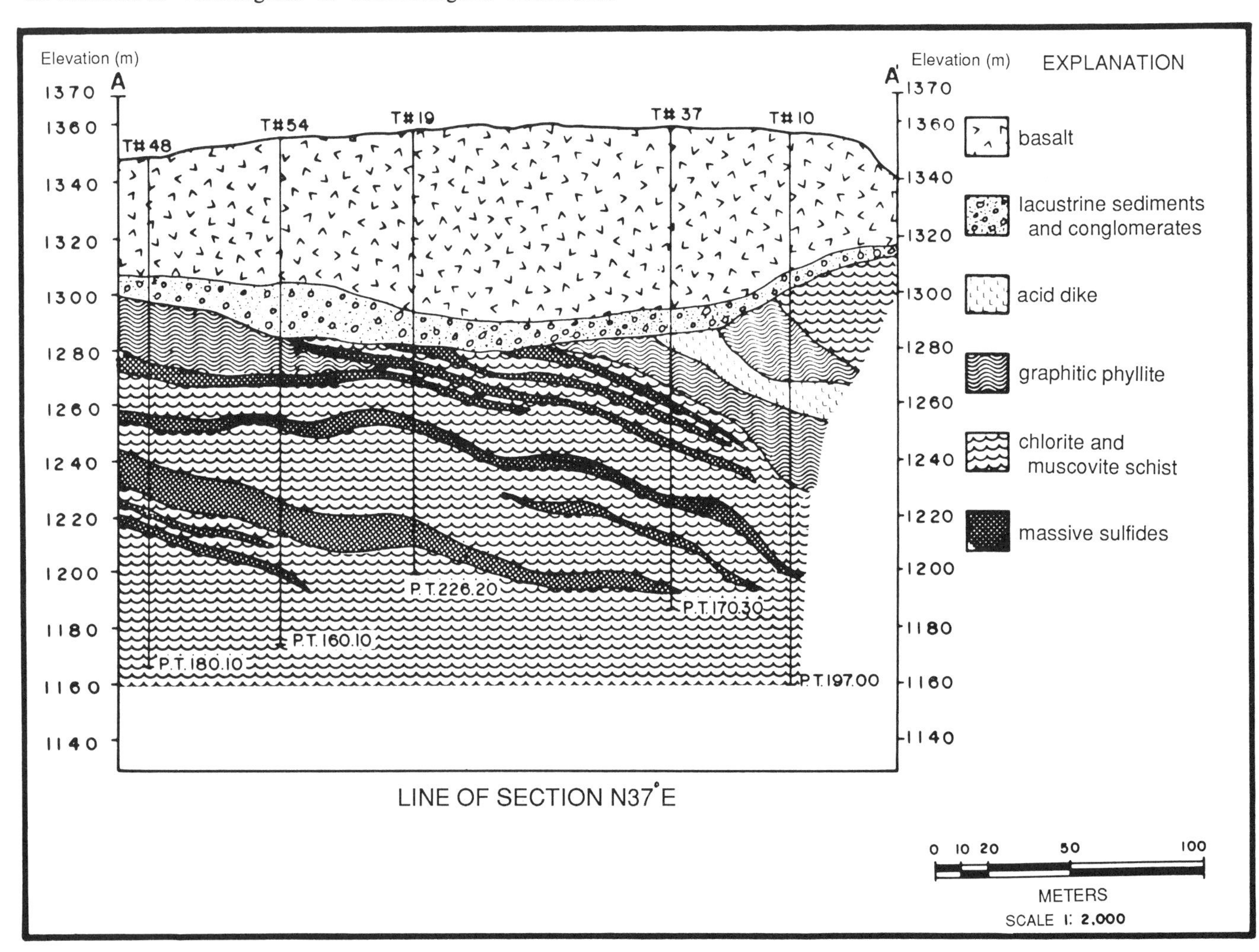

Figure 3. Representative geologic cross section of the Tizapa ore deposit.

ECONOMIC IMPORTANCE OF THE REGION

Mexico's metamorphic rocks have not been sufficiently studied and are therefore not well known; their commercial ore prospects remain underestimated. Because metamorphic events mask and often destroy previous elements that permit identification of the original rocks and environments, the study of metamorphic rocks is traditionally demanding and frequently speculative, besides requiring more sophisticated field and laboratory techniques and equipment than usual, which partly explains the lack of experience in Mexico up to now in this field.

Nevertheless, it is now well known that numerous originally syngenetic (i.e., massive sulfide) deposits in metamorphic rocks are significant worldwide producers of large volumes of silver, lead, zinc, copper, and barite. Also, they occur as groups of several lenses not necessarily at the same stratigraphic level, and they evolved in very active submarine hydrothermal environments, commonly in association with volcanism.

Such environments are usually related to, and associated with, subduction zones and at one time or another become involved in the deformations and tectonic transport inherent to these, which explains the metamorphic nature of most massive sulfide deposits.

In Mexico, regionally extensive volcanic-sedimentary rocks developed in the appropriate geologic environments for volcanogenic and sedimentogenic massive sulfides to evolve: near Tizapa, the Santa Rosa deposit and several folded and metamorphosed massive sulfide lenses in the Almoloya de Las Granadas area; farther north in the state of Guerrero, the Rey de Plata and Campo Morado deposits in volcanic-sedimentary rocks of low metamorphic grade are possibly contemporaneous with the Tizapa sequence; La Dicha, in the Ixcuinatoyac area west of Ocotito (Guerrero) is similar, with a series of massive sulfide bodies in metasedimentary rocks and probably tuffaceous horizons of the late Paleozoic(?) Ixcuinatoyac Formation.

The Tizapa deposit is about 300 m structurally higher than the augengneiss, and the Santa Rosa massive sulfide occurs approximately 1,000 m above the Tizapa. In theory this implies a metamorphosed volcanic-sedimentary thickness of at least 1,000 m and more than 10 km long that shows prospects of containing polymetallic massive sulfides. Additional hydrothermal mineralization appears as small, isolated quartz veins of low gold-silver, lead, copper, and zinc content filling narrow fractures, which however, do not persist along strike nor at depth. The upper volcanic-sedimentary sequence also shows indications of hypogene mineralization.

MINERALIZATION

The orebodies consist of sulfides only and are therefore very regular, at least mineralogically, allowing uniformity in the metallurgical treatment. Grades are also more or less regular, although some sulfides with low Zn, Pb, and Ag content and more than 1 percent Cu may require separate treatment or mixture with other sulfides to homogenize for milling.

Lateral continuity and thicknesses of the orebodies remain unclear in some cases; additional drilling is required for improved correlation of orebodies drilled to date. The continuity of the deposit has been verified over approximately 400 m north-south and 400 m east-west.

CONCLUSIONS

a. The Tizapa low- to medium-grade metamorphic sequence contains the following parent rocks: basal plutonic granodiorite unconformably underlying arkosic sandstone that grades upward to carbonaceous pelitic sediments with intercalated intermediate tuffs, graywackes, pelitic sediments, rhyolites, and limestones at the top.

b. Igneous rocks include a hornblende diorite stock, small isolated rhyolites with flow structure, and numerous acid to intermediate dikes and basaltic flows, unconformably covering a large part of the metamorphic rocks of the region.

c. The Tizapa area exhibits one extensional and three compressional deformations, two progressive metamorphic episodes syntectonic with the first two deformations, and regressive metamorphism.

d. The Tizapa deposit is made up of a series of lenticular, stratiform, polymetallic massive sulfide orebodies conformable with the host rocks (graphitic phyllites, muscovite schists, and chlorite-muscovite schists).

e. The true dimensions of the deposit are yet to be defined. One or more orebodies show promise of continuing laterally west, northwest, and northeast. There is a real possibility of finding one or more orebodies at depth.

f. Geochemical mercury vapor anomalies suggest the possible continuation toward the southwest of some orebodies or the presence of others following the structural trend of those already known.

g. Tizapa is tentatively classified as a Zn-Pb-Cu "sedimentogenic massive sulfide" with Ag, Au, and Cd.

h. Both the deposit and the host rocks are provisionally assigned to the Permo-Triassic on the basis of stratigraphic and structural positions, degrees of deformation, and metamorphism.

REFERENCES

De Cserna, Z., 1978, Notas sobre la geología de la región comprendida entre Iguala, Ciudad Altamirano y Temascaltepec; Estados de Guerrero y México, *in* Sociedad Geologica Mexicana, Field guide of the geological excursion to Tierra Caliente, States of Guerrero and México, p. 1–25.

Franklin, J. M., Sangster, D. M., and Lydon, J. W., 1981, Volcanic-associated massive sulfide deposits, *in* Skinner, B. J., ed., Economic Geology; 75th Anniversary Volume, p. 485–627.

Salas, G. P., 1975, Carta y provincias metalogenéticas de la República Mexicana: México, D.F., Consejo de Recursos Minerales Publication 21-E, 242 p.

Unpublished sources of information

Nieto O., J., Cruz, R., Colorado, D., Figueroa, M., and González, P., 1977, Elementos tectónicos y metalogenéticos para considerar el potencial económico minero de la región comprendida entre Zacualpan y El Oro, México: México, D.F., Consejo de Recursos Minerales, VI Seminario Interno Sobre Exploración Geológico Minera, p. 111–128.

Parga, P.J.J., 1981, Geología del área de Tizapa, Municipio de Zacazonapan, México [M.S. thesis]: México, D.F., Universidad Nacional Autonoma de México, Facultad des Ciencas (Geología), 135 p.

Printed in U.S.A.

The Geology of North America
Vol. P-3, Economic Geology, Mexico
The Geological Society of America, 1991

Chapter 49

Taxco Mining District, state of Guerrero

Guillermo P. Salas
Geologo Consultor, Asesor del Consejo de Recursos Minerales, Centro Minero Nacional, Carretera México-Pachuca, Pachuca, Hgo., Mexico, D.F.

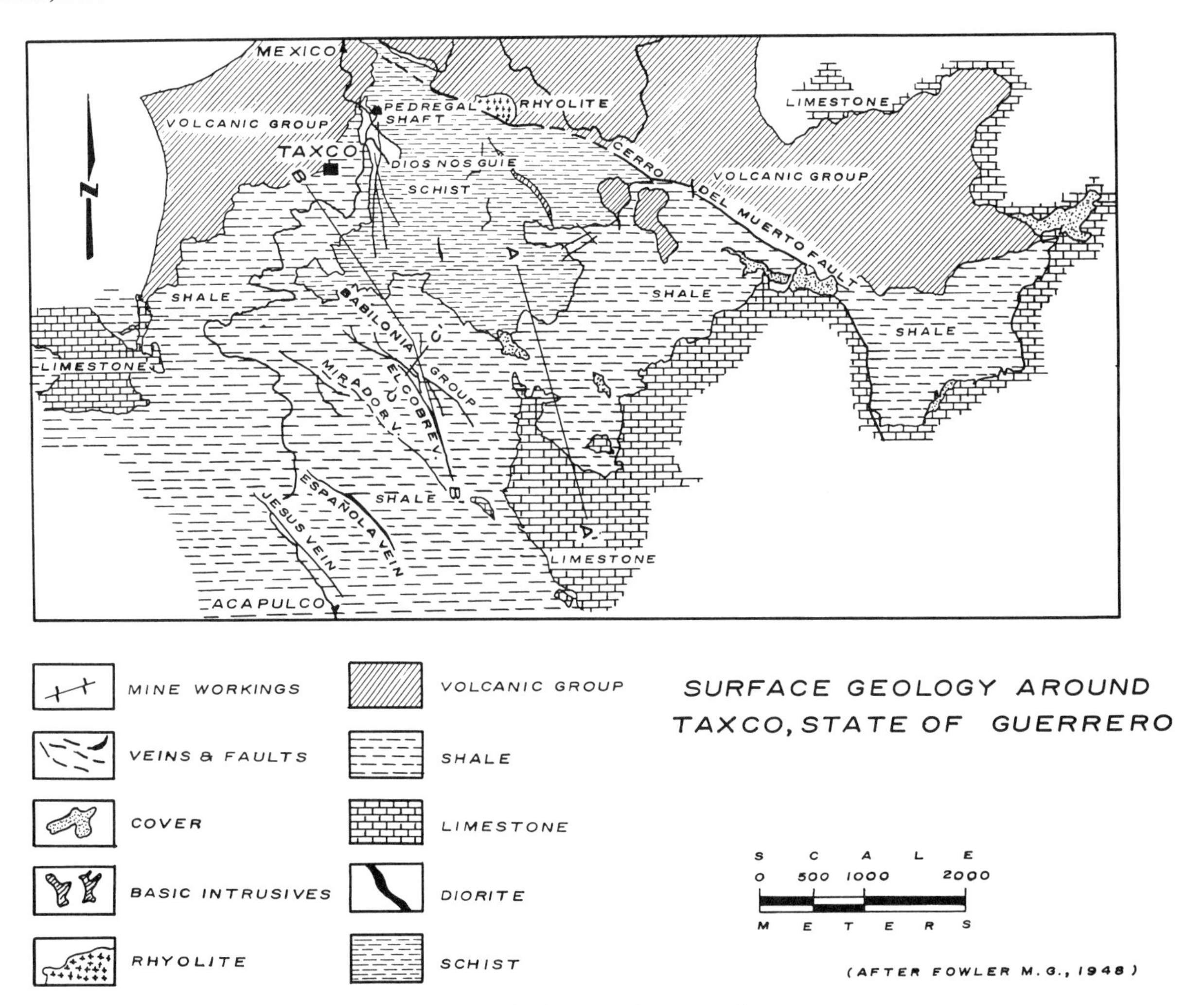

Figure 1. Geologic map of the Taxco Mine area.

INTRODUCTION

The Taxco Mining District is located in the Neovolcanic Axis Metallogenetic Province in northern Guerrero State, at an elevation of 1,700 m (Fig. 1).

This district includes the San Antonio, Guerrero, Remedios, Guadalupe, and Golondrina Mines. In 1534, Conquistadors Juan de La Cabra, Juan Salcedo, Muriel and Hernán Cortés mined and processed silver from the Socavón del Rey (now known as Mina del Pedregal), which was worked sporadically until 1747. In 1802, José de la Borda and José Vicente de Anza discovered a large silver bonanza in Tehuilotepec; José de la Estaca found another in the Juliántra area; these mines were almost inactive from 1802 to 1880. A flotation plant was installed in 1920 by the

Salas, G. P., 1991, Taxco Mining District, state of Guerrero, *in* Salas, G. P., ed., Economic Geology, Mexico: Boulder, Colorado, Geological Society of America, The Geology of North America, v. P-3.

American Smelting and Refining Company. Ores were processed by amalgamation. Industrial Minera México, S.A. (IMMSA), owner of all the properties in the district, which employs 1,298 workers at the IMMSA mines and Taxco District plant, was producing 3,300 t/d in 1984.

REGIONAL GEOLOGY

This district lies in the northern part of the Balsas-Mezcala basin on the flank of the Paleozoic Taxco-Zitácuaro Massif. Basement schists with northeast foliation underlie Lower Cretaceous limestones and Upper Cretaceous shales. This marine section is overlain by thick continental conglomerates, which together with Tertiary and Quaternary pyroclastic and volcanic flows, form the upper part of the sequence. Dikes and bodies of diverse composition and ages intrude the above section, which is also faulted. Faults have been mineralized by hydrothermal solutions filling metasomatic fissures or veins. Subsequent faults cross the generally north-south–trending axes of the folded structures. The rocks range in age from middle Paleozoic to Quaternary.

GEOLOGY OF THE DISTRICT

The generally N30°W to north-south–trending, 40° to 80° northeast- and southeast-dipping orebody rests on Taxco schists (Fig. 2) that may be sericitic, chloritic, or talc-bearing. Dolomites and limestones overlie the basement unconformably and are also unconformably overlain by flysch-type thinly bedded shales and sandstones. These, in turn, conformably underlie a section of continental clastics, lacustrine gypsum, calcareous tuffs, and volcanic conglomerates (Fig. 1).

Three types of intrusives are present: diorite, restricted to the Taxco schists; felsitic dikes, cross-cutting the Mesozoic section as well as some veins; and hornblende diabase, affecting the entire sequence as dikes and small bodies that displace the veins.

Fissure veins throughout the stratigraphic section of Mesozoic schists and sedimentary rocks contain the most important orebodies, such as Veta El Cobre (Fig. 2). Metasomatic veins occur as part of the fissure fills; also, some veins seem derived from distant skarn.

The fissure veins contain silver-bearing galena and sphalerite; secondary polybasite, proustite, and pyrargirite; and pyrite, chalcopyrite, fluorite, stibnite, jamesonite, arsenopyrite, marcasite, quartz, and calcite as gangue.

Stockwork deposits also occur where fault veins are intercepted by secondary faults and/or dikes, which are common in the schists.

The mineralization is of hydrothermal origin; minerals correspond to temperatures of 200 to 300° C in the epithermal phase. It has been suggested that the mineralization was derived from a large igneous body not yet found, and that low-grade lead and zinc disseminated in the schists could have been remobilized by greenschist metamorphism and subsequent intrusions.

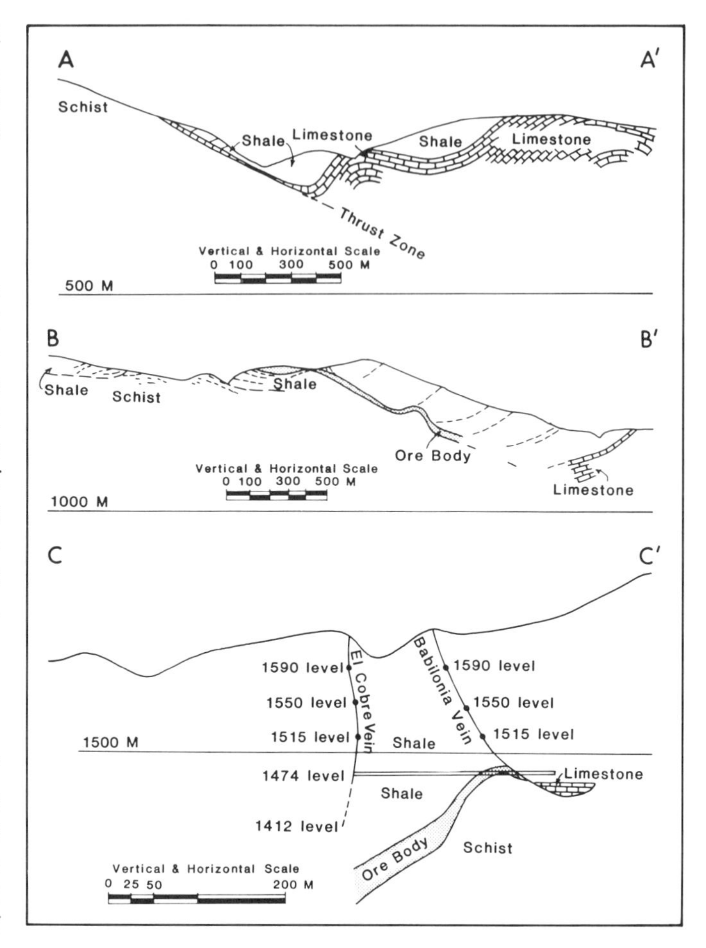

Figure 2. Cross sections of Taxco Mine area; locations shown on Figure 1.

SILVER PRODUCTION

The IMMSA ores, of which the presently most important are from the El Cobre vein, are processed in the abovementioned 3,300 t/d flotation plant. The IMMSA Taxco Unit produced 1,392,826 kg of silver from 1942 to 1980, and 212,533 kg of silver during 1981 and 1982.

REFERENCES CITED

Fowler, G. M., Hernon, R. M., and Stone, E. A., 1948, The Taxco Mining District, Guerrero [Mexico], *in* Dunham, K. C., ed., Symposium on the geology, paragenesis and reserves of lead and zinc: London, 18th International Geological Congress, p. 107–116.

Printed in U.S.A.

The Geology of North America
Vol. P-3, Economic Geology, Mexico
The Geological Society of America, 1991

Chapter 50

Central Mesa Metallogenic Province

Guillermo P. Salas
Geologo Consultor, Asesor del Consejo de Recursos Minerales, Centro Minero Nacional, Carretera México-Pachuca, Pachuca, Hgo., Mexico, D.F.

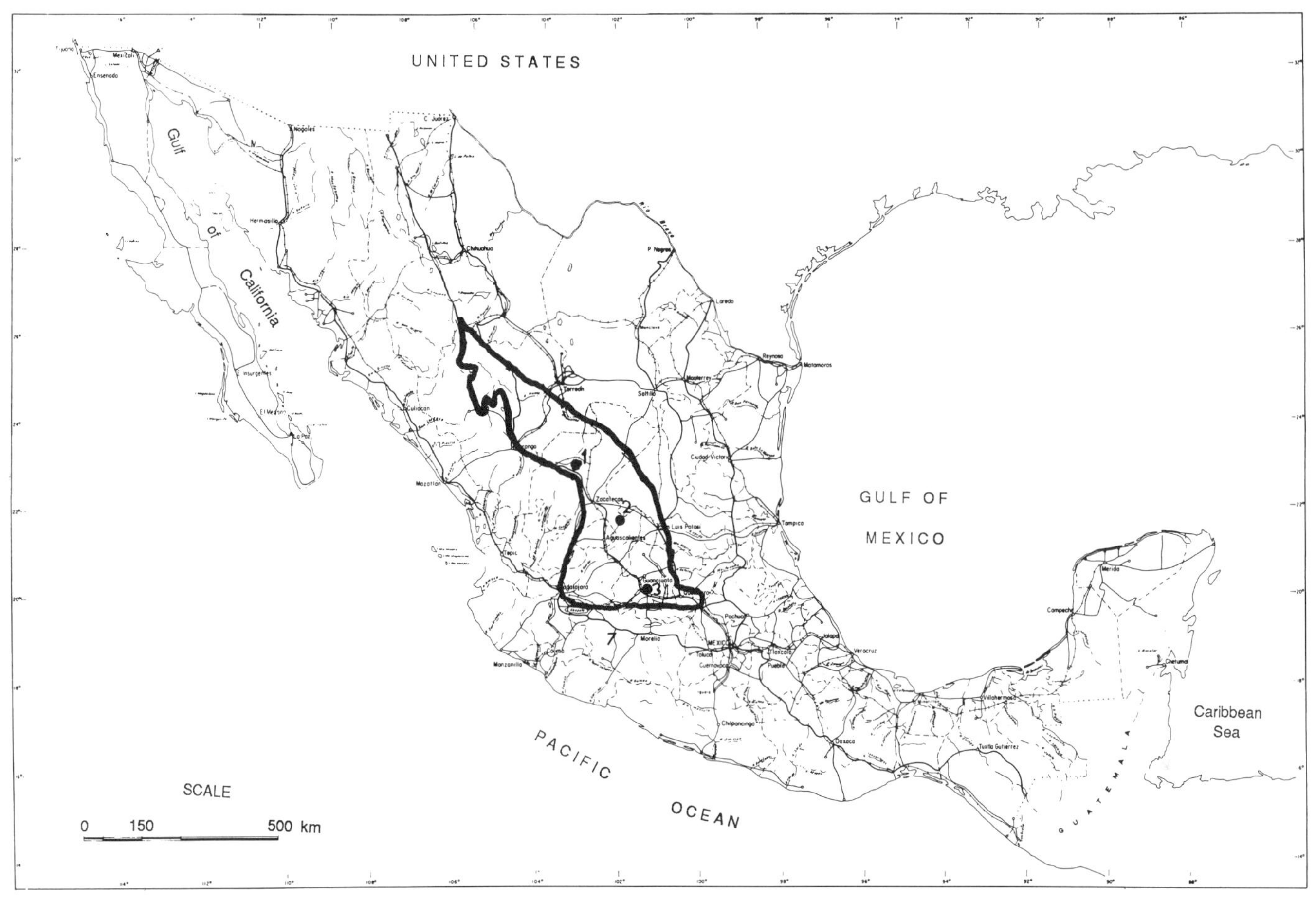

Figure 1. Map showing locations of mineral deposits of the Mesa Central Metallogenic Province discussed in the following chapters: 1, Fresnillo; 2, Real de Angeles; 3, Guanajuato.

INTRODUCTION

The Central Mesa Metallogenic Province does not necessarily fit the established definitions of a metallogenic province. Physiographically, it occupies a plateau between the Sierra Madre Oriental and Sierra Madre Occidental that covers approximately 117,000 km^2. It is bounded on the south by the Neovolcanic Axis province and extends northward to the Zacatecas and Chihuahua borders (Fig. 1).

Metallogenically, this is a remarkable province with abundant epigenetic ore deposits: mercury and fluorite in Querétaro and San Luis Potosí, manganese in Zacatecas and San Luis Potosí, tin in Guanajuato and Durango, and polymetallic deposits in San Luis Potosí and at Fresnillo in Zacatecas. CRM (1980) discusses the geology and some mine districts in detail; only general aspects are outlined here.

Salas, G. P., 1991, Central Mesa Metallogenic Province, *in* Salas, G. P., ed., Economic Geology, Mexico: Boulder, Colorado, Geological Society of America, The Geology of North America, v. P-3.

The geology of the province encloses part of the Sierra Madre Oriental folded limestones, which most probably underlie the province's predominant volcanic sequence. A possibly noteworthy observation is that the andesitic lava flows in the south become more felsic north of Querétaro with more extensive and abundant rhyolites.

Mines or mine districts considered typical of the province are Fresnillo, in Zacatecas; Real de Angeles, in Zacatecas; and Guanajuato, in Guanajuato; these are discussed in Chapters 51 to 53.

REFERENCE CITED

CRM, 1980, Carta y provincias metalogeneticos de la republica Mexicana: México, D.F., Consejo de Recursos Minerales, Boletín 21-E (2nd edition), 200 p.

Printed in U.S.A.

The Geology of North America
Vol. P-3, Economic Geology, Mexico
The Geological Society of America, 1991

Chapter 51

Geology of the Fresnillo Mining District, Zacatecas

E. García M., F. Querol S., and G. K. Lowther
Cía. Fresnillo, S.A. de C.V., Río de la Plata 48 esq. Lerma, Col. Cuauhtemos, 06500 Mexico, D.F.

INTRODUCTION

The Fresnillo Mining District is located in central Zacatecas State about 60 km northwest of the town of Zacatecas at geographic coordinates 23°10′29″ and 102°52′39″ and an average elevation of 2,200 m (Fig. 1). Fresnillo is connected directly with Mexico City by Federal Highway 45 (Panamerican) and the Mexico–Ciudad Juárez railroad; Zacatecas airport is 35 km southeast of the district.

The state of Zacatecas has been renowned for its mines since the time of the Spaniards; expeditions under Juan de Tolosa in 1546 discovered innumerable veins in the area of today's town of Zacatecas. Francisco de Ibarra discovered Fresnillo farther north in 1554, naming it "Aguas del Fresnillo" (Isunza, 1981). Shortly after, he explored the area called Cerro Proaño where Diego Fernández de Proaño had already located some veins. These discoveries were followed by many others in the state, including Sombrerete, and Concepción del Oro (Fig. 1).

According to a Murguía (in Stone and MacCarthy, 1942), Fresnillo ores were mined normally from 1717 to 1751 by numerous small-scale prospectors; operations were suspended in 1757 due mainly to economic problems caused by mine drainage difficulties. Fresnillo mines became government property by state congressional decree in 1830; the governor at the time, Francisco García Salinas, having studied records in Mexico City, reopened mine operations using convicts as mineworkers. Lack of capital and labor in addition to a cholera epidemic that decimated the population in 1833, brought mine drainage and production to a stop. The governor then applied unsuccessfully to London for financial aid to purchase two steam boilers and pressurizing equipment. In late 1833 the state congress offered the Fresnillo mines to English capital for development (Auld and Bychan, 1834). The English-based Compañía Zacatecano-Mexicana ran the mines from 1835 to 1872; Manager José González Echavarría acquired two Cornish made steam-boilers to drain the Fresnillo Beleña and San Francisco Mines, which at the time took 300 men to work the hoists. The first of these engines began work at one side of the Beleña Shaft on December 19, 1836. Two Cornish engines installed at Hacienda Grande (today's Hacienda Proaño) built in 1842 allowed smelting of 140 tons of ore per day. From 1835 to 1872, mineral was hauled through 35 shafts, and mineworks reached the −425-m-level.

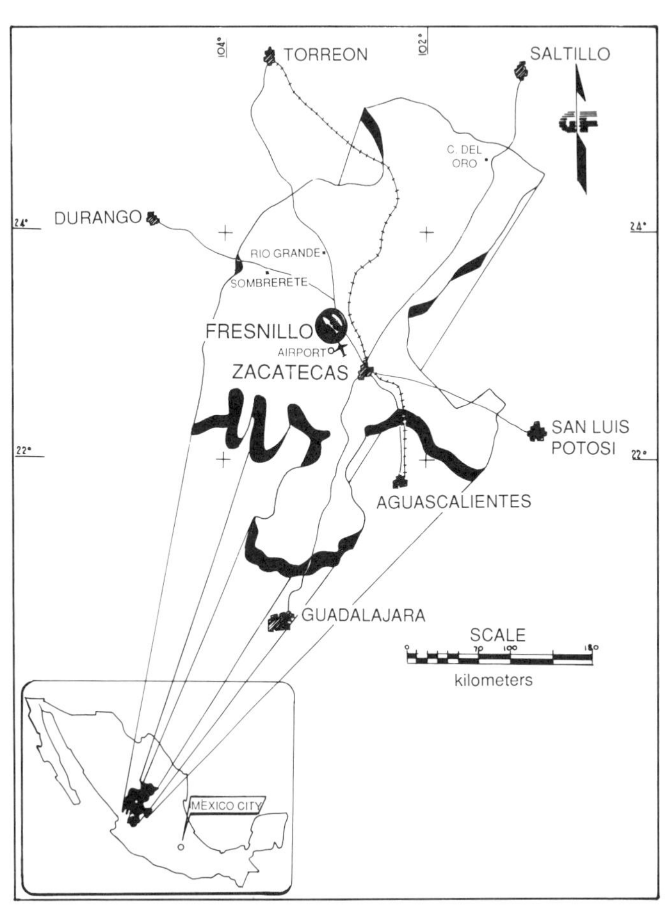

Figure 1. Map showing location of the Fresnillo Mining District.

The company abandoned some of its mines the year after operations were suspended by the war with the United States in 1847; the so-called Reform War and the French occupation depressed Proaño Mine activities, and the company finally left in 1872. In 1900, the New York–based Robert Towne interests acquired 2 million tons of tailings from the patio (yard) process

García M., E., Querol S., F., and Lowther, G. K., 1991, Geology of the Fresnillo Mining District, Zacatecas, *in* Salas, G. P., ed., Economic Geology, Mexico: Boulder, Colorado, Geological Society of America, The Geology of North America, v. P-3.

remaining north of the hacienda. The Fresnillo Company of New York was founded in 1910 after the successful operation of a hyposulfite-leaching plant, and a 400-t/day cyanide-process plant for oxidized ores began to operate in 1912. The Mexican Revolution suspended operations from 1913 to 1919.

In 1919 the Fresnillo Company leased its properties to the English-owned Mexican Corporation who erected a 2,200 ton/day cyanide-process plant, later increased to 3,000 tons/day, that ran from 1921 to 1943. Treatment of sulfide ores since 1926 continues today. In 1929 the Fresnillo Company and Mexican Corporation merged as Fresnillo Company of New York until September 1961, when the business was Mexicanized under the new mining legislation and named Compañía Fresnillo, S.A.

REGIONAL GEOLOGY

Introduction

The Fresnillo Mining District lies on the south edge of the Mesa Central physiographic province, bordering the Sierra Madre Occidental on the west and consequently showing geomorphologic similarities with both provinces (Álvarez, 1969).

Fresnillo is on a slightly northwest-tilted plain covered by detrital material that is nonconsolidated in some places, strongly cemented by caliche in others, and locally up to 30 m thick (Fig. 2). This cover is dissected by numerous intermittent streams that rise in the surrounding high sierras and carry abundant debris during rainy seasons. One of the most conspicuous local elevations is the mature, gently sloping Cerro Proaño, 2,300 m above sea level, covered on its north flank by younger but strongly eroded conglomerates and rhyolite tuffs.

Regional stratigraphy

The oldest outcrops in the region, west of the town of Zacatecas, comprise Triassic metasediments, a Cretaceous intrusion, and Tertiary intrusive and extrusive rocks. The metasediments, mostly phyllites containing conformable quartzite and marble lenses and fine-grained, ammonite-bearing quartzites, all in the greenschist facies, underwent deformation and metamorphism in

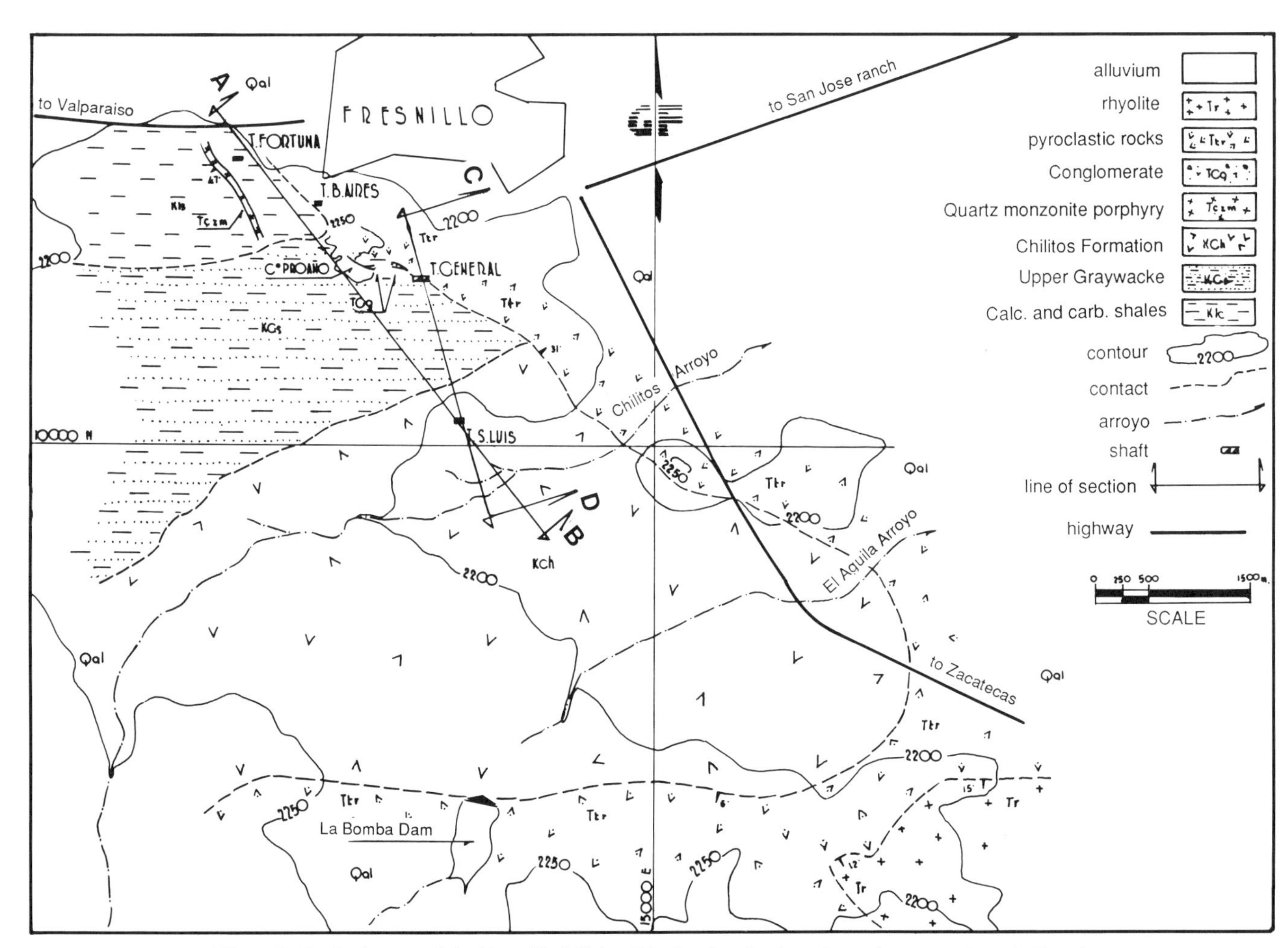

Figure 2. Geologic map of the Fresnillo Mining District showing locations of cross sections A–B and C–D (see Figs. 4 and 7).

Early Jurassic and Late Cretaceous times. The final metamorphic episode has been dated at 73 to 75 Ma. The metasediments are intruded by a shallow hypabyssal diorite dated 74 to 75 Ma (i.e., similar in age to the tectonic event). Subsequent erosion of the metasediments and diorite caused the accumulation of conglomerates cemented by calcite and hematite. Magmatism continued in late Tertiary time with the intrusion of rhyolite dikes, tuffs, flows, and volcanic breccias. Hydrothermal mineralization took place in the metasediments, diorite, conglomerates, and rhyolites shortly after the rhyolite dikes were emplaced (Ranson and others, 1982). As discussed below, apart from the volcanic rocks, no correlation to other areas, either chronological or lithological, is possible with the rocks exposed at Fresnillo. A hitherto unreported tectonic-stratigraphic discontinuity appears to exist between Fresnillo and Zacatecas.

Regional tectonics

The Fresnillo area reflects several tectonic-magmatic episodes since the Triassic, when the older stratigraphic interval was subjected to north-south forces; a second period of strong deformation and incipient metamorphism at the end of the Cretaceous (Campanian-Maastrichtian) coincides with the northeast trend of the Laramide orogeny, which produced the present-day folds and considerable thrusting (López Ramos, 1965); most of the regional structures trend north-south. Plutonic, volcanic, and subvolcanic magmatism with associated hydrothermal activity is evident at every stage from the Jurassic to the late Tertiary.

STRATIGRAPHY

Cretaceous, Tertiary, and Quaternary sedimentary and volcanic rocks crop out in the Fresnillo Mining District. The pronounced lateral variations of the rocks, the absence of key-horizons, strong cross-bedding, and tight folding hinder the definition at the mine area of formal units as recommended by the rules of stratigraphic terminology. The informal units or formations discussed here are used in previous surface and mine geological reports.

Cretaceous

The Early Cretaceous Proaño Group is divided into three members (Stone and MacCarthy, 1942; see Fig. 3).

Lower Graywacke (KGi). This nonoutcropping lower member of the Proaño Group, known only within the mineworks, is a series of graywackes with intercalated shales and limestone lenses. The unit may exceed 700 m in thickness, and its lower part is unknown; the upper contact is conformable with the overlying member.

Calcareous and carbonaceous shales (KLc). This forms the middle member of the Proaño Group. It is composed of laminar dark-gray shales and intercalated limestones in the southeast and also intercalated graywackes toward the northwest; the interval is ill-defined owing to lateral changes that take place over short distances. The unit's contacts at the surface seem to outline a generally northwest-trending, southeast-dipping anticline. Placement of the upper contact is based on diamond-drillholes and minework projections at different levels. The unit's maximum thickness is estimated at approximately 300 m. There seems to be no unconformity either with the underlying Lower Graywacke or the overlying Upper Graywacke.

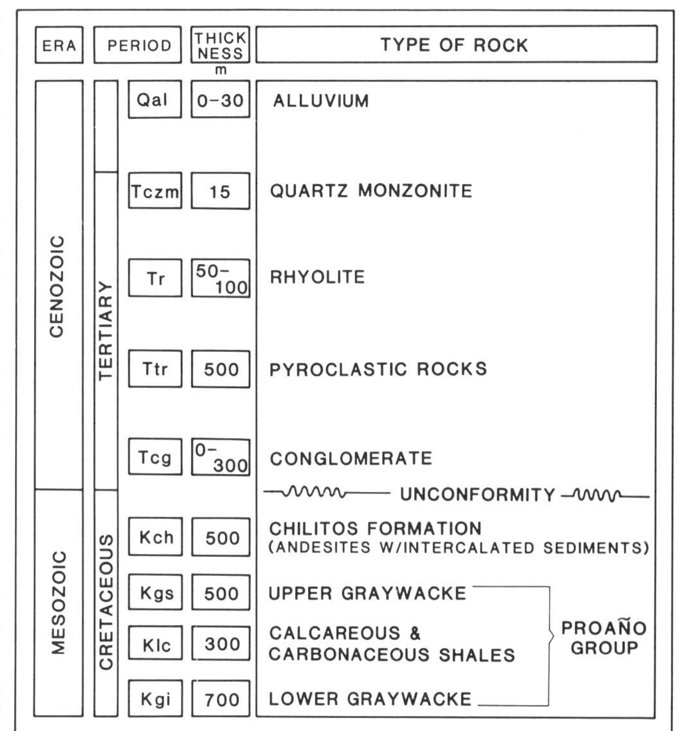

Figure 3. Stratigraphic column for the Fresnillo Mining District.

Upper Graywacke (KGs). This upper unit of the Proaño Group crops out in the southern part of Cerro Proaño; an estimated 500-m-thick interval of fine- to coarse-grained, dark greenish gray sandstones, trending generally N70°E and dipping 35° southeast, conformably overlies the calcareous and carbonaceous shales. Within the mine it underlies the Chilitos Formation with apparent conformity. Micropaleontological sampling of all three units within the mineworks revealed partially calcified radiolaria in the upper graywacke only; the most significant *Patanellium* sp., *Pseudodictyomira* sp., *Mirifusus* sp., and *Archaeodictyomira* sp.) confirm its Early Cretaceous age as no younger than Valanginian (Dávila A., 1981; Fig. 3).

Chilitos Formation (KCh). This formation, well exposed along the Arroyo Chilitos, comprises spilitized basalt agglomerates and flows with intercalated siltstones, graywackes, and marl, and limestone and shale lenses, which indicates the temporal and syngenetic relation of all these rocks. The Chilitos Formation is a marine volcanic sequence typical of hydrothermally metamorphosed oceanic crust. Dark-gray to light-green basalts and dolerites with reddish brown tones exhibit pillow-lava structures and porphyritic texture showing plagioclases and some ferromagnesian minerals, with quartz-, calcite-, and celadonite-filled vacuoles; the mesostasis is usually chloritized and sometimes silicified. The unit is conformable above the Upper Graywacke both at the surface and underground. Toward the Cerros del Pópulo and El Cristo area it is unconformably overlain by conglomerate and Tertiary rhyolite tuffs. Although the upper contact is unknown, it exceeds 500 m in thickness.

Macrofossils from the Arroyo Chilitos sediments indicate upper Valanginian–lower Hauterivian (i.e., Early Cretaceous age; Cantú, 1974).

Tertiary

Conglomerate (TCg). This unit is exposed at the surface in Cerro Proaño and encountered in both surface and underground diamond-drilling. It is composed mostly of angular to subangular fragments, up to 6 cm in size, of limestone, calcareous shale, and graywacke, and has been more appropriately described as sedimentary breccia by several previous authors. Petrographic studies report rock and quartz fragments over 1 cm in size as sedarenite with quartz and sublithoarenite with a chert groundmass. The rock fragments, in order of decreasing abundance, are siltstone, chert, and arkosic sandstone. The 0 to 300-m-thick unit overlies the Cretaceous rocks with angular unconformity (known locally as the San Pedro fault) and is normally overlain by pyroclastic rocks.

Pyroclastic rocks (Ttr). This thick unit outcrops extensively in the district (Fig. 3) and has been mapped mostly southeast of Fresnillo. Ignimbrites, volcanic breccia, rhyolite tuffs, and lithic tuffs identified on the basis of distinctive textures are grouped as a single 500-m-thick unit, dipping generally up to 30°S, that overlies the conglomerate conformably and the Cretaceous rocks with angular unconformity. Recent K/Ar dating of the ignimbrites indicates 37.6 Ma (J. J. Sawkins, personal communication, 1985), practically coinciding with similar rocks (36.8 Ma) dated near the town of Zacatecas (Clark and others, 1981; Fig. 4).

Rhyolite (Tr). Good outcrops of these rocks at and to the south of Loma del Puerto consist of quartz, biotite, and some feldspar phenocrysts in an aphanitic groundmass, showing flow texture and abundant chalcedony, both colloform and in druses.

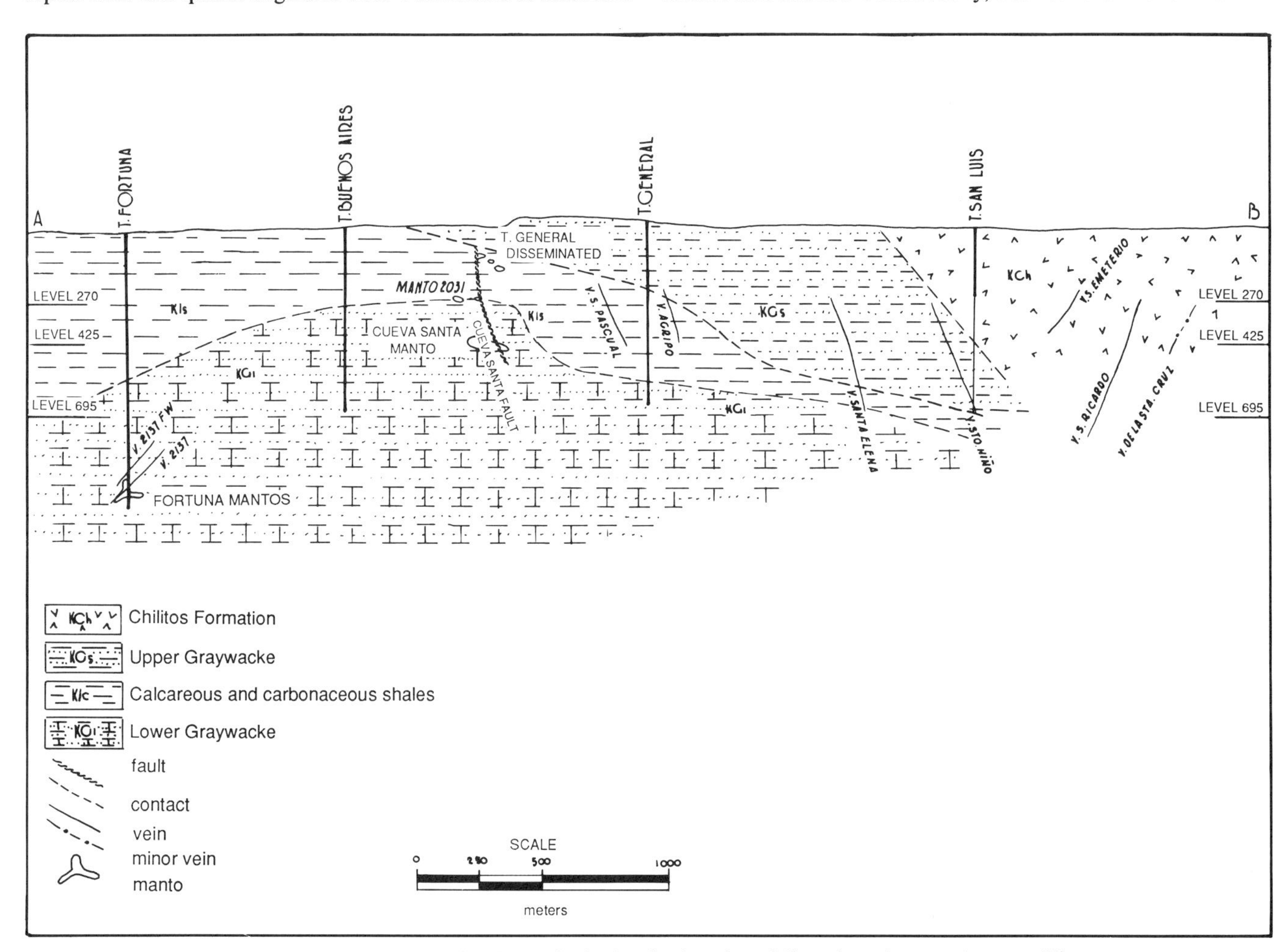

Figure 4. Geologic cross section A–B (Fig. 2) showing location of disseminated ores and mantos. View to the northeast.

Two directions of banding, south and northeast, are present. Vitrophyres also present in the area consist of abundant glass with some plagioclase and ferromagnesian (including biotite) phenocrysts, ranging in texture from vitreous to fine-grained porphyritic with quartz, biotite, and K-feldspar phenocrysts in an aphanitic groundmass.

Quartz-monzonite porphyry (Tczm). This intrusive rock exposed 150 m southwest of the Fortuna Shaft as a 15-m-thick, N50°W–trending, generally vertical dike, and is described petrographically as yellowish to greenish brown–weathering porphyry with quartz and altered feldspar phenocrysts in an aphanitic groundmass. The body is well defined underground, intruding the Lower Graywacke and Calcareous–Carbonaceous Shale members of the Proaño Group. Clark and others (1981) report an age of 75.7 ± 1.5 Ma (K/Ar) in a sample of the quartz-monzonite stock within the Fortuna area mine. Recent K/Ar determinations by Lang (J. J. Sawkins, personal communication, 1985) indicate 29 Ma for the Fresnillo intrusive and 30.8 Ma for similar bodies exposed in the Plateros area, concurring with both the geological observations and the relation between the mineralization and this igneous body, and thus invalidating Clark's age determination.

Quaternary

Conglomerate and alluvium. Intensive regional weathering mostly eroded the positive (mainly volcanic) parts of the area, and the resulting sediments accumulated in the existing ample valleys. Thus, in low-lying areas a perceptibly horizontal, 0- to 30-m-thick, oligomictic conglomerate of rhyolites and conglomeratic sandstones is almost completely covered by soil.

STRUCTURAL GEOLOGY

Poor-quality outcrops and lack of chronological data on some units hamper the structural interpretation of Fresnillo District rocks. The main structure is evidently an anticline (Fig. 4) described as an anticlinorium trending N10°E and plunging to the south. It is affected by numerous diversely oriented faults and fractures, among which the N45°E and east-northeast to east-southeast systems stand out as hydrothermal fluid conduits. Both systems dip generally south, with some exceptions such as the north-dipping 2270 and San Ricardo fault veins. The strikes of mineralized faults and fractures appear to be parallel to formational contacts (see Fig. 6); many of them develop parabolically around Cerro Proaño. However, in the direction of dip the almost vertically dipping faults intersect the 20 to 40° southeast and southwest-dipping formations.

TYPES OF DEPOSITS

Stockworks, disseminated ores, mantos, chimneys, and veins have been commercially developed in the district. Throughout the history of the mines, most production has come from narrow veins.

Stockworks

The Cerro Proaño stockwork or massive deposit (locally termed Glory Hole after the formerly used mining method), comprises an area of 700 by 200 m with an average depth of 60 m and five minor Glory Holes: Catillas, San Pascual, San Nicholás, Espíritu Santo, and Deseada (Fig. 5).

This type of deposit is derived from the principal outcropping veins: San Pascual, Espíritu Santo, Providencia, Rosario, Amarillas, El Pilar, and El Refugio, which near the upper levels, branch out into innumerable variably trending smaller veins and veinlets.

The very irregularly shaped orebodies usually adopt the shape of elongated funnels along a main structure, 150 m wide in the upper parts and 10 to 15 m wide in the lower parts, that extends up to 300 m along strike.

Quartz, feldspar, montmorillonite, and kaolin, with lesser amounts of pyrite, sphalerite, galena, cryptomelane, iron oxides, and some calcite form the gangue. Ore minerals are silver sulfosalts, native silver, cerargyrite, and native gold. The deposit is developed in the upper graywacke member and the Chilitos Formation. Colored argillaceous alteration and manganese oxides are mineralogical markers that indicate increasing gold and silver grades as the tones deepen. Although all the richer veins and zones have been mined, approximately 7 million metric tons of low-grade mineralized zones with 0.23 ppm Au and 85 ppm Ag still remain.

Mantos and chimneys

Mantos normally develop in breccia or stockwork zones, at vein walls, and along favorable calcareous beds in calcareous shale and lower graywacke members of the Proaño Group. These are mostly tabular, irregularly shaped stockworks in which mineralization fills fracture veinlets and forms bands replacing parts of the favorable beds.

Replacement occurs in calcareous shale beds, mostly tending to occupy fold crests. This type of orebody is locally called disseminates in the Tiro General area and mantos in the Fortuna area. Fracture fills predominate in the Tiro General orebodies nearer the surface, and selective replacement of bands in the beds of the Fortuna area at the 800- to 1,100-m depth interval. Both are described below (Figs. 6 to 8).

Tiro General area. A disseminated deposit is concentrated toward the upper, more calcareous, portion of the shales between levels 70 to 150 and 215 to 470 in the Tiro General area. The Cueva Santa, Espíritu Santo, Providencia, San Pascual, Cueva Santa Branch, 2200, and 2031 veins are genetically and spatially related to these orebodies (Figs. 6, 7). The upper-level mantos occur in the calcareous shale member, and the lower level in calcareous units of the lower graywacke member. Ore grades are usually higher nearer the source veins in the more disseminated bodies, which also show a pyritic aureole that grades to propylitic alteration. The bodies extend no more than 25 m from the veins

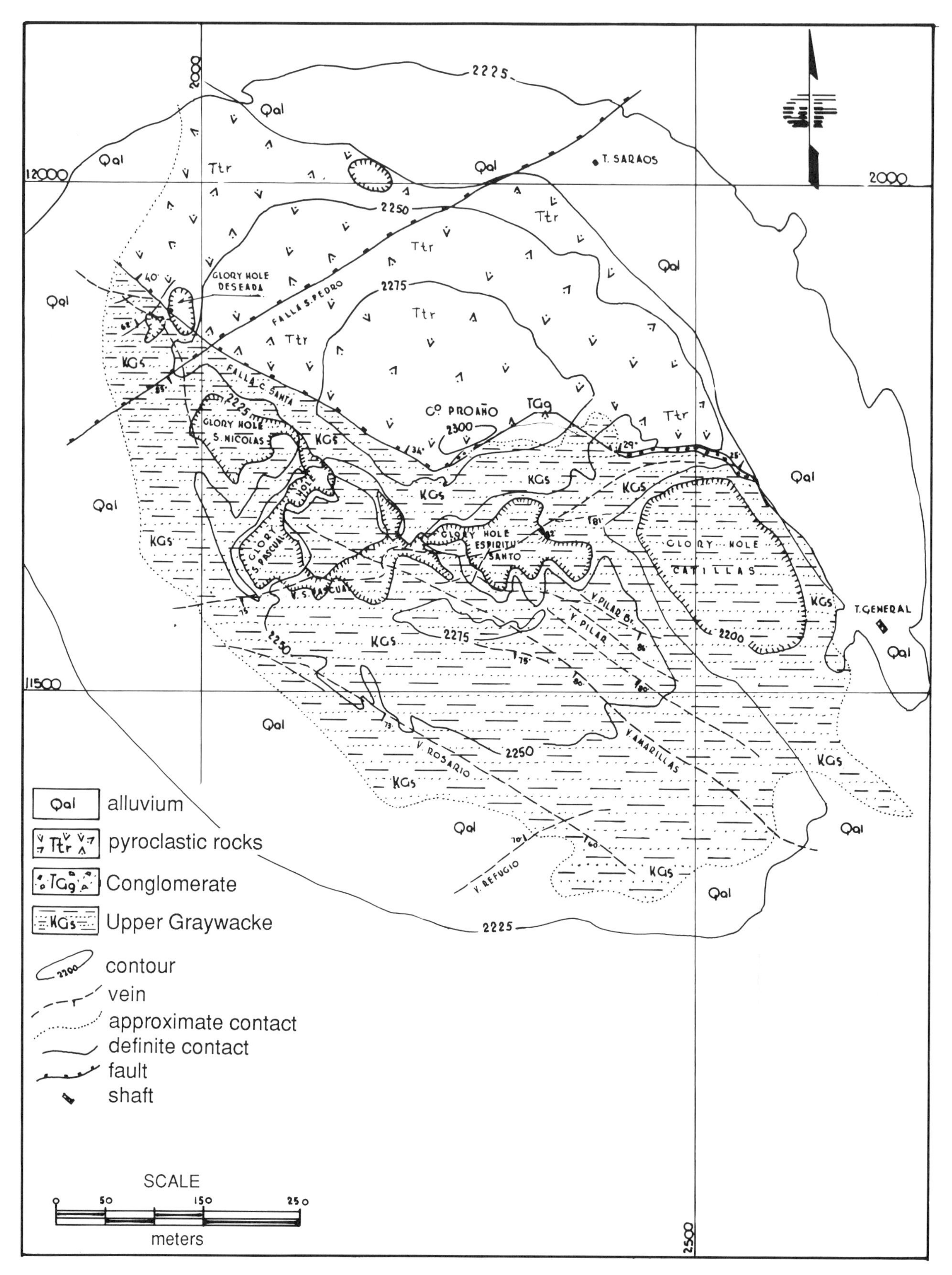

Figure 5. Geologic map of Cerro Proaño showing the principal veins.

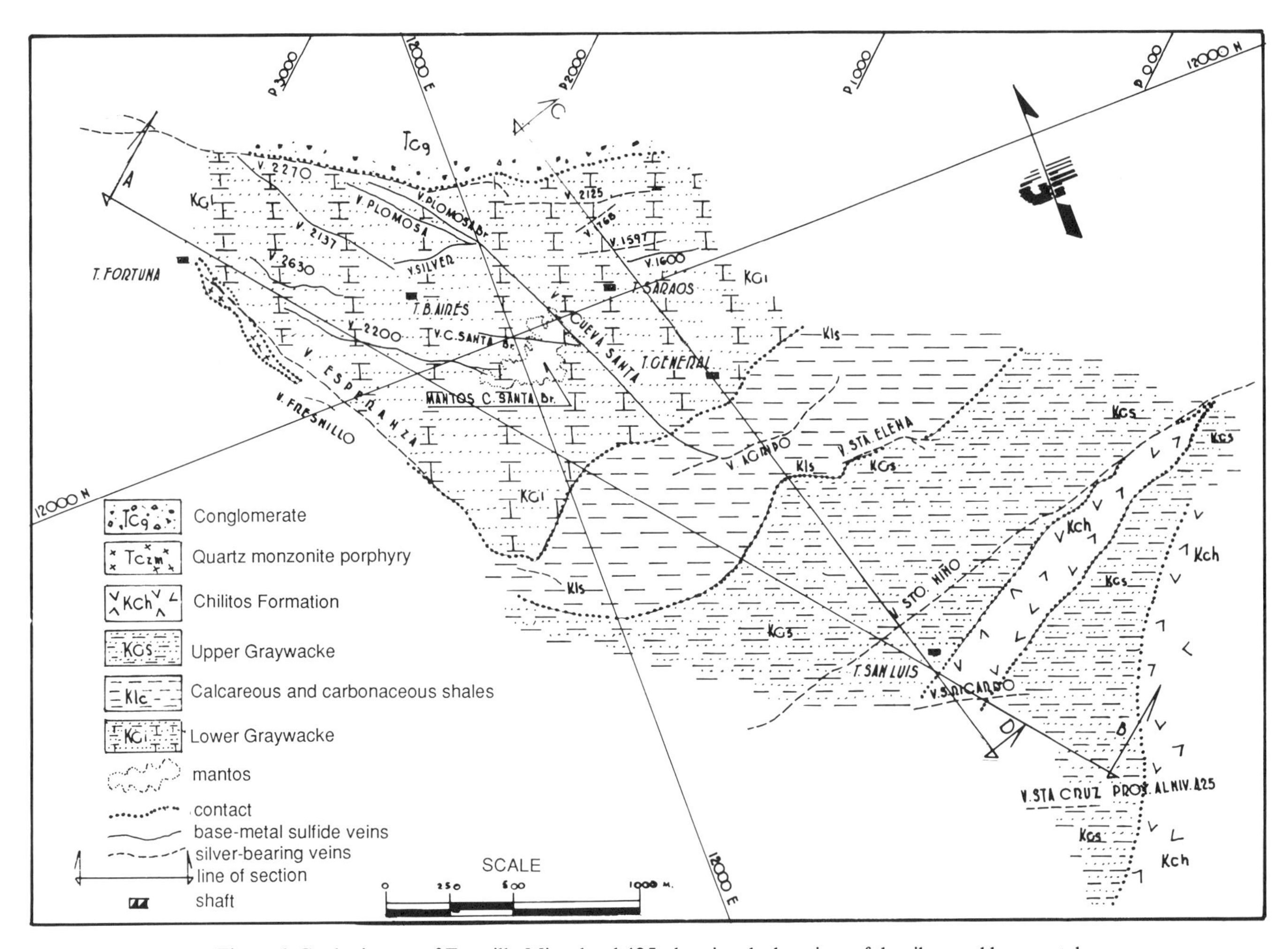

Figure 6. Geologic map of Fresnillo Mine, level 425, showing the locations of the silver and base-metal sulfide veins.

and may be 100 m in length along them. The exceptional Cueva Santa Branch manto, largest of the Tiro General orebodies, extends for 400 m along a calcareous shale horizon in the lower graywacke member with an average thickness of 8 m, from the Cueva Santa to the Cueva Santa Branch and thence to the 2200 vein (Fig. 6). This manto produced more than one million tons of ore with 65 ppm Ag, 2.2 percent Pb, and 3.3 percent Zn. Other smaller mantos, including Cueva Santa, Espíritu Santo, Providencia, and San Pascual (Fig. 5), contain sphalerite, galena, pyrite, arsenopyrite, chalcopyrite, pyrrhotite, tennantite, and tetrahedrite. MacDonald (1978) reports formation temperatures of 185 to 348°C (average 270°C) for the Cueva Santa Branch manto, and also that gangue and ore minerals were formed at practically the same temperature.

Fortuna area. The Fortuna mantos occur between depths of approximately 800 and 1,100 m (Fig. 8); all are developed in the Lower Graywacke and are spatially related to a quartz-monzonite intrusion. Based on their position and structural relations, Kreczmer (1977) indicates that the mineralization was the final event associated with the igneous emplacement. These orebodies were in 1951 after developing Vein 2137 toward the northwest at level 875, where drilling uncovered Manto 1981 and subsequently the Manto Inferior. Mantos discovered to date are the Manto Inferior, Manto Superior, and Mantos 3060, 2981, 2939, and 3004 (Fig. 8).

The Manto Inferior, located east of the Manto Superior, is the largest and most extensive; its upper part wedges out approximately at level 905, and diamond-drilling has traced it down to the 1070 depth-level, with an average length of 80 m along strike and a thickness that occasionally reaches 10 m. The Manto Superior, known from levels 845 to 980, extends for approximately 45 m along strike and is 4 to 6 m thick. The 3060 Mantos are a series of short aligned mantos at the Manto Superior footwall, with the highest Pb, Zn, and Ag values. The 2981 Manto in the northeasternmost Fortuna area is known from level 875 to below level 965 and shows the lowest silver grades. The 2939 Manto, between levels 920 and 965, is considered a faulted portion of the Manto Inferior. The 3004 Manto is a short, thin mineralized horizon at level 875 between the Manto Inferior and the Chimney.

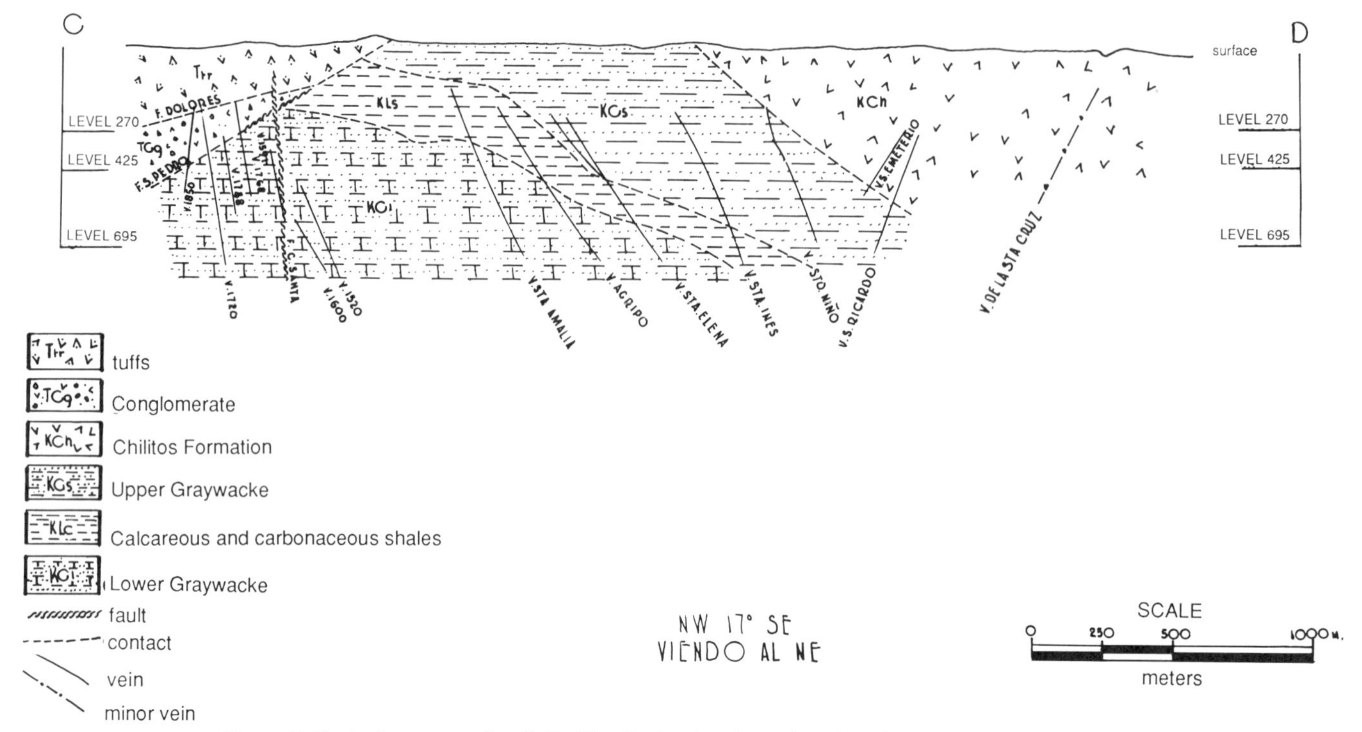

Figure 7. Geologic cross section C–D (Fig. 2) showing the main veins of the southeastern area. View towards the northeast.

The hydrothermal alteration is axinite, concentrated with preference in the shales. Hedenbergite and epidote replace the graywacke; usually the hanging wall of the manto is a ceiling of shales with axinite, and the footwall is graywackes with hedenbergite (Kreczmer, 1977).

The sulfides of the Fortuna mantos form massive lenses, commonly in stratified bodies that follow and mostly replace the shale horizons. Sulfides, in order of decreasing abundance, are marmatite, pyrite, galena, pyrrhotite, chalcopyrite, argentite, arsenopyrite, tetrahedrite, magnetite, matildite, and pavonite. Gangue is mainly quartz and axinite, with lesser calcite, hedenbergite, chlorite, epidote, siderite, and carbonaceous material.

The chimneys are tabular bodies rising with preference from the hanging wall of some mantos near the intrusive body, which suggests that the latter paved the way for their emplacement by deforming and brecciating the host rocks. Chimneys identified to date are the 2912, 2990 and 2907 (Fig. 8).

Chimney 2907, northeast of the quartz monzonite, rises from the Manto Inferior hanging wall approximately at Level 980, with an average 20-m diameter, and has been explored to above level 875. Chimney 2912, in the southern part of the intrusive body, rises from the Manto Superior hanging wall near Level 980 with an approximate 15-m diameter, wedging out near Level 920. Chimney 2990, barely 10 m in diameter, is known from Levels 875 to 830 at the northernmost end of the intrusive.

Mineralization in the chimneys is similar to that of mantos, which however, are notably higher in silver values (+20 ppm). The chimneys at their upper levels are strongly brecciated; quartz, axinite, epidote, calcite, chlorite, and hedenbergite (in order of decreasing abundance), together with sulfides, form the groundmass. The fragments are strongly silicified at the upper levels, less so deeper down where brecciation diminishes and banded mineralized structures appear. The fragments show the same alteration minerals as the groundmass.

A narrow fringe of grossularite and almandine skarn develops around the intrusive body; the lower part of this zone displays anomalous values of finely disseminated silver due to a high matildite content.

Production to date from the Fortuna area mantos and chimneys is 21 million metric tons of 67 ppm Ag, 2.6 percent Pb, 5.02 percent Zn, and 0.12 percent Cu.

Homogenization temperature determinations on fluid calcite and quartz inclusions associated with the mineralization indicate deposition temperatures of 250 to 350°C for the mantos and 300 to 350°C for the chimneys; salinities range from 4 to 12 percent in NaCl-equivalent weight (Kreczmer, 1977).

Veins

Fissure veins led to the original discovery of the district and have now revived it. The first mines worked veins exposed at Cerro Proaño; explorations discovered over 100 others in the central portion of the area, now abandoned, where only small pillars remain.

Veins are developed mostly in Proaño Group graywackes and shales; however, they also occur within the Conglomerate

(Tcg) and persist for up to 100 m within the overlying tuffs. Both the Chilitos Formation and the Proaño Group are favorable for vein emplacement in the new Eastern part of the district (Fig. 7).

Average vein thickness is 1.5 m, but narrower veins have also been mined; the small thickness is compensated by the oreshoots' horizontal extension of up to 1,100 m along some of the veins; vertical extensions range usually from 180 to 750 m below the present surface, but some veins—for instance, Cueva Santa and 2137—persist down to 1,010 and 920 m, respectively.

Most veins trend generally northwest or almost east-west, varying toward both the north and south and dipping usually 55 to 80° south, with the important exceptions of the 2270 and San Ricardo silver veins, which dip 40 and 75° north, respectively.

The Fresnillo District veins have been divided into two groups according to their mineralization and silver/basic metals contents, described below.

Silver- or light sulfide-bearing veins. These are quartz and calcite veins with variable amounts of pyrite and some galena, sphalerite, and chalcopyrite. Pyrargirite-proustite is the principal silver mineral, with minor polibasite and acanthite. In the most productive zones these veins commonly contain an average 770 ppm Ag with less than 2 percent of combined Pb-Zn. At depth the long, continuous upper-level oreshoots gradually shorten and occasionally branch out into more than one, with gradually decreasing silver contents, until they finally become uncommercial.

Most of the veins exhibit vertical zoning of the silver, lead, and zinc content, the combined Pb-Zn content reaching 5 or 6 percent at depth. The silver-bearing veins have little vertical extension (105 to 470 m) and are most productive between 240 and 425 m.

The most important are 2270, Santo Niño, San Ricardo, Santa Elena, 2125, Agripo, and the Esperanza system (Fig. 7).

Drilling in the southeast part of the district has detected other as yet undeveloped veins. Although large silver volumes have been mined from Vein 2270, the richest in the district, remaining reserves are 193 million metric tons of 0.34 ppm Au, 849 ppm Ag, 0.75 percent Pb, and 1.4 percent Zn.

The N70 to 80°E–trending, 65 to 80°SE–dipping, and 2.8-m-thick Santo Niño Vein has been mined at levels 270 and 425 for more than 2.5 km. The vein lodges in graywackes and greenstones of the upper Proaño Group and Chilitos Formation, re-

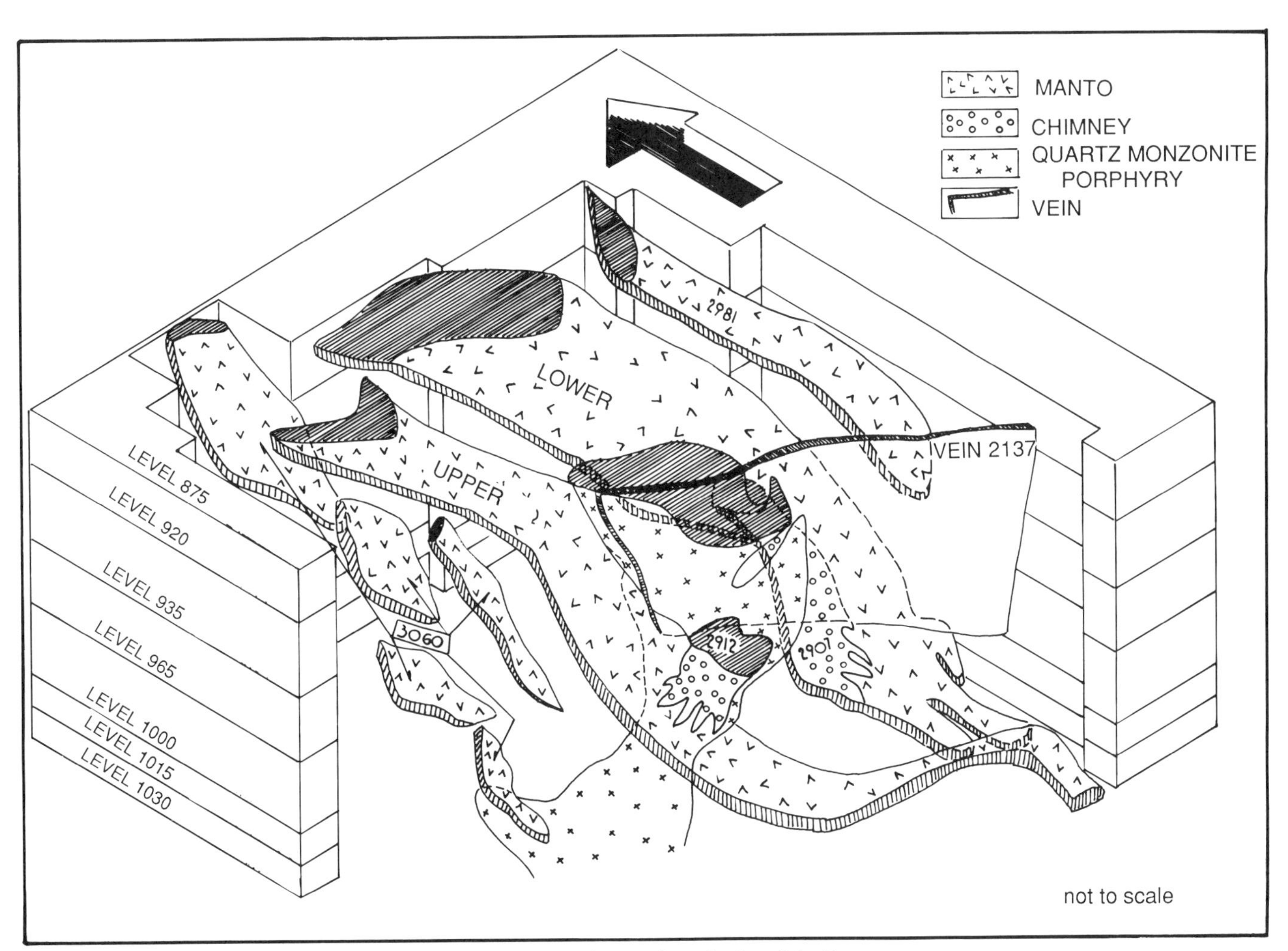

Figure 8. Stereodiagram of the main orebodies of the Fortuna area.

spectively, and at the contact along some stretches. To date, 2,861,000 metric tons of ore have been quantified, averaging 0.52 ppm Au, 730 ppm Ag, 0.55 percent Pb, and 1.06 percent Zn, of which a total 293.5 million mt of 0.56 ppm Au, 836 ppm Ag, 0.48 percent Pb, and 0.88 percent Zn have been extracted.

The mineralogy of the Santo Niño Vein is presently under detailed study; preliminary information (A. Gemmell, personal communication, 1985) indicates pyrite, arsenopyrite, sphalerite, galena, chalcopyrite, pyrargirite-proustite, polybasite-arsenopolybasite, tetrahedrite, acanthite, marcasite, and pyrrhotite. Oxidation products in the upper part of the vein are limonite, hematite, malachite, azurite, and native silver.

Pb-Zn or heavy sulfide veins. Base-metal veins have higher sulfide contents than the light sulfide veins described above. Main sulfides, in order of decreasing abundance, are pyrite, sphalerite, galena, pyrrhotite, chalcopyrite, and arsenopyrite; the sphalerite contains 5 to 12 percent Fe. Silver content is considerable, mostly included in galena as polybasite, silver-bearing tetrahedrite, and pyrargirite; the latter appears in the upper portions of some basic metals oreshoots. Matildite was reported in the Cueva Santa and 2137 veins by Donald C. Harris (De Cserna and others, 1977). The most abundant gangue minerals are axinite and hedenbergite.

In contrast to the light sulfide veins, these are vertically more extensive and more persistent at depth. The most important veins trend northwest, dipping 50 to 65°south at depth but following more vertical fractures in the upper parts. Cueva Santa and 2137, the most commercial, extend up to 1,130 m (Cueva Santa), with variable vertical extensions as shown in Table 1.

Stone and MacCarthy (1942) report little change in the Pb, Zn, and Ag distribution of the Cueva Santa Vein, although lead content seems to increase upward; silver values increase both below (presence of Bi mineralization) and above (light sulfide mineralization).

Vein 2137 has been studied in greater detail. Two types of mineralization are present: basic sulfides below and light sulfides above. Stone and MacCarthy (1942) locate the change or overlap of both types between Levels 340 and 425; Koch and Link (1967) place it between levels 385 and 470. The separation between both types is clearly evident in the contrasting Ag/Pb ratios, with maximum frequency of around 735 gm Ag/1 percent Pb in the light sulfides and 30 gm Ag/1 percent Pb in the basic sulfides. Other veins (2200, 2630, and 2137 footwall) show similar distributions; thus, all veins of this type exhibit a steep gradient in the silver values from high in the upper part to low at depth. A third type of deep mineralization with considerably increased silver contents has been found to coincide with the presence of bismuth sulfosalts, but is not present in all the veins. Bismuth sulfosalts have also been reported at greater depths in the Fortuna area mantos.

TABLE 1. VERTICAL DIMENSIONS AND DISTRIBUTION OF BASE-METALS VEINS

Vein	Vertical Distance (m)	Level (m)
Cueva Santa	960	105 to 1,010
2137	705	215 to 920
2630	390	305 to 965
2200	390	215 to 605
1600	315	470 to 785

Investigations on fluid inclusions in veins

Although the data are not directly correlatable, conclusions are that silver-bearing veins formed at lower temperatures than base metal veins, and also that salinities in the silver veins and in the upper levels of the ore deposits are very low (Table 2).

Isotopic studies of sulfur contained in galena and sphalerites of several Fresnillo Mine orebodies (Kreczmer, 1975; González and others, 1984) coincide with a general rank of $-\delta^{34}S/^{00}$. The numerous natural variables affecting these values (pH, fO_2, temperature, boiling-point and decompression of solutions, and the corresponding distribution coefficients) have led to widely contradictory interpretations of the results.

Isotopic analyses of lead in ores of one of the Cueva Santa Vein mantos show identical isotopic compositions in both orebodies; lead is of J type (i.e., crustal origin), a product of uranium disintegration.

GENESIS OF THE DEPOSIT

The available information leads to the following conclusions:

1. The Fresnillo mineral deposits, of hydrothermal origin, was generated by a heat source related to quartz-monzonite intrusion, of which only small apophyses crop out; this was certainly the last magmatic event recorded in the region, which for 37 m.y. was subjected to strong acid volcanism and probably associated hydrothermal activity.

2. The main pathways for the upward-flowing solutions were mostly tectonic but occasionally were sedimentary contacts or bedding planes.

3. Preliminary, still unpublished, biochemical data confirm the location of the main source of Fresnillo mineralization around Cerro Proaño, and generally oriented following the main northwest-trending fault pattern to the west, and the east-west fault pattern to the east, of Cerro Proaño.

4. The deformations that caused folds, fractures, and faults in the sedimentary rocks and provided the necessary permeability for the hydrothermal processes preceded the magmatism and may be ascribed to the uppermost Cretaceous/lowermost Tertiary Laramide Revolution. However, movements also may have occurred during the volcanic and hydrothermal episodes.

TABLE 2. COMPARISON OF TEMPERATURE AND SALINITY OF SILVER-BEARING VEINS

Vein	Level	Highest Temperature (°C)	Mineral	Salinity	Author
2200	385-470	260-310	qtz	....	MacDonald, 1978
Cueva Santa 2137	875-965	300-360	lstn qtz	....	Kreczmer, 1975
Santa Elena	270-425	207-298	qtz lstn	0–1.1	Ruvalcaba, written communication, 1980
Santo Niño	270 (W and Center)	220-250	qtz lstn	0–3	Simmons, personal communication
Santo Niño	270 (E)	190-210	qtz lstn	0–3	Simmons, personal communication
Santo Niño	425	220-300	qtz lstn	4–12	Simmons, personal communication
Santa Elena	425	210-241	qtz lstn	0.3–12.5	Albinson, personal communication

5. The deposit displays both vertical and horizontal zoning; the central deep part of the district is rich in base metals, and the outer, upper portion is rich in silver and poor in base metals.

6. Homogenization temperatures in gangue minerals, although varying in time and space, indicate a geothermal system of less than 400°C, averaging generally 250 to 300°C.

7. Judging by the low salinities reported in fluid inclusions, which are less than 12 percent NaCl equivalent, the geothermal system was mainly derived from meteoric fluids.

8. There is no definitive indication of the origin of the metals and sulfur; some authors suggest a syngenetic origin in the sedimentary rocks and subsequent remobilization by Oligocene hydrothermal activity.

9. The mineralization process changed with time, producing several stages of mineral precipitation. Boiling played an important role in the silver veins and surely in the reopening of the veins.

10. The age of the mineralizing hydrothermal process is restricted by that of the igneous bodies intimately related to the mineralization, although perhaps somewhat older (29 Ma and 30.8 Ma) and by that of the alteration of a Plateros District vein (30.0 Ma). The Fresnillo mineralization is thus inferred to have occurred in mid-Oligocene time (29 to 30 Ma).

AVERAGE TONNAGES AND GRADES

Reliable figures on the total production from the district are unavailable in view of the many companies that operated in the course of its long history. Church (1906, in Stone and MacCarthy, 1942) estimated 1,333 metric tons of silver production

TABLE 3. PRODUCTION OF OXIDE ORES FROM 1921 TO 1984

Periods	Metric Tons	Assays Au (g/ton)	Ag (g/ton)
1921–1943	13,044,400	0.30	190

Sulfide Ores

Periods	Metric Tons	Au (g/ton)	Ag (g/ton)	Pb (g/ton)	Zn (%)	Cu (%)
1926–1981	22,037,200	0.65	269	3.1	4.1	0.33
1982	302,700	0.29	407	0.9	1.9	0.06
1983	345,100	0.25	385	0.9	1.7	0.06
1984	397,900	0.26	367	0.8	1.6	0.06
Total and average 1926–1984	23,082,900	0.63	274	3.0	4.0	0.32

TABLE 4. ORE RESERVES IN FRESNILLO MINE

	Metric Tons	Assays Au (g/ton)	Ag (g/ton)	Pb (%)	Zn (%)	Cu (%)
Total proved ore	1,939,400	0.38	563	0.87	1.57	0.05
Total probable ore	1,576,100	0.48	722	0.56	1.06	0.03
Proved and probable Total	3,515,500	0.42	634	0.73	1.34	0.04

from 1832 to 1903, which according to Livingston (1932, *in* Stone and MacCarthy, 1942), was 43 percent more (i.e., 2,480 mt). Stone and MacCarthy (1942) compiled data on oxide zone productions from 1921 to 1941 and on the sulfide zone production from 1926 to 1941. Updated information from the Mexican Corporation, the Fresnillo Company, and Compañía Fresnillo, S.A. de C.V., files appears on the following tables, showing production during 1982 to 1984, production increases, and comparative annual average grades (see Table 3).

The latest estimated Fresnillo Unit reserves as of September 30, 1984, are shown on Table 4.

Ore blocks and pillars in abandoned areas (i.e., the central nonproducing part of the mine) have been omitted. Quantified reserves, possible reserves, potential based on tons per meter of depth, structures detected only by drilling, and other information, give the mine 14.3 years of remaining useful life according to future production programs.

TABLE 5. AVERAGE MONTHLY METALLURGICAL RESULTS

	Metric Tons/ Month Processed or Produced	Au (ppm)	Ag (ppm)	Assays Pb (%)	Zn (%)	Cu (%)
Millheads	33,200	0.26	367	0.85	1.58	0.06
Pb concentrates	1,145	4.54	8.964	22.47	7.69	1.21
Zn concentrates	595	0.70	856	0.31	51.39	0.70
Recoveries		Au (%)	Ag (%)	Pb (%)	Zn (%)	Cu (%)
Pb concentrate		59.24	84.36	91.55	16.81	65.30
Zn concentrate		4.74	4.19	0.65	58.34	19.48

REFERENCES

Auld, R. O., and Bychan, J. H., 1834, Notice of the silver mines of Fresnillo: London.

Cantú Chapa, C. M., 1974, Una nueva localidad del Cretacico Inferior en México: Instituto Mexicano del Petróleo Revista, v. 6, p. 51–55.

Clark, K. F., Damon, P. E., Shafiquilla, M., Ponce, B. F., and Cárdenas, D., 1981, Sección geológica-estructural a través de la parte sur de la Sierra Madre Occidental, entre Fresnillo y la costa de Nayarít: AIMMGM National Convention Memoirs, v. 14, p. 74–103.

Dávila A., V. M., 1981, Radiolarios del Cretácico Inferior de la Formación Plateros, Distrito Minero de Fresnillo, Zacatecas: Instituto de Geologia Revista, v. 5, no. 1, p. 119–120.

De Cserna, Z., Delevauz, M. H., and Harris, D. C., 1977, Datos isotópicos mineralógicos y modelo genético propuesto para los yacimientos de plomo, zinc y plata de Fresnillo, Zacatecas: Instituto de Geología Revista v. 1, no. 1, p. 110–116.

González P., E., Arnold, E. M., and Acosta A., E., 1984, Análisis metalogenético preliminar del Distrito Minero de Fresnillo, Zacatecas, sobre la base de 50 medidas isotópicas J34S°/100: Geomimet, no. 129, p. 27–34.

Isunza E., A., 1981, Fresnillo, monografia e historia grafica, no. 1 & 2: Fresnillo, Imprenta Mignon.

Koch, G. S., Jr., and Link, R. F., 1967, Geometry of metal distribution in five veins of the Fresnillo Mine, Zacatecas, Mexico: U.S. Bureau of Mines Report of Investigations 6919, 64 p.

Ranson, W. A., Fernández, L., Sommons, W. B., Jr., and Enciso de la Vega, S., 1982, Petrology of the metamorphic rocks of Zacatecas, Mexico: Sociedad de Geología Mexicana Boletín, v. 43, no. 1, p. 37–59.

Stone, J. B. and MacCarthy, J. C., 1942, Mineral and metal variations in the veins of Fresnillo, Zacatecas, Mexico: American Institute of Mining and Metallurgical Engineers Technical Publication 1500, Mining Technology, v. 6, no. 5, 16 p.

Unpublished sources of information

Álvarez, M., Jr., 1969, Apuntes de geología de México: Universidad Nacional Autonoma de México.

Kreczmer, M. J., 1975, The geology and geochemistry of the Fortuna mineralization, Fresnillo, Zacatecas, Mexico [M.S. thesis]: University of Toronto, 155 p.

López-Ramos, E., 1965, Recorrido geologico de los Estados de San Luis Potosí y Zacatecas: México, D.F., Petroleos Mexicana, Ingenieria Geologia Centro No. 86, 18 p.

MacDonald, J. A., 1978, The geology and genesis of the Cueva Santa Branch silver, lead, zinc manto orebody, Fresnillo Mine, Mexico [M.S. thesis]: University of Toronto.

Printed in U.S.A.

The Geology of North America
Vol. P-3, Economic Geology, Mexico
The Geological Society of America, 1991

Chapter 52

Geology of the Real de Angeles deposit, Noria de Angeles municipality, Zacatecas

J. Bravo N.
Cía. Minera Real de Angeles, S.A. de C.V., Ave de la Convención Nte. 1302-2o. piso, Aguasacalientes, Ags., Mexico

INTRODUCTION

The Real de Angeles deposit is the result of geologic events causing a mass of marine sedimentary rocks approximately 400 m long by 400 m wide and 400 m of known depth to become more or less uniformly mineralized in silver, lead, zinc, and other elements.

Exploration and quantification of the deposit's potential, and subsequent planning, financing, and construction of the necessary installations have resulted in its successful development.

In 1985, opencast mining produced 11,400,000 troy ounces of silver, 37,800 tons of lead, and 30,600 tons of zinc, approximately equivalent to 19 percent of the silver, 21 percent of the lead, and 16 percent of the zinc produced in Mexico.

Minera Real de Angeles has countered the drastic price reductions by increasing treated ore volumes from the original 10,000 tons/day capacity to 14,500 tons/day, reducing costs and increasing productivity indexes; 85 percent of the financial commitments will have been covered by the second half of 1986.

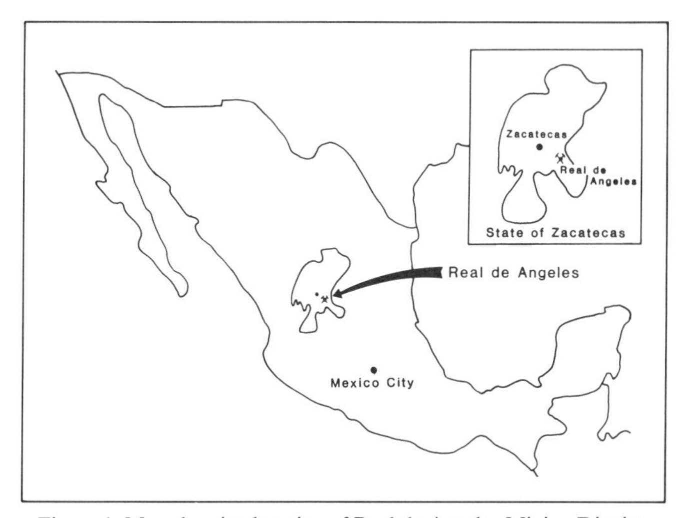

Figure 1. Map showing location of Real de Angeles Mining District.

Mina Real de Angeles is located in Noria de Angeles Municipality, eastern Zacatecas State in central Mexico, at geographic coordinates 22°25′N and 101°54′W (Fig. 1), at the center of a triangle formed by the towns of San Luis Potosí, Zacatecas, and Aguascalientes, to which it is connected by roads; the Aguascalientes–San Luis Potosí railroad stops at Estación La Honda, 10 km east of the mine.

History

The date of initial mining activity in the area was probably in the late sixteenth century; the earliest records are from 1705. Captains José Antonio Palencia and Anselmo Antonio García developed two mines from 1760 to 1777, which Agustín Ruiz de La Peña continued to work from 1798 to 1812; Rafael Carrera reactivated mining activities from 1840 to 1860. Lead-silver concentrates were first produced from 1890 to 1910 when operations temporarily came to a close.

Mining activity during this period focused mainly on silver-rich oxidized ores in vein structures. Possibly owing to its isolation and moderate production, Real de Angeles was never as renowned as other districts such as Zacatecas, Guanajuato, or Tepezalá-Asientos. Nevertheless, mining activities brought about a village settlement over the deposit and small haciendas where the ores were treated, traces of which still remain. Real de Angeles economic activity was always associated with the town of San Luis Potosí where the successive proprietors were based.

In 1970 the Gamma, C.A., Company contracted with mine property concessionaires, until 1972 when the contract was cancelled, to carry out exploration that included 20 diamond-drill holes. The properties were then offered to Compañía Explomín, S.A. de C.V., in 1973 for exploration activities, which closed in September 1975 with a feasibility study. The construction phase, decided in November 1979, took from January 1980 to July 1982 when production began.

GEOLOGY

The Real de Angeles deposit is part of the silver belt enclosing the important mining districts of Pachuca, Guanajuato,

Bravo N., J., 1991, Geology of the Real de Angeles deposit, Noria de Angeles municipality, Zacatecas, *in* Salas, G. P., ed., Economic Geology, Mexico: Boulder, Colorado, Geological Society of America, The Geology of North America, v. P-3.

Tepezalá-Asientos, Zacatecas, Fresnillo, Sombrerete, San Martín, and other mine districts.

Despite its long-lived mining activity, published information on Real de Angeles is scarce. Most information is in unpublished reports by companies that worked the area; published papers are mainly university theses and conference memoirs. The present geological information is derived from Bravo (1986).

Geomorphology

Real de Angeles Mine occupies the middle part of the Mesa Central Province, at an elevation of 2,300 m. Local geomorphology is characterized by rounded hills, small ranges, and low-lying mesas; shallow arroyos drain the area during the rainy season (June to September) into closed, flat-bottomed endorheic basins, generating short-lived shallow lagoons. Real de Angeles is situated between the Villa Hidalgo basin on the east and the Maravillas basin on the west (Fig. 2), both of which were investigated as potential water supplies for the mining operations.

The deposit occupies the central part of a circular domal structure, having a radius of approximately 5 km, at the northern border of a small east-west range that forms the south flank of an anticlinal fold.

Stratigraphy

Sedimentary, igneous, and metamorphic outcrops occur within a 50-km^2 area surrounding the mine.

Greenstone. The oldest outcrops belong to a volcanic-sedimentary series of regionally metamorphosed, poorly bedded, green andesite tuffs, quartzites, and clays, very similar to the Triassic unit known as Zacatecas greenstone. The unit forms a belt 15 km northwest of the mine and also north of Villa Hidalgo.

Flysch. Also present in the area is a marine flysch-type sequence of alternating well-bedded, occasionally weakly metamorphosed, greenish brown weathering, black siltstones and sandstones, which serve as the host rocks for the Real de Angeles deposit; they are exposed north and east of the mine (Fig. 2). The

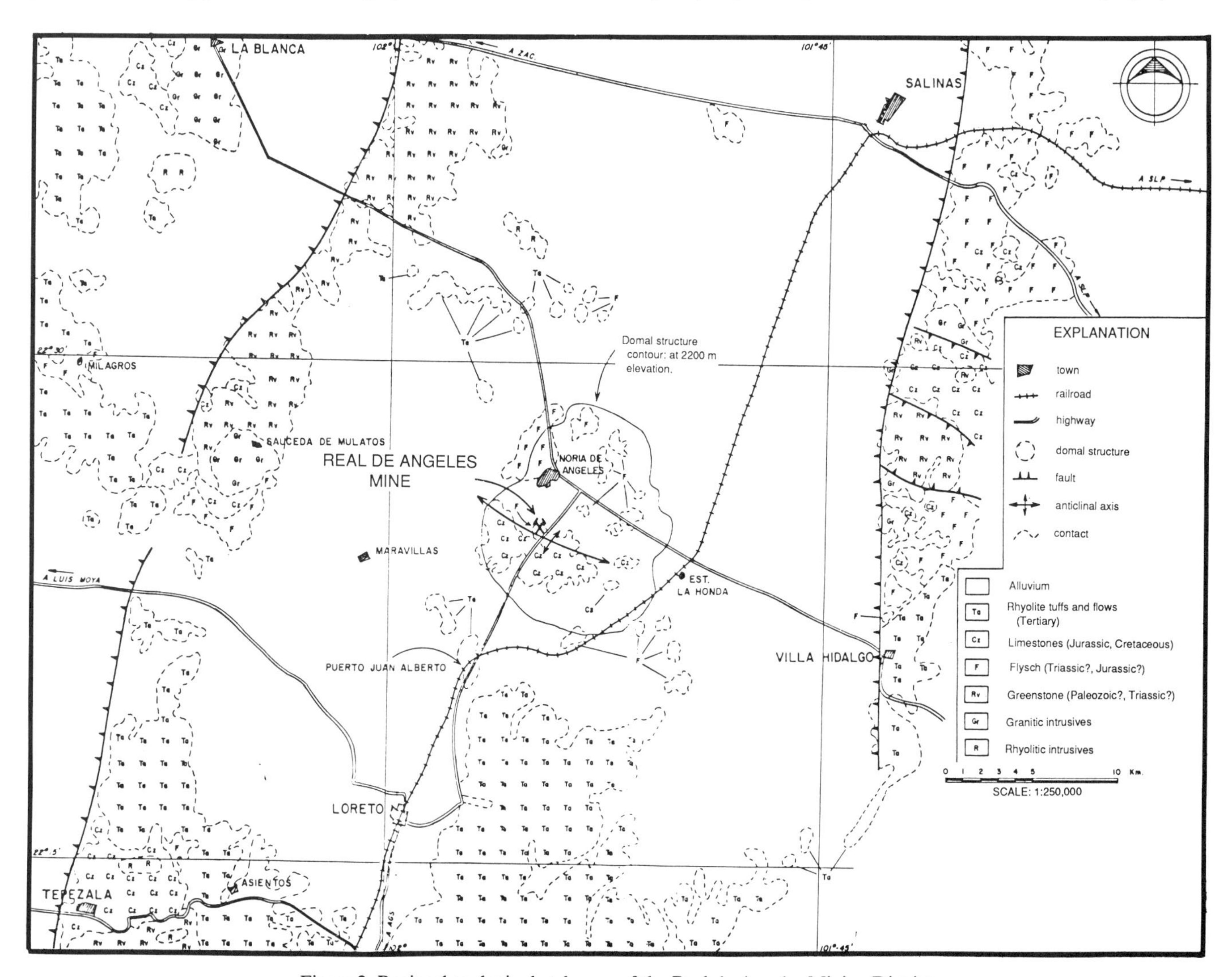

Figure 2. Regional geologic sketch map of the Real de Angeles Mining District.

unit is generally barren of fossils, although rare, poorly preserved ammonites have been reported recently; identified pollen and spores indicate a Triassic to Jurassic age. The unit extends regionally northward for several tens of kilometers, forming the core of the Charcas anticline, and may be the host rock for the Fresnillo Cerro Proaño deposit (see the preceding chapter in this volume).

Weathered outcrops of this Jurassic sequence are very similar to the Upper Cretaceous Caracol Formation and are occasionally mistaken for it. Lenticular horizons of siliceous rocks, 2 to 20 m thick, microscopically identified as radiolarite, are found in the upper part of the sequence, are commonly mineralized, and bear manganese oxides occasionally with associated silver values. These rocks crop out south of the mineworks, in small ranges north of the mine, and east of Juan Alberto. Regionally they may be correlative with outcrops to the southeast, in Pozos (Guanajuato), around Villa de Cos and Montaña de Manganeso, 50 to 100 km northeast of Zacatecas, and west of Fresnillo, about 50 km to the northwest.

Limestones. A rapid transition from mostly clastic to calcareous sediments overlies the Jurassic flysch radiolarite immediately south of the mine. Pink argillaceous limestones grading to purer gray limestones with black chert lenses and beds, assigned to the Lower Cretaceous (Albian-Cenomanian) on the basis of their abundant ammonite content, are therefore correlatable with the Cuesta del Cura Formation. As early as 1930, Burckhardt had correctly dated these limestones as Albian-Cenomanian.

The numerous Jurassic and Cretaceous limestone localities in the vicinity of the mine include Tepezalá-Asientos and Sauceda de Mulatos, north and south of La Blanca, and the Benito Juárez area south of Salinas.

Rhyolite tuffs and flows. Horizontal mesas formed of rhyolite tuffs and flows overlying older rocks are conspicuous south of the mine between Loreto and the Juan Alberto harbor; these Tertiary pyroclastic and volcanic rocks also occur as isolated erosional remnants north and east of the mine.

Alluvium. Water wells drilled for the treatment plant in the Maravillas area intersected volcanic rhyolite flows intercalated between 100-m-thick underlying and overlying alluvium intervals, proving the prolonged accumulation of thick alluvium deposits, at times exceeding 250 m, in endorheic basins.

Igneous intrusive rocks

Two periods of igneous activity have been identified in the region. The first, possibly Laramide in age, generated the La Blanca–La Tesorera and Sauceda de Mulatos granite intrusions northwest and west of the mine, as well as Peñón Blanco 25 km northeast, with associated gold, silver, copper, wollastonite, fluorite, and radioactive mineralizations.

The second (Tertiary) intrusive period gave rise to rhyolite dikes and small stocks in the Pinos area, Tepezalá-Asientos, El Morro, and Cerro San Miguel, 15 km north of the mine, with associated gold, silver, zinc, copper, and phosphorite mineralizations.

Regional structures

The region's sedimentary rocks evidently underwent several periods of deformation, but folding orientations are difficult to define.

Locally the deposit occupies the central part of an extensive west-northwest–trending anticline; the well-preserved south flank shows south-dipping flysch and limestone beds; the eroded north flank shows only gently north-dipping beds on the north side of the mine cut.

Segments of an old east-west fault system have been identified; a regional younger extensional normal fault system crosscutting the former generated troughs and uplifted blocks that formed north-northeast–trending parallel ridges and valleys.

Regional geologic history

Possibly Paleozoic metamorphic schistose outcrops north of the town of Zacatecas form what is considered to be the regional basement. Triassic seas invading the region deposited the volcanic-sedimentary sequence of the Zacatecas Formation. The Paleozoic rocks and the Zacatecas Formation were partially uplifted possibly during the Jurassic, emerging as islands aligned with the Coahuila Peninsula, which provided the source for clastic sediments, deposited as Jurassic flysch, within certain basins; thin Jurassic limestones were laid down in other basins. Basin subsidence in the Lower Cretaceous caused the islands to disappear, and deposition became essentially calcareous (Cuesta del Cura Formation).

Clastic deposition returned in the Upper Cretaceous (Indidura Formation); renewed uplift of Paleozoic and Triassic rocks and their consequent erosion also possibly provided the source for the clastic deposits of the Caracol Formation.

The Laramide orogeny caused folding and granitic intrusions, mostly at anticlinal cores, which favored the formation of ore deposits associated with metamorphic haloes, such as Peñón Blanco, La Blanca–La Tesorera, and Sauceda de Mulatos.

Another Eocene-Oligocene intrusive rhyolitic episode caused pyroclastic and lava deposition as well as mineralization, at the Pinos, Tepezalá-Asientos, El Morro, and Milagros areas. The Real de Angeles mineralization apparently corresponds to this event; isotopic determinations in galena indicate a probable age of less than 25 Ma.

ORE DEPOSIT

The geometry of the Real de Angeles orebody is that of a southwest-tilted inverted cone (Figs. 3 and 4) with an outcropping horizontal cross section 450 by 400 m in diameter and a known depth of 400 m. The greater resistance to erosion of the silicified rocks of the deposit developed a peculiarly shaped hill 40 m high in the original outcrop. The lateral boundaries of the orebody are sharply faulted. Sulfide content diminishes rapidly toward the periphery of the deposit where veinlets are filled only with calcite.

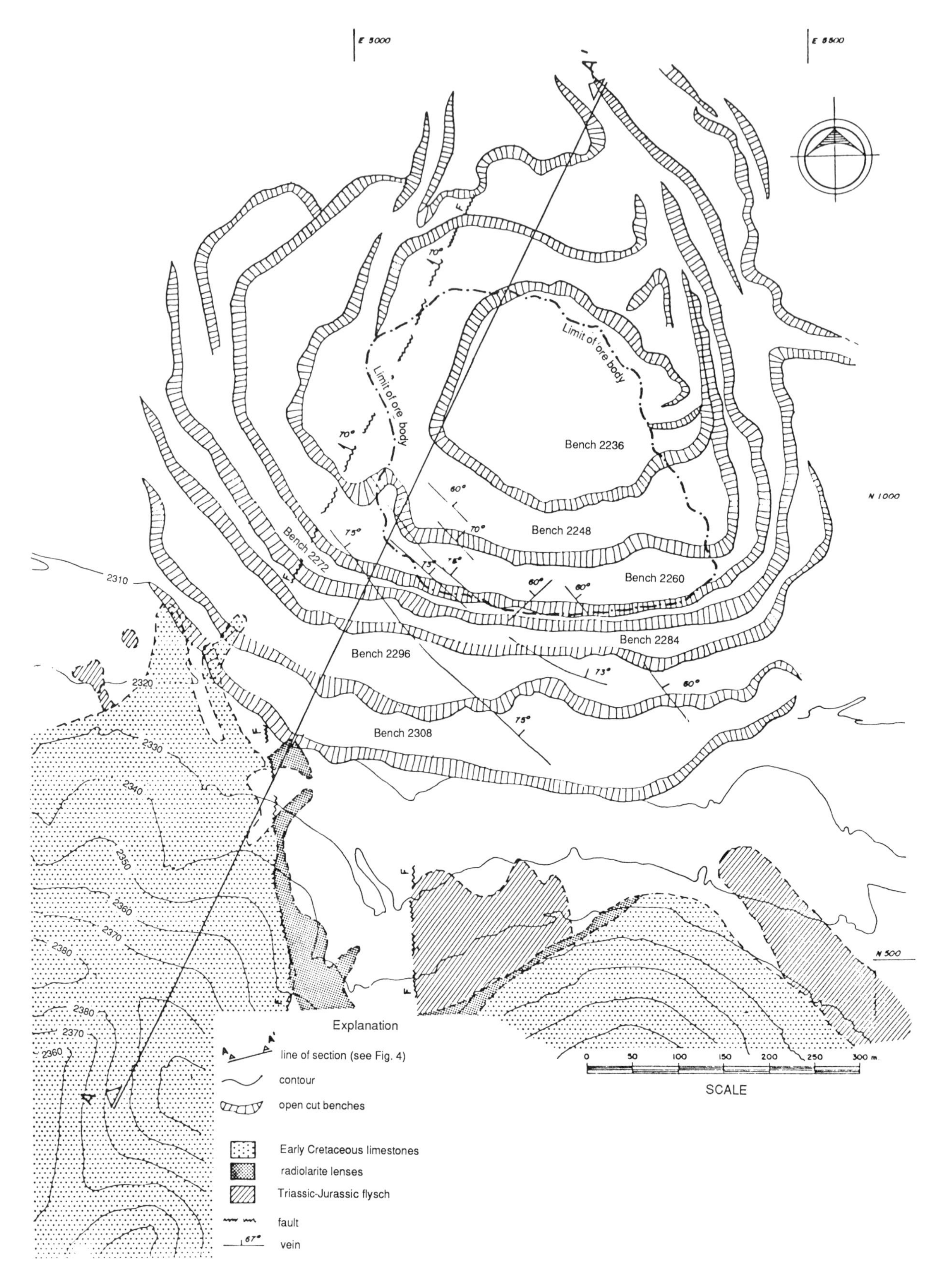

Figure 3. Geologic map of opencast area, Real de Angeles Mine.

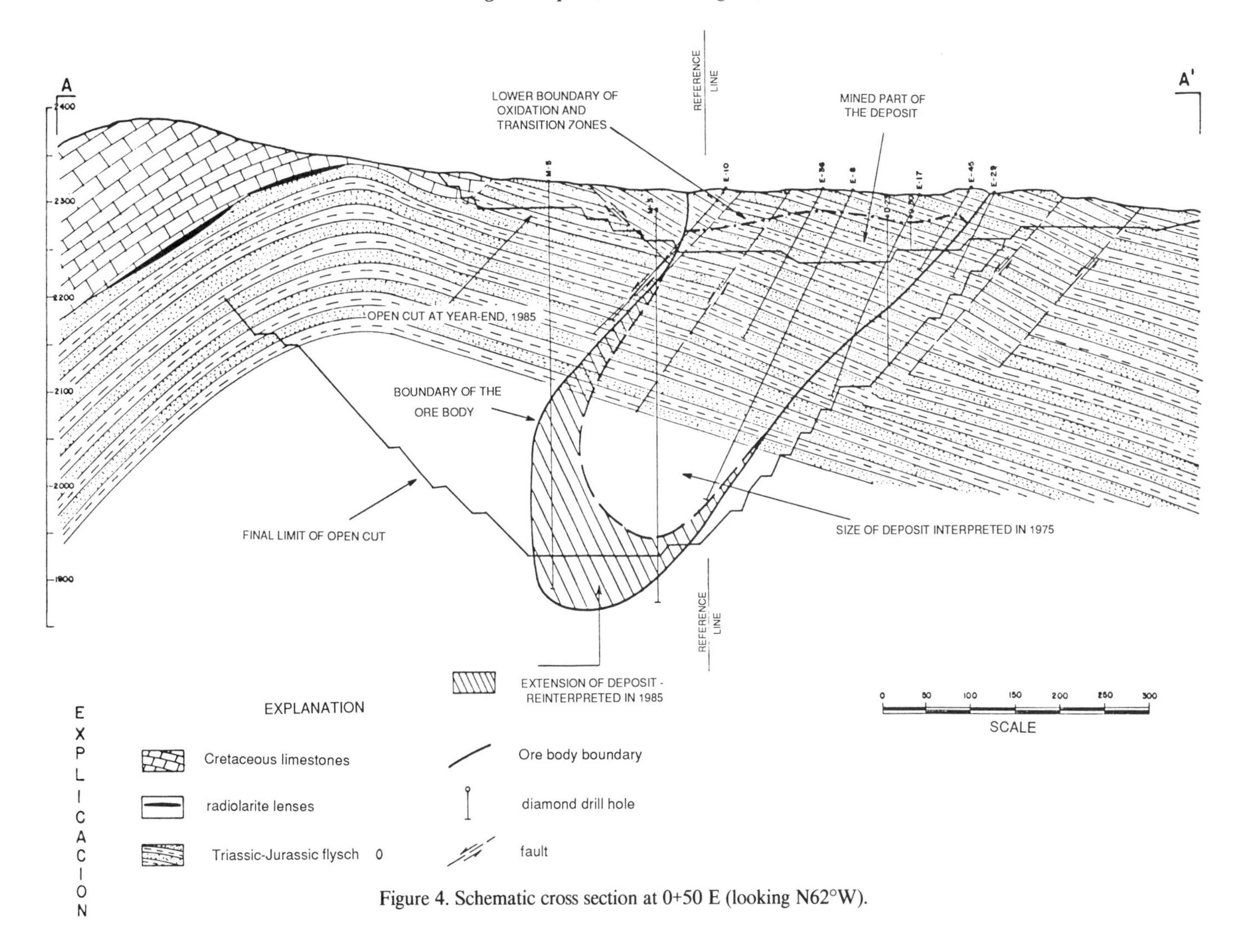

Figure 4. Schematic cross section at 0+50 E (looking N62°W).

The host rock is a possibly Triassic to Jurassic shallow marine clastic sequence consisting of sandstones, siltstones, and argillites—a flysch-type stratigraphy that commonly displays lenticularity, cross-bedding, and slumping, and is strongly affected by folding, fracturing, hydrothermal alteration, and weathering near the surface.

The mineralization is evidently controlled by the stratigraphy; an estimated 80 percent or more of the disseminated and stockwork silver-lead-zinc values occur in sandstones. Certain stratigraphic arrangements such as interbedded medium-thick (0.20 to 0.60 m) sandstones and thin siltstones are the most favorable.

The commercial mineralization in general occurs disseminated and as stockwork, veins, and massive aggregates. The stockwork is a swarm of veinlets less than 1 mm to several centimeters thick, developed preferentially within individual sandstone beds, and commonly perpendicular to the bedding. Disseminations appear as diminutive mineral grains scattered in the intergranular spaces of the sandstone groundmass. Veins occupy fault zones or areas of strong parallel fractures cross-cutting the sedimentary interval. Some veins occur within the disseminated body but mostly toward the south edge. Vein thicknesses range from 0.5 to 1.5 m. The most important system trends N70°W and dips 70°NE; smaller systems trend east-west and northeast-southwest. Massive sulfide aggregates occur commonly as 0.15- to 0.60-m-thick beds or lenses parallel to the bedding planes.

Reserves

The deposit shows reserves in the range of 65 million tons of 85 g/ton Ag, 1 percent Pb, and 0.95 percent zinc. Equivalent silver values for the next reserves calculations and a new deep-drilling program for increased reserves are presently being considered.

Mineralogy

Two different styles of mineralization are present; one is clearly syngenetic, extending outward from the boundaries of the

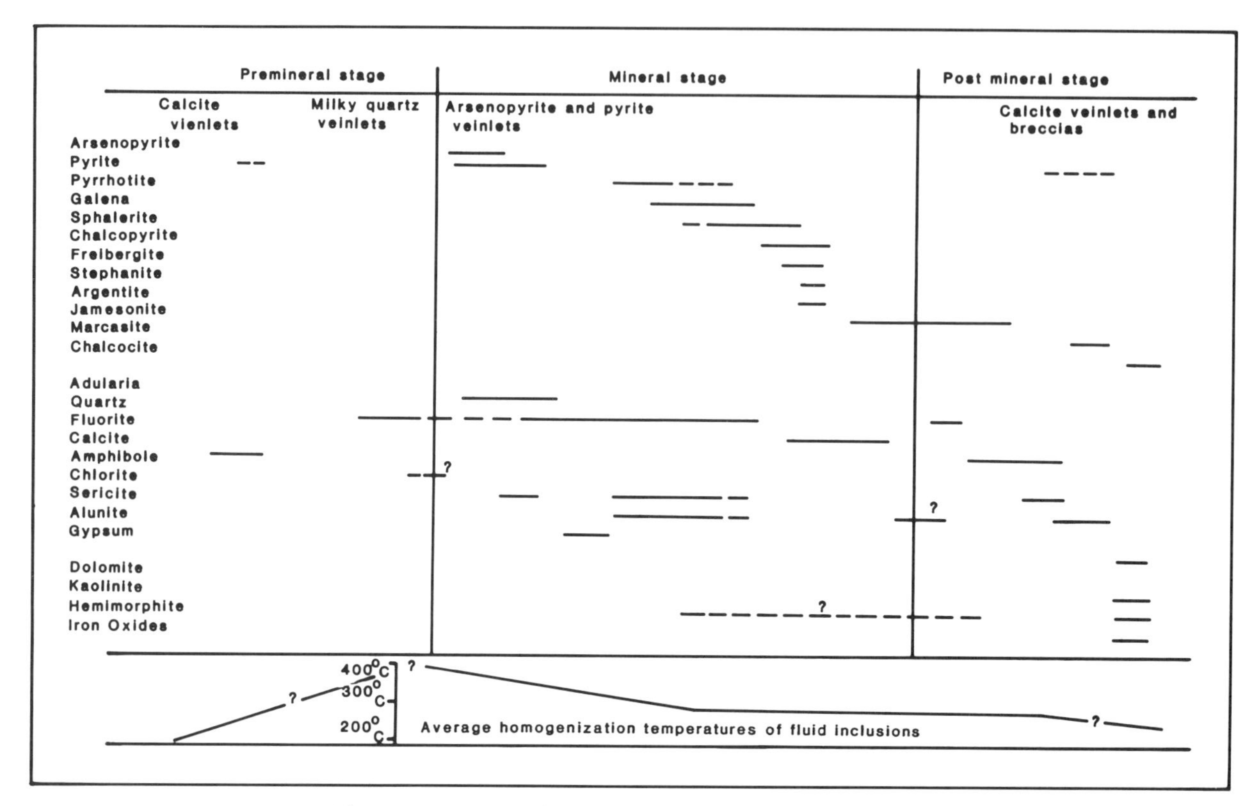

Figure 5. Paragenesis of Real de Angeles ore and gangue minerals.

commercial deposit and containing interbedded small layers and lenses of pyrite and pyrrhotite in siltstone beds. Ore mineralization occurs mainly in clearly epigenetic structures, such as veins and veinlets closely associated with hydrothermal alteration, silicification, and chloritization minerals.

Ores have been classified as sulfides, transitional, and oxides, according to the effects of different degrees of weathering on the metallurgy; to date the latter two types have already been mined. The ores are described below.

Sulfide zone. This zone forms the largest part of the deposit. Ore minerals are freibergite, tetrahedrite, silver-bearing galena, and marmatite; pyrrhotite, pyrite, and arsenopyrite form the gangue, together with quartz, calcite, and fluorite. The zone extends from level 2260 to the bottom of the known deposit (level 1866). The sulfides display vertical mineralogical zoning, presently under study; lead, silver, and pyrite contents decrease slightly, and zinc, copper, and arsenopyrite increase with depth. Sulfide aggregates occur discretely, which simplifies their release and metallurgical concentration by differential flotation.

Transition zone. This zone, 25 m thick in the central part of the deposit and thinning to a few meters at the periphery, comprises a variable mixture of mineralogical species of the sulfide and oxide zones; supergene enrichment has slightly increased the silver values.

Oxidation zone. Secondary minerals in this zone are silver halide, cerargyrite, and bromyrite, as well as jarosite and argentojarosite; oxidized lead and zinc minerals such as cerussite, anglesite, hydrozincite, and hemimorphite are accompanied by iron and manganese oxides.

In contrast to the transition zone, the thickness of the oxidation zone is minimum (2 m) at the center of the deposit and increases to 30 m toward the periphery.

Paragenesis

Field evidence (structures and alterations) and laboratory information (salinities, formation temperatures, ore, gangue, and alteration mineral textures) indicate a hydrothermal deposit model.

The evolution of a highly permeable system adequate for mineralization is explained by the folding episode. The preparation of the host rock included the opening of main channels (veins) and generation of microfractures perpendicular to bedding in response to the different degrees of competence between the sandstone beds (fracturing) and the siltstones (foliation). Also, the primary porosity of the sandstone beds contributed available spaces for deposition of minerals.

A deep-seated intrusive body, suggested by the present domal structure, is proposed as the source of the mineralizing solutions.

According to Mark Pearson, Real Angeles paragenesis took place in three stages—premineral, mineral, and postmineral—and most sulfides were deposited sequentially in time at one stage only (Fig. 5).

Classification of the deposit

Fluid inclusion measurements indicate that ore and gangue minerals precipitated from solutions of low to high salinity. Formation temperatures, restricted within a range of 285 to 298°C, show this to be a mesothermal deposit. High pressures, leading to mineral deposition predominantly in free spaces within fractures, microfractures, and between grains, have also been proposed.

Exploration

Surface mineralization and alteration, and the remains of old mineworks, clearly indicated the presence of a mineralized zone. Exploration activity was initiated in 1971 aimed at evaluating this low-grade, large-volume deposit and defining its economic feasibility, as well as locating water supplies for an eventual treatment plant and nonmineralized areas for the mine installations.

During production, the exploratory effort focused on extending the deposit at depth toward the south-southwest, defining grade distributions, and solving specific mining and metallurgical problems.

Geologic mapping and sampling. An intensive surface and underground mapping program, as well as sampling for metallurgical assays, was carried out in 1974 and 1975; subsequent production control uses rotary-drill ditch samples.

Diamond-drilling. Four diamond-drill programs for a total of 91 drillholes and over 17,400 m of cores to date defined the deposit and quantified reserves and grades. Campañía Gamma, S.A., drilled 20 holes in 1971 and the beginnings of 1972. Fifty-eight holes drilled by Explomín from 1974 to mid-1975 led to preliminary estimates of reserves (manual by polygons, and by computer) for the economic feasibility study. Four holes drilled by Explomín at year-end 1975 proved the deposit's extension at depth toward the southwest, and mine holes drilled by Minera Real de Angeles in 1984 increased ore reserves southwest of the deposit.

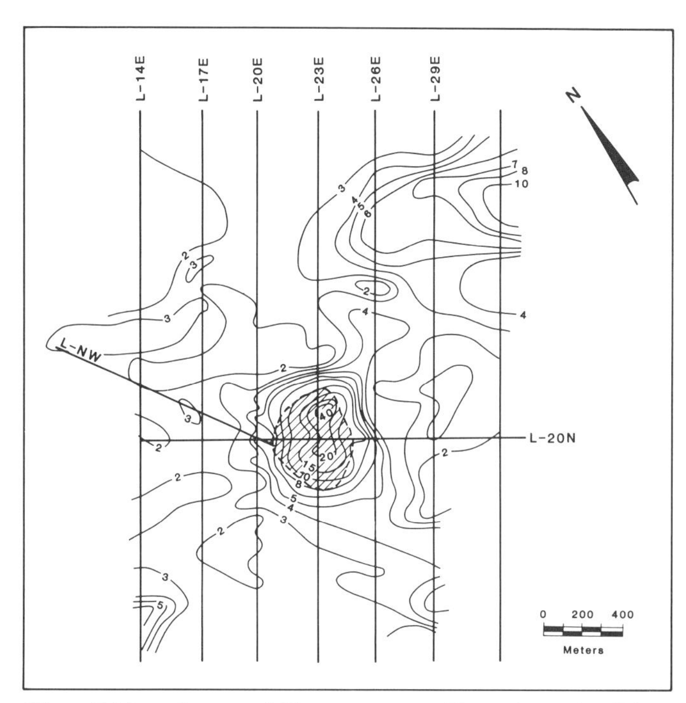

Figure 7. Magnetic susceptibility contour map. Electrode spacing 100 m. Contours in milliseconds. Shaded area indicates the Real de Angeles ore deposit.

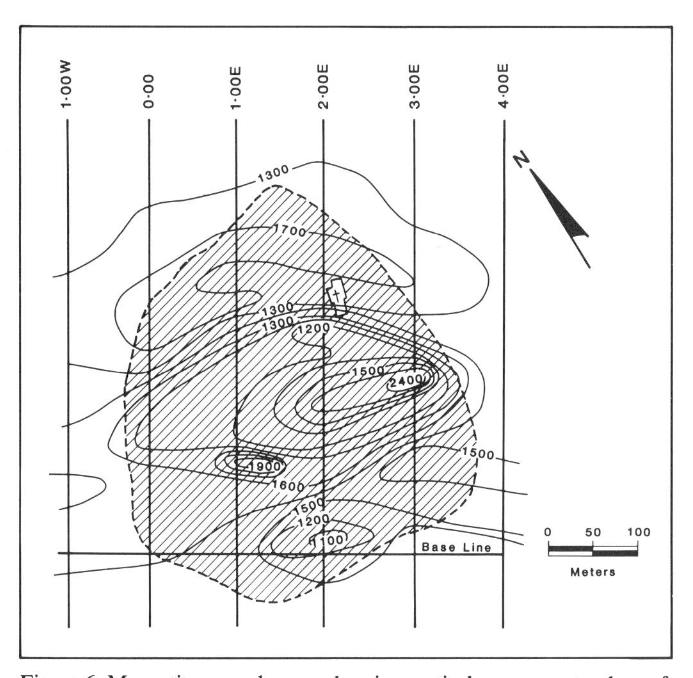

Figure 6. Magnetic anomaly map showing vertical component values of the magnetic field. Contour interval 100 gammas. Shaded area indicates the Real de Angeles ore deposit.

Drillholes are spaced approximately 50 m apart. Except for nine vertical drillholes, the rest have 208° azimuths and variable (45 to 75°) inclinations. The drill plan sought to intersect the main mineralization controls (i.e., north-dipping stratification and northwest-trending, northeast-dipping vein system) at right angles. Average drillhole depth is 150 m; some reach 400 m.

Geophysics. Magnetometry and induced polarization were applied over the deposit and surrounding area in 1975 to demark the deposit's lateral boundaries, aid diamond-drilling orientations, and investigate mineral potential at sites preselected for plant installations and for dump and bin areas.

In general, the ore minerals do not respond to electric susceptibility and have no natural magnetism; however, the associated gangue sulfides are highly susceptible (pyrite) and magnetic (pyrrhotite).

The deposit was clearly defined by its high susceptibility (over 10 milliseconds), low resistivity (under 75 ohm-m), and strong magnetism; the results show this to be an excellent prospecting tool over alluvium-covered areas.

Figures 6, 7, and 8 show magnetic anomaly, susceptibility and apparent resistivity maps, respectively.

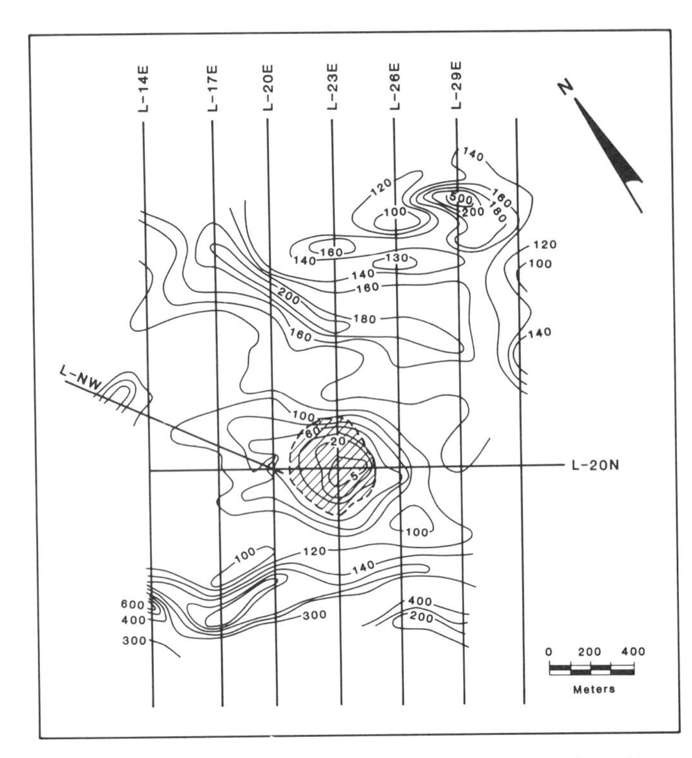

Figure 8. Apparent resistivity contour map. Electrode spacing 100 m. Contours in ohm-m. Shaded area indicates the Real de Angeles ore deposit.

MINING-METALLURGICAL OPERATION

Mine

The Real de Angeles deposit is opencut-mined. The estimated final cut will be a 400-m-deep truncated cone with a 1,200-m upper diameter, 45° slopes, 12-m-high benches, and hauling ramps with a 10 percent gradient. The mining operation includes drilling, blasting, loading, and hauling. Table 1 shows increases obtained at the cut since the start of operations.

Smelter

Ores are treated by selective flotation, producing a lead-silver concentrate of approximately 55 percent lead and 4.5 kgs Ag per ton and an approximately 48 percent grade zinc concentrate. The metallurgical process comprises seven main stages: (a) three crushing processes (primary, secondary, and tertiary), (b) two milling processes (rod- and ball-milling), (c) two lead flotation processes (primary and exhaustive with three clean-ups), (d) two zinc flotation processes (primary and exhaustive with three clean-ups), (e) thickening, (f) filtering, and (g) handling of concentrates.

Smelter productivity for the Real de Angeles deposit is shown on Table 2. The lead-silver concentrates are sold to the Peñoles foundry at Torreón, Coahuila. The zinc concentrate is transported to Tampico harbor, Tamaulipas, and exported.

TABLE 1. MINE PRODUCTION DATA

	1982*	1983
Mineral extracted	1,208,000 tons 6,700 tons/day	3,734,000 tons 10,372 tons/day
Waste	1,422,000 tons 7,900 tons/day	5,789,000 tons 16,080 tons/day
	1984	**1985**
Mineral extracted	4,355,000 tons 12,097 tons/day	4,991,000 tons 13,864 tons/day
Waste	9,513,000 tons 26,425 tons/day	14,100,000 tons 39,167 tons/day

*Approximately 6 month's production.

TABLE 2. SMELTER PRODUCTIVITY

	1982*	1983	1984	1985
Products				
Mill tonnage	1,208,000	3,734,054	4,355,180	4,904,000
Ag production (oz)	2,126,700	7,600,000	9,000,000	11,400,000
Pb production (tons)	7,179	27,823	30,467	37,800
Zn production (tons)	5,444	21,914	22,270	30,688
Recoveries		(%)	(%)	(%)
Silver		65	68.5	78
Lead		78	78.3	82
Zinc		61	65.0	65
Smelter availability	84.95	92.2	94	

*Approximate 6 month's production.

REFERENCES CITED

Bravo N., J., 1986, Real de Angeles, *in* Ordoñez C., J. E., ed., Minas Mexicanas, Vol. 1: Mexico, D.F., American Institute of Mining and Metallurgical Engineers and Society of Economic Geologists, Mexican Chapter, p. 185–209.

Stoiser L., R., and Bravo N., J., 1979, Geophysical response of gangue minerals as a tool in the exploration of the Real de Angeles silver-lead-zinc Mineral District, Zacatecas, Mexico: Geological Survey of Canada Economic Geology Report 31, p. 727–734.

Unpublished sources of information

Avila, P. C., 1976, Calculo de reservas minerales por el método de poligonos en el Proyecto Real de Angeles [professional thesis]: Universidad Autónoma de San Luis Potosí.

Bravo N., J., 1975, Volumen de Geología, correspondiente al estudio de viabilidad economica del Proyecto Real de Angeles: Minera Real de Angeles report.

Gutierrez B., J., 1985, Conferencía sobre la Minera Real de Angeles: American Institute of Mining and Metallurgical Engineers, Tucson, Arizona.

Pearson, M. F., 1985, Geology, mineralogy and isotopic studies of the Real de Angeles silver deposit, Zacatecas, Mexico [M.S. thesis]: El Paso, University of Texas.

Posadas H., A., 1973, Geología y yacimientos minerales de Real de Angeles [professional thesis]: Universidad Autónoma de San Luis Potosí.

Serrano G., J., 1984, Historia del Real de Minas de Angeles, 1700–1984; Reporte final de Investigación: Minera Real de Angeles report.

Printed in U.S.A.

The Geology of North America
Vol. P-3, Economic Geology, Mexico
The Geological Society of America, 1991

Chapter 53

Mineral deposits of the Guanajuato Mining District, Guanajuato

F. Querol S., G. K. Lowther, and E. Navarro
Cía. Fresnillo, S.A. de C.V., Río de la Plata 48 esq. Lerma, Col. Cuauhtemoc, 06500 Mexico, D.F.

INTRODUCTION

The Guanajuato Mining District covers an area 20 km long by an average of 16 km wide located 475 km northwest of Mexico City in the approximate center of Guanajuato State (Fig. 1). It includes the La Luz area on the west, Peregrina–El Cubo on the east, and Santa Rosa to the northeast (Fig. 2).

The town of Guanajuato occupies a narrow valley almost at the center of the district, at 21°01′01″N and 101°15′20″W.

The mining district connects with Highway 57 by way of Dolores Hidalgo and San Luis de La Paz; and with Highway 45 by way of Silao. The Guanajuato-Silao railroad branch connects the district with the main Mexico–Ciudad Juárez line. A small-craft airport (Los Infantes) is about 18 km south of Guanajuato on the road to Irapuato; commercial airlines fly to Mexico City from the town of León, on Highway 45, 62 km west from Guanajuato. A 3.2-km all-weather dirt road from La Olla Dam in Guanajuato City to Mineral del Cubo leads to the Torres-Cedros Mine, the largest producer in the district.

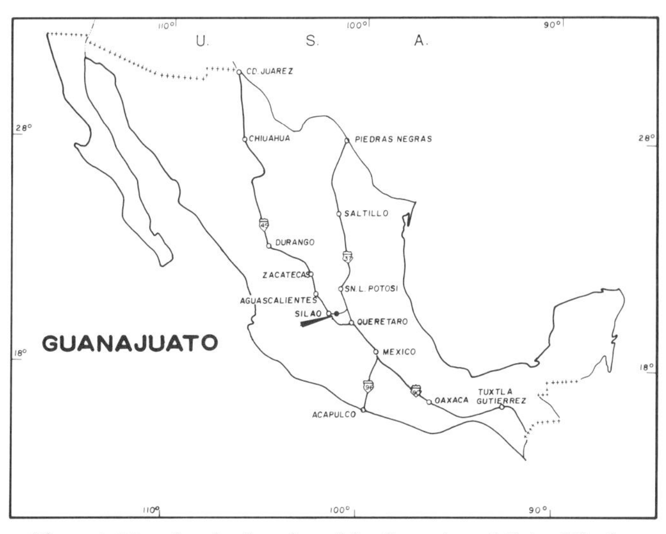

Figure 1. Map showing location of the Guanajuato Mining District.

History of the district

Information on the history of the district comes from Antúnez (1964), Terrazas (1968), and Wandke and Martínez (1928).

The Guanajuato District area is situated in the southern part of the pre-Columbian Chichimeca Empire, finally conquered by the Spaniards under Nuño de Guzmán in 1529. In 1548 a mule-train drover on the way to Mexico from the recently discovered Zacatecas Mines discovered melted silver around a campfire near the San Bernabé Vein (today's La Luz Vein) in the La Luz District northwest of Guanajuato. This man, Juan de Rayas, remained in the area and discovered the crest of the Guanajuato Veta Madre (Mother Vein) in what is now the Rayas Mine (Fig. 3) in 1550. Further discoveries until 1558 along the Veta Madre would become the La Valenciana, Tepeyac, Mellado, Cata, Sirena, and other mines. Once the local tribes had been pacified, the district made quick progress and in 1619 was granted the title of Noble y Leal Villa de Santa Fe de Guanajuato by Philip III of Spain. The use of gunpowder was introduced in 1726 by Don José de Sardaneta y Legaspi and increased production considerably. Soon the construction of the Tiro de Rayas was begun. From 1760 to 1770, Antonio Obregón y Alcocer, later Count of Valenciana, carried out numerous explorations that led to the discovery and development of the Valenciana oreshoot; mineworks at El Cedro probably began that same year (1770) and subsequently in Peregrina, El Monte, San Nicolás, Santa Rosa, and El Cubo. Large volumes of rich ores produced in Guanajuato from 1775 to 1810 led to its being ranked as the city of Guanajuato by Philip V of Spain.

The bonanza continued until 1821 when insurrectionists burned all mine installations including the recently erected Tiro Valenciana. Guanajuato became a refuge for criminals and thieves, and all attempts to reactivate the mines failed. English capital was first introduced in 1823 but met with no success until 1868 when the Valenciana mineworks were finally renewed; the mine was drained using four English-made steam hoists and leather buckets.

The La Luz District was a large bonanza from 1845 to 1854. Impure nitroglycerine (dynamite) introduced in 1872 fur-

Querol S., F., Lowther, G. K., and Navarro, E., 1991, Mineral deposits of the Guanajuato Mining District, Guanajuato, *in* Salas, G. P., ed., Economic Geology, Mexico: Boulder, Colorado, Geological Society of America, The Geology of North America, v. P-3.

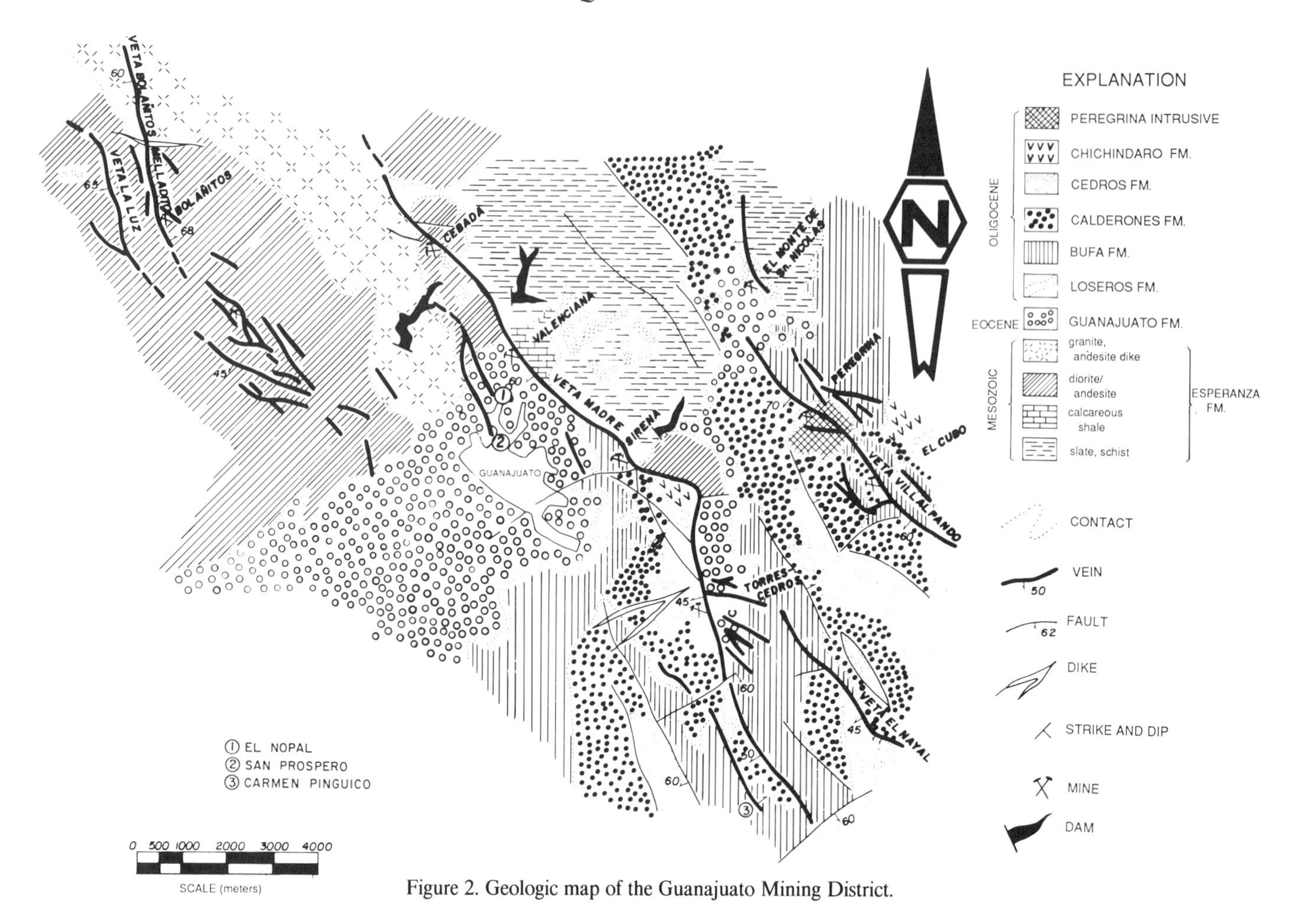

Figure 2. Geologic map of the Guanajuato Mining District.

thered mining operations. Guanajuato continued producing successfully until 1888 when vast floods led to the loss of many lives and mining activities were suspended. Activity continued on a smaller scale in the following years until 1898 when the mines were closed down because of intolerable ventilation conditions that cost many lives.

All the cores produced until then were treated by the patio process, which Bartolomé de Medina had invented in Pachuca in 1555. This process was basically an empirical method of amalgamation involving first, ore crushing by "arrastras" (enormous animal-hauled rocks), then mixing of the fine crushed product with salt, iron, copper-bearing pyrites, lime, and vegetable ash, and the addition of mercury to recover gold and silver as amalgam; processing took place in large patios, with the mixture being moved continuously by horse or mule traction.

At the end of the nineteenth and beginning of the twentieth century, North Americans, attracted by the historic Spanish and English bonanzas, promoted the Guanajuato mines and installed cyanide-process plants mostly for ores that had not been patio-processed; gold and silver were extracted mainly from tailings, fills, and pillars, but no new exploration was done. In 1935, workers demanded improved salaries and benefits, and the companies departed, leaving the mines to the workers in lieu of unpaid salaries. The Sociedad Cooperativa Minero-Metalúrgica de Santa Fe de Guanajuato was thus founded and initiated formal operations in 1939. Government aid since 1947 helped to overcome the lack of reserves and capital; the cooperative drained the mines and progressively increased production to the present, becoming one of the district's principal producers (Terrazas, 1968).

Meanwhile, mining went on at La Peregrina within the Sierra group of veins. The small Peregrina Mine (as compared to Veta Madre) was discovered in 1804 and was worked intermittently during the past century. Compañía Minera de Guadalupe, S.A., initiated activities in 1933, extracted old fills and installed a 70 ton/day treatment plant, remaining active until 1944. The Peregrina Mine and Plant were improved in 1958, and mining activities continued until 1968 when they were acquired by Negociación Minera Santa Lucía, S.A. de C.V.

The Bolañitos Mine was unimportant until several ore shoots were discovered in 1887; by 1893 and 1894 these had produced 50 tons/day of 2,000 ppm Ag. In 1972 and 1973, Compañía Minera Las Torres, S.A. de C.V., evaluated approximately 200,000 tons of 1.92 ppm Au and 242 ppm Ag in the mine area, based on its own and previous drilling information.

An extensive exploration program by Compañía Fresnillo in 1968, which included 36 diamond-drillholes for a total 18,000 m, led to the discovery of the Torres-Cedros orebody (today's Las Torres Mine), comprising 2,375,000 tons of 2.2 ppm Au and 353 ppm Ag. Construction of Tiro Guanajuato, the orebody extraction and services center, began in 1970.

The Cebada Mine was explored intermittently by diamond-drilling between 1944 and 1948 and remained under contract to the Fresnillo Company until 1961 when the contract was transferred to the recently nationalized Compañía Fresnillo, S.A. de C.V. The properties were acquired from the owner in 1968 and incorporated to Negociación Minera Santa Lucía, S.A. de C.V.

At present the Guanajuato Group—Compañía Minera Las Torres, S.A. de C.V., Negociación Minera Santa Lucía, S.A. de C.V., and Compañía Minera Cedros, S.A. de C.V., under management by Compañía Fresnillo, S.A. de C.V., the major shareholder of the Las Torres and Santa Lucía Companies—directs exploration and development of the Las Torres–Cedros, Bolañitos, Cebada, and Peregrina Mines in the Guanajuato District.

Production of the district

Precise Guanajuato District production figures in 437 years of mining history are unavailable; however, Terrazas (1968) estimates that 30,596 tons of silver and 133 tons of gold were produced up to 1968. The present daily company production is as follows: Sociedad Cooperative Minero-Metalúrgica Santa Fe de Guanajuato, 750 tons; Compañía Minera El Cubo, S.A., 200 tons; and Grupo Guanajuato, 2,000 tons.

Table 1 shows total production figures of the Guanajuato Group mines; from 1976 to 1984 the group's production was approximately 157 tons of Ag per year in concentrates.

TABLE 1. GUANAJUATO GROUP TOTAL PRODUCTION

Mine	Tonnage	Grades Au (ppm)	Grades Ag (ppm)
Torres	2,265,177	1.09	260
Cedros	1,192,097	1.49	274
Peregrina	645,922	4.33	306
Cebada	415,368	4.84	453
Bolañitos	379,262	3.08	288

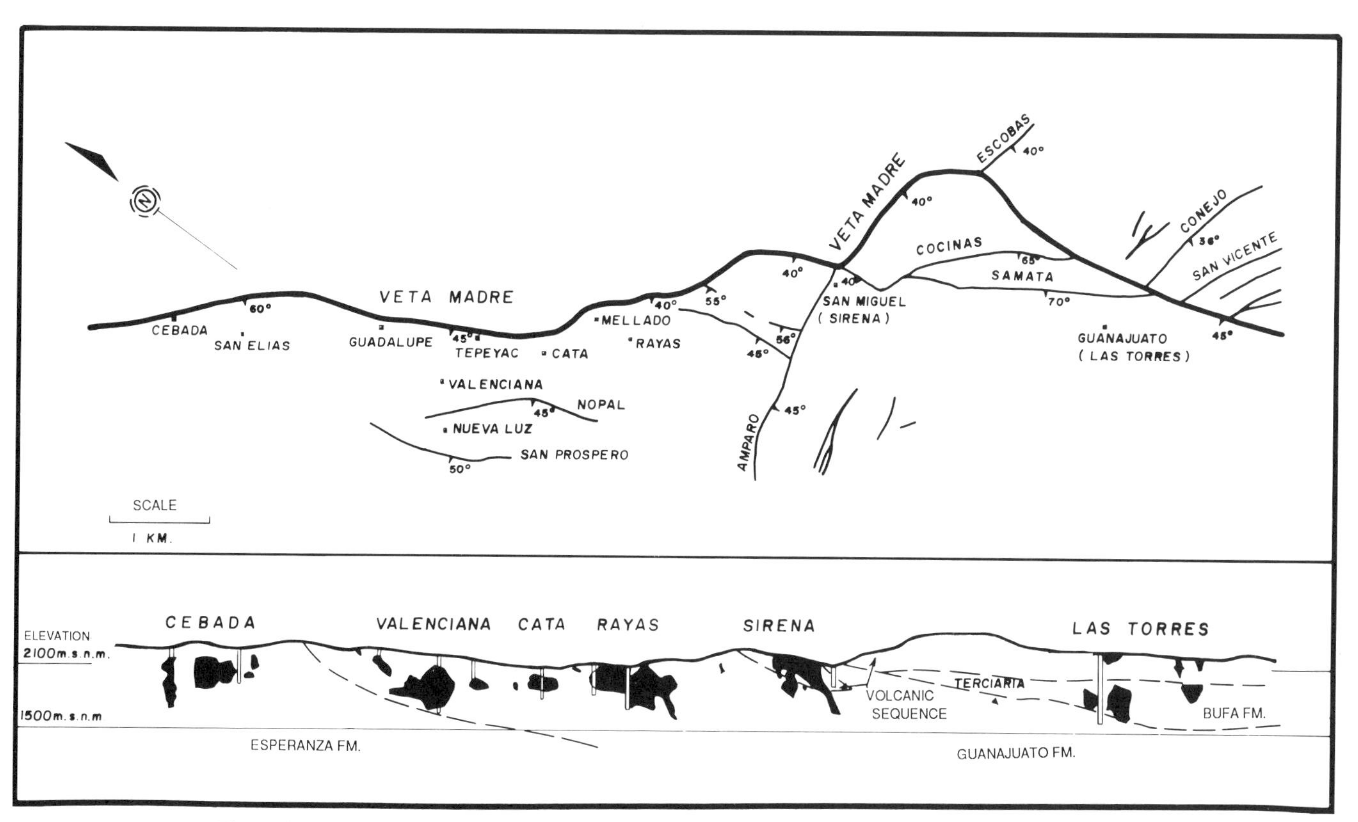

Figure 3. Map and longitudinal cross section of Veta Madre showing the most important mines and formations in the hanging wall.

GEOLOGY

The Guanajuato District lies in the southern part of the Meseta Central physiographic province, a semidesertic plateau averaging 2,000 m in elevation with local heights exceeding 3,000 m. The district area is part of the eastern flank of an extensive northwest-trending anticline that forms the Sierra de Guanajuato and Sierra Gorda; the anticline is affected by numerous staggered faults semiparallel to its axis and by almost perpendicular fractures.

The basement contains Mesozoic metamorphosed igneous and sedimentary rocks unconformably overlain by Tertiary conglomeratic red beds followed by a thick Oligocene volcanic sequence. The youngest volcanic evidence appears in the northwest part of the district where possibly Pleistocene basaltic lava flows—seemingly much younger than the mineralization—cover the heights of the neighboring mountains (Gómez de la Rosa, 1961) and are probably related to Neovolcanic Axis basalts.

Stratigraphy

Outcrops in the region range from Mesozoic to Recent in age. Buchanan's (1980) description of the stratigraphy of the district is briefly summarized on Figure 4.

Esperanza Formation. This is the oldest exposed unit; it includes gray to black carbonaceous and fossiliferous shale, calcilutite, and thin-bedded sandstones with intercalated basalt and andesite lavas, all slightly metamorphosed to phyllitic shales, slates, and marbles with local tactites and metasomatic marbles. Although the base is unknown, the formation is over 500 m thick, according to information from the deepest Valenciana mineworks (Echegoyen and others, 1970). No fossils have been identified; however, Edwards (1955) reports Late Jurassic–Early Cretaceous corals in limestone fragments from the overlying conglomerate, which are possibly derived from the Esperanza. The formation is exposed at the Veta Madre footwall north, northwest, and west of Guanajuato.

Guanajuato Formation. A series of thin- to thick-bedded red beds unconformably overlying the Esperanza Formation consists of a poorly sorted conglomerate of shale and sandstone pebbles and cobbles in an arenaceous groundmass. Near the base, the clasts are interbedded with thin volcanoclastic sandstones and andesite flows; this lower part of the formation includes fine-grained shale, sandstones, and cobble conglomerate, separated by an unconformity from an overlying pebble conglomerate that grades upward to cobble conglomerate again. The formation is restricted to the Veta Madre hanging wall and is covered by younger volcanic rocks in fault contact with the Esperanza Formation on the west. The Guanajuato Formation's estimated thickness is 1,500 to 2,000 m. Edwards (1955) assigned it to the early Tertiary—probably late Eocene or early Oligocene—on the basis of its fossil vertebrates.

Loseros Formation. Conformably overlying the Guanajuato conglomerate, the Loseros Formation consists of thin- to medium-bedded, fine-grained (subrounded to angular), green, red, and purple volcanoclastic sandstones, 10 to 25 m thick over most of the district; in the vicinity of Las Torres Mine it becomes less platy and coarser grained, and is up to 52 m thick (Edwards, 1955). Assigned to the Oligocene by its stratigraphic position, some authors (Cepeda, 1967) consider it to be the lower member of the Bufa Formation.

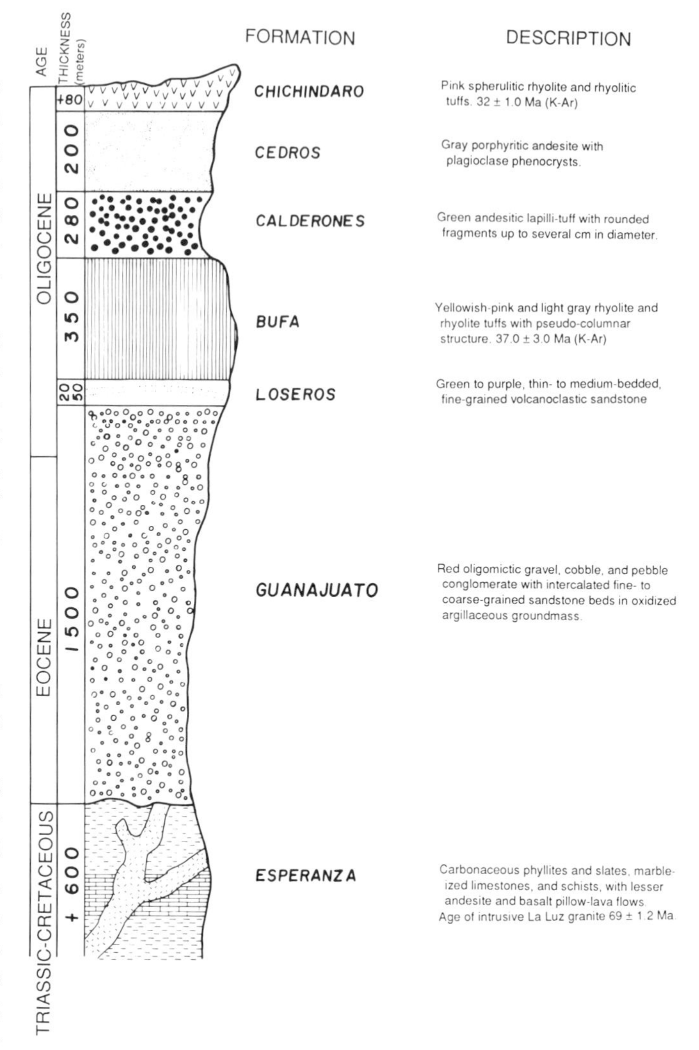

Figure 4. Stratigraphic column of the Guanajuato Mining District.

Bufa Formation. Unconformably overlying the Loseros Formation is the Bufa Formation, an air-fall deposit of yellow, pink, and white lithic lapilli and vitreous rhyolite tuff, which occasionally displays weak columnar jointing and poorly welded glass fragments toward the center of the formation. Quartz, sanidine, and scarce plagioclase crystals form 15 to 25 percent of the rock volume.

Rhyolite and andesite fragments occur sporadically, and some fragments consist of shales; the tuffaceous groundmass makes up 5 to 80 percent of the volumetric whole. Gross (1975) reports an isotopic age determination of 37.0 ± 3.0 Ma (K/Ar) (i.e., early Oligocene). The formation varies widely in thickness, averaging 360 m. The Bufa Formation is the Veta Madre hanging-wall host rock in Las Torres Mine, and is at the footwall southeast of the mine. Most of the Sierra de Guanajuato ores are formed in these rocks.

Calderones Formation. Unconformably overlying the La Bufa Formation is a green lithic lapilli and crystalline andesite tuff named the Calderones Formation. Crystal fragments consist of plagioclase (An_{24}), augite, hypersthene, and some quartz; most lithic fragments are chloritized; angular porphyritic and aphanitic andesite fragments at times form 75 percent of the rock. The formation is typically medium- to thick-bedded except at the base, which includes laminar volcanoclastic sandstones and shales of andesitic composition and fluvial origin, present only in eroded paleochannels within the formation.

Locally, beds within the formation display dessication, rain, and current traces as well as graded bedding. The formation is the most extensively exposed in the district, ranges 200 to 250 m in thickness, and is assigned to the Oligocene on the basis of its stratigraphic position (Gross, 1975). The upper 5 m of the unit is an interval of finely laminated crystalline andesite tuff grading imperceptibly to the overlying Cedros Formation.

Cedros Formation. This unit consists of lava flows of gray to black porphyritic hypersthene andesite with intercalated gray andesite-lapilli tuffs. Rock volumes consist of plagioclase (An_{29}) phenocrysts (10 to 20 percent), augite phenocrysts (2 to 3 percent), hypersthene phenocrysts (7 to 8 percent), and an aphanitic groundmass (50 percent; Taylor, 1972). Widely varying thickness estimates range from 250 to 640 m. The formation is well exposed in the Veta Madre hanging-wall host rock and, though not isotopically dated, has been assigned to the middle Oligocene (Buchanan, 1980).

Chichíndaro Formation. Unconformably overlying the Cedros Formation, this unit consists of massively bedded, poorly sorted crystalline and vitreous ignimbrites with irregular lenses of flow breccia. At the type locality it is 100 m thick, but its original thickness is unknown, because it is the youngest unit in the area and is not covered by any other unit. Gross (1975) determined the K/Ar age of this premineralization unit as 32.0 ± 1.0 Ma. It is displaced by the Veta Madre and strongly altered adjacent to the vein.

Intrusive rocks

La Luz Complex. The La Luz Complex is an extensive granite stock bordering the western boundary of the district, with possibly genetically related rhyolite dikes up to 3 m wide. Other dikes of unknown age in relation to the granite are quartz-monzonite, monzonite, and diorite. These intermediate and siliceous dikes intrude a diorite and gabbro near La Luz (Echegoyen and others, 1970), which is seen to have assimilated shale xenoliths of the Esperanza Formation, altering them to laminated black tactites. The La Luz Complex is restricted to the north and west portions of the district, in fault contact with the Guanajuato Formation. It is assigned a Late Cretaceous–early Tertiary age but is not yet isotopically dated (Buchanan, 1980).

Calderones and Cedros dikes. A series of andesite dikes invading the Calderones and Cedros Formations has been proposed as the source for the Cedros andesite. These are porphyritic and microporphyritic, fine-grained, variably colored andesites with phenocrysts up to 10 mm in diameter, similar in composition to the Cedros Formation, that occur as normally vertical tabular bodies up to 20 m wide and 1 km in length.

Chichindaro dikes. White rocks of rhyolitic composition and porphyritic texture with aphanitic groundmass, in which sanidine and quartz phenocrysts are visible, occur in the eastern part of the area, cross-cutting the Cedros Formation as short (less than 250 m) and narrow (less than 10 m wide) dikes.

Structures

The Guanajuato Mine District occupies the northeast flank of a northwest-trending regional anticline; sediments and lavas dip an average 10 to 20° north and northeast. The Esperanza Formation, which has a history of earlier deformation, displays more complex structures. The Tertiary volcanic sediments are locally faulted; the faults in the area may be grouped (Buchanan, 1980) into three series, listed below from older to younger:

1. Older: northeast to east, premineral;
2. Intermediate: northwest, premineral and contemporaneous with mineralization, comprising three main systems: (a) La Sierra system, (b) Veta Madre system, and (c) La Luz system;
3. Youngest: northeast, postmineralization.

The northeast-to-east group dips both north and south with slight displacements. Near Las Torres Mine they apparently precede or are contemporaneous with the Calderones Formation and are frequently occupied by andesite dikes feeding the Cedros Formation; on occasion they are slightly mineralized and the dikes are propylitized.

Although the northwest series trends predominantly northwest, variations trend north-northeast. The northeasternmost La Sierra system contains subparallel, mostly 40 to 80° southeast- to sometimes northeast-dipping faults; the Peregrina and El Cubo Mines produce from veins lodged in this system.

The Veta Madre fault system, 4 km southwest of the La Sierra system, 4 km southwest of the La Sierra system, is the longest of the northwest-trending group; the Veta Madre dips consistently 35 to 55° southwest and has been traced for over 20 km at the surface. According to Buchanan (1980), although it is not completely mapped, it may be as long as 100 km. The system affects all the rocks in the district; total displacements range from 1,200 m in the southeast to 1,700 m at La Valenciana Mine. Parallel short faults are common, particularly at the Veta Madre hanging wall. Cymoid slumping on both sides of the fault is common where changes in strike take place; the fault hosts the Veta Madre deposit, which ranges in thickness from veinlet width to 30 m at depth.

The La Luz system is the most widely varying in attitude of the northwest-trending group. Many of its faults dip 40 to 80° northeast, and others dip 40 to 80° southwest; strikes in that part of the system are generally northwest, curving more toward the west-northwest at the southeastern end of the system, with numerous bifurcations. The degrees of displacement in these faults are unknown.

The northeast group is the youngest. Faults are rare, and their displacements are generally less than 20 m. The known faults are not mineralized and are assumed to be post-mineralization.

MINERAL DEPOSITS

Introduction

The Guanajuato District is a gold and silver producer; basic metals and pyrite appear as traces except for isolated pockets. Gold and silver occur in veins and stockworks consisting essentially of quartz, calcite, and adularia in the northwest-trending group of normal faults described above. Veins occur in all three systems, mostly filling fractures, together with some replacement; the most important is the traditional Veta Madre in the fault of the same name. Many secondary fractures, occasionally containing secondary mineralization, arise from the Veta Madre hanging and footwalls. Other extensive and ore-rich structures approximately parallel to the Veta Madre are La Luz (7 km), Bolañitos-Melladito (6 km), Carmen-Pinguico (3 km), San Próspero (2 km), and El Nopal (2 km) at the hanging wall; and Villalpando (6 km), El Nayal (3 km), and El Conejo (1.5 km) at the footwall (Figs. 2, 3).

Description of the orebodies

Guanajuato orebodies display two main geometries: tabular veins and stockworks. The veins have well-defined walls and crustified texture originating from preexisting fissure fill. Replacement at the walls sometimes results in lenticular cross sections for a vein. Veins are usually small, except in the Valenciana, La Luz, Bolañitos, and Cebada orebodies where they form true lodes.

Stockworks known to date occur at the Veta Madre hanging wall in the Guanajuato Formation (Sirena and Rayas bodies) and in rhyolite of the Bufa Formation (Las Torres Mine). In contrast with the veins, the commercial ore boundary is gradual and ill-defined, made up of a lattice of interconnected mineralized veins and veinlets. Veinlets may be a few millimeters to 30 cm wide and become less frequent away from the main vein. Adjacent to the stockworks, the Veta Madre is frequently sterile and is left behind at the stope walls or floors. The stockworks are the most important orebodies in the district by their tonnage (Valenciana, Rayas, Sirena, and Las Torres). Veinlet systems in the stockworks normally follow north-northwest and east-west directions at 45° from the Veta Madre fault plane.

Individual ore shoots vary from centimeters to 40 m in width and from meters to 450 m along strike and dip. The largest, such as Valenciana, has produced several million tons of ore. Mineralization is under no direct lithologic control, its position depending primarily on structures pre- or contemporaneous with mineralization and on depths from the paleosurface.

La Sierra vein system. Ore shoots of this system occur at different elevations within quartz veins in La Sierra structures. The Peregrina and El Cubo Mines are the most productive to date. Two systems of parallel veins composed of quartz, carbonates, and adularia occur at Peregrina. The Mina Vieja system (Gross, 1975) crops out and contains ore shoots less than 100 m long extending vertically from the surface down to 300 m (2,000 m above sea level; asl); vein material (15 ppm Ag, 0.1 ppm Au) continues downward below the deepest mineworks. The Mina Nueva system does not crop out, ore shoots are usually over 100 m long, and extend from 2,200 m to 1,700 m asl. Gross (1975) terms the latter "lower ore" and the former "upper ore." La Sierra veins are characterized by a larger Au/Ag ratio, having 1.5 to 4 gm Au/100 gm Ag (Wandke and Martínez, 1928). Villalpando Vein shoots extend continuously from 1,850 to 2,350 m asl; however, toward the southwest the shoot is nearer the surface and therefore wedges out downward at lesser depths.

La Luz vein system. This system is located about 11 km to the northwest of the Veta Madre hanging wall. Ore deposits extend for about 8 km along the La Luz, Bolañitos-Melladito, and several smaller veins; strikes range N 10-60°W with 45°-80° southwest dips, except for the northeast-dipping Bolañitos Vein. Bolañitos ore shoots extend continuously from elevations of 2,080 m to 2,300 m at the surface where the ore crops out.

Veta Madre system. Figure 3 shows the distribution of ore shoots along this imposing vein. According to Gross (1975), the ores follow a vertical pattern, and ore shoots may be separated into the Upper Ore, from 2,100 m asl to the surface; the Lower Ore, from 1,700 to 2,200 m asl; and the Deep Ore, below 1,800 m. No clear separation is found, however, between the Upper and Lower ore shoots in several mines (e.g., Torres and Valenciana), besides which the Deep Ore has not been found outside of Rayas. What is really evident in most Guanajuato veins, and even more so in the Veta Madre, is that the stockwork zone is confined to what Gross terms Lower Ore (i.e., 1,700 to 2,200 m asl), and that the Upper Ore is confined to the tabular structure of the vein itself. The interval between the Lower and Upper Ores, rated as sterile by Gross, is not always so, representing merely the disappearance of the stockwork. Las Torres contains four main shoots: 5020, 3 N, Cedros, and 4100 (Figs. 5 to 7). The largest is 5020 (Figs. 5 and 6), developed at the Veta Madre hanging wall, where it is mineralized and varies from 30 cm to 3 m in width. The rest of the 5020 shoot is a chimney-shaped stockwork-mineralized brecciated structure, in which stockwork veinlets range from a few millimeters to 30 cm in thickness and vary in density, which decreases away from the Veta Madre. The shoot is thickest at 1,800 m asl where it is 225 m long, narrowing downward until it wedges out at 1,675 m asl. It separates upward into two lesser shoots at 1,900 m asl, and disappears at 2,000 m asl where only the vein remains. Its width also decreases toward the ends from a maximum 40 m in the central parts. The approximate tonnage in the shoot, including produced ore and total reserves, is 2,917,800 tons of 1.4 ppm Au and 287 ppm Ag (K. Lowther, personal communication, 1984).

Alteration

Host-rock alteration depends mostly on the nature of the rock itself. Away from the veins the fresh rock is found grading at first slightly, and later strongly, to propylitized rock near the fractures, and at some places to potassic alteration adjacent to the veins. Phyllic and argillic alteration in transverse fractures sometimes is found superimposed on this zonal pattern and at other places in formerly fresh rock. Alterations are briefly described below.

Argillic alteration occurs very near to, and closely associated with, the ores, and consists of kaolinite, halloysite, and montmorillonite.

Phyllic alteration follows argillic alteration spatially but is much more developed. It is found above and at the sides of mineralized zones, where it is assumed to be closely related to boiling-solution processes; the alteration was apparently caused

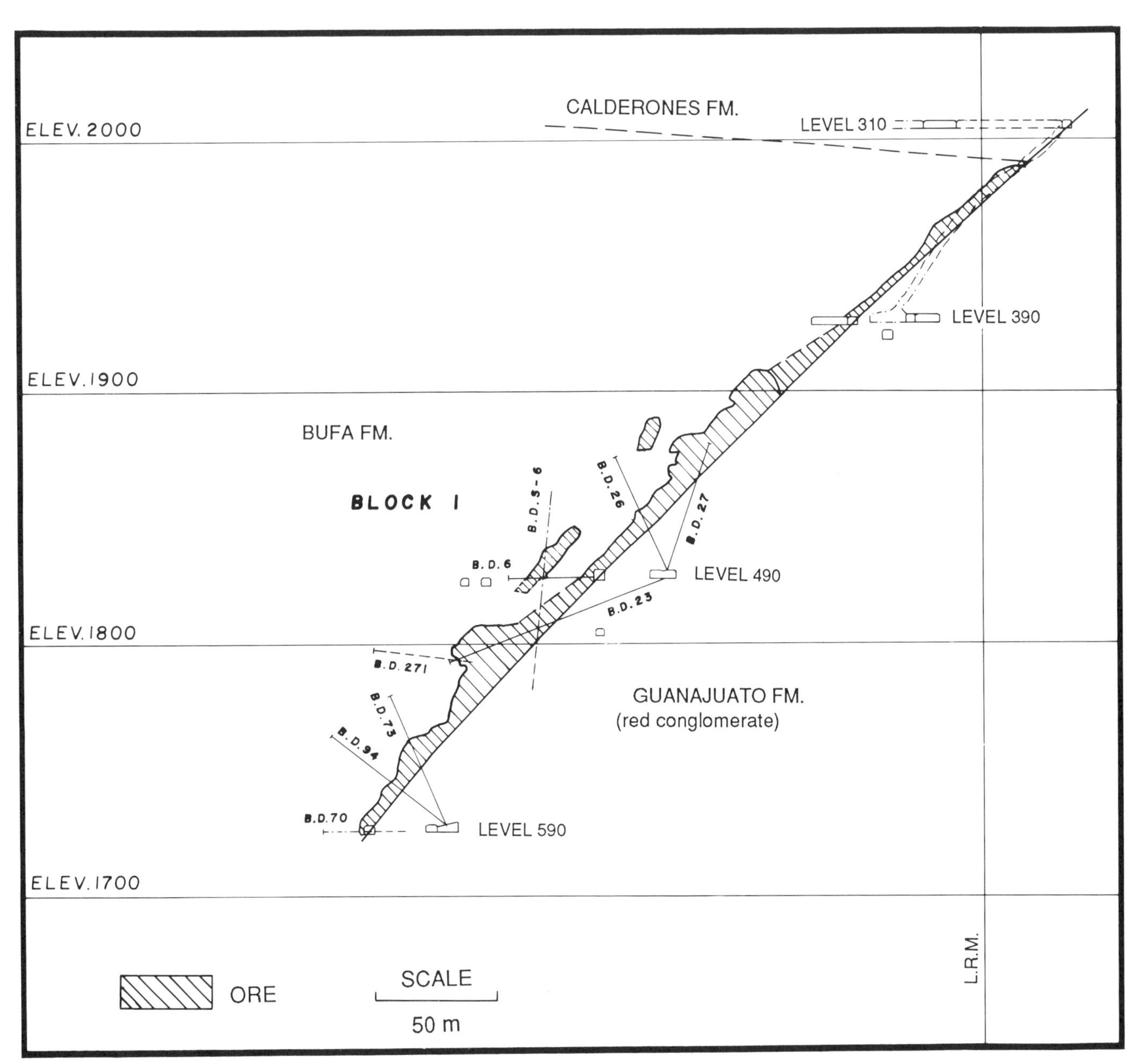

Figure 5. Cross section of the 5020 oreshoot on parallel 5120, looking northwest.

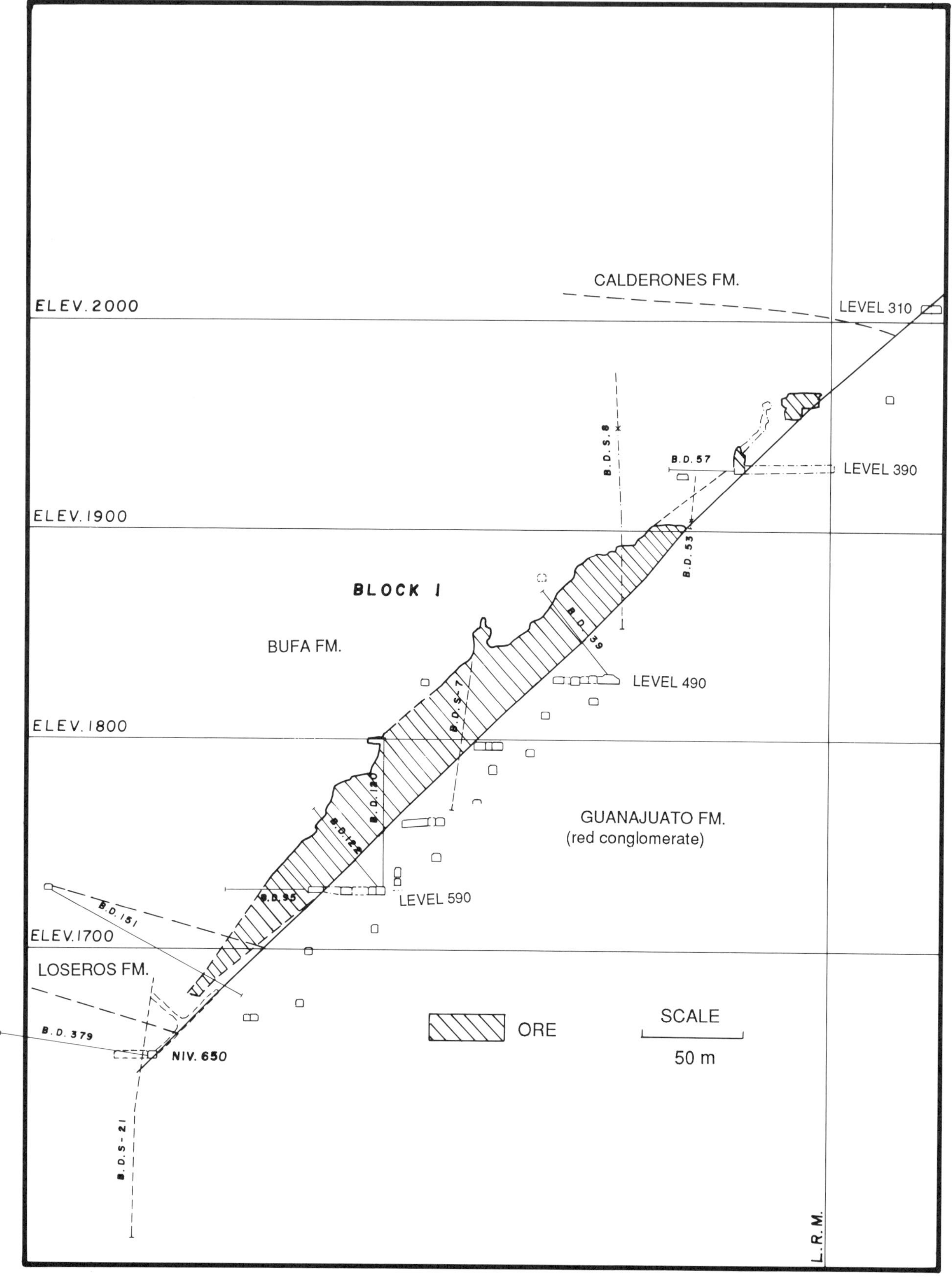

Figure 6. Cross section of the 5020 oreshoot on parallel 4980, looking northwest.

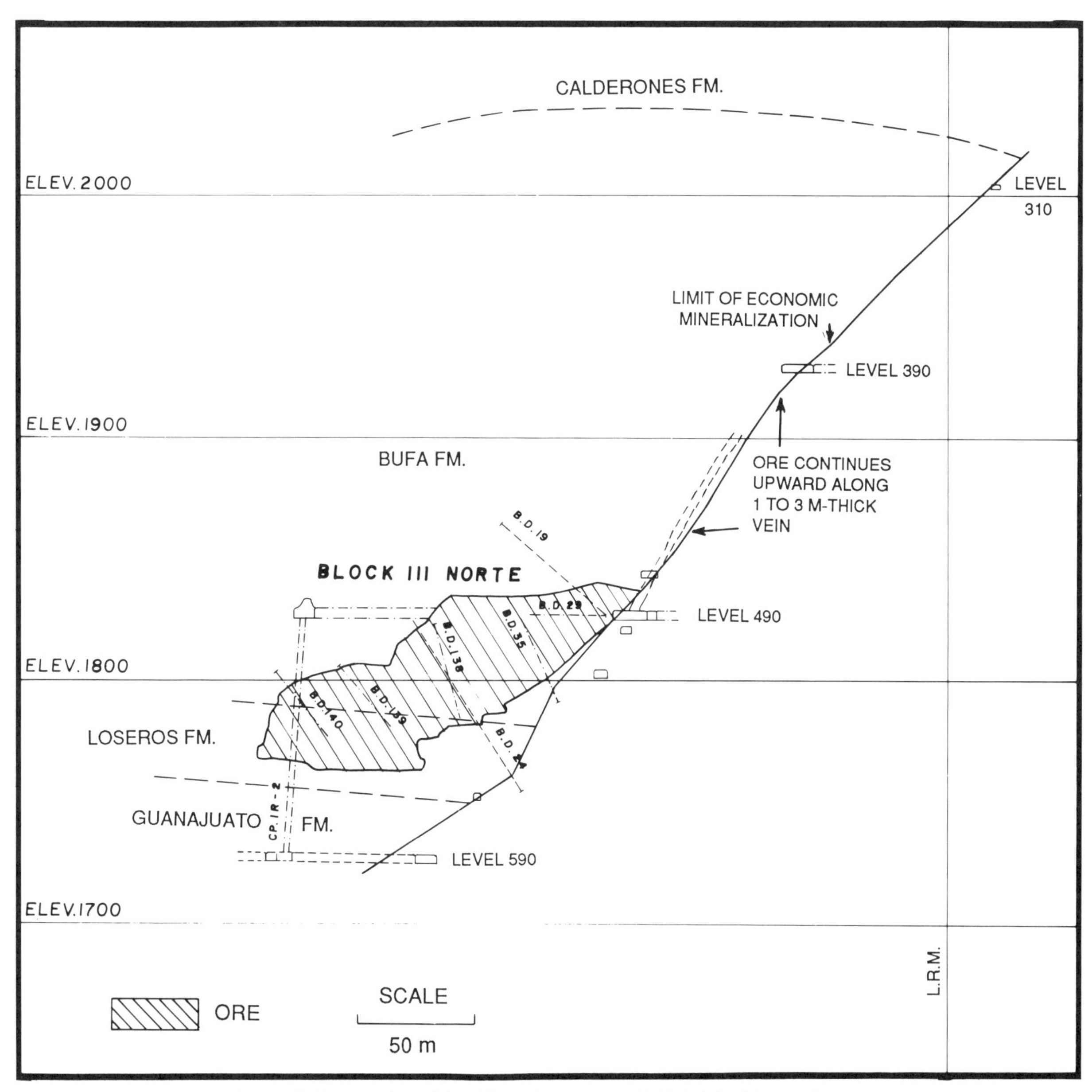

Figure 7. Cross section of the 3N oreshoot on parallel 5280, looking northwest.

by vapors such as HCl and H_2S released during boiling and transported in the volatile phase. Phyllic alteration is characterized by sericite, illite, and pyrite.

Potassic alteration is characterized mostly by adularia and sericite; it may contain lesser amounts of illite, quartz, and calcite; and it occurs in close association with oreshoots only where rock is very fractured.

Propylitic alteration occurs extensively over the area and is useful only at the regional level; it is strongest in rocks originally rich in Mg and Fe, and includes chlorite, clays, calcite, and occasional epidote; it is controlled by the density of fractures in the rock.

Silicification, though not considered as such by many, is closely related to the preparation of the ground together with the potassic alteration, most probably influencing the mechanical behavior of the rock during the formation of the deposit by making it fragile, more easily fractured, and therefore more permeable to mineralizing solutions in the area of the present chimneys. Silicification, being intimately associated with the mineralization process and easily identifiable in the field, is an important type of alteration, usually accompanied by low-grade base-metal sulfides, some gold, and as much as 70 ppm Ag. It may extend 150 m above the Upper Ore shoots; it both precedes and accompanies the ore, and is unknown above 2,350 m asl.

The various rock types in the district reacted differently to hydrothermal alteration, and the alterations described above do not follow simple spatial patterns. Schists and phyllites show preferential silicification and sericitization, whereas the conglomerate tends toward chlorite, adularia, and some sericite. Sericite was observed neither in the Bolañitos area (La Luz) nor in Cebada (northwest of Veta Madre) and is therefore of little use in surface exploration.

Although lithology is not an important element in the location of ore shoots, the rock type evidently determined the types of structures and the area widths favorable to mineralization.

Mineralogy

Ore mineralogy in the Guanajuato District is relatively simple as to its essential minerals despite the diversity of veins, alteration, and host rocks.

An outstanding feature of Guanajuato ores is the abundance of selenium minerals, which distinguish the ores from those of other mine districts. Petruck and Owens (1974) identified the following minerals in samples from the Peregrina, Cebada, Valenciana, and Torres Mines:

- sulfides: pyrite, marcasite, acanthite (pseudomorphs after argentite), polybasite, pearceite, sphalerite, chalcopyrite, galena, pyrargirite, tetrahedrite, arsenopyrite, pyrrhotite;
- metals: gold, silver, and electrum;
- selenides: aguilarite and naumannite;
- silicates: quartz, adularia, chlorite, montmorillonite, and nontronites;
- carbonates: calcite, dolomite, and siderite.

Wandke and Martínez (1928) report a variety of oxidation-zone minerals. Taylor (1972) reports argentite, stephanite, guanajuatite, paraguanajuatite, and proustite, none of which were mentioned by Petruck and Owens (1974).

The most abundant minerals are pyrite, argentite-acanthite, aguilarite, polybasite, electrum, galena, sphalerite, and chalcopyrite. Sulfosalts are much more abundant in the La Luz system (Taylor, 1972): principally proustite and minor amounts of polybasite, stephanite, and pyrargirite.

Petruck and Owens (1974) classify three different types of Guanajuato ores: type 1, with high chalcopyrite, sphalerite, and galena contents, low pyrite and silver sulfide contents, and a low acanthite:polybasite ratio; type 2, abundant pyrite, with low to moderate proportions of silver sulfides, sphalerite, chalcopyrite, and galena, and an acanthite: polybasite ratio of 0.25 to 0.30; and type 3, large amounts of pyrite, low quantities of sphalerite, chalcopyrite, and galena relative to silver sulfides, an acanthite: polybasite ratio of 1.5 to 6.0, with significant amounts of pyrargirite.

Textural relations indicate two distinct assemblages of ore minerals: (a) pyrite, sphalerite, galena, chalcopyrite, and tetrahedrite; and (b) pyrite, acanthite, polybasite, pyrargirite, chalcopyrite, electrum, silver, and gold. The paragenetic sequence indicates that assemblage a was deposited before b, where selenium is concentrated. Ore type 1 is rich in assemblage a minerals and ore types 2 and 3 are rich in assemblage b minerals, although they contain some assemblage a minerals (Petruck and Owens, 1974).

Guanajuatite and paraguanajuatite were reported by Taylor (1972) in the area of the Sierra and Veta Madre veins. They are found in close association and are the last sulfoselenides to crystallize; selenium contents are variable in the district and even in veins; they are apparently more abundant southeast of Veta Madre and in the Villalpando vein of La Sierra.

Geochemistry

Study of fluid inclusions in transport minerals of the ores, as well as the distribution of elements along the veins (Gross, 1975; Vasallo, 1980; Buchanan, 1980) has led to the following conclusions:

1. Selenium is not uniformly distributed either in the ores or the disseminates; a considerable gradient seems to exist along which values diminish, becoming very low at depth.

2. With respect to the distribution of elements in hydrothermal fluids:

- The great abundance of potassium is also reflected in the ore mineralogy; additionally, potassium content is very high in the host rock itself. Taylor (1972) points out the high potassium content in all the volcanic rocks of the district and shows they are consanguineous differentiation products of the same potassium-rich magma, such as an alumina-rich basalt. It is therefore suggested that the potassium in the hydrothermal fluids was leached from the Tertiary volcanic rocks.
- Antimony, copper, and lead are abundant in both pre-ore solutions and in the ores.
- Lithium, bismuth, arsenic, and magnesium are abundant in hydrothermal solutions associated with the mineralization.
- Silver content is somewhat high, but not a considerable anomaly (11 ppm).
- The solutions included in vein minerals geographically above the Deep Ore show high Cu, Pb, Se, and F values.

Isotopic studies of sulfur by Taylor (1972) and Gross (1975) conclude that the origin of the sulfur in the Guanajuato sulfides is unclear and difficult to explain.

Thermometry

Isotopic data, gold content in adularia, and Taylor's (1972) distribution of Ag/Au ratios indicate vertical and horizontal temperature gradients in the mineralization process of the Guanajuato veins; temperatures are greater at depth southeast of Veta Madre and in the Sierra veins. Fluid inclusion studies by Gross (1975) established a vertical temperature gradient of 10° C/100 m at Veta Madre and 8° C/100 m in the Sierra veins. Nevertheless, both Vasallo (1980) and Buchanan (1980), also based on fluid inclusion studies, deny the presence of a vertical gradient; Buchanan (1980) proposes a 12°C/km horizontal gradient, opposed to Taylor's, that diminishes from Rayas to Las Torres. Table 2 shows the diversity of homogenization temperature values obtained in the Guanajuato District, which however, tend

TABLE 2. DIVERSITY OF HOMOGENIZATION TEMPERATURES, GUANAJUATO MINING DISTRICT

Place	Depth	Temperature (°C)	Mineral	Source*
Peregrina Mine				
Upper ore	2,200	285	Quartz + cc.	1
	2,300	260	Quartz + cc.	1
Peregrina Mine				
Lower ore	2,200	280	Quartz + cc.	1
	2,000	290	Quartz + cc.	1
	1,800	300	Quartz + cc.	1
Rayas-Sirena	1,600	330-335	Quartz + cc.	1
	1,950	290	Quartz + cc.	1
	2,150	270	Quartz + cc.	1
Veta Madre SE	1,700	330	Quartz + cc.	1
	1,860	302	Quartz + cc.	1
	1,950	280	Quartz + cc.	1
	2,150	266	Quartz + cc.	1
	2,400	263	Quartz + cc.	1
–Torres (ore)	....	225-235	Quartz + cc.	2
(pre-ore)	....	231	Quartz + cc.	2
–Rayas	1,705	261-385	Quartz + cc.	2
Deep ore	174	284	Quartz + cc.	2
Lower ore	1,875	323-350	Quartz + cc.	2
Lower ore	2,090	290-360	Quartz + cc.	2
Lower ore	2,100	329-360	Quartz + cc.	2
Upper ore	2,130	258-261	Quartz + cc.	2
–Veta Madre SE		260-270	Quartz	3
Veta Madre SE		230	Quartz + mineral	3
Veta Madre SE		210	Quartz + mineral	3
Post-ore		160	Calcite	3

*1 = Gross (1975); 2 = Buchanan (1980); 3 = Vasallo (1980).

to indicate that mineralization precipitated in the Lower, or main, Ore zone within a temperature range of 210 to 300°C, the average being 230°C (Buchanan, 1980). All authors agree that mineralization took place by stages.

Isotopic geology

Geochronology. The age of the mineralization is well established by the ages of both host rocks and gangue minerals. The lower member of the volcanic series is 37 ± 3 Ma and the upper 32 ± 1.0 Ma (Gross, 1975). Radiometric dating of Veta Madre adularia ranges from 27.4 ± 0.4 to 29.2 ± 2 Ma; three determinations in Sierra veins show 28.3 ± 5.0 to 30.7 ± 3 Ma (Gross, 1975; Taylor, 1972). Although the wide age range would imply an active hydrothermal system for over 3 m.y., the mineralization evidently took place immediately after the last volcanic emissions in the area, possibly in relation to the geothermal gradient imposed by the magmatism and simultaneous deformation.

Isotopic ratios. As mentioned previously, sulfur isotopic ratios do not indicate the source of the sulfur in Guanajuato ores with any certainty, but do show they are not exclusively magmatic. Lead isotope studies (Gross, 1975; Taylor, 1972) identify the lead as J type, which points to its origin in the crust as a disintegration product of uranium; Rb/Sr ratios suggest the lavas to be differentiation products of subcrustal magma. All indications are that both the metals and most of the sulfur may have been leached from the prevolcanic sequence, and Gross (1975) proposes the Esperanza Formation as the source of the metals.

Origin of the deposit

Several authors have theorized about the origin of the Guanajuato deposit; the following stages are considered of importance in this respect:

1. Deposition, lithification, deformation, and intrusion by siliceous rocks, of a volcanic-sedimentary sequence (now Esperanza Formation).
2. Early Tertiary erosion and subsequent deposition of the Guanajuato Conglomerate, followed in the final stages by Oligocene volcanism.
3. Deformation and faulting caused by the volcanism, followed by more volcanism.
4. Normal faulting and development of the geothermal system in the mid-Oligocene. L. O. Buchanan (written communication, 1977) proposes six stages of simultaneous fracturing and hydrothermal alteration (summarized on Fig. 8). As described previously, the system is characterized basically by:

- potassium-rich and silica-rich alkaline waters with very diluted salts at maximum temperatures of 350°C;
- mineralization in at least three stages (the intermediate being the commercially important), and consequent deposition of banded quartz with very fine-grained mineralization;
- generally low sulfur fugacities, particularly at low temperatures or in higher parts of the system, and high selenium fugacity;
- commercial mineralization concentrated in two well-defined vertical zones or ranges, explained by various authors as caused either by decrease in temperature (Taylor, 1972; Gross, 1975), or by normal boiling and flashing (Buchanan, 1980).

GUANAJUATO GROUP METALLURGY

Present smelter capacity is 2,250 mt/day of ores derived from the various mines in the following proportions: 61 percent Torres-Cedros Mine, 1.35 percent Peregrina Mine, 8.5 percent Cebada Mine, 15 percent Bolañitos, and more recently 2 percent of purchased ore from mines outside of the district.

Average grades per mine are as follows:

Mine	*Gms/Ton Au*	*Gms/Ton Ag*
Torres	1.20	230
Cedros	1.10	170
Peregrina	4.50	350
Cebada	3.00	270
Bolañitos	2.40	220

Average monthly recoveries are 96 percent for gold and 83 percent for silver.

Concentration is by bulk primary flotation followed by three clean-ups, producing a concentrate with the following average grades: 28 percent pyrite, 110 ppm Au, 11,300 ppm Ag.

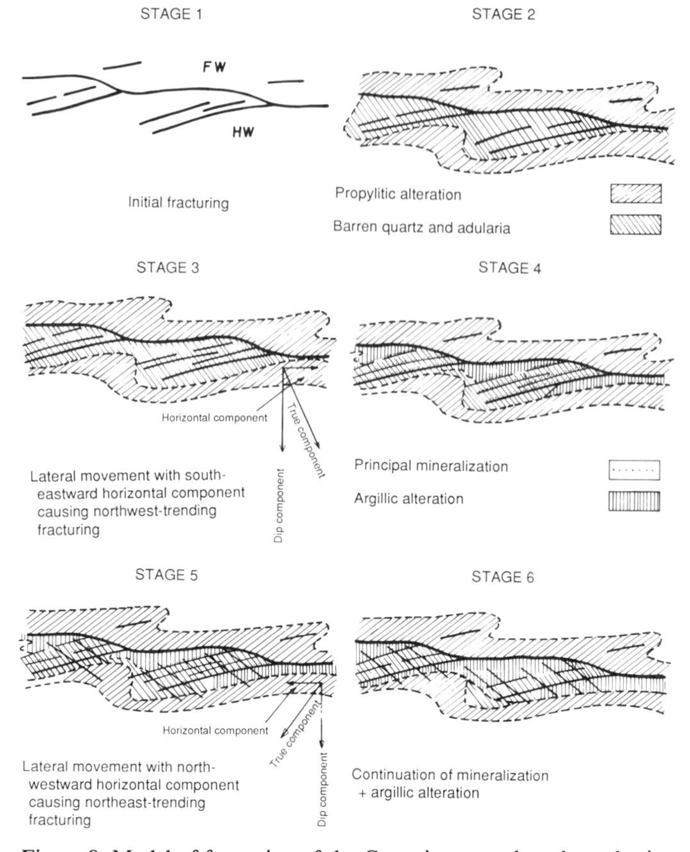

Figure 8. Model of formation of the Guanajuato stockworks and mineralization (after L. J. Buchanan, written communication, 1977).

REFERENCES

Autúnez E., F., 1964, Monografia histórica y minera sobre el Distrito de Guanajuato, México, D.F.: México D.F., Consejo de Recursos Naturales No Renovables Publication 17-E, 590 p.

Cepeda D., L. J., 1967, Estudio petrológico y mineralógico de la region "El Cubo," Minicipio de Guanajuato, Gto.: Assocíación Mexicana de Geólogos Petroleros Boletín, v. 19, p. 39–107.

Echegoyen S., J., Romero M., S., and Velásquez S., S., 1970, Geología y yacimientos minerales de la parte central del Distrito Minero de Guanajuato, México, D.F.: Consejo de Recursas Naturales No Renovables, Boletín 75, 36 p.

Edwards, J. D., 1955, Studies of some early Tertiary red conglomerates of central Mexico: U.S. Geological Survey Professional Paper 264-H, p. 153–185.

Gross, W. H., 1975, New ore discovery and source of silver-gold veins, Guanajuato, Mexico: Economic Geology, v. 70, p. 1175–1189.

Petruck, W., and Owens, D., 1974, Some mineralogical characteristics of the silver deposits in the Guanajuato Mining District, Mexico: Economic Geology, v. 69, p. 1078–1085.

Wandke, A. and Martínez, J., 1928, The Guanajuato Mining District of Mexico: Economic Geology, v. 23, p. 1–44.

Unpublished sources of information

Buchanan, L. J., 1980, Ore controls of vertically stacked deposits, Guanajuato, Mexico: Society of Mining Engineers preprint no's 80–82.

Cervantes, R. A., 1979, Aspectos estructurales de la Veta Madre en el área de Chichindaro, Guanajuato [professional thesis]: Universidad Autonoma de San Luis Potosí.

Gómez de la Rosa, R., 1961, Estudio geológico-minero del Distrito de La Luz, Guanajuato, Gto. [professional thesis]: Mexico, D.F., Universidad Nacional Autonoma de México.

Taylor, P. S., 1972, Mineral variations in the silver veins of Guanajuato [Ph.D. thesis]: Dartmouth College, 139 p.

Terrazas, A., 1968, Brief notes on the Guanajuato Mining District (1522–1968): Santa Fe de Guanajuato, Sociedad Cooperativo de Minero-Metalurgico.

Vasallo, L. F., 1980, Acerca de la composición quimica y la temperatura del sistema acantita-aguilarita-naumannita del depósito de Torres, Guanajuato, México: Moscow State University.

Printed in U.S.A.

The Geology of North America
Vol. P-3, Economic Geology, Mexico
The Geological Society of America, 1991

Chapter 54

History of exploration for sulfur in southeast Mexico

Guillermo P. Salas
Geologo Consultor, Asesor del Consejo de Recursos Minerales, Centro Minero Nacional, Carretera México-Pachuca, Pachuca, Hgo., Mexico, D.F.

INTRODUCTION

Mexico's demographic growth (3.5 percent per year) demands substantially increased food production in the next decades; to a greater or lesser degree this is also true for the rest of the world. Sulfur is essential for the fertilizer industry and is extremely important in general heavy industry as well. The value of the better geological understanding of southeast Mexico's Salt Basin domes, where cap-rock sulfur deposits are found, is here stressed in view of this element's weight in the future economic development of the country and in the worldwide fertilizer industry.

Although sulfur has been known to exist in the western part of the southeastern Salt Basin (Figs. 1, 2) since the beginning of this century, exploration on a commercial basis was not undertaken until 1948. Commercial production of sulfur using the Frasch process began in 1955. The Jáltipan dome was drilled by the (independent) Panamerican Sulphur Company in 1955; the San Cristóbal dome (now paid out and abandoned) was worked by Gulf Sulphur Company in 1956; the Amexquite Norte dome (northeast of Salinas) has been worked by Companía Azufrera Veracruzana since 1956, and the Texistepec Este dome by Companía Central Minera, S.A. de C.V., which suspended operations temporarily in 1958. Production in the Nopalapa dome, discov-

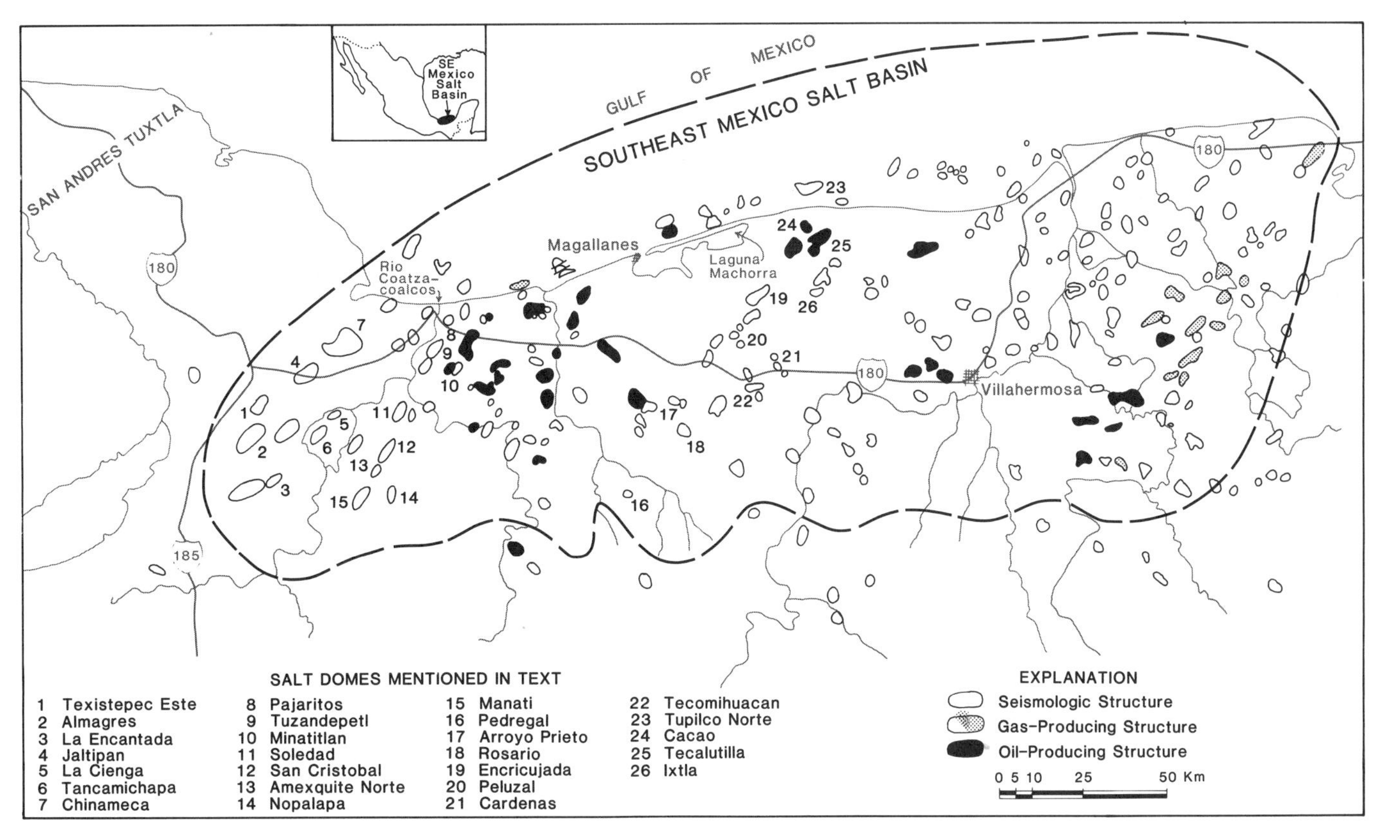

Figure 1. Index map for the southeast Mexico Salt Basin.

Salas, G. P., 1991, History of exploration for sulfur in southeast Mexico, *in* Salas, G. P., ed., Economic Geology, Mexico: Boulder, Colorado, Geological Society of America, The Geology of North America, v. P-3.

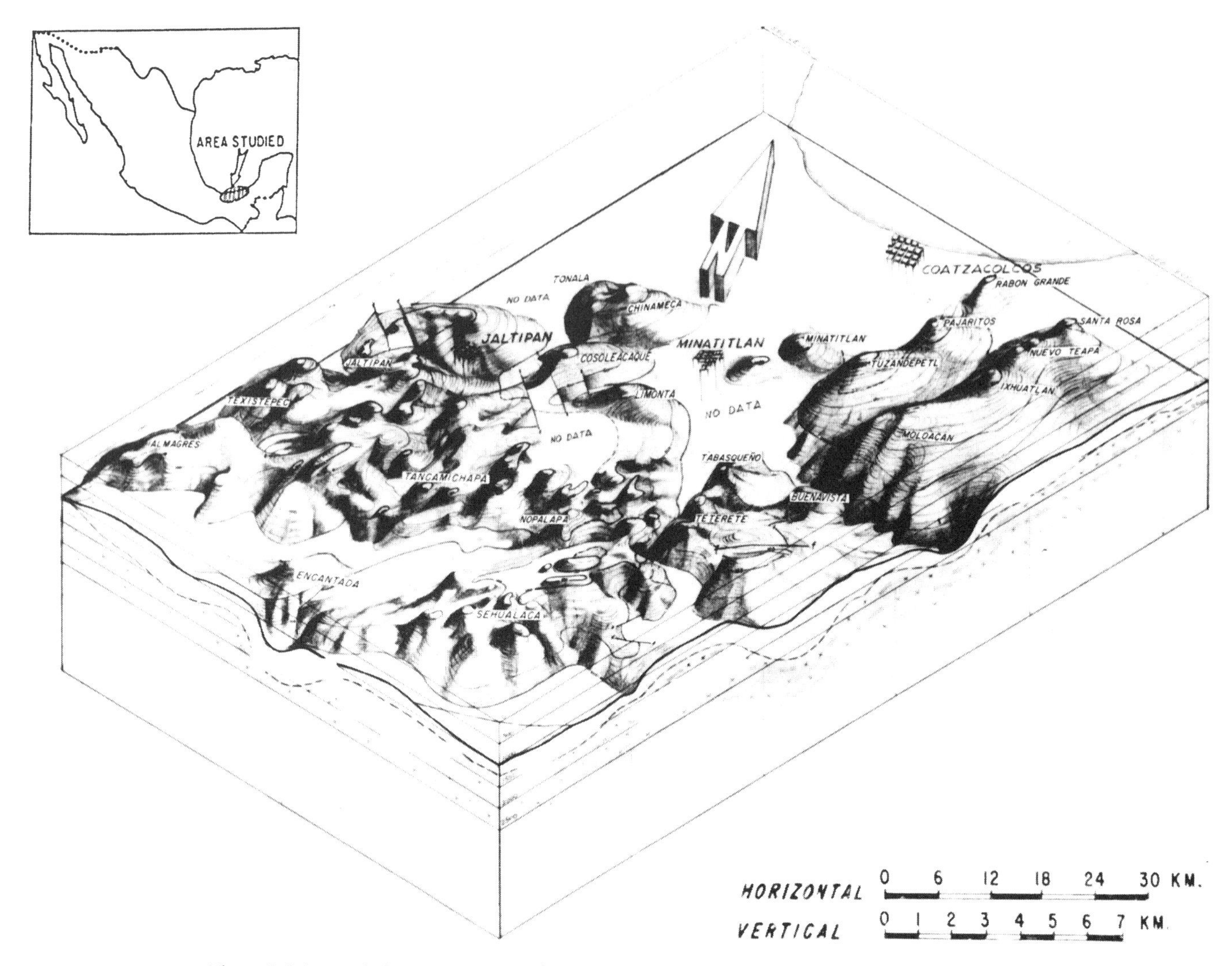

Figure 2. Salt morphology, western part of the southeast Mexico Salt Basin (after PEMEX and CRNNR seismic maps).

ered and worked by Companía Exploradora del Istmo (Texas Gulf Sulphur Company), was also suspended in 1959 owing to internal legal problems but is expected to be resumed.

GENERAL GEOLOGY OF THE SOUTHEAST MEXICO SALT BASIN

Figure 1 shows the new conception, geographic extension, and approximate area of southeast Mexico's Salt Basin. Its northern boundary in the southern Gulf of Mexico continental shelf is still unknown. Ewing and Antoine (1966) suggest the possible connection of the Sigsbee Knolls with the Salt Basin domes. The writer disagrees with this idea in view of the widely different geologic histories of both areas, but concurs that more exploration is needed to clarify whether or not a connection exists between the Salt Basin on the continent and the Sigsbee Knolls at the central Gulf of Mexico sea floor.

The basin is well demarked on the west by the San Andrés Tuxtla Volcanic Massif; its eastern boundary is still somewhat undefined. The Salinas Formation seems to wedge out toward the east through a facies change from salt to anhydrite, gypsum, and dolomite in the Campeche Bay and Yucatán areas. Deep wells drilled in Yucatán encounter only gypsum and anhydrite, with no clastics eastward from Champotón, Campeche (see Fig. 4).

The Salt Basin area has been covered by more or less detailed geological and geophysical surveys, including photogeology, surface and subsurface geology, gravity, magnetometry, and both refraction and reflection seismics; in addition, over 2,000 wells have been drilled in the basin. However, additional and more thorough geological exploration needs to be done in every field. The available information permits description of the general geology of the area and some structures to illustrate the oil and sulfur environment in the basin.

SALT DOMES

The Salt Basin domes display marked differences from those occurring in the Texas and Louisiana Gulf Coast of the United States or the Zechstein Salt Basin in Germany. The southeast Mexico domes are shallow, and usually ovoid in cross section (Fig. 2) and more of a protruding type; their elevation above the adjacent synclines is often equal to their cross section, probably because they were never deeply buried. In contrast, the United States and German domes are typically deep-seated penetration or intrusive domes of almost circular cross section and great columnar vertical development (Fig. 3), with overhanging cap rocks that spill out onto the flanks.

The salt domes of southeast Mexico are in fact apophyses protruding from a salt mass that forms a large subsurface salt platform. The domes are very shallow in the west and southwest part of the basin and deeper toward the northeast (Figs. 2 and 4), generally showing more similarity to the Romanian and Iranian domes than to those in Texas, Louisiana, and Germany. Seemingly the Mexican salt deposits were never buried at depths that would allow geothermal heat or high pressures to enhance the plasticity of the salt. Salt hydroactivity sufficient to increase its plasticity by acting as a lubricant appears to have acted only in relatively recent geologic time.

The whole salt mass from which the dome apophyses are derived appears to have been down-tilted toward the north and east and uplifted toward the southwest, to such an extent that, at least in the Amexquite Norte, Texistepec, and other domes (all in the southwest part of the basin; Fig. 1), the cap rock is exposed, and the sulfur previously lodged in the pores is oxidized. These features are considerably important in sulfur prospecting; domes with sulfur-bearing cap rocks have been found only in the southwest and west part of the basin.

AGE OF THE SALT

There has been much speculation in the past as to the age of the salt in the Salt Basin. That it is pre-Jurassic seems clearly demonstrated: the mid-Jurassic ammonite-bearing Chinameca Limestone was found overlying the salt in exploration wells drilled by Companía Central Minera north of the Hibueras railroad station on the northeast flank of the Chinameca dome, north of Minatitlán (Veracruz; Fig. 5); in addition, Benavides (1956, p. 507) found the red beds of the Salina Formation underlying the Kimmeridgian Chinameca Limestone.

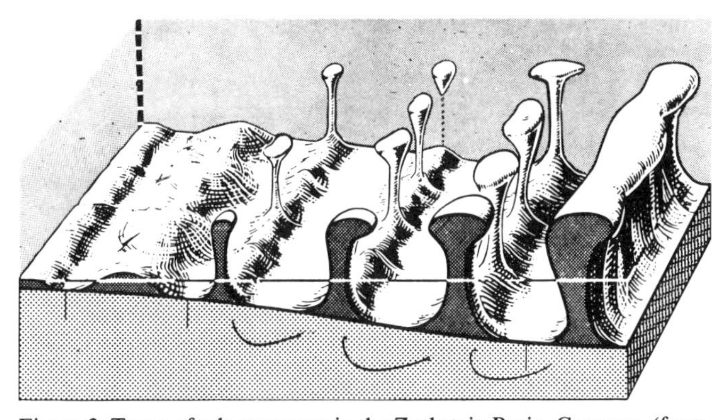

Figure 3. Types of salt structures in the Zechstein Basin, Germany (from Murray, 1966).

Both indicate that the salt is pre-Kimmeridgian. Recently recorded (as yet unpublished) palynological evidence establishes its age as Lias-Rhaetian, and therefore correlatable with the Louann Salt Formation of northeast Mexico and Texas, according to Ewing and Antoine (1966, p. 482) and Murray (1961, p. 262).

MOVEMENTS OF THE SALT

The salt ascended and intruded apparently in post-Kimmeridgian time and continued to be mobile and plastic for all or most of the Lower and mid-Cretaceous (Salas, 1949). The area subsided slightly during the Late Cretaceous and reemerged at the end of the Mesozoic and early Tertiary; Eocene and Oligocene deposits over the domes are either very thin or absent. It may be assumed that salt ascension began before the Eocene and could have caused some upper Mesozoic and lower Tertiary sediments to slump down toward the synclines or their flanks with the advance of the salt intrusion into the overlying sedimentary section.

According to Borchert and Muir (1964), salt rises at the rate of 1 or 2 mm per year. This is assumed to have occurred throughout the Tertiary to the present and would explain the absence or very slight thickness of upper Cretaceous or Eocene sediments over the anticlinal axial areas caused by the salt intrusions; the section thickens in the synclinal areas or dome flanks. Other arguments might be advanced to explain sedimentary thinning over axial areas of the dome, but the writer deems that the diapir existed and was active during late Mesozoic and Tertiary time, and that the topographically high domes at that time did not allow deposition at their crests. The subsidence of the Salt Basin and the immersion of this mass during the Oligocene and Miocene caused over 5,000 m of sediments to be deposited in the eastern part, whereas Oligocene and Miocene formations overlying the salt platforms are much thinner or absent on the west.

The salt continues to rise even now; present-day differential erosion develops mounds, hills, and mountains—at times several hundred meters high—and rejuvenates drainage throughout the Gulf Coastal Plain, which was eroded to base level and peneplained at the end of the Pleistocene. Streams as important as the Coatzacoalcos have deviated from their erratic courses due to uplift of the salt mass; this is the explanation for Tancamichapa Island 35 km south-southwest of Minatitlán (Fig. 6, inset map), where the Coatzacoalcos divides to form the Río Chiquito at the southern tip of the island and merges again several tens of kilometers farther north.

A detailed topographic map of the Salt Basin, as yet unavailable, would show that the topographic highs are generally

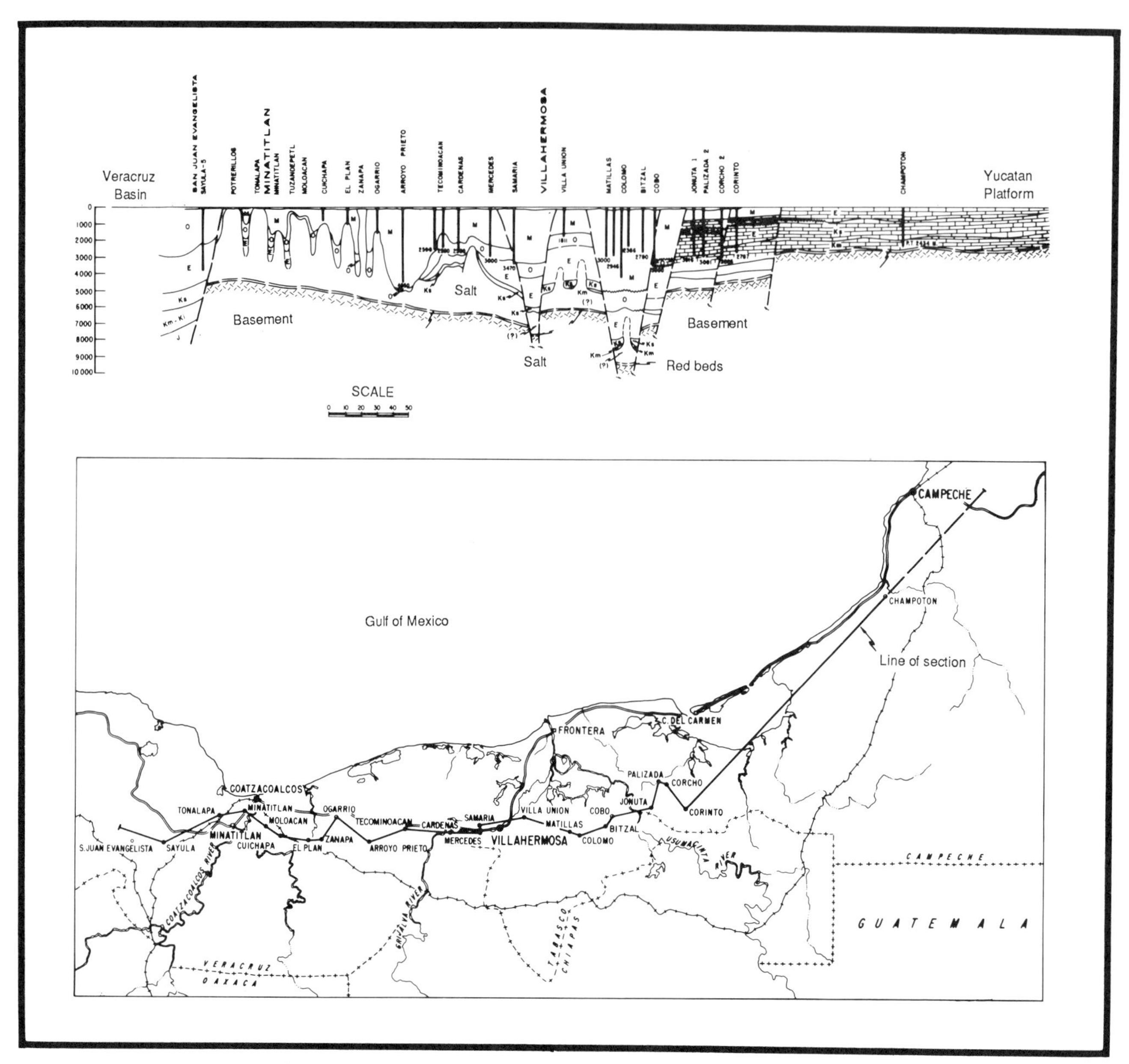

Figure 4. Geological cross section through the southeast Mexico Salt Basin.

located over the domes. Low-lying dome axial areas, occasionally forming (at times salt-water) marshes or lagoons owing to differential dissolution of the dome crest and the consequent sedimentary subsidence at the axial area, occur only where the salt is shallow, as in Chinameca, La Encantada, and others (Fig. 2).

SALT TECTONICS

The similarity of the Salt Basin domes with those of Iran and Romania disappears when their tectonic histories are compared. Overturned folds such as those found in the Near East and European domes, or reverse faults, are unknown in the Salt Basin. Considering that the effects of the early Tertiary (Laramide) orogeny—transmitted to the salt mass through mostly incompetent early Tertiary sediments—were never sufficiently intensive after the late Miocene, the clastic sediments overlying the salt domes remained unaffected by tangential forces, and overturned folds or reverse faults did not develop. Thus, the marked similarity of the rhythmic folding of the salt in the Salt Basin to that in the Zechstein Basin of Germany also disappears, because the salt in the German domes is intrusive or penetrating, akin to that of the Texas and Louisiana basins. This "needle" characteristic is absent in the Mexican domes that intrude the Oligocene and Miocene formations.

The southeast Mexico Salt Basin exhibits definite orientation trends in the way that the apophyses arose from the salt mass. There is a decided northwest trend of the folds within the rigid framework along the south boundary of the Tertiary basin (Sierra

Madre Oriental–type folding) that contrasts with the definite right-angle northeast trend of the alignment of the domal structures within the Salt Basin. The axial orientation of the domes and of the folds that developed over them appears parallel to the western limit of the Salt Basin's western half. For instance, the Chinameca, Jáltipan, Almagres, and other domes along the Salt Basin's western limit trend northeast-southwest, parallel to that western boundary. Fold axes farther west, outside the salt mass and nearer to the San Andrés Tuxtla Massif, are also northeast-southwest. It seems evident that the massif exerted southeast compression extending within the salt that caused northeast-southwest folds and salt walls, but insufficient for overturned folds, reverse faults, or "schuppen" structures to develop. Since San Andrés Tuxtla Massif volcanic and/or intrusive rocks are post-Oligocene, and most of the salt intrusions into younger beds fractured and folded the Miocene sediments, it appears reasonable to assume that the San Andrés Tuxtla Massif caused the northeast-southwest orientation of the salt folds and that of the apophyses in the western Salt Basin. Miocene formations show some northeast-southwest folding east of the volcano and outside the Salt Basin.

Consequently, it also seems clear that the early Tertiary (Laramide) orogeny had little influence on salt tectonics, apart perhaps from triggering the ascent and penetration of plastic salt at the end of the Mesozoic. As mentioned above, the salt domes display rhythmic alignment. Figures 1 and 2 show several of these northeast-southwest-trending domes in the western part of the basin: Chinameca, Jáltipan, Almagres, and others; approximately 10 km farther east is the alignment of Minatitlán, Texistepec, La Ciénaga, Tancamichapa, and La Encantada, followed 12 km to the east by an equally northeast-southwest line from the Pajaritos dome through the Tuzandepetl, Soledad, San Cristóbal, and Manatí domes (Figs. 1, 2, 6); this northeast-southwest alignment trend continues farther east. There would appear to be rhythmic folding of the salt that affects the shallowest part of the salt domes in the western Salt Basin; these apophyses alignments are spaced 10 to 12 km apart between fold axes. The easternmost domes that show such a trend are Tupilco Norte, Cacao, Encrucijada, Arroyo Prieto, Rosario, and Pedregal (Fig. 1).

No salt structures have been mapped between Ixtla (south of Tecolutilla) to the Cárdenas-Tecomihuacán line, or farther east to the Samaria structure (Fig. 1). No structural evidence is found within several square kilometers of the area southwest of Laguna Machorra between Magallanes and Peluzal, where according to the rhythmic fold reasoning, structural alignments similar to the above should be present; the writer suggests lack of exploration or misinterpretation in the area as a possible explanation.

The northeast-southwest alignment of dome groups along the east and west boundaries of the Salt Basin is not encountered in structures of the sedimentary section overlying the salt domes in the western part of the basin, where fold axes and principal faults display random orientations.

The dome areas show transverse and radial faults similar to those of the Texas and Louisiana domes. Other faults display a

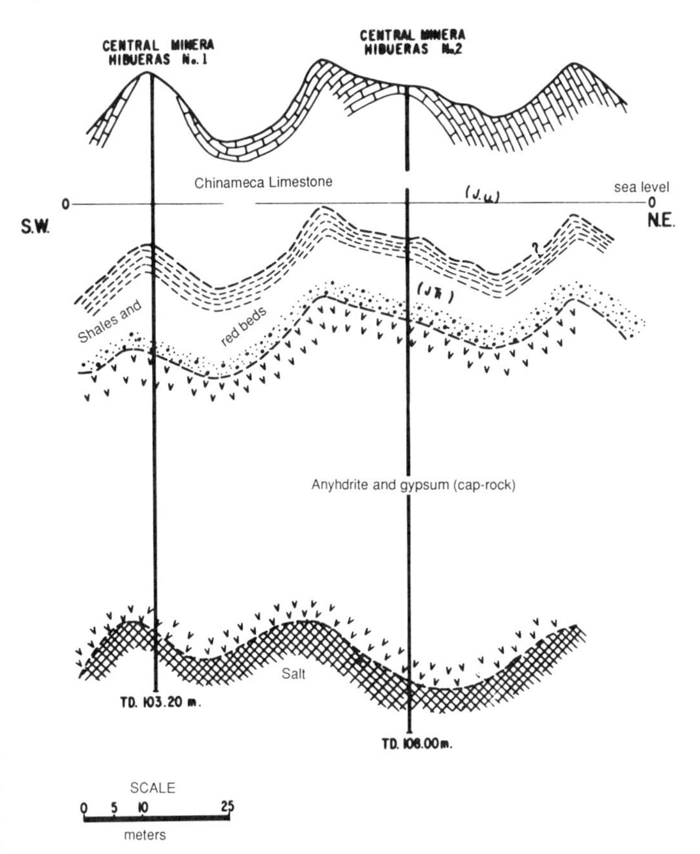

Figure 5. Cross section of the northeast flank of the Chinameca dome (from unpublished reports, CRNNR).

tendency to run parallel to the northeast-southwest salt structure. On the whole, it would seem that the mechanics of salt intrusion acted differentially on clastic sedimentary sections of different textures. Perhaps salt tectonics behaved differently in the sedimentary section, owing to some control exerted by the position of the domes within the basin (i.e., the behavior of folding and tectonic forces may have varied according to the depths of the domes within the basin).

Detailed gravity and seismic reflection studies have obviously revealed a great number of structures in these areas, and should detect others as yet undiscovered. These structures are believed to be of directly diapiric origin.

THE SULFUR CASE

Origin of the salt domes

Barton (1925) and DeGolyer (1925) suggested that certain stresses coinciding simultaneously within a salt basin would cause plastic salt to move upward, giving rise to such as those in the Gulf Coast of the United States. Since then many investigators have uncovered a great number of factors regarding salt history and characterizations that bear on the origin of salt domes.

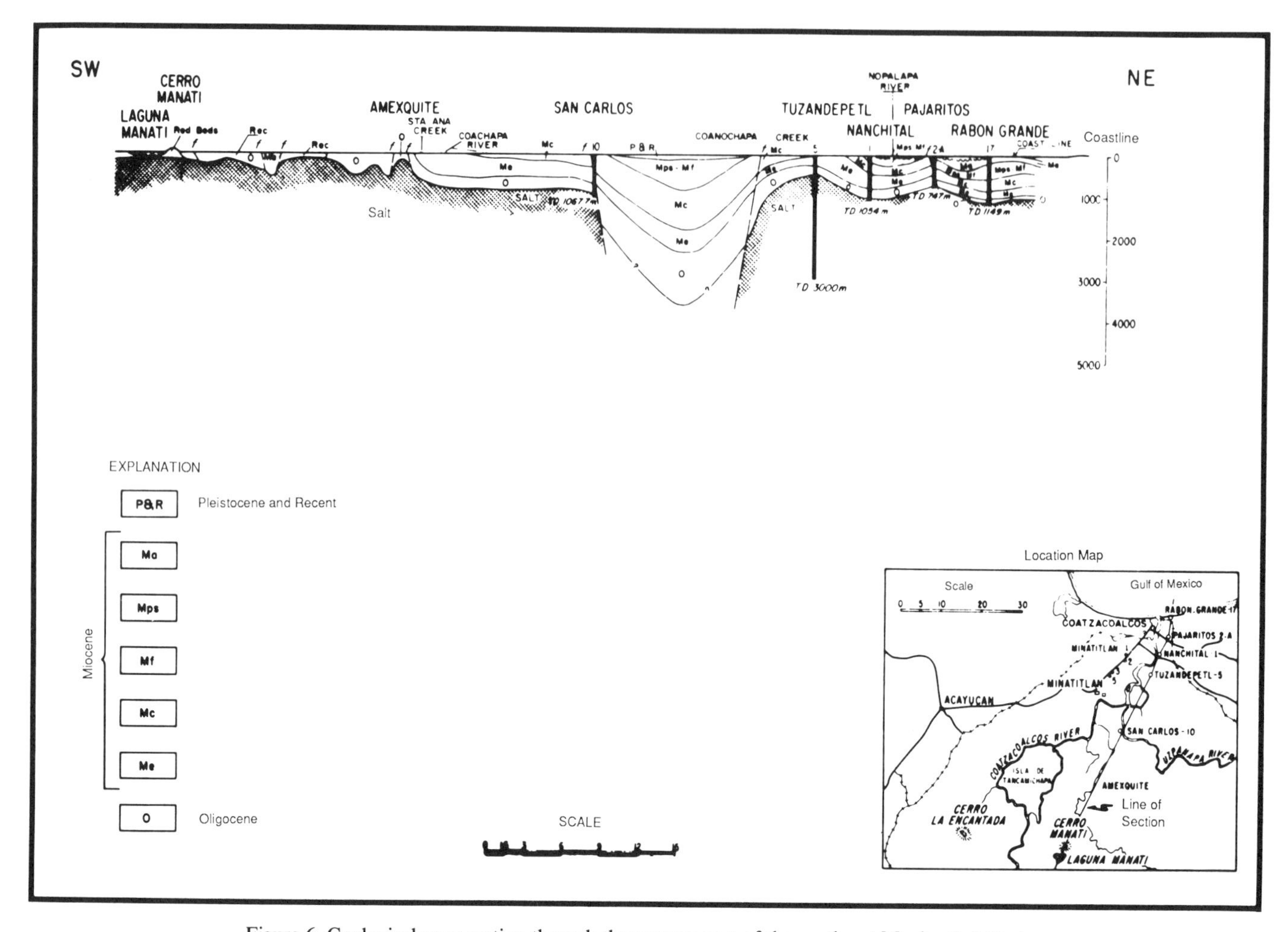

Figure 6. Geological cross section through the western part of the southeast Mexico Salt Basin.

1. Salt becomes mobile or plastic under appropriate pressures (in this case the weight of the overburden, under abnormal temperatures (the geothermal gradient or higher temperatures), through the effect of shear forces, and by water lubrication, which increases its mobility.

2. The salt forming the domes is sedimentary in its origin, derived from the evaporation of sea water.

3. Salt and other evaporites accumulate by cycles, as shown by the cyclic layers or laminae of insoluble impurities within evaporite masses (Fig. 7). (Some authors mention parallelism between 11-layer sequences of insoluble material within the salt mass and the 11 years of the sunspot cycle, which seem to coincide with these impurities in several parts of the world.) Such impurities make up 1 to 10 percent of the salt. The insoluble material usually consists of 99 percent $CaSO_4$ (anhydrite) and 1 percent SiO_2, Fe_2O_3, $SrSO_4$, MnS, $CaCO_3$, or pyrite (Taylor, 1937; Hungsberg, 1960; Enciso de la Vega, 1963).

4. Potassium, the final crystallization product of marine brines, appears in lower percentages in evaporite deposits than in sea water, which indicates that evaporation rarely came to completion. In addition, evidence (in present deposits) shows that potassium is found where evaporation is more complete, usually where the basins are shallowest; the presence of potassium salts—even though comparatively scarce—may therefore be interpreted as indicating the proximity of old coastlines in the Salt Basin.

5. Present-day salt occurs worldwide over extensive areas of high evaporation levels. Within the widespread North American evaporation zone, southeast Mexico's Salt Basin may very possibly be correlative in time with the Louann of Texas, Louisiana, and northeast Mexico as mentioned earlier. The salt in the salt basin of northeast Mexico (Cuchillo Parado, Chihuahua) is believed to be older, possibly of Permian (Ochoan) age.

6. Salt wells up and rises due to differential gravitational pressures over originally high areas. Dome groups that develop under some type of structural control, unfortunately poorly understood to date, occasionally display cyclic folding, as in the Zechstein Salt Basin of Germany, or parallel arrangements of domal structures, as shown by the writer (Fig. 2).

7. The cap rock (Taylor, 1937; Enciso de la Vega, 1963) is formed by differential solution of originally impure salt. Insoluble residues accumulate and harden to cap rock composed mostly of $CaSO_4$ (as either anhydrite or gypsum).

Salt domes were originally viewed as an exclusively geological problem. Today, however, geologists have also combined information provided by chemists, physicists, biologists, mathematicians, and engineers, to arrive at the present concept of their origin.

Origin of the sulfur

Based on the above and additional investigations, it has been shown that anhydrite may alter biochemically to limestone and sulfur when sulfur bacteria and hydrocarbons are present in a reducing environment. The possible reactions would be as follows:

$$CaSO_4 + CH_4 = CaS + CO_2 + 2H_2O$$
$$CaS + CO_2 + H_2O = CaCO_3 + H_2S$$
$$3H_2S + SO_4^{=} = 4\,S + 2H_2O + 2\,OH^-.$$

Several authors agree that sulfur bacteria are needed to accelerate the reaction, which otherwise would be simply a chemical reduction with extremely slow development of sulfur in the cap rock. Three species of bacteria have been found in sulfur-bearing cap rocks: *Desulfovibrio desulfuricans* (wide spectrum of temperatures), *Desulfovibrio orientis* (medium temperature), and *Clostridium nigrificans* (thermophilic, high temperature); only the first (*D. desulfuricans*) has been found in Mexican salt domes.

On the basis of the H_2S and $CaCO_3$ as final products of the biochemical reaction, the Consejo de Recursos Naturales No Renovables (now Consejo de Recursos Minerales—CRM) launched a research program in an attempt to define the carbonate/sulfate and sulfur/carbonate ratios in the cap rock by quantitative analysis of cores with abundant, medium, scarce, or no sulfur contents. Results show abundant calcium carbonate in sulfur-rich cores, and that the carbonate-sulfate ratio varies in proportion to the sulfur content. This could be an important contribution to sulfur exploration by drilling, which until now has included continuous coring of the cap rock from top to base at considerable drilling costs.

On the other hand, reliance on the lithological description of ditch samples only, without coring, is less expensive and time consuming when no $CaCO_3$ is found in the cap rock. When calcium carbonate is present in any part of the cap rock or dome crest, sulfur is normally found in the immediate vicinity; if this exploration method is used, sulfur and calcium carbonate traces must be thoroughly searched for in the ditch samples.

Other minerals sometimes form 5 to 10 percent of the cap rock; for instance, where sulfur is present, pyrite and marcasite are found in such percentages and could be used as guides to the presence of sulfur (Table 1). On occasion other minerals appear even in pure anhydrite when the cap rock has no calcium carbonate, but they usually make up less than 1 percent of the rock.

Figure 7. Photographs of evaporites in cores from wells on the Tancamichapa salt dome (see Fig. 2 for location). Left and center, laminated and deformed anhydrite; right, laminated anhydrite in recrystallized salt, showing vertical dip and redeposition.

Sulfur in the cap rock. Commercial amounts of sulfur are found at depths of less than 300 m in five domes in the west and southwest part of southeast Mexico's Salt Basin. It usually occurs either in cavities, or scattered within more or less porous or semiporous crystalline limestone that may be a few to more than 100 m thick, even in the same dome. The position of the sulfur in the cap rock varies from one dome to another, appearing in the structurally highest part of the dome, at the highest part of the cap rock, or at its flanks, and is therefore unpredictable. In every case the sulfur occurs within the dome's axial area. In the Amexquite Norte dome, cap rock containing sulfur has been found interbedded within the anhydrite (Fig. 8); in this case the cap rock was almost certainly strongly faulted, showing considerable perturbation and sulfur-filled fault zones.

In sulfur-bearing domes, the limestone cap rock is commonly in direct contact with overlying Miocene or Oligocene shales. An anhydrite bed of variable thickness (few to several m, generally less than 30 or 40 m) underlies the sulfur-bearing limestone and directly overlies the salt. The anhydrite/salt contact is occasionally transitional, marked by the appearance of salt layers, which gradually become thinner, within the anhydrite beds until only salt with anhydrite or gypsum laminae is present.

In some Amexquite Norte and Nopalapa dome wells the salt sometimes contains intercalated shales probably contemporaneous with those overlying the dome or appearing at the flanks. It is not clear whether these interbedded shales are part of the sequence, or intruded the plastic salt, or whether the salt intruded laterally along the shale bedding planes. The writer suggests that thin shales form part of the sequence; when shales are very thick, the well may be drilling through tectonically interbedded salt and shale in a plastic sequence. At times the shale appears also in the sulfur-bearing cap rock (some Amexquite Norte and Nopalapa wells), a probably unique occurrence in the world, which has posed production problems in these prolific domes.

The salt, shales, and limestone cap rock display evidence of oil and/or sulfur impregnation.

The fact that the cap rock in commercial sulfur-producing domes is limestone and occurs at shallow depths may be evidence that the genetic process necessarily involves shallowness by reason of the salt intrusion, which in turn causes upward arching of possible gas or oil reservoir beds and allows hydrocarbon migration into the faulted cap rock, bringing about the reducing environment required for sulfur bacteria to act and cause sulfur to precipitate as the final product.

Where the cap rock has been cored in salt domes drilled for oil by PEMEX (Petróleos Mexicanos), it is always definitely anhydrite; none of the deep domes (more than about 800 to 1,000 m) have shown limestone in the cap rock, and certainly no sulfur. A possible exception is a core in one of the La Venta oilfield wells, which is rumored (unconfirmed) to show a calcareous cap rock with sulfur.

It would therefore seem that cap-rock depth in the domes is also a significant factor in sulfur exploration. This should also be investigated, because more than 1,000 wells have reached cap rock at shallower depths both north and south of the Río Uzoanapa in the western part (Veracruz) of the southeast Mexico Salt Basin without encountering calcareous cap rock or sulfur. However, the writer considers it very important that depth should not exceed 300 m—at most—when searching for sulfur in salt dome cap rock, even though it may vary somewhat. Dome axial area tectonics and hydrocarbon evidence, as possible guides to the presence of sulfur, should also be thoroughly examined prior to drilling.

TABLE 1. PETROLOGIC PROPORTIONS IN THE CAP ROCK

Types of Cap rocks	CaCO	CaSo	S	FeS
Cap rocks with **good** S content	Abundant	Traces	Abundant	1 to 5 percent
Cap rocks with **medium** S content	Abundant	Abundant	Less abundant	5 to 10 percent
Cap rocks with **poor** S content	Scarce or trace	Abundant	Trace	Trace
Cap rocks **barren**	Absent	100 percent	Absent	Absent

Other sulfur exploration techniques. Since $SO_4^=$ to $SO_3^=$ reduction is an exothermic reaction, the search for present thermal anomalies using aerial infrared exploration has been proposed.

a. The CRM, jointly with the Instituto de Geología, has begun a series of field observations in a preliminary attempt to establish infrared emission coefficients of different rock types within the Salt Basin, including cap rock with or without sulfur; although a number of samples have been studied, it would be premature to speculate on the results. Given the premises that sulfur genesis in salt dome cap rocks requires shallow subsurface depths and that the biochemical reaction is exothermic, generating H_2S before becoming sulfur, aerial infrared remote sensing to map thermal anomalies seems advisable.

b. Pursuing simplified surface exploration procedures prior to drilling, sulfur isotope studies are being carried out by the Instituto de Geología to establish $^{32}S/^{34}S$ ratios in surface and subsurface samples from areas directly above sulfur-producing salt domes.

Feely and Kulp (1957) found that sulfur isotopes derived from calcareous cap rock in salt domes always show $^{32}S/^{34}S$ ratios within a range of 21.85 to 22.45, very similar to that of volcanic sulfur; the type of sulfur is of course known by the sampling environment, thereby avoiding errors in isotope data interpretation.

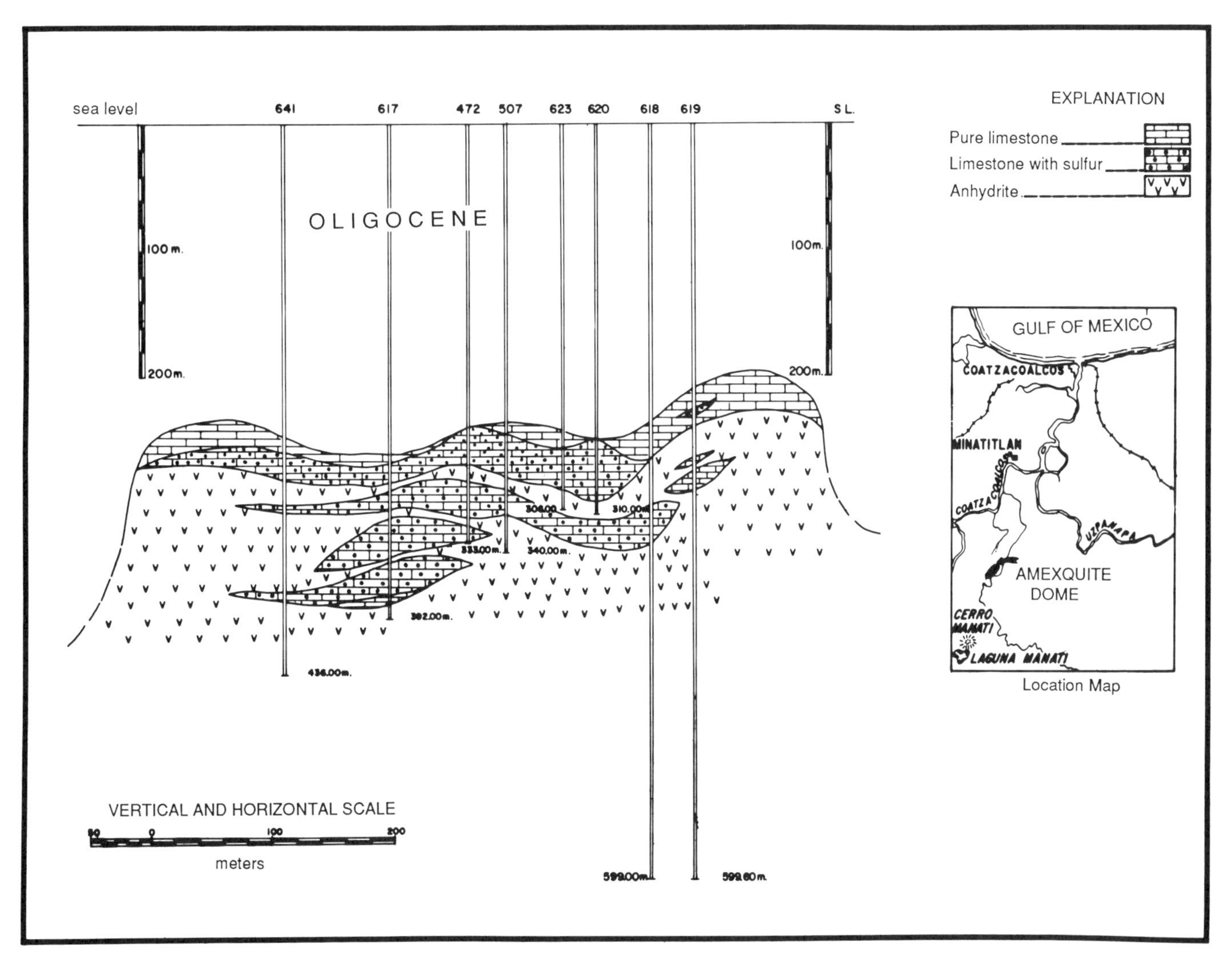

Figure 8. Cross section of Amexquite dome showing distribution of sulfur-bearing cap rocks.

c. Although analysis of hydrocarbons in soils overlying or surrounding salt domes (to map contours over the structure) has not been attempted, such a procedure would appear recommendable, given that hydrocarbons are essential in the reduction of $SO_4^=$ to $SO_3^=$ and O_2.

To summarize, this chapter establishes that the only commercial sulfur-bearing salt domes in Mexico—Jáltipan, Amexquite Norte, Texistepec, and Nopalapa—produce from calcareous cap rock at depths less than 300 m. Commercial sulfur has not been found in deeper domes. Consequently, the cap-rock depth seems essential in the formation of sulfur, and exploration for new deposits should therefore focus on domes where the cap rock is known to be less than 400 or 500 m deep; the writer considers that exploration in deeper cap rock will have negative results.

Because sulfur companies will drill hundreds or thousands of wells in search of sulfur in the high parts of all these domes, it would be highly desirable that the PEMEX Exploration Management commission specialize personnel to search for oil-related information provided by the sulfur-exploration programs.

The above conclusions and recommendations are worth testing; it should be clearly understood that sulfur will not be found by any method if drilling is not done to prove its presence in commercial quantities, similar to oil, which is only known to be there once it is drilled for and tested.

REFERENCES CITED

Barton, D. C., 1925, The salt domes of south Texas: American Association of Petroleum Geologists Bulletin, v. 9, p. 536–589.

Benavides, G. L., 1956, Notas sobre la geología petrolera de Mexico, *in* Guzman J., E. J., ed., Symposium sobre yacimientos de petroleo y gas; v. 3, Mexico, D. F., 20th International Geological Congress, p. 351–562.

Borchert, H., and Muir, R. O., 1964, Salt deposits; The origin, metamorphism and deformation of evaporites: London, D. Van Nostrand Co., Ltd., 338 p.

DeGolyer, E. L., 1925, Origin of North American salt domes: American Association of Petroleum Geologists Bulletin, v. 9, p. 831–874.

Enciso de la Vega, S., 1963, Estudio mineralogico y petrografico de algunos domos salinos del Istmo de Tehuantepec: México, D.F., Universidad Nacional Autonoma de México, Instituto de Geología Boletín 65, 48 p.

Ewing, M., and Antoine, J., 1966, New seismic data concerning sediments and diapiric structures in Sigsbee Deep and upper continental slope, Gulf of Mexico: American Association of Petroleum Geologists Bulletin, v. 50, p. 479–504.

Feely, H. W., and Kulp, J. L., 1957, Origin of Gulf Coast salt-dome sulphur deposits: American Association of Petroleum Geologists Bulletin, v. 41, p. 1802–1853.

Hungsberg, U., 1960, Origen del azufre en el casquete de los domos salinos, cuenca salina del Istmo de Tehuantepec: México, D.F., Consejo Recursos Naturales No Renovables Boletín 51, 96 p.

Murray, G. E., 1961, Geology of the Atlantic and Gulf coastal province of North America: New York, Harper and Brothers, 692 p.

——, 1966, Salt structures of Gulf of Mexico Basin; A review: American Association of Petroleum Geologists Bulletin, v. 50, p. 439–478.

Salas, G. P., 1949, El Cretacico do la Cuenca de Macuspana Tabasco y su correlación: Sociedad Geologico Mexicana Boletín, v. 14, p. 47–65.

Taylor, R. E., 1937, Water-insoluble residue in rock salt of Louisiana salt plugs: American Association of Petroleum Geologists Bulletin, v. 21, p. 1268–1310.

Printed in U.S.A.

Index

[Italic page numbers indicate major references]

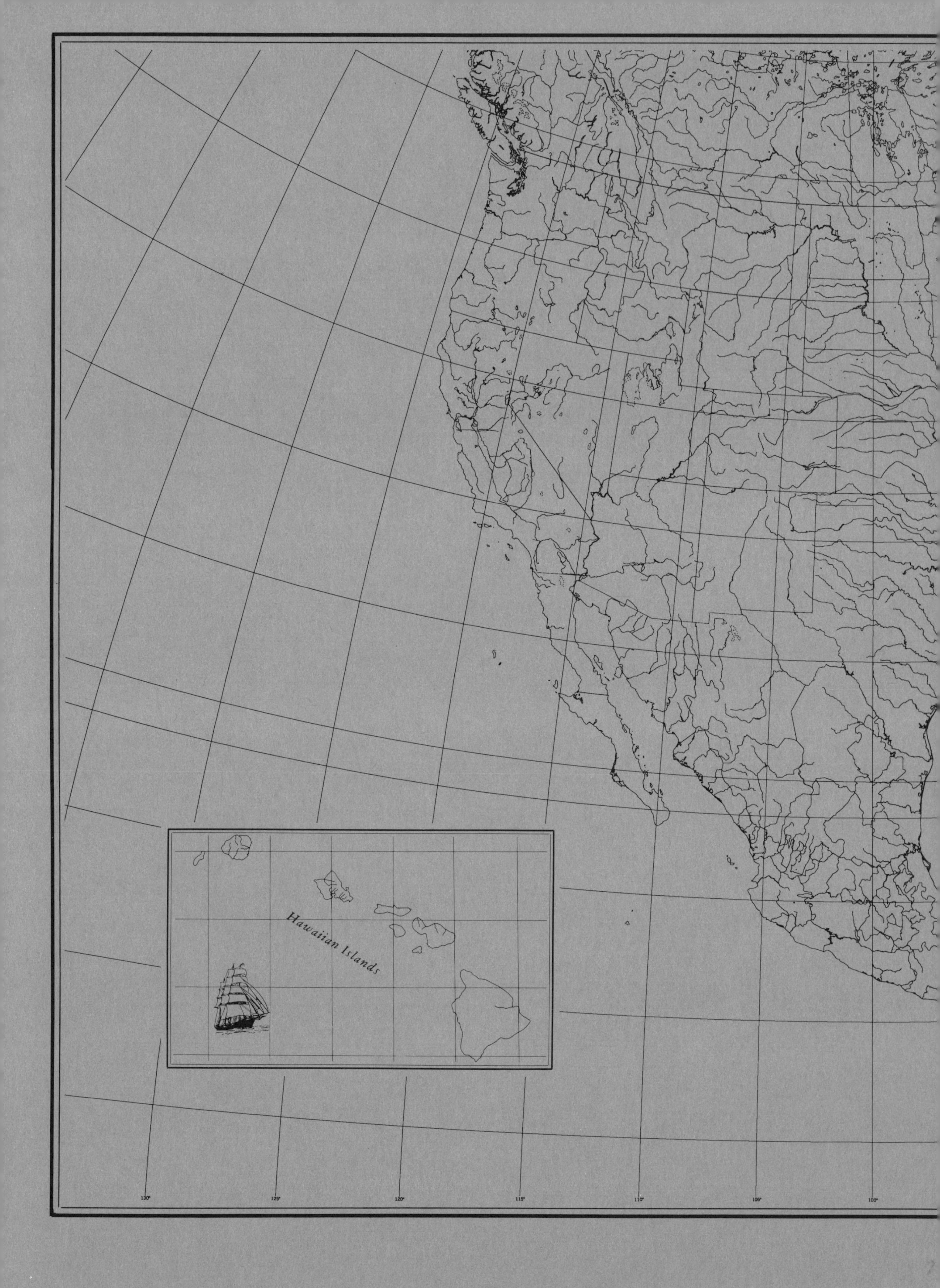
Hawaiian Islands
130°
125°
120°
115°
110°
105°
100°